# BIG IDEAS MATH®
## Algebra 1
### A Common Core Curriculum

CALIFORNIA EDITION

**Ron Larson**
**Laurie Boswell**

Erie, Pennsylvania
BigIdeasLearning.com

Big Ideas Learning, LLC
1762 Norcross Road
Erie, PA 16510-3838
USA

For product information and customer support, contact Big Ideas Learning at **1-877-552-7766** or visit us at ***BigIdeasLearning.com***.

About the Cover
The cover images on the *Big Ideas Math* series illustrate the advancements in aviation from the hot-air balloon to spacecraft. This progression symbolizes the launch of a student's successful journey in mathematics. The sunrise in the background is representative of the dawn of the Common Core era in math education, while the cradle signifies the balanced instruction that is a pillar of the *Big Ideas Math* series.

Copyright © 2015 by Big Ideas Learning, LLC. All rights reserved.

No part of this work may be reproduced or transmitted in any form or by any means, electronic or mechanical, including, but not limited to, photocopying and recording, or by any information storage or retrieval system, without prior written permission of Big Ideas Learning, LLC unless such copying is expressly permitted by copyright law. Address inquiries to Permissions, Big Ideas Learning, LLC, 1762 Norcross Road, Erie, PA 16510.

*Big Ideas Learning* and *Big Ideas Math* are registered trademarks of Larson Texts, Inc.

Common Core State Standards: © Copyright 2010. National Governors Association Center for Best Practices and Council of Chief State School Officers. All rights reserved.

Printed in the U.S.A.

ISBN 13: 978-1-60840-675-3
ISBN 10: 1-60840-675-X

3 4 5 6 7 8 9 10 WEB 17 16 15 14

# AUTHORS

**Ron Larson** is a professor of mathematics at Penn State Erie, The Behrend College, where he has taught since receiving his Ph.D. in mathematics from the University of Colorado. Dr. Larson is well known as the lead author of a comprehensive program for mathematics that spans middle school, high school, and college courses. His high school and Advanced Placement books are published by Houghton Mifflin Harcourt. Ron's numerous professional activities keep him in constant touch with the needs of students, teachers, and supervisors. Ron and Laurie Boswell began writing together in 1992. Since that time, they have authored over two dozen textbooks. In their collaboration, Ron is primarily responsible for the pupil edition and Laurie is primarily responsible for the teaching edition of the text.

**Laurie Boswell** is the Head of School and a mathematics teacher at the Riverside School in Lyndonville, Vermont. Dr. Boswell received her Ed.D. from the University of Vermont in 2010. She is a recipient of the Presidential Award for Excellence in Mathematics Teaching. Laurie has taught math to students at all levels, elementary through college. In addition, Laurie was a Tandy Technology Scholar, and served on the NCTM Board of Directors from 2002 to 2005. She currently serves on the board of NCSM, and is a popular national speaker. Along with Ron, Laurie has co-authored numerous math programs.

# ABOUT THE BOOK

*Big Ideas Math Algebra 1* is the final book in the *Big Ideas Math* series. *Big Ideas Math Algebra 1* uses the same research-based strategy of a balanced approach to instruction that made the *Big Ideas Math* series so successful. This approach opens doors to abstract thought, reasoning, and inquiry as students persevere to answer the Essential Questions that drive instruction. The foundation of the program is the Common Core Standards for Mathematical Content and Standards for Mathematical Practice.

The *Big Ideas Math* series exposes students to highly motivating and relevant problems. Woven throughout the series are the depth and rigor students need to prepare for Calculus and other college-level courses. The *Big Ideas Math Algebra 1* book completes the compacted and advanced pathways for middle school students.

We consider *Big Ideas Math* to be the crowning jewel of 30 years of achievement in writing educational materials.

*Ron Larson*

*Laurie Boswell*

# TEACHER REVIEWERS

- Lisa Amspacher
  Milton Hershey School
  Hershey, PA

- Mary Ballerina
  Orange County Public Schools
  Orlando, FL

- Lisa Bubello
  School District of Palm
    Beach County
  Lake Worth, FL

- Sam Coffman
  North East School District
  North East, PA

- Kristen Karbon
  Troy School District
  Rochester Hills, MI

- Laurie Mallis
  Westglades Middle School
  Coral Springs, FL

- Dave Morris
  Union City Area
    School District
  Union City, PA

- Bonnie Pendergast
  Tolleson Union High
    School District
  Tolleson, AZ

- Valerie Sullivan
  Lamoille South
    Supervisory Union
  Morrisville, VT

- Becky Walker
  Appleton Area School District
  Appleton, WI

- Zena Wiltshire
  Dade County Public Schools
  Miami, FL

# STUDENT REVIEWERS

- Mike Carter
- Matthew Cauley
- Amelia Davis
- Wisdom Dowds
- John Flatley
- Nick Ganger

- Hannah Iadeluca
- Paige Lavine
- Emma Louie
- David Nichols
- Mikala Parnell
- Jordan Pashupathi

- Stephen Piglowski
- Robby Quinn
- Michael Rawlings
- Garrett Sample
- Andrew Samuels
- Addie Sedelmyer
- Tyler Steffy
- Erin Taylor
- Reid Wilson

# CONSULTANTS

- **Patsy Davis**
  Educational Consultant
  Knoxville, Tennessee

- **Bob Fulenwider**
  Mathematics Consultant
  Bakersfield, California

- **Linda Hall**
  Mathematics Assessment Consultant
  Norman, Oklahoma

- **Ryan Keating**
  Special Education Advisor
  Gilbert, Arizona

- **Michael McDowell**
  Project-Based Instruction Specialist
  Fairfax, California

- **Sean McKeighan**
  Interdisciplinary Advisor
  Norman, Oklahoma

- **Bonnie Spence**
  Differentiated Instruction Consultant
  Missoula, Montana

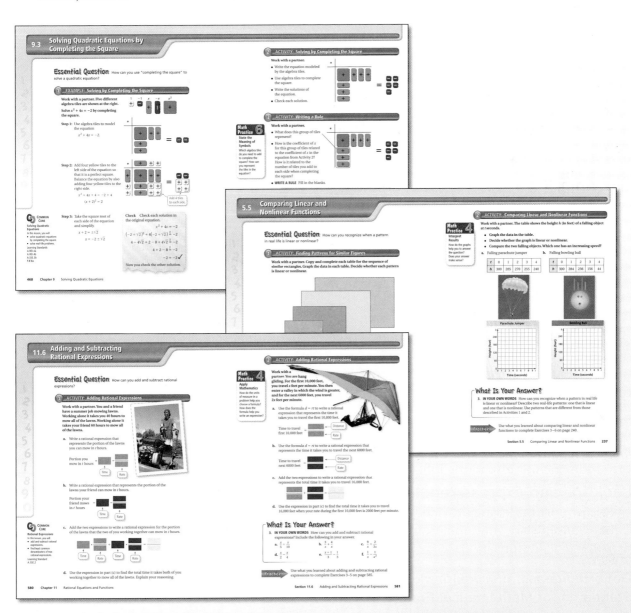

# Common Core State Standards for Mathematical Practice

**Make sense of problems and persevere in solving them.**
- Multiple representations are presented to help students move from concrete to representative and into abstract thinking
- *Essential Questions* help students focus and analyze
- *In Your Own Words* provide opportunities for students to look for meaning and entry points to a problem

**Reason abstractly and quantitatively.**
- Visual problem solving models help students create a coherent representation of the problem
- Opportunities for students to decontextualize and contextualize problems are presented in every lesson

**Construct viable arguments and critique the reasoning of others.**
- *Error Analysis*; *Different Words, Same Question*; and *Which One Doesn't Belong* features provide students the opportunity to construct arguments and critique the reasoning of others
- *Inductive Reasoning* activities help students make conjectures and build a logical progression of statements to explore their conjecture

**Model with mathematics.**
- Real-life situations are translated into diagrams, tables, equations, and graphs to help students analyze relations and to draw conclusions
- Real-life problems are provided to help students learn to apply the mathematics that they are learning to everyday life

**Use appropriate tools strategically.**
- *Graphic Organizers* support the thought process of what, when, and how to solve problems
- A variety of tool papers, such as graph paper, number lines, and manipulatives, are available as students consider how to approach a problem
- Opportunities to use the web, graphing calculators, and spreadsheets support student learning

**Attend to precision.**
- *On Your Own* questions encourage students to formulate consistent and appropriate reasoning
- Cooperative learning opportunities support precise communication

**Look for and make use of structure.**
- *Inductive Reasoning* activities provide students the opportunity to see patterns and structure in mathematics
- Real-world problems help students use the structure of mathematics to break down and solve more difficult problems

**Look for and express regularity in repeated reasoning.**
- Opportunities are provided to help students make generalizations
- Students are continually encouraged to check for reasonableness in their solutions

Go to *BigIdeasMath.com* for more information on the Common Core State Standards for Mathematical Practice.

# Common Core State Standards for Mathematical Content for Algebra 1

## Chapter Coverage for Standards

1—2—3—4—5—6—7—8—9—10—11—12

### Conceptual Category: Number and Quantity
- The Real Number System
- Quantities

1—2—3—4—5—6—7—8—9—10—11—12

### Conceptual Category: Algebra
- Seeing Structure in Expressions
- Arithmetic with Polynomials and Rational Expressions
- Creating Equations
- Reasoning with Equations and Inequalities

1—2—3—4—5—6—7—8—9—10—11—12

### Conceptual Category: Functions
- Interpreting Functions
- Building Functions
- Linear, Quadratic, and Exponential Models

1—2—3—4—5—6—7—8—9—10—11—12

### Conceptual Category: Geometry
- Geometric Measurement and Dimension

1—2—3—4—5—6—7—8—9—10—11—12

### Conceptual Category: Statistics and Probability
- Interpreting Categorical and Quantitative Data

Go to *BigIdeasMath.com* for more information on the Common Core State Standards for Mathematical Content.

# 1 Solving Linear Equations

"I like talking about math, and working with a partner allows me to do that."

| | |
|---|---|
| What You Learned Before | 1 |
| **Section 1.1 Solving Simple Equations** | |
| Activity | 2 |
| Lesson | 4 |
| **Section 1.2 Solving Multi-Step Equations** | |
| Activity | 10 |
| Lesson | 12 |
| Study Help/Graphic Organizer | 16 |
| 1.1–1.2 Quiz | 17 |
| **Section 1.3 Solving Equations with Variables on Both Sides** | |
| Activity | 18 |
| Lesson | 20 |
| Extension: Solving Absolute Value Equations | 24 |
| **Section 1.4 Rewriting Equations and Formulas** | |
| Activity | 26 |
| Lesson | 28 |
| 1.3–1.4 Quiz | 32 |
| Chapter Review | 33 |
| Chapter Test | 36 |
| Standards Assessment | 37 |

# Graphing and Writing Linear Equations

| | | |
|---|---|---|
| | What You Learned Before | 41 |
| Section 2.1 | **Graphing Linear Equations** | |
| | Activity | 42 |
| | Lesson | 44 |
| Section 2.2 | **Slope of a Line** | |
| | Activity | 48 |
| | Lesson | 50 |
| | Extension: Slopes of Parallel and Perpendicular Lines | 56 |
| Section 2.3 | **Graphing Linear Equations in Slope-Intercept Form** | |
| | Activity | 58 |
| | Lesson | 60 |
| Section 2.4 | **Graphing Linear Equations in Standard Form** | |
| | Activity | 64 |
| | Lesson | 66 |
| | Study Help/Graphic Organizer | 70 |
| | 2.1–2.4 Quiz | 71 |
| Section 2.5 | **Writing Equations in Slope-Intercept Form** | |
| | Activity | 72 |
| | Lesson | 74 |
| Section 2.6 | **Writing Equations in Point-Slope Form** | |
| | Activity | 78 |
| | Lesson | 80 |
| | Extension: Writing Equations of Parallel and Perpendicular Lines | 84 |
| Section 2.7 | **Solving Real-Life Problems** | |
| | Activity | 86 |
| | Lesson | 88 |
| | 2.5–2.7 Quiz | 92 |
| | Chapter Review | 93 |
| | Chapter Test | 98 |
| | Standards Assessment | 99 |

"*With my eBook, I get to decide when I use technology and when I use print.*"

# 3 Solving Linear Inequalities

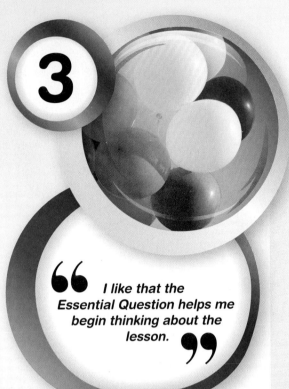

"I like that the Essential Question helps me begin thinking about the lesson."

|  | What You Learned Before | 103 |
| --- | --- | --- |
| Section 3.1 | **Writing and Graphing Inequalities** | |
|  | Activity | 104 |
|  | Lesson | 106 |
| Section 3.2 | **Solving Inequalities Using Addition or Subtraction** | |
|  | Activity | 110 |
|  | Lesson | 112 |
| Section 3.3 | **Solving Inequalities Using Multiplication or Division** | |
|  | Activity | 116 |
|  | Lesson | 118 |
|  | Study Help/Graphic Organizer | 124 |
|  | 3.1–3.3 Quiz | 125 |
| Section 3.4 | **Solving Multi-Step Inequalities** | |
|  | Activity | 126 |
|  | Lesson | 128 |
|  | Extension: Solving Compound Inequalities | 132 |
| Section 3.5 | **Graphing Linear Inequalities in Two Variables** | |
|  | Activity | 136 |
|  | Lesson | 138 |
|  | 3.4–3.5 Quiz | 144 |
|  | Chapter Review | 145 |
|  | Chapter Test | 148 |
|  | Standards Assessment | 149 |

# Solving Systems of Linear Equations

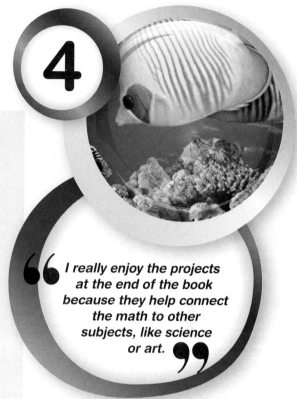

|  |  |  |
|---|---|---|
|  | What You Learned Before | 153 |
| Section 4.1 | **Solving Systems of Linear Equations by Graphing** | |
|  | Activity | 154 |
|  | Lesson | 156 |
| Section 4.2 | **Solving Systems of Linear Equations by Substitution** | |
|  | Activity | 160 |
|  | Lesson | 162 |
|  | Study Help/Graphic Organizer | 166 |
|  | 4.1–4.2 Quiz | 167 |
| Section 4.3 | **Solving Systems of Linear Equations by Elimination** | |
|  | Activity | 168 |
|  | Lesson | 170 |
| Section 4.4 | **Solving Special Systems of Linear Equations** | |
|  | Activity | 176 |
|  | Lesson | 178 |
|  | Extension: Solving Linear Equations by Graphing | 182 |
| Section 4.5 | **Systems of Linear Inequalities** | |
|  | Activity | 184 |
|  | Lesson | 186 |
|  | 4.3–4.4 Quiz | 192 |
|  | Chapter Review | 193 |
|  | Chapter Test | 196 |
|  | Standards Assessment | 197 |

"I really enjoy the projects at the end of the book because they help connect the math to other subjects, like science or art."

# 5 Linear Functions

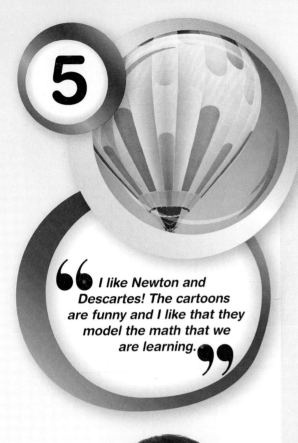

*I like Newton and Descartes! The cartoons are funny and I like that they model the math that we are learning.*

| | What You Learned Before | 201 |

**Section 5.1 Domain and Range of a Function**
- Activity ............ 202
- Lesson ............ 204
- Extension: Relations and Functions ............ 208

**Section 5.2 Discrete and Continuous Domains**
- Activity ............ 210
- Lesson ............ 212

**Section 5.3 Linear Function Patterns**
- Activity ............ 216
- Lesson ............ 218
- Study Help/Graphic Organizer ............ 222
- 5.1–5.3 Quiz ............ 223

**Section 5.4 Function Notation**
- Activity ............ 224
- Lesson ............ 226
- Extension: Special Functions ............ 232

**Section 5.5 Comparing Linear and Nonlinear Functions**
- Activity ............ 236
- Lesson ............ 238

**Section 5.6 Arithmetic Sequences**
- Activity ............ 242
- Lesson ............ 244
- 5.4–5.6 Quiz ............ 250
- Chapter Review ............ 251
- Chapter Test ............ 254
- Standards Assessment ............ 255

# Exponential Equations and Functions

## 6

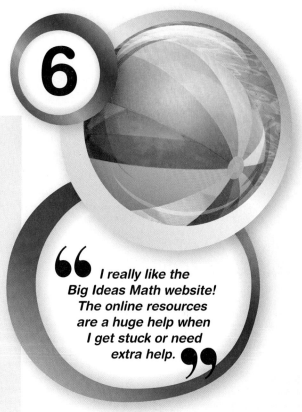

| | What You Learned Before | 259 |
|---|---|---|
| Section 6.1 | **Properties of Square Roots** | |
| | Activity | 260 |
| | Lesson | 262 |
| | Extension: Real Number Operations | 266 |
| Section 6.2 | **Properties of Exponents** | |
| | Activity | 268 |
| | Lesson | 270 |
| Section 6.3 | **Radicals and Rational Exponents** | |
| | Activity | 276 |
| | Lesson | 278 |
| | Study Help/Graphic Organizer | 282 |
| | 6.1–6.3 Quiz | 283 |
| Section 6.4 | **Exponential Functions** | |
| | Activity | 284 |
| | Lesson | 286 |
| | Extension: Solving Exponential Equations | 292 |
| Section 6.5 | **Exponential Growth** | |
| | Activity | 294 |
| | Lesson | 296 |
| Section 6.6 | **Exponential Decay** | |
| | Activity | 300 |
| | Lesson | 302 |
| Section 6.7 | **Geometric Sequences** | |
| | Activity | 306 |
| | Lesson | 308 |
| | Extension: Recursively Defined Sequences | 312 |
| | 6.4–6.7 Quiz | 316 |
| | Chapter Review | 317 |
| | Chapter Test | 322 |
| | Standards Assessment | 323 |

*"I really like the Big Ideas Math website! The online resources are a huge help when I get stuck or need extra help."*

# 7 Polynomial Equations and Factoring

> *I like the real-life application exercises because they show me how I can use the math in my own life.*

|  |  |
|---|---|
| What You Learned Before | 327 |
| **Section 7.1 Polynomials** | |
| Activity | 328 |
| Lesson | 330 |
| **Section 7.2 Adding and Subtracting Polynomials** | |
| Activity | 334 |
| Lesson | 336 |
| **Section 7.3 Multiplying Polynomials** | |
| Activity | 340 |
| Lesson | 342 |
| **Section 7.4 Special Products of Polynomials** | |
| Activity | 348 |
| Lesson | 350 |
| Study Help/Graphic Organizer | 354 |
| 7.1–7.4 Quiz | 355 |
| **Section 7.5 Solving Polynomial Equations in Factored Form** | |
| Activity | 356 |
| Lesson | 358 |
| **Section 7.6 Factoring Polynomials Using the GCF** | |
| Activity | 362 |
| Lesson | 364 |
| **Section 7.7 Factoring $x^2 + bx + c$** | |
| Activity | 368 |
| Lesson | 370 |
| **Section 7.8 Factoring $ax^2 + bx + c$** | |
| Activity | 376 |
| Lesson | 378 |
| **Section 7.9 Factoring Special Products** | |
| Activity | 382 |
| Lesson | 384 |
| Extension: Factoring Polynomials Completely | 388 |
| 7.5–7.9 Quiz | 390 |
| Chapter Review | 391 |
| Chapter Test | 396 |
| Standards Assessment | 397 |

# Graphing Quadratic Functions

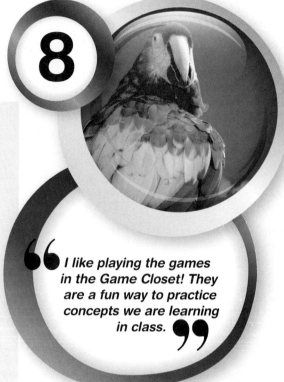

| | | |
|---|---|---|
| | What You Learned Before | 401 |
| Section 8.1 | **Graphing $y = ax^2$** | |
| | Activity | 402 |
| | Lesson | 404 |
| Section 8.2 | **Focus of a Parabola** | |
| | Activity | 410 |
| | Lesson | 412 |
| Section 8.3 | **Graphing $y = ax^2 + c$** | |
| | Activity | 416 |
| | Lesson | 418 |
| | Study Help/Graphic Organizer | 422 |
| | 8.1–8.3 Quiz | 423 |
| Section 8.4 | **Graphing $y = ax^2 + bx + c$** | |
| | Activity | 424 |
| | Lesson | 426 |
| | Extension: Graphing $y = a(x - h)^2 + k$ | 432 |
| Section 8.5 | **Comparing Linear, Exponential, and Quadratic Functions** | |
| | Activity | 434 |
| | Lesson | 436 |
| | Extension: Comparing Graphs of Functions | 442 |
| | 8.4–8.5 Quiz | 444 |
| | Chapter Review | 445 |
| | Chapter Test | 448 |
| | Standards Assessment | 449 |

"I like playing the games in the Game Closet! They are a fun way to practice concepts we are learning in class."

# 9 Solving Quadratic Equations

"With the BigIdeasMath.com website I don't have to worry if I forget my book or my workbook at school."

What You Learned Before .................. 453

**Section 9.1 Solving Quadratic Equations by Graphing**
Activity .................. 454
Lesson .................. 456

**Section 9.2 Solving Quadratic Equations Using Square Roots**
Activity .................. 462
Lesson .................. 464

**Section 9.3 Solving Quadratic Equations by Completing the Square**
Activity .................. 468
Lesson .................. 470
Study Help/Graphic Organizer .................. 474
9.1–9.3 Quiz .................. 475

**Section 9.4 Solving Quadratic Equations Using the Quadratic Formula**
Activity .................. 476
Lesson .................. 478
Extension: Choosing a Solution Method .................. 484

**Section 9.5 Solving Systems of Linear and Quadratic Equations**
Activity .................. 486
Lesson .................. 488
9.4–9.5 Quiz .................. 492
Chapter Review .................. 493
Chapter Test .................. 496
Standards Assessment .................. 497

# Square Root Functions and Geometry

## 10

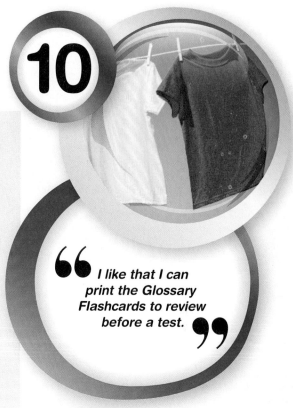

> *I like that I can print the Glossary Flashcards to review before a test.*

|  |  |  |
|---|---|---|
|  | What You Learned Before | 501 |
| Section 10.1 | **Graphing Square Root Functions** | |
|  | Activity | 502 |
|  | Lesson | 504 |
|  | Extension: Rationalizing the Denominator | 508 |
| Section 10.2 | **Solving Square Root Equations** | |
|  | Activity | 510 |
|  | Lesson | 512 |
|  | Study Help/Graphic Organizer | 518 |
|  | 10.1–10.2 Quiz | 519 |
| Section 10.3 | **The Pythagorean Theorem** | |
|  | Activity | 520 |
|  | Lesson | 522 |
| Section 10.4 | **Using the Pythagorean Theorem** | |
|  | Activity | 526 |
|  | Lesson | 528 |
|  | 10.3–10.4 Quiz | 532 |
|  | Chapter Review | 533 |
|  | Chapter Test | 536 |
|  | Standards Assessment | 537 |

# Rational Equations and Functions

|  |  |
|---|---|
|  | What You Learned Before ............... 541 |
| Section 11.1 | **Direct and Inverse Variation** |
|  | Activity ............................................. 542 |
|  | Lesson ............................................... 544 |
| Section 11.2 | **Graphing Rational Functions** |
|  | Activity ............................................. 550 |
|  | Lesson ............................................... 552 |
|  | Extension: Inverse of a Function .... 558 |
| Section 11.3 | **Simplifying Rational Expressions** |
|  | Activity ............................................. 560 |
|  | Lesson ............................................... 562 |
|  | Study Help/Graphic Organizer ....... 566 |
|  | 11.1–11.3 Quiz ................................. 567 |
| Section 11.4 | **Multiplying and Dividing Rational Expressions** |
|  | Activity ............................................. 568 |
|  | Lesson ............................................... 570 |
| Section 11.5 | **Dividing Polynomials** |
|  | Activity ............................................. 574 |
|  | Lesson ............................................... 576 |
| Section 11.6 | **Adding and Subtracting Rational Expressions** |
|  | Activity ............................................. 580 |
|  | Lesson ............................................... 582 |
| Section 11.7 | **Solving Rational Equations** |
|  | Activity ............................................. 588 |
|  | Lesson ............................................... 590 |
|  | 11.4–11.7 Quiz ................................. 594 |
|  | Chapter Review ................................ 595 |
|  | Chapter Test .................................... 600 |
|  | Standards Assessment .................... 601 |

"Before my school had Big Ideas Math I would always lose test points because I left units off my answers. Now I see why they are so important."

# Data Analysis and Displays

|  |  |  |
|---|---|---|
|  | What You Learned Before | 605 |
| Section 12.1 | **Measures of Central Tendency** |  |
|  | Activity | 606 |
|  | Lesson | 608 |
| Section 12.2 | **Measures of Dispersion** |  |
|  | Activity | 612 |
|  | Lesson | 614 |
| Section 12.3 | **Box-and-Whisker Plots** |  |
|  | Activity | 618 |
|  | Lesson | 620 |
| Section 12.4 | **Shapes of Distributions** |  |
|  | Activity | 626 |
|  | Lesson | 628 |
|  | Study Help/Graphic Organizer | 634 |
|  | 12.1–12.4 Quiz | 635 |
| Section 12.5 | **Scatter Plots and Lines of Fit** |  |
|  | Activity | 636 |
|  | Lesson | 638 |
| Section 12.6 | **Analyzing Lines of Fit** |  |
|  | Activity | 644 |
|  | Lesson | 646 |
| Section 12.7 | **Two-Way Tables** |  |
|  | Activity | 652 |
|  | Lesson | 654 |
| Section 12.8 | **Choosing a Data Display** |  |
|  | Activity | 658 |
|  | Lesson | 660 |
|  | 12.5–12.8 Quiz | 664 |
|  | Chapter Review | 665 |
|  | Chapter Test | 670 |
|  | Standards Assessment | 671 |

> "Word problems used to confuse me. Now I understand how to look for patterns and what the question is asking!"

## Appendix A: My Big Ideas Projects

| Section A.1 | **Literature Project** | A2 |
|---|---|---|
| Section A.2 | **History Project** | A4 |
| Section A.3 | **Art Project** | A6 |
| Section A.4 | **Science Project** | A8 |

| **Selected Answers** | A10 |
|---|---|
| **Key Vocabulary Index** | A65 |
| **Student Index** | A66 |
| **Common Core State Standards** | A82 |
| **Mathematics Reference Sheet** | B1 |

# How to Use Your Math Book

- Read the **Essential Question** in the activity.

  Discuss the **Math Practice** question with your partner.

  Work with a partner to decide **What Is Your Answer?**

  Now you are ready to do the **Practice** problems.

- Find the **Key Vocabulary** words, **highlighted in yellow**.

  Read their definitions. Study the concepts in each **Key Idea**.

  If you forget a definition, you can look it up online in the

  **Multi-Language Glossary at BigIdeasMath.com**.

- After you study each **EXAMPLE**, do the exercises in the **On Your Own**.

  **Now You're Ready** to do the exercises that correspond to the example.

  As you study, look for a **Study Tip** or a **Common Error**.

- The exercises are divided into 3 parts.

  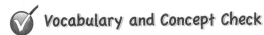 Vocabulary and Concept Check

   Practice and Problem Solving

   Fair Game Review

  If an exercise has a ① next to it, look back at Example 1 for help with that exercise.

  More help is available at **Check It Out Lesson Tutorials BigIdeasMath.com**.

- To help study for your test, use the following.

  **Quiz**  **Study Help**

  **Chapter Review**  **Chapter Test**

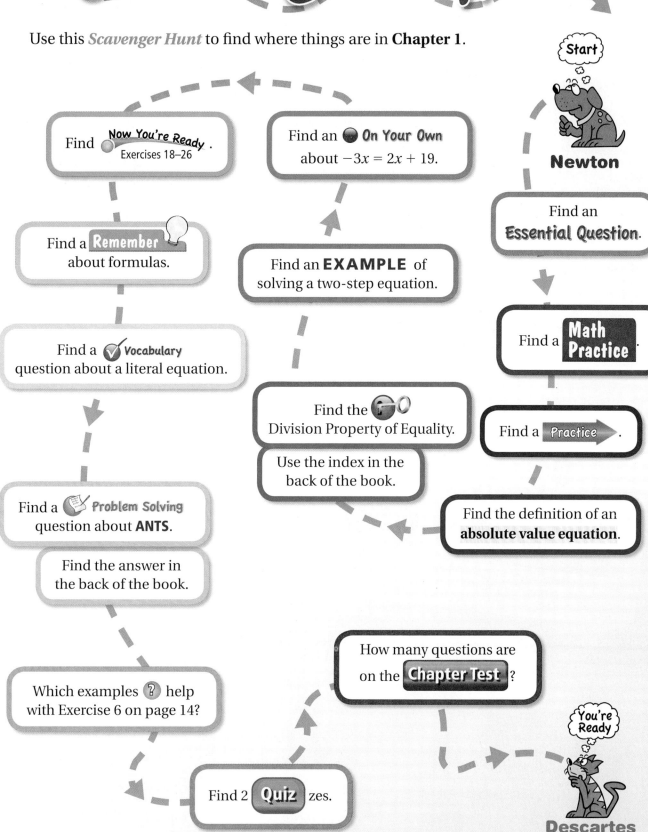

# 1 Solving Linear Equations

- **1.1** Solving Simple Equations
- **1.2** Solving Multi-Step Equations
- **1.3** Solving Equations with Variables on Both Sides
- **1.4** Rewriting Equations and Formulas

"Dear Sir: Here is my suggestion for a good math problem."

"A box contains a total of 30 dog and cat treats. There are 5 times more dog treats than cat treats."

"How many of each type of treat are there?"

"Push faster, Descartes! According to the formula $R = D \div T$, the time needs to be 10 minutes or less to break our all-time speed record!"

# What You Learned Before

● **Adding and Subtracting Integers** (7.NS.1d)

**Example 1** Find $4 + (-12)$.

$$4 + (-12) = -8$$

**Example 2** Find $-7 - (-16)$.

$$-7 - (-16) = -7 + 16 \quad \text{Add the opposite of } -16.$$
$$= 9 \quad \text{Add.}$$

### Try It Yourself
**Add or subtract.**

1. $-5 + (-2)$
2. $0 + (-13)$
3. $-6 + 14$
4. $19 - (-13)$
5. $-1 - 6$
6. $-5 - (-7)$

● **Multiplying and Dividing Integers** (7.NS.2c)

**Example 3** Find $-3 \cdot (-5)$.

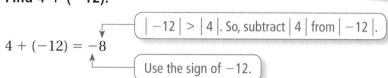

$$-3 \cdot (-5) = 15$$

**Example 4** Find $15 \div (-3)$.

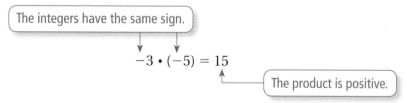

$$15 \div (-3) = -5$$

### Try It Yourself
**Multiply or divide.**

7. $-3(8)$
8. $-7 \cdot (-9)$
9. $4 \cdot (-7)$
10. $-24 \div (-6)$
11. $-16 \div 2$
12. $12 \div (-3)$

# 1.1 Solving Simple Equations

**Essential Question** How can you use inductive reasoning to discover rules in mathematics? How can you test a rule?

### 1 ACTIVITY: Sum of the Angles of a Triangle

Work with a partner. Copy the triangles. Use a protractor to measure the angles of each triangle. Copy and complete the table to organize your results.

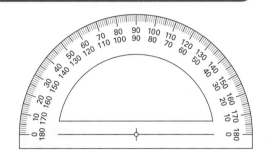

a.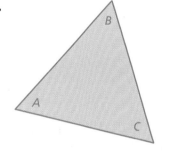

b.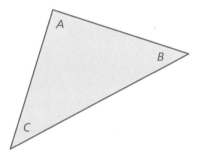

c.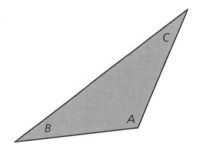

d.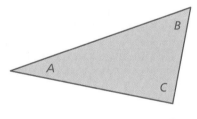

**COMMON CORE**

**Solving Equations**
In this lesson, you will
- write and solve simple equations.
- solve real-life problems.

Learning Standards
A.CED.1
A.REI.1
A.REI.3

| Triangle | Angle A (degrees) | Angle B (degrees) | Angle C (degrees) | A + B + C |
|---|---|---|---|---|
| a. | | | | |
| b. | | | | |
| c. | | | | |
| d. | | | | |

## Math Practice 3

**Construct Arguments**

How can you use results from the previous activity to write a rule?

### 2  ACTIVITY: Writing a Rule

**Work with a partner. Use inductive reasoning to write and test a rule.**

a. Use the completed table in Activity 1 to write a rule about the sum of the angle measures of a triangle.

b. **TEST YOUR RULE** Draw four triangles that are different from those in Activity 1. Measure the angles of each triangle. Organize your results in a table. Find the sum of the angle measures of each triangle.

### 3  ACTIVITY: Applying Your Rule

**Work with a partner. Use the rule you wrote in Activity 2 to write an equation for each triangle. Then, solve the equation to find the value of $x$. Use a protractor to check the reasonableness of your answer.**

a.
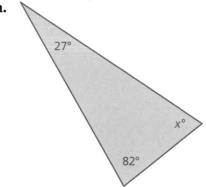

b.
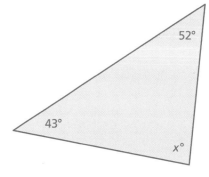

c.
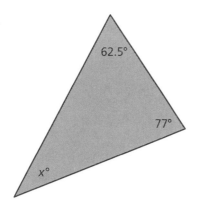

d.
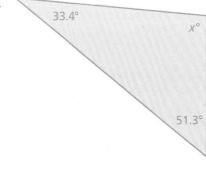

## What Is Your Answer?

4. **IN YOUR OWN WORDS** How can you use inductive reasoning to discover rules in mathematics? How can you test a rule? How can you use a rule to solve problems in mathematics?

**Practice**

Use what you learned about solving simple equations to complete Exercises 4–6 on page 7.

Section 1.1   Solving Simple Equations   3

# 1.1 Lesson

## Key Ideas

**Remember**
Addition and subtraction are inverse operations.

**Addition Property of Equality**

**Words** Adding the same number to each side of an equation produces an equivalent equation.

**Algebra** If $a = b$, then $a + c = b + c$.

**Subtraction Property of Equality**

**Words** Subtracting the same number from each side of an equation produces an equivalent equation.

**Algebra** If $a = b$, then $a - c = b - c$.

### EXAMPLE 1  Solving Equations Using Addition or Subtraction

a. Solve $x - 7 = -6$.

$$\begin{aligned} x - 7 &= -6 &&\text{Write the equation.} \\ +7 &\phantom{=}+7 &&\text{Add 7 to each side.} \\ x &= 1 &&\text{Simplify.} \end{aligned}$$

Undo the subtraction.

∴ The solution is $x = 1$.

**Check**
$x - 7 = -6$
$1 - 7 \stackrel{?}{=} -6$
$-6 = -6$ ✓

b. Solve $y + 3.4 = 0.5$.

$$\begin{aligned} y + 3.4 &= 0.5 &&\text{Write the equation.} \\ -3.4 &\phantom{=}-3.4 &&\text{Subtract 3.4 from each side.} \\ y &= -2.9 &&\text{Simplify.} \end{aligned}$$

Undo the addition.

∴ The solution is $y = -2.9$.

**Check**
$y + 3.4 = 0.5$
$-2.9 + 3.4 \stackrel{?}{=} 0.5$
$0.5 = 0.5$ ✓

c. Solve $h + 2\pi = 3\pi$.

$$\begin{aligned} h + 2\pi &= 3\pi &&\text{Write the equation.} \\ -2\pi &\phantom{=}-2\pi &&\text{Subtract } 2\pi \text{ from each side.} \\ h &= \pi &&\text{Simplify.} \end{aligned}$$

Undo the addition.

∴ The solution is $h = \pi$.

### On Your Own

**Solve the equation. Check your solution.**

1. $b + 2 = -5$
2. $g - 1.7 = -0.9$
3. $-3 = k + 3$
4. $r - \pi = \pi$
5. $t - \dfrac{1}{4} = -\dfrac{3}{4}$
6. $5.6 + z = -8$

## Key Ideas

**Multiplication Property of Equality**

**Words** Multiplying each side of an equation by the same number produces an equivalent equation.

**Algebra** If $a = b$, then $a \cdot c = b \cdot c$.

**Division Property of Equality**

**Words** Dividing each side of an equation by the same number produces an equivalent equation.

**Algebra** If $a = b$, then $a \div c = b \div c, c \neq 0$.

**Remember**
Multiplication and division are inverse operations.

### EXAMPLE 2 — Solving Equations Using Multiplication or Division

a. Solve $-\dfrac{3}{4}n = -2$.

$$-\dfrac{3}{4}n = -2 \quad \text{Write the equation.}$$

$$-\dfrac{4}{3} \cdot \left(-\dfrac{3}{4}n\right) = -\dfrac{4}{3} \cdot (-2) \quad \text{Multiply each side by } -\dfrac{4}{3}, \text{ the reciprocal of } -\dfrac{3}{4}.$$

$$n = \dfrac{8}{3} \quad \text{Simplify.}$$

*Use the reciprocal.*

∴ The solution is $n = \dfrac{8}{3}$.

b. Solve $\pi x = 3\pi$.

$$\pi x = 3\pi \quad \text{Write the equation.}$$

$$\dfrac{\pi x}{\pi} = \dfrac{3\pi}{\pi} \quad \text{Divide each side by } \pi.$$

$$x = 3 \quad \text{Simplify.}$$

*Undo the multiplication.*

∴ The solution is $x = 3$.

**Check**
$\pi x = 3\pi$
$\pi(3) \stackrel{?}{=} 3\pi$
$3\pi = 3\pi$ ✓

### On Your Own

**Solve the equation. Check your solution.**

7. $\dfrac{y}{4} = -7$
8. $6\pi = \pi x$
9. $0.09w = 1.8$

### EXAMPLE 3  Solving an Equation

What value of $k$ makes the equation $k + 4 \div 0.2 = 5$ true?

  Ⓐ $-15$  Ⓑ $-5$  Ⓒ $-3$  Ⓓ $1.5$

$$k + 4 \div 0.2 = 5 \quad \text{Write the equation.}$$
$$k + 20 = 5 \quad \text{Divide 4 by 0.2.}$$
$$\underline{-20 \quad -20} \quad \text{Subtract 20 from each side.}$$
$$k = -15 \quad \text{Simplify.}$$

∴ The correct answer is Ⓐ.

### EXAMPLE 4  Real-Life Application

The melting point of bromine is $-7°C$.

The *melting point* of a solid is the temperature at which the solid becomes a liquid. The melting point of bromine is $\frac{1}{30}$ of the melting point of nitrogen. Write and solve an equation to find the melting point of nitrogen.

**Words**  The melting point of bromine   is  $\frac{1}{30}$  of  the melting point of nitrogen.

**Variable**  Let $n$ be the melting point of nitrogen.

**Equation**  $-7 = \frac{1}{30} \cdot n$

$$-7 = \frac{1}{30}n \quad \text{Write the equation.}$$
$$30 \cdot (-7) = 30 \cdot \left(\frac{1}{30}n\right) \quad \text{Multiply each side by 30.}$$
$$-210 = n \quad \text{Simplify.}$$

∴ The melting point of nitrogen is $-210°C$.

### On Your Own

Exercises 33–38

10. Solve $p - 8 \div \frac{1}{2} = -3$.

11. Solve $q + |-10| = 2$.

12. The melting point of mercury is about $\frac{1}{4}$ of the melting point of krypton. The melting point of mercury is $-39°C$. Write and solve an equation to find the melting point of krypton.

## 1.1 Exercises

### Vocabulary and Concept Check

1. **VOCABULARY** Which of the operations +, −, ×, and ÷ are inverses of each other?

2. **VOCABULARY** Are the equations $3x = -9$ and $4x = -12$ equivalent? Explain.

3. **WHICH ONE DOESN'T BELONG?** Which equation does *not* belong with the other three? Explain your reasoning.

   | $x - 2 = 4$ | $x - 3 = 6$ | $x - 5 = 1$ | $x - 6 = 0$ |

### Practice and Problem Solving

**CHOOSE TOOLS** Find the value of *x*. Check the reasonableness of your answer.

4.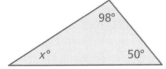

5.

6.

**Solve the equation. Check your solution.**

7. $x + 12 = 7$

8. $g - 16 = 8$

9. $-9 + p = 12$

10. $0.7 + y = -1.34$

11. $x - 8\pi = \pi$

12. $4\pi = w - 6\pi$

13. $\dfrac{5}{6} = \dfrac{1}{3} + d$

14. $\dfrac{3}{8} = r + \dfrac{2}{3}$

15. $n - 1.4 = -6.3$

16. **CONCERT** A discounted concert ticket is $14.50 less than the original price *p*. You pay $53 for a discounted ticket. Write and solve an equation to find the original price.

17. **BOWLING** Your friend's final bowling score is 105. Your final bowling score is 14 pins less than your friend's final score.

   a. Write and solve an equation to find your final score.

   b. Your friend made a spare in the tenth frame. Did you? Explain.

Section 1.1  Solving Simple Equations  7

**Solve the equation. Check your solution.**

② 18. $7x = 35$
19. $4 = -0.8n$
20. $6 = -\dfrac{w}{8}$

21. $\dfrac{m}{\pi} = 7.3$
22. $-4.3g = 25.8$
23. $\dfrac{3}{2} = \dfrac{9}{10}k$

24. $-7.8x = -1.56$
25. $-2 = \dfrac{6}{7}p$
26. $3\pi d = 12\pi$

27. **ERROR ANALYSIS** Describe and correct the error in solving the equation.

$$-1.5 + k = 8.2$$
$$k = 8.2 + (-1.5)$$
$$k = 6.7$$

28. **TENNIS** A gym teacher orders 42 tennis balls. Each package contains 3 tennis balls. Which of the following equations represents the number $x$ of packages?

$x + 3 = 42$  $\quad$  $3x = 42$  $\quad$  $\dfrac{x}{3} = 42$  $\quad$  $x = \dfrac{3}{42}$

**MODELING** In Exercises 29–32, write and solve an equation to answer the question.

29. **PARK** You clean a community park for 6.5 hours. You earn $42.25. How much do you earn per hour?

30. **ROCKET LAUNCH** A rocket is scheduled to launch from a command center in 3.75 hours. What time is it now?

Launch Time 11:20 A.M.

31. **BANKING** After earning interest, the balance of an account is $420. The new balance is $\dfrac{7}{6}$ of the original balance. How much interest was earned?

**Tallest Coasters at Cedar Point**

| Roller Coaster | Height (feet) |
| --- | --- |
| Top Thrill Dragster | 420 |
| Millennium Force | 310 |
| Magnum XL-200 | 205 |
| Mantis | ? |

32. **ROLLER COASTER** Cedar Point amusement park has some of the tallest roller coasters in the United States. The Mantis is 165 feet shorter than the Millennium Force. What is the height of the Mantis?

**Solve the equation. Check your solution.**

**33.** $-3 = h + 8 \div 2$

**34.** $12 = w - |-7|$

**35.** $q + |6.4| = 9.6$

**36.** $d - 2.8 \div 0.2 = -14$

**37.** $\dfrac{8}{9} = x + \dfrac{1}{3}(7)$

**38.** $p - \dfrac{1}{4} \cdot 3 = -\dfrac{5}{6}$

**39. LOGIC** Without solving, is the solution of $-2x = -15$ *greater than* or *less than* $-15$? Explain.

**40. OPEN-ENDED** Write a subtraction equation and a division equation that each has a solution of $-2$.

**41. ANTS** Some ant species can carry 50 times their body weight. It takes 32 ants to carry the cherry. About how much does each ant weigh?

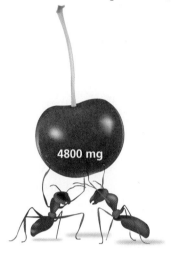

**42. REASONING** One-fourth of the girls and one-eighth of the boys in a class retake their school pictures. The photographer retakes pictures for 16 girls and 7 boys. How many students are in the class?

**43. VOLUME** The volume $V$ of the cylinder is $72\pi$ cubic inches. Use the formula $V = Bh$ to find the height $h$ of the cylinder.

**44.**  A neighbor pays you and two friends $90 to paint her garage. The money is divided three ways in the ratio $2:3:5$.

  **a.** How much does each person receive?
  **b.** What is one possible reason the money is not divided evenly?

## Fair Game Review  What you learned in previous grades & lessons

**Simplify the expression.**  *(Skills Review Handbook)*

**45.** $2(x - 2) + 5x$

**46.** $0.4b - 3.2 + 1.2b$

**47.** $\dfrac{1}{4}g + 6g - \dfrac{2}{3}$

**48. MULTIPLE CHOICE** The temperature at 4 P.M. was $-12\,°C$. By 11 P.M. the temperature had dropped 14 degrees. What was the temperature at 11 P.M.? *(Skills Review Handbook)*

  **Ⓐ** $-26\,°C$    **Ⓑ** $-2\,°C$    **Ⓒ** $2\,°C$    **Ⓓ** $26\,°C$

# 1.2 Solving Multi-Step Equations

**Essential Question** How can you solve a multi-step equation? How can you check the reasonableness of your solution?

### 1 ACTIVITY: Solving for the Angles of a Triangle

Work with a partner. Write an equation for each triangle. Solve the equation to find the value of the variable. Then find the angle measures of each triangle. Use a protractor to check the reasonableness of your answer.

a.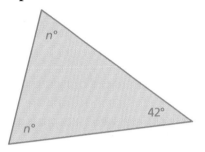

b.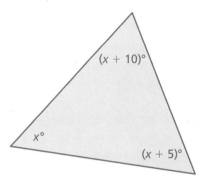

c.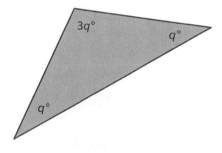

d.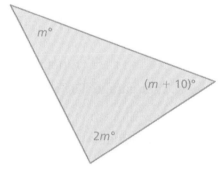

**COMMON CORE**

**Solving Equations**

In this lesson, you will
- write and solve multi-step equations.
- solve real-life problems.

Learning Standards
A.CED.1
A.REI.1
A.REI.3

e.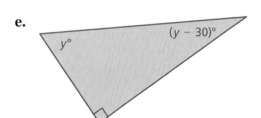

f.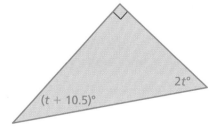

10   Chapter 1   Solving Linear Equations

### 2 ACTIVITY: Problem-Solving Strategy

**Math Practice**

**Consider Similar Problems**
What do you know about triangles? How can you use the methods from the previous activity to help you find angle measures?

Work with a partner.

The six triangles form a rectangle.

Find the angle measures of each triangle. Use a protractor to check the reasonableness of your answers.

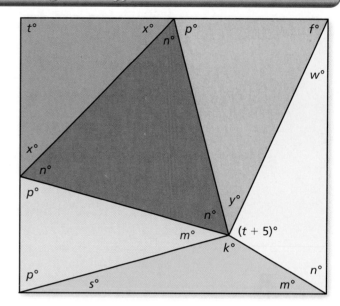

### 3 ACTIVITY: Puzzle

Work with a partner. A survey asked 200 people to name their favorite weekday. The results are shown in the circle graph.

a. How many degrees are in each part of the circle graph?
b. What percent of the people chose each day?
c. How many people chose each day?
d. Organize your results in a table.

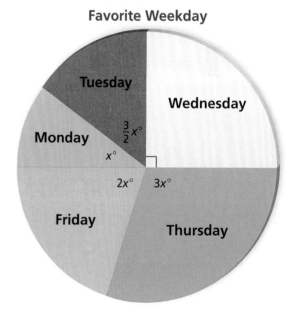

Favorite Weekday

## What Is Your Answer?

4. **IN YOUR OWN WORDS** How can you solve a multi-step equation? How can you check the reasonableness of your solution?

**Practice**

Use what you learned about solving multi-step equations to complete Exercises 3–5 on page 14.

Section 1.2  Solving Multi-Step Equations  11

## 1.2 Lesson

### Key Idea

**Solving Multi-Step Equations**

To solve multi-step equations, use inverse operations to isolate the variable.

**EXAMPLE 1** **Solving a Two-Step Equation**

The height (in feet) of a tree after $x$ years is $1.5x + 15$. After how many years is the tree 24 feet tall?

| | | |
|---|---|---|
| | $1.5x + 15 = 24$ | Write an equation. |
| Undo the addition. → | $\underline{-15 \quad -15}$ | Subtract 15 from each side. |
| | $1.5x = 9$ | Simplify. |
| Undo the multiplication. → | $\dfrac{1.5x}{1.5} = \dfrac{9}{1.5}$ | Divide each side by 1.5. |
| | $x = 6$ | Simplify. |

∴ The tree is 24 feet tall after 6 years.

**EXAMPLE 2** **Combining Like Terms to Solve an Equation**

Solve $8x - 6x - 25 = -35$.

| | | |
|---|---|---|
| | $8x - 6x - 25 = -35$ | Write the equation. |
| | $2x - 25 = -35$ | Combine like terms. |
| Undo the subtraction. → | $\underline{+25 \quad +25}$ | Add 25 to each side. |
| | $2x = -10$ | Simplify. |
| Undo the multiplication. → | $\dfrac{2x}{2} = \dfrac{-10}{2}$ | Divide each side by 2. |
| | $x = -5$ | Simplify. |

∴ The solution is $x = -5$.

### On Your Own

**Solve the equation. Check your solution.**

Exercises 6–9

1. $-3z + 1 = 7$
2. $\dfrac{1}{2}x - 9 = -25$
3. $-4n - 8n + 17 = 23$

## EXAMPLE 3 — Using the Distributive Property to Solve an Equation

Solve $2(1 - 5x) + 4 = -8$.

$2(1 - 5x) + 4 = -8$    Write the equation.
$2(1) - 2(5x) + 4 = -8$    Use Distributive Property.
$2 - 10x + 4 = -8$    Multiply.
$-10x + 6 = -8$    Combine like terms.
$\underline{\phantom{-10x} - 6 \phantom{=} - 6}$    Subtract 6 from each side.
$-10x = -14$    Simplify.
$\dfrac{-10x}{-10} = \dfrac{-14}{-10}$    Divide each side by $-10$.
$x = 1.4$    Simplify.

**Study Tip**

Here is another way to solve the equation in Example 3.

$2(1 - 5x) + 4 = -8$
$2(1 - 5x) = -12$
$1 - 5x = -6$
$-5x = -7$
$x = 1.4$

## EXAMPLE 4 — Real-Life Application

Use the table to find the number of miles $x$ you need to run on Friday so that the mean number of miles run per day is 1.5.

| Day | Miles |
|---|---|
| Monday | 2 |
| Tuesday | 0 |
| Wednesday | 1.5 |
| Thursday | 0 |
| Friday | $x$ |

Write an equation using the definition of mean.

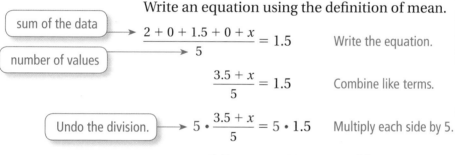

sum of the data → $\dfrac{2 + 0 + 1.5 + 0 + x}{5} = 1.5$    Write the equation.
number of values →

$\dfrac{3.5 + x}{5} = 1.5$    Combine like terms.

Undo the division. → $5 \cdot \dfrac{3.5 + x}{5} = 5 \cdot 1.5$    Multiply each side by 5.

$3.5 + x = 7.5$    Simplify.

Undo the addition. → $\underline{\phantom{3.5 + x =} - 3.5 \phantom{=} - 3.5}$    Subtract 3.5 from each side.

$x = 4$    Simplify.

∴ You need to run 4 miles on Friday.

### On Your Own

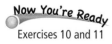

Exercises 10 and 11

Solve the equation. Check your solution.

4. $-3(x + 2) + 5x = -9$
5. $5 + 1.5(2d - 1) = 0.5$

6. You scored 88, 92, and 87 on three tests. Write and solve an equation to find the score you need on the fourth test so that your mean test score is 90.

# 1.2 Exercises

## Vocabulary and Concept Check

1. **WRITING** Write the verbal statement as an equation. Then solve.

   2 more than 3 times a number is 17.

2. **OPEN-ENDED** Explain how to solve the equation $2(4x - 11) + 9 = 19$.

## Practice and Problem Solving

**CHOOSE TOOLS** Find the value of the variable. Then find the angle measures of the polygon. Check the reasonableness of your answer.

3.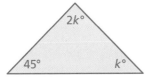

   Sum of angle measures: 180°

4.

   Sum of angle measures: 360°

5.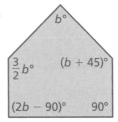

   Sum of angle measures: 540°

**Solve the equation. Check your solution.**

6. $10x + 2 = 32$

7. $19 - 4c = 17$

8. $1.1x + 1.2x - 5.4 = -10$

9. $\frac{2}{3}h - \frac{1}{3}h + 11 = 8$

10. $6(5 - 8v) + 12 = -54$

11. $21(2 - x) + 12x = 44$

12. **ERROR ANALYSIS** Describe and correct the error in solving the equation.

    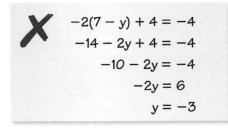

13. **WATCHES** The cost $C$ (in dollars) of making $n$ watches is represented by $C = 15n + 85$. How many watches are made when the cost is $385?

14. **HOUSE** The height of the house is 26 feet. What is the height $x$ of each story?

14  Chapter 1  Solving Linear Equations

**In Exercises 15–17, write and solve an equation to answer the question.**

15. **POSTCARD** The area of the postcard is 24 square inches. What is the width $b$ of the message (in inches)?

16. **BREAKFAST** You order two servings of pancakes and a fruit cup. The cost of the fruit cup is $1.50. You leave a 15% tip. Your total bill is $11.50. How much does one serving of pancakes cost?

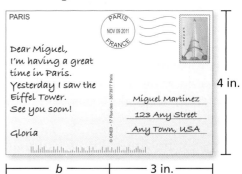

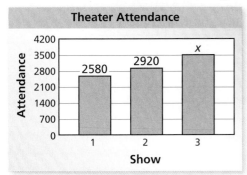

17. **PROBLEM SOLVING** How many people must attend the third show so that the average attendance for the three shows is 3000?

**The letters $a$, $b$, and $c$ represent constants. Solve the equation for $x$.**

18. $x + a = \dfrac{3}{4}$

19. $bx = -7$

20. $2bx - bx = -8$

21. $4cx - b = 5b$

22. $ax - b = 12.5$

23. $ax + b = c$

24. **DIVING** Divers in a competition are scored by an international panel of judges. The highest and lowest scores are dropped. The total of the remaining scores is multiplied by the degree of difficulty of the dive. This product is multiplied by 0.6 to determine the final score.

   a. A diver's final score is 77.7. What is the degree of difficulty of the dive?

| Judge | Russia | China | Mexico | Germany | Italy | Japan | Brazil |
|---|---|---|---|---|---|---|---|
| Score | 7.5 | 8.0 | 6.5 | 8.5 | 7.0 | 7.5 | 7.0 |

   b. **Critical Thinking** The degree of difficulty of a dive is 4.0. The diver's final score is 97.2. Judges award half or whole points from 0 to 10. What scores could the judges have given the diver?

**Fair Game Review** What you learned in previous grades & lessons

**Let $a = 3$ and $b = -2$. Copy and complete the statement using <, >, or =.**
*(Skills Review Handbook)*

25. $-5a \quad \underline{\phantom{xx}} \quad 4$

26. $5 \quad \underline{\phantom{xx}} \quad b + 7$

27. $a - 4 \quad \underline{\phantom{xx}} \quad 10b + 8$

28. **MULTIPLE CHOICE** What value of $x$ makes the equation $x + 5 = 2x$ true?
   *(Skills Review Handbook)*

   Ⓐ $-1$   Ⓑ $0$   Ⓒ $3$   Ⓓ $5$

# 1 Study Help

You can use a **Y chart** to compare two topics. List differences in the branches and similarities in the base of the Y. Here is an example of a Y chart that compares solving simple equations using addition to solving simple equations using subtraction.

**Solving Simple Equations Using Addition**
- Add the same number to each side of the equation.

**Solving Simple Equations Using Subtraction**
- Subtract the same number from each side of the equation.

- You can solve the equation in one step.
- You produce an equivalent equation.
- The variable can be on either side of the equation.
- It is always a good idea to check your solution.

## On Your Own

**Make Y charts to help you study and compare these topics.**

1. solving simple equations using multiplication and solving simple equations using division

2. solving simple equations and solving multi-step equations

**After you complete this chapter, make Y charts for the following topics.**

3. solving equations with the variable on one side and solving equations with variables on both sides

4. solving multi-step equations and solving equations with variables on both sides

5. solving multi-step equations and rewriting literal equations

6. solving multi-step equations and solving absolute value equations

"I made a Y chart to compare and contrast Fluffy's characteristics with yours."

16   Chapter 1   Solving Linear Equations

## 1.1–1.2 Quiz

**Solve the equation. Check your solution.** *(Section 1.1)*

1. $-\dfrac{1}{2} = y - 1$

2. $-3\pi + w = 2\pi$

3. $1.2m = 0.6$

4. $q + 2.7 = -0.9$

**Solve the equation. Check your solution.** *(Section 1.2)*

5. $-4k + 17 = 1$

6. $\dfrac{1}{4}z + 8 = 12$

7. $-3(2n + 1) + 7 = -5$

8. $2.5(t - 2) - 6 = 9$

**Find the value of $x$. Then find the angle measures of the polygon.** *(Section 1.2)*

9.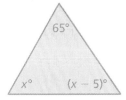

Sum of angle measures: 180°

10.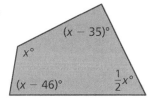

Sum of angle measures: 360°

11. **JEWELER** The equation $P = 2.5m + 35$ represents the price $P$ (in dollars) of a bracelet, where $m$ is the cost of the materials (in dollars). The price of a bracelet is $115. What is the cost of the materials? *(Section 1.2)*

12. **PASTURE** A 455-foot fence encloses a pasture. What is the length of each side of the pasture? *(Section 1.2)*

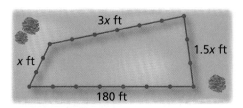

13. **POSTERS** A machine prints 230 movie posters each hour. Write and solve an equation to find the number of hours it takes the machine to print 1265 posters. *(Section 1.1)*

14. **BASKETBALL** Use the table to write and solve an equation to find the number of points $p$ you need to score in the fourth game so that the mean number of points is 20? *(Section 1.2)*

| Game | Points |
|---|---|
| 1 | 25 |
| 2 | 15 |
| 3 | 18 |
| 4 | $p$ |

# 1.3 Solving Equations with Variables on Both Sides

**Essential Question** How can you solve an equation that has variables on both sides?

### 1  ACTIVITY: Perimeter and Area

Work with a partner. Each figure has the unusual property that the value of its perimeter (in feet) is equal to the value of its area (in square feet).

- Write an equation (value of perimeter = value of area) for each figure.
- Solve each equation for $x$.
- Use the value of $x$ to find the perimeter and area of each figure.
- Check your solution by comparing the value of the perimeter and the value of the area of each figure.

a.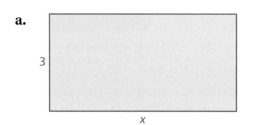

b.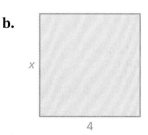

c.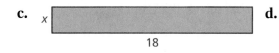

d.

e.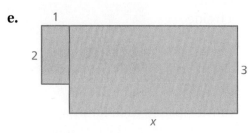

f.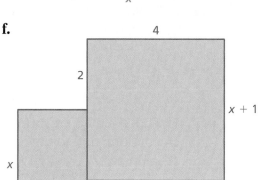

g.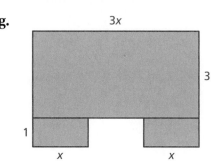

**COMMON CORE**

**Solving Equations**

In this lesson, you will
- write and solve equations with variables on both sides of the equal sign.
- solve real-life problems.

Learning Standards
A.CED.1
A.REI.1
A.REI.3

## 2  ACTIVITY: Surface Area and Volume

**Math Practice**

**Understand Quantities**
What is the value of each quantity? What does each quantity represent?

Work with a partner. Each solid has the unusual property that the value of its surface area (in square inches) is equal to the value of its volume (in cubic inches).

- Write an equation (value of surface area = value of volume) for each solid.
- Solve each equation for $x$.
- Use the value of $x$ to find the surface area and volume of each solid.
- Check your solution by comparing the value of the surface area and the value of the volume of each solid.

a.
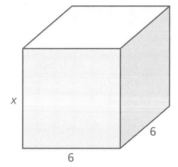

b.
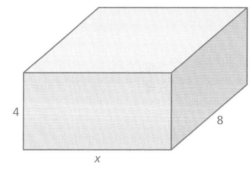

## 3  ACTIVITY: Puzzle

Work with a partner. The two triangles are similar. The perimeter of the larger triangle is 150% of the perimeter of the smaller triangle. Find the dimensions of each triangle.

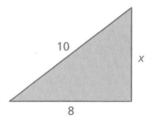
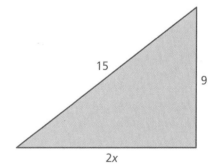

## What Is Your Answer?

4. **IN YOUR OWN WORDS** How can you solve an equation that has variables on both sides? Write an equation that has variables on both sides. Solve the equation.

**Practice**

Use what you learned about solving equations with variables on both sides to complete Exercises 3–5 on page 22.

## 1.3 Lesson

### Key Idea

**Solving Equations with Variables on Both Sides**

To solve equations with variables on both sides, collect the variable terms on one side and the constant terms on the other side.

**EXAMPLE 1** **Solving an Equation with Variables on Both Sides**

Solve $15 - 2x = -7x$. Check your solution.

| | | |
|---|---|---|
| | $15 - 2x = -7x$ | Write the equation. |
| Undo the subtraction. | $+2x \quad +2x$ | Add $2x$ to each side. |
| | $15 = -5x$ | Simplify. |
| Undo the multiplication. | $\dfrac{15}{-5} = \dfrac{-5x}{-5}$ | Divide each side by $-5$. |
| | $-3 = x$ | Simplify. |

**Check**
$15 - 2x = -7x$
$15 - 2(-3) \stackrel{?}{=} -7(-3)$
$21 = 21$ ✓

∴ The solution is $x = -3$.

**EXAMPLE 2** **Solving Equations with Variables on Both Sides**

a. Solve $3(5x + 2) = 15x$.

$3(5x + 2) = 15x$
$15x + 6 = 15x$
$\underline{-15x \qquad -15x}$
$6 = 0$ ✗

∴ The equation $6 = 0$ is never true. So, the equation has no solution.

b. Solve $-2(4y + 1) = -8y - 2$.

$-2(4y + 1) = -8y - 2$
$-8y - 2 = -8y - 2$
$\underline{+8y \qquad +8y}$
$-2 = -2$

∴ The equation $-2 = -2$ is always true. So, the equation has infinitely many solutions.

**Remember**

When solving a linear equation that has no solution, you will obtain an equivalent equation that is not true for any value of $x$. When the equation has infinitely many solutions, you will obtain an equivalent equation that is true for all values of $x$.

### On Your Own

Now You're Ready
Exercises 6–14

Solve the equation. Check your solution, if possible.

1. $-3x = 2x + 19$
2. $4(1 - p) = -4p + 4$
3. $6m - m = \dfrac{5}{6}(6m - 10)$
4. $10k + 7 = -3 - 10k$

## EXAMPLE 3 Writing and Solving an Equation

**The circles are identical. What is the area of each circle?**

Ⓐ 2   Ⓑ 4   Ⓒ $16\pi$   Ⓓ $64\pi$

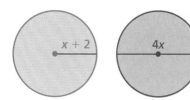

The circles are identical, so the radius of each circle is the same.

| | | |
|---|---|---|
| $x + 2 =$ | $2x$ | Write an equation. The radius of the purple circle is $2x$. |
| $-x$ | $-x$ | Subtract $x$ from each side. |
| $2 =$ | $x$ | Simplify. |

∴ The area of each circle is $\pi r^2 = \pi(4)^2 = 16\pi$. So, the correct answer is Ⓒ.

## EXAMPLE 4 Real-Life Application

**A boat travels $x$ miles per hour upstream on the Mississippi River. On the return trip, the boat travels 2 miles per hour faster. How far does the boat travel upstream?**

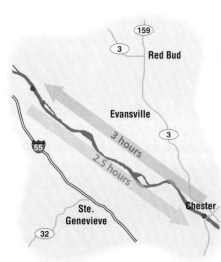

The speed of the boat on the return trip is $(x + 2)$ miles per hour.

Distance upstream = Distance of return trip

| | | |
|---|---|---|
| $3x =$ | $2.5(x + 2)$ | Write an equation. |
| $3x =$ | $2.5x + 5$ | Use Distributive Property. |
| $-2.5x$ | $-2.5x$ | Subtract $2.5x$ from each side. |
| $0.5x =$ | $5$ | Simplify. |
| $\dfrac{0.5x}{0.5} =$ | $\dfrac{5}{0.5}$ | Divide each side by 0.5. |
| $x =$ | $10$ | Simplify. |

∴ The boat travels 10 miles per hour for 3 hours upstream. So, it travels 30 miles upstream.

### On Your Own

**5. WHAT IF?** In Example 3, the diameter of the purple circle is $3x$. What is the area of each circle?

**6.** A boat travels $x$ miles per hour from one island to another island in 2.5 hours. The boat travels 5 miles per hour faster on the return trip of 2 hours. What is the distance between the islands?

## 1.3 Exercises

### Vocabulary and Concept Check

1. **WRITING** Is $x = 3$ a solution of the equation $3x - 5 = 4x - 9$? Explain.
2. **OPEN-ENDED** Write an equation that has variables on both sides and has a solution of $-3$.

### Practice and Problem Solving

The value of the solid's surface area is equal to the value of the solid's volume. Find the value of $x$.

3.
   11 in.   3 in.

4.   2.5 cm

   $x$

5.   6 in.

   5 in.   $x$

Solve the equation. Check your solution, if possible.

6. $m - 4 = 2m$
7. $3k - 1 = 7k + 2$
8. $-2x + 10 = -2(x + 5)$
9. $-24 - \frac{1}{8}p = \frac{3}{8}p$
10. $5(4w - 20) = 4(5w - 25)$
11. $\frac{3}{2}(16n + 3) = 24n$
12. $3(4z - 7) = -21 + 12z$
13. $0.1x = 0.2(x + 2)$
14. $\frac{1}{6}d + \frac{2}{3} = \frac{1}{4}(d - 2)$

15. **ERROR ANALYSIS** Describe and correct the error in solving the equation.

    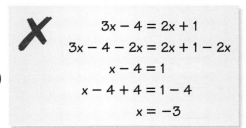

16. **TRAIL MIX** The equation $4.05p + 14.40 = 4.50(p + 3)$ represents the number $p$ of pounds of peanuts you need to make trail mix. How many pounds of peanuts do you need for the trail mix?

17. **CARS** Write and solve an equation to find the number of miles you must drive to have the same cost for each of the car rentals.

    $15 plus $0.50 per mile       $25 plus $0.25 per mile

**A polygon is *regular* if each of its sides has the same length. Find the perimeter of the regular polygon.**

18.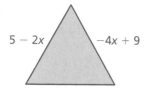
$5 - 2x$    $-4x + 9$

19.
$3(x - 1)$
$5x - 6$

20. $x + 7$
$\frac{4}{3}x - \frac{1}{3}$

21. **WRITING** Write a linear equation that has (a) no solution and (b) infinitely many solutions. Justify your answers.

22. **PRECISION** Sending a DVD in an express delivery service envelope costs the same as sending the DVD in a priority service box. What is the weight of the DVD with its packing material?

| Packing Material | Priority | Express |
|---|---|---|
| Box | $2.25 | $2.50 per lb | $8.50 per lb |
| Envelope | $1.10 | $2.50 per lb | $8.50 per lb |

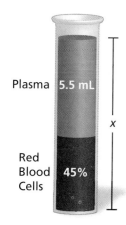

23. **STRUCTURE** Would you solve the equation $0.25x + 7 = \frac{1}{3}x - 8$ using fractions or decimals? Explain.

24. **BLOOD SAMPLE** The amount of red blood cells in a blood sample is equal to the total amount in the sample minus the amount of plasma. What is the total amount $x$ of blood drawn?

25. **NUTRITION** One serving of oatmeal provides 16% of the fiber you need daily. You must get the remaining 21 grams of fiber from other sources. How many grams of fiber should you consume daily?

26. **Geometry** The perimeter of the square is equal to the perimeter of the triangle. What are the side lengths of each figure?

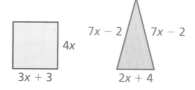

## Fair Game Review  *What you learned in previous grades & lessons*

**Find the volume of the figure. Use 3.14 for $\pi$.** *(Skills Review Handbook)*

27.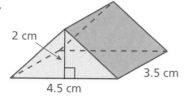
2 cm, 3.5 cm, 4.5 cm

28.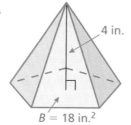
4 in., $B = 18$ in.²

29.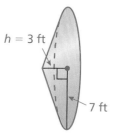
$h = 3$ ft, 7 ft

30. **MULTIPLE CHOICE** A car travels 480 miles on 15 gallons of gasoline. How many miles does the car travel per gallon? *(Section 1.1)*

  Ⓐ 28 mi/gal    Ⓑ 30 mi/gal    Ⓒ 32 mi/gal    Ⓓ 35 mi/gal

# Extension 1.3 Solving Absolute Value Equations

**Key Vocabulary**
absolute value equation, p. 24

An **absolute value equation** is an equation that contains an absolute value expression. Here are three examples.

$$|x| = 2 \qquad |x+1| = 5 \qquad 3|2x+1| = 6$$

You can solve these types of equations by solving two related linear equations.

## Key Idea

**Solving Absolute Value Equations**

To solve $|ax+b| = c$ for $c \geq 0$, solve the related linear equations

$$ax + b = c \quad \text{or} \quad ax + b = -c.$$

### EXAMPLE 1 — Solving Absolute Value Equations

**a.** Solve $|x - 4| = 6$. Graph the solutions.

Write two related linear equations for $|x - 4| = 6$. Then solve.

| | | | | |
|---|---|---|---|---|
| $x - 4 =$ | $6$ | or | $x - 4 = -6$ | Write related linear equations. |
| $+ 4$ | $+4$ | | $+4 \quad +4$ | Add 4 to each side. |
| $x =$ | $10$ | or | $x = -2$ | Simplify. |

∴ The solutions are $x = -2$ and $x = 10$.

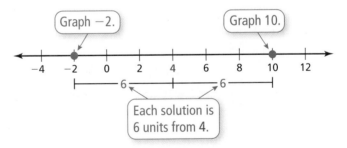

Each solution is 6 units from 4.

**Check**

$|x - 4| = 6$
$|-2 - 4| \stackrel{?}{=} 6$
$|-6| \stackrel{?}{=} 6$
$6 = 6$ ✓

$|x - 4| = 6$
$|10 - 4| \stackrel{?}{=} 6$
$|6| \stackrel{?}{=} 6$
$6 = 6$ ✓

**b.** Solve $|3x + 1| = -5$.

The absolute value of an expression must be greater than or equal to 0. The expression $|3x + 1|$ cannot equal $-5$.

∴ So, the equation has no solution.

## Practice

Solve the equation. Graph the solutions, if possible.

1. $|x| = 10$
2. $|x - 1| = 4$
3. $|3 + x| = -3$
4. $|4x - 5| = 8$

# EXAMPLE 2 — Solving an Absolute Value Equation

**Solving Equations**

In this extension, you will
- write and solve absolute value equations.

**Learning Standards**
A.CED.1
A.REI.1
A.REI.3

Solve $|3x + 9| - 10 = -4$.

$$|3x + 9| - 10 = -4 \quad \text{Write the equation.}$$
$$\phantom{|3x+9|}\ +10 \ +10 \quad \text{Add 10 to each side.}$$
$$|3x + 9| = 6 \quad \text{Simplify.}$$

Write two related linear equations for $|3x + 9| = 6$. Then solve.

| | | | | |
|---|---|---|---|---|
| $3x + 9 = 6$ | or | $3x + 9 = -6$ | | Write related linear equations. |
| $-9 \ -9$ | | $-9 \ -9$ | | Subtract 9 from each side. |
| $3x = -3$ | or | $3x = -15$ | | Simplify. |
| $\dfrac{3x}{3} = \dfrac{-3}{3}$ | | $\dfrac{3x}{3} = \dfrac{-15}{3}$ | | Divide each side by 3. |
| $x = -1$ | or | $x = -5$ | | Simplify. |

# EXAMPLE 3 — Real-Life Application

In a cheerleading competition, the minimum length of a routine is 4 minutes. The maximum length of a routine is 5 minutes. Write an absolute value equation that has these minimum and maximum lengths as its solutions.

**Step 1:** Graph the minimum and maximum lengths on a number line. Then find the point that is halfway between the lengths.

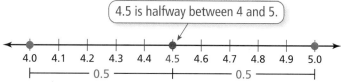

4.5 is halfway between 4 and 5.

**Step 2:** Write the equation. Each solution is 0.5 unit from 4.5.

Halfway point | Distance from halfway point

$$|x - 4.5| = 0.5$$

∴ The equation is $|x - 4.5| = 0.5$.

## Practice

**Solve the equation. Check your solutions.**

5. $|x - 2| + 5 = 9$
6. $4|2x + 7| = 16$
7. $-2|5x - 1| - 3 = -11$

8. **WRITING** Write an absolute value equation that has 5 and 15 as its solutions.

9. **POEM CONTEST** For a poem contest, the minimum length of a poem is 16 lines. The maximum length is 32 lines. Write an absolute value equation that has these minimum and maximum lengths as its solutions.

# 1.4 Rewriting Equations and Formulas

**Essential Question** How can you use a formula for one measurement to write a formula for a different measurement?

### 1 ACTIVITY: Using Perimeter and Area Formulas

**Work with a partner.**

a.
- Write a formula for the perimeter $P$ of a rectangle.
- Solve the formula for $w$.
- Use the new formula to find the width of the rectangle.

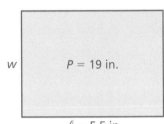

b.
- Write a formula for the area $A$ of a triangle.
- Solve the formula for $h$.
- Use the new formula to find the height of the triangle.

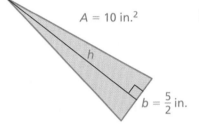

c.
- Write a formula for the circumference $C$ of a circle.
- Solve the formula for $r$.
- Use the new formula to find the radius of the circle.

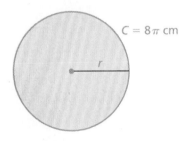

**COMMON CORE**

**Rewriting Equations**

In this lesson, you will
- rewrite equations to solve for one variable in terms of the other variable(s).

Learning Standard
A.CED.4

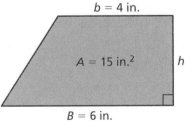

d.
- Write a formula for the area $A$ of a trapezoid.
- Solve the formula for $h$.
- Use the new formula to find the height of the trapezoid.

e.
- Write a formula for the area $A$ of a parallelogram.
- Solve the formula for $h$.
- Use the new formula to find the height of the parallelogram.

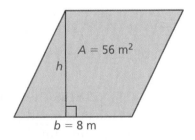

**26** Chapter 1 Solving Linear Equations

## 2 ACTIVITY: Using Volume Formulas

**Math Practice 7**

**Look for Structure**
What values in the formula do you know? What value are you trying to find?

**Work with a partner.**

a. 
- Write a formula for the volume $V$ of a prism.
- Solve the formula for $h$.
- Use the new formula to find the height of the prism.

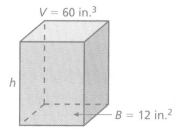

b. 
- Write a formula for the volume $V$ of a pyramid.
- Solve the formula for $B$.
- Use the new formula to find the area of the base of the pyramid.

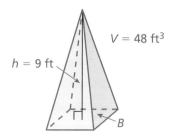

c. 
- Write a formula for the volume $V$ of a cylinder.
- Solve the formula for $B$.
- Use the new formula to find the area of the base of the cylinder.

d. 
- Write a formula for the volume $V$ of a cone.
- Solve the formula for $h$.
- Use the new formula to find the height of the cone.

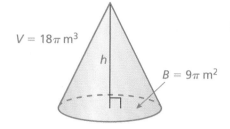

## What Is Your Answer?

3. **IN YOUR OWN WORDS** How can you use a formula for one measurement to write a formula for a different measurement? Give an example that is different from the examples on these two pages.

Use what you learned about rewriting equations and formulas to complete Exercises 3 and 4 on page 30.

# 1.4 Lesson

**Key Vocabulary**
literal equation, p. 28

An equation that has two or more variables is called a **literal equation**. To rewrite a literal equation, solve for one variable in terms of the other variable(s).

## EXAMPLE 1 Rewriting an Equation

**Solve the equation $2y + 5x = 6$ for $y$.**

| | | |
|---|---|---|
| | $2y + 5x = 6$ | Write the equation. |
| Undo the addition. → | $2y + 5x - 5x = 6 - 5x$ | Subtract $5x$ from each side. |
| | $2y = 6 - 5x$ | Simplify. |
| Undo the multiplication. → | $\dfrac{2y}{2} = \dfrac{6 - 5x}{2}$ | Divide each side by 2. |
| | $y = 3 - \dfrac{5}{2}x$ | Simplify. |

### On Your Own

**Now You're Ready**
Exercises 5–10

**Solve the equation for $y$.**

1. $5y - x = 10$
2. $4x - 4y = 1$
3. $12 = 6x + 3y$

## EXAMPLE 2 Rewriting a Formula

**The formula for the surface area $S$ of a cone is $S = \pi r^2 + \pi r \ell$. Solve the formula for the slant height $\ell$.**

**Remember**
A *formula* shows how one variable is related to one or more other variables. A formula is a type of literal equation.

| | |
|---|---|
| $S = \pi r^2 + \pi r \ell$ | Write the equation. |
| $S - \pi r^2 = \pi r^2 - \pi r^2 + \pi r \ell$ | Subtract $\pi r^2$ from each side. |
| $S - \pi r^2 = \pi r \ell$ | Simplify. |
| $\dfrac{S - \pi r^2}{\pi r} = \dfrac{\pi r \ell}{\pi r}$ | Divide each side by $\pi r$. |
| $\dfrac{S - \pi r^2}{\pi r} = \ell$ | Simplify. |

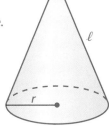

### On Your Own

**Now You're Ready**
Exercises 14–19

**Solve the formula for the red variable.**

4. Area of rectangle: $A = bh$
5. Simple interest: $I = Prt$
6. Surface area of cylinder: $S = 2\pi r^2 + 2\pi r h$

28 Chapter 1 Solving Linear Equations

### Key Idea

**Temperature Conversion**

A formula for converting from degrees Fahrenheit $F$ to degrees Celsius $C$ is

$$C = \frac{5}{9}(F - 32).$$

**EXAMPLE 3** **Rewriting the Temperature Formula**

**Solve the temperature formula for $F$.**

$C = \dfrac{5}{9}(F - 32)$     Write the temperature formula.

Use the reciprocal. → $\dfrac{9}{5} \cdot C = \dfrac{9}{5} \cdot \dfrac{5}{9}(F - 32)$     Multiply each side by $\dfrac{9}{5}$, the reciprocal of $\dfrac{5}{9}$.

$\dfrac{9}{5}C = F - 32$     Simplify.

Undo the subtraction. → $\dfrac{9}{5}C + 32 = F - 32 + 32$     Add 32 to each side.

$\dfrac{9}{5}C + 32 = F$     Simplify.

∴ The rewritten formula is $F = \dfrac{9}{5}C + 32$.

**EXAMPLE 4** **Real-Life Application**

Sun 11,000°F

Lightning 30,000°C

**Which has the greater temperature?**

Convert the Celsius temperature of lightning to Fahrenheit.

$F = \dfrac{9}{5}C + 32$     Write the rewritten formula from Example 3.

$= \dfrac{9}{5}(30,000) + 32$     Substitute 30,000 for $C$.

$= 54,032$     Simplify.

∴ Because 54,032°F is greater than 11,000°F, lightning has the greater temperature.

### On Your Own

**7.** Room temperature is considered to be 70°F. Suppose the temperature is 23°C. Is this greater than or less than room temperature?

## 1.4 Exercises

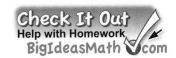

### Vocabulary and Concept Check

1. **VOCABULARY** Is $-2x = \dfrac{3}{8}$ a literal equation? Explain.

2. **DIFFERENT WORDS, SAME QUESTION** Which is different? Find "both" answers.

   | Solve $4x - 2y = 6$ for $y$. | Solve $6 = 4x - 2y$ for $y$. |
   |---|---|
   | Solve $4x - 2y = 6$ for $y$ in terms of $x$. | Solve $4x - 2y = 6$ for $x$ in terms of $y$. |

### Practice and Problem Solving

3. a. Write a formula for the area $A$ of a triangle.
   b. Solve the formula for $b$.
   c. Use the new formula to find the base of the triangle.

4. a. Write a formula for the volume $V$ of a prism.
   b. Solve the formula for $B$.
   c. Use the new formula to find the area of the base of the prism.

$A = 36$ mm²,  $h = 6$ mm

$V = 36$ in.³,  $h = 6$ in.

**Solve the equation for $y$.**

5. $\dfrac{1}{3}x + y = 4$

6. $3x + \dfrac{1}{5}y = 7$

7. $6 = 4x + 9y$

8. $\pi = 7x - 2y$

9. $4.2x - 1.4y = 2.1$

10. $6y - 1.5x = 8$

11. **ERROR ANALYSIS** Describe and correct the error in rewriting the equation.

    $2x - y = 5$
    $y = -2x + 5$

12. **TEMPERATURE** The formula $K = C + 273.15$ converts temperatures from Celsius $C$ to Kelvin $K$.

    a. Solve the formula for $C$.
    b. Convert 300 $K$ to Celsius.

13. **INTEREST** The formula for simple interest is $I = Prt$.

    a. Solve the formula for $t$.
    b. Use the new formula to find the value of $t$ in the table.

    | | |
    |---|---|
    | $I$ | $75 |
    | $P$ | $500 |
    | $r$ | 5% |
    | $t$ | |

30  Chapter 1  Solving Linear Equations

**Solve the equation for the red variable.**

**14.** $d = rt$

**15.** $e = mc^2$

**16.** $R - C = P$

**17.** $A = \dfrac{1}{2}\pi w^2 + 2\ell w$

**18.** $B = 3\dfrac{V}{h}$

**19.** $g = \dfrac{1}{6}(w + 40)$

**20. LOGIC** Why is it useful to rewrite a formula in terms of another variable?

**21. REASONING** The formula $K = \dfrac{5}{9}(F - 32) + 273.15$ converts temperatures from Fahrenheit $F$ to Kelvin $K$.

   **a.** Solve the formula for $F$.

   **b.** The freezing point of water is 273.15 Kelvin. What is this temperature in Fahrenheit?

   **c.** The temperature of dry ice is $-78.5\,°C$. Which is colder, dry ice or liquid nitrogen?

Liquid nitrogen
77.35 K

Navy Pier Ferris Wheel
$C = 439.6$ ft

**22. FERRIS WHEEL** The Navy Pier Ferris Wheel in Chicago has a circumference that is 56% of the circumference of the first Ferris wheel built in 1893.

   **a.** What is the radius of the Navy Pier Ferris Wheel?

   **b.** What was the radius of the first Ferris wheel?

   **c.** The first Ferris wheel took 9 minutes to make a complete revolution. How fast was the wheel moving?

**23. Repeated Reasoning** The formula for the volume of a sphere is $V = \dfrac{4}{3}\pi r^3$. Solve the formula for $r^3$. Use guess, check, and revise to find the radius of the sphere.

$V = 381.51$ in.$^3$   ⊢— $r$ —⊣

# Fair Game Review  *What you learned in previous grades & lessons*

**Multiply.** *(Skills Review Handbook)*

**24.** $5 \times \dfrac{3}{4}$

**25.** $2.4 \times \dfrac{8}{3}$

**26.** $\dfrac{1}{4} \times \dfrac{3}{2} \times \dfrac{8}{9}$

**27.** $25 \times \dfrac{3}{5} \times \dfrac{1}{12}$

**28. MULTIPLE CHOICE** Which of the following is not equivalent to $\dfrac{3}{4}$? *(Skills Review Handbook)*

   **Ⓐ** 0.75     **Ⓑ** 3 : 4     **Ⓒ** 75%     **Ⓓ** 4 : 3

# 1.3–1.4 Quiz

**Solve the equation. Check your solution.** *(Section 1.3)*

1. $2(x + 4) = -5x + 1$
2. $\frac{1}{2}s = 4s - 21$
3. $8.3z = 4.1z + 10.5$
4. $3(b + 5) = 4(2b - 5)$

**Solve the equation. Graph the solutions, if possible.** *(Section 1.3)*

5. $|d + 10| = 6$
6. $-4|w - 1| = -8$

**Solve the equation for y.** *(Section 1.4)*

7. $6x - 3y = 9$
8. $8 = 2y - 10x$

**Solve the formula for the red variable.** *(Section 1.4)*

9. Volume of a cylinder: $V = \pi r^2 h$
10. Area of a trapezoid: $A = \frac{1}{2}h(b + B)$

11. **TEMPERATURE** In which city is the water temperature higher? *(Section 1.4)*

12. **INTEREST** The formula for simple interest $I$ is $I = Prt$. Solve the formula for the interest rate $r$. What is the interest rate $r$ if the principal $P$ is $1500, the time $t$ is 2 years, and the interest earned $I$ is $90? *(Section 1.4)*

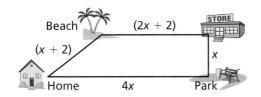

13. **ROUTES** From your home, the route to the store that passes the beach is 2 miles shorter than the route to the store that passes the park. What is the length of each route? *(Section 1.3)*

14. **PERIMETER** Use the triangle shown. *(Section 1.4)*

    a. Write a formula for the perimeter $P$ of the triangle.

    b. Solve the formula for $b$.

    c. Use the new formula to find $b$ when $a$ is 10 feet and $c$ is 17 feet.

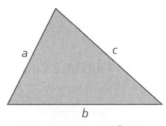

Perimeter = 42 feet

# 1 Chapter Review

## Review Key Vocabulary

absolute value equation, *p. 24*        literal equation, *p. 28*

## Review Examples and Exercises

### 1.1 Solving Simple Equations *(pp. 2–9)*

The *boiling point* of a liquid is the temperature at which the liquid becomes a gas. The boiling point of mercury is about $\frac{41}{200}$ of the boiling point of lead. Write and solve an equation to find the boiling point of lead.

Let $x$ be the boiling point of lead.

$\frac{41}{200}x = 357$      Write the equation.

$\frac{200}{41} \cdot \left(\frac{41}{200}x\right) = \frac{200}{41} \cdot 357$      Multiply each side by $\frac{200}{41}$.

$x \approx 1741$      Simplify.

∴ The boiling point of lead is about 1741°C.

Mercury 357°C

#### Exercises

Solve the equation. Check your solution.

**1.** $y + 8 = -11$      **2.** $3.2 = -0.4n$      **3.** $-\frac{t}{4} = -3\pi$

### 1.2 Solving Multi-Step Equations *(pp. 10–15)*

a. Solve $-4p - 9 = 3$.

$$-4p - 9 = 3$$
$$\underline{+9 \quad +9}$$
$$-4p = 12$$
$$\frac{-4p}{-4} = \frac{12}{-4}$$
$$p = -3$$

∴ The solution is $p = -3$.

b. Solve $-14x + 28 + 6x = -44$.

$$-14x + 28 + 6x = -44$$
$$-8x + 28 = -44$$
$$\underline{-28 \quad -28}$$
$$-8x = -72$$
$$\frac{-8x}{-8} = \frac{-72}{-8}$$
$$x = 9$$

∴ The solution is $x = 9$.

## Exercises

**Solve the equation. Check your solution.**

4. $7y + 15 = -27$
5. $8 - \dfrac{3}{2}b = 11$
6. $-2(3z + 1) - 10 = 4$
7. $-3n - 2n + 9 = 29$
8. $2.5(4x - 6) - 5 = 10$
9. $\dfrac{2}{5}w + \dfrac{4}{5}w - 4 = 1$

**Find the value of $x$. Then find the angle measures of the polygon.**

10.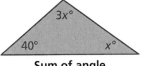
Sum of angle measures: 180°

11.
Sum of angle measures: 360°

12.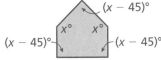
Sum of angle measures: 540°

## 1.3 Solving Equations with Variables on Both Sides (pp. 18–25)

**a. Solve $3n - 2 = 11n + 18$.**

$$
\begin{aligned}
3n - 2 &= 11n + 18 &&\text{Write the equation.} \\
-11n &\phantom{=} -11n &&\text{Subtract } 11n \text{ from each side.} \\
-8n - 2 &= 18 &&\text{Simplify.} \\
+2 &\phantom{=} +2 &&\text{Add 2 to each side.} \\
-8n &= 20 &&\text{Simplify.} \\
\dfrac{-8n}{-8} &= \dfrac{20}{-8} &&\text{Divide each side by } -8. \\
n &= -\dfrac{5}{2} &&\text{Simplify.}
\end{aligned}
$$

∴ The solution is $n = -\dfrac{5}{2}$.

**b. Solve $|x - 7| = 3$.**

$|x - 7| = 3$      Write the equation.

$x - 7 = 3$    or    $x - 7 = -3$      Write two related linear equations.

$+7 \phantom{==} +7$          $+7 \phantom{==} +7$      Add 7 to each side.

$x = 10$    or    $x = 4$      Simplify.

∴ The solutions are $x = 4$ and $x = 10$.

## Exercises

**Solve the equation. Check your solution, if possible.**

**13.** $5m - 1 = 4m + 5$

**14.** $3(5p - 3) = 5(p - 1)$

**15.** $\dfrac{2}{5}n + \dfrac{1}{10} = \dfrac{1}{2}(n + 4)$

**Solve the equation. Check your solutions, if possible.**

**16.** $|x + 5| = 17$

**17.** $|2w - 9| = 1$

**18.** $-3|6y - 7| + 10 = -8$

### 1.4 Rewriting Equations and Formulas (pp. 26–31)

The equation for a line in slope-intercept form is $y = mx + b$.
Solve the equation for $x$.

$$y = mx + b \qquad \text{Write the equation.}$$
$$y - b = mx + b - b \qquad \text{Subtract } b \text{ from each side.}$$
$$y - b = mx \qquad \text{Simplify.}$$
$$\dfrac{y - b}{m} = \dfrac{mx}{m} \qquad \text{Divide each side by } m.$$
$$\dfrac{y - b}{m} = x \qquad \text{Simplify.}$$

So, $x = \dfrac{y - b}{m}$.

## Exercises

**Solve the equation for $y$.**

**19.** $5x - 5y = 30$

**20.** $14 = 8x + 2y$

**21.** $1 - 2y = -x$

**22. a.** The formula $F = \dfrac{9}{5}(K - 273.15) + 32$ converts a temperature from Kelvin $K$ to Fahrenheit $F$. Solve the formula for $K$.

  **b.** Convert 240°F to Kelvin $K$. Round your answer to the nearest hundredth.

**23. a.** Write the formula for the area $A$ of a trapezoid.

  **b.** Solve the formula for $h$.

  **c.** Use the new formula to find the height $h$ of the trapezoid.

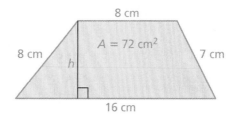

# 1 Chapter Test

**Solve the equation. Check your solution.**

1. $4 + y = 9.5$
2. $-\dfrac{x}{9} = -8$
3. $z - \dfrac{2}{3} = \dfrac{1}{8}$
4. $r - |-4| = 11$
5. $3.8n - 13 = 1.4n + 5$
6. $9(8d - 5) + 13 = 12d - 2$

**Find the value of $x$. Then find the angle measures of the polygon.**

7.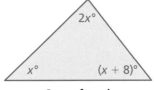
Sum of angle measures: 180°

8.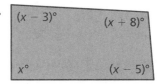
Sum of angle measures: 360°

**Solve the equation. Graph the solutions, if possible.**

9. $|2p - 3| = 7$
10. $5|3v - 8| = -10$

**Solve the equation for $y$.**

11. $1.2x - 4y = 28$
12. $0.5 = 0.4y - 0.25x$

**Solve the formula for the red variable.**

13. Perimeter of a rectangle: $P = 2\ell + 2w$
14. Distance formula: $d = rt$

15. **BASKETBALL** Your basketball team wins a game by 13 points. The opposing team scores 72 points. Explain how to find your team's score.

16. **CYCLING** You are biking at a speed of 18 miles per hour. You are 3 miles behind your friend who is biking at a speed of 12 miles per hour. Write and solve an equation to find the amount of time it takes for you to catch up to your friend.

17. **VOLCANOES** Two scientists are measuring lava temperatures. One scientist records a temperature of 1725°F. The other scientist records a temperature of 950°C. Which is the greater temperature? $\left(\text{Use } C = \dfrac{5}{9}(F - 32).\right)$

18. **JOBS** Your profit for mowing lawns this week is $24. You are paid $8 per hour and you paid $40 for gas for the lawnmower. How many hours did you work this week?

# Standards Assessment

1. Which value of $x$ makes the equation true? *(A.REI.3)*

$$4x = 32$$

   A. 8      C. 36

   B. 28      D. 128

**Test-Taking Strategy**
**Solve Directly or Eliminate Choices**

"When a cat wakes up, it's grumpy for $x$ hours, where $2x - 5x = x - 4$. What's $x$?
 Ⓐ 0 Ⓑ 1 Ⓒ 2 Ⓓ -3"

"Don't talk to me until I've had my morning milk."

"You can eliminate A and D. Then, solve directly to determine that the correct answer is B."

2. A taxi ride costs $3 plus $2 for each mile driven. When you rode in a taxi, the total cost was $39. This can be modeled by the equation below, where $m$ represents the number of miles driven.

$$2m + 3 = 39$$

   How long was your taxi ride? *(A.REI.3)*

   F. 72 mi      H. 21 mi

   G. 34 mi      I. 18 mi

3. A car traveling at a speed of 65 miles per hour is 5 miles behind a truck traveling at a speed of 55 miles per hour. Which equation can be used to find the amount of time $t$ it takes for the car to catch up to the truck? *(A.CED.1)*

   A. $65t + 5 = 55t$      C. $65t + 55t = 5$

   B. $65t = 5 + 55t$      D. $65(t - 5) = 55t$

4. What is the perimeter of the square? *(A.CED.3, A.REI.3)*

   $2(x + 5)$

   $3x + 2$

5. The formula below relates distance, rate, and time.

$$d = rt$$

   Solve this formula for $t$. *(A.CED.4)*

   F. $t = dr$      H. $t = d - r$

   G. $t = \dfrac{d}{r}$      I. $t = \dfrac{r}{d}$

6. What could be the first step to solve the equation shown below? *(A.REI.1)*

$$3x + 5 = 2(x + 7)$$

   A. Combine $3x$ and 5.

   B. Multiply $x$ by 2 and 7 by 2.

   C. Subtract $x$ from $3x$.

   D. Subtract 5 from 7.

7. You work as a sales representative. You earn $400 per week plus 5% of your total sales for the week. *(A.CED.1, A.REI.3)*

   **Part A** Last week, you had total sales of $5000. Find your total earnings. Show your work.

   **Part B** One week, you earned $1350. Let $s$ represent your total sales that week. Write an equation that could be used to find $s$.

   **Part C** Using your equation from Part B, find $s$. Explain all steps.

8. In 10 years, Maria will be 39 years old. Let $m$ represent Maria's age today. Which equation can be used to find $m$? *(A.CED.1)*

   F. $m = 39 + 10$

   G. $m - 10 = 39$

   H. $m + 10 = 39$

   I. $10m = 39$

9. Which value of $y$ makes the equation below true? *(A.REI.3)*

$$3y + 8 = 7y + 11$$

   A. $-4.75$

   B. $-0.75$

   C. $0.75$

   D. $4.75$

10. The equation below is used to convert a Celsius temperature $C$ to its equivalent Fahrenheit temperature $F$.

$$F = \frac{9}{5}C + 32$$

   Which formula can be used to convert a Fahrenheit temperature to its equivalent Celsius temperature? *(A.CED.4)*

   F. $C = \frac{9}{5}(F - 32)$

   G. $C = \frac{5}{9}F - 32$

   H. $C = \frac{5}{9}(F + 32)$

   I. $C = \frac{5}{9}(F - 32)$

**11.**  You have already saved $35 for a new cell phone. You need $175 in all. You think you can save $10 per week. At this rate, how many more weeks will you need to save money before you can buy the new cell phone? *(A.CED.1, A.REI.3)*

**12.** Solve $-8x - x + 5 = -6x + 5 - 3x$. *(A.REI.3)*

 **A.** $x = 0$ **C.** Infinitely many solutions

 **B.** $x = 5$ **D.** No solution

**13.** Solve $3|2x + 1| = 12$. *(A.REI.3)*

 **F.** The solutions are $-\dfrac{5}{2}$ and $\dfrac{3}{2}$. **H.** The solutions are $-\dfrac{5}{2}$ and $-\dfrac{3}{2}$.

 **G.** The solutions are $\dfrac{5}{2}$ and $\dfrac{3}{2}$. **I.** No solution

**14.** Which value of $x$ makes the equation below true? *(A.REI.3)*

$$6(x - 3) = 4x - 7$$

 **A.** $-5.5$ **C.** $1.1$

 **B.** $-2$ **D.** $5.5$

**15.** The drawing below shows equal weights on two sides of a balance scale. *(A.CED.1)*

What can you conclude from the drawing?

 **F.** A mug weighs one-third as much as a trophy.

 **G.** A mug weighs one-half as much as a trophy.

 **H.** A mug weighs twice as much as a trophy.

 **I.** A mug weighs three times as much as a trophy.

# 2 Graphing and Writing Linear Equations

- **2.1** Graphing Linear Equations
- **2.2** Slope of a Line
- **2.3** Graphing Linear Equations in Slope-Intercept Form
- **2.4** Graphing Linear Equations in Standard Form
- **2.5** Writing Equations in Slope-Intercept Form
- **2.6** Writing Equations in Point-Slope Form
- **2.7** Solving Real-Life Problems

"Okay Descartes, stand on the *y*-axis and try to intercept the pass when I throw."

"Here's an easy example of a line with a slope of 1."

"You eat one mouse treat the first day. Two treats the second day. And so on. Get it?"

# What You Learned Before

● **Evaluating Expressions Using Order of Operations** (6.EE.2c)

**Example 1** Evaluate $2xy + 3(x + y)$ when $x = 4$ and $y = 7$.

$2xy + 3(x + y) = 2(4)(7) + 3(4 + 7)$     Substitute 4 for $x$ and 7 for $y$.
$\qquad\qquad\quad = 8(7) + 3(4 + 7)$     Use order of operations.
$\qquad\qquad\quad = 56 + 3(11)$     Simplify.
$\qquad\qquad\quad = 56 + 33$     Multiply.
$\qquad\qquad\quad = 89$     Add.

**Try It Yourself**

Evaluate the expression when $a = \dfrac{1}{4}$ and $b = 6$.

1. $-8ab$
2. $16a^2 - 4b$
3. $\dfrac{5b}{32a^2}$
4. $12a + (b - a - 4)$

● **Plotting Points** (6.NS.6c)

**Example 2** Write the ordered pair that corresponds to Point $U$.

Point $U$ is 3 units to the left of the origin and 4 units down. So, the $x$-coordinate is $-3$ and the $y$-coordinate is $-4$.

∴ The ordered pair $(-3, -4)$ corresponds to Point $U$.

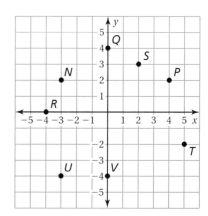

**Example 3** Which point is located at $(5, -2)$?

Start at the origin. Move 5 units right and 2 units down.

∴ Point $T$ is located at $(5, -2)$.

**Try It Yourself**

**Use the graph to answer the question.**

5. Write the ordered pair that corresponds to Point $Q$.
6. Write the ordered pair that corresponds to Point $P$.
7. Which point is located at $(-4, 0)$?
8. Which point is located in Quadrant II?

# 2.1 Graphing Linear Equations

**Essential Question** How can you recognize a linear equation? How can you draw its graph?

### 1 ACTIVITY: Graphing a Linear Equation

**Work with a partner.**

a. Use the equation $y = \frac{1}{2}x + 1$ to complete the table. (Choose any two $x$-values and find the $y$-values.)

| x | Solution Points | |
|---|---|---|
| $y = \frac{1}{2}x + 1$ | | |

b. Write the two ordered pairs given by the table. These are called **solution points** of the equation.

c. **PRECISION** Plot the two solution points. Draw a line *exactly* through the two points.

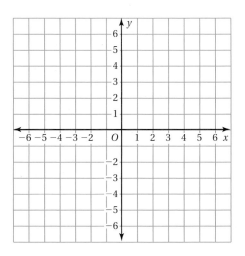

d. Find a different point on the line. Check that this point is a solution point of the equation $y = \frac{1}{2}x + 1$.

e. **LOGIC** Do you think it is true that *any* point on the line is a solution point of the equation $y = \frac{1}{2}x + 1$? Explain.

f. Choose five additional $x$-values for the table. (Choose positive and negative $x$-values.) Plot the five corresponding solution points. Does each point lie on the line?

| x | Solution Points | | | | |
|---|---|---|---|---|---|
| $y = \frac{1}{2}x + 1$ | | | | | |

g. **LOGIC** Do you think it is true that *any* solution point of the equation $y = \frac{1}{2}x + 1$ is a point on the line? Explain.

h. **THE MEANING OF A WORD** Why is $y = ax + b$ called a *linear equation*?

---

**COMMON CORE**

**Graphing Equations**

In this lesson, you will
- understand that lines represent solutions of linear equations.
- graph linear equations.

Learning Standards
A.CED.2
A.REI.10

### 2 ACTIVITY: Using a Graphing Calculator

**Math Practice 5**

**Recognize Usefulness of Tools**

What are some advantages and disadvantages of using a graphing calculator to graph a linear equation?

Use a graphing calculator to graph $y = 2x + 5$.

a. Enter the equation $y = 2x + 5$ into your calculator.

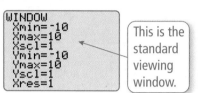

b. Check the settings of the *viewing window*. The boundaries of the graph are set by the minimum and maximum $x$- and $y$-values. The number of units between the tick marks are set by the $x$- and $y$-scales.

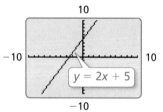

This is the standard viewing window.

c. Graph $y = 2x + 5$ on your calculator.

d. Change the settings of the viewing window to match those shown.

Compare the two graphs.

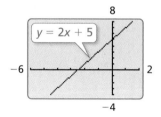

## What Is Your Answer?

3. **IN YOUR OWN WORDS** How can you recognize a linear equation? How can you draw its graph? Write an equation that is linear. Write an equation that is *not* linear.

4. Use a graphing calculator to graph $y = 5x - 12$ in the standard viewing window.

   a. Can you tell where the line crosses the $x$-axis? Can you tell where the line crosses the $y$-axis?

   b. How can you adjust the viewing window so that you can determine where the line crosses the $x$- and $y$-axes?

5. **CHOOSE TOOLS** You want to graph $y = 2.5x - 3.8$. Would you graph it by hand or using a graphing calculator? Why?

**Practice**

Use what you learned about graphing linear equations to complete Exercises 3 and 4 on page 46.

Section 2.1   Graphing Linear Equations

# 2.1 Lesson

**Key Vocabulary**
linear equation, *p. 44*
solution of a linear equation, *p. 44*

## Key Idea

**Linear Equations**

A **linear equation** is an equation whose graph is a line. The points on the line are **solutions** of the equation.

You can use a graph to show the solutions of a linear equation. The graph below is for the equation $y = x + 1$.

| x | y | (x, y) |
|---|---|--------|
| −1 | 0 | (−1, 0) |
| 0 | 1 | (0, 1) |
| 2 | 3 | (2, 3) |

**Remember**
An ordered pair (x, y) is used to locate a point in a coordinate plane.

### EXAMPLE 1  Graphing a Linear Equation

Graph $y = -2x + 1$.

**Step 1:** Make a table of values.

| x | y = −2x + 1 | y | (x, y) |
|---|-------------|---|--------|
| −1 | y = −2(−1) + 1 | 3 | (−1, 3) |
| 0 | y = −2(0) + 1 | 1 | (0, 1) |
| 2 | y = −2(2) + 1 | −3 | (2, −3) |

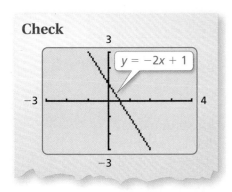

**Check**

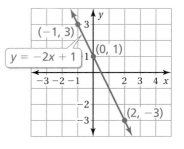

**Step 2:** Plot the ordered pairs.

**Step 3:** Draw a line through the points.

## Key Idea

**Graphing Horizontal and Vertical Lines**

The graph of $y = b$ is a horizontal line passing through (0, b).

The graph of $x = a$ is a vertical line passing through (a, 0).

44   Chapter 2   Graphing and Writing Linear Equations

### EXAMPLE 2  Graphing a Horizontal Line and a Vertical Line

a. Graph $y = -3$.

The graph of $y = -3$ is a horizontal line passing through $(0, -3)$. Draw a horizontal line through this point.

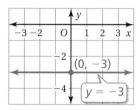

b. Graph $x = 2$.

The graph of $x = 2$ is a vertical line passing through $(2, 0)$. Draw a vertical line through this point.

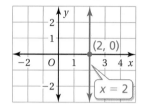

#### On Your Own

*Now You're Ready*
*Exercises 5–16*

Graph the linear equation. Use a graphing calculator to check your graph, if possible.

1. $y = 3x$
2. $y = -\frac{1}{2}x + 2$
3. $x = -4$
4. $y = -1.5$

### EXAMPLE 3  Real-Life Application

The wind speed $y$ (in miles per hour) of a tropical storm is $y = 2x + 66$, where $x$ is the number of hours after the storm enters the Gulf of Mexico.

a. Graph the equation.
b. When does the storm become a hurricane?

A tropical storm becomes a hurricane when wind speeds are at least 74 miles per hour.

a. Make a table of values.

| x | y = 2x + 66 | y | (x, y) |
|---|---|---|---|
| 0 | y = 2(0) + 66 | 66 | (0, 66) |
| 1 | y = 2(1) + 66 | 68 | (1, 68) |
| 2 | y = 2(2) + 66 | 70 | (2, 70) |
| 3 | y = 2(3) + 66 | 72 | (3, 72) |

Plot the ordered pairs and draw a line through the points.

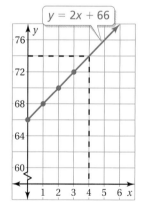

b. From the graph, you can see that $y = 74$ when $x = 4$. So, the storm becomes a hurricane 4 hours after it enters the Gulf of Mexico.

#### On Your Own

5. **WHAT IF?** In Example 3, the wind speed of the storm is $y = 1.5x + 62$. When does the storm become a hurricane?

Section 2.1  Graphing Linear Equations

# 2.1 Exercises

## Vocabulary and Concept Check

1. **VOCABULARY** What type of graph represents the solutions of the equation $y = 2x + 3$?

2. **WHICH ONE DOESN'T BELONG?** Which equation does *not* belong with the other three? Explain your reasoning.

   $y = 0.5x - 0.2$  $\quad$  $4x + 3 = y$  $\quad$  $y = x^2 + 6$  $\quad$  $\frac{3}{4}x + \frac{1}{3} = y$

## Practice and Problem Solving

**PRECISION** Copy and complete the table. Plot the two solution points and draw a line *exactly* through the two points. Find a different solution point on the line.

3. 
| x | | |
|---|---|---|
| $y = 3x - 1$ | | |

4. 
| x | | |
|---|---|---|
| $y = \frac{1}{3}x + 2$ | | |

**Graph the linear equation. Use a graphing calculator to check your graph, if possible.**

5. $y = -5x$
6. $y = \frac{1}{4}x$
7. $y = 5$
8. $x = -6$

9. $y = x - 3$
10. $y = -7x - 1$
11. $y = -\frac{x}{3} + 4$
12. $y = \frac{3}{4}x - \frac{1}{2}$

13. $y = -\frac{2}{3}$
14. $y = 6.75$
15. $x = -0.5$
16. $x = \frac{1}{4}$

17. **ERROR ANALYSIS** Describe and correct the error in graphing the equation.

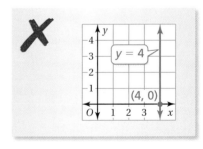

18. **MESSAGING** You sign up for an unlimited text messaging plan for your cell phone. The equation $y = 20$ represents the cost $y$ (in dollars) for sending $x$ text messages. Graph the equation. What does the graph tell you?

19. **MAIL** The equation $y = 2x + 3$ represents the cost $y$ (in dollars) of mailing a package that weighs $x$ pounds.

   a. Graph the equation.
   b. Use the graph to estimate how much it costs to mail the package.
   c. Use the equation to find exactly how much it costs to mail the package.

**Solve for y. Then graph the equation. Use a graphing calculator to check your graph.**

**20.** $y - 3x = 1$  **21.** $5x + 2y = 4$

**22.** $-\dfrac{1}{3}y + 4x = 3$  **23.** $x + 0.5y = 1.5$

**24. SAVINGS** You have $100 in your savings account and plan to deposit $12.50 each month.

   **a.** Write and graph a linear equation that represents the balance in your account.

   **b.** How many months will it take you to save enough money to buy 10 acres of land on Mars?

**25. CAMERA** One second of video on your digital camera uses the same amount of memory as two pictures. Your camera can store 250 pictures.

   **a.** Write and graph a linear equation that represents the number $y$ of pictures your camera can store if you take $x$ seconds of video.

   **b.** How many pictures can your camera store after you take the video shown?

**26. PROBLEM SOLVING** Along the U.S. Atlantic Coast, the sea level is rising about 2 millimeters per year. How many millimeters has sea level risen since you were born? How do you know? Use a linear equation and a graph to justify your answer.

**27. Geometry** The sum $S$ of the measures of the angles of a polygon is $S = (n - 2) \cdot 180°$, where $n$ is the number of sides of the polygon.

   **a.** Plot four points $(n, S)$ that satisfy the equation. Do the points lie on a line? Explain your reasoning.

   **b.** Does the value $n = 3.5$ make sense in the context of the problem? Explain your reasoning.

 **Fair Game Review**  *What you learned in previous grades & lessons*

**Write the ordered pair corresponding to the point.**
*(Skills Review Handbook)*

**28.** Point $A$  **29.** Point $B$

**30.** Point $C$  **31.** Point $D$

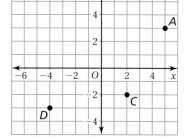

**32. MULTIPLE CHOICE** A debate team has 15 female members. The ratio of females to males is 3 : 2. How many males are on the debate team? *(Skills Review Handbook)*

   Ⓐ 6  Ⓑ 10  Ⓒ 22  Ⓓ 25

# 2.2 Slope of a Line

**Essential Question** How can the slope of a line be used to describe the line?

**Slope** is the rate of change between any two points on a line. It is the measure of the *steepness* of the line.

To find the slope of a line, find the ratio of the change in $y$ (vertical change) to the change in $x$ (horizontal change).

$$\text{slope} = \frac{\text{change in } y}{\text{change in } x}$$

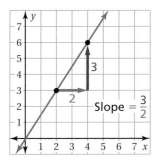

Slope = $\frac{3}{2}$

## 1 ACTIVITY: Finding the Slope of a Line

Work with a partner. Find the slope of each line using two methods.

Method 1: Use the two black points.
Method 2: Use the two pink points.

Do you get the same slope using each method? Why do you think this happens?

a.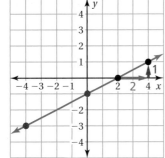

b.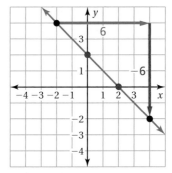

**Slope of a Line**

In this lesson, you will
- find slopes of lines using two points.
- find slopes of lines from tables.

Preparing for Standards
F.IF.4
F.IF.6

c.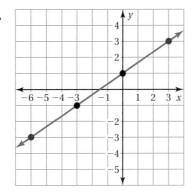

d.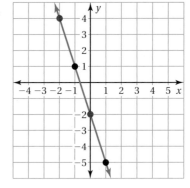

48    Chapter 2    Graphing and Writing Linear Equations

## 2 ACTIVITY: Drawing Lines with Given Slopes

**Math Practice**

**Make Conjectures**
What does the slope tell you about the graph of the line? Explain.

**Work with a partner.**
- Draw a line through the black point using the given slope.
- Draw a line through the pink point using the given slope.
- What do you notice about the two lines?

**a.** Slope = $\dfrac{3}{4}$

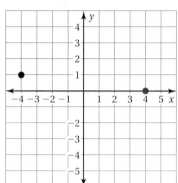

**b.** Slope = $-\dfrac{4}{3}$

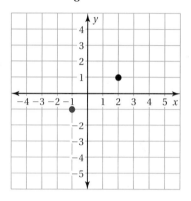

## 3 ACTIVITY: Drawing Lines with Given Slopes

**Work with a partner.**

- Examine the lines drawn through the black points in parts (a) and (b) of Activity 2. Draw these two lines in the same coordinate plane.

- Describe the angle formed by the two lines. What do you notice about the product of the slopes of the two lines?

### What Is Your Answer?

4. **IN YOUR OWN WORDS** How can the slope of a line be used to describe the line?

5. Based on your results in Activity 2, make a conjecture about two different nonvertical lines in the same plane that have the same slope.

6. **REPEATED REASONING** Repeat Activity 3 for the lines drawn through the pink points in Activity 2. Based on your results, make a conjecture about two lines in the same plane whose slopes have a product of −1.

**Practice**

Use what you learned about the slope of a line to complete Exercises 4–6 on page 53.

## 2.2 Lesson

**Key Vocabulary**
slope, p. 50
rise, p. 50
run, p. 50

### Key Idea

**Slope**

The **slope** of a line is a ratio of the change in $y$ (the **rise**) to the change in $x$ (the **run**) between any two points, $(x_1, y_1)$ and $(x_2, y_2)$, on the line.

$$\text{slope} = \frac{\text{rise}}{\text{run}} = \frac{\text{change in } y}{\text{change in } x} = \frac{y_2 - y_1}{x_2 - x_1}$$

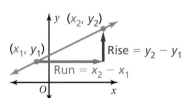

*Positive slope*      *Negative slope*

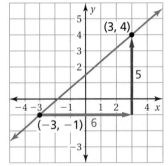

The line rises from left to right.     The line falls from left to right.

**Reading**

In the slope formula, $x_1$ is read as "x sub one" and $y_2$ is read as "y sub two". The numbers 1 and 2 in $x_1$ and $y_2$ are called *subscripts*.

### EXAMPLE 1   Finding the Slope of a Line

**Describe the slope of the line. Then find the slope.**

a.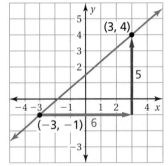

b.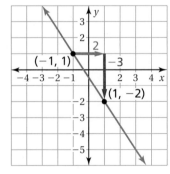

The line rises from left to right. So, the slope is positive. Let $(x_1, y_1) = (-3, -1)$ and $(x_2, y_2) = (3, 4)$.

$$\text{slope} = \frac{y_2 - y_1}{x_2 - x_1}$$

$$= \frac{4 - (-1)}{3 - (-3)}$$

$$= \frac{5}{6}$$

The line falls from left to right. So, the slope is negative. Let $(x_1, y_1) = (-1, 1)$ and $(x_2, y_2) = (1, -2)$.

$$\text{slope} = \frac{y_2 - y_1}{x_2 - x_1}$$

$$= \frac{-2 - 1}{1 - (-1)}$$

$$= \frac{-3}{2}, \text{ or } -\frac{3}{2}$$

**Study Tip**

When finding slope, you can label either point as $(x_1, y_1)$ and the other point as $(x_2, y_2)$.

● **On Your Own**

**Find the slope of the line.**

1.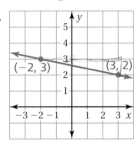
2.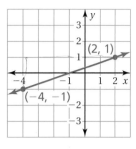
3.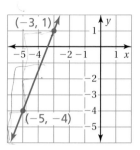

Exercises 7–9

---

**EXAMPLE 2** **Finding the Slope of a Horizontal Line**

**Find the slope of the line.**

There is no change in *y*. So, the change in *y* is 0.

$$\text{slope} = \frac{y_2 - y_1}{x_2 - x_1}$$

$$= \frac{5 - 5}{6 - (-1)}$$

$$= \frac{0}{7}, \text{ or } 0$$

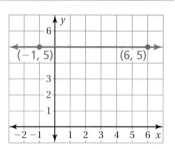

∴ The slope is 0.

---

**EXAMPLE 3** **Finding the Slope of a Vertical Line**

**Find the slope of the line.**

There is no change in *x*. So, the change in *x* is 0.

$$\text{slope} = \frac{y_2 - y_1}{x_2 - x_1}$$

$$= \frac{6 - 2}{4 - 4}$$

$$= \frac{4}{0} \; ✗$$

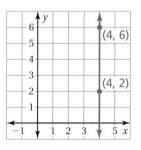

∴ Because division by zero is undefined, the slope of the line is undefined.

---

● **On Your Own**

**Find the slope of the line through the given points.**

Exercises 13–15

4. (1, −2), (7, −2)
5. (−3, −3), (−3, −5)
6. (0, 8), (0, 0)

7. How do you know that the slope of every horizontal line is 0? How do you know that the slope of every vertical line is undefined?

# EXAMPLE 4 Finding Slope from a Table

The points in the table lie on a line. How can you find the slope of the line from the table? What is the slope?

| x | 1 | 4 | 7 | 10 |
|---|---|---|---|----|
| y | 8 | 6 | 4 | 2  |

Choose any two points from the table and use the slope formula.

Use the points $(x_1, y_1) = (1, 8)$ and $(x_2, y_2) = (4, 6)$.

$$\text{slope} = \frac{y_2 - y_1}{x_2 - x_1}$$

$$= \frac{6 - 8}{4 - 1}$$

$$= \frac{-2}{3}$$

∴ The slope is $-\frac{2}{3}$.

**Check**

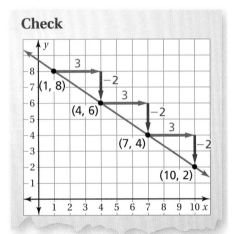

## On Your Own

**Now You're Ready**
Exercises 21–24

The points in the table lie on a line. How can you find the slope of the line from the table? What is the slope?

8.
| x | 1 | 3 | 5 | 7  |
|---|---|---|---|----|
| y | 2 | 5 | 8 | 11 |

9.
| x | −3 | −2 | −1 | 0 |
|---|----|----|----|---|
| y | 6  | 4  | 2  | 0 |

## Summary

**Slope**

*Positive slope*

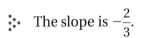

*Negative slope*

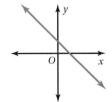

*Slope of 0*

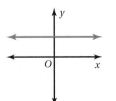

*Undefined slope*

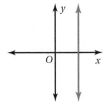

The line rises from left to right.

The line falls from left to right.

The line is horizontal.

The line is vertical.

## 2.2 Exercises

###  Vocabulary and Concept Check

1. **CRITICAL THINKING** Refer to the graph.
   a. Which lines have positive slopes?
   b. Which line has the steepest slope?
   c. Do any of the lines have undefined slope? Explain.

2. **OPEN-ENDED** Describe a real-life situation in which you need to know the slope.

3. **REASONING** The slope of a line is 0. What do you know about the line?

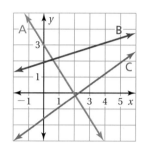

###  Practice and Problem Solving

**Draw a line through each point using the given slope. What do you notice about the two lines?**

4. Slope = 1

5. Slope = −3

6. Slope = $\dfrac{1}{4}$

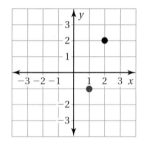 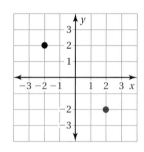 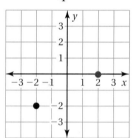

**Find the slope of the line.**

 7.

8.

9.

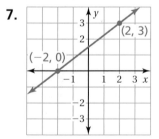 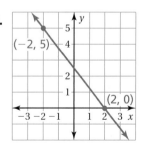 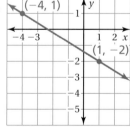

 10.

11.

12.

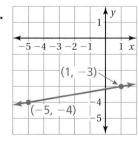 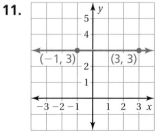 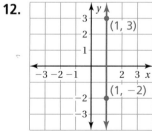

Section 2.2  Slope of a Line  53

**Find the slope of the line through the given points.**

**13.** $(4, -1), (-2, -1)$  **14.** $(5, -3), (5, 8)$  **15.** $(-7, 0), (-7, -6)$

**16.** $(-3, 1), (-1, 5)$  **17.** $(10, 4), (4, 15)$  **18.** $(-3, 6), (2, 6)$

**19. ERROR ANALYSIS** Describe and correct the error in finding the slope of the line.

**20. CRITICAL THINKING** Is it more difficult to walk up the ramp or the hill? Explain.

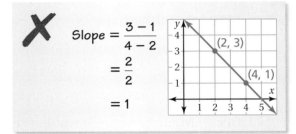

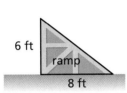
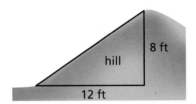

**The points in the table lie on a line. How can you find the slope of the line from the table? What is the slope?**

**21.**

| x | 1 | 3 | 5 | 7 |
|---|---|---|---|---|
| y | 2 | 10 | 18 | 26 |

**22.**

| x | -3 | 2 | 7 | 12 |
|---|---|---|---|---|
| y | 0 | 2 | 4 | 6 |

**23.**

| x | -6 | -2 | 2 | 6 |
|---|---|---|---|---|
| y | 8 | 5 | 2 | -1 |

**24.**

| x | -8 | -2 | 4 | 10 |
|---|---|---|---|---|
| y | 8 | 1 | -6 | -13 |

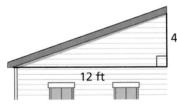

**25. PITCH** Carpenters refer to the slope of a roof as the *pitch* of the roof. Find the pitch of the roof.

**26. PROJECT** The guidelines for a wheelchair ramp suggest that the ratio of the rise to the run be no greater than $1:12$.

  **a. CHOOSE TOOLS** Find a wheelchair ramp in your school or neighborhood. Measure its slope. Does the ramp follow the guidelines?

  **b.** Design a wheelchair ramp that provides access to a building with a front door that is 2.5 feet higher than the sidewalk. Illustrate your design.

**Use an equation to find the value of $k$ so that the line that passes through the given points has the given slope.**

**27.** $(1, 3), (5, k)$; slope $= 2$

**28.** $(-2, k), (2, 0)$; slope $= -1$

**29.** $(-4, k), (6, -7)$; slope $= -\dfrac{1}{5}$

**30.** $(4, -4), (k, -1)$; slope $= \dfrac{3}{4}$

31. **TURNPIKE TRAVEL** The graph shows the cost of traveling by car on a turnpike.

    a. Find the slope of the line.
    b. Explain the meaning of the slope as a rate of change.

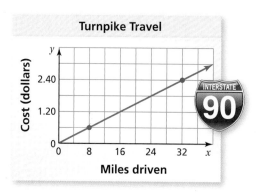

32. **BOAT RAMP** Which is steeper: the boat ramp or a road with a 12% grade? Explain. (*Note:* Road grade is the vertical increase divided by the horizontal distance.)

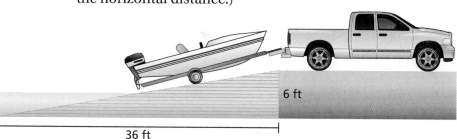

33. **REASONING** Do the points $A(-2, -1)$, $B(1, 5)$, and $C(4, 11)$ lie on the same line? Without using a graph, how do you know?

34. **BUSINESS** A small business earns a profit of $6500 in January and $17,500 in May. What is the rate of change in profit for this time period?

35. **STRUCTURE** Choose two points in the coordinate plane. Use the slope formula to find the slope of the line that passes through the two points. Then find the slope using the formula $\dfrac{y_1 - y_2}{x_1 - x_2}$. Are your results the same? Explain.

36. **Critical Thinking** The top and bottom of the slide are level with the ground, which has a slope of 0.

    a. What is the slope of the main portion of the slide?
    b. How does the slope change if the bottom of the slide is only 12 inches above the ground? Is the slide steeper? Explain.

 **Fair Game Review** *What you learned in previous grades & lessons*

**Graph the linear equation.** (*Section 2.1*)

37. $y = -\dfrac{1}{2}x$

38. $y = 3x - \dfrac{3}{4}$

39. $y = -\dfrac{x}{3} - \dfrac{3}{2}$

40. **MULTIPLE CHOICE** What is the prime factorization of 84? (*Skills Review Handbook*)

    Ⓐ $2 \times 3 \times 7$  Ⓑ $2^2 \times 3 \times 7$  Ⓒ $2 \times 3^2 \times 7$  Ⓓ $2^2 \times 21$

# Extension 2.2 Slopes of Parallel and Perpendicular Lines

**Key Vocabulary**
perpendicular lines, p. 57

**Study Tip**
Vertical lines have undefined slopes.

## Key Idea

**Parallel Lines and Slopes**

Two different lines in the same plane that never intersect are parallel lines. Nonvertical parallel lines have the same slope.

All vertical lines are parallel.

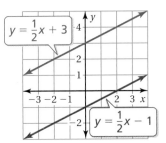

### EXAMPLE 1 Identifying Parallel Lines

**Which two lines are parallel? How do you know?**

Find the slope of each line.

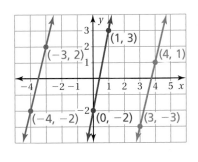

**Blue Line**

$$\text{slope} = \frac{y_2 - y_1}{x_2 - x_1}$$

$$= \frac{-2 - 2}{-4 - (-3)}$$

$$= \frac{-4}{-1}, \text{ or } 4$$

**Red Line**

$$\text{slope} = \frac{y_2 - y_1}{x_2 - x_1}$$

$$= \frac{-2 - 3}{0 - 1}$$

$$= \frac{-5}{-1}, \text{ or } 5$$

**Green Line**

$$\text{slope} = \frac{y_2 - y_1}{x_2 - x_1}$$

$$= \frac{-3 - 1}{3 - 4}$$

$$= \frac{-4}{-1}, \text{ or } 4$$

The slope of the blue and green lines is 4. The slope of the red line is 5.

∴ The blue and green lines have the same slope, so they are parallel.

## Practice

**Which lines are parallel? How do you know?**

1.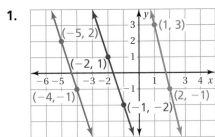

2.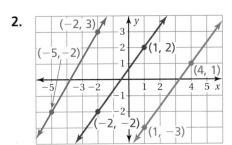

**Are the given lines parallel? Explain your reasoning.**

3. $y = -5, y = 3$

4. $y = 0, x = 0$

5. $x = -4, x = 1$

6. **GEOMETRY** The vertices of a quadrilateral are $A(-5, 3)$, $B(2, 2)$, $C(4, -3)$, and $D(-2, -2)$. How can you use slope to determine whether the quadrilateral is a parallelogram? Is it a parallelogram? Justify your answer.

**Common Core**

**Parallel and Perpendicular Lines**

In this extension, you will
- identify parallel and perpendicular lines.

Preparing for Standards
F.IF.4
F.IF.6

### Key Idea

**Perpendicular Lines and Slope**

Two lines in the same plane that intersect to form right angles are **perpendicular lines**. Two nonvertical lines are perpendicular if and only if the product of their slopes is $-1$.

Vertical lines are perpendicular to horizontal lines.

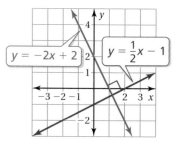

### EXAMPLE 2 Identifying Perpendicular Lines

**Which two lines are perpendicular? How do you know?**

Find the slope of each line.

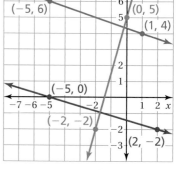

*Blue Line*

$\text{slope} = \dfrac{y_2 - y_1}{x_2 - x_1}$

$= \dfrac{4 - 6}{1 - (-5)}$

$= \dfrac{-2}{6}$, or $-\dfrac{1}{3}$

*Red Line*

$\text{slope} = \dfrac{y_2 - y_1}{x_2 - x_1}$

$= \dfrac{-2 - 0}{2 - (-5)}$

$= -\dfrac{2}{7}$

*Green Line*

$\text{slope} = \dfrac{y_2 - y_1}{x_2 - x_1}$

$= \dfrac{5 - (-2)}{0 - (-2)}$

$= \dfrac{7}{2}$

The slope of the red line is $-\dfrac{2}{7}$. The slope of the green line is $\dfrac{7}{2}$.

∴ Because $-\dfrac{2}{7} \cdot \dfrac{7}{2} = -1$, the red and green lines are perpendicular.

### Practice

**Which lines are perpendicular? How do you know?**

7.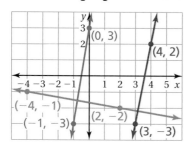

8.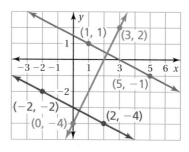

**Are the given lines perpendicular? Explain your reasoning.**

9. $x = -2, y = 8$

10. $x = -8, x = 7$

11. $y = 0, x = 0$

12. **GEOMETRY** The vertices of a parallelogram are $J(-5, 0)$, $K(1, 4)$, $L(3, 1)$, and $M(-3, -3)$. How can you use slope to determine whether the parallelogram is a rectangle? Is it a rectangle? Justify your answer.

Extension 2.2  Slopes of Parallel and Perpendicular Lines  57

# 2.3 Graphing Linear Equations in Slope-Intercept Form

**Essential Question** How can you describe the graph of the equation $y = mx + b$?

## 1 ACTIVITY: Finding Slopes and y-Intercepts

**Work with a partner.**

- Graph the equation.
- Find the slope of the line.
- Find the point where the line crosses the y-axis.

a. $y = -\dfrac{1}{2}x + 1$

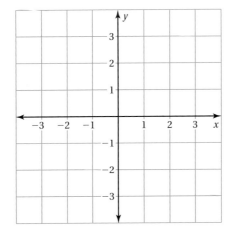

b. $y = -x + 2$

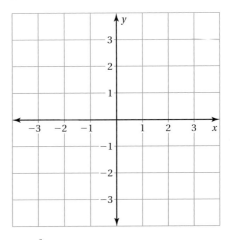

c. $y = -x - 2$

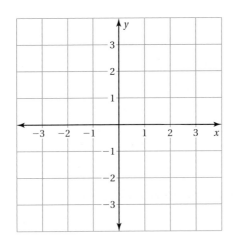

d. $y = \dfrac{1}{2}x + 1$

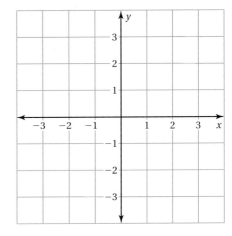

**COMMON CORE**

**Graphing Equations**

In this lesson, you will
- find slopes and y-intercepts of graphs of linear equations.
- graph linear equations written in slope-intercept form.

Learning Standards
A.CED.2
A.REI.10
F.IF.4

## Inductive Reasoning

**Math Practice 7**

**Look for Patterns**
What patterns do you notice in the table? What does this tell you about the graph of the equation?

Work with a partner. Graph each equation. Then copy and complete the table.

| | Equation | Description of Graph | Slope of Graph | Point of Intersection with y-axis |
|---|---|---|---|---|
| 1a | 2. $y = -\frac{1}{2}x + 1$ | Line | $-\frac{1}{2}$ | (0, 1) |
| 1b | 3. $y = -x + 2$ | | | |
| 1c | 4. $y = -x - 2$ | | | |
| 1d | 5. $y = \frac{1}{2}x + 1$ | | | |
| | 6. $y = x + 2$ | | | |
| | 7. $y = x - 2$ | | | |
| | 8. $y = \frac{1}{2}x - 1$ | | | |
| | 9. $y = -\frac{1}{2}x - 1$ | | | |
| | 10. $y = 3x + 2$ | | | |
| | 11. $y = 3x - 2$ | | | |
| | 12. $y = -2x + 3$ | | | |

## What Is Your Answer?

13. **IN YOUR OWN WORDS** How can you describe the graph of the equation $y = mx + b$?

    a. How does the value of $m$ affect the graph of the equation?
    b. How does the value of $b$ affect the graph of the equation?
    c. Check your answers to parts (a) and (b) with three equations that are not in the table.

14. **LOGIC** Why do you think $y = mx + b$ is called the "slope-intercept" form of the equation of a line? Use drawings or diagrams to support your answer.

Use what you learned about graphing linear equations in slope-intercept form to complete Exercises 4–6 on page 62.

Section 2.3  Graphing Linear Equations in Slope-Intercept Form

## 2.3 Lesson

Check It Out
Lesson Tutorials
BigIdeasMath.com

**Key Vocabulary**
*x*-intercept, p. 60
*y*-intercept, p. 60
slope-intercept form, p. 60

 **Key Ideas**

### Intercepts

The **x-intercept** of a line is the *x*-coordinate of the point where the line crosses the *x*-axis. It occurs when $y = 0$.

The **y-intercept** of a line is the *y*-coordinate of the point where the line crosses the *y*-axis. It occurs when $x = 0$.

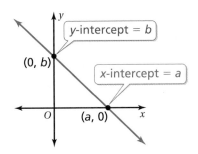

### Slope-Intercept Form

**Words** A linear equation written in the form $y = mx + b$ is in **slope-intercept form**. The slope of the line is $m$ and the *y*-intercept of the line is $b$.

**Algebra** $y = mx + b$
↑ ↑
slope  y-intercept

---

**EXAMPLE 1  Identifying Slopes and y-Intercepts**

**Find the slope and y-intercept of the graph of each linear equation.**

**a.** $y = -4x - 2$

$y = -4x + (-2)$   Write in slope-intercept form.

∴ The slope is $-4$ and the *y*-intercept is $-2$.

**b.** $y - 5 = \dfrac{3}{2}x$

$y = \dfrac{3}{2}x + 5$   Add 5 to each side.

∴ The slope is $\dfrac{3}{2}$ and the *y*-intercept is 5.

---

**On Your Own**

Now You're Ready
Exercises 7–15

**Find the slope and y-intercept of the graph of the linear equation.**

**1.** $y = 3x - 7$

**2.** $y - 1 = -\dfrac{2}{3}x$

---

60  Chapter 2   Graphing and Writing Linear Equations

**EXAMPLE 2  Graphing a Linear Equation in Slope-Intercept Form**

Graph $y = -3x + 3$. Identify the $x$-intercept.

**Step 1:** Find the slope and $y$-intercept.

$$y = -3x + 3$$

slope ↑      ↑ $y$-intercept

**Step 2:** The $y$-intercept is 3. So, plot (0, 3).

**Step 3:** Use the slope to find another point and draw the line.

$$\text{slope} = \frac{\text{rise}}{\text{run}} = \frac{-3}{1}$$

Plot the point that is 1 unit right and 3 units down from (0, 3). Draw a line through the two points.

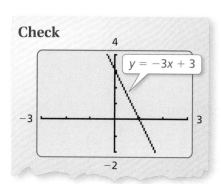

Check

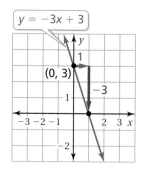

∴ The line crosses the $x$-axis at (1, 0). So, the $x$-intercept is 1.

**EXAMPLE 3  Real-Life Application**

The cost $y$ (in dollars) of taking a taxi $x$ miles is $y = 2.5x + 2$.
(a) Graph the equation. (b) Interpret the $y$-intercept and slope.

**a.** The slope of the line is $2.5 = \frac{5}{2}$. Use the slope and $y$-intercept to graph the equation.

The $y$-intercept is 2. So, plot (0, 2).

Use the slope to plot another point, (2, 7). Draw a line through the points.

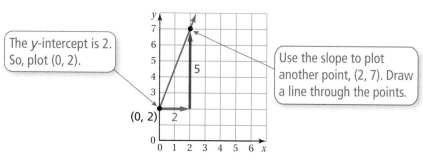

**b.** The slope is 2.5. So, the cost per mile is $2.50. The $y$-intercept is 2. So, there is an initial fee of $2 to take the taxi.

### On Your Own

Exercises 18–23

Graph the linear equation. Identify the $x$-intercept. Use a graphing calculator to check your answer.

**3.** $y = x - 4$

**4.** $y = -\frac{1}{2}x + 1$

**5.** In Example 3, the cost $y$ (in dollars) of taking a different taxi $x$ miles is $y = 2x + 1.5$. Interpret the $y$-intercept and slope.

# 2.3 Exercises

## Vocabulary and Concept Check

1. **VOCABULARY** How can you find the *x*-intercept of the graph of $2x + 3y = 6$?

2. **CRITICAL THINKING** Is the equation $y = 3x$ in slope-intercept form? Explain.

3. **OPEN-ENDED** Describe a real-life situation that can be modeled by a linear equation. Write the equation. Interpret the *y*-intercept and slope.

## Practice and Problem Solving

**Match the equation with its graph. Identify the slope and *y*-intercept.**

4. $y = 2x + 1$

5. $y = \frac{1}{3}x - 2$

6. $y = -\frac{2}{3}x + 1$

A.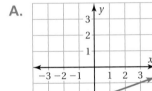

B.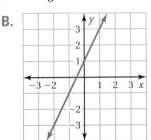

C.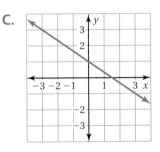

**Find the slope and *y*-intercept of the graph of the linear equation.**

7. $y = 4x - 5$

8. $y = -7x + 12$

9. $y = -\frac{4}{5}x - 2$

10. $y = 2.25x + 3$

11. $y + 1 = \frac{4}{3}x$

12. $y - 6 = \frac{3}{8}x$

13. $y - 3.5 = -2x$

14. $y + 5 = -\frac{1}{2}x$

15. $y = 1.5x + 11$

16. **ERROR ANALYSIS** Describe and correct the error in finding the slope and *y*-intercept of the graph of the linear equation.

> ✗ $y = 4x - 3$
> The slope is 4 and the *y*-intercept is 3.

17. **SKYDIVING** A skydiver parachutes to the ground. The height *y* (in feet) of the skydiver after *x* seconds is $y = -10x + 3000$.

   a. Graph the equation.

   b. Interpret the *x*-intercept and slope.

**Graph the linear equation. Identify the x-intercept. Use a graphing calculator to check your answer.**

**18.** $y = \dfrac{1}{5}x + 3$     **19.** $y = 6x - 7$     **20.** $y = -\dfrac{8}{3}x + 9$

**21.** $y = -1.4x - 1$     **22.** $y + 9 = -3x$     **23.** $y - 4 = -\dfrac{3}{5}x$

**24. PHONES** The cost $y$ (in dollars) of making a long distance phone call for $x$ minutes is $y = 0.25x + 2$.

  **a.** Graph the equation.
  **b.** Interpret the slope and $y$-intercept.

**25. APPLES** Write a linear equation that models the cost $y$ of picking $x$ pounds of apples. Graph the equation.

Admission: $5.00
Apples: $0.75 per lb

**26. ELEVATOR** The basement of a building is 40 feet below ground level. The elevator rises at a rate of 5 feet per second. You enter the elevator in the basement. Write an equation that represents the height $y$ (in feet) of the elevator after $x$ seconds. Graph the equation.

**27. REASONING** You work in an electronics store. You earn a fixed amount of $35 per day, plus a 15% bonus on the merchandise you sell. Write an equation that models the amount $y$ (in dollars) you earn for selling $x$ dollars of merchandise in one day. Graph the equation.

**28. Critical Thinking** Six friends create a website. The website earns money by selling banner ads. The site has five banner ads. It costs $120 a month to operate the website.

  **a.** A banner ad earns $0.005 per click. Write a linear equation that represents the monthly income $y$ (in dollars) for $x$ clicks.
  **b.** Draw a graph of the equation in part (a). On the graph, label the number of clicks needed for the friends to start making a profit.

## Fair Game Review  *What you learned in previous grades & lessons*

**Solve the equation for y.** *(Section 1.4)*

**29.** $y - 2x = 3$     **30.** $4x + 5y = 13$     **31.** $2x - 3y = 6$     **32.** $7x + 4y = 8$

**33. MULTIPLE CHOICE** Which point is a solution of the equation $3x - 8y = 11$? *(Section 2.1)*

  **Ⓐ** $(1, 1)$     **Ⓑ** $(1, -1)$     **Ⓒ** $(-1, 1)$     **Ⓓ** $(-1, -1)$

## 2.4 Graphing Linear Equations in Standard Form

**Essential Question** How can you describe the graph of the equation $ax + by = c$?

### 1 ACTIVITY: Using a Table to Plot Points

Work with a partner. You sold a total of $16 worth of tickets to a school concert. You lost track of how many of each type of ticket you sold.

$$\frac{\$4}{\text{Adult}} \cdot \text{Number of Adult Tickets} + \frac{\$2}{\text{Child}} \cdot \text{Number of Child Tickets} = \$16$$

a. Let $x$ represent the number of adult tickets.
   Let $y$ represent the number of child tickets.
   Write an equation that relates $x$ and $y$.

b. Copy and complete the table showing the different combinations of tickets you might have sold.

| Number of Adult Tickets, x |  |  |  |  |  |
|---|---|---|---|---|---|
| Number of Child Tickets, y |  |  |  |  |  |

c. Plot the points from the table. Describe the pattern formed by the points.

d. If you remember how many adult tickets you sold, can you determine how many child tickets you sold? Explain your reasoning.

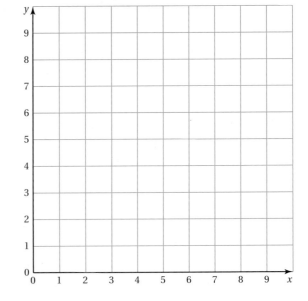

**Common Core**

**Graphing Equations**
In this lesson, you will
- graph linear equations written in standard form.

Learning Standards
A.CED.2
A.REI.10
F.IF.4

## 2 ACTIVITY: Rewriting an Equation

**Math Practice**

**Understand Quantities**
What do the equation and the graph represent? How can you use this information to solve the problem?

Work with a partner. You sold a total of $16 worth of cheese. You forgot how many pounds of each type of cheese you sold.

**CHEESE FOR SALE**
Swiss: $4/lb   Cheddar: $2/lb

$$\frac{\$4}{\text{lb}} \cdot \text{Pounds of Swiss} + \frac{\$2}{\text{lb}} \cdot \text{Pounds of Cheddar} = \$16$$

a. Let $x$ represent the number of pounds of Swiss cheese.

   Let $y$ represent the number of pounds of Cheddar cheese.

   Write an equation that relates $x$ and $y$.

b. Rewrite the equation in slope-intercept form. Then graph the equation.

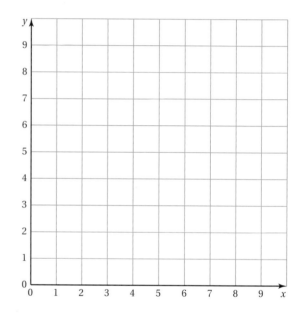

## What Is Your Answer?

3. **IN YOUR OWN WORDS** How can you describe the graph of the equation $ax + by = c$?

4. Activities 1 and 2 show two different methods for graphing $ax + by = c$. Describe the two methods. Which method do you prefer? Explain.

5. Write a real-life problem that is similar to those shown in Activities 1 and 2.

6. Why do you think it might be easier to graph $x + y = 10$ using standard form instead of rewriting it in slope-intercept form and then graphing?

**Practice** — Use what you learned about graphing linear equations in standard form to complete Exercises 3 and 4 on page 68.

Section 2.4  Graphing Linear Equations in Standard Form

# 2.4 Lesson

**Key Vocabulary**
standard form, *p. 66*

**Standard Form of a Linear Equation**

The **standard form** of a linear equation is

$$ax + by = c$$

where $a$ and $b$ are not both zero.

**Study Tip**
Any linear equation can be written in standard form.

### EXAMPLE 1 Graphing a Linear Equation in Standard Form

Graph $-2x + 3y = -6$.

**Step 1:** Write the equation in slope-intercept form.

| | |
|---|---|
| $-2x + 3y = -6$ | Write the equation. |
| $3y = 2x - 6$ | Add $2x$ to each side. |
| $y = \dfrac{2}{3}x - 2$ | Divide each side by 3. |

**Step 2:** Use the slope and $y$-intercept to graph the equation.

$$y = \tfrac{2}{3}x + (-2)$$

slope      $y$-intercept

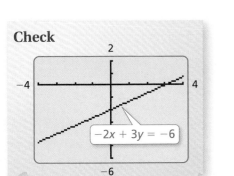

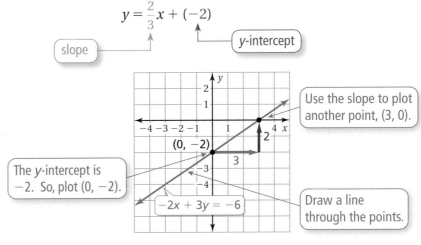

The $y$-intercept is $-2$. So, plot $(0, -2)$.

Use the slope to plot another point, $(3, 0)$.

Draw a line through the points.

### On Your Own

**Now You're Ready**
Exercises 5–10

Graph the linear equation. Use a graphing calculator to check your graph.

1. $x + y = -2$
2. $-\dfrac{1}{2}x + 2y = 6$
3. $-\dfrac{2}{3}x + y = 0$
4. $2x + y = 5$

Chapter 2    Graphing and Writing Linear Equations

### EXAMPLE 2 — Graphing a Linear Equation in Standard Form

Graph $x + 3y = -3$ using intercepts.

**Step 1:** To find the $x$-intercept, substitute 0 for $y$.

$$x + 3y = -3$$
$$x + 3(0) = -3$$
$$x = -3$$

To find the $y$-intercept, substitute 0 for $x$.

$$x + 3y = -3$$
$$0 + 3y = -3$$
$$y = -1$$

**Step 2:** Graph the equation.

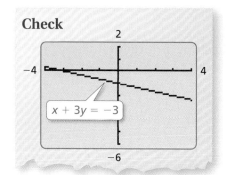

Check

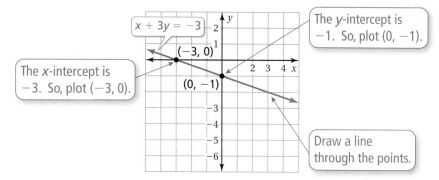

The $x$-intercept is $-3$. So, plot $(-3, 0)$.

The $y$-intercept is $-1$. So, plot $(0, -1)$.

Draw a line through the points.

### EXAMPLE 3 — Real-Life Application

Bananas $0.60/pound
Apples $1.50/pound

You have $6 to spend on apples and bananas. (a) Graph the equation $1.5x + 0.6y = 6$, where $x$ is the number of pounds of apples and $y$ is the number of pounds of bananas. (b) Interpret the intercepts.

a. Find the intercepts and graph the equation.

| $x$-intercept | $y$-intercept |
|---|---|
| $1.5x + 0.6y = 6$ | $1.5x + 0.6y = 6$ |
| $1.5x + 0.6(0) = 6$ | $1.5(0) + 0.6y = 6$ |
| $x = 4$ | $y = 10$ |

b. The $x$-intercept shows that you can buy 4 pounds of apples if you don't buy any bananas. The $y$-intercept shows that you can buy 10 pounds of bananas if you don't buy any apples.

### On Your Own

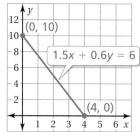
Now You're Ready
Exercises 16–18

Graph the linear equation using intercepts. Use a graphing calculator to check your graph.

**5.** $2x - y = 8$     **6.** $x + 3y = 6$

**7. WHAT IF?** In Example 3, you buy $y$ pounds of oranges instead of bananas. Oranges cost $1.20 per pound. Graph the equation $1.5x + 1.2y = 6$. Interpret the intercepts.

Section 2.4  Graphing Linear Equations in Standard Form  67

## 2.4 Exercises

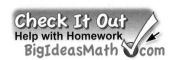

###  Vocabulary and Concept Check

1. **VOCABULARY** Is the equation $y = -2x + 5$ in standard form? Explain.

2. **REASONING** Does the graph represent a linear equation? Explain.

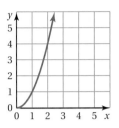

###  Practice and Problem Solving

**Define two variables for the verbal model. Write an equation in slope-intercept form that relates the variables. Graph the equation.**

3. $\dfrac{\$2.00}{\text{pound}} \cdot \text{Pounds of peaches} + \dfrac{\$1.50}{\text{pound}} \cdot \text{Pounds of apples} = \$15$

4. $\dfrac{16 \text{ miles}}{\text{hour}} \cdot \text{Hours biked} + \dfrac{2 \text{ miles}}{\text{hour}} \cdot \text{Hours walked} = 32 \text{ miles}$

**Write the linear equation in slope-intercept form.**

① 5. $2x + y = 17$

6. $5x - y = \dfrac{1}{4}$

7. $-\dfrac{1}{2}x + y = 10$

**Graph the linear equation. Use a graphing calculator to check your graph.**

8. $-18x + 9y = 72$

9. $16x - 4y = 2$

10. $\dfrac{1}{4}x + \dfrac{3}{4}y = 1$

**Use the graph to find the *x*- and *y*-intercepts.**

11.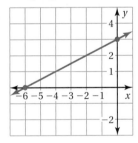

12.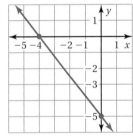

13.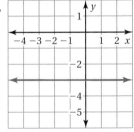

14. **ERROR ANALYSIS** Describe and correct the error in finding the *x*-intercept.

15. **BRACELET** A charm bracelet costs $65, plus $25 for each charm.

   a. Write an equation in standard form that represents the total cost of the bracelet.

   b. How much does the bracelet shown cost?

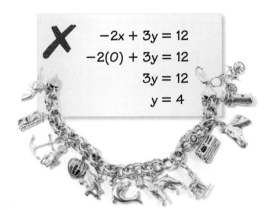

68 Chapter 2 Graphing and Writing Linear Equations

**Graph the linear equation using intercepts. Use a graphing calculator to check your graph.**

**16.** $3x - 4y = -12$  **17.** $2x + y = 8$  **18.** $\frac{1}{3}x - \frac{1}{6}y = -\frac{2}{3}$

**19. SHOPPING** The amount of money you spend on $x$ CDs and $y$ DVDs is given by the equation $14x + 18y = 126$. Find the intercepts and graph the equation.

**20. SCUBA** Five friends go scuba diving. They rent a boat for $x$ days and scuba gear for $y$ days. The total spent is $1000.

Boat: $250/day
Gear: $50/day

 a. Write an equation in standard form that represents the situation.
 b. Graph the equation and interpret the intercepts.

**21. MODELING** You work at a restaurant as a host and a server. You earn $9.45 for each hour you work as a host and $7.65 for each hour you work as a server.

 a. Write an equation in standard form that models your earnings.
 b. Graph the equation.

**Basic Information**
Pay to the Order of:
.................... John Doe
# of hours worked as
................. host: $x$
# of hours worked as
................. server: $y$
Earnings for this pay
......... period: $160.65

**22. LOGIC** Does the graph of every linear equation have an $x$-intercept? Explain your reasoning. Include an example.

**23. Critical Thinking** For a house call, a veterinarian charges $70, plus $40 an hour.

 a. Write an equation that represents the total fee $y$ charged by the veterinarian for a visit lasting $x$ hours.
 b. Find the $x$-intercept. Will this point appear on the graph of the equation? Explain your reasoning.
 c. Graph the equation.

 **Fair Game Review** *What you learned in previous grades & lessons*

**Copy and complete the table of values.** *(Skills Review Handbook)*

**24.**

| $x$ | $-2$ | $-1$ | $0$ | $1$ | $2$ |
|---|---|---|---|---|---|
| $2x + 5$ | | | | | |

**25.**

| $x$ | $-2$ | $-1$ | $0$ | $1$ | $2$ |
|---|---|---|---|---|---|
| $-5 - 3x$ | | | | | |

**26. MULTIPLE CHOICE** Which value of $x$ makes the equation $4x - 12 = 3x - 9$ true? *(Section 1.3)*

  Ⓐ $-1$   Ⓑ $0$   Ⓒ $1$   Ⓓ $3$

# 2 Study Help

Check It Out
Graphic Organizer
BigIdeasMath.com

You can use a **process diagram** to show the steps involved in a procedure. Here is an example of a process diagram for graphing a linear equation.

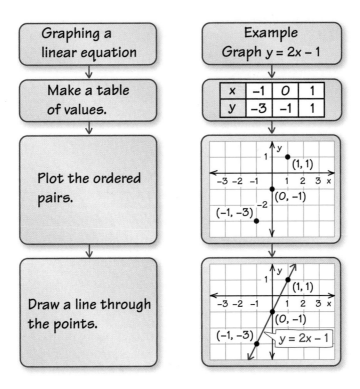

## On Your Own

**Make process diagrams with examples to help you study these topics.**

1. finding the slope of a line
2. graphing a linear equation using
   a. slope and *y*-intercept
   b. *x*- and *y*-intercepts

**After you complete this chapter, make process diagrams for the following topics.**

3. writing equations in slope-intercept form
4. writing equations in point-slope form
5. writing equations of parallel lines
6. writing equations of perpendicular lines

"Here is a process diagram with suggestions for what to do if a hyena knocks on your door."

# 2.1–2.4 Quiz

**Graph the linear equation using a table.** *(Section 2.1)*

1. $y = -x + 8$
2. $y = \dfrac{x}{3} - 4$
3. $x = -1$
4. $y = 3.5$

**Find the slope of the line.** *(Section 2.2)*

5.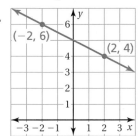
6.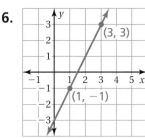
7.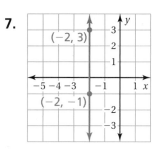

8. What is the slope of a line that is parallel to the line in Exercise 5? What is the slope of a line that is perpendicular to the line in Exercise 5? *(Section 2.2)*

9. Are the lines $y = -1$ and $x = 1$ parallel? Are they perpendicular? Justify your answer. *(Section 2.2)*

**Find the slope and *y*-intercept of the graph of the linear equation.** *(Section 2.3)*

10. $y = \dfrac{1}{4}x - 8$
11. $y = -x + 3$

**Find the *x*- and *y*-intercepts of the graph of the equation.** *(Section 2.4)*

12. $3x - 2y = 12$
13. $x + 5y = 15$

14. **BANKING** A bank charges $3 each time you use an out-of-network ATM. At the beginning of the month, you have $1500 in your bank account. You withdraw $60 from your bank account each time you use an out-of-network ATM. Write and graph a linear equation that represents the balance in your account after you use an out-of-network ATM *x* times. *(Section 2.1)*

15. **STATE FAIR** Write a linear equation that models the cost *y* of one person going on *x* rides at the fair. Graph the equation. *(Section 2.3)*

16. **PAINTING** You used $90 worth of paint for a school float. *(Section 2.4)*

   a. Graph the equation $18x + 15y = 90$, where *x* is the number of gallons of blue paint and *y* is the number of gallons of white paint.

   b. Interpret the intercepts.

# 2.5 Writing Equations in Slope-Intercept Form

**Essential Question** How can you write an equation of a line when you are given the slope and *y*-intercept of the line?

## 1 ACTIVITY: Writing Equations of Lines

Work with a partner.
- Find the slope of each line.
- Find the *y*-intercept of each line.
- Write an equation for each line.
- What do the three lines have in common?

a.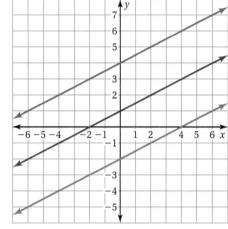

b.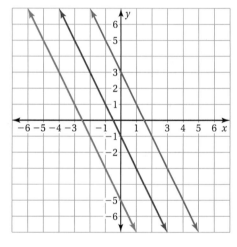

c.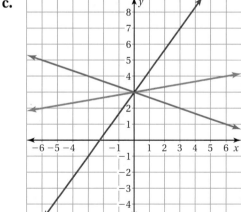

d.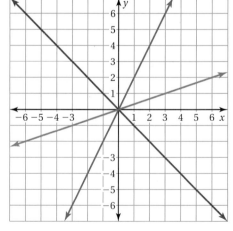

**COMMON CORE**

**Graphing Equations**

In this lesson, you will
- write equations of lines in slope-intercept form.

Learning Standards
8.F.3
A.CED.2
A.CED.3

72  Chapter 2  Graphing and Writing Linear Equations

### 2 ACTIVITY: Describing a Parallelogram

**Math Practice**

**Analyze Givens**
What do you need to know in order to write an equation?

Work with a partner.
- Find the area of each parallelogram.
- Write an equation for each side of each parallelogram.
- What do you notice about the slopes of the opposite sides of each parallelogram?

a.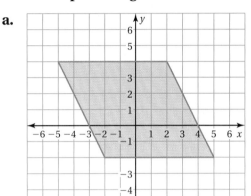

b.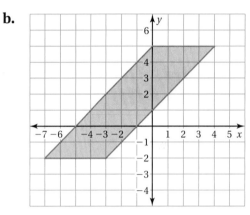

### 3 ACTIVITY: Interpreting the Slope and y-Intercept

Work with a partner. The graph shows a trip taken by a car where $t$ is the time (in hours) and $y$ is the distance (in miles) from Phoenix.

a. How far from Phoenix was the car at the beginning of the trip?
b. What was the car's speed?
c. How long did the trip last?
d. How far from Phoenix was the car at the end of the trip?

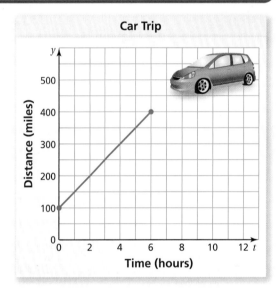

## What Is Your Answer?

4. **IN YOUR OWN WORDS** How can you write an equation of a line when you are given the slope and y-intercept of the line? Give an example that is different from those in Activities 1, 2, and 3.

**Practice** → Use what you learned about writing equations in slope-intercept form to complete Exercises 3 and 4 on page 76.

Section 2.5 Writing Equations in Slope-Intercept Form

## 2.5 Lesson

### EXAMPLE 1 — Writing Equations in Slope-Intercept Form

**Write an equation of the line in slope-intercept form.**

a.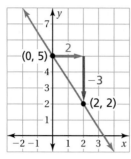

Find the slope and y-intercept.

$$\text{slope} = \frac{y_2 - y_1}{x_2 - x_1}$$

$$= \frac{2 - 5}{2 - 0}$$

$$= \frac{-3}{2}, \text{ or } -\frac{3}{2}$$

**Study Tip**

After writing an equation, check that the given points are solutions of the equation.

Because the line crosses the y-axis at (0, 5), the y-intercept is 5.

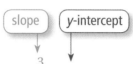

So, the equation is $y = -\frac{3}{2}x + 5$.

b.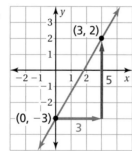

Find the slope and y-intercept.

$$\text{slope} = \frac{y_2 - y_1}{x_2 - x_1}$$

$$= \frac{-3 - 2}{0 - 3}$$

$$= \frac{-5}{-3}, \text{ or } \frac{5}{3}$$

Because the line crosses the y-axis at (0, −3), the y-intercept is −3.

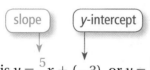

So, the equation is $y = \frac{5}{3}x + (-3)$, or $y = \frac{5}{3}x - 3$.

### On Your Own

**Now You're Ready**
Exercises 5–10

**Write an equation of the line in slope-intercept form.**

1.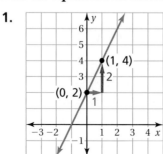

2.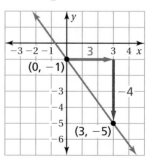

### EXAMPLE 2  Writing an Equation

**Which equation is shown in the graph?**

Ⓐ $y = -4$   Ⓑ $y = -3$
Ⓒ $y = 0$    Ⓓ $y = -3x$

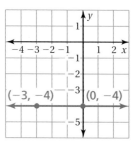

**Remember**
The graph of $y = a$ is a horizontal line that passes through $(0, a)$.

Find the slope and y-intercept.

The line is horizontal, so the change in y is 0.

$$\text{slope} = \frac{\text{change in } y}{\text{change in } x} = \frac{0}{3} = 0$$

Because the line crosses the y-axis at $(0, -4)$, the y-intercept is $-4$.

∴ So, the equation is $y = 0x + (-4)$, or $y = -4$. The correct answer is Ⓐ.

### EXAMPLE 3  Real-Life Application

The graph shows the distance remaining to complete a tunnel.
(a) Write an equation that represents the distance y (in feet) remaining after x months. (b) How much time does it take to complete the tunnel?

Engineers used tunnel boring machines like the ones shown above to dig an extension of the Metro Gold Line in Los Angeles. The new tunnels are 1.7 miles long and 21 feet wide.

a.  Find the slope and y-intercept.

$$\text{slope} = \frac{\text{change in } y}{\text{change in } x} = \frac{-2000}{4} = -500$$

Because the line crosses the y-axis at $(0, 3500)$, the y-intercept is 3500.

∴ So, the equation is $y = -500x + 3500$.

b.  The tunnel is complete when the distance remaining is 0 feet. So, find the value of x when $y = 0$.

| | |
|---|---|
| $y = -500x + 3500$ | Write the equation. |
| $0 = -500x + 3500$ | Substitute 0 for y. |
| $-3500 = -500x$ | Subtract 3500 from each side. |
| $7 = x$ | Solve for x. |

∴ It takes 7 months to complete the tunnel.

### On Your Own

**Now You're Ready**
Exercises 13–15

3. Write an equation of the line that passes through $(0, 5)$ and $(4, 5)$.

4. **WHAT IF?** In Example 3, the points are $(0, 3500)$ and $(5, 1500)$. How long does it take to complete the tunnel?

# 2.5 Exercises

## Vocabulary and Concept Check

1. **PRECISION** Explain how to find the slope of a line given the intercepts of the line.
2. **WRITING** Explain how to write an equation of a line using its graph.

## Practice and Problem Solving

**Write an equation for each side of the figure.**

3.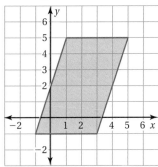

4.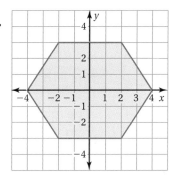

**Write an equation of the line in slope-intercept form.**

5.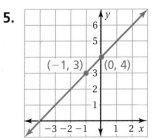

6.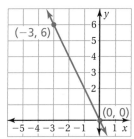

7.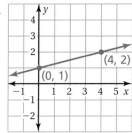

8.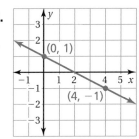

9.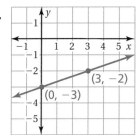

10.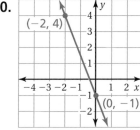

11. **ERROR ANALYSIS** Describe and correct the error in writing the equation of the line.

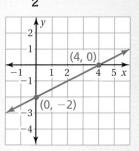

12. **BOA** A boa constrictor is 18 inches long at birth and grows 8 inches per year. Write an equation that represents the length $y$ (in feet) of a boa constrictor that is $x$ years old.

**76** Chapter 2 Graphing and Writing Linear Equations

**Write an equation of the line that passes through the points.**

**13.** (2, 5), (0, 5)      **14.** (−3, 0), (0, 0)      **15.** (0, −2), (4, −2)

**16. WALKATHON** One of your friends gives you $10 for a charity walkathon. Another friend gives you an amount per mile. After 5 miles, you have raised $13.50 total. Write an equation that represents the amount $y$ of money you have raised after $x$ miles.

**17. BRAKING TIME** During each second of braking, an automobile slows by about 10 miles per hour.

   **a.** Plot the points (0, 60) and (6, 0). What do the points represent?

   **b.** Draw a line through the points. What does the line represent?

   **c.** Write an equation of the line.

**18. PAPER** You have 500 sheets of notebook paper. After 1 week, you have 72% of the sheets left. You use the same number of sheets each week. Write an equation that represents the number $y$ of pages remaining after $x$ weeks.

**19. Critical Thinking** The palm tree on the left is 10 years old. The palm tree on the right is 8 years old. The trees grow at the same rate.

   **a.** Estimate the height $y$ (in feet) of each tree.

   **b.** Plot the two points $(x, y)$, where $x$ is the age of each tree and $y$ is the height of each tree.

   **c.** What is the rate of growth of the trees?

   **d.** Write an equation that represents the height of a palm tree in terms of its age.

 **Fair Game Review** *What you learned in previous grades & lessons*

**Plot the ordered pair in a coordinate plane.** *(Skills Review Handbook)*

**20.** (1, 4)      **21.** (−1, −2)      **22.** (0, 1)      **23.** (2, 7)

**24. MULTIPLE CHOICE** Which of the following statements is true? *(Section 2.3)*

   **Ⓐ** The $x$-intercept is 5.

   **Ⓑ** The $x$-intercept is −2.

   **Ⓒ** The $y$-intercept is 5.

   **Ⓓ** The $y$-intercept is −2.

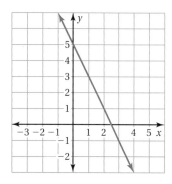

## 2.6 Writing Equations in Point-Slope Form

**Essential Question** How can you write an equation of a line when you are given the slope and a point on the line?

### 1 ACTIVITY: Writing Equations of Lines

Work with a partner.
- Sketch the line that has the given slope and passes through the given point.
- Find the *y*-intercept of the line.
- Write an equation of the line.

a. $m = -2$

b. $m = \dfrac{1}{3}$

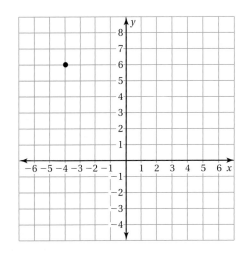
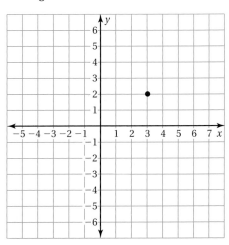

c. $m = -\dfrac{2}{3}$

d. $m = \dfrac{5}{2}$

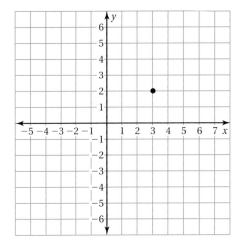
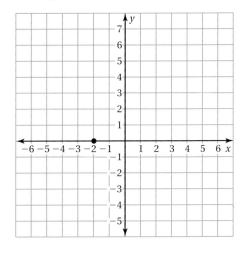

**COMMON CORE**

**Writing Equations**
In this lesson, you will
- write equations of lines using a slope and a point.
- write equations of lines using two points.

Learning Standards
A.CED.2
A.REI.10
F.IF.4
F.IF.6

Chapter 2  Graphing and Writing Linear Equations

## 2 ACTIVITY: Developing a Formula

**Math Practice**

**Construct Arguments**
How does a graph help you develop a formula?

**Work with a partner.**

a. Draw a nonvertical line that passes through the point $(x_1, y_1)$.

b. Plot another point on your line. Label this point as $(x, y)$. This point represents any other point on the line.

c. Label the rise and run of the line through the points $(x_1, y_1)$ and $(x, y)$.

d. The rise can be written as $y - y_1$. The run can be written as $x - x_1$. Explain why this is true.

e. Write an equation for the slope $m$ of the line using the expressions from part (d).

f. Multiply each side of the equation by the expression in the denominator. Write your result. What does this result represent?

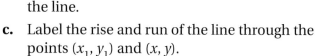

## 3 ACTIVITY: Writing an Equation

**Work with a partner.**

For 4 months, you have saved $25 a month. You now have $175 in your savings account.

- Draw a graph that shows the balance in your account after $t$ months.

- Use your result from Activity 2 to write an equation that represents the balance $A$ after $t$ months.

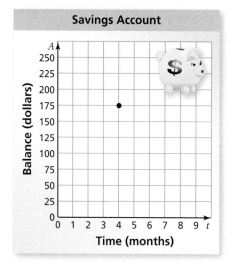

### What Is Your Answer?

4. Redo Activity 1 using the formula you found in Activity 2. Compare the results. What do you notice?

5. The formula you found in Activity 2, $y - y_1 = m(x - x_1)$, is called the *point-slope form* of the equation of a line. Why is $y - y_1 = m(x - x_1)$ called the "point-slope" form? Why do you think it is important?

6. **IN YOUR OWN WORDS** How can you write an equation of a line when you are given the slope and a point on the line? Give an example that is different from those in Activity 1.

**Practice**
Use what you learned about writing equations using a slope and a point to complete Exercises 3–5 on page 82.

## 2.6 Lesson

**Key Vocabulary**
point-slope form, p. 80

### Key Idea

**Point-Slope Form**

**Words** A linear equation written in the form $y - y_1 = m(x - x_1)$ is in **point-slope form.** The line passes through the point $(x_1, y_1)$ and the slope of the line is $m$.

**Algebra**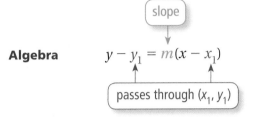

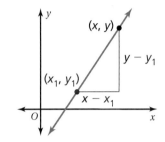

### EXAMPLE 1 — Writing an Equation Using a Slope and a Point

Write in point-slope form an equation of the line that passes through the point $(-6, 1)$ with slope $\frac{2}{3}$.

$y - y_1 = m(x - x_1)$    Write the point-slope form.

$y - 1 = \frac{2}{3}[x - (-6)]$    Substitute $\frac{2}{3}$ for $m$, $-6$ for $x_1$, and 1 for $y_1$.

$y - 1 = \frac{2}{3}(x + 6)$    Simplify.

∴ So, the equation is $y - 1 = \frac{2}{3}(x + 6)$.

**Check** Check that $(-6, 1)$ is a solution of the equation.

$y - 1 = \frac{2}{3}(x + 6)$    Write the equation.

$1 - 1 \stackrel{?}{=} \frac{2}{3}(-6 + 6)$    Substitute.

$0 = 0$ ✓    Simplify.

### On Your Own

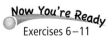
Exercises 6–11

Write in point-slope form an equation of the line that passes through the given point and has the given slope.

1. $(1, 2)$; $m = -4$    2. $(7, 0)$; $m = 1$    3. $(-8, -5)$; $m = -\frac{3}{4}$

**80**   Chapter 2   Graphing and Writing Linear Equations

## EXAMPLE 2 Writing an Equation Using Two Points

Write in slope-intercept form an equation of the line that passes through the points (2, 4) and (5, −2).

Find the slope: $m = \dfrac{y_2 - y_1}{x_2 - x_1} = \dfrac{-2 - 4}{5 - 2} = \dfrac{-6}{3} = -2$

Then use the slope $m = -2$ and the point (2, 4) to write an equation of the line.

| | |
|---|---|
| $y - y_1 = m(x - x_1)$ | Write the point-slope form. |
| $y - 4 = -2(x - 2)$ | Substitute −2 for $m$, 2 for $x_1$, and 4 for $y_1$. |
| $y - 4 = -2x + 4$ | Use Distributive Property. |
| $y = -2x + 8$ | Write in slope-intercept form. |

**Study Tip**

You can use either of the given points to write the equation of the line.
Use $m = -2$ and (5, −2).
$y - (-2) = -2(x - 5)$
$y + 2 = -2x + 10$
$y = -2x + 8$ ✓

## EXAMPLE 3 Real-Life Application

You finish parasailing and are being pulled back to the boat. After 2 seconds, you are 25 feet above the boat. (a) Write and graph an equation that represents your height $y$ (in feet) above the boat after $x$ seconds. (b) At what height were you parasailing?

a. You are being pulled down at the rate of 10 feet per second. So, the slope is −10. You are 25 feet above the boat after 2 seconds. So, the line passes through (2, 25). Use the point-slope form.

| | |
|---|---|
| $y - 25 = -10(x - 2)$ | Substitute for $m$, $x_1$, and $y_1$. |
| $y - 25 = -10x + 20$ | Use Distributive Property. |
| $y = -10x + 45$ | Write in slope-intercept form. |

∴ So, the equation is $y = -10x + 45$.

b. You start descending when $x = 0$. The $y$-intercept is 45. So, you were parasailing at a height of 45 feet.

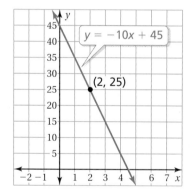

### On Your Own

Write in slope-intercept form an equation of the line that passes through the given points.

**4.** (−2, 1), (3, −4)    **5.** (−5, −5), (−3, 3)    **6.** (−8, 6), (−2, 9)

**7. WHAT IF?** In Example 3, you are 35 feet above the boat after 2 seconds. Write and graph an equation that represents your height $y$ (in feet) above the boat after $x$ seconds.

# 2.6 Exercises

## Vocabulary and Concept Check

1. **VOCABULARY** From the equation $y - 3 = -2(x + 1)$, identify the slope and a point on the line.

2. **WRITING** Describe how to write an equation of a line using (a) its slope and a point on the line, and (b) two points on the line.

## Practice and Problem Solving

Use the point-slope form to write an equation of the line with the given slope that passes through the given point.

3. $m = \dfrac{1}{2}$

4. $m = -\dfrac{3}{4}$

5. $m = -3$

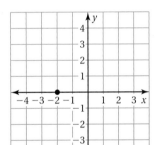
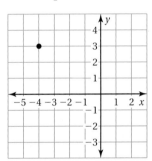
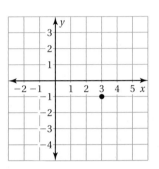

Write in point-slope form an equation of the line that passes through the given point and has the given slope.

6. $(3, 0);\ m = -\dfrac{2}{3}$

7. $(4, 8);\ m = \dfrac{3}{4}$

8. $(1, -3);\ m = 4$

9. $(7, -5);\ m = -\dfrac{1}{7}$

10. $(3, 3);\ m = \dfrac{5}{3}$

11. $(-1, -4);\ m = -2$

Write in slope-intercept form an equation of the line that passes through the given points.

12. $(-1, -1),\ (1, 5)$

13. $(2, 4),\ (3, 6)$

14. $(-2, 3),\ (2, 7)$

15. $(4, 1),\ (8, 2)$

16. $(-9, 5),\ (-3, 3)$

17. $(1, 2),\ (-2, -1)$

18. **CHEMISTRY** At 0°C, the volume of a gas is 22 liters. For each degree the temperature $T$ (in degrees Celsius) increases, the volume $V$ (in liters) of the gas increases by $\dfrac{2}{25}$. Write an equation that represents the volume of the gas in terms of the temperature.

19. **CARS** After it is purchased, the value of a new car decreases $4000 each year. After 3 years, the car is worth $18,000.

   a. Write an equation that represents the value $V$ (in dollars) of the car $x$ years after it is purchased.

   b. What was the original value of the car?

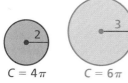

20. **CIRCUMFERENCE** Consider the circles shown.

   a. Plot the points $(2, 4\pi)$ and $(3, 6\pi)$.

   b. Write an equation of the line that passes through the two points.

21. **CRICKETS** According to Dolbear's Law, you can predict the temperature $T$ (in degrees Fahrenheit) by counting the number $x$ of chirps made by a snowy tree cricket in 1 minute. For each rise in temperature of 0.25°F, the cricket makes an additional chirp each minute.

   a. A cricket chirps 40 times in 1 minute when the temperature is 50°F. Write an equation that represents the temperature in terms of the number of chirps in 1 minute.

   b. You count 100 chirps in 1 minute. What is the temperature?

   c. The temperature is 96°F. How many chirps would you expect the cricket to make?

Leaning Tower of Pisa

22. **WATERING CAN** You water the plants in your classroom at a constant rate. After 5 seconds, your watering can contains 58 ounces of water. Fifteen seconds later, the can contains 28 ounces of water.

   a. Write an equation that represents the amount $y$ (in ounces) of water in the can after $x$ seconds.

   b. How much water was in the can when you started watering the plants?

   c. When is the watering can empty?

23.  **Problem Solving** The Leaning Tower of Pisa in Italy was built between 1173 and 1350.

   a. Write an equation for the yellow line.

   b. The tower is 56 meters tall. How far off center is the top of the tower?

## Fair Game Review  *What you learned in previous grades & lessons*

**Find the percent of the number.**  *(Skills Review Handbook)*

24. 15% of 300

25. 140% of 125

26. 6% of $-75$

27. **MULTIPLE CHOICE** What is the $x$-intercept of the equation $3x + 5y = 30$? *(Section 2.4)*

   Ⓐ $-10$   Ⓑ $-6$   Ⓒ 6   Ⓓ 10

# Extension 2.6 Writing Equations of Parallel and Perpendicular Lines

You can use the slope-intercept form or the point-slope form to write equations of parallel and perpendicular lines.

### EXAMPLE 1  Writing an Equation of a Parallel Line

Write an equation of the line that passes through $(6, -2)$ and is parallel to the line $y = \frac{1}{2}x + 3$.

**Remember**
Lines that are parallel have the same slope. Lines that are perpendicular have slopes whose product is $-1$.

**Step 1:** Find the slope of the parallel line.

The slope of $y = \frac{1}{2}x + 3$ is $\frac{1}{2}$. So, the parallel line that passes through $(6, -2)$ has the same slope, $\frac{1}{2}$.

**Step 2:** Use the slope $\frac{1}{2}$ and the slope-intercept form to find the $y$-intercept of the parallel line that passes through $(6, -2)$.

$y = mx + b$   Write the slope-intercept form.

$-2 = \frac{1}{2}(6) + b$   Substitute $\frac{1}{2}$ for $m$, 6 for $x$, and $-2$ for $y$.

$-2 = 3 + b$   Multiply.

$-5 = b$   Subtract 3 from each side.

The parallel line has a slope of $\frac{1}{2}$ and a $y$-intercept of $-5$.

∴ So, an equation of the parallel line is $y = \frac{1}{2}x + (-5)$, or $y = \frac{1}{2}x - 5$.

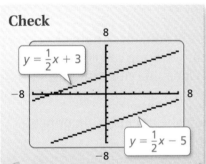

## Practice

Write an equation of the line that passes through the given point and is parallel to the given line. Use a graphing calculator to check your answer.

1. $(-2, 1); y = 3x - 4$
2. $(6, -3); y = -\frac{2}{3}x + 5$
3. $(-4, -5); y = -4x - 1$

Write an equation of the line that passes through the given point and is parallel to the line shown in the graph.

4. $(2, -3)$
5. $(0, 0)$
6. $(-5, 8)$
7. $(-1, -7)$

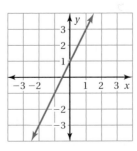

# EXAMPLE 2  Writing an Equation of a Perpendicular Line

**Writing Equations**
In this extension, you will
- write equations of parallel and perpendicular lines.

Learning Standards
A.CED.2
A.REI.10
F.IF.4
F.IF.6

Write an equation of the line that passes through $(-3, 1)$ and is perpendicular to the line shown in the graph.

**Step 1:** Find the slope of the line in the graph.

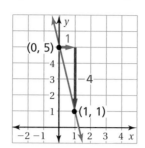

$$\text{slope} = \frac{\text{change in } y}{\text{change in } x} = \frac{-4}{1} = -4$$

**Step 2:** Find the slope of the perpendicular line.

The slope of the line in the graph is $-4$. Because $-4 \cdot \frac{1}{4} = -1$, the slope of the perpendicular line is $\frac{1}{4}$.

**Step 3:** Use the slope $m = \frac{1}{4}$ and the point-slope form to write an equation of the perpendicular line that passes through $(-3, 1)$.

| | |
|---|---|
| $y - y_1 = m(x - x_1)$ | Write the point-slope form. |
| $y - 1 = \frac{1}{4}[x - (-3)]$ | Substitute $\frac{1}{4}$ for $m$, $-3$ for $x_1$, and 1 for $y_1$. |
| $y - 1 = \frac{1}{4}x + \frac{3}{4}$ | Simplify. |
| $y = \frac{1}{4}x + \frac{7}{4}$ | Add 1 to each side. |

So, an equation of the perpendicular line is $y = \frac{1}{4}x + \frac{7}{4}$.

## Practice

Write an equation of the line that passes through the given point and is perpendicular to the given line.

**8.** $(4, 2); y = -\frac{1}{3}x + 1$  **9.** $(0, -1); y = x - 6$  **10.** $(-3, 7); y = -2x - 5$

Write an equation of the line that passes through the given point and is perpendicular to the line shown in the graph.

**11.** $(1, 4)$  **12.** $(0, 6)$

**13.** $(-5, -2)$  **14.** $(3, -3)$

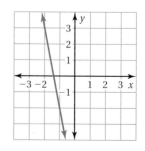

**15. REASONING** Rework Example 1 using the point-slope form. Rework Example 2 using the slope-intercept form. Which method do you prefer? Explain your reasoning.

# 2.7 Solving Real-Life Problems

**Essential Question** How can you use a linear equation in two variables to model and solve a real-life problem?

### 1 EXAMPLE: Writing a Story

Write a story that uses the graph at the right.

- In your story, interpret the slope of the line, the *y*-intercept, and the *x*-intercept.
- Make a table that shows data from the graph.
- Label the axes of the graph with units.
- Draw pictures for your story.

There are many possible stories. Here is one about a reef tank.

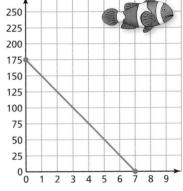

Tom works at an aquarium shop on Saturdays. One Saturday, when Tom gets to work, he is asked to clean a 175-gallon reef tank.

His first job is to drain the tank. He puts a hose into the tank and starts a siphon. Tom wonders if the tank will finish draining before he leaves work.

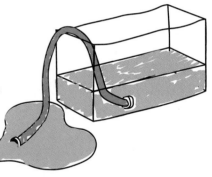

He measures the amount of water that is draining out and finds that 12.5 gallons drain out in 30 minutes. So, he figures that the rate is 25 gallons per hour. To see when the tank will be empty, Tom makes a table and draws a graph.

**COMMON CORE**

**Writing Equations**

In this lesson, you will
- solve real-life problems involving linear equations.

Applying Standards
8.F.4
A.CED.2
F.IF.4

x-intercept: number of hours to empty the tank

| x | 0 | 1 | 2 | 3 | 4 | 5 | 6 | 7 |
|---|---|---|---|---|---|---|---|---|
| y | 175 | 150 | 125 | 100 | 75 | 50 | 25 | 0 |

y-intercept: amount of water in full tank

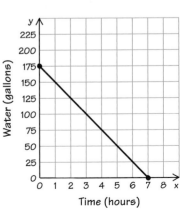

From the table and also from the graph, Tom sees that the tank will be empty after 7 hours. This will give him 1 hour to wash the tank before going home.

**86 Chapter 2** Graphing and Writing Linear Equations

## 2 ACTIVITY: Writing a Story

**Math Practice 6**

**Label Axes**
What information is needed to label the axes? How do you know where to place the labels?

Work with a partner. Write a story that uses the graph of a line.

- In your story, interpret the slope of the line, the *y*-intercept, and the *x*-intercept.
- Make a table that shows data from the graph.
- Label the axes of the graph with units.
- Draw pictures for your story.

## 3 ACTIVITY: Drawing Graphs

Work with a partner. Describe a real-life problem that has the given rate and intercepts. Draw a line that represents the problem.

a. Rate: −30 feet per second
   *y*-intercept: 150 feet
   *x*-intercept: 5 seconds

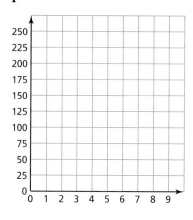

b. Rate: −25 dollars per month
   *y*-intercept: $200
   *x*-intercept: 8 months

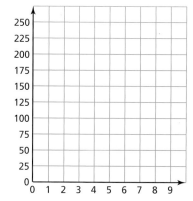

## What Is Your Answer?

4. **IN YOUR OWN WORDS** How can you use a linear equation in two variables to model and solve a real-life problem? List three different rates that can be represented by slopes in real-life problems.

**Practice** — Use what you learned about solving real-life problems to complete Exercises 4 and 5 on page 90.

## 2.7 Lesson

### EXAMPLE 1 Real-Life Application

The percent $y$ (in decimal form) of battery power remaining $x$ hours after you turn on a laptop computer is $y = -0.2x + 1$. (a) Graph the equation. (b) Interpret the $x$- and $y$-intercepts. (c) After how many hours is the battery power at 75%?

**a.** Use the slope and the $y$-intercept to graph the equation.

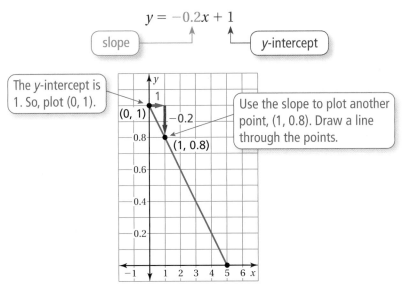

**b.** To find the $x$-intercept, substitute 0 for $y$ in the equation.

| | |
|---|---|
| $y = -0.2x + 1$ | Write the equation. |
| $0 = -0.2x + 1$ | Substitute 0 for $y$. |
| $5 = x$ | Solve for $x$. |

∴ The $x$-intercept is 5. So, the battery lasts 5 hours. The $y$-intercept is 1. So, the battery power is at 100% when you turn on the laptop.

**c.** Find the value of $x$ when $y = 0.75$.

| | |
|---|---|
| $y = -0.2x + 1$ | Write the equation. |
| $0.75 = -0.2x + 1$ | Substitute 0.75 for $y$. |
| $1.25 = x$ | Solve for $x$. |

75% Remaining

∴ The battery power is at 75% after 1.25 hours.

### On Your Own

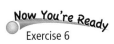
Exercise 6

**1.** The amount $y$ (in gallons) of gasoline remaining in a gas tank after driving $x$ hours is $y = -2x + 12$. (a) Graph the equation. (b) Interpret the $x$- and $y$-intercepts. (c) After how many hours are there 5 gallons left?

### EXAMPLE 2 Real-Life Application

The graph relates temperatures $y$ (in degrees Fahrenheit) to temperatures $x$ (in degrees Celsius). (a) Find the slope and $y$-intercept. (b) Write an equation of the line. (c) What is the mean temperature of Earth in degrees Fahrenheit?

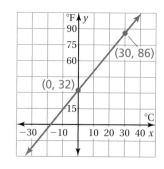

Mean Temperature: 15°C

a. slope = $\dfrac{\text{change in } y}{\text{change in } x} = \dfrac{54}{30} = \dfrac{9}{5}$

The line crosses the $y$-axis at $(0, 32)$. So, the $y$-intercept is 32.

∴ The slope is $\dfrac{9}{5}$ and the $y$-intercept is 32.

b. Use the slope and $y$-intercept to write an equation.

∴ The equation is $y = \dfrac{9}{5}x + 32$.

c. In degrees Celsius, the mean temperature of Earth is 15°. To find the mean temperature in degrees Fahrenheit, find the value of $y$ when $x = 15$.

$y = \dfrac{9}{5}x + 32$     Write the equation.

$= \dfrac{9}{5}(15) + 32$     Substitute 15 for $x$.

$= 59$     Simplify.

∴ The mean temperature of Earth is 59°F.

### On Your Own

**Now You're Ready**
Exercise 7

2. The graph shows the height $y$ (in feet) of a flag $x$ seconds after you start raising it up a flagpole.

   a. Find and interpret the slope.
   b. Write an equation of the line.
   c. What is the height of the flag after 9 seconds?

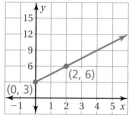

Section 2.7  Solving Real-Life Problems

# 2.7 Exercises

## Vocabulary and Concept Check

1. **REASONING** Explain how to find the slope, *y*-intercept, and *x*-intercept of the line shown.

2. **OPEN-ENDED** Describe a real-life situation that uses a negative slope.

3. **REASONING** In a real-life situation, what does the slope of a line represent?

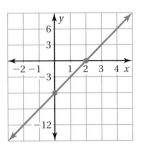

## Practice and Problem Solving

**Describe a real-life problem that has the given rate and intercepts. Draw a line that represents the problem.**

4. Rate: −1.6 gallons per hour
   *y*-intercept: 16 gallons
   *x*-intercept: 10 hours

5. Rate: −3°F per hour
   *y*-intercept: 21°F
   *x*-intercept: 7 hours

6. **DOWNLOAD** You are downloading a song. The percent *y* (in decimal form) of megabytes remaining to download after *x* seconds is $y = -0.1x + 1$.

   a. Graph the equation.
   b. Interpret the *x*- and *y*-intercepts.
   c. After how many seconds is the download 50% complete?

7. **HIKING** The graph relates temperature *y* (in degrees Fahrenheit) to altitude *x* (in thousands of feet).

   a. Find the slope and *y*-intercept.
   b. Write an equation of the line.
   c. What is the temperature at sea level?

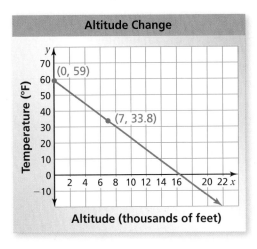

90   Chapter 2   Graphing and Writing Linear Equations

8. **REASONING** Your family is driving from Cincinnati to St. Louis. The graph relates your distance from St. Louis $y$ (in miles) and travel time $x$ (in hours).

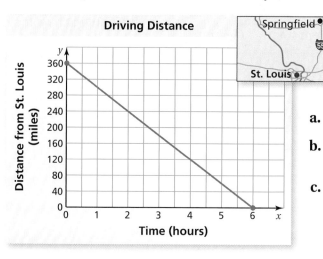

a. Interpret the $x$- and $y$-intercepts.

b. What is the slope? What does the slope represent in this situation?

c. Write an equation of the line. How would the graph and the equation change if you were able to travel in a straight line?

9. **PROJECT** Use a map or the Internet to find the latitude and longitude of your school to the nearest whole number. Then find the latitudes and longitudes of: Antananarivo, Madagascar; Denver, Colorado; Brasilia, Brazil; London, England; and Beijing, China.

a. Plot a point for each of the cities in the same coordinate plane. Let the positive $y$-axis represent north and the positive $x$-axis represent east.

b. Write an equation of the line that passes through Denver and Beijing.

c. In part (b), what geographic location does the $y$-intercept represent?

10. **Reasoning** A band is performing at an auditorium for a fee of $1500. In addition to this fee, the band receives 30% of each $20 ticket sold. The maximum capacity of the auditorium is 800 people.

a. Write an equation that represents the band's revenue $R$ when $x$ tickets are sold.

b. The band needs $5000 for new equipment. How many tickets must be sold for the band to earn enough money to buy the new equipment?

## Fair Game Review What you learned in previous grades & lessons

**Solve the equation. Check your solution.** *(Section 1.2)*

11. $-h - 7h + 13 = 3$

12. $4(k - 10) - 4 = 12$

13. $9 + 2.5(2q - 3) = -10$

14. **MULTIPLE CHOICE** Which equation is the slope-intercept form of $24x - 8y = 56$? *(Section 2.4)*

Ⓐ $y = -3x + 7$   Ⓑ $y = 3x - 7$   Ⓒ $y = -3x - 7$   Ⓓ $y = 3x + 7$

## 2.5–2.7 Quiz

**Write an equation of the line in slope-intercept form.** *(Section 2.5)*

1.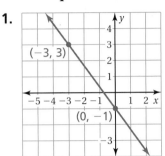

2.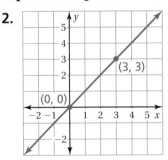

3.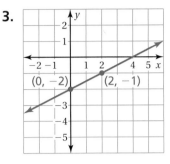

**Write in point-slope form an equation of the line that passes through the given point and has the given slope.** *(Section 2.6)*

4. $(1, 3)$; $m = 2$

5. $(-3, -2)$; $m = \dfrac{1}{3}$

6. $(-1, 4)$; $m = -1$

7. $(8, -5)$; $m = -\dfrac{1}{8}$

**Write in slope-intercept form an equation of the line that passes through the given points.** *(Section 2.6)*

8. $\left(0, -\dfrac{2}{3}\right), \left(-3, -\dfrac{2}{3}\right)$

9. $(4, 0), (0, 4)$

10. Write an equation of the line that passes through $(2, -5)$ and is (a) parallel to and (b) perpendicular to the line $y = \dfrac{1}{3}x + 4$. *(Section 2.6)*

11. **CONSTRUCTION** A construction crew is extending a highway sound barrier that is 13 miles long. The crew builds $\dfrac{1}{2}$ mile per week. Write an equation for the length $y$ (in miles) of the barrier after $x$ weeks. *(Section 2.5)*

12. **FISH POND** You are draining a fish pond. The amount $y$ (in liters) of water remaining after $x$ hours is $y = -60x + 480$. (a) Graph the equation. (b) Interpret the $x$- and $y$-intercepts. *(Section 2.7)*

13. **WATER** A recreation department bought bottled water to sell at a fair. The graph shows the number $y$ of bottles remaining after each hour $x$. *(Section 2.7)*

    a. Find the slope and $y$-intercept.
    b. Write an equation of the line.
    c. The fair started at 10 A.M. When did the recreation department run out of bottled water?

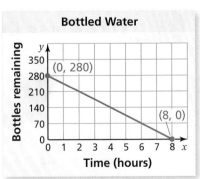

# 2 Chapter Review

## Review Key Vocabulary

linear equation, *p. 44*
solution of a linear equation, *p. 44*
slope, *p. 50*
rise, *p. 50*
run, *p. 50*
perpendicular lines, *p. 57*

*x*-intercept, *p. 60*
*y*-intercept, *p. 60*
slope-intercept form, *p. 60*
standard form, *p. 66*
point-slope form, *p. 80*

## Review Examples and Exercises

### 2.1 Graphing Linear Equations (pp. 42–47)

Graph $y = 3x - 1$.

**Step 1:** Make a table of values.

| x | y = 3x − 1 | y | (x, y) |
|---|---|---|---|
| −2 | y = 3(−2) − 1 | −7 | (−2, −7) |
| −1 | y = 3(−1) − 1 | −4 | (−1, −4) |
| 0 | y = 3(0) − 1 | −1 | (0, −1) |
| 1 | y = 3(1) − 1 | 2 | (1, 2) |

**Step 2:** Plot the ordered pairs.   **Step 3:** Draw a line through the points.

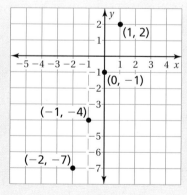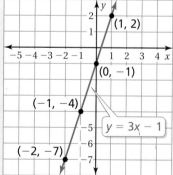

### Exercises

**Graph the linear equation.**

1. $y = \dfrac{3}{5}x$
2. $y = -2$
3. $y = 9 - x$
4. $y = 1$
5. $y = \dfrac{2}{3}x + 2$
6. $x = -5$

## 2.2 Slope of a Line (pp. 48–57)

**Find the slope of each line in the graph.**

Red Line: slope $= \dfrac{y_2 - y_1}{x_2 - x_1} = \dfrac{5 - (-3)}{2 - 2} = \dfrac{8}{0}$

∴ The slope of the red line is undefined.

Blue Line: slope $= \dfrac{y_2 - y_1}{x_2 - x_1} = \dfrac{-1 - 2}{4 - (-3)} = \dfrac{-3}{7}$, or $-\dfrac{3}{7}$

Green Line: slope $= \dfrac{y_2 - y_1}{x_2 - x_1} = \dfrac{4 - 4}{5 - 0} = \dfrac{0}{5}$, or 0

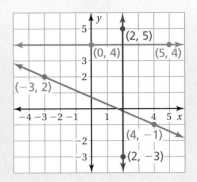

### Exercises

**The points in the table lie on a line. How can you find the slope of the line from the table? What is the slope?**

7.
| x | 0 | 1 | 2 | 3 |
|---|---|---|---|---|
| y | −1 | 0 | 1 | 2 |

8.
| x | −2 | 0 | 2 | 4 |
|---|---|---|---|---|
| y | 3 | 4 | 5 | 6 |

9. Are the lines $x = 2$ and $y = 4$ parallel? Are they perpendicular? Explain.

## 2.3 Graphing Linear Equations in Slope-Intercept Form (pp. 58–63)

**Graph $y = 0.5x - 3$. Identify the x-intercept.**

**Step 1:** Find the slope and y-intercept.

$y = 0.5x + (-3)$

↑ slope   ↑ y-intercept

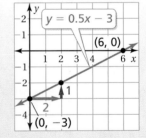

**Step 2:** The y-intercept is −3. So, plot (0, −3).

**Step 3:** Use the slope to find another point and draw the line.

slope $= \dfrac{\text{rise}}{\text{run}} = \dfrac{1}{2}$

Plot the point that is 2 units right and 1 unit up from (0, −3). Draw a line through the two points.

∴ The line crosses the x-axis at (6, 0). So, the x-intercept is 6.

### Exercises

**Graph the linear equation. Identify the x-intercept. Use a graphing calculator to check your answer.**

10. $y = 2x - 6$

11. $y = -4x + 8$

12. $y = -x - 8$

## 2.4 Graphing Linear Equations in Standard Form (pp. 64–69)

**Graph $8x + 4y = 16$.**

**Step 1:** Write the equation in slope-intercept form.

$8x + 4y = 16$  Write the equation.
$4y = -8x + 16$  Subtract $8x$ from each side.
$y = -2x + 4$  Divide each side by 4.

**Step 2:** Use the slope and y-intercept to plot two points.

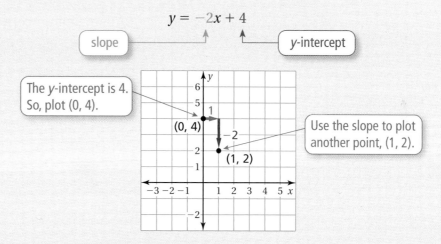

**Step 3:** Draw a line through the points.

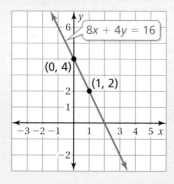

### Exercises

**Graph the linear equation.**

13. $\frac{1}{4}x + y = 3$

14. $-4x + 2y = 8$

15. $x + 5y = 10$

16. $-\frac{1}{2}x + \frac{1}{8}y = \frac{3}{4}$

17. A dog kennel charges $30 per night to board your dog and $6 for each hour of play time. The amount of money you spend is given by $30x + 6y = 180$, where $x$ is the number of nights and $y$ is the number of hours of play time. Graph the equation and interpret the intercepts.

## 2.5 Writing Equations in Slope-Intercept Form (pp. 72–77)

**Write an equation of the line in slope-intercept form.**

a. 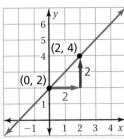 Find the slope and y-intercept.

$$\text{slope} = \frac{y_2 - y_1}{x_2 - x_1} = \frac{4 - 2}{2 - 0} = \frac{2}{2}, \text{ or } 1$$

Because the line crosses the y-axis at (0, 2), the y-intercept is 2.

∴ So, the equation is $y = 1x + 2$, or $y = x + 2$.

b. 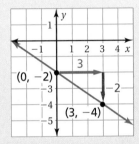 Find the slope and y-intercept.

$$\text{slope} = \frac{y_2 - y_1}{x_2 - x_1} = \frac{-4 - (-2)}{3 - 0} = \frac{-2}{3}, \text{ or } -\frac{2}{3}$$

Because the line crosses the y-axis at (0, −2), the y-intercept is −2.

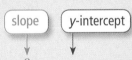

∴ So, the equation is $y = -\frac{2}{3}x + (-2)$, or $y = -\frac{2}{3}x - 2$.

### Exercises

**Write an equation of the line in slope-intercept form.**

18.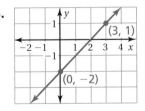

19.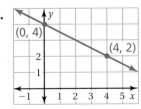

20. Write an equation of the line that passes through (0, 8) and (6, 8).

21. Write an equation of the line that passes through (0, −5) and (−5, −5).

## 2.6 Writing Equations in Point-Slope Form (pp. 78–85)

**Write in slope-intercept form an equation of the line that passes through the points (2, 1) and (3, 5).**

Find the slope.

$$m = \frac{y_2 - y_1}{x_2 - x_1} = \frac{5 - 1}{3 - 2} = \frac{4}{1}, \text{ or } 4$$

Then use the slope and one of the given points to write an equation of the line.

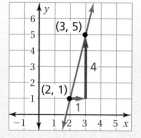

Use $m = 4$ and (2, 1).

| | |
|---|---|
| $y - y_1 = m(x - x_1)$ | Write the point-slope form. |
| $y - 1 = 4(x - 2)$ | Substitute 4 for $m$, 2 for $x_1$, and 1 for $y_1$. |
| $y - 1 = 4x - 8$ | Use Distributive Property. |
| $y = 4x - 7$ | Write in slope-intercept form. |

∴ So, the equation is $y = 4x - 7$.

### Exercises

22. Write in point-slope form an equation of the line that passes through the point (4, 4) with slope 3.

23. Write in slope-intercept form an equation of the line that passes through the points (−4, 2) and (6, −3).

## 2.7 Solving Real-Life Problems (pp. 86–91)

**The amount $y$ (in dollars) of money you have left after playing $x$ games at a carnival is $y = -0.75x + 10$. How much money do you have after playing eight games?**

| | |
|---|---|
| $y = -0.75x + 10$ | Write the equation. |
| $= -0.75(8) + 10$ | Substitute 8 for $x$. |
| $= 4$ | Simplify. |

∴ You have $4 left after playing 8 games.

### Exercises

24. **HAY** The amount $y$ (in bales) of hay remaining after feeding cows for $x$ days is $y = -3.5x + 105$. (a) Graph the equation. (b) Interpret the $x$- and $y$-intercepts. (c) How many bales are left after 10 days?

# 2 Chapter Test

**Find the slope and y-intercept of the graph of the linear equation.**

1. $y = 6x - 5$
2. $y = 20x + 15$
3. $y = -5x - 16$
4. $y - 1 = 3x + 8.4$
5. $y + 4.3 = 0.1x$
6. $-\frac{1}{2}x + 2y = 7$

**Graph the linear equation.**

7. $y = 2x + 4$
8. $y = -\frac{1}{2}x - 5$
9. $-3x + 6y = 12$

10. Which lines are parallel? Which lines are perpendicular? Explain.

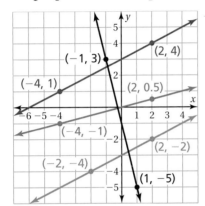

11. The points in the table lie on a line. How can you find the slope of the line from the table? What is the slope?

| x | y |
|---|---|
| -1 | -4 |
| 0 | -1 |
| 1 | 2 |
| 2 | 5 |

**Write an equation of the line in slope-intercept form.**

12.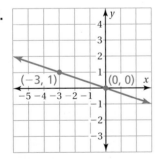

13.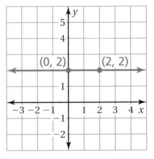

**Write in slope-intercept form an equation of the line that passes through the given points.**

14. $(-1, 5), (3, -3)$
15. $(-4, 1), (4, 3)$
16. $(-2, 5), (-1, 1)$

17. **BRAILLE** Because of its size and detail, Braille takes longer to read than text. A person reading Braille reads at 25% the rate of a person reading text.

    a. Write and graph an equation that represents the average rate $y$ of a Braille reader in terms of the average rate $x$ of a text reader.

    b. Interpret the solution (180, 45).

    c. What happens to $y$ as $x$ increases? Explain.

98  Chapter 2  Graphing and Writing Linear Equations

# 2 Standards Assessment

1. Which equation matches the line shown in the graph? *(A.REI.10)*

   A. $y = 2x - 2$

   B. $y = 2x + 1$

   C. $y = x - 2$

   D. $y = x + 1$

   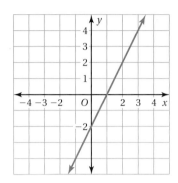

**Test-Taking Strategy**
**Estimate the Answer**

"Using estimation you can see that there are about 300 hairs. So, it's got to be C."

2. A line has a slope of 4 and passes through the point $(a, b)$. Which point must also lie on this line? *(F.IF.6)*

   F. $(a, b + 4)$

   G. $(2a, 8b)$

   H. $(a + 1, b + 4)$

   I. $(2a, 5b)$

3. A car's value depreciates at a rate of $2,500 per year. Three years after it was purchased, the car's value was $21,000. Which equation can be used to find $v$, its value in dollars, $n$ years after it was purchased? *(A.CED.2)*

   A. $v = 28,500 - 2,500n$

   B. $v = 21,000 - 2,500n$

   C. $v = 18,500 - 2,500n$

   D. $v = 18,500 - n$

4. The equation $6x - 5y = 14$ is written in standard form. Which point lies on the graph of this equation? *(A.REI.10)*

   F. $(-4, -1)$

   G. $(-2, 4)$

   H. $(-1, -4)$

   I. $(4, -2)$

**5.** The line shown in the graph below has a slope of −3. What is the equation of the line? *(A.REI.10)*

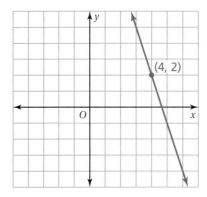

**A.** $y = 3x - 10$

**B.** $y = -3x + 10$

**C.** $y = -3x + 14$

**D.** $y = -3x - 14$

**6.** A cell phone plan costs $10 per month plus $0.10 for each minute used. Last month, you spent $18.50 using this plan. This can be modeled by the equation below, where $m$ represents the number of minutes used.

$$0.1m + 10 = 18.5$$

How many minutes did you use last month? *(A.REI.3)*

**F.** 8.4 min

**G.** 85 min

**H.** 185 min

**I.** 285 min

**7.** What is the slope of the line that passes through the points (2, −2) and (8, 1)? *(F.IF.6)*

**8.** It costs $40 to rent a car for one day. In addition, the rental agency charges you for each mile driven, as shown in the graph. *(F.IF.6)*

*Part A* Determine the slope of the line joining the points on the graph.

*Part B* Explain what the slope represents.

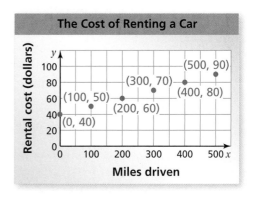

**9.** Which value of $x$ makes the equation below true?   *(A.REI.3)*

$$7 + 2x = 4x - 5$$

**A.** $-2$

**B.** 1

**C.** 2

**D.** 6

**10.** Which line has a slope of 0?   *(F.IF.6)*

**F.**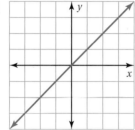

**H.**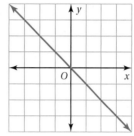

**G.**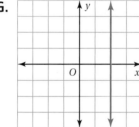

**I.**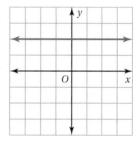

**11.** Solve the formula $K = 3M - 7$ for $M$.   *(A.CED.4)*

**A.** $M = K + 7$

**B.** $M = \dfrac{K+7}{3}$

**C.** $M = \dfrac{K}{3} + 7$

**D.** $M = \dfrac{K-7}{3}$

**12.** The linear equation $5x + 2y = 10$ is written in standard form. What is the slope of the graph of this equation?   *(F.IF.6)*

**F.** 5

**G.** 2.5

**H.** $-2.5$

**I.** $-5$

**13.** Solve $4\pi x = 16\pi$.   *(A.REI.3)*

**A.** $x = 4$

**B.** $x = 2\pi$

**C.** $x = 4\pi$

**D.** $x = 12\pi$

# 3 Solving Linear Inequalities

- 3.1 Writing and Graphing Inequalities
- 3.2 Solving Inequalities Using Addition or Subtraction
- 3.3 Solving Inequalities Using Multiplication or Division
- 3.4 Solving Multi-Step Inequalities
- 3.5 Graphing Linear Inequalities in Two Variables

"Here is a math quiz, Descartes. Tell me about these symbols."

"That's easy. One just means I am happy."

"The other means that I have a piece of spaghetti stuck between my fangs."

"Just think of the Addition Property of Inequality in this way. If Fluffy has more cat treats than you have ..."

"... and you each get 2 more cat treats, then Fluffy will STILL have more cat treats than you have!"

"This guy really knows how to hurt a cat, doesn't he?"

# What You Learned Before

- ## Comparing Real Numbers (8.NS.1)

    Complete the number sentence with <, >, or =.

    **Example 1**  $\dfrac{1}{3}$ ⬚ 0.3

    $\dfrac{1}{3} = \dfrac{10}{30}$, $0.3 = \dfrac{3}{10} = \dfrac{9}{30}$

    Because $\dfrac{10}{30}$ is greater than $\dfrac{9}{30}$, $\dfrac{1}{3}$ is greater than 0.3.

    ∴ So, $\dfrac{1}{3} > 0.3$.

    **Example 2**  $\sqrt{6}$ ⬚ 2.5

    Use a calculator to estimate $\sqrt{6}$.

    $\sqrt{6} \approx 2.45$

    Because 2.45 is less than 2.5, $\sqrt{6}$ is less than 2.5.

    ∴ So, $\sqrt{6} < 2.5$.

    ### Try It Yourself
    Complete the number sentence with <, >, or =.

    1. $\dfrac{1}{4}$ ⬚ 0.25
    2. 0.1 ⬚ $\dfrac{1}{9}$
    3. $\pi$ ⬚ $\sqrt{10}$

- ## Graphing Inequalities (6.EE.5)

    **Example 3**  Graph $x \geq 3$.

    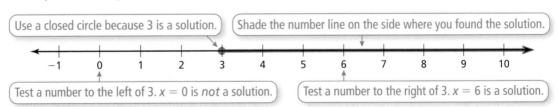

    **Example 4**  Graph $x < 2$.

    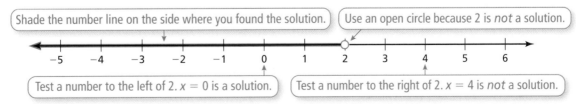

    ### Try It Yourself
    Graph the inequality.

    4. $x \geq 0$
    5. $x < 6$
    6. $x \leq 4$
    7. $x > 10$

# 3.1 Writing and Graphing Inequalities

**Essential Question** How can you use an inequality to describe a real-life statement?

### 1 ACTIVITY: Writing and Graphing Inequalities

**Work with a partner. Write an inequality for the statement. Then sketch the graph of all the numbers that make the inequality true.**

a. **Statement:** The temperature $t$ in Minot, North Dakota has never been below $-36°$F.

   **Inequality:** _____

   **Graph:**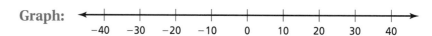

b. **Statement:** The elevation $e$ in Wisconsin is at most 1951.5 feet above sea level.

   **Inequality:** _____

   **Graph:**

### 2 ACTIVITY: Writing and Graphing Inequalities

**Work with a partner. Write an inequality for the graph. Then, in words, describe all the values of $x$ that make the inequality true.**

a.

b.

c.

d.

**Common Core**

**Writing Inequalities**
In this lesson, you will
- write and graph inequalities.

Learning Standards
A.CED.1
A.CED.3

### 3 ACTIVITY: Triangle Inequality

**Math Practice 3**

**Construct Arguments**

How can you use results from this activity to write a rule?

**Work with a partner. Use 8 to 10 pieces of spaghetti.**

- Break one piece of spaghetti into three parts that can be used to form a triangle.

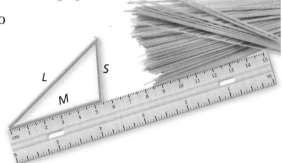

- Form a triangle and use a centimeter ruler to measure each side. Round the side lengths to the nearest tenth.

- Record the side lengths in a table.

| Side Lengths That Form a Triangle |||| 
|---|---|---|---|
| Small | Medium | Large | S + M |
|  |  |  |  |
|  |  |  |  |
|  |  |  |  |

- Repeat the process with two other pieces of spaghetti.

- Repeat the experiment by breaking pieces of spaghetti into three pieces that *do not* form a triangle. Record the lengths in a table.

| Side Lengths That Do Not Form a Triangle ||||
|---|---|---|---|
| Small | Medium | Large | S + M |
|  |  |  |  |
|  |  |  |  |
|  |  |  |  |

- **INDUCTIVE REASONING** Write a rule that uses an inequality to compare the lengths of three sides of a triangle.

- Use your rule to decide whether the following triangles are possible. Explain.

a. 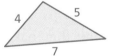   b. 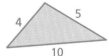   c.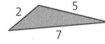

## What Is Your Answer?

4. **IN YOUR OWN WORDS** How can you use an inequality to describe a real-life statement? Give two examples of real-life statements that can be represented by inequalities.

**Practice** — Use what you learned about writing and graphing inequalities to complete Exercises 4 and 5 on page 108.

# 3.1 Lesson

**Key Vocabulary**
inequality, *p. 106*
solution of an inequality, *p. 106*
solution set, *p. 106*
graph of an inequality, *p. 107*

An **inequality** is a mathematical sentence that compares expressions. It contains the symbol <, >, ≤, or ≥. To write an inequality, look for the following phrases to determine where to place the inequality symbol.

| | Inequality Symbols | | | |
|---|---|---|---|---|
| Symbol | < | > | ≤ | ≥ |
| Key Phrases | • is less than<br>• is fewer than | • is greater than<br>• is more than | • is less than or equal to<br>• is at most<br>• is no more than | • is greater than or equal to<br>• is at least<br>• is no less than |

### EXAMPLE 1  Writing an Inequality

**A number $w$ minus 3.5 is less than or equal to $-2$. Write this sentence as an inequality.**

A number $w$ minus 3.5   is less than or equal to   $-2$.
$\qquad\quad w - 3.5 \qquad\qquad\qquad \leq \qquad\qquad\quad -2$

∴ An inequality is $w - 3.5 \leq -2$.

### On Your Own

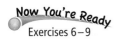
Exercises 6–9

**Write the word sentence as an inequality.**

1. A number $b$ is fewer than 30.4.
2. Twice a number $k$ is at least $-\dfrac{7}{10}$.

---

A **solution of an inequality** is a value that makes the inequality true. An inequality can have more than one solution. The set of all solutions of an inequality is called the **solution set**.

| Value of $x$ | $x + 5 \geq -2$ | Is the inequality true? |
|---|---|---|
| $-6$ | $-6 + 5 \stackrel{?}{\geq} -2$<br>$-1 \geq -2$ ✓ | yes |
| $-7$ | $-7 + 5 \stackrel{?}{\geq} -2$<br>$-2 \geq -2$ ✓ | yes |
| $-8$ | $-8 + 5 \stackrel{?}{\geq} -2$<br>$-3 \not\geq -2$ ✗ | no |

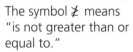

**Reading**
The symbol $\not\geq$ means "is not greater than or equal to."

# EXAMPLE 2 Checking Solutions

**Tell whether −4 is a solution of each inequality.**

a. $x + 8 < -3$

$x + 8 < -3$    Write the inequality.

$-4 + 8 \stackrel{?}{<} -3$    Substitute −4 for x.

$4 \not< -3$ ✗    Simplify.

4 is *not* less than −3.

∴ So, −4 is *not* a solution of the inequality.

b. $-4.5x > -21$

$-4.5x > -21$

$-4.5(-4) \stackrel{?}{>} -21$

$18 > -21$ ✓

18 is greater than −21.

∴ So, −4 is a solution of the inequality.

### On Your Own

*Now You're Ready*
Exercises 11–16

**Tell whether −6 is a solution of the inequality.**

3. $c + 4 < -1$
4. $5 - m \leq 10$
5. $21 \div x \geq -3.5$

---

The **graph of an inequality** shows all of the solutions of the inequality on a number line. An open circle ○ is used when a number is *not* a solution. A closed circle ● is used when a number is a solution. An arrow to the left or right shows that the graph continues in that direction.

# EXAMPLE 3 Graphing an Inequality

**Graph $y \leq -3$.**

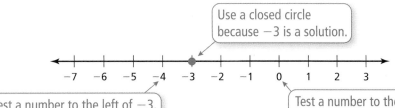

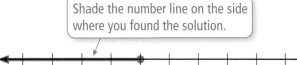

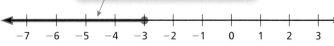

### On Your Own

*Now You're Ready*
Exercises 17–20

**Graph the inequality on a number line.**

6. $b > -8$
7. $g \leq 1.4$
8. $r < -\dfrac{1}{2}$
9. $v \geq \sqrt{36}$

# 3.1 Exercises

## Vocabulary and Concept Check

1. **VOCABULARY** Would an open circle or a closed circle be used in the graph of the inequality $k < 250$? Explain.

2. **DIFFERENT WORDS, SAME QUESTION** Which is different? Write "both" inequalities.

   | $w$ is greater than or equal to $-7$. | $w$ is no less than $-7$. |

   | $w$ is no more than $-7$. | $w$ is at least $-7$. |

3. **REASONING** Do $x \geq -9$ and $-9 \geq x$ represent the same inequality? Explain.

## Practice and Problem Solving

**Write an inequality for the graph. Then, in words, describe all the values of $x$ that make the inequality true.**

4. (number line with closed circle at 9, arrow left; marks $-3, 0, 3, 6, 9, 12, 15, 18$)

5. (number line with open circle at $-3$, arrow left; marks $-7, -6, -5, -4, -3, -2, -1$)

**Write the word sentence as an inequality.**

6. A number $x$ is no less than $-4$.

7. A number $y$ added to 5.2 is less than 23.

8. A number $b$ multiplied by $-5$ is at most $-\dfrac{3}{4}$.

9. A number $k$ minus 8.3 is greater than 48.

10. **ERROR ANALYSIS** Describe and correct the error in writing the word sentence as an inequality.

    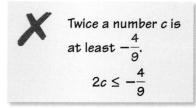

**Tell whether the given value is a solution of the inequality.**

11. $s + 6 \leq 12$; $s = 4$

12. $15n > -3$; $n = -2$

13. $a - 2.5 \leq 1.6$; $a = 4.1$

14. $-3.3q > -13$; $q = 4.6$

15. $\dfrac{4}{5}h \geq -4$; $h = -15$

16. $\dfrac{1}{12} - p < \dfrac{1}{3}$; $p = \dfrac{1}{6}$

**Graph the inequality on a number line.**

17. $g \geq -6$

18. $q > 1.25$

19. $z < 11\dfrac{1}{4}$

20. $w \leq -\sqrt{64}$

21. **DRIVING** When you are driving with a learner's license, a licensed driver who is 21 years of age or older must be with you. Write an inequality that represents this situation.

**Tell whether the given value is a solution of the inequality.**

**22.** $3p > 5 + p$; $p = 4$

**23.** $\dfrac{y}{2} \geq y - 11$; $y = 18$

**24. LOGIC** Each video game rating is matched with the inequality that represents the suggested ages of players. Your friend is old enough to play "E 10+" games. Is your friend old enough to play "T" games? Explain.

$x \geq 3$    $x \geq 6$    $x \geq 10$    $x \geq 13$    $x \geq 17$

The ESRB rating icons are registered trademarks of the Entertainment Software Association.

**25. SCUBA DIVING** Three requirements for a scuba diving training course are shown.

   **a.** Write and graph three inequalities that represent the requirements.

   **b.** You can swim 10 lengths of a 25-yard pool. Do you satisfy the swimming requirement of the course? Justify your answer.

**26. REPEATED REASONING** On an airplane, the maximum sum of the length, width, and height of a carry-on bag is 45 inches. Find three different sets of dimensions that are reasonable for a carry-on bag. Use a diagram to justify your answer.

**27. Critical Thinking** A number $m$ is less than another number $n$. The number $n$ is less than or equal to a third number $p$.

   **a.** Write two inequalities representing these relationships.

   **b.** Describe the relationship between $m$ and $p$.

   **c.** Can $m$ be equal to $p$? Explain.

## Fair Game Review  What you learned in previous grades & lessons

**Solve the equation. Check your solution.** *(Section 1.1)*

**28.** $r - 12 = 3$     **29.** $4.2 + p = 2.5$     **30.** $n - 3\pi = 7\pi$

**31. MULTIPLE CHOICE** Which of the following is the equation of the line in slope-intercept form? *(Section 2.5)*

   Ⓐ $y = -2x + 1$     Ⓑ $y = -x - 1$
   Ⓒ $y = x + 1$       Ⓓ $y = -x + 1$

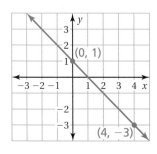

# 3.2 Solving Inequalities Using Addition or Subtraction

**Essential Question** How can you use addition or subtraction to solve an inequality?

## 1 ACTIVITY: Quarterback Passing Efficiency

**Work with a partner.** The National Collegiate Athletic Association (NCAA) uses the following formula to rank the passing efficiency $P$ of quarterbacks.

$$P = \frac{8.4Y + 100C + 330T - 200N}{A}$$

$Y$ = total length of all completed passes (in Yards)
$C$ = Completed passes
$T$ = passes resulting in a Touchdown
$N$ = iNtercepted passes
$A$ = Attempted passes
$M$ = incoMplete passes

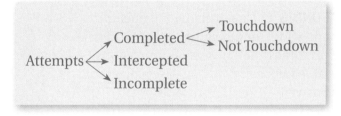

Which of the following equations or inequalities are true relationships among the variables? Explain your reasoning.

a. $C + N < A$  b. $C + N \leq A$  c. $T < C$  d. $T \leq C$
e. $N < A$  f. $A > T$  g. $A - C \geq M$  h. $A = C + N + M$

## 2 ACTIVITY: Quarterback Passing Efficiency

**Common Core**

**Solving Inequalities**
In this lesson, you will
- write and solve inequalities using addition or subtraction.
- solve real-life problems.

Learning Standards
A.CED.1
A.CED.3
A.REI.3

**Work with a partner.** Which of the following quarterbacks has a passing efficiency rating that satisfies the inequality $P > 100$? Show your work.

| Player | Attempts | Completions | Yards | Touchdowns | Interceptions |
|---|---|---|---|---|---|
| A | 149 | 88 | 1065 | 7 | 9 |
| B | 400 | 205 | 2000 | 10 | 3 |
| C | 426 | 244 | 3105 | 30 | 9 |
| D | 188 | 89 | 1167 | 6 | 15 |

### 3 ACTIVITY: Finding Solutions of Inequalities

**Math Practice**

**Find General Methods**
What method did you use to choose the values for the formula? Why?

Work with a partner. Use the passing efficiency formula to create a passing record that makes the inequality true. Then describe the values of $P$ that make the inequality true.

**a.** $P < 0$

| Attempts | Completions | Yards | Touchdowns | Interceptions |
|---|---|---|---|---|
|  |  |  |  |  |

**b.** $P + 100 \geq 250$

| Attempts | Completions | Yards | Touchdowns | Interceptions |
|---|---|---|---|---|
|  |  |  |  |  |

**c.** $180 < P - 50$

| Attempts | Completions | Yards | Touchdowns | Interceptions |
|---|---|---|---|---|
|  |  |  |  |  |

**d.** $P + 30 \geq 120$

| Attempts | Completions | Yards | Touchdowns | Interceptions |
|---|---|---|---|---|
|  |  |  |  |  |

**e.** $P - 250 > -80$

| Attempts | Completions | Yards | Touchdowns | Interceptions |
|---|---|---|---|---|
|  |  |  |  |  |

### What Is Your Answer?

**4.** Write a rule that describes how to solve inequalities like those in Activity 3. Then use your rule to solve each of the inequalities in Activity 3.

**5.** **IN YOUR OWN WORDS** How can you use addition or subtraction to solve an inequality?

**6.** How is solving the inequality $x + 3 < 4$ similar to solving the equation $x + 3 = 4$? How is it different?

**Practice**

Use what you learned about solving inequalities using addition or subtraction to complete Exercises 3–5 on page 114.

Section 3.2 Solving Inequalities Using Addition or Subtraction

## 3.2 Lesson

### Key Ideas

**Addition Property of Inequality**

**Words** If you add the same number to each side of an inequality, the inequality remains true.

**Numbers**
$$-3 < 2$$
$$+4 \phantom{<} +4$$
$$\phantom{-}1 < 6$$

**Algebra**
$$x - 3 > -10$$
$$+3 \phantom{>} +3$$
$$x > -7$$

**Subtraction Property of Inequality**

**Words** If you subtract the same number from each side of an inequality, the inequality remains true.

**Numbers**
$$-3 < 1$$
$$-5 \phantom{<} -5$$
$$-8 < -4$$

**Algebra**
$$x + 7 > -20$$
$$-7 \phantom{>} -7$$
$$x > -27$$

These properties are also true for ≤ and ≥.

**Study Tip:** You can solve inequalities the same way you solve equations. Use inverse operations to get the variable by itself.

---

### EXAMPLE 1  Solving an Inequality Using Addition

Solve $x - 6 \geq -10$. Graph the solution.

$$x - 6 \geq -10 \qquad \text{Write the inequality.}$$
$$+6 \phantom{\geq} +6 \qquad \text{Add 6 to each side.}$$
$$x \geq -4 \qquad \text{Simplify.}$$

Undo the subtraction.

∴ The solution is $x \geq -4$.

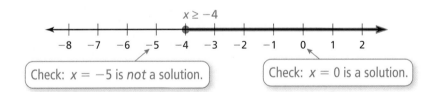

Check: $x = -5$ is *not* a solution.  Check: $x = 0$ is a solution.

**Study Tip:** To check a solution, you check some numbers that are solutions and some that are not.

### On Your Own

**Solve the inequality. Graph the solution.**

1. $b - 2 > -9$
2. $m - 3.8 \leq 5$
3. $\dfrac{1}{4} > y - \dfrac{1}{4}$

### EXAMPLE 2  Solving an Inequality Using Subtraction

Solve $-8 > 1.4 + x$. Graph the solution.

$$-8 > 1.4 + x \quad \text{Write the inequality.}$$
$$\underline{-1.4 \quad -1.4} \quad \text{Subtract 1.4 from each side.}$$
$$-9.4 > x \quad \text{Simplify.}$$

Undo the addition.

**Reading**
The inequality $-9.4 > x$ is the same as $x < -9.4$.

∴ The solution is $x < -9.4$.

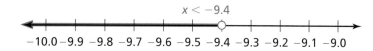

### On Your Own

**Now You're Ready**
Exercises 6–17

Solve the inequality. Graph the solution.

4. $k + 5 \leq -3$
5. $\dfrac{5}{6} \leq z + \dfrac{2}{3}$
6. $p + 0.7 > -2.3$

---

### EXAMPLE 3  Real-Life Application

**On a train, carry-on bags can weigh no more than 50 pounds. Your bag weighs 24.8 pounds. Write and solve an inequality that represents the amount of weight you can add to your bag.**

**Words**   Weight of your bag   plus   amount of weight you can add   is no more than   the weight limit.

**Variable**   Let $w$ be the possible weight you can add.

**Inequality**   24.8   +   $w$   ≤   50

$$24.8 + w \leq 50 \quad \text{Write the inequality.}$$
$$\underline{-24.8 \quad -24.8} \quad \text{Subtract 24.8 from each side.}$$
$$w \leq 25.2 \quad \text{Simplify.}$$

∴ You can add no more than 25.2 pounds to your bag.

### On Your Own

7. **WHAT IF?** Your carry-on bag weighs 32.5 pounds. Write and solve an inequality that represents the possible weight you can add to your bag.

Section 3.2  Solving Inequalities Using Addition or Subtraction

# 3.2 Exercises

## Vocabulary and Concept Check

1. **REASONING** Is the inequality $r - 5 \leq 8$ the same as $8 \leq r - 5$? Explain.

2. **WHICH ONE DOESN'T BELONG?** Which inequality does *not* belong with the other three? Explain your reasoning.

   $c + \dfrac{7}{2} \leq \dfrac{3}{2}$     $c + \dfrac{7}{2} \geq \dfrac{3}{2}$     $\dfrac{3}{2} \geq c + \dfrac{7}{2}$     $c - \dfrac{3}{2} \leq -\dfrac{7}{2}$

## Practice and Problem Solving

**Use the formula in Activity 1 to create a passing record that makes the inequality true.**

3. $P \geq 180$
4. $P + 40 < 110$
5. $280 \leq P - 20$

**Solve the inequality. Graph the solution.**

6. $y - 3 \geq 7$
7. $t - 8 > -4$
8. $n + 11 \leq 20$

9. $a + 7 > -1$
10. $5 < v - \dfrac{1}{2}$
11. $\dfrac{1}{5} > d + \dfrac{4}{5}$

12. $-\dfrac{2}{3} \leq g - \dfrac{1}{3}$
13. $m + \dfrac{7}{4} \leq \dfrac{11}{4}$
14. $11.2 \leq k + 9.8$

15. $h - 1.7 < -3.2$
16. $0 > s + \pi$
17. $5 \geq u - 4.5$

18. **ERROR ANALYSIS** Describe and correct the error in graphing the solution of the inequality.

    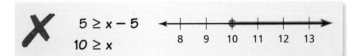

19. **PROBLEM SOLVING** The maximum volume of a great white pelican's bill is about 700 cubic inches.

    a. A pelican scoops up 100 cubic inches of water. Write and solve an inequality that represents the additional volume the pelican's bill can contain.

    b. A pelican's stomach can contain about one-third the maximum amount that its bill can contain. Write an inequality that represents the volume of the pelican's stomach.

**Write and solve an inequality that represents the value of *x*.**

20. The perimeter is less than 16 feet.

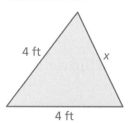

21. The base is greater than the height.

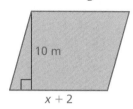

22. The perimeter is less than or equal to 5 feet.

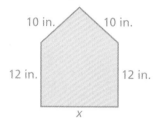

23. **REASONING** The solution of $w + c \leq 8$ is $w \leq 3$. What is the value of $c$?

24. **FENCE** The hole for a fence post is 2 feet deep. The top of the fence post needs to be at least 4 feet above the ground. Write and solve an inequality that represents the required length of the fence post.

25. **VIDEO GAME** You need at least 12,000 points to advance to the next level of a video game.

   a. Write and solve an inequality that represents the number of points you need to advance.

   b. You find a treasure chest that increases your score by 60%. Explain how this changes the inequality.

26. **MODELING** A circuit overloads at 1800 watts of electricity. A microwave that uses 1100 watts of electricity is plugged into the circuit.

   a. Use a model to write and solve an inequality that represents the additional number of watts you can plug in without overloading the circuit.

   b. In addition to the microwave, what two appliances in the table can you plug in without overloading the circuit? Explain.

| Appliance | Watts |
|---|---|
| Clock radio | 50 |
| Blender | 300 |
| Hot plate | 1200 |
| Toaster | 800 |

27. **Critical Thinking** The maximum surface area of the solid is $15\pi$ square millimeters. Write and solve an inequality that represents the height of the cylinder.

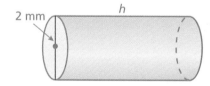

## Fair Game Review  *What you learned in previous grades & lessons*

**Solve the equation.** *(Section 1.1)*

28. $6 = 3x$

29. $\dfrac{r}{5} = 2$

30. $4c = 15$

31. $8 = \dfrac{2}{3}b$

**Find the square root.** *(Skills Review Handbook)*

32. $\sqrt{49}$

33. $\sqrt{0.25}$

34. $\sqrt{\dfrac{4}{9}}$

35. $\sqrt{12}$

# 3.3 Solving Inequalities Using Multiplication or Division

**Essential Question** How can you use multiplication or division to solve an inequality?

### 1 ACTIVITY: Using a Table to Solve an Inequality

Work with a partner.
- Copy and complete the table.
- Decide which graph represents the solution of the inequality.
- Write the solution of the inequality.

a. $3x \leq 6$

| x | −1 | 0 | 1 | 2 | 3 | 4 | 5 |
|---|---|---|---|---|---|---|---|
| 3x | | | | | | | |
| 3x $\overset{?}{\leq}$ 6 | | | | | | | |

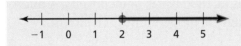 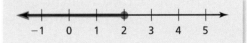

b. $-2x > 4$

| x | −5 | −4 | −3 | −2 | −1 | 0 | 1 |
|---|---|---|---|---|---|---|---|
| −2x | | | | | | | |
| −2x $\overset{?}{>}$ 4 | | | | | | | |

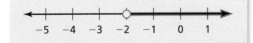 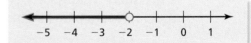

**COMMON CORE**

**Solving Inequalities**

In this lesson, you will
- solve inequalities using multiplication or division.
- solve real-life problems.

Learning Standards
A.CED.1
A.CED.3
A.REI.3

### 2 ACTIVITY: Writing a Rule

Work with a partner. Use a table to solve each inequality.

a. $3x > 3$   b. $4x \leq 4$   c. $-2x \geq 6$   d. $-5x < 10$

Write a rule that describes how to solve inequalities like those in Activity 1. Then use your rule to solve each of the four inequalities above.

116   Chapter 3   Solving Linear Inequalities

## 3  ACTIVITY: Using a Table to Solve an Inequality

**Math Practice**

**Look for Patterns**
How do the patterns help you complete this activity?

Work with a partner.

- Copy and complete the table.
- Decide which graph represents the solution of the inequality.
- Write the solution of the inequality.

a. $\dfrac{x}{2} \geq 1$

| x | −1 | 0 | 1 | 2 | 3 | 4 | 5 |
|---|---|---|---|---|---|---|---|
| $\dfrac{x}{2}$ | | | | | | | |
| $\dfrac{x}{2} \overset{?}{\geq} 1$ | | | | | | | |

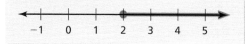   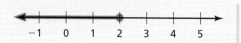

b. $\dfrac{x}{-3} < \dfrac{2}{3}$

| x | −5 | −4 | −3 | −2 | −1 | 0 | 1 |
|---|---|---|---|---|---|---|---|
| $\dfrac{x}{-3}$ | | | | | | | |
| $\dfrac{x}{-3} \overset{?}{<} \dfrac{2}{3}$ | | | | | | | |

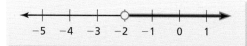   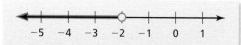

## 4  ACTIVITY: Writing a Rule

**Work with a partner. Use a table to solve each inequality.**

a. $\dfrac{x}{4} \geq 1$   b. $\dfrac{x}{2} < \dfrac{3}{2}$   c. $\dfrac{x}{-2} > 2$   d. $\dfrac{x}{-5} \leq \dfrac{1}{5}$

Write a rule that describes how to solve inequalities like those in Activity 3. Then use your rule to solve each of the four inequalities above.

### What Is Your Answer?

5. **IN YOUR OWN WORDS** How can you use multiplication or division to solve an inequality?

**Practice**  Use what you learned about solving inequalities using multiplication or division to complete Exercises 4–9 on page 121.

Section 3.3  Solving Inequalities Using Multiplication or Division

# 3.3 Lesson

## Key Idea

**Multiplication and Division Properties of Inequality (Case 1)**

**Words** If you multiply or divide each side of an inequality by the same *positive* number, the inequality remains true.

**Numbers**

$-6 < 8$  $\quad\quad\quad\quad\quad$  $6 > -8$

$2 \cdot (-6) < 2 \cdot 8$  $\quad\quad$  $\dfrac{6}{2} > \dfrac{-8}{2}$

$-12 < 16$  $\quad\quad\quad\quad$  $3 > -4$

**Algebra**

$\dfrac{x}{2} < -9$  $\quad\quad\quad\quad$  $4x > -12$

$2 \cdot \dfrac{x}{2} < 2 \cdot (-9)$  $\quad\quad$  $\dfrac{4x}{4} > \dfrac{-12}{4}$

$x < -18$  $\quad\quad\quad\quad$  $x > -3$

These properties are also true for ≤ and ≥.

**Remember**

Multiplication and division are inverse operations.

### EXAMPLE 1  Solving an Inequality Using Multiplication

Solve $\dfrac{x}{8} > -5$. Graph the solution.

$\dfrac{x}{8} > -5$  $\quad\quad$ Write the inequality.

Undo the division. $\longrightarrow$ $8 \cdot \dfrac{x}{8} > 8 \cdot (-5)$  $\quad\quad$ Multiply each side by 8.

$x > -40$  $\quad\quad$ Simplify.

 The solution is $x > -40$.

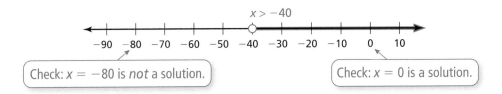

### On Your Own

**Solve the inequality. Graph the solution.**

1. $a \div 2 < 4$  $\quad\quad$ 2. $\dfrac{n}{7} \geq -1$  $\quad\quad$ 3. $-6.4 \geq \dfrac{w}{5}$

# EXAMPLE 2 Solving an Inequality Using Division

Solve $3x \leq -24$. Graph the solution.

$3x \leq -24$     Write the inequality.

Undo the multiplication. → $\dfrac{3x}{3} \leq \dfrac{-24}{3}$     Divide each side by 3.

$x \leq -8$     Simplify.

∴ The solution is $x \leq -8$.

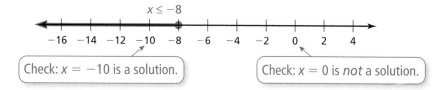

Check: $x = -10$ is a solution.  Check: $x = 0$ is *not* a solution.

## On Your Own

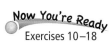
Exercises 10–18

Solve the inequality. Graph the solution.

**4.** $4b \geq 36$     **5.** $2k > -10$     **6.** $-18 > 1.5q$

## Key Idea

**Multiplication and Division Properties of Inequality (Case 2)**

**Words**    If you multiply or divide each side of an inequality by the same *negative* number, the direction of the inequality symbol must be reversed for the inequality to remain true.

**Numbers**
$$-6 < 8 \qquad\qquad 6 > -8$$
$$(-2) \cdot (-6) \;\boxed{>}\; (-2) \cdot 8 \qquad \dfrac{6}{-2} \;\boxed{<}\; \dfrac{-8}{-2}$$
$$12 > -16 \qquad\qquad -3 < 4$$

**Algebra**
$$\dfrac{x}{-6} < 3 \qquad\qquad -5x > 30$$
$$-6 \cdot \dfrac{x}{-6} \;\boxed{>}\; -6 \cdot 3 \qquad \dfrac{-5x}{-5} \;\boxed{<}\; \dfrac{30}{-5}$$
$$x > -18 \qquad\qquad x < -6$$

These properties are also true for $\leq$ and $\geq$.

**Common Error**

A negative sign in an inequality does not necessarily mean you must reverse the inequality symbol.

Only reverse the inequality symbol when you multiply or divide both sides by a negative number.

Section 3.3    Solving Inequalities Using Multiplication or Division

**EXAMPLE 3** **Solving an Inequality Using Multiplication**

Solve $\dfrac{y}{-3} > 2$. Graph the solution.

$\dfrac{y}{-3} > 2$    Write the inequality.

Undo the division. → $-3 \cdot \dfrac{y}{-3}\ \boxed{<}\ -3 \cdot 2$    Multiply each side by $-3$. Reverse the inequality symbol.

$y < -6$    Simplify.

∴ The solution is $y < -6$.

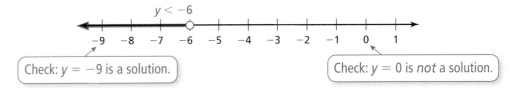

Check: $y = -9$ is a solution.    Check: $y = 0$ is *not* a solution.

**EXAMPLE 4** **Solving an Inequality Using Division**

Solve $-7y \le -35$. Graph the solution.

$-7y \le -35$    Write the inequality.

Undo the multiplication. → $\dfrac{-7y}{-7} \ge \dfrac{-35}{-7}$    Divide each side by $-7$. Reverse the inequality symbol.

$y \ge 5$    Simplify.

∴ The solution is $y \ge 5$.

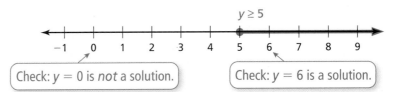

Check: $y = 0$ is *not* a solution.    Check: $y = 6$ is a solution.

**On Your Own**

**Now You're Ready**
Exercises 27–35

Solve the inequality. Graph the solution.

7. $\dfrac{p}{-4} < 7$

8. $\dfrac{x}{-5} \le -5$

9. $1 \ge -\dfrac{1}{10}z$

10. $-9m > 63$

11. $-2r \ge -22$

12. $-0.4y \ge -12$

# 3.3 Exercises

## Vocabulary and Concept Check

1. **VOCABULARY** Explain how to solve $\frac{x}{6} < -5$.

2. **WRITING** Explain how solving $2x < -8$ is different from solving $-2x < 8$.

3. **OPEN-ENDED** Write an inequality that is solved using the Division Property of Inequality where the inequality symbol needs to be reversed.

## Practice and Problem Solving

**Use a table to solve the inequality.**

4. $4x < 4$
5. $-2x \leq 2$
6. $-5x > 15$
7. $\frac{x}{-3} \geq 1$
8. $\frac{x}{-2} > \frac{5}{2}$
9. $\frac{x}{4} \leq \frac{3}{8}$

**Solve the inequality. Graph the solution.**

 10. $3n > 18$
11. $\frac{c}{4} \leq -9$
12. $1.2m < 12$

13. $-14 > x \div 2$
14. $\frac{w}{5} \geq -2.6$
15. $5 < 2.5k$

16. $4x \leq -\frac{3}{2}$
17. $2.6y \leq -10.4$
18. $10.2 > \frac{b}{3.4}$

19. **ERROR ANALYSIS** Describe and correct the error in solving the inequality.

**Write the word sentence as an inequality. Then solve the inequality.**

20. The quotient of a number and 3 is at most 4.

21. A number divided by 8 is less than $-2$.

22. Four times a number is at least $-12$.

23. The product of 5 and a number is greater than 20.

24. **CAMERA** You earn $9.50 per hour at your summer job. Write and solve an inequality that represents the number of hours you need to work in order to buy a digital camera that costs $247.

25. **COPIES** You have $3.65 to make copies. Write and solve an inequality that represents the number of copies you can make.

26. **SPEED LIMIT** The maximum speed limit for a school bus is 55 miles per hour. Write and solve an inequality that represents the number of hours it takes to travel 165 miles in a school bus.

**Solve the inequality. Graph the solution.**

③ ④ 27. $-2n \leq 10$

28. $-5w > 30$

29. $\dfrac{h}{-6} \geq 7$

30. $-8 < -\dfrac{1}{3}x$

31. $-2y < -11$

32. $-7d \geq 56$

33. $2.4 > -\dfrac{m}{5}$

34. $\dfrac{k}{-0.5} \leq 18$

35. $-2.5 > \dfrac{b}{-1.6}$

36. **ERROR ANALYSIS** Describe and correct the error in solving the inequality.

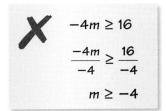

37. **CRITICAL THINKING** Are all numbers greater than zero solutions of $-x > 0$? Explain.

38. **TRUCKING** In many states, the maximum height (including freight) of a vehicle is 13.5 feet.

   a. Five crates are stacked vertically on the bed of the truck. Is this legal? Explain.

   b. Write and solve an inequality to justify your answer to part (a).

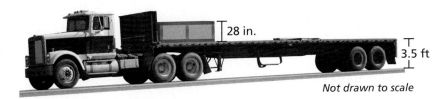

*Not drawn to scale*

**Write and solve an inequality that represents the value of $x$.**

39. Area $\geq 102$ cm$^2$

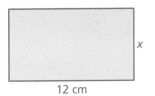

40. Area $< 30$ ft$^2$

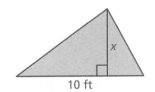

41. **TRIP** You and three friends are planning a trip. You want to keep the cost below $80 per person. Write and solve an inequality that represents the total cost of the trip.

42. **PRECISION** Explain why the direction of the inequality symbol must be reversed when multiplying or dividing by the same negative number.

43. **PROJECT** Choose two musical artists to research.

    a. Use the Internet or a magazine to complete the table.

    b. Find and compare the average number of copies sold per month for each CD. Which CD do you consider to be the most successful? Explain.

    c. Assume each CD continues to sell at the average rate. Write and solve an inequality that represents the number of months it will take for the total number of copies sold to exceed twice the current number sold.

| Artist | Name of CD | Release Date | Current Number of Copies Sold |
|---|---|---|---|
| 1. | | | |
| 2. | | | |

**Structure** Describe all numbers that satisfy *both* inequalities. Include a graph with your description.

44. $3m > -12$ and $2m < 12$

45. $\dfrac{n}{2} \geq -3$ and $\dfrac{n}{-4} \geq 1$

46. $2x \geq -4$ and $2x \geq 4$

47. $\dfrac{m}{-4} > -5$ and $\dfrac{m}{4} < 10$

## Fair Game Review  What you learned in previous grades & lessons

**Solve the equation.** *(Section 1.2)*

48. $-4w + 5 = -11$

49. $4(x - 3) = 21$

50. $\dfrac{v}{6} - 7 = 4$

51. $\dfrac{m + 300}{4} = 96$

52. **MULTIPLE CHOICE** Which of the following is *not* a solution of $p - 3.9 \geq 0.8$? *(Section 3.2)*

    Ⓐ $p = -4.5$   Ⓑ $p = 4.7$   Ⓒ $p = 4.75$   Ⓓ $p = 5$

# 3 Study Help

You can use a **four square** to organize information about a topic. Each of the four squares can be a category, such as *definition, vocabulary, example, non-example, words, algebra, table, numbers, visual, graph,* or *equation*. Here is an example of a four square for an inequality.

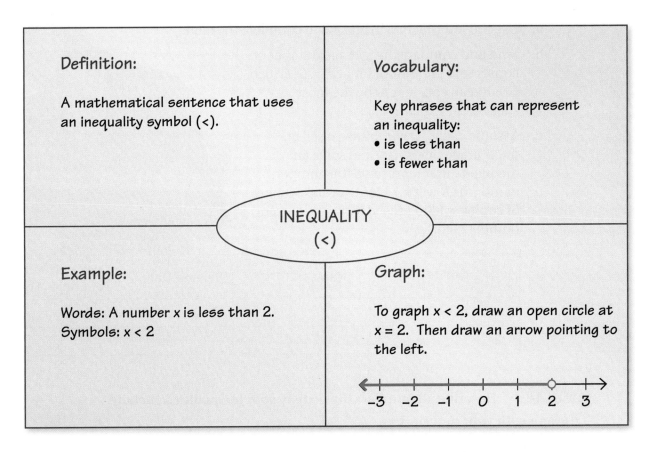

## On Your Own

**Make four squares to help you study these topics.**

1. inequality ($\geq$)
2. solving an inequality using addition
3. solving an inequality using subtraction
4. solving an inequality using multiplication
5. solving an inequality using division

**After you complete this chapter, make four squares for the following topics.**

6. solving a compound inequality
7. graphing an inequality in two variables

"Sorry, but I have limited space in my four square. I needed pet names with only three letters."

# 3.1–3.3 Quiz

**Write the word sentence as an inequality.** *(Section 3.1)*

1. A number $x$ plus 1 is less than $-13$.
2. A number $t$ minus 1.6 is at most 9.

**Tell whether the given value is a solution of the inequality.** *(Section 3.1)*

3. $12n < -2$; $n = -1$
4. $y + 4 < -3$; $y = -7$

**Graph the inequality on a number line.** *(Section 3.1)*

5. $x > -10$
6. $w < 6.8$

**Solve the inequality. Graph the solution.** *(Section 3.2 and Section 3.3)*

7. $x - 2 < 4$
8. $g + 14 \geq 30$
9. $h - 1 \leq -9$
10. $\dfrac{3}{2} < p + \dfrac{1}{2}$
11. $\dfrac{n}{-6} \geq -2$
12. $-4y \geq 60$

**Write the word sentence as an inequality. Then solve the inequality.** *(Section 3.3)*

13. The quotient of a number and 6 is more than 9.
14. Five times a number is at most $-10$.

**LIFEGUARDS NEEDED**
**Take Our Training Course NOW!!!**
Lifeguard Training Requirements
- Swim at least 100 yards
- Tread water for at least 5 minutes
- Swim 10 yards or more underwater without taking a breath

15. **LIFEGUARD** Three requirements for a lifeguard training course are shown. *(Section 3.1)*

    a. Write and graph three inequalities that represent the requirements.

    b. You can swim 350 feet. Do you satisfy the swimming requirement of the course? Explain.

16. **REASONING** The solution of $x - a > 4$ is $x > 11$. What is the value of $a$? *(Section 3.2)*

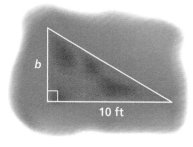

17. **GARDEN** The area of the triangular garden must be less than 35 square feet. Write and solve an inequality that represents the value of $b$. *(Section 3.3)*

# 3.4 Solving Multi-Step Inequalities

**Essential Question** How can you use an inequality to describe the area and perimeter of a composite figure?

## 1 ACTIVITY: Areas and Perimeters of Composite Figures

**Work with a partner.**

a. For what values of $x$ will the area of the blue region be greater than 12 square units?

b. For what values of $x$ will the sum of the inner and outer perimeters of the blue region be greater than 20 units?

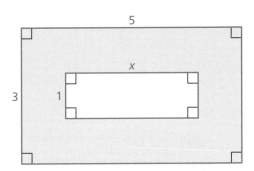

c. For what values of $y$ will the area of the trapezoid be less than or equal to 10 square units?

d. For what values of $y$ will the perimeter of the trapezoid be less than or equal to 16 units?

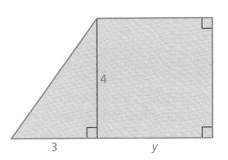

e. For what values of $w$ will the area of the red region be greater than or equal to 36 square units?

f. For what values of $w$ will the sum of the inner and outer perimeters of the red region be greater than 47 units?

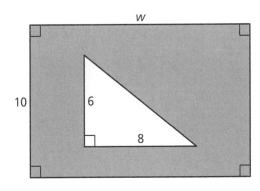

**COMMON CORE**

**Solving Inequalities**

In this lesson, you will
- write and solve multi-step inequalities.
- solve real-life problems.

Learning Standards
A.CED.1
A.CED.3
A.REI.3

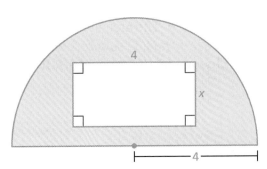

g. For what values of $x$ will the area of the yellow region be less than $4\pi$ square units?

h. For what values of $x$ will the sum of the inner and outer perimeters of the yellow region be less than $4\pi + 20$ units?

126    Chapter 3    Solving Linear Inequalities

## 2 ACTIVITY: Volume and Surface Area of a Composite Solid

**Math Practice**

**Use Operations**
Which operations will you use to find the volume and surface area of the composite solid?

**Work with a partner.**

a. For what values of $x$ will the volume of the solid be greater than or equal to 42 cubic units?

b. For what values of $x$ will the surface area of the solid be greater than 72 square units?

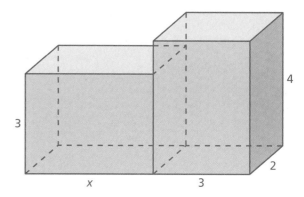

## 3 ACTIVITY: Planning a Budget

**Work with a partner.**

**You are building a patio. You want to cover the patio with Spanish tile that costs $5 per square foot. Your budget for the tile is $1700. How wide can you make the patio without going over your budget?**

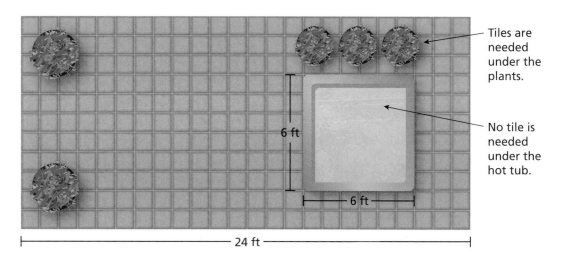

- Tiles are needed under the plants.
- No tile is needed under the hot tub.

## What Is Your Answer?

4. **IN YOUR OWN WORDS** How can you use an inequality to describe the area and perimeter of a composite figure? Give an example. Include a diagram with your example.

**Practice** — Use what you learned about solving multi-step inequalities to complete Exercises 3 and 4 on page 130.

## 3.4 Lesson

You can use the properties of inequality to solve multi-step inequalities the same way you use the properties of equality to solve multi-step equations.

### EXAMPLE 1 Solving a Multi-Step Inequality

Solve $\dfrac{y}{-6} + 7 < 9$. Graph the solution.

| | |
|---|---|
| $\dfrac{y}{-6} + 7 < 9$ | Write the inequality. |
| Undo the addition. $\quad \dfrac{-7 \quad -7}{}$ | Subtract 7 from each side. |
| $\dfrac{y}{-6} < 2$ | Simplify. |
| Undo the division. $\quad -6 \cdot \dfrac{y}{-6} > -6 \cdot 2$ | Multiply each side by $-6$. Reverse the inequality symbol. |
| $y > -12$ | Simplify. |

∴ The solution is $y > -12$.

### On Your Own

Now You're Ready
Exercises 5–10

Solve the inequality. Graph the solution.

1. $4b - 1 < 7$
2. $8 + 9c \geq -28$
3. $\dfrac{n}{-2} + 11 > 12$

When solving an inequality, if you obtain an inequality that is true, such as $-5 < 0$, then the solution is the set of *all real numbers*. If you obtain an inequality that is false, such as $3 \leq -2$, then the inequality has *no solutions*.

### EXAMPLE 2 Solving an Inequality with No Solution

Solve $8x - 3 > 4(2x + 3)$.

| | |
|---|---|
| $8x - 3 > 4(2x + 3)$ | Write the inequality. |
| $8x - 3 > 8x + 12$ | Distributive Property |
| $\underline{-8x \qquad -8x}$ | Subtract $8x$ from each side. |
| $-3 \not> 12$ ✗ | Simplify. |

∴ The inequality $-3 > 12$ is false. So, there are no solutions.

128    Chapter 3    Solving Linear Inequalities

# EXAMPLE 3 Solving an Inequality with Infinitely Many Solutions

Which graph represents the solution of $2(5x - 1) \leq 7 + 10x$?

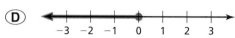

**Study Tip**

The graph of the set of all real numbers is the entire number line.

$$2(5x - 1) \leq 7 + 10x \quad \text{Write the inequality.}$$
$$10x - 2 \leq 7 + 10x \quad \text{Distributive Property}$$
$$\underline{-10x \qquad\qquad -10x} \quad \text{Subtract 10x from each side.}$$
$$-2 \leq 7 \quad \text{Simplify.}$$

∴ The inequality $-2 \leq 7$ is true. So, the solution is the set of all real numbers. The correct answer is Ⓑ.

# EXAMPLE 4 Real-Life Application

**You need a mean score of at least 90 to advance to the next round of the trivia game. What score do you need on the fifth game to advance?**

Use the definition of mean to write and solve an inequality. Let $x$ be the score on the fifth game.

$$\frac{95 + 91 + 77 + 89 + x}{5} \geq 90$$

The meaning of the phrase "at least" is greater than or equal to.

$$\frac{352 + x}{5} \geq 90 \quad \text{Simplify.}$$
$$5 \cdot \frac{352 + x}{5} \geq 5 \cdot 90 \quad \text{Multiply each side by 5.}$$
$$352 + x \geq 450 \quad \text{Simplify.}$$
$$\underline{-352 \qquad\qquad -352} \quad \text{Subtract 352 from each side.}$$
$$x \geq 98 \quad \text{Simplify.}$$

**Remember**

The mean in Example 4 is equal to the sum of the game scores divided by the number of games.

∴ You need at least 98 points to advance to the next round.

## On Your Own

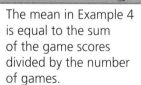

Now You're Ready
Exercises 12–20

**Solve the inequality, if possible.**

4. $2(k - 5) < 2k + 5$
5. $-4(3n - 1) > -12n + 5.2$

6. **WHAT IF?** In Example 4, you need a mean score of at least 88 to advance to the next round of the trivia game. What score do you need on the fifth game to advance?

Section 3.4 Solving Multi-Step Inequalities 129

## 3.4 Exercises

### Vocabulary and Concept Check

1. **WRITING** Compare and contrast solving multi-step inequalities and solving multi-step equations.

2. **WRITING** How do you know when an inequality has no solutions? How do you know when the solution of an inequality is the set of all real numbers?

### Practice and Problem Solving

3. For what values of $k$ will the perimeter of the octagon be less than or equal to 64 units?

4. For what values of $h$ will the surface area of the solid be greater than 46 square units?

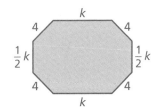
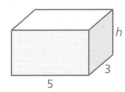

**Solve the inequality. Graph the solution.**

5. $7b + 4 \geq 11$

6. $2v - 4 < 8$

7. $1 - \dfrac{m}{3} \leq 6$

8. $\dfrac{4}{5} < 3w - \dfrac{11}{5}$

9. $1.8 < 0.5 - 1.3p$

10. $-2.4r + 9.6 \geq 4.8$

11. **ERROR ANALYSIS** Describe and correct the error in solving the inequality.

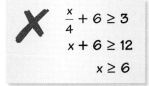

**Solve the inequality, if possible.**

12. $6(g + 2) \leq 18$

13. $4(y - 2) \geq 4y - 9$

14. $-10 \geq \dfrac{5}{3}(h - 3)$

15. $-\dfrac{1}{3}(u + 2) > 5$

16. $2.7 > 0.9(n - 1.7)$

17. $10 > -2.5(z - 3.1)$

18. $5(w + 4) \leq 5w + 20$

19. $-(6 - x) < x - 7.5$

20. $12c - 5 > 3(4c + 1)$

21. **ATM** Write and solve an inequality that represents the number of $20 bills you can withdraw from the account without going below the minimum balance.

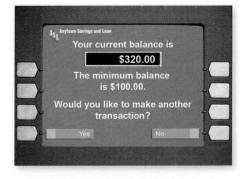

**Solve the inequality. Graph the solution.**

**22.** $5x - 2x + 7 \leq 15 + 10$

**23.** $7b - 12b + 1.4 > 8.4 - 22$

**24. TYPING** One line of text on a page uses about $\frac{3}{16}$ of an inch. There are 1-inch margins at the top and bottom of a page. Write and solve an inequality to find the number of lines that can be typed on a page that is 11 inches long.

**25. WOODWORKING** A woodworker builds a cabinet in 20 hours. The cabinet is sold at a store for $500. Write and solve an inequality that represents the hourly wage the store can pay the woodworker and still make a profit of at least $100.

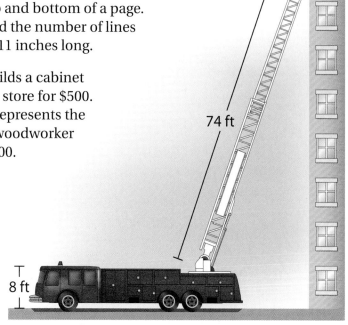

**26. FIRE TRUCK** The height of one story of a building is about 10 feet. The bottom of the ladder on the fire truck must be at least 24 feet away from the building. Write and solve an inequality to find the number of stories the ladder can reach.

**27. REASONING** A drive-in movie theater charges $3.50 per car.

  **a.** The drive-in has already admitted 100 cars. Write and solve an inequality to find how many more cars the drive-in needs to admit to earn at least $500.

  **b.** The theater increases the price by $1 per car. How does this affect the total number of cars needed to earn $500? Explain.

**28. Challenge** For what values of $r$ will the area of the shaded region be greater than or equal to $9(\pi - 2)$?

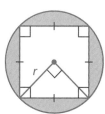

 **Fair Game Review** *What you learned in previous grades & lessons*

**Graph the linear equation.** *(Section 2.1)*

**29.** $y = 4x - 1$   **30.** $y = -4$   **31.** $x = 5$   **32.** $y = -\frac{1}{2}x + 3$

**33. MULTIPLE CHOICE** Which of the following is shown in the graph? *(Section 2.4)*

  **A** $3x + 4y = -12$   **B** $3x - 4y = -12$
  **C** $3x + 4y = 12$   **D** $3x - 4y = 12$

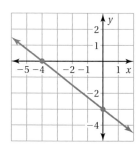

# Extension 3.4 Solving Compound Inequalities

**Key Vocabulary**
compound inequality, p. 132
absolute value inequality, p. 134

A **compound inequality** is an inequality formed by joining two inequalities with the word "and" or the word "or."

Solutions of a compound inequality with "and" consist of numbers that are solutions of both inequalities.

Solutions of a compound inequality with "or" consist of numbers that are solutions of at least one of the inequalities.

$x \geq 2$

$y \leq -2$

$x < 5$

$y > 1$

$2 \leq x$ and $x < 5$
$2 \leq x < 5$

$y \leq -2$ or $y > 1$

### EXAMPLE 1 Writing and Graphing Compound Inequalities

Write each word sentence as an inequality. Graph the inequality.

**a. A number $x$ is greater than $-8$ and less than or equal to 4.**

A number $x$ is greater than $-8$ and less than or equal to 4.

$x > -8$    and    $x \leq 4$

**Study Tip**
A compound inequality with "and" can be written as a single inequality. For example, you can write $x > -8$ and $x \leq 4$ as $-8 < x \leq 4$.

**b. A number $y$ is at most 0 or at least 7.**

A number $y$ is at most 0 or at least 7.

$y \leq 0$    or    $y \geq 7$

## Practice

**In Exercises 1–4, write the word sentence as an inequality. Graph the inequality.**

1. A number $k$ is more than 3 and less than 9.

2. A number $n$ is greater than or equal to 6 and no more than 11.

3. A number $w$ is fewer than $-10$ or no less than $-6$.

4. A number $z$ is less than or equal to $-5$ or more than 4.

5. Write an inequality to describe the graph.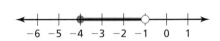

6. The world's longest human life span is 122 years. Write and graph a compound inequality that describes the ages of all humans.

You can solve compound inequalities by solving two inequalities separately. When a compound inequality with "and" is written as a single inequality, you can solve the inequality by performing the same operation on each expression.

## EXAMPLE 2  Solving a Compound Inequality with "And"

Solve $-3 < -2x + 1 \leq 9$. Graph the solution.

$$-3 < -2x + 1 \leq 9 \quad \text{Write the inequality.}$$
$$\underline{-1} \qquad \underline{-1} \quad \underline{-1} \quad \text{Subtract 1 from each expression.}$$
$$-4 < -2x \qquad \leq 8 \quad \text{Simplify.}$$
$$\frac{-4}{-2} > \frac{-2x}{-2} \qquad \geq \frac{8}{-2} \quad \text{Divide each expression by } -2. \text{ Reverse the inequality symbols.}$$
$$2 > \quad x \quad \geq -4 \quad \text{Simplify.}$$

∴ The solution is $-4 \leq x < 2$.

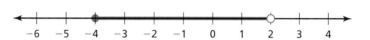

**Study Tip**

You can also solve the inequality in Example 2 by solving the inequalities
$-3 < -2x + 1$
and
$-2x + 1 \leq 9$
separately.

## EXAMPLE 3  Solving a Compound Inequality with "Or"

Solve $3x - 5 < -8$ or $2x - 1 > 5$. Graph the solution.

$$3x - 5 < -8 \quad \text{or} \quad 2x - 1 > 5 \quad \text{Write the inequality.}$$
$$\underline{+5} \quad \underline{+5} \qquad \underline{+1} \quad \underline{+1} \quad \text{Addition Property of Inequality}$$
$$3x < -3 \quad \text{or} \quad 2x > 6 \quad \text{Simplify.}$$
$$\frac{3x}{3} < \frac{-3}{3} \quad \text{or} \quad \frac{2x}{2} > \frac{6}{2} \quad \text{Division Property of Inequality}$$
$$x < -1 \quad \text{or} \quad x > 3 \quad \text{Simplify.}$$

∴ The solution is $x < -1$ or $x > 3$.

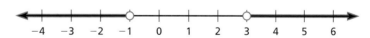

**Solving Inequalities**

In this extension, you will
- write, solve, and graph compound inequalities.
- write, solve, and graph absolute value inequalities.

Applying Standards
A.CED.1
A.CED.3
A.REI.3

## Practice

**Solve the inequality. Graph the solution.**

**7.** $4 < x - 5 < 7$

**8.** $-1 \leq 2x + 3 < 7$

**9.** $15 > -3x + 9 \geq 0$

**10.** $4x + 1 \leq -11$ or $3x - 4 \geq 5$

**11.** $-2x - 7 < 5$ or $-5x + 6 \geq 41$

**Study Tip**

When an absolute value expression is on the left side of an inequality, use an "and" statement for < and ≤, and an "or" statement for > and ≥.

An **absolute value inequality** is an inequality that contains an absolute value expression. For example, $|x| < 2$ and $|x| > 2$ are absolute value inequalities.

The distance between $x$ and 0 is less than 2.

$$|x| < 2$$

The graph of $|x| < 2$ is $x > -2$ and $x < 2$.

The distance between $x$ and 0 is greater than 2.

$$|x| > 2$$

The graph of $|x| > 2$ is $x < -2$ or $x > 2$.

You can solve these types of inequalities by solving a compound inequality.

## Key Idea

**Solving Absolute Value Inequalities**

To solve $|ax + b| < c$ for $c > 0$, solve the compound inequality

$$ax + b > -c \quad \text{and} \quad ax + b < c.$$

To solve $|ax + b| > c$ for $c > 0$, solve the compound inequality

$$ax + b < -c \quad \text{or} \quad ax + b > c.$$

In the inequalities above, you can replace < with ≤ and > with ≥.

### EXAMPLE 4 Solving Absolute Value Inequalities

**a. Solve $|x + 7| \leq 2$. Graph the solution.**

Use $|x + 7| \leq 2$ to write a compound inequality. Then solve.

| $x + 7 \geq -2$ | and | $x + 7 \leq 2$ | Write compound inequality. |
| $-7 \quad -7$ | | $-7 \quad -7$ | Subtract 7 from each side. |
| $x \geq -9$ | and | $x \leq -5$ | Simplify. |

∴ The solution is $x \geq -9$ and $x \leq -5$.

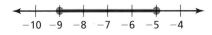

**b. Solve $|8x - 11| < 0$.**

The absolute value of an expression must be greater than or equal to 0. The expression $|8x - 11|$ cannot be less than 0.

∴ So, the inequality has no solution.

### EXAMPLE 5 Solving an Absolute Value Inequality

Solve $4|2x - 5| + 1 > 29$.

$4|2x - 5| + 1 > 29$    Write the inequality.

$|2x - 5| > 7$    Isolate the absolute value expression.

Use $|2x - 5| > 7$ to write a compond inequality. Then solve.

| | | | |
|---|---|---|---|
| $2x - 5 < -7$ | or | $2x - 5 > 7$ | Write compound inequality. |
| $+5 \quad +5$ | | $+5 \quad +5$ | Add 5 to each side. |
| $2x < -2$ | or | $2x > 12$ | Simplify. |
| $\dfrac{2x}{2} < \dfrac{-2}{2}$ | or | $\dfrac{2x}{2} > \dfrac{12}{2}$ | Divide each side by 2. |
| $x < -1$ | or | $x > 6$ | Simplify. |

### EXAMPLE 6 Real-Life Application

In a poll, 47% of voters say they plan to reelect the mayor. The poll has a margin of error of ±2 percentage points. Write and solve an absolute value inequality to find the least and greatest percents of voters who plan to reelect the mayor.

**Words**    Actual percent of voters    minus    percent of voters in poll    is less than or equal to    the margin of error.

**Variable**    Let $x$ represent the actual percent of voters who plan on reelecting the mayor.

**Inequality**    $|x - 47| \leq 2$

| | | | |
|---|---|---|---|
| $x - 47 \geq -2$ | and | $x - 47 \leq 2$ | Write compound inequality. |
| $+47 \quad +47$ | | $+47 \quad +47$ | Add 47 to each side. |
| $x \geq 45$ | and | $x \leq 49$ | Simplify. |

∴ The least percent of voters who plan to reelect the mayor is 45%. The greatest percent of voters who plan to reelect the mayor is 49%.

### Practice

**Solve the inequality. Graph the solution, if possible.**

12. $|x - 3| \geq 4$
13. $|x + 7| < 1$
14. $11 \geq |4x - 5|$
15. $|8x - 9| < 0$
16. $3|2x + 5| - 8 \geq 19$
17. $-2|x - 10| + 1 > -7$

18. **NUMBER SENSE** What is the solution of $|4x - 2| \geq -6$? Explain.

19. **MODELING** In Example 6, 44% of the voters say they plan to reelect the mayor. The poll has a margin of error of ±3 percentage points. Use a model to write and solve an absolute value inequality to find the least and greatest percents of voters who plan to reelect the mayor.

## 3.5 Graphing Linear Inequalities in Two Variables

**Essential Question** How can you use a coordinate plane to solve problems involving linear inequalities?

### 1 ACTIVITY: Graphing Inequalities

**Work with a partner.**

a. Graph $y = x + 1$ in a coordinate plane.

b. Choose three points that lie above the graph of $y = x + 1$. Substitute the values of $x$ and $y$ of each point in the inequality $y > x + 1$. If the substitutions result in true statements, plot the points on the graph.

c. Choose three points that lie below the graph of $y = x + 1$. Substitute the values of $x$ and $y$ of each point in the inequality $y > x + 1$. If the substitutions result in true statements, plot the points on the graph.

d. To graph $y > x + 1$, would you choose points above or below $y = x + 1$?

e. Choose a point that lies on the graph of $y = x + 1$. Substitute the values of $x$ and $y$ in the inequality $y > x + 1$. What do you notice? Do you think the graph of $y > x + 1$ includes the points that lie on the graph of $y = x + 1$? Explain your reasoning.

f. Explain how you could change the inequality so that it includes the points that lie on the graph of $y = x + 1$.

### 2 ACTIVITY: Writing and Graphing Inequalities

**COMMON CORE**

**Graphing Inequalities**

In this lesson, you will
- graph linear inequalities in two variables.

Learning Standard
A.REI.12

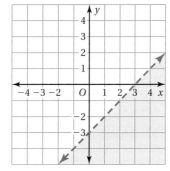

Work with a partner. The graph of a linear inequality in two variables shows all the solutions of the inequality in a coordinate plane. An ordered pair $(x, y)$ is a solution of an inequality if the inequality is true when the values of $x$ and $y$ are substituted in the inequality.

a. Write an equation for the graph of the dashed blue line.

b. The solutions of an inequality are represented by the shaded region. In words, describe the solutions of the inequality.

c. Write an inequality for the graph. Which inequality symbol did you use? Explain your reasoning.

**136** Chapter 3 Solving Linear Inequalities

### 3 EXAMPLE: Using a Graphing Calculator

**Math Practice 5**

**Recognize Usefulness of Tools**

When do you think it would be useful to use a graphing calculator?

Use a graphing calculator to graph $y \geq \frac{1}{4}x - 3$.

a. Enter the equation $y = \frac{1}{4}x - 3$ into your calculator.

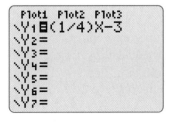

b. The inequality contains the symbol $\geq$. So, the region to be shaded is above the graph of $y = \frac{1}{4}x - 3$. Adjust your graphing calculator so that the region above the graph will be shaded.

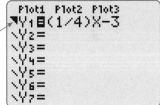

For some calculators, this icon represents the region above the graph.

c. Graph $y \geq \frac{1}{4}x - 3$ on your calculator.

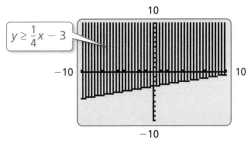

Some graphing calculators always use a solid line when graphing inequalities. In this case, you will have to decide whether the line should be dashed or solid.

### What Is Your Answer?

4. Use a graphing calculator to graph each inequality in a standard viewing window.

   a. $y > x + 5$
   b. $y \leq -\frac{1}{2}x + 1$
   c. $y \geq -x - 4$

5. **IN YOUR OWN WORDS** How can you use a coordinate plane to solve problems involving linear inequalities? Give an example of a real-life problem that can be represented by an inequality in two variables.

**Practice** → Use what you learned about writing and graphing inequalities to complete Exercises 8–10 on page 141.

Section 3.5  Graphing Linear Inequalities in Two Variables  137

# 3.5 Lesson

**Key Vocabulary**
linear inequality in two variables, *p. 138*
solution of a linear inequality, *p. 138*
graph of a linear inequality, *p. 138*
half-planes, *p. 138*

A **linear inequality in two variables** $x$ and $y$ can be written as

$$ax + by < c \quad ax + by \le c \quad ax + by > c \quad ax + by \ge c$$

where $a$, $b$, and $c$ are real numbers. A **solution of a linear inequality** in two variables is an ordered pair $(x, y)$ that makes the inequality true.

### EXAMPLE 1   Checking Solutions of a Linear Inequality

**Tell whether the ordered pair is a solution of the inequality.**

**a.** $2x + y < -3;\ (-1, 9)$

$$2x + y < -3 \qquad \text{Write the inequality.}$$
$$2(-1) + 9 \stackrel{?}{<} -3 \qquad \text{Substitute } -1 \text{ for } x \text{ and } 9 \text{ for } y.$$
$$7 \not< -3 \quad ✗ \qquad \text{Simplify. 7 is } not \text{ less than } -3.$$

∴ So, $(-1, 9)$ is *not* a solution of the inequality.

**b.** $x - 3y \ge 8;\ (2, -2)$

$$x - 3y \ge 8 \qquad \text{Write the inequality.}$$
$$2 - 3(-2) \stackrel{?}{\ge} 8 \qquad \text{Substitute 2 for } x \text{ and } -2 \text{ for } y.$$
$$8 \ge 8 \quad ✓ \qquad \text{Simplify. 8 is equal to 8.}$$

∴ So, $(2, -2)$ is a solution of the inequality.

### On Your Own

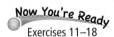
Exercises 11–18

**Tell whether the ordered pair is a solution of the inequality.**

1. $x + y > 0;\ (-2, 2)$
2. $4x - y \ge 5;\ (0, 0)$
3. $5x - 2y \le -1;\ (-4, -1)$
4. $-2x - 3y < 15;\ (5, -7)$

---

**Reading**
A dashed boundary line means that points on the line are *not* solutions. A solid boundary line means that points on the line are solutions.

The **graph of a linear inequality** in two variables shows all of the solutions of the inequality in a coordinate plane.

All solutions of $y < 2x$ lie on one side of the boundary line $y = 2x$.

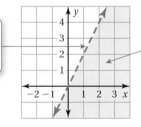

The boundary line divides the coordinate plane into two **half-planes**. The shaded half-plane is the graph of $y < 2x$.

## Key Idea

**Graphing a Linear Inequality in Two Variables**

**Step 1** Graph the boundary line for the inequality. Use a dashed line for < or >. Use a solid line for ≤ or ≥.

**Step 2** Test a point that is not on the boundary line to determine if it is a solution of the inequality.

**Step 3** If the test point is a solution, shade the half-plane that contains the point. If the test point is *not* a solution, shade the half-plane that does *not* contain the point.

It is convenient to use the origin as a test point because it is easily substituted. However, you must choose a different test point if the origin is on the boundary line.

### EXAMPLE 2  Graphing Linear Inequalities in One Variable

**a. Graph $y \leq 2$ in a coordinate plane.**

**Step 1:** Graph $y = 2$. Use a solid line because the inequality symbol is ≤.

**Step 2:** Test (0, 0).

$y \leq 2$     Write the inequality.

$0 \leq 2$ ✓     Substitute.

**Step 3:** Because (0, 0) is a solution, shade the half-plane that contains (0, 0).

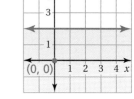

**b. Graph $x > 1$ in a coordinate plane.**

**Step 1:** Graph $x = 1$. Use a dashed line because the inequality symbol is >.

**Step 2:** Test (0, 0).

$x > 1$     Write the inequality.

$0 \not> 1$ ✗     Substitute.

**Step 3:** Because (0, 0) is *not* a solution, shade the half-plane that does *not* contain (0, 0).

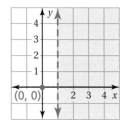

### On Your Own

**Now You're Ready**
Exercises 32–34

Graph the inequality in a coordinate plane.

**5.** $y > -1$

**6.** $y \geq -5$

**7.** $x \leq -4$

**8.** $3.5 > x$

## EXAMPLE 3 Graphing Linear Inequalities in Two Variables

Graph $-x + 2y > 2$ in a coordinate plane.

**Step 1:** Graph $-x + 2y = 2$, or $y = \frac{1}{2}x + 1$.

Use a dashed line because the inequality symbol is >.

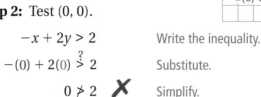

**Step 2:** Test (0, 0).

$-x + 2y > 2$   Write the inequality.

$-(0) + 2(0) \stackrel{?}{>} 2$   Substitute.

$0 \not> 2$ ✗   Simplify.

**Step 3:** Because (0, 0) is *not* a solution, shade the half-plane that does *not* contain (0, 0).

## EXAMPLE 4 Real-Life Application

You can spend at most $10 on grapes and apples for a fruit salad. Grapes cost $2.50 per pound and apples cost $1 per pound. Write and graph an inequality for the amounts of grapes and apples you can buy. Identify and interpret two solutions of the inequality.

**Words**  Cost per pound of grapes · times · Pounds of grapes · plus · Cost per pound of apples · times · Pounds of apples · is at most · Amount you can spend

**Variables**  Let $x$ be pounds of grapes and $y$ be pounds of apples.

**Inequality**  2.50 · $x$ + 1 · $y$ ≤ 10

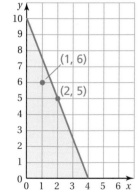

**Step 1:** Graph $2.5x + y = 10$, or $y = -2.5x + 10$. Use a solid line because the inequality symbol is ≤.

**Step 2:** Test (0, 0).

$2.5x + y \le 10$   Write the inequality.

$2.5(0) + 0 \stackrel{?}{\le} 10$   Substitute.

$0 \le 10$ ✓   Simplify.

**Step 3:** Because (0, 0) is a solution, shade the half-plane that contains (0, 0).

∴ Two possible solutions are (1, 6) and (2, 5). So, you can buy 1 pound of grapes and 6 pounds of apples, or 2 pounds of grapes and 5 pounds of apples.

### On Your Own

Now You're Ready
Exercises 35–40

Graph the inequality in a coordinate plane.

**9.** $x + y \le -4$   **10.** $x - 2y < 0$   **11.** $2x + 2y \ge 3$

## 3.5 Exercises

### Vocabulary and Concept Check

1. **VOCABULARY** How can you tell whether an ordered pair is a solution of an inequality?

2. **OPEN-ENDED** Write an example of an inequality in two variables.

3. **WRITING** Compare the graph of a linear inequality in two variables with the graph of a linear equation in two variables.

4. **REASONING** Why do you only need to test one point when graphing a linear inequality?

**Match the inequality with its graph.**

5. $x > -1$  
6. $y > 1$  
7. $x < 1$

A. 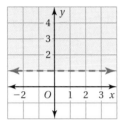   B. 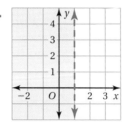   C.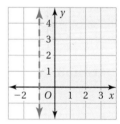

### Practice and Problem Solving

**In words, describe the solutions of the inequality.**

8. $y > x - 1$
9. $y \geq -x + 5$
10. $y < x - 2$

**Tell whether the ordered pair is a solution of the inequality.**

11. $x + y < 7$; $(6, -1)$
12. $2x - y \leq 0$; $(-2, -5)$
13. $x + 3y \geq -2$; $(-4, -2)$
14. $3x + 2y > -6$; $(0, 0)$
15. $-6x + 4y \leq 5$; $(3, -5)$
16. $3x - 5y \geq -8$; $(-1, 1)$
17. $-x - 6y > 12$; $(-8, 2)$
18. $-4x - 8y < -15$; $(-6, 3)$

**Tell whether the ordered pair is a solution of the inequality whose graph is shown.**

19. $(0, 4)$
20. $(0, 0)$
21. $(-1, -2)$
22. $(-1, 3)$
23. $(3, 3)$
24. $(-2, -1)$

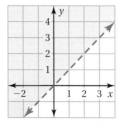

25. **FABRIC** You can spend at most $60 on lace. Cotton lace is $2 per yard and linen lace is $3 per yard. Write an inequality for the amounts of lace you can buy. Can you buy 12 yards of cotton lace and 15 yards of linen lace? Explain.

Section 3.5   Graphing Linear Inequalities in Two Variables   141

**In Exercises 26–28, use the inequality $2x + y < -1$.**

**26.** Write the equation of the boundary line in slope-intercept form.

**27.** Tell whether you would use a solid line or a dashed line to graph the boundary line. Then graph the boundary line.

**28.** Test the point (0, 0) in the inequality. Is the test point a solution? If so, shade the half-plane that contains the point. If not, shade the half-plane that does *not* contain the point.

**Match the inequality with its graph.**

**29.** $3x - 2y \leq 6$     **30.** $3x - 2y < 6$     **31.** $3x - 2y \geq 6$

A. 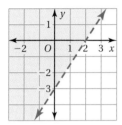     B. 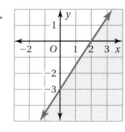     C.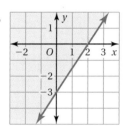

**Graph the inequality in a coordinate plane.**

**32.** $y < 5$     **33.** $x \geq -3$     **34.** $x < 2$

**35.** $y \leq 3x - 1$     **36.** $-2x + y > -4$     **37.** $3x - 2y \geq 0$

**38.** $5x - 2y \leq 6$     **39.** $2x - y < -3$     **40.** $-x + 4y > -2$

**ERROR ANALYSIS** Describe and correct the error in graphing the inequality.

**41.** $y < -x + 1$

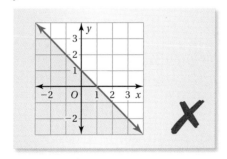

**42.** $y \leq 3x - 2$

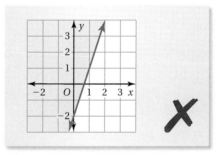

**43. CRITICAL THINKING** When graphing a linear inequality in two variables, why must you choose a test point that is *not* on the boundary line?

**44. MODELING** In order for the drama club to cover the expenses of producing a play, at least $1500 worth of tickets must be sold.

   **a.** Use a model to write an inequality that represents this situation.

   **b.** Graph the inequality.

   **c.** Eighty adults and 110 students attend the play. Does the drama club cover its expenses? Explain.

School Play
Adults: $10
Students: $6

**Tell whether the ordered pair is a solution of the inequality.**

**45.** $y < \frac{1}{3}x + \frac{1}{4}$; (6, 2)

**46.** $2.5 - y \le 1.8x$; (0.5, 1.5)

**47.** $0.2x + 1.6y \ge -1$; (10, -2.2)

**48.** $2x - \frac{2}{3}y > -5$; $\left(\frac{3}{4}, 4\right)$

**Write an inequality that represents the graph.**

**49.**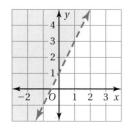

**50.**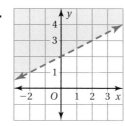

**51.**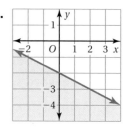

**52. REASONING** How many solutions does the inequality $2x + y \ge 5$ have?

**53. PROBLEM SOLVING** After buying your admission ticket, you have $9 to spend at the movies. Arcade games cost $0.75 per game and drinks cost $2.25.

**a.** Write and graph an inequality that represents the numbers of arcade games you can play and drinks you can buy.

**b.** Identify and interpret two solutions of the inequality.

**54. Critical Thinking** Large boxes weigh 75 pounds and small boxes weigh 40 pounds.

Weight Limit 2000 lb

**a.** Write and graph an inequality that represents the numbers of large and small boxes a 200-pound delivery person can take on the elevator.

**b.** Identify and interpret two solutions of the inequality that are on the boundary line.

**c.** Explain why the solutions in part (b) might not be practical in real life.

 **Fair Game Review** What you learned in previous grades & lessons

**Multiply.** *(Skills Review Handbook)*

**55.** $4 \cdot 4 \cdot 4 \cdot 4$

**56.** $(-2) \cdot (-2) \cdot (-2)$

**57.** $3 \cdot 3 \cdot 3 \cdot 3 \cdot 3$

**58. MULTIPLE CHOICE** Which graph represents the solution of $-5(x - 9) \ge -35$? *(Section 3.4)*

Ⓐ

Ⓑ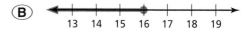

Ⓒ

Ⓓ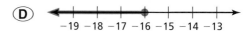

# 3.4–3.5 Quiz

**Solve the inequality. Graph the solution.** *(Section 3.4)*

1. $2m + 1 \geq 7$
2. $\dfrac{n}{6} - 8 \leq 2$
3. $2 - \dfrac{j}{5} > 7$
4. $\dfrac{5}{4} > -3w - \dfrac{7}{4}$

**Write the word sentence as an inequality. Graph the inequality.** *(Section 3.4)*

5. A number $h$ is greater than 1 and less than 6.
6. A number $q$ is less than or equal to $-3$ or at least 2.

**Solve the inequality. Graph the solution, if possible.** *(Section 3.4)*

7. $7 > -2y + 5 > -3$
8. $3z + 2 \leq -10$ or $z - 7 \geq -5$
9. $|2b - 1| \leq 3$
10. $-4|r - 1| + 7 < -9$

**Graph the inequality in a coordinate plane.** *(Section 3.5)*

11. $y \geq -8$
12. $x < 6$
13. $x + y > 5$
14. $4x - 4y \leq 8$

15. **PARTY** You buy lunch for guests at a party. You can spend no more than $100. You will spend $20 on beverages and $10 per guest on sandwiches. Write and solve an inequality to find the number of guests you can invite to the party. *(Section 3.4)*

16. **BOOKS** You have a gift card worth $50. You want to buy several paperback books that cost $6 each. Write and solve an inequality to find the number of books you can buy and still have at least $20 on the gift card. *(Section 3.4)*

17. **SUPPLIES** You have $6 to spend on pens and notebooks. Pens cost $0.75 each and notebooks cost $1.50 each. Write and graph an inequality that represents the numbers of pens and notebooks you can buy. Identify and interpret a solution of the inequality. *(Section 3.5)*

# 3 Chapter Review

## Review Key Vocabulary

inequality, *p. 106*
solution of an inequality, *p. 106*
solution set, *p. 106*
graph of an inequality, *p. 107*
compound inequality, *p. 132*
absolute value inequality, *p. 134*
linear inequality in two variables, *p. 138*
solution of a linear inequality, *p. 138*
graph of a linear inequality, *p. 138*
half-planes, *p. 138*

## Review Examples and Exercises

### 3.1 Writing and Graphing Inequalities (pp. 104–109)

a. Four plus a number $w$ is at least $-\frac{1}{2}$. Write this sentence as an inequality.

$$\underbrace{\text{Four plus a number } w}_{4 + w} \quad \underbrace{\text{is at least}}_{\geq} \quad -\frac{1}{2}.$$

∴ An inequality is $4 + w \geq -\frac{1}{2}$.

b. Graph $m > 4$.

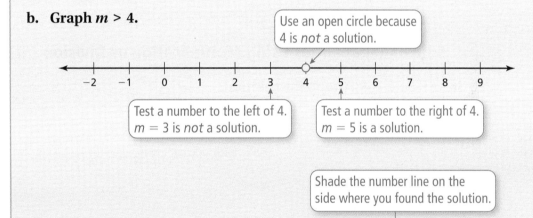

Use an open circle because 4 is *not* a solution.

Test a number to the left of 4. $m = 3$ is *not* a solution.

Test a number to the right of 4. $m = 5$ is a solution.

Shade the number line on the side where you found the solution.

### Exercises

**Write the word sentence as an inequality.**

1. A number $v$ is less than $-2$.
2. A number $x$ minus $\frac{1}{4}$ is no more than $-\frac{3}{4}$.

**Tell whether the given value is a solution of the inequality.**

3. $10 - q < 3$; $q = 6$
4. $12 \div m \geq -4$; $m = -3$

**Graph the inequality on a number line.**

5. $p < 1.2$
6. $n > 10\frac{1}{4}$

## 3.2 Solving Inequalities Using Addition or Subtraction (pp. 110–115)

Solve $-4 < n - 3$. Graph the solution.

$$-4 < n - 3 \quad \text{Write the inequality.}$$

Undo the subtraction. → $+3 \quad +3 \quad$ Add 3 to each side.

$$-1 < n \quad \text{Simplify.}$$

∴ The solution is $n > -1$.

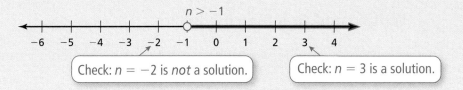

Check: $n = -2$ is *not* a solution.   Check: $n = 3$ is a solution.

### Exercises

Solve the inequality. Graph the solution.

7. $b + 13 < 18$
8. $x - 3 \leq 10$
9. $y + 1 \geq -2$

## 3.3 Solving Inequalities Using Multiplication or Division (pp. 116–123)

Solve $-8a \geq -48$. Graph the solution.

$$-8a \geq -48 \quad \text{Write the inequality.}$$

Undo the multiplication. → $\dfrac{-8a}{-8} \leq \dfrac{-48}{-8} \quad$ Divide each side by $-8$. Reverse the inequality symbol.

$$a \leq 6 \quad \text{Simplify.}$$

∴ The solution is $a \leq 6$.

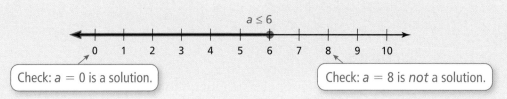

Check: $a = 0$ is a solution.   Check: $a = 8$ is *not* a solution.

### Exercises

Solve the inequality. Graph the solution.

10. $\dfrac{x}{2} \geq 4$
11. $4z < -44$
12. $-2q \geq -18$

## 3.4 Solving Multi-Step Inequalities (pp. 126–135)

Solve $2x - 3 \leq -9$. Graph the solution.

| | | |
|---|---|---|
| | $2x - 3 \leq -9$ | Write the inequality. |
| Step 1: Undo the subtraction. | $\underline{+3 \quad +3}$ | Add 3 to each side. |
| | $2x \leq -6$ | Simplify. |
| Step 2: Undo the multiplication. | $\dfrac{2x}{2} \leq \dfrac{-6}{2}$ | Divide each side by 2. |
| | $x \leq -3$ | Simplify. |

∴ The solution is $x \leq -3$.

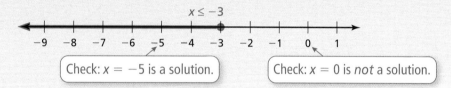

Check: $x = -5$ is a solution.    Check: $x = 0$ is *not* a solution.

### Exercises

Solve the inequality. Graph the solution.

**13.** $4x + 3 < 11$  **14.** $\dfrac{z}{-4} - 3 \leq 1$  **15.** $-3w - 4 > 8$

**16.** $4 > x - 7 > -6$  **17.** $2x + 2 \leq 4 \text{ or } x + 2 \geq 5$  **18.** $|x - 3| > 1$

## 3.5 Graphing Linear Inequalities in Two Variables (pp. 136–143)

Graph $4x + 2y \geq -6$ in a coordinate plane.

**Step 1:** Graph $4x + 2y = -6$, or $y = -2x - 3$. Use a solid line because the inequality symbol is $\geq$.

**Step 2:** Test $(0, 0)$.

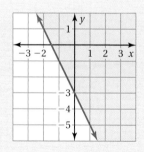

$$4x + 2y \geq -6 \quad \text{Write the inequality.}$$
$$4(0) + 2(0) \stackrel{?}{\geq} -6 \quad \text{Substitute.}$$
$$0 \geq -6 \checkmark \quad \text{Simplify.}$$

**Step 3:** Because $(0, 0)$ is a solution, shade the half-plane that contains $(0, 0)$.

### Exercises

Graph the inequality in a coordinate plane.

**19.** $-9x + 3y > 3$  **20.** $-2x + 2y \leq 4$  **21.** $5x + 10y < 40$

Chapter Review

# 3 Chapter Test

**Write the word sentence as an inequality.**

1. A number $j$ plus 20.5 is greater than or equal to 50.

2. A number $r$ multiplied by $\frac{1}{7}$ is less than $-14$.

**Tell whether the given value is a solution of the inequality.**

3. $v - 2 \leq 7$; $v = 9$

4. $\frac{3}{10} p < 0$; $p = 10$

5. $-3n \geq 6$; $n = -3$

**Solve the inequality. Graph the solution.**

6. $n - 3 > -3$

7. $x - \frac{7}{8} \leq \frac{9}{8}$

8. $-6b \geq -30$

9. $\frac{y}{-4} \geq 13$

10. $3v - 7 \geq -13.3$

11. $-5(t + 11) < -60$

12. $3 \leq x + 5 \leq 9$

13. $3x - 2 \leq 4$ or $x - 4 \geq 6$

14. $|x + 5| < 12$

**Graph the inequality in a coordinate plane.**

15. $x > -6$

16. $y < 2$

17. $x \geq -1$

18. $3x + y \geq 7$

19. $4x + 2y \leq 8$

20. $3x - 9y > 18$

21. **VOTING** U.S. citizens must be at least 18 years of age on Election Day to vote. Write an inequality that represents this situation.

22. **GARAGE** The vertical clearance for a hotel parking garage is 10 feet. Write and solve an inequality that represents the height (in feet) of the vehicle.

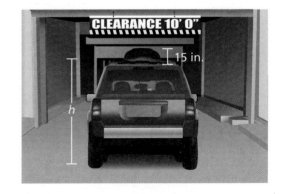

23. **TRADING CARDS** You have $25 to buy trading cards online. Each pack of cards costs $4.50. Shipping costs $2.95. Write and solve an inequality to find the number of packs of trading cards you can buy.

24. **SCIENCE QUIZZES** The table shows your scores on four science quizzes. What score do you need on the fifth quiz to have a mean score of at least 80?

| Quiz | 1 | 2 | 3 | 4 | 5 |
|---|---|---|---|---|---|
| Score (%) | 76 | 87 | 73 | 72 | ? |

# 3 Standards Assessment

1. A line contains the points $(-3, 5)$ and $(6, 8)$. What is the equation of the line? *(8.F.4, F.LE.2)*

   A. $y = \dfrac{1}{3}x$
   
   B. $y = \dfrac{1}{3}x + 6$
   
   C. $y = 3x - 10$
   
   D. $y = 3x + 14$

2. Two lines have the same $y$-intercept. The slope of one line is 1 and the slope of the other line is $-1$. What can you conclude? *(F.IF.4)*

   F. The lines are parallel.
   
   G. The lines meet at exactly one point.
   
   H. The lines meet at more than one point.
   
   I. The situation described is impossible.

**Test-Taking Strategy**
**After Answering Easy Questions, Relax**

"After answering the easy questions, relax and try the harder ones. For this, $x = \$946$. So, it's D."

3. What value of $x$ makes the equation below true? *(A.REI.3)*

   $$4x - 11 = -4$$

4. The perimeter of the triangle shown below is greater than 50 centimeters. Which inequality represents this algebraically? *(A.CED.1)*

   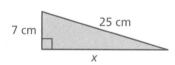

   A. $\dfrac{1}{2}(7x) < 50$
   
   B. $x + 32 < 50$
   
   C. $\dfrac{1}{2}(7x) > 50$
   
   D. $x + 32 > 50$

5. Which value is a solution of $x - 2 \geq -3$? *(A.REI.3)*

   F. $-6$
   
   G. $-5$
   
   H. $-\dfrac{3}{2}$
   
   I. $-1$

6. Water is leaking from a jug at a constant rate. After leaking for 2 hours, the jug contains 48 fluid ounces of water. After leaking for 5 hours, the jug contains 42 fluid ounces of water.   *(8.F.4)*

   *Part A* Find the rate at which water is leaking from the jug.

   *Part B* Find how many fluid ounces of water were in the jug before it started leaking. Show your work and explain your reasoning.

   *Part C* Write an equation that shows how many fluid ounces $y$ of water are left in the jug after it has been leaking for $h$ hours.

   *Part D* Find how many hours it will take the jug to empty entirely. Show your work and explain your reasoning.

7. Which graph represents the inequality below?   *(A.REI.12)*

   $$3x + 6y > 6$$

   **A.**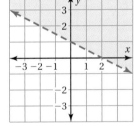

   **C.**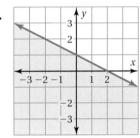

   **B.**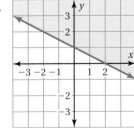

   **D.**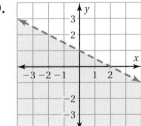

8. Solve $-5x - 2 \geq 8$.   *(A.REI.3)*

   **F.** $x < -2$

   **H.** $x \leq -\dfrac{6}{5}$

   **G.** $x \leq -2$

   **I.** $x \geq -2$

9. Which graph represents the inequality below? (A.REI.3)

$$-2x + 3 < 1$$

A. ⟵—|—|—|—○—|—|—|—|—⟶ $x$
   −4 −3 −2 −1  0  1  2  3  4

C. ⟵—|—|—|—|—|—○—|—|—|—⟶ $x$
   −4 −3 −2 −1  0  1  2  3  4

B. ⟵—|—|—|—○—|—|—|—|—⟶ $x$
   −4 −3 −2 −1  0  1  2  3  4

D. ⟵—|—|—|—|—○—|—|—|—⟶ $x$
   −4 −3 −2 −1  0  1  2  3  4

**The graph below shows how many calories $c$ are burned during $m$ minutes of playing basketball. Use the graph for Exercises 10 and 11.**

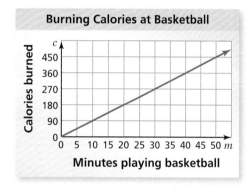

10. How many calories are burned in 25 minutes? (A.CED.3)

11. Which equation represents the graph? (A.CED.2)

   F. $c = 9m$
   G. $c = 90m$
   H. $c = m + 80$
   I. $m = 9c$

12. Which word sentence represents the inequality below? (A.REI.3)

$$3 \leq x \leq 8$$

   A. A number $x$ is fewer than 8 or no less than 3.

   B. A number $x$ is less than or equal to 3 or more than 8.

   C. A number $x$ is more than 3 and less than 8.

   D. A number $x$ is greater than or equal to 3 and no more than 8.

Standards Assessment    151

# 4 Solving Systems of Linear Equations

- **4.1** Solving Systems of Linear Equations by Graphing
- **4.2** Solving Systems of Linear Equations by Substitution
- **4.3** Solving Systems of Linear Equations by Elimination
- **4.4** Solving Special Systems of Linear Equations
- **4.5** Systems of Linear Inequalities

"Can you graph a system of linear equations that shows the number of biscuits and treats that I am going to share with you?"

"Hey look over here. Can you estimate the solution of the system of linear equations that I made with these cattails?"

# What You Learned Before

"Hold your tail a bit lower."

## ● Solving Multi-Step Equations (A.REI.3)

**Example 1** Solve $4x - 2(3x + 1) = 16$.

$$4x - 2(3x + 1) = 16 \quad \text{Write the equation.}$$
$$4x - 6x - 2 = 16 \quad \text{Use Distributive Property.}$$
$$-2x - 2 = 16 \quad \text{Combine like terms.}$$
$$-2x = 18 \quad \text{Add 2 to each side.}$$
$$x = -9 \quad \text{Divide each side by } -2.$$

∴ The solution is $x = -9$.

### Try It Yourself
Solve the equation. Check your solution.

1. $-5x + 8 = -7$
2. $7w + w - 15 = 17$
3. $-3(z - 8) + 10 = -5$
4. $2 = 10c - 4(2c - 9)$

## ● Graphing Linear Inequalities (A.REI.12)

**Example 2** Graph $4x + y \geq -2$ in a coordinate plane.

**Step 1:** Graph $4x + y = -2$. Use a solid line because the inequality symbol is $\geq$.

**Step 2:** Test $(0, 0)$.

$$4x + y \geq -2 \quad \text{Write the inequality.}$$
$$4(0) + 0 \stackrel{?}{\geq} -2 \quad \text{Substitute.}$$
$$0 \geq -2 \checkmark \quad \text{Simplify.}$$

**Step 3:** Because $(0, 0)$ is a solution, shade the half-plane that contains $(0, 0)$.

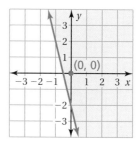

### Try It Yourself
Graph the inequality in a coordinate plane.

5. $x < 5$
6. $y \leq -3$
7. $x + y > -8$
8. $x - 2y \geq 6$

# 4.1 Solving Systems of Linear Equations by Graphing

**Essential Question** How can you solve a system of linear equations?

### 1 ACTIVITY: Writing a System of Linear Equations

Work with a partner.

Your family starts a bed-and-breakfast. They spend $500 fixing up a bedroom to rent. The cost for food and utilities is $10 per night. Your family charges $60 per night to rent the bedroom.

**a.** Write an equation that represents the costs.

$$\text{Cost, } C \text{ (in dollars)} = \$10 \text{ per night} \cdot \text{Number of nights, } x + \$500$$

**b.** Write an equation that represents the revenue (income).

$$\text{Revenue, } R \text{ (in dollars)} = \$60 \text{ per night} \cdot \text{Number of nights, } x$$

**c.** A set of two (or more) linear equations is called a **system of linear equations.** Write the system of linear equations for this problem.

### 2 ACTIVITY: Using a Table to Solve a System

**COMMON CORE**

**Systems of Equations**

In this lesson, you will
- write and solve systems of linear equations by graphing.
- solve real-life problems.

Learning Standards
8.EE.8a
8.EE.8b
8.EE.8c
A.CED.3
A.REI.6

Use the cost and revenue equations from Activity 1 to find how many nights your family needs to rent the bedroom before recovering the cost of fixing up the bedroom. This is the *break-even point*.

**a.** Copy and complete the table.

| x | 0 | 1 | 2 | 3 | 4 | 5 | 6 | 7 | 8 | 9 | 10 | 11 |
|---|---|---|---|---|---|---|---|---|---|---|----|----|
| C |   |   |   |   |   |   |   |   |   |   |    |    |
| R |   |   |   |   |   |   |   |   |   |   |    |    |

**b.** How many nights does your family need to rent the bedroom before breaking even?

### 3 ACTIVITY: Using a Graph to Solve a System

a. Graph the cost equation from Activity 1.

b. In the same coordinate plane, graph the revenue equation from Activity 1.

c. Find the point of intersection of the two graphs. What does this point represent? How does this compare to the break-even point in Activity 2? Explain.

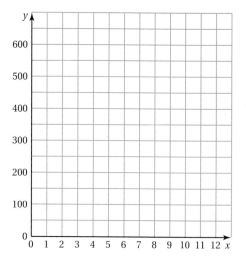

### 4 EXAMPLE: Using a Graphing Calculator

Use a graphing calculator to solve the system.

$y = 10x + 500$    Equation 1
$y = 60x$    Equation 2

**Math Practice 5**

**Use Technology to Explore**
How do you decide the values for the viewing window of your calculator? What other viewing windows could you use?

a. Enter the equations into your calculator. Then graph the equations in an appropriate window.

b. To find the solution, use the *intersect* feature to find the point of intersection. The solution is (10, 600).

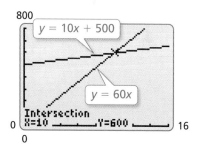

### What Is Your Answer?

5. **IN YOUR OWN WORDS** How can you solve a system of linear equations? How can you check your solution?

6. Solve one of the systems by using a table, another system by sketching a graph, and the remaining system by using a graphing calculator. Explain why you chose each method.

   a. $y = 4.3x + 1.2$
      $y = -1.7x - 2.4$

   b. $y = x$
      $y = -2x + 9$

   c. $y = -x - 5$
      $y = 3x + 1$

**Practice** — Use what you learned about systems of linear equations to complete Exercises 4–6 on page 158.

Section 4.1    Solving Systems of Linear Equations by Graphing

## 4.1 Lesson

Check It Out
Lesson Tutorials
BigIdeasMath com

**Key Vocabulary**
system of linear equations, *p. 156*
solution of a system of linear equations, *p. 156*

A **system of linear equations** is a set of two or more linear equations in the same variables. An example is shown below.

$y = x + 1$   Equation 1
$y = 2x - 7$   Equation 2

A **solution of a system of linear equations** in two variables is an ordered pair that is a solution of each equation in the system. The solution of a system of linear equations is the point of intersection of the graphs of the equations.

###  Key Idea

**Solving a System of Linear Equations by Graphing**

**Step 1** Graph each equation in the same coordinate plane.
**Step 2** Estimate the point of intersection.
**Step 3** Check the point from Step 2 by substituting for $x$ and $y$ in each equation of the original system.

**Reading**
A system of linear equations is also called a *linear system*.

---

**EXAMPLE 1** **Solving a System of Linear Equations by Graphing**

Solve the system by graphing.   $y = 2x + 5$   Equation 1
                                 $y = -4x - 1$  Equation 2

**Step 1:** Graph each equation.

**Step 2:** Estimate the point of intersection. The graphs appear to intersect at $(-1, 3)$.

**Step 3:** Check the point from Step 2.

Equation 1            Equation 2
$y = 2x + 5$          $y = -4x - 1$
$3 \stackrel{?}{=} 2(-1) + 5$    $3 \stackrel{?}{=} -4(-1) - 1$
$3 = 3$ ✓             $3 = 3$ ✓

∴ The solution is $(-1, 3)$.

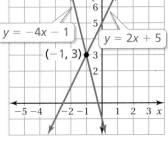

**Check**
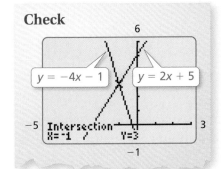

### On Your Own

Now You're Ready
Exercises 10–12

**Solve the system of linear equations by graphing.**

1. $y = x - 1$
   $y = -x + 3$

2. $y = -5x + 14$
   $y = x - 10$

3. $y = x$
   $y = 2x + 1$

# EXAMPLE 2 Real-Life Application

A kicker on a football team scores 1 point for making an extra point and 3 points for making a field goal. The kicker makes a total of 8 extra points and field goals in a game and scores 12 points. Write and solve a system of linear equations to find the number $x$ of extra points and the number $y$ of field goals.

Use a verbal model to write a system of linear equations.

| Number of extra points, $x$ | + | Number of field goals, $y$ | = | Total number of kicks |

| Points per extra point | · | Number of extra points, $x$ | + | Points per field goal | · | Number of field goals, $y$ | = | Total number of points |

The system is: $x + y = 8$    Equation 1
$x + 3y = 12$    Equation 2

**Step 1:** Graph each equation.

**Step 2:** Estimate the point of intersection. The graphs appear to intersect at (6, 2).

**Step 3:** Check your point from Step 2.

Equation 1      Equation 2

$x + y = 8$      $x + 3y = 12$

$6 + 2 \stackrel{?}{=} 8$      $6 + 3(2) \stackrel{?}{=} 12$

$8 = 8$ ✓      $12 = 12$ ✓

∴ The solution is (6, 2). So, the kicker made 6 extra points and 2 field goals.

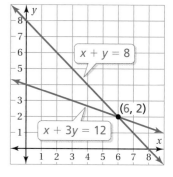

**Check**

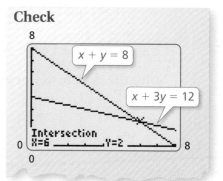

**Study Tip**

It may be easier to graph the equations in a system by rewriting the equations in slope-intercept form.

## On Your Own

**Now You're Ready**
Exercises 13–15

Solve the system of linear equations by graphing.

4. $y = -4x - 7$
    $x + y = 2$

5. $x - y = 5$
    $-3x + y = -1$

6. $\frac{1}{2}x + y = -6$
    $6x + 2y = 8$

7. **WHAT IF?** In Example 2, the kicker makes a total of 7 extra points and field goals and scores 17 points. Write and solve a system of linear equations to find the numbers of extra points and field goals.

Section 4.1    Solving Systems of Linear Equations by Graphing

# 4.1 Exercises

##  Vocabulary and Concept Check

1. **VOCABULARY** Do the equations $4x - 3y = 5$ and $7y + 2x = -8$ form a system of linear equations? Explain.

2. **WRITING** What does it mean to solve a system of equations?

3. **WRITING** You graph a system of linear equations and the solution appears to be (3, 4). How can you verify that the solution is (3, 4)?

##  Practice and Problem Solving

**Use a table to find the break-even point. Check your solution.**

4. $C = 15x + 150$
   $R = 45x$

5. $C = 24x + 80$
   $R = 44x$

6. $C = 36x + 200$
   $R = 76x$

**Match the system of linear equations with the corresponding graph. Use the graph to estimate the solution. Check your solution.**

7. $y = 1.5x - 2$
   $y = -x + 13$

8. $y = x + 4$
   $y = 3x - 1$

9. $y = \frac{2}{3}x - 3$
   $y = -2x + 5$

A.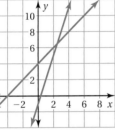
B.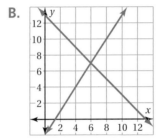
C.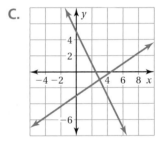

**Solve the system of linear equations by graphing.**

10. $y = 2x + 9$
    $y = 6 - x$

11. $y = -x - 4$
    $y = \frac{3}{5}x + 4$

12. $y = 2x + 5$
    $y = \frac{1}{2}x - 1$

13. $x + y = 27$
    $y = x + 3$

14. $y - x = 17$
    $y = 4x + 2$

15. $x - y = 7$
    $0.5x + y = 5$

16. **CARRIAGE RIDES** The cost $C$ (in dollars) for the care and maintenance of a horse and carriage is $C = 15x + 2000$, where $x$ is the number of rides.

    a. Write an equation for the revenue $R$ in terms of the number of rides.
    
    b. How many rides are needed to break even?

**Use a graphing calculator to solve the system of linear equations.**

17. $2.2x + y = 12.5$
    $1.4x - 4y = 1$

18. $2.1x + 4.2y = 14.7$
    $-5.7x - 1.9y = -11.4$

19. $-1.1x - 5.5y = -4.4$
    $0.8x - 3.2y = -11.2$

20. **ERROR ANALYSIS** Describe and correct the error in solving the system of linear equations.

21. **REASONING** Is it possible for a system of two linear equations to have exactly two solutions? Explain your reasoning.

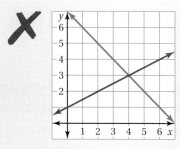

The solution of the linear system $y = 0.5x + 1$ and $y = -x + 7$ is $x = 4$.

22. **MODELING** You have a total of 42 math and science problems for homework. You have 10 more math problems than science problems. How many problems do you have in each subject? Use a system of linear equations to justify your answer.

23. **CANOE RACE** You and your friend are in a canoe race. Your friend is a half mile in front of you and paddling 3 miles per hour. You are paddling 3.4 miles per hour.

    a. You are 8.5 miles from the finish line. How long will it take you to catch up to your friend?

    b. You both maintain your paddling rates for the remainder of the race. How far ahead of your friend will you be when you cross the finish line?

24. **Critical Thinking** Your friend is trying to grow her hair as long as her cousin's hair. The table shows their hair lengths (in inches) in different months.

| Month | Friend's Hair (in.) | Cousin's Hair (in.) |
|---|---|---|
| 3 | 4 | 7 |
| 8 | 6.5 | 9 |

   a. Write a system of linear equations that represents this situation.

   b. Will your friend's hair ever be as long as her cousin's hair? If so, in what month?

**Fair Game Review** What you learned in previous grades & lessons

**Solve the equation. Check your solution.** *(Section 1.2)*

25. $\dfrac{3}{4}c - \dfrac{1}{4}c + 3 = 7$

26. $5(2 - y) + y = -6$

27. $6x - 3(x + 8) = 9$

28. **MULTIPLE CHOICE** The graph of which equation is perpendicular to the graph of $y = 2x + 1$? *(Section 2.2)*

    Ⓐ $y = -2x - 1$   Ⓑ $y = -\dfrac{1}{2}x + 2$   Ⓒ $y = \dfrac{1}{2}x - 1$   Ⓓ $y = 2x + 2$

# 4.2 Solving Systems of Linear Equations by Substitution

**Essential Question** How can you use substitution to solve a system of linear equations?

### 1 ACTIVITY: Using Substitution to Solve a System

**Work with a partner.** Solve each system of linear equations using two methods.

**Method 1:** Solve for $x$ first.

Solve for $x$ in one of the equations. Use the expression for $x$ to find the solution of the system. Explain how you did it.

**Method 2:** Solve for $y$ first.

Solve for $y$ in one of the equations. Use the expression for $y$ to find the solution of the system. Explain how you did it.

Is the solution the same using both methods?

a. $6x - y = 11$
   $2x + 3y = 7$

b. $2x - 3y = -1$
   $x - y = 1$

c. $3x + y = 5$
   $5x - 4y = -3$

d. $5x - y = 2$
   $3x - 6y = 12$

e. $x + y = -1$
   $5x + y = -13$

f. $2x - 6y = -6$
   $7x - 8y = 5$

### 2 ACTIVITY: Writing and Solving a System of Equations

**Common Core**

**Systems of Equations**

In this lesson, you will
- write and solve systems of linear equations by substitution.
- solve real-life problems.

Learning Standards
8.EE.8b
8.EE.8c
A.CED.3
A.REI.6

**Work with a partner.**

a. Roll a pair of number cubes that have different colors. Then write the ordered pair shown by the number cubes. The ordered pair at the right is (3, 4).

b. Write a system of linear equations that has this ordered pair as its solution.

c. Exchange systems with your partner and use one of the methods from Activity 1 to solve the system.

$x$-value

$y$-value

### 3 ACTIVITY: Solving a Secret Code

**Math Practice**

**Check Progress**
As you complete each system of equations, how do you know your answer is correct?

Work with a partner. Decode the quote by Archimedes.

$\overline{-8}\,\overline{-7}\,\overline{7}\;\overline{-5}\;\;\overline{-4}\,\overline{-5}\;\overline{-3}\;\;\overline{-2}\,\overline{-1}\,\overline{-3}\;\overline{0}\;\overline{-5}\;\;\overline{1}\;\overline{2}\;\;\overline{3}\;\overline{1}\;\overline{-3}\;\overline{4}\;\overline{5}$ ,

$\overline{-3}\;\overline{4}\;\overline{5}\;\;\overline{-7}\;\;\overline{6}\;\overline{-7}\,\overline{-1}\,\overline{-1}\;\;\overline{-4}\;\overline{2}\;\overline{7}\;\overline{-5}\;\;\overline{1}\;\overline{8}\;\overline{-5}\;\;\overline{-5}\,\overline{-3}\;\overline{9}\;\overline{1}\;\overline{8}$ .

(A, C)  $x + y = -3$
        $x - y = -3$

(D, E)  $x + y = 0$
        $x - y = 10$

(G, H)  $x + y = 0$
        $x - y = -16$

(I, L)  $x + 2y = -9$
        $2x - y = -13$

(M, N)  $x + 2y = 4$
        $2x - y = -12$

(O, P)  $x + 2y = -2$
        $2x - y = 6$

(R, S)  $2x + y = 21$
        $x - y = 6$

(T, U)  $2x + y = -7$
        $x - y = 10$

(V, W)  $2x + y = 20$
        $x - y = 1$

## What Is Your Answer?

4. **IN YOUR OWN WORDS** How can you use substitution to solve a system of linear equations?

**Practice**  Use what you learned about systems of linear equations to complete Exercises 4–6 on page 164.

## 4.2 Lesson

Another way to solve systems of linear equations is to use substitution.

### Key Idea

**Solving a System of Linear Equations by Substitution**

**Step 1** Solve one of the equations for one of the variables.

**Step 2** Substitute the expression from Step 1 into the other equation and solve for the other variable.

**Step 3** Substitute the value from Step 2 into one of the original equations and solve.

### EXAMPLE 1  Solving a System of Linear Equations by Substitution

Solve the system by substitution.   $y = 2x - 4$   Equation 1
$\phantom{Solve the system by substitution.\ \ \ }$ $7x - 2y = 5$   Equation 2

**Step 1:** Equation 1 is already solved for $y$.

**Step 2:** Substitute $2x - 4$ for $y$ in Equation 2.

| | |
|---|---|
| $7x - 2y = 5$ | Equation 2 |
| $7x - 2(2x - 4) = 5$ | Substitute $2x - 4$ for $y$. |
| $7x - 4x + 8 = 5$ | Use the Distributive Property. |
| $3x + 8 = 5$ | Combine like terms. |
| $3x = -3$ | Subtract 8 from each side. |
| $x = -1$ | Divide each side by 3. |

**Check**

Equation 1
$y = 2x - 4$
$-6 \stackrel{?}{=} 2(-1) - 4$
$-6 = -6$ ✓

Equation 2
$7x - 2y = 5$
$7(-1) - 2(-6) \stackrel{?}{=} 5$
$5 = 5$ ✓

**Step 3:** Substitute $-1$ for $x$ in Equation 1 and solve for $y$.

| | |
|---|---|
| $y = 2x - 4$ | Equation 1 |
| $= 2(-1) - 4$ | Substitute $-1$ for $x$. |
| $= -2 - 4$ | Multiply. |
| $= -6$ | Subtract. |

∴ The solution is $(-1, -6)$.

### On Your Own

Exercises 10–15

Solve the system of linear equations by substitution. Check your solution.

1. $y = 2x + 3$
   $y = 5x$

2. $4x + 2y = 0$
   $y = \frac{1}{2}x - 5$

3. $x = 5y + 3$
   $2x + 4y = -1$

# EXAMPLE 2  Real-Life Application

You buy a total of 50 turkey burgers and veggie burgers for $90. You pay $2 per turkey burger and $1.50 per veggie burger. Write and solve a system of linear equations to find the number $x$ of turkey burgers and the number $y$ of veggie burgers you buy.

Use a verbal model to write a system of linear equations.

$$\boxed{\text{Number of turkey burgers, } x} + \boxed{\text{Number of veggie burgers, } y} = \boxed{\text{Total number of burgers}}$$

$$\boxed{\text{Cost per turkey burger}} \cdot \boxed{\text{Number of turkey burgers, } x} + \boxed{\text{Cost per veggie burger}} \cdot \boxed{\text{Number of veggie burgers, } y} = \boxed{\text{Total cost}}$$

The system is: $x + y = 50$  Equation 1
$2x + 1.5y = 90$  Equation 2

**Step 1:** Solve Equation 1 for $x$.

$x + y = 50$   Equation 1
$x = 50 - y$   Subtract $y$ from each side.

> **Study Tip**
> It is easiest to solve for a variable that has a coefficient of 1 or −1.

**Step 2:** Substitute $50 - y$ for $x$ in Equation 2.

$2x + 1.5y = 90$   Equation 2
$2(50 - y) + 1.5y = 90$   Substitute $50 - y$ for $x$.
$100 - 2y + 1.5y = 90$   Use the Distributive Property.
$-0.5y = -10$   Simplify.
$y = 20$   Divide each side by $-0.5$.

**Step 3:** Substitute 20 for $y$ in Equation 1 and solve for $x$.

$x + y = 50$   Equation 1
$x + 20 = 50$   Substitute 20 for $y$.
$x = 30$   Subtract 20 from each side.

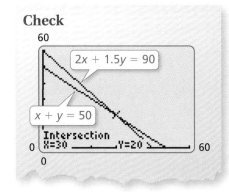

∴ You buy 30 turkey burgers and 20 veggie burgers.

## On Your Own

Now You're Ready
Exercises 18–20

**4.** A juice stand sells lemonade for $2 per cup and orange juice for $3 per cup. The juice stand sells a total of 100 cups of juice for $240. Write and solve a system of linear equations to find the number of cups of lemonade and the number of cups of orange juice sold.

## 4.2 Exercises

### Vocabulary and Concept Check

1. **WRITING** Describe how to solve a system of linear equations by substitution.
2. **NUMBER SENSE** When solving a system of linear equations by substitution, how do you decide which variable to solve for in Step 1?
3. **REASONING** Does solving a system of linear equations by graphing give the same solution as solving by substitution? Explain your reasoning.

### Practice and Problem Solving

**Write a system of linear equations that has the ordered pair as its solution. Use a method from Activity 1 to solve the system.**

4.   5.   6.

**Tell which equation you would use in Step 1 when solving the system by substitution. Explain your reasoning.**

7. $2x + 3y = 5$
   $4x - y = 3$

8. $\frac{2}{3}x + 5y = -1$
   $x + 6y = 0$

9. $2x + 10y = 14$
   $5x - 9y = 1$

**Solve the system of linear equations by substitution. Check your solution.**

10. $y = x - 4$
    $y = 4x - 10$

11. $y = 2x + 5$
    $y = 3x - 1$

12. $x = 2y + 7$
    $3x - 2y = 3$

13. $4x - 2y = 14$
    $y = \frac{1}{2}x - 1$

14. $2x = y - 10$
    $x + 7 = y$

15. $8x - \frac{1}{3}y = 0$
    $12x + 3 = y$

16. **SCHOOL CLUBS** There are a total of 64 students in a drama club and a yearbook club. The drama club has 10 more students than the yearbook club.

    a. Write a system of linear equations that represents this situation.
    b. How many students are in the drama club? the yearbook club?

17. **THEATER** A drama club earns $1040 from a production. A total of 64 adult tickets and 132 student tickets are sold. An adult ticket costs twice as much as a student ticket.

    a. Write a system of linear equations that represents this situation.
    b. What is the cost of each ticket?

164 Chapter 4 Solving Systems of Linear Equations

**Solve the system of linear equations by substitution. Check your solution.**

2️⃣ **18.** $y - x = 0$
$2x - 5y = 9$

**19.** $x + 4y = 14$
$3x + 7y = 22$

**20.** $-2x - 5y = 3$
$3x + 8y = -6$

**21. ERROR ANALYSIS** Describe and correct the error in solving the system of linear equations.

✗ $2x + y = 5$    Equation 1
$3x - 2y = 4$    Equation 2

Step 1:
$2x + y = 5$
$y = -2x + 5$

Step 2:
$2x + (-2x + 5) = 5$
$2x - 2x + 5 = 5$
$5 = 5$

**22. STRUCTURE** The measure of the obtuse angle in the isosceles triangle is two and a half times the measure of one base angle. Write and solve a system of linear equations to find the measures of all the angles.

**23. ANIMAL SHELTER** An animal shelter has a total of 65 abandoned cats and dogs. The ratio of cats to dogs is 6 : 7. How many cats are in the shelter? How many dogs are in the shelter? Justify your answers.

**24. NUMBER SENSE** The sum of the digits of a two-digit number is 8. When the digits are reversed, the number increases by 36. Find the original number.

**25.** *Repeated Reasoning* A DJ has a total of 1075 dance, rock, and country songs on her system. The dance selection is three times the size of the rock selection. The country selection has 105 more songs than the rock selection. How many songs on the system are dance? rock? country?

 **Fair Game Review** *What you learned in previous grades & lessons*

**Write the equation in standard form.** *(Section 2.4)*

**26.** $3x - 9 = 7y$

**27.** $8 - 5y = -2x$

**28.** $6x = y + 3$

**29. MULTIPLE CHOICE** Use the figure to find the measure of ∠2. *(Skills Review Handbook)*

Ⓐ 17°     Ⓑ 73°
Ⓒ 83°     Ⓓ 107°

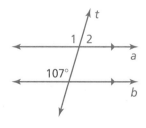

# 4 Study Help

You can use a **notetaking organizer** to write notes, vocabulary, and questions about a topic. Here is an example of a notetaking organizer for solving systems of linear equations by graphing.

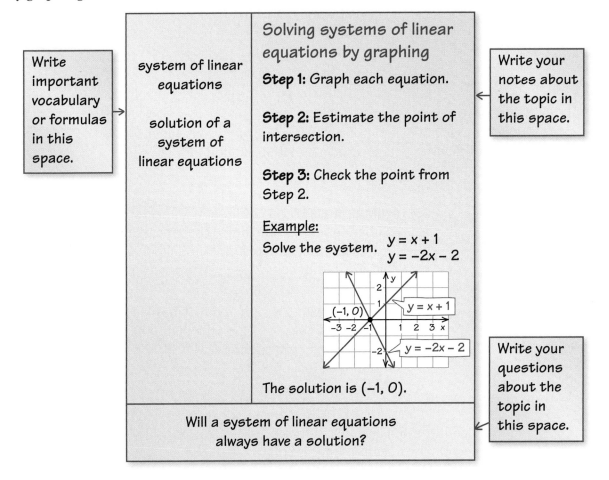

## On Your Own

**Make a notetaking organizer to help you study this topic.**

1. solving systems of linear equations by substitution

**After you complete this chapter, make notetaking organizers for the following topics.**

2. solving systems of linear equations by elimination

3. graphing systems of linear inequalities

"My notetaking organizer has me thinking about retirement when I won't have to fetch sticks anymore."

## 4.1–4.2 Quiz

**Match the system of linear equations with the corresponding graph. Use the graph to estimate the solution. Check your solution.** *(Section 4.1)*

1. $y = x - 2$
   $y = -2x + 1$

2. $y = x - 3$
   $y = -\frac{1}{3}x + 1$

3. $y = \frac{1}{2}x - 2$
   $y = 4x + 5$

A.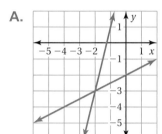
B.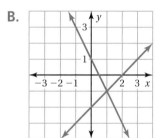
C.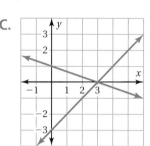

**Solve the system of linear equations by graphing.** *(Section 4.1)*

4. $y = 2x - 3$
   $y = -x + 9$

5. $6x + y = -2$
   $y = -3x + 1$

6. $4x + 2y = 2$
   $3x = 4 - y$

**Solve the system of linear equations by substitution. Check your solution.** *(Section 4.2)*

7. $y = x - 8$
   $y = 2x - 14$

8. $x = 2y + 2$
   $2x - 5y = 1$

9. $x - 5y = 1$
   $-2x + 9y = -1$

10. **MOVIE CLUB** Members of a movie rental club pay a $15 annual membership fee and $2 for new release movies. Nonmembers pay $3 for new release movies. *(Section 4.1)*

    a. Write a system of linear equations that represents this situation.

    b. When is it beneficial to have a membership?

11. **NUMBER SENSE** The sum of two numbers is 38. The greater number is 8 more than the other number. Find each number. Use a system of linear equations to justify your answer. *(Section 4.1)*

12. **VOLLEYBALL** The length of a sand volleyball court is twice its width. The perimeter is 180 feet. Find the length and width of the sand volleyball court. *(Section 4.2)*

13. **MEDICAL STAFF** A hospital employs a total of 77 nurses and doctors. The ratio of nurses to doctors is 9 : 2. How many nurses are employed at the hospital? How many doctors are employed at the hospital? *(Section 4.2)*

# 4.3 Solving Systems of Linear Equations by Elimination

**Essential Question** How can you use elimination to solve a system of linear equations?

### 1 ACTIVITY: Using Elimination to Solve a System

**Work with a partner. Solve each system of linear equations using two methods.**

**Method 1:** Subtract.

Subtract Equation 2 from Equation 1. What is the result? Explain how you can use the result to solve the system of equations.

**Method 2:** Add.

Add the two equations. What is the result? Explain how you can use the result to solve the system of equations.

Is the solution the same using both methods?

a.  $2x + y = 4$
    $2x - y = 0$

b.  $3x - y = 4$
    $3x + y = 2$

c.  $x + 2y = 7$
    $x - 2y = -5$

### 2 ACTIVITY: Using Elimination to Solve a System

**Work with a partner.**

$2x + y = 2$   Equation 1
$x + 5y = 1$   Equation 2

**Systems of Equations**
In this lesson, you will
- write and solve systems of linear equations by elimination.
- solve real-life problems.

Learning Standards
8.EE.8b
8.EE.8c
A.CED.3
A.REI.5
A.REI.6

a. Can you add or subtract the equations to solve the system of linear equations? Explain.

b. Explain what property you can apply to Equation 1 in the system so that the $y$ coefficients are the same.

c. Explain what property you can apply to Equation 2 in the system so that the $x$ coefficients are the same.

d. You solve the system in part (b). Your partner solves the system in part (c). Compare your solutions.

e. Use a graphing calculator to check your solution.

### 3 ACTIVITY: Solving a Secret Code

**Math Practice**

**Find Entry Points**
What is the first thing you do to solve a system of linear equations by elimination? Why?

Work with a partner. Solve the puzzle to find the name of a famous mathematician who lived in Egypt around 350 A.D.

|     | −3 | −2 | −1 | 0 | 1 | 2 | 3 | 4 |
|-----|----|----|----|----|----|----|----|----|
| 4   | B | W | R | M | F | Y | K | N |
| 3   | O | J | A | S | I | D | X | Z |
| 2   | Q | P | C | E | G | B | T | J |
| 1   | M | R | C | Z | N | O | U | W |
| 0   | K | X | U | H | L | Y | S | Q |
| −1  | F | E | A | S | W | K | R | M |
| −2  | G | J | Z | N | H | V | D | G |
| −3  | E | L | X | L | F | Q | O | B |

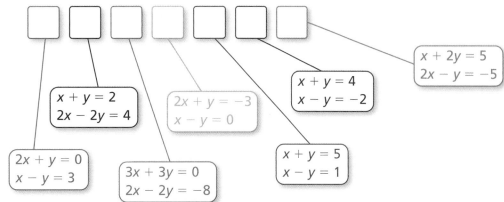

- $2x + y = 0$
- $x - y = 3$

- $x + y = 2$
- $2x - 2y = 4$

- $3x + 3y = 0$
- $2x - 2y = -8$

- $2x + y = -3$
- $x - y = 0$

- $x + y = 5$
- $x - y = 1$

- $x + y = 4$
- $x - y = -2$

- $x + 2y = 5$
- $2x - y = -5$

## What Is Your Answer?

**4. IN YOUR OWN WORDS** How can you use elimination to solve a system of linear equations?

**5.** When can you add or subtract equations in a system to solve the system? When do you have to multiply first? Justify your answers with examples.

**6.** In Activity 2, why can you multiply equations in the system by a constant and not change the solution of the system? Explain your reasoning.

**Practice**

Use what you learned about systems of linear equations to complete Exercises 4–6 on page 173.

## 4.3 Lesson

### Key Idea

**Solving a System of Linear Equations by Elimination**

**Step 1** Multiply, if necessary, one or both equations by a constant so at least one pair of like terms has the same or opposite coefficients.

**Step 2** Add or subtract the equations to eliminate one of the variables.

**Step 3** Solve the resulting equation for the remaining variable.

**Step 4** Substitute the value from Step 3 into one of the original equations and solve.

### EXAMPLE 1  Solving a System of Linear Equations by Elimination

Solve the system by elimination.   $x + 3y = -2$   Equation 1
$x - 3y = 16$   Equation 2

**Study Tip**

Because the coefficients of $x$ are the same, you can also solve the system by subtracting in Step 2.

$x + 3y = -2$
$x - 3y = 16$
$\overline{\phantom{xx}6y = -18}$

So, $y = -3$.

**Step 1:** The coefficients of the $y$-terms are already opposites.

**Step 2:** Add the equations.

$x + 3y = -2$   Equation 1
$x - 3y = 16$   Equation 2
$\overline{2x \phantom{xxxx} = 14}$   Add the equations.

**Step 3:** Solve for $x$.

$2x = 14$   Equation from Step 2
$x = 7$   Divide each side by 2.

**Step 4:** Substitute 7 for $x$ in one of the original equations and solve for $y$.

$x + 3y = -2$   Equation 1
$7 + 3y = -2$   Substitute 7 for $x$.
$3y = -9$   Subtract 7 from each side.
$y = -3$   Divide each side by 3.

∴ The solution is $(7, -3)$.

**Check**

Equation 1
$x + 3y = -2$
$7 + 3(-3) \stackrel{?}{=} -2$
$-2 = -2$ ✓

Equation 2
$x - 3y = 16$
$7 - 3(-3) \stackrel{?}{=} 16$
$16 = 16$ ✓

### On Your Own

Now You're Ready
Exercises 7–12

Solve the system of linear equations by elimination. Check your solution.

1. $2x - y = 9$
   $4x + y = 21$

2. $-5x + 2y = 13$
   $5x + y = -1$

3. $3x + 4y = -6$
   $7x + 4y = -14$

# EXAMPLE 2  Solving a System of Linear Equations by Elimination

**Solve the system by elimination.**  $-6x + 5y = 25$  Equation 1
$-2x - 4y = 14$  Equation 2

**Step 1:** Multiply Equation 2 by 3.

$-6x + 5y = 25$         $-6x + 5y = 25$   Equation 1
$-2x - 4y = 14$ **Multiply by 3.** $-6x - 12y = 42$   Revised Equation 2

### Study Tip
In Example 2, notice that you can also multiply Equation 2 by $-3$ and then add the equations.

**Step 2:** Subtract the equations.

$-6x + 5y = 25$   Equation 1
$-6x - 12y = 42$   Revised Equation 2
$\overline{\phantom{-6x - }17y = -17}$   Subtract the equations.

**Step 3:** Solve for $y$.

$17y = -17$   Equation from Step 2
$y = -1$   Divide each side by 17.

**Step 4:** Substitute $-1$ for $y$ in one of the original equations and solve for $x$.

$-2x - 4y = 14$   Equation 2
$-2x - 4(-1) = 14$   Substitute $-1$ for $y$.
$-2x + 4 = 14$   Multiply.
$-2x = 10$   Subtract 4 from each side.
$x = -5$   Divide each side by $-2$.

 The solution is $(-5, -1)$.

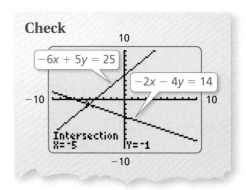

## On Your Own

**Solve the system of linear equations by elimination. Check your solution.**

**4.** $3x + y = 11$
$6x + 3y = 24$

**5.** $4x - 5y = -19$
$-x - 2y = 8$

**6.** $5y = 15 - 5x$
$y = -2x + 3$

## EXAMPLE 3 Real-Life Application

**You buy 8 hostas and 15 daylilies for $193. Your friend buys 3 hostas and 12 daylilies for $117. Write and solve a system of linear equations to find the cost of each daylily.**

Use a verbal model to write a system of linear equations.

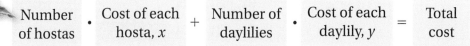

$$\text{Number of hostas} \cdot \text{Cost of each hosta, } x + \text{Number of daylilies} \cdot \text{Cost of each daylily, } y = \text{Total cost}$$

The system is:   $8x + 15y = 193$   Equation 1 (You)
$\phantom{The system is: }$ $3x + 12y = 117$   Equation 2 (Your friend)

**Step 1:** To find the cost $y$ of each daylily, eliminate the $x$-terms. Multiply Equation 1 by 3. Multiply Equation 2 by 8.

$8x + 15y = 193$   **Multiply by 3.**   $24x + 45y = 579$   Revised Equation 1

$3x + 12y = 117$   **Multiply by 8.**   $24x + 96y = 936$   Revised Equation 2

**Step 2:** Subtract the revised equations.

$$\begin{array}{r} 24x + 45y = 579 \\ \underline{24x + 96y = 936} \\ -51y = -357 \end{array}$$

Revised Equation 1
Revised Equation 2
Subtract the equations.

**Step 3:** Solving the equation $-51y = -357$ gives $y = 7$.

∴ Each daylily costs $7.

### On Your Own

Exercises 16–21

7. A landscaper buys 4 peonies and 9 geraniums for $190. Another landscaper buys 5 peonies and 6 geraniums for $185. Write and solve a system of linear equations to find the cost of each peony.

 Summary

**Methods for Solving Systems of Linear Equations**

| Method | When to Use |
| --- | --- |
| Graphing (Lesson 4.1) | To estimate solutions |
| Substitution (Lesson 4.2) | When one of the variables in one of the equations has a coefficient of 1 or $-1$ |
| Elimination (Lesson 4.3) | When at least one pair of like terms has the same or opposite coefficients |
| Elimination (Multiply First) (Lesson 4.3) | When one of the variables cannot be eliminated by adding or subtracting the equations |

# 4.3 Exercises

## Vocabulary and Concept Check

1. **WRITING** Describe how to solve a system of linear equations by elimination.

2. **NUMBER SENSE** When should you use multiplication to solve a system of linear equations by elimination?

3. **WHICH ONE DOESN'T BELONG?** Which system of equations does *not* belong with the other three? Explain your reasoning.

   | $3x + 3y = 3$ | $-2x + y = 6$ | $2x + 3y = 11$ | $x + y = 5$ |
   |---|---|---|---|
   | $2x - 3y = 7$ | $2x - 3y = -10$ | $3x - 2y = 10$ | $3x - y = 3$ |

## Practice and Problem Solving

**Use a method from Activity 1 to solve the system.**

4. $x + y = 3$
   $x - y = 1$

5. $-x + 3y = 0$
   $x + 3y = 12$

6. $3x + 2y = 3$
   $3x - 2y = -9$

**Solve the system of linear equations by elimination. Check your solution.**

7. $x + 3y = 5$
   $-x - y = -3$

8. $x - 2y = -7$
   $3x + 2y = 3$

9. $4x + 3y = -5$
   $-x + 3y = -10$

10. $2x + 7y = 1$
    $2x - 4y = 12$

11. $2x + 5y = 16$
    $3x - 5y = -1$

12. $3x - 2y = 4$
    $6x - 2y = -2$

13. **ERROR ANALYSIS** Describe and correct the error in solving the system of linear equations.

    ✗  $5x + 2y = 9$    Equation 1
       $3x - 2y = -1$    Equation 2
       ―――――――
       $2x\phantom{+2y} = 10$
       $x = 5$
       The solution is $(5, -8)$.

14. **RAFFLE TICKETS** You and your friend are selling raffle tickets for a new laptop. You sell 14 more tickets than your friend sells. Together, you and your friend sell 58 tickets.

    a. Write a system of linear equations that represents this situation.

    b. How many tickets do each of you sell?

15. **JOGGING** You can jog around your block twice and the park once in 10 minutes. You can jog around your block twice and the park 3 times in 22 minutes.

    a. Write a system of linear equations that represents this situation.

    b. How long does it take you to jog around the park?

**Solve the system of linear equations by elimination. Check your solution.**

16. $2x - y = 0$
    $3x - 2y = -3$

17. $x + 4y = 1$
    $3x + 5y = 10$

18. $-2x + 3y = 7$
    $5x + 8y = -2$

19. $3x + 3 = 3y$
    $2x - 6y = 2$

20. $2x - 6 = 4y$
    $7y = -3x + 9$

21. $5x = 4y + 8$
    $3y = 3x - 3$

22. **ERROR ANALYSIS** Describe and correct the error in solving the system of linear equations.

$x + y = 1$     Equation 1    Multiply by −5.    $-5x + 5y = -5$
$5x + 3y = -3$  Equation 2                        $5x + 3y = -3$
                                                  ─────────────
                                                  $8y = -8$
                                                  $y = -1$

The solution is $(2, -1)$.

23. **REASONING** For what values of $a$ and $b$ should you solve the system by elimination?

   a. $4x - y = 3$
      $ax + 10y = 6$

   b. $x - 7y = 6$
      $-6x + by = 9$

24. **AIRPLANES** Two airplanes are flying to the same airport. Their positions are shown in the graph. Write a system of linear equations that represents this situation. Solve the system by elimination to justify your answer.

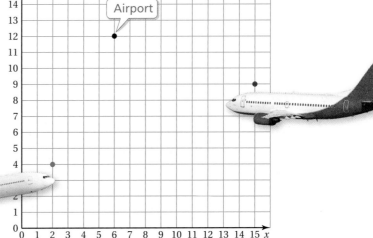

25. **TEST PRACTICE** The table shows the number of correct answers on a practice standardized test. You score 86 points on the test and your friend scores 76 points.

| | You | Your Friend |
|---|---|---|
| Multiple Choice | 23 | 28 |
| Short Response | 10 | 5 |

   a. Write a system of linear equations that represents this situation.

   b. How many points is each type of question worth?

26. **LOGIC** You solve a system of equations in which $x$ represents the number of adult tickets sold and $y$ represents the number of student tickets sold. Can $(-6, 24)$ be the solution of the system? Explain your reasoning.

27. **VACATION** The table shows the activities of two tourists at a vacation resort. You want to go parasailing for one hour and horseback riding for two hours. How much do you expect to pay?

|  | Parasailing | Horseback Riding | Total Cost |
|---|---|---|---|
| Tourist 1 | 2 hours | 5 hours | $205 |
| Tourist 2 | 3 hours | 3 hours | $240 |

28. **REASONING** The solution of a system of linear equations is $(2, -4)$. One equation in the system is $2x + y = 0$. Explain how you could find a second equation for the system. Then find a second equation. Solve the system by elimination to justify your answer.

29. **JEWELER** A metal alloy is a mixture of two or more metals. A jeweler wants to make 8 grams of 18-carat gold, which is 75% gold. The jeweler has an alloy that is 90% gold and an alloy that is 50% gold. How much of each alloy should the jeweler use?

30. **PROBLEM SOLVING** A power boat takes 30 minutes to travel 10 miles downstream. The return trip takes 50 minutes. What is the speed of the current?

31. **Critical Thinking** Solve the system of equations by elimination.

$2x - y + 3z = -1$
$x + 2y - 4z = -1$
$y - 2z = 0$

### Fair Game Review *What you learned in previous grades & lessons*

**Decide whether the two equations are equivalent.** *(Section 1.2 and Section 1.3)*

32. $4n + 1 = n - 8$
    $3n = -9$

33. $2a + 6 = 12$
    $a + 3 = 6$

34. $7v - \dfrac{3}{2} = 5$
    $14v - 3 = 15$

35. **MULTIPLE CHOICE** Which line has the same slope as $y = \dfrac{1}{2}x - 3$? *(Section 2.3)*

Ⓐ $y = -2x + 4$    Ⓑ $y = 2x + 3$    Ⓒ $y - 2x = 5$    Ⓓ $2y - x = 7$

# 4.4 Solving Special Systems of Linear Equations

**Essential Question** Can a system of linear equations have no solution? Can a system of linear equations have many solutions?

### 1 ACTIVITY: Writing a System of Linear Equations

Work with a partner. Your cousin is 3 years older than you. Your ages can be represented by two linear equations.

$y = t$     Your age

$y = t + 3$     Your cousin's age

a. Graph both equations in the same coordinate plane.

b. What is the vertical distance between the two graphs? What does this distance represent?

c. Do the two graphs intersect? If not, what does this mean in terms of your age and your cousin's age?

### 2 ACTIVITY: Using a Table to Solve a System

Work with a partner. You invest $500 for equipment to make dog backpacks. Each backpack costs you $15 for materials. You sell each backpack for $15.

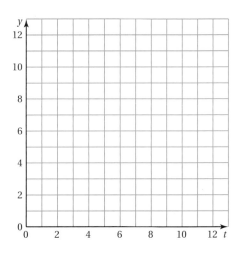

**COMMON CORE**

**Systems of Equations**

In this lesson, you will
- solve systems of linear equations having no solution or infinitely many solutions.

Learning Standards
8.EE.8a
8.EE.8b
8.EE.8c
A.CED.3
A.REI.6

a. Copy and complete the table for your cost $C$ and your revenue $R$.

| x | 0 | 1 | 2 | 3 | 4 | 5 | 6 | 7 | 8 | 9 | 10 |
|---|---|---|---|---|---|---|---|---|---|---|----|
| C |   |   |   |   |   |   |   |   |   |   |    |
| R |   |   |   |   |   |   |   |   |   |   |    |

b. When will your company break even? What is wrong?

176    Chapter 4    Solving Systems of Linear Equations

## 3  ACTIVITY: Using a Graph to Solve a Puzzle

**Math Practice 4**

**Analyze Relationships**
What do you know about the graphs of the two equations? How does this relate to the number of solutions?

Work with a partner. Let $x$ and $y$ be two numbers. Here are two clues about the values of $x$ and $y$.

|  | **Words** | **Equation** |
|---|---|---|
| **Clue 1:** | $y$ is 4 more than twice the value of $x$. | $y = 2x + 4$ |
| **Clue 2:** | The difference of $3y$ and $6x$ is 12. | $3y - 6x = 12$ |

a. Graph both equations in the same coordinate plane.

b. Do the two lines intersect? Explain.

c. What is the solution of the puzzle?

d. Use the equation $y = 2x + 4$ to complete the table.

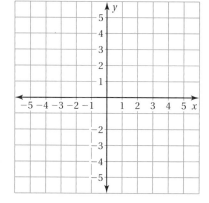

| x | 0 | 1 | 2 | 3 | 4 | 5 | 6 | 7 | 8 | 9 | 10 |
|---|---|---|---|---|---|---|---|---|---|---|---|
| y |   |   |   |   |   |   |   |   |   |   |   |

e. Does each solution in the table satisfy *both* clues?

f. What can you conclude? How many solutions does the puzzle have? How can you describe them?

## What Is Your Answer?

4. **IN YOUR OWN WORDS** Can a system of linear equations have no solution? Can a system of linear equations have many solutions? Give examples to support your answers.

**Practice**  Use what you learned about special systems of linear equations to complete Exercises 3 and 4 on page 180.

Section 4.4  Solving Special Systems of Linear Equations

## 4.4 Lesson

### Key Idea

**Solutions of Systems of Linear Equations**

A system of linear equations can have *one solution*, *no solution*, or *infinitely many solutions*.

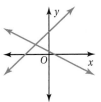 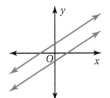 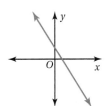

**One solution**　　　**No solution**　　　**Infinitely many solutions**

The lines intersect.　The lines are parallel.　The lines are the same.

### EXAMPLE 1　Solving a System: No Solution

**Solve the system.**　　$y = 3x + 1$　　Equation 1
　　　　　　　　　　　$y = 3x - 5$　　Equation 2

**Method 1:** Solve by graphing.

Graph each equation.

The lines have the same slope and different $y$-intercepts. So, the lines are parallel.

Because parallel lines do not intersect, there is no point that is a solution of both equations.

∴ So, the system of linear equations has no solution.

**Method 2:** Solve by substitution.

Substitute $3x - 5$ for $y$ in Equation 1.

$$y = 3x + 1 \quad \text{Equation 1}$$
$$3x - 5 = 3x + 1 \quad \text{Substitute } 3x - 5 \text{ for } y.$$
$$-5 \neq 1 \quad \text{✗} \quad \text{Subtract } 3x \text{ from each side.}$$

∴ The equation $-5 = 1$ is never true. So, the system of linear equations has no solution.

### On Your Own

**Now You're Ready**
Exercises 8–10

Solve the system of linear equations. Check your solution.

**1.** $y = -x + 3$　　**2.** $y = -5x - 2$　　**3.** $x = 2y + 10$
　　$y = -x + 5$　　　　$5x + y = 0$　　　　　$2x + 3y = -1$

EXAMPLE 2  **Solving a System: Infinitely Many Solutions**

Rectangle A

4y

2x

Rectangle B

12y

6x

The perimeter of Rectangle A is 36 units. The perimeter of Rectangle B is 108 units. Write and solve a system of linear equations to find the values of $x$ and $y$.

*Perimeter of Rectangle A*

$2(2x) + 2(4y) = 36$

$4x + 8y = 36$    Equation 1

*Perimeter of Rectangle B*

$2(6x) + 2(12y) = 108$

$12x + 24y = 108$    Equation 2

The system is:  $4x + 8y = 36$    Equation 1
$12x + 24y = 108$    Equation 2

**Method 1:** Solve by graphing.

Graph each equation.

The lines have the same slope and the same $y$-intercept. So, the lines are the same.

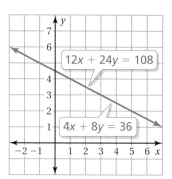

∴ Because the lines are the same, all the points on the line are solutions of both equations. So, the system of linear equations has infinitely many solutions.

**Method 2:** Solve by elimination.

Multiply Equation 1 by 3 and subtract the equations.

$4x + 8y = 36$    **Multiply by 3.**    $12x + 24y = 108$    Revised Equation 1
$12x + 24y = 108$                        $12x + 24y = 108$    Equation 2
                                         $0 = 0$              Subtract.

∴ The equation $0 = 0$ is always true. So, the solutions are all the points on the line $4x + 8y = 36$. The system of linear equations has infinitely many solutions.

**On Your Own**

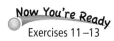

Solve the system of linear equations. Check your solution.

4.  $x + y = 3$
    $x - y = -3$

5.  $2x + y = 5$
    $4x + 2y = 0$

6.  $2x - 4y = 10$
    $-12x + 24y = -60$

7.  **WHAT IF?** What happens to the solution in Example 2 if the perimeter of Rectangle A is 54 units? Explain.

# 4.4 Exercises

## Vocabulary and Concept Check

1. **WRITING** Describe the difference between the graph of a system of linear equations that has *no solution* and the graph of a system of linear equations that has *infinitely many solutions*.

2. **REASONING** When solving a system of linear equations algebraically, how do you know when the system has *no solution*? *infinitely many solutions*?

## Practice and Problem Solving

Let *x* and *y* be two numbers. Find the solution of the puzzle.

3. $y$ is $\frac{1}{3}$ more than 4 times the value of $x$.

   The difference of $3y$ and $12x$ is 1.

4. $\frac{1}{2}$ of $x$ plus 3 is equal to $y$.

   $x$ is 6 more than twice the value of $y$.

Without graphing, determine whether the system of linear equations has *one solution*, *infinitely many solutions*, or *no solution*. Explain your reasoning.

5. $y = 5x - 9$
   $y = 5x + 9$

6. $y = 6x + 2$
   $y = 3x + 1$

7. $y = 8x - 2$
   $y - 8x = -2$

Solve the system of linear equations. Check your solution.

8. $y = 2x - 2$
   $y = 2x + 9$

9. $y = 3x + 1$
   $-x + 2y = -3$

10. $y = \frac{\pi}{3}x + \pi$
    $-\pi x + 3y = -6\pi$

11. $y = -\frac{1}{6}x + 5$
    $x + 6y = 30$

12. $\frac{1}{3}x + y = 1$
    $2x + 6y = 6$

13. $-2x + y = 1.3$
    $2(0.5x - y) = 4.6$

14. **ERROR ANALYSIS** Describe and correct the error in solving the system of linear equations.

    $y = -2x + 4$
    $y = -2x + 6$
    The lines have the same slope so there are infinitely many solutions.

15. **PIG RACE** In a pig race, your pig gets a head start of 3 feet and is running at a rate of 2 feet per second. Your friend's pig is also running at a rate of 2 feet per second. A system of linear equations that represents this situation is $y = 2x + 3$ and $y = 2x$. Will your friend's pig catch up to your pig? Explain.

**16. REASONING** One equation in a system of linear equations has a slope of −3. The other equation has a slope of 4. How many solutions does the system have? Explain.

**17. LOGIC** How can you use the slopes and *y*-intercepts of equations in a system of linear equations to determine whether the system has *one solution, infinitely many solutions,* or *no solution*? Explain your reasoning.

$4x + 8y = 64$
$8x + 16y = 128$

**18. MONEY** You and a friend both work two different jobs. The system of linear equations represents the total earnings for *x* hours worked at the first job and *y* hours worked at the second job. Your friend earns twice as much as you.

  **a.** One week, both of you work 4 hours at the first job. How many hours do you and your friend work at the second job?

  **b.** Both of you work the same number of hours at the second job. Compare the number of hours you each work at the first job.

**19. DOWNLOADS** You download a digital album for $10. Then you and your friend download the same number of individual songs for $0.99 each. Write a system of linear equations that represents this situation. Will you and your friend spend the same amount of money? Explain.

**20. REASONING** Does the system shown *always, sometimes,* or *never* have no solution when $a = b$? $a \geq b$? $a < b$? Explain your reasoning.

$y = ax + 1$
$y = bx + 4$

**21. SKIING** The table shows the number of lift tickets and ski rentals sold to two different groups. Is it possible to determine how much each lift ticket costs? Justify your answer.

**22.** ⟐Precision⟐ Find the values of *a* and *b* so the system shown has the solution (2, 3). Does the system have any other solutions? Explain.

| Group | 1 | 2 |
|---|---|---|
| Number of Lift Tickets | 36 | 24 |
| Number of Ski Rentals | 18 | 12 |
| Total Cost (dollars) | 684 | 456 |

$12x - 2by = 12$
$3ax - by = 6$

**Fair Game Review** What you learned in previous grades & lessons

**Graph the inequality in a coordinate plane.** *(Section 3.5)*

**23.** $3x + y \geq 6$     **24.** $-3x - 4y \geq 4$     **25.** $-4x + 3y < -12$

**26. MULTIPLE CHOICE** What is the solution of $-2(y + 5) \leq 16$? *(Section 3.4)*

  Ⓐ $y \leq -13$     Ⓑ $y \geq -13$     Ⓒ $y \leq -3$     Ⓓ $y \geq -3$

# Extension 4.4 Solving Linear Equations by Graphing

**COMMON CORE**

**Systems of Equations**

In this extension, you will
- solve linear equations by graphing a system of linear equations.

Learning Standards
8.EE.8a
8.EE.8b
8.EE.8c
A.CED.3
A.REI.6

## Key Idea

**Solving Equations Using Graphs**

**Step 1:** To solve the equation $ax + b = cx + d$, write two linear equations.

$$ax + b = cx + d$$

$y = ax + b$ and $y = cx + d$

**Step 2:** Graph the system of linear equations. The $x$-value of the solution of the system of linear equations is the solution of the equation $ax + b = cx + d$.

### EXAMPLE 1 Solving an Equation Using a Graph

Solve $x - 2 = -\dfrac{1}{2}x + 1$ using a graph. Check your solution.

**Step 1:** Write a system of linear equations using each side of the equation.

$$x - 2 = -\dfrac{1}{2}x + 1$$

$y = x - 2$ and $y = -\dfrac{1}{2}x + 1$

**Check**

$x - 2 = -\dfrac{1}{2}x + 1$

$2 - 2 \stackrel{?}{=} -\dfrac{1}{2}(2) + 1$

$0 = 0$ ✓

**Step 2:** Graph the system.

$y = x - 2$

$y = -\dfrac{1}{2}x + 1$

The graphs intersect at $(2, 0)$.

∴ So, the solution of the equation is $x = 2$.

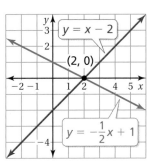

## Practice

Use a graph to solve the equation. Check your solution.

1. $2x + 3 = 4$
2. $2x = x - 3$
3. $3x + 1 = 3x + 2$
4. $\dfrac{1}{3}x = x + 8$
5. $1.5x + 2 = 11 - 3x$
6. $3 - 2x = -2x + 3$

7. **STRUCTURE** Write an equation with variables on both sides that has no solution. How can you change the equation so that it has infinitely many solutions?

## EXAMPLE 2  Real-Life Application

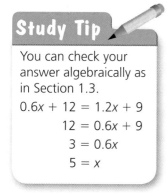

Plant A

Plant B (9 in.)

Plant A (12 in.)

Plant A grows 0.6 inch per month. Plant B grows twice as fast.

a. Use the model to write an equation.

b. After how many months $x$ are the plants the same height?

$$\text{Growth rate} \cdot \text{Months, } x + \text{Original height} = \text{Growth rate} \cdot \text{Months, } x + \text{Original height}$$

a. The equation is $0.6x + 12 = 1.2x + 9$.

b. Write a system of linear equations using each side of the equation. Then use a graphing calculator to graph the system.

$$0.6x + 12 = 1.2x + 9$$

$y = 0.6x + 12$  $y = 1.2x + 9$

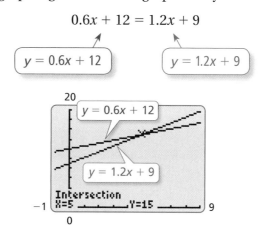

**Study Tip**

You can check your answer algebraically as in Section 1.3.

$0.6x + 12 = 1.2x + 9$
$12 = 0.6x + 9$
$3 = 0.6x$
$5 = x$

The solution of the system is (5, 15).

∴ So, the plants are both 15 inches tall after 5 months.

## Practice

**Use a graph to solve the equation. Check your solution.**

8. $6x - 2 = x + 11$

9. $\frac{4}{3}x - 1 = \frac{2}{3}x + 6$

10. $1.75x = 2.25x + 10.25$

11. **WHAT IF?** In Example 2, the growth rate of Plant A is 0.5 inch per month. After how many months $x$ are the plants the same height?

# 4.5 Systems of Linear Inequalities

**Essential Question** How can you sketch the graph of a system of linear inequalities?

### 1  ACTIVITY: Graphing Linear Inequalities

Work with a partner. Match the linear inequality with its graph.

$$2x + y \leq 4 \qquad \text{Inequality 1}$$
$$2x - y \leq 0 \qquad \text{Inequality 2}$$

a.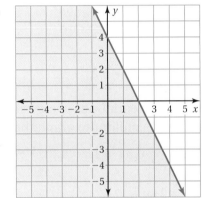

b.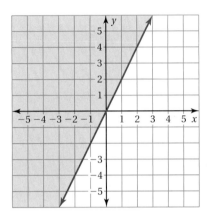

### 2  ACTIVITY: Graphing a System of Linear Inequalities

Work with a partner. Consider the system of linear inequalities given in Activity 1.

$$2x + y \leq 4 \qquad \text{Inequality 1}$$
$$2x - y \leq 0 \qquad \text{Inequality 2}$$

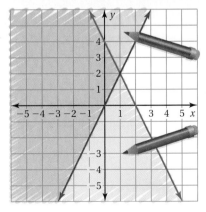

**COMMON CORE**

**Systems of Inequalities**
In this lesson, you will
- write and graph systems of linear inequalities in two variables.
- solve real-life problems.

Learning Standards
A.CED.3
A.REI.12

Use colored pencils to shade the solutions of the two linear inequalities. When you graph both inequalities in the same coordinate plane, what do you get?

Describe each of the shaded regions in the graph at the right. What does the unshaded region represent?

184   Chapter 4   Solving Systems of Linear Equations

### 3 ACTIVITY: Writing a System of Linear Inequalities

Work with a partner. Write a system of 4 linear inequalities whose solution is the traffic sign at the right.

(1) $x + y \leq$ 

(2) $x + y \geq$ 

(3) $x - y \leq$ 

(4) $x - y \geq$ 

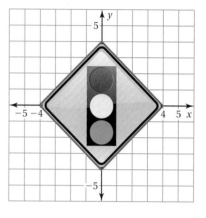

### 4 ACTIVITY: Representing a State by a Linear System

**Math Practice 6**

**Use Clear Definitions**

What is a system of linear inequalities? How can this definition help you identify the states?

Two states can be represented as the graph of a system of linear inequalities. Identify the two states. Explain your reasoning.

## What Is Your Answer?

5. **IN YOUR OWN WORDS** How can you sketch the graph of a system of linear inequalities?

6. When graphing a system of linear inequalities, which region represents the solution of the system? Do you think all systems have a solution? Explain.

**Practice** — Use what you learned about systems of linear inequalities to complete Exercises 7–9 on page 189.

# 4.5 Lesson

**Key Vocabulary**
system of linear inequalities, *p. 186*
solution of a system of linear inequalities, *p. 186*
graph of a system of linear inequalities, *p. 186*

A **system of linear inequalities** is a set of two or more linear inequalities in the same variables. An example is shown below.

$$y < x + 2 \quad \text{Inequality 1}$$
$$y \geq 2x - 1 \quad \text{Inequality 2}$$

A **solution of a system of linear inequalities** in two variables is an ordered pair that is a solution of each inequality in the system.

### EXAMPLE 1  Checking Solutions

**Tell whether each ordered pair is a solution of the system.**

$$y < 2x \quad \text{Inequality 1}$$
$$y \geq x + 1 \quad \text{Inequality 2}$$

**a.** $(3, 5)$

| Inequality 1 | Inequality 2 |
|---|---|
| $y < 2x$ | $y \geq x + 1$ |
| $5 \stackrel{?}{<} 2(3)$ | $5 \stackrel{?}{\geq} 3 + 1$ |
| $5 < 6$ ✓ | $5 \geq 4$ ✓ |

∴ $(3, 5)$ is a solution of both inequalities. So, it is a solution of the system.

**b.** $(-2, 0)$

| Inequality 1 | Inequality 2 |
|---|---|
| $y < 2x$ | $y \geq x + 1$ |
| $0 \stackrel{?}{<} 2(-2)$ | $0 \stackrel{?}{\geq} -2 + 1$ |
| $0 \not< -4$ ✗ | $0 \geq -1$ ✓ |

∴ $(-2, 0)$ is not a solution of both inequalities. So, it is not a solution of the system.

### On Your Own

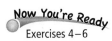
Exercises 4–6

**Tell whether the ordered pair is a solution of the system of linear inequalities.**

1. $y < 5$
   $y > x - 4$; $(-1, 5)$

2. $y \leq -2x + 5$
   $y < x + 3$; $(0, -1)$

The **graph of a system of linear inequalities** is the graph of all of the solutions of the system.

 **Key Idea**

**Graphing a System of Linear Inequalities**

**Step 1** Graph each inequality in the same coordinate plane.

**Step 2** Find the intersection of the half-planes. This intersection is the graph of the system.

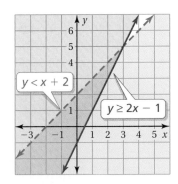

**186** Chapter 4  Solving Systems of Linear Equations

# EXAMPLE 2 Graphing a System of Linear Inequalities

**Graph the system.**

$y \leq 3$   Inequality 1
$y > x + 2$   Inequality 2

**Step 1:** Graph each inequality.

**Step 2:** Find the intersection of the half-planes. One solution is $(-3, 1)$.

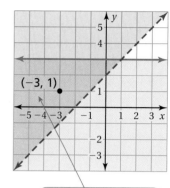

The solution is the purple shaded region.

**Check**

Verify that $(-3, 1)$ is a solution of each inequality.

Inequality 1
$y \leq 3$
$1 \leq 3$ ✓

Inequality 2
$y > x + 2$
$1 \stackrel{?}{>} -3 + 2$
$1 > -1$ ✓

# EXAMPLE 3 Graphing a System of Linear Inequalities: No Solution

**Graph the system.**

$2x + y < -1$   Inequality 1
$2x + y > 3$   Inequality 2

**Step 1:** Graph each inequality.

**Step 2:** Find the intersection of the half-planes.

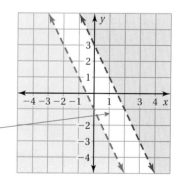

The lines are parallel and the half-planes do not intersect.

So, the system has no solution.

### On Your Own

**Graph the system of linear inequalities.**

3. $y \geq -x + 4$
   $x + y \leq 0$

4. $y > 2x - 3$
   $y \geq \frac{1}{2}x + 1$

5. $-2x + y < 4$
   $2x + y > 4$

## EXAMPLE 4  Writing a System of Linear Inequalities

**Write a system of linear inequalities represented by the graph.**

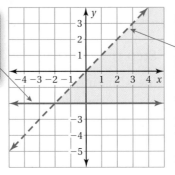

The horizontal boundary line passes through (0, −2). So, an equation of the line is $y = -2$.

The slope of the other boundary line is 1 and the $y$-intercept is 0. So, an equation of the line is $y = x$.

The shaded region is *above* the *solid* boundary line, so the inequality is $y \geq -2$.

The shaded region is *below* the *dashed* boundary line, so the inequality is $y < x$.

∴ The system is $y \geq -2$ and $y < x$.

## EXAMPLE 5  Real-Life Application

**You have at most 8 hours to spend at the mall and at the beach. You want to spend at least 2 hours at the mall and more than 4 hours at the beach. Write and graph a system that represents the situation. How much time could you spend at each location?**

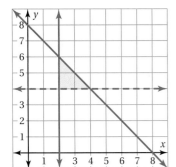

Use the constraints to write a system of linear inequalities. Let $x$ be the number of hours at the mall and let $y$ be the number of hours at the beach.

| | |
|---|---|
| $x + y \leq 8$ | at most 8 hours at the mall and at the beach |
| $x \geq 2$ | at least 2 hours at the mall |
| $y > 4$ | more than 4 hours at the beach |

Graph the system. One ordered pair in the solution region is (2.5, 5).

∴ So, you could spend 2.5 hours at the mall and 5 hours at the beach.

### On Your Own

**Now You're Ready**
Exercises 24–26

Write a system of linear inequalities represented by the graph.

6.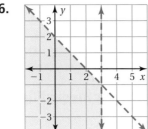

7.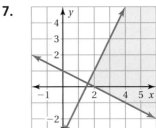

8. **WHAT IF?** In Example 5, you want to spend at least 3 hours at the mall. How does this change the system? Is (2.5, 5) still a solution? Explain.

# 4.5 Exercises

## Vocabulary and Concept Check

1. **VOCABULARY** How can you verify that an ordered pair is a solution of a system of linear inequalities?

2. **WRITING** How are solving systems of linear inequalities and systems of linear equations similar? How are they different?

3. **REASONING** Is the point shown a solution of the system of linear inequalities? Explain.

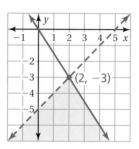

## Practice and Problem Solving

**Tell whether the ordered pair is a solution of the system of linear inequalities.**

4. $y < 4$
   $y > x + 3$; $(-5, 2)$

5. $y > -2$
   $y \leq x - 5$; $(1, -1)$

6. $y \leq x + 7$
   $y \geq 2x + 3$; $(0, 0)$

**Graph the system of linear inequalities.**

7. $y < -3$
   $y \geq 5x$

8. $y > -x + 3$
   $-2x + y \geq 0$

9. $x + y > 1$
   $-x - y < -3$

10. $y < -2$
    $y > 2$

11. $y \geq -5$
    $y < 3x + 1$

12. $x + y > 4$
    $y > \dfrac{3}{2}x - 9$

13. $-x + y < -1$
    $-x - 1 \geq -y$

14. $2x + y \leq 5$
    $y + 2 \geq -2x$

15. $-2x - 5y < 15$
    $-4x > 10y + 60$

16. **MUFFINS** You can spend at most $21 on fruit. Blueberries cost $4 per pound and strawberries cost $3 per pound. You need at least 3 pounds to make muffins.

    a. Define the variables.

    b. Write a system of linear inequalities that represents this situation.

    c. Graph the system of linear inequalities.

    d. Is it possible to buy 4 pounds of blueberries and 1 pound of strawberries in this situation? Justify your answer.

Section 4.5  Systems of Linear Inequalities  189

**ERROR ANALYSIS** Describe and correct the error in graphing the system of linear inequalities.

**17.** $y \geq x + 3$

$y < -x - 2$

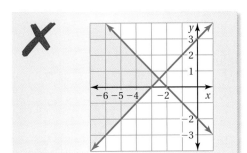

**18.** $y \leq 3x + 4$

$y > \dfrac{1}{2}x + 2$

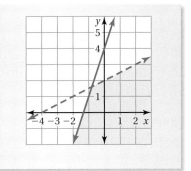

**Match the graph with the corresponding system of linear inequalities.**

**19.**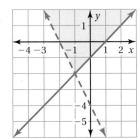

**20.**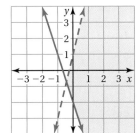

**21.**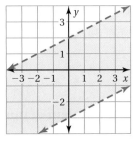

**A.** $y < 4x + 1$

$y \geq -3x - 2$

**B.** $-x + y \geq -1$

$2x + y > -4$

**C.** $-\dfrac{1}{2}x + y < 2$

$-2x + 4y > -12$

**22. REASONING** Describe the intersection of the half-planes of the system shown.

$$x - y \leq 4$$
$$x - y \geq 4$$

**23. JOBS** You earn $12 per hour working as a manager at a grocery store. You also coach a soccer team for $10 per hour. You need to earn at least $110 per week, but you do not want to work more than 20 hours per week.

**a.** Write and graph a system of linear inequalities that represents this situation.

**b.** Identify and interpret one solution of the system.

190  Chapter 4  Solving Systems of Linear Equations

**Write a system of linear inequalities represented by the graph.**

24. 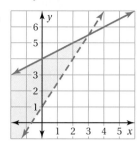  25. 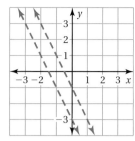  26.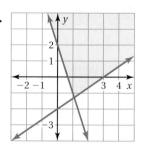

**Graph the system of linear inequalities.**

27. $y > 1$
    $x \geq 2$
    $y > x - 1$

28. $y \leq 5x - 6$
    $y > 0.5x - 4$
    $y < -x + 7$

29. $-4x + 2y < 12$
    $6x + y \leq 9$
    $-9x + 3y \geq -15$

30. **STRUCTURE** Write a system of linear inequalities that is equivalent to $|y| < x$ where $x > 0$. Graph the system.

31. **REPEATED REASONING** One inequality in a system is $-4x + 2y > 6$. Write another inequality so the system has (a) *no solution* and (b) *infinitely many solutions*.

32. **AMUSEMENT PARK** You have at most 8 hours to spend at an amusement park. You want to spend less than 3 hours playing games and at least 4 hours on rides. How much time can you spend on each activity?

33. **ROAD TRIP** On a road trip, you drive about 70 miles per hour and your friend drives about 60 miles per hour. The plan is to drive less than 15 hours and at least 600 miles each day. Your friend will drive more hours than you. Identify and interpret one solution of this situation.

34. **Geometry** The following points are the vertices of a triangle.

$$(2, 5), (6, -3), (-2, -3)$$

a. Write a system of linear inequalities that represents the triangle.
b. Find the area of the triangle.

**Fair Game Review** What you learned in previous grades & lessons

**Evaluate the expression when $a = -2$, $b = 3$, and $c = -1$.** *(Skills Review Handbook)*

35. $4a - bc$

36. $ab + c^2$

37. $-3c - ac$

38. **MULTIPLE CHOICE** What is the solution of $2(x - 4) = -(-x + 3)$? *(Section 1.3)*

  Ⓐ $x = -5$   Ⓑ $x = 2$   Ⓒ $x = 5$   Ⓓ $x = 7$

# 4.3–4.5 Quiz

**Solve the system of linear equations by elimination. Check your solution.** *(Section 4.3)*

1. $x + 2y = 4$
   $-x - y = 2$

2. $2x - y = 1$
   $x + 3y - 4 = 0$

3. $3x = -4y + 10$
   $4x + 3y = 11$

**Solve the system of linear equations. Check your solution.** *(Section 4.4)*

4. $3x - 2y = 16$
   $6x - 4y = 32$

5. $4y = x - 8$
   $-\dfrac{1}{4}x + y = -1$

6. $-2x + y = -2$
   $3x + y = 3$

**Use a graph to solve the equation. Check your solution.** *(Section 4.4)*

7. $4x - 1 = 2x$

8. $-\dfrac{1}{2}x + 1 = -x + 1$

9. $1 - 3x = -3x + 2$

**Graph the system of linear inequalities.** *(Section 4.5)*

10. $y \le \dfrac{1}{2}x + 1$
    $y > -x - 1$

11. $2x + y \ge -3$
    $2x < -y - 4$

12. $-5x + y + 1 > 0$
    $\dfrac{3}{4}x + y \ge -2$

**Write a system of linear inequalities represented by the graph.** *(Section 4.5)*

13.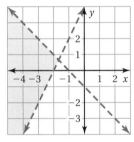

14.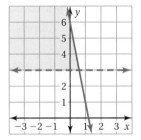

15. **RENTALS** A business rents bicycles and in-line skates. Bicycle rentals cost $25 per day and in-line skate rentals cost $20 per day. The business has 20 rentals today and makes $455. *(Section 4.3)*

    a. Write a system of linear equations that represents this situation.

    b. How many bicycle rentals and in-line skate rentals did the business have today?

16. **JOBS** You earn $11 per hour delivering pizzas. You also work part-time at a convenience store where you earn $9 per hour. You want to earn at least $150 per week, but you can only work 25 hours per week. How many hours can you work at each job? *(Section 4.5)*

192   Chapter 4   Solving Systems of Linear Equations

# 4 Chapter Review

## Review Key Vocabulary

system of linear equations, *p. 156*

solution of a system of linear equations, *p. 156*

system of linear inequalities, *p. 186*

solution of a system of linear inequalities, *p. 186*

graph of a system of linear inequalities, *p. 186*

## Review Examples and Exercises

### 4.1 Solving Systems of Linear Equations by Graphing (pp. 154–159)

**Solve the system by graphing.**  $y = -2x$   Equation 1
$y = 3x + 5$   Equation 2

**Step 1:** Graph each equation.

**Step 2:** Estimate the point of intersection. The graphs appear to intersect at $(-1, 2)$.

**Step 3:** Check the point from Step 2.

$y = -2x$ $\qquad$ $y = 3x + 5$
$2 \stackrel{?}{=} -2(-1)$ $\qquad$ $2 \stackrel{?}{=} 3(-1) + 5$
$2 = 2$ ✓ $\qquad$ $2 = 2$ ✓

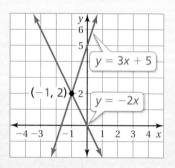

∴ The solution is $(-1, 2)$.

#### Exercises

**Solve the system of linear equations by graphing.**

1. $y = 2x - 3$
   $y = x + 2$

2. $y = -x + 4$
   $x + 3y = 0$

3. $x - y = -2$
   $2x - 3y = -2$

### 4.2 Solving Systems of Linear Equations by Substitution (pp. 160–165)

**Solve the system by substitution.**  $x = 1 + y$   Equation 1
$x + 3y = 13$   Equation 2

**Step 1:** Equation 1 is already solved for $x$.

**Step 2:** Substitute $1 + y$ for $x$ in Equation 2.

$1 + y + 3y = 13$ $\qquad$ Substitute $1 + y$ for $x$.
$y = 3$ $\qquad$ Solve for $y$.

**Step 3:** Substituting 3 for $y$ in Equation 1 gives $x = 4$.

∴ The solution is $(4, 3)$.

### Exercises

**Solve the system of linear equations by substitution. Check your solution.**

**4.** $y = -3x - 7$

$y = x + 9$

**5.** $\frac{1}{2}x + y = -4$

$y = 2x + 16$

**6.** $-x + 5y = 28$

$x + 3y = 20$

## 4.3 Solving Systems of Linear Equations by Elimination (pp. 168–175)

You have a total of 5 quarters and dimes in your pocket. The value of the coins is $0.80. Write and solve a system of linear equations to find the number $x$ of dimes and the number $y$ of quarters in your pocket.

Use a verbal model to write a system of linear equations.

$$\boxed{\text{Number of dimes, } x} + \boxed{\text{Number of quarters, } y} = \boxed{\text{Number of coins}}$$

$$\boxed{\text{Value of a dime}} \cdot \boxed{\text{Number of dimes, } x} + \boxed{\text{Value of a quarter}} \cdot \boxed{\text{Number of quarters, } y} = \boxed{\text{Total value}}$$

The system is $x + y = 5$ and $0.1x + 0.25y = 0.8$.

**Step 1:** Multiply Equation 2 by 10.

$x + y = 5$                               $x + y = 5$           Equation 1

$0.1x + 0.25y = 0.8$   **Multiply by 10.**   $x + 2.5y = 8$       Revised Equation 2

**Step 2:** Subtract the equations.

$\phantom{-}x + \phantom{2.5}y = \phantom{-}5$     Equation 1
$\underline{\phantom{-}x + 2.5y = \phantom{-}8}$     Revised Equation 2
$\phantom{-x + }-1.5y = -3$     Subtract the equations.

**Step 3:** Solving the equation $-1.5y = -3$ gives $y = 2$.

**Step 4:** Substitute 2 for $y$ in one of the original equations and solve for $x$.

$x + y = 5$     Equation 1

$x + 2 = 5$     Substitute 2 for $y$.

$x = 3$     Subtract 2 from each side.

∴ So, you have 3 dimes and 2 quarters in your pocket.

### Exercises

**7. GIFT BASKET** A gift basket that contains jars of jam and packages of bread mix costs $45. There are 8 items in the basket. Jars of jam cost $6 each and packages of bread mix cost $5 each. Write and solve a system of linear equations to find the number of jars of jam and the number of packages of bread mix in the gift basket.

## 4.4 Solving Special Systems of Linear Equations (pp. 176–183)

**Solve the system.**   $y = -5x - 8$   Equation 1
$y = -5x + 4$   Equation 2

Solve by substitution. Substitute $-5x + 4$ for $y$ in Equation 1.

$y = -5x - 8$   Equation 1
$-5x + 4 = -5x - 8$   Substitute $-5x + 4$ for $y$.
$4 \neq -8$ ✗   Add $5x$ to each side.

∴ The equation $4 = -8$ is never true. So, the system of linear equations has no solution.

### Exercises

**Solve the system of linear equations. Check your solution.**

8. $x + 2y = -5$
   $x - 2y = -5$

9. $3x - 2y = 1$
   $9x - 6y = 3$

10. $8x - 2y = 16$
    $-4x + y = 8$

11. Use a graph to solve $2x - 9 = 7x + 11$. Check your solution.

## 4.5 Systems of Linear Inequalities (pp. 184–191)

**Graph the system.**   $y < x - 2$   Inequality 1
$y \geq 2x - 4$   Inequality 2

**Step 1:** Graph each inequality.

**Step 2:** Find the intersection of the half-planes. One solution is $(0, -3)$.

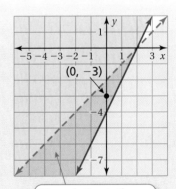

The solution is the purple shaded region.

**Check**
Verify that $(0, -3)$ is a solution of each inequality.

Inequality 1
$y < x - 2$
$-3 \overset{?}{<} 0 - 2$
$-3 < -2$ ✓

Inequality 2
$y \geq 2x - 4$
$-3 \overset{?}{\geq} 2(0) - 4$
$-3 \geq -4$ ✓

### Exercises

**Graph the system of linear inequalities.**

12. $y \leq x - 3$
    $y \geq x + 1$

13. $y > -2x + 3$
    $y \geq \frac{1}{4}x - 1$

14. $x + 2y > 4$
    $2x + y < 4$

# 4 Chapter Test

**Solve the system of linear equations by graphing.**

1. $y = 4 - x$
   $y = x - 4$

2. $y = \frac{1}{2}x + 10$
   $y = 4x - 4$

3. $y + x = 0$
   $3y + 6x = -9$

**Solve the system of linear equations by substitution. Check your solution.**

4. $-3x + y = 2$
   $-x + y - 4 = 0$

5. $x + y = 20$
   $y = 2x - 1$

6. $x - y = 3$
   $x + 2y = -6$

**Solve the system of linear equations by elimination. Check your solution.**

7. $2x + y = 3$
   $x - y = 3$

8. $x + y = 12$
   $3x = 2y + 6$

9. $-2x + y + 3 = 0$
   $3x + 4y = -1$

**Without graphing, determine whether the system of linear equations has *one solution*, *infinitely many solutions*, or *no solution*. Explain your reasoning.**

10. $y = 4x + 8$
    $y = 5x + 1$

11. $2y = 16x - 2$
    $y = 8x - 1$

12. $y = -3x + 2$
    $6x + 2y = 10$

**Use a graph to solve the equation. Check your solution.**

13. $\frac{1}{4}x - 4 = \frac{3}{4}x + 2$

14. $8x - 14 = -2x - 4$

**Graph the system of linear inequalities.**

15. $y > \frac{1}{2}x + 4$
    $2y \leq x + 4$

16. $y \geq -\frac{2}{3}x + 1$
    $-3x + y > -2$

17. $x + y < 1$
    $5x + y > 4$

18. **BOUQUET** A bouquet of lilies and tulips has 12 flowers. Lilies cost $3 each and tulips cost $2 each. The bouquet costs $32. Write and solve a system of linear equations to find the number of lilies and tulips in the bouquet.

19. **DINNER** How much does it cost for two specials and two glasses of milk?

20. **SHOPPING** You have $110 to spend at the mall. You want to buy at most 6 articles of clothing. A clothing store sells shirts for $12 and pairs of pants for $18. You want to have at least $20 left over for food.

    a. Write and graph a system of linear inequalities that represents this situation.

    b. How many shirts and pairs of pants can you buy at the store?

# 4 Standards Assessment

1. What is the solution of the system of equations shown below? *(A.REI.6)*

   $$y = -\frac{2}{3}x - 1$$
   $$4x + 6y = -6$$

   **A.** $\left(-\frac{3}{2}, 0\right)$   **C.** No solution

   **B.** $(0, -1)$   **D.** Infinitely many solutions

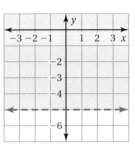

**Test-Taking Strategy**
**Read Question Before Answering**

"Take your time and read the question carefully before choosing your answer."

2. Which inequality is shown in the coordinate plane? *(A.REI.12)*

   **F.** $y > -5$

   **G.** $y < -5$

   **H.** $y \geq -5$

   **I.** $y \leq -5$

3. What is the slope of a line that is perpendicular to the line $y = -0.25x + 3$? *(F.IF.6)*

4. Which graph shows the solution of $-4x + y > -3$? *(A.REI.12)*

   **A.**
   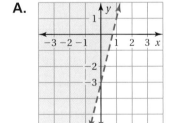

   **B.**
   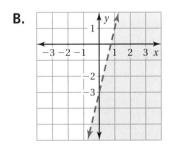

   **C.**
   (graph)

   **D.**
   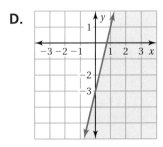

5. Which point is a solution of the system of equations shown below? (A.REI.6)

$$x + 3y = 10$$
$$x = 2y - 5$$

F. (1, 3)

G. (3, 1)

H. (55, −15)

I. (−35, −15)

6. A system of two linear equations has no solution. What can you conclude about the graphs of the two equations? (8.EE.8b)

A. The lines have the same slope and the same $y$-intercept.

B. The lines have the same slope and different $y$-intercepts.

C. The lines have different slopes and the same $y$-intercept.

D. The lines have different slopes and different $y$-intercepts.

7. A scenic train ride has one price for adults and one price for children. One family of two adults and two children pays $62 for the train ride. Another family of one adult and four children pays $70. Which system of linear equations can be used to find the price $x$ for an adult and the price $y$ for a child? (8.EE.8c)

F. $2x + 2y = 70$
   $x + 4y = 62$

G. $x + y = 62$
   $x + y = 70$

H. $2x + 2y = 62$
   $4x + y = 70$

I. $2x + 2y = 62$
   $x + 4y = 70$

8. Which graph shows the solution of $-\dfrac{x}{4} - 10 > -18$? (A.REI.3)

A.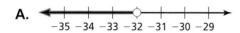
    −35 −34 −33 −32 −31 −30 −29

C. 
    −35 −34 −33 −32 −31 −30 −29

B. 
    29  30  31  32  33  34  35

D. 
    29  30  31  32  33  34  35

9. What value of $w$ makes the equation below true? (A.REI.3)

$$7w - 3w = 2(3w + 11)$$

10. The graph of which equation is parallel to the line that passes through the points $(-1, 5)$ and $(4, 7)$? *(F.IF.6)*

   F. $y = \frac{2}{3}x + 6$     H. $y = \frac{2}{5}x + 1$

   G. $y = -\frac{5}{2}x + 4$     I. $y = \frac{5}{2}x - 1$

11. Which of the following is true for the system of inequalities shown below? *(A.REI.12)*

   $$y < -2$$
   $$y > 3x + 5$$

   A. The graph of the system is located in Quadrants I, II, and III.

   B. The graph of the system is located in Quadrant II only.

   C. The graph of the system is located in Quadrant III only.

   D. The graph of the system is located in Quadrants I, II, III, and IV.

12. You buy 3 T-shirts and 2 pairs of shorts for $42.50. Your friend buys 5 T-shirts and 3 pairs of shorts for $67.50. Use a system of linear equations to find the cost of each T-shirt. Show your work and explain your reasoning. *(8.EE.8c)*

   *Think Solve Explain*

13. The two figures have the same area. What is the value of $y$? *(A.CED.1)*

   F. $\frac{1}{4}$     H. $3$

   G. $\frac{15}{8}$     I. $8$

    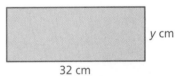

14. The sum of one-third of a number and 10 is equal to 13. What is the number? *(A.CED.1)*

   A. $\frac{8}{3}$     B. $9$     C. $29$     D. $69$

15. Solve the equation $4x + 7y = 16$ for $x$. *(A.CED.4)*

   F. $x = 4 + \frac{7}{4}y$     H. $x = 4 + \frac{4}{7}y$

   G. $x = 4 - \frac{7}{4}y$     I. $x = 16 - 7y$

# 5 Linear Functions

**5.1** Domain and Range of a Function
**5.2** Discrete and Continuous Domains
**5.3** Linear Function Patterns
**5.4** Function Notation
**5.5** Comparing Linear and Nonlinear Functions
**5.6** Arithmetic Sequences

"Here's how I remember that the range is the y-values."

"I draw a cabin on the y-axis. Then, I hum 'Home, Home on the range'."

"It is my treat-converter function machine. However many cat treats I input, the machine outputs TWICE that many dog biscuits. Isn't that cool?"

# What You Learned Before

● **Recognizing Patterns** (6.EE.9)

Describe the pattern of inputs and outputs.

**Example 1**

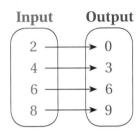

∴ As the input increases by 2, the output increases by 3.

**Example 2**

| Input, x | 6 | 1 | −4 | −9 | −14 |
|---|---|---|---|---|---|
| Output, y | 7 | 8 | 9 | 10 | 11 |

∴ As the input x decreases by 5, the output y increases by 1.

**Example 3** Draw a mapping diagram for the graph. Then describe the pattern of inputs and outputs.

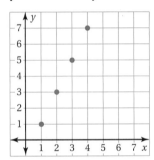
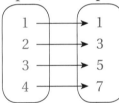

∴ As the input increases by 1, the output increases by 2.

**Try It Yourself**

Describe the pattern of inputs x and outputs y.

**1.**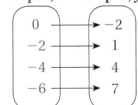

**2.**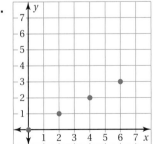

**3.**

| Input, x | 0 | −1 | −2 | −3 | −4 |
|---|---|---|---|---|---|
| Output, y | 7 | 3.5 | 0 | −3.5 | −7 |

"Do you think the stripes in this shirt make me look too linear?"

## 5.1 Domain and Range of a Function

**Essential Question** How can you find the domain and range of a function?

### 1 ACTIVITY: The Domain and Range of a Function

Work with a partner. In Activity 1 in Section 2.4, you completed the table shown below. The table shows the number of adult and child tickets sold for a school concert.

input → 
| Number of Adult Tickets, $x$ | 0 | 1 | 2 | 3 | 4 |
|---|---|---|---|---|---|
| Number of Child Tickets, $y$ | 8 | 6 | 4 | 2 | 0 |
 ← output

The variables $x$ and $y$ are related by the linear equation $4x + 2y = 16$.

**a.** Write the equation in *function form* by solving for $y$.

**b.** The **domain** of a function is the set of all input values. Find the domain of the function.

Domain = 

Why is $x = 5$ not in the domain of the function?

Why is $x = \frac{1}{2}$ not in the domain of the function?

**c.** The **range** of a function is the set of all output values. Find the range of the function.

Range = 

**d.** Functions can be described in many ways.
- by an equation
- by an input-output table
- in words
- by a graph
- as a set of ordered pairs

Use the graph to write the function as a set of ordered pairs.

( , ), ( , ), ( , ),
( , ), ( , )

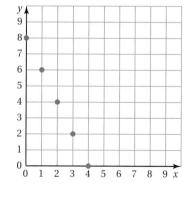

**COMMON CORE**

**Functions**
In this lesson, you will
- find the domain and range of functions from graphs or tables.

Learning Standards
8.F.1
F.IF.1
F.IF.5

202  Chapter 5  Linear Functions

## 2 ACTIVITY: Finding Domains and Ranges

**Math Practice**

**Use Definitions**
What does the domain of a function represent? What does the range represent?

Work with a partner.
- Copy and complete each input-output table.
- Find the domain and range of the function represented by the table.

a. $y = -3x + 4$

| x | −2 | −1 | 0 | 1 | 2 |
|---|---|---|---|---|---|
| y |  |  |  |  |  |

b. $y = \frac{1}{2}x - 6$

| x | 0 | 1 | 2 | 3 | 4 |
|---|---|---|---|---|---|
| y |  |  |  |  |  |

c.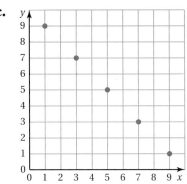

| x |  |  |  |  |  |
|---|---|---|---|---|---|
| y |  |  |  |  |  |

d.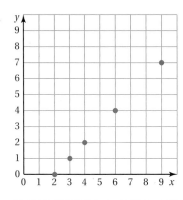

| x |  |  |  |  |  |
|---|---|---|---|---|---|
| y |  |  |  |  |  |

### What Is Your Answer?

3. **IN YOUR OWN WORDS** How can you find the domain and range of a function?

4. The following are general rules for finding a person's foot length.

To find the length y (in inches) of a woman's foot, divide her shoe size x by 3 and add 7.

To find the length y (in inches) of a man's foot, divide his shoe size x by 3 and add 7.3.

© 2013 Zappos.com, Inc.

a. Write an equation for one of the statements.
b. Make an input-output table for the function in part (a). Use shoe sizes $5\frac{1}{2}$ to 12.
c. Label the domain and range of the function on the table.

**Practice**

Use what you learned about the domain and range of a function to complete Exercise 3 on page 206.

Section 5.1 Domain and Range of a Function

# 5.1 Lesson

## Key Vocabulary
function, *p. 204*
domain, *p. 204*
range, *p. 204*
independent variable, *p. 204*
dependent variable, *p. 204*

## Key Idea

**Functions**

A **function** is a relationship that pairs each *input* with exactly one *output*. The **domain** is the set of all possible input values. The **range** is the set of all possible output values.

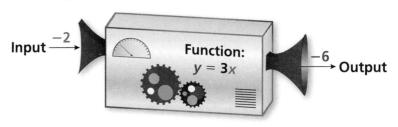

### EXAMPLE 1  Finding Domain and Range from a Graph

**Find the domain and range of the function represented by the graph.**

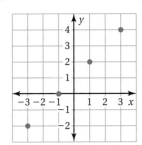

Write the ordered pairs. Identify the inputs and outputs.

$$(-3, -2), (-1, 0), (1, 2), (3, 4)$$

with inputs above and outputs below.

∴ The domain is $-3, -1, 1,$ and $3$. The range is $-2, 0, 2,$ and $4$.

## On Your Own

**Now You're Ready** Exercises 4–6

**Find the domain and range of the function represented by the graph.**

1.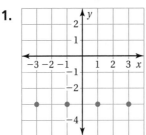

2.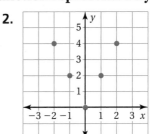

When an equation represents a function, the variable that represents input values is the **independent variable** because it can be *any* value in the domain. The variable that represents output values is the **dependent variable** because it *depends* on the value of the independent variable.

204  Chapter 5  Linear Functions

### EXAMPLE 2 — Finding the Range of a Function

The function $y = -3x + 12$ gives the amount $y$ (in fluid ounces) of juice remaining in a bottle after you take $x$ gulps. (a) Identify the independent and dependent variables. (b) The domain is 0, 1, 2, 3, and 4. What is the range?

a. Because the amount $y$ remaining depends on the number $x$ of gulps, $y$ is the dependent variable and $x$ is the independent variable.

b. Make an input-output table to find the range.

| Input, x | $-3x + 12$ | Output, y |
|---|---|---|
| 0 | $-3(0) + 12$ | 12 |
| 1 | $-3(1) + 12$ | 9 |
| 2 | $-3(2) + 12$ | 6 |
| 3 | $-3(3) + 12$ | 3 |
| 4 | $-3(4) + 12$ | 0 |

∴ The range is 12, 9, 6, 3, and 0.

### EXAMPLE 3 — Real-Life Application

The table shows the percent $y$ (in decimal form) of the moon that was visible at midnight $x$ days after May 19, 2014. (a) Interpret the domain and range. (b) What percent of the moon was visible on May 21, 2014?

| x | y |
|---|---|
| 0 | 0.76 |
| 1 | 0.65 |
| 2 | 0.54 |
| 3 | 0.43 |
| 4 | 0.32 |

a. Zero days after May 19 is May 19. One day after May 19 is May 20. So, the domain of 0, 1, 2, 3, and 4 represents May 19, 20, 21, 22, and 23.

The range is 0.76, 0.65, 0.54, 0.43, and 0.32. These amounts are decreasing, so the moon was less visible each day.

b. May 21, 2014 corresponds to the input $x = 2$. When $x = 2$, $y = 0.54$. So, 0.54, or 54% of the moon was visible on May 21, 2014.

### On Your Own

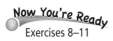
Exercises 8–11

3. The function $y = -4x + 14$ gives the number $y$ of avocados you have left after making $x$ batches of guacamole.

   a. Identify the independent and dependent variables.
   b. The domain is 0, 1, 2, and 3. What is the range?

4. The table shows the percent $y$ (in decimal form) of the moon that was visible at midnight $x$ days after March 24, 2015.

   | x | 0 | 1 | 2 | 3 | 4 |
   |---|---|---|---|---|---|
   | y | 0.19 | 0.29 | 0.39 | 0.49 | 0.59 |

   a. Interpret the domain and range.
   b. What percent of the moon was visible on March 28, 2015?

# 5.1 Exercises

## Vocabulary and Concept Check

1. **VOCABULARY** How are independent variables and dependent variables different?
2. **DIFFERENT WORDS, SAME QUESTION** Which is different? Find "both" answers.

| | |
|---|---|
| Find the range of the function represented by the table. | Find the inputs of the function represented by the table. |
| Find the $x$-values of the function represented by $(2, 7)$, $(4, 5)$, and $(6, -1)$. | Find the domain of the function represented by $(2, 7)$, $(4, 5)$, and $(6, -1)$. |

| x | 2 | 4 | 6 |
|---|---|---|---|
| y | 7 | 5 | -1 |

## Practice and Problem Solving

3. The number of earrings and headbands you can buy with $24 is represented by the equation $8x + 4y = 24$. The table shows the numbers of earrings and headbands.

   a. Write the equation in function form.
   b. Find the domain and range.
   c. Why is $x = 6$ not in the domain of the function?

| Earrings, x | 0 | 1 | 2 | 3 |
|---|---|---|---|---|
| Headbands, y | 6 | 4 | 2 | 0 |

**Find the domain and range of the function represented by the graph.**

4.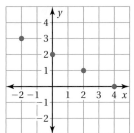

5.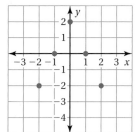

6.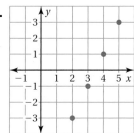

7. **ERROR ANALYSIS** Describe and correct the error in finding the domain and range of the function represented by the graph.

8. **PARKING METER** The number of quarters you put into a parking meter affects the amount of time on the meter. Identify the independent and dependent variables.

 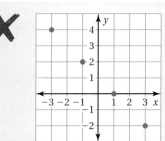

The domain is $-2, 0, 2,$ and $4$.

The range is $-3, -1, 1, 3$.

206  Chapter 5   Linear Functions

Copy and complete the input-output table for the function. Then find the domain and range of the function represented by the table.

**9.** $y = 6x + 2$

| x | −1 | 0 | 1 | 2 |
|---|---|---|---|---|
| y | | | | |

**10.** $y = -\dfrac{1}{4}x - 2$

| x | 0 | 4 | 8 | 12 |
|---|---|---|---|---|
| y | | | | |

**11.** $y = 1.5x + 3$

| x | −1 | 0 | 1 | 2 |
|---|---|---|---|---|
| y | | | | |

**12. VAULTING** In the sport of vaulting, a vaulter performs a routine while on a moving horse. For each round $x$ of competition, the vaulter receives a score $y$ from 1 to 10.

| x | y |
|---|---|
| 1 | 6.856 |
| 2 | 7.923 |
| 3 | 8.135 |

a. Find the domain and range of the function represented by the table.

b. Interpret the domain and range.

c. What is the mean score of the vaulter?

**13. MANATEE** A manatee eats the equivalent of about 12% of its body weight each day.

a. Write an equation that represents the amount $y$ (in pounds) of food a manatee eats each day for its weight $x$. Identify the independent variable and the dependent variable.

b. Make an input-output table for the equation in part (a). Use the inputs 150, 300, 450, 600, 750, and 900.

c. Find the domain and range of the function represented by the table.

d. The weights of three manatees are 300 pounds, 750 pounds, and 1050 pounds. What is the total amount of food that these three manatees eat in a day? in a week?

**14.** **Precision** Describe the domain and range of the function.

a. $y = |x|$  b. $y = -|x|$  c. $y = |x| - 6$  d. $y = -|x| + 4$

**Fair Game Review** What you learned in previous grades & lessons

**Graph the linear equation.** *(Section 2.1)*

**15.** $y = 2x + 8$  **16.** $5x + 6y = 12$  **17.** $-x - 3y = 2$  **18.** $y = 7x - 5$

**19. MULTIPLE CHOICE** The minimum number of people needed for a group rate at an amusement park is 8. Which inequality represents the number of people needed to get the group rate? *(Section 3.1)*

Ⓐ $x \leq 8$   Ⓑ $x > 8$   Ⓒ $x < 8$   Ⓓ $x \geq 8$

# Extension 5.1 Relations and Functions

**Key Vocabulary**
relation, p. 208
Vertical Line Test, p. 209

A **relation** pairs inputs with outputs. A relation that pairs each input with *exactly one* output is a function.

### EXAMPLE 1 Determining Whether Relations are Functions

**Functions**
In this extension, you will
• determine whether relations are functions.
• use the vertical line test to determine whether a graph represents a function.

Learning Standards:
8.F.1
F.IF.1
F.IF.5

Determine whether each relation is a function.

a. $(-2, 2), (-1, 2), (0, 2), (1, 0), (2, 0)$

Every input has exactly one output.

∴ So, the relation is a function.

b. 
| Input | -2 | -1 | 0 | 0 | 1 | 2 |
|---|---|---|---|---|---|---|
| Output | 3 | 4 | 5 | 6 | 7 | 8 |

The input 0 has two outputs, 5 and 6.

∴ So, the relation is *not* a function.

c.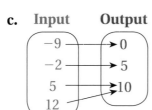

Every input has exactly one output.

∴ So, the relation is a function.

## Practice

**Determine whether the relation is a function.**

1. $(-5, 0), (0, 0), (5, 0), (5, 10), (10, 10)$

2. 
| Input | Output |
|---|---|
| 2 | 2.6 |
| 4 | 5.2 |
| 6 | 7.8 |

3.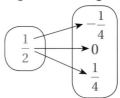

**Determine whether the statement is *true* or *false*. Explain your reasoning.**

4. Every function is a relation.

5. Every relation is a function.

6. When you switch the inputs and outputs of any function, the resulting relation is a function.

7. **REASONING** You record the number $x$ of runs scored by the winning team and the number $y$ of runs scored by the losing team for each softball game in a team's season. Does the relation necessarily represent a function? Explain.

You can use a vertical line test to determine whether a graph represents a function.

> **Vertical Line Test**
>
> **Words** A graph represents a function when no vertical line passes through more than one point on the graph.
>
> **Examples**   Function                    Not a function
>
>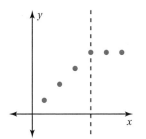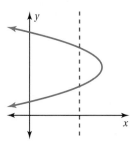

**EXAMPLE 2   Using the Vertical Line Test**

Determine whether each graph represents a function.

a. 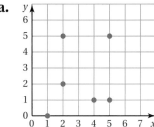      b.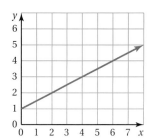

You can draw a vertical line through (2, 2) and (2, 5).

∴ So, the graph does *not* represent a function.

No vertical line can be drawn through two points on the graph.

∴ So, the graph represents a function.

● **Practice**

Determine whether the graph represents a function.

8. 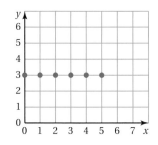   9. 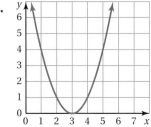   10.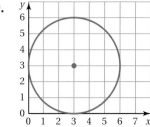

11. **REASONING** You studied linear equations in Chapter 2. Do all linear equations represent functions? Explain your reasoning.

Extension 5.1   Relations and Functions   209

## 5.2 Discrete and Continuous Domains

**Essential Question** How can you decide whether the domain of a function is discrete or continuous?

### 1 EXAMPLE: Discrete and Continuous Domains

In Activities 1 and 2 in Section 2.4, you studied two real-life problems represented by the same equation.

$$4x + 2y = 16 \quad \text{or} \quad y = -2x + 8$$

**a.**

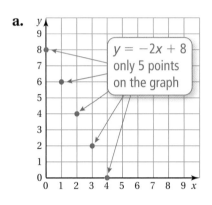

$y = -2x + 8$
only 5 points on the graph

Domain (*x*-values): 0, 1, 2, 3, 4

Range (*y*-values): 8, 6, 4, 2, 0

The domain is **discrete** because it consists of only the numbers 0, 1, 2, 3, and 4.

**b.**

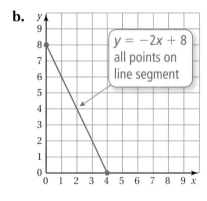

$y = -2x + 8$
all points on line segment

CHEESE FOR SALE
Swiss: $4/lb   Cheddar: $2/lb

Domain (*x*-values): $0 \le x \le 4$

Range (*y*-values): $0 \le y \le 8$

The domain is **continuous** because it consists of all numbers from 0 to 4 on the number line.

**Functions**

In this lesson, you will
- graph discrete and continuous data.
- determine whether functions have a discrete or continuous domain.

Learning Standards
8.F.1
F.IF.1
F.IF.5

## 2 ACTIVITY: Discrete and Continuous Domains

**Math Practice 4**

**Apply Mathematics**
How can you use mathematics to represent and solve each problem?

Work with a partner.
- Write a function to represent each problem.
- Graph each function.
- Describe the domain and range of each function. Is the domain discrete or continuous?

a. You are in charge of reserving hotel rooms for a youth soccer team. Each room costs $69, plus $6 tax, per night. You need each room for two nights. You need 10 to 16 rooms. Write a function for the total hotel cost.

b. The airline you are using for the soccer trip needs an estimate of the total weight of the team's luggage. You determine that there will be 36 pieces of luggage and each piece will weigh from 25 to 45 pounds. Write a function for the total weight of the luggage.

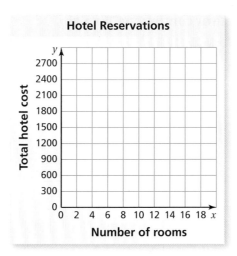

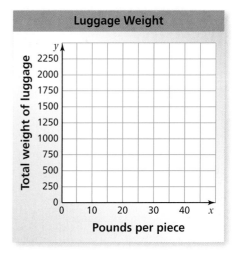

## What Is Your Answer?

3. **IN YOUR OWN WORDS** How can you decide whether the domain of a function is discrete or continuous? Describe two real-life examples of functions: one with a discrete domain and one with a continuous domain.

**Practice**

Use what you learned about discrete and continuous domains to complete Exercises 3 and 4 on page 214.

Section 5.2   Discrete and Continuous Domains   211

## 5.2 Lesson

**Key Vocabulary**
discrete domain, p. 212
continuous domain, p. 212

### Key Idea

**Discrete and Continuous Domains**

A **discrete domain** is a set of input values that consists of only certain numbers in an interval.

**Example:** Integers from 1 to 5

A **continuous domain** is a set of input values that consists of all numbers in an interval.

**Example:** All numbers from 1 to 5

### EXAMPLE 1 — Graphing Discrete Data

The function $y = 15.95x$ represents the cost $y$ (in dollars) of $x$ tickets for a museum. Graph the function using a domain of 0, 1, 2, 3, and 4. Is the domain discrete or continuous? Explain.

Make an input-output table.

| Input, $x$ | $15.95x$ | Output, $y$ | Ordered Pair, $(x, y)$ |
|---|---|---|---|
| 0 | 15.95(0) | 0 | (0, 0) |
| 1 | 15.95(1) | 15.95 | (1, 15.95) |
| 2 | 15.95(2) | 31.9 | (2, 31.9) |
| 3 | 15.95(3) | 47.85 | (3, 47.85) |
| 4 | 15.95(4) | 63.8 | (4, 63.8) |

Plot the ordered pairs. Because you cannot buy part of a ticket, the graph consists of individual points.

∴ So, the domain is discrete.

### On Your Own

1. The function $m = 50 - 9d$ represents the amount of money $m$ (in dollars) you have after buying $d$ DVDs. Graph the function. Is the domain discrete or continuous? Explain.

# EXAMPLE 2  Graphing Continuous Data

A cereal bar contains 130 calories. The number $c$ of calories consumed is a function of the number $b$ of bars eaten. Graph the function. Is the domain discrete or continuous?

Make an input-output table.

| Input, $b$ | Output, $c$ | Ordered Pair, $(b, c)$ |
|---|---|---|
| 0 | 0 | (0, 0) |
| 1 | 130 | (1, 130) |
| 2 | 260 | (2, 260) |
| 3 | 390 | (3, 390) |
| 4 | 520 | (4, 520) |

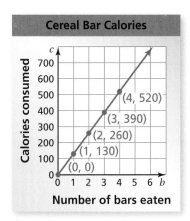

Plot the ordered pairs. Because you can eat part of a cereal bar, $b$ can be any value greater than or equal to 0. Draw a line through the points.

∴ So, the domain is continuous.

# EXAMPLE 3  Real-Life Application

You conduct an experiment on the speed of sound waves in dry air at 86°F. You record your data in a table. Which of the following is true?

| Input Time, $t$ (seconds) | Output Distance, $d$ (miles) |
|---|---|
| 2 | 0.434 |
| 4 | 0.868 |
| 6 | 1.302 |
| 8 | 1.736 |
| 10 | 2.170 |

**A** The domain is $2 \leq t \leq 10$ and it is discrete.

**B** The domain is $2 \leq t \leq 10$ and it is continuous.

**C** The domain is $0.434 \leq d \leq 2.17$ and it is discrete.

**D** The domain is $0.434 \leq d \leq 2.17$ and it is continuous.

The domain is the set of possible input values, or the time $t$. The time $t$ can be any value from 2 to 10. So, the domain is continuous.

∴ The correct answer is **B**.

## On Your Own

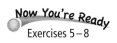
Exercises 5–8

2. A 20-gallon bathtub is draining at a rate of 2.5 gallons per minute. The number $g$ of gallons remaining is a function of the number $m$ of minutes. Graph the function. Is the domain discrete or continuous?

3. Is the domain discrete or continuous? Explain.

| Input Number of Stories | 1 | 2 | 3 |
|---|---|---|---|
| Output Height of Building (feet) | 12 | 24 | 36 |

Section 5.2  Discrete and Continuous Domains  213

## 5.2 Exercises

###  Vocabulary and Concept Check

1. **VOCABULARY** Explain how continuous domains and discrete domains are different.
2. **WRITING** Describe how you can use a graph to determine whether a domain is discrete or continuous.

###  Practice and Problem Solving

**Describe the domain and range of the function. Is the domain discrete or continuous?**

3.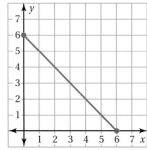

4.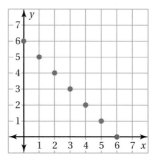

**Graph the function. Is the domain discrete or continuous?**

 5.

| Input Bags, x | Output Marbles, y |
|---|---|
| 2 | 20 |
| 4 | 40 |
| 6 | 60 |

6.

| Input Years, x | Output Height of a Tree, y (feet) |
|---|---|
| 0 | 3 |
| 1 | 6 |
| 2 | 9 |

7.

| Input Width, x (inches) | Output Volume, y (cubic inches) |
|---|---|
| 5 | 50 |
| 10 | 100 |
| 15 | 150 |

8.

| Input Hats, x | Output Cost, y (dollars) |
|---|---|
| 0 | 0 |
| 1 | 8.45 |
| 2 | 16.9 |

9. **ERROR ANALYSIS** Describe and correct the error made in the statement about the domain.

10. **YARN** The function $m = 40 - 8.5b$ represents the amount $m$ of money (in dollars) that you have after buying $b$ balls of yarn. Graph the function using a domain of 0, 1, 2, and 3. Is the domain discrete or continuous?

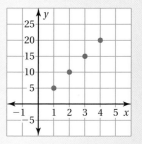

2.5 is in the domain.

214 Chapter 5 Linear Functions

11. **TICKETS** The number $t$ of tickets sold at a concert is a function of the ticket cost $c$.

   a. Which variable is independent? dependent?
   b. Is the domain discrete or continuous?

12. **DISTANCE** The function $y = 3.28x$ converts length from $x$ meters to $y$ feet.

   a. Graph the function. Which variable is independent? dependent?
   b. Is the domain discrete or continuous?

13. **LOGIC** The area $A$ of the triangle is a function of the height $h$. Your friend says the domain is discrete. Is he correct? Explain.

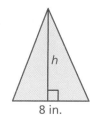

14. **PACKING** You are packing books into a box. The box can hold at most 10 books. The function $y = 5.2x$ represents the weight $y$ (in pounds) of $x$ books.

   a. Is 52 in the range? Explain.
   b. Is 15 in the domain? Explain.
   c. Graph the function. Is the domain discrete or continuous?

15. **Reasoning** Describe a real-world situation for the given constraints.

   a. A negative number in the domain and the domain is continuous
   b. A negative number in the range and the domain is discrete

## Fair Game Review  What you learned in previous grades & lessons

**Find the slope of the line.** *(Section 2.2)*

16.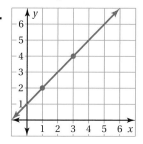

17.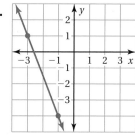

18.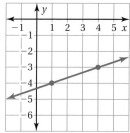

19. **MULTIPLE CHOICE** What is the $y$-intercept of the graph of the linear equation? *(Section 2.3)*

   Ⓐ −4    Ⓑ −2
   Ⓒ 2     Ⓓ 4

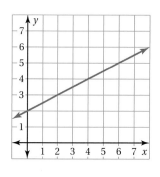

## 5.3 Linear Function Patterns

**Essential Question** How can you use a linear function to describe a linear pattern?

### 1 ACTIVITY: Finding Linear Patterns

**Work with a partner.**
- Plot the points from the table in a coordinate plane.
- Write a linear equation for the function represented by the graph.

a.
| x | 0 | 2 | 4 | 6 | 8 |
|---|---|---|---|---|---|
| y | 150 | 125 | 100 | 75 | 50 |

b.
| x | 4 | 6 | 8 | 10 | 12 |
|---|---|---|---|---|---|
| y | 15 | 20 | 25 | 30 | 35 |

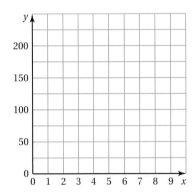

c.
| x | −4 | −2 | 0 | 2 | 4 |
|---|---|---|---|---|---|
| y | 4 | 6 | 8 | 10 | 12 |

d.
| x | −4 | −2 | 0 | 2 | 4 |
|---|---|---|---|---|---|
| y | 1 | 0 | −1 | −2 | −3 |

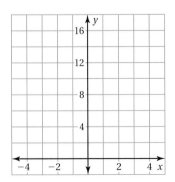
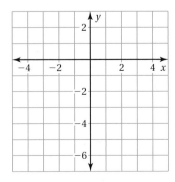

**COMMON CORE**

**Functions**
In this lesson, you will
- write linear functions from graphs or tables.
- solve linear functions.
- solve real-life problems.

Learning Standards
8.F.3
8.F.4
F.BF.1a
F.LE.2

**216** Chapter 5   Linear Functions

## 2 ACTIVITY: Finding Linear Patterns

**Math Practice 4**

**Analyze Relationships**
What is the relationship between the variables? How does this help you write a linear function?

Work with a partner. The table shows a familiar linear pattern from geometry.
- Write a linear function that relates $y$ to $x$.
- What do the variables $x$ and $y$ represent?
- Graph the linear function.

a.

| x | 1 | 2 | 3 | 4 | 5 |
|---|---|---|---|---|---|
| y | $2\pi$ | $4\pi$ | $6\pi$ | $8\pi$ | $10\pi$ |

b.

| x | 1 | 2 | 3 | 4 | 5 |
|---|---|---|---|---|---|
| y | 10 | 12 | 14 | 16 | 18 |

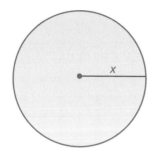

c.

| x | 1 | 2 | 3 | 4 | 5 |
|---|---|---|---|---|---|
| y | 5 | 6 | 7 | 8 | 9 |

d.

| x | 1 | 2 | 3 | 4 | 5 |
|---|---|---|---|---|---|
| y | 28 | 40 | 52 | 64 | 76 |

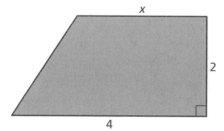

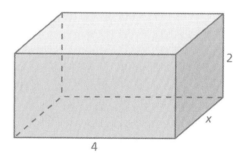

### What Is Your Answer?

3. **IN YOUR OWN WORDS** How can you use a linear function to describe a linear pattern?

4. Describe the strategy you used to find the linear functions in Activities 1 and 2.

**Practice**  Use what you learned about linear function patterns to complete Exercises 4 and 5 on page 220.

Section 5.3  Linear Function Patterns  **217**

## 5.3 Lesson

**Key Vocabulary**
linear function, p. 218

A **linear function** is a function whose graph is a nonvertical line. A linear function can be written in the form $y = mx + b$.

### EXAMPLE 1  Finding a Linear Function Using a Graph

**Use the graph to write a linear function that relates $y$ to $x$.**

The points lie on a line. Find the slope and $y$-intercept of the line.

$$\text{slope} = \frac{\text{change in } y}{\text{change in } x} = \frac{3 - 0}{4 - 2} = \frac{3}{2}$$

Because the line crosses the $y$-axis at $(0, -3)$, the $y$-intercept is $-3$.

∴ So, the linear function is $y = \dfrac{3}{2}x - 3$.

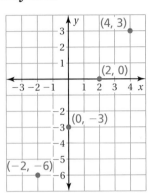

### EXAMPLE 2  Finding a Linear Function Using a Table

**Use the table to write a linear function that relates $y$ to $x$.**

| x | −3 | −2 | −1 | 0 |
|---|----|----|----|---|
| y | 9  | 7  | 5  | 3 |

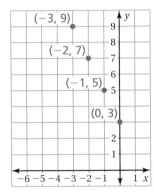

Plot the points in the table.

The points lie on a line. Find the slope and $y$-intercept of the line.

$$\text{slope} = \frac{\text{change in } y}{\text{change in } x} = \frac{9 - 7}{-3 - (-2)} = \frac{2}{-1} = -2$$

Because the line crosses the $y$-axis at $(0, 3)$, the $y$-intercept is 3.

∴ So, the linear function is $y = -2x + 3$.

### On Your Own

Now You're Ready
Exercises 6–11

**Use the graph or table to write a linear function that relates $y$ to $x$.**

1.

2.

| x | −2 | −1 | 0 | 1 |
|---|----|----|---|---|
| y | 2  | 2  | 2 | 2 |

**218  Chapter 5  Linear Functions**

## EXAMPLE 3 Real-Life Application

| Hours Kayaking, x | Calories Burned, y |
|---|---|
| 2 | 600 |
| 4 | 1200 |
| 6 | 1800 |
| 8 | 2400 |

Graph the data in the table. (a) Is the domain discrete or continuous? (b) Write a linear function that relates y to x. (c) How many calories do you burn in 4.5 hours?

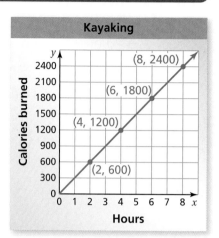

a. Plot the points. Time can represent any value greater than or equal to 0, so the domain is continuous. Draw a line through the points.

b. The y-intercept is 0 and the slope is $\frac{1200 - 600}{4 - 2} = \frac{600}{2} = 300$.

∴ So, the linear function is $y = 300x$.

c. Find the value of y when $x = 4.5$.

$y = 300x$     Write the equation.

$\phantom{y} = 300(4.5)$     Substitute 4.5 for x.

$\phantom{y} = 1350$     Multiply.

∴ You burn 1350 calories in 4.5 hours of kayaking.

### On Your Own

| Hours Rock Climbing, x | Calories Burned, y |
|---|---|
| 3 | 1950 |
| 6 | 3900 |
| 9 | 5850 |
| 12 | 7800 |

3. Graph the data in the table.
   a. Is the domain discrete or continuous?
   b. Write a linear function that relates y to x.
   c. How many calories do you burn in 5.5 hours?

## Summary

**Representing a Function**

**Words** An output is 2 more than the input.

**Equation** $y = x + 2$

**Input-Output Table**

| Input, x | −1 | 0 | 1 | 2 |
|---|---|---|---|---|
| Output, y | 1 | 2 | 3 | 4 |

**Graph**

Section 5.3    Linear Function Patterns

## 5.3 Exercises

### Vocabulary and Concept Check

1. **VOCABULARY** Describe four ways to represent a function.
2. **VOCABULARY** Does the graph represent a linear function? Explain.
3. **REASONING** Do all linear functions have a $y$-intercept? Explain.

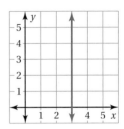

### Practice and Problem Solving

The table shows a familiar linear pattern from geometry. Write a linear function that relates $y$ to $x$. What do the variables $x$ and $y$ represent? Graph the linear function.

4.
| x | 1 | 2 | 3 | 4 | 5 |
|---|---|---|---|---|---|
| y | $\pi$ | $2\pi$ | $3\pi$ | $4\pi$ | $5\pi$ |

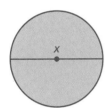

5.
| x | 1 | 2 | 3 | 4 | 5 |
|---|---|---|---|---|---|
| y | 2 | 4 | 6 | 8 | 10 |

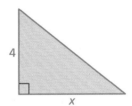

Use the graph or table to write a linear function that relates $y$ to $x$.

6.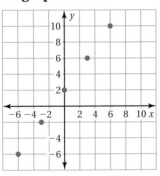

7.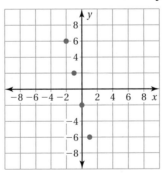

8.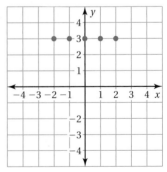

9.
| x | −2 | −1 | 0 | 1 |
|---|---|---|---|---|
| y | −4 | −2 | 0 | 2 |

10.
| x | −8 | −4 | 0 | 4 |
|---|---|---|---|---|
| y | 2 | 1 | 0 | −1 |

11.
| x | −3 | 0 | 3 | 6 |
|---|---|---|---|---|
| y | 3 | 5 | 7 | 9 |

12. **MOVIES** The table shows the cost $y$ (in dollars) of renting $x$ movies.

   a. Which variable is independent? dependent?
   b. Graph the data. Is the domain discrete or continuous?
   c. Write a function that relates $y$ to $x$.
   d. How much does it cost to rent three movies?

| Number of Movies, x | 0 | 1 | 2 | 4 |
|---|---|---|---|---|
| Cost, y | 0 | 3 | 6 | 12 |

220 Chapter 5 Linear Functions

13. **BIKE JUMPS** A bunny hop is a bike trick in which the rider brings both tires off the ground without using a ramp. The table shows the height $y$ (in inches) of a bunny hop on a bike that weighs $x$ pounds.

| Weight, x | 19 | 21 | 23 |
|---|---|---|---|
| Height, y | 10.2 | 9.8 | 9.4 |

   a. Graph the data. Then describe the pattern.
   b. Write a linear function that relates the height of a bunny hop to the weight of the bike.
   c. What is the height of a bunny hop on a bike that weighs 21.5 pounds?

14. **REASONING** Can the graph of a function be a horizontal line? Explain your reasoning.

| Years of Education, x | Annual Salary, y |
|---|---|
| 0 | 28 |
| 2 | 40 |
| 4 | 52 |
| 6 | 64 |
| 10 | 88 |

15. **SALARY** The table shows a person's annual salary $y$ (in thousands of dollars) after $x$ years of education beyond high school.

   a. Graph the data. Then describe the pattern.
   b. What is the annual salary of the person after 8 years of education beyond high school?

16. **Problem Solving** The Heat Index is calculated using the relative humidity and the temperature. For every 1 degree increase in the temperature from 94°F to 98°F at 75% relative humidity, the Heat Index rises 4°F.

   a. On a summer day, the relative humidity is 75%, the temperature is 94°F, and the Heat Index is 122°F. Construct a table that relates the temperature $t$ to the Heat Index $H$. Start the table at 94°F and end it at 98°F.
   b. Identify the independent and dependent variables.
   c. Write a linear function that represents this situation.
   d. Estimate the Heat Index when the temperature is 100°F.

## Fair Game Review   What you learned in previous grades & lessons

**Evaluate the expression when $x = -2, 0,$ and $3$.**   *(Skills Review Handbook)*

17. $x - 2$       18. $-3x + 2$       19. $0.5x - 0.25$

20. **MULTIPLE CHOICE** Which expression has a value less than 1?   *(Skills Review Handbook)*

   Ⓐ $\dfrac{1}{5^{-2}}$       Ⓑ $5^{-2}$       Ⓒ $5^0$       Ⓓ $5^2$

# 5 Study Help

Check It Out
Graphic Organizer
BigIdeasMath.com

You can use a **comparison chart** to compare two topics. Here is an example of a comparison chart for domain and range.

|  | Domain | Range |
|---|---|---|
| Definition | the set of all possible input values | the set of all possible output values |
| Algebra<br>Example: y = mx + b | x-values | corresponding y-values |
| Ordered pairs<br>Example: (−4, 0), (−3, 1),<br>(−2, 2), (−1, 3) | −4, −3, −2, −1 | 0, 1, 2, 3 |
| Table<br>Example:<br>x: −1, 0, 2, 3<br>y: 1, 0, 4, 9 | −1, 0, 2, 3 | 0, 1, 4, 9 |
| Graph<br>Example: | −3, −1, 2, 3 | −1, 1, 2 |

## On Your Own

**Make comparison charts to help you study and compare these topics.**

1. independent variable and dependent variable
2. discrete domain and continuous domain
3. linear functions with positive slopes and linear functions with negative slopes

**After you complete this chapter, make a comparison chart for the following topics.**

4. linear functions and nonlinear functions

"Creating a comparison chart causes canines to crystalize concepts."

# 5.1–5.3 Quiz

**Find the domain and range of the function represented by the graph.** *(Section 5.1)*

1.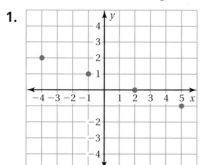
2.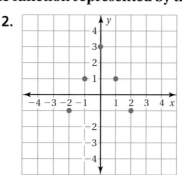
3.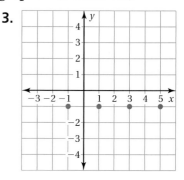

**Graph the function. Is the domain discrete or continuous?** *(Section 5.2)*

4. 
| Minutes, x | 0 | 10 | 20 | 30 |
|---|---|---|---|---|
| Height, y | 40 | 35 | 30 | 25 |

5. 
| Relay Teams, x | 2 | 4 | 6 | 8 |
|---|---|---|---|---|
| Athletes, y | 8 | 16 | 24 | 32 |

**Use the graph or table to write a linear function that relates y to x.** *(Section 5.3)*

6. (graph)

7. 
| x | y |
|---|---|
| −3 | −3 |
| 0 | −1 |
| 3 | 1 |
| 6 | 3 |

8. **VIDEO GAME** The function $m = 30 - 3r$ represents the amount $m$ (in dollars) of money you have after renting $r$ video games. Graph the function using a domain of 0, 1, 2, 3, and 4. Is the domain discrete or continuous? *(Section 5.2)*

9. **ADVERTISING** The table shows the revenue $R$ (in millions of dollars) of a company when it spends $A$ (in millions of dollars) on advertising. *(Section 5.3)*

   | Advertising, A | Revenue, R |
   |---|---|
   | 0 | 2 |
   | 2 | 6 |
   | 4 | 10 |
   | 6 | 14 |
   | 8 | 18 |

   a. Write a linear function that relates the revenue to the advertising cost.
   b. What is the revenue of the company when it spends $10 million on advertising?

10. **WATER** Water accounts for about 60% of a person's body weight. *(Section 5.1)*

    a. Write an equation that represents the water weight $y$ of a person who weighs $x$ pounds. Identify the independent variable and the dependent variable.
    b. Make an input-output table for the equation in part (a). Use the inputs 100, 120, 140, and 160.
    c. Find the domain and range of the function represented by the table.

# 5.4 Function Notation

**Essential Question** How can you use function notation to represent a function?

By naming a function $f$, you can write the function using **function notation**.

$f(x) = 2x - 3$    Function notation

This is read as "$f$ of $x$ equals $2x$ minus 3." The notation $f(x)$ is another name for $y$. When function notation is used, the parentheses do not imply multiplication. You can use letters other than $f$ to name a function. The letters $g$, $h$, $j$, and $k$ are often used to name functions.

### 1 ACTIVITY: Matching Functions with Their Graphs

**Work with a partner. Match each function with its graph.**

**a.** $f(x) = 2x - 3$  
**b.** $g(x) = -x + 2$  
**c.** $h(x) = x^2 - 1$  
**d.** $j(x) = 2x^2 - 3$

A.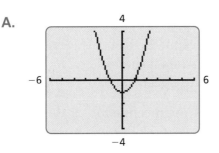

B.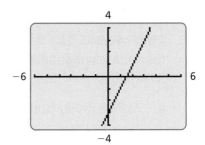

C.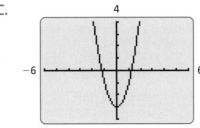

D.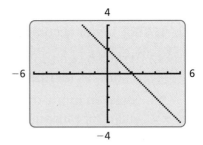

**COMMON CORE**

**Functions**

In this lesson, you will
- evaluate, solve, and graph functions written in function notation.
- compare graphs of linear functions.

Learning Standards
F.BF.3
F.IF.1
F.IF.2
F.IF.7b

224  Chapter 5  Linear Functions

## 2 ACTIVITY: Evaluating a Function

**Evaluate Results**
Does your answer seem reasonable? How can you check your answer?

Work with a partner. Consider the function

$$f(x) = -x + 3.$$

Locate the points $(x, f(x))$ on the graph. Explain how you found each point.

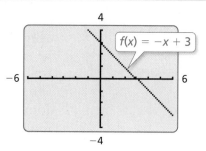

a. $(-1, f(-1))$

b. $(0, f(0))$

c. $(1, f(1))$

d. $(2, f(2))$

## 3 ACTIVITY: Comparing Graphs of Functions

Work with a partner. The graph of a function from trigonometry is shown at the right. Use the graph to sketch the graph of each function. Explain your reasoning.

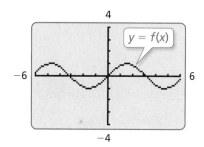

a. $g(x) = f(x) + 2$

b. $g(x) = f(x) + 1$

c. $g(x) = f(x) - 1$

d. $g(x) = f(x) - 2$

### What Is Your Answer?

4. **IN YOUR OWN WORDS** How can you use function notation to represent a function? How are standard notation and function notation similar? How are they different?

| Standard Notation | Function Notation |
| --- | --- |
| $y = 2x + 5$ | $f(x) = 2x + 5$ |

5. Use what you discovered in Activity 3 to write a general observation that compares the graphs of

$$y = f(x) \quad \text{and} \quad y = f(x) + c.$$

**Practice** Use what you learned about function notation to complete Exercises 4–6 on page 229.

Section 5.4  Function Notation  225

## 5.4 Lesson

Check It Out
Lesson Tutorials
BigIdeasMath.com

**Key Vocabulary**
function notation, p. 226

In Section 5.3, you learned that you can write a linear function in the form $y = mx + b$. By naming a linear function $f$, you can also write the function using **function notation.**

$$f(x) = mx + b \qquad \text{Function notation}$$

The notation $f(x)$ is another name for $y$. If $f$ is a function and $x$ is in its domain, then $f(x)$ represents the output of $f$ corresponding to the input $x$. You can use letters other than $f$ to name a function, such as $g$ or $h$.

### EXAMPLE 1 Evaluating a Function

**Reading**
The notation $f(x)$ is read as "the value of $f$ at $x$" or "$f$ of $x$." It does not mean "$f$ times $x$."

Evaluate $f(x) = -4x + 7$ when $x = 2$.

| | |
|---|---|
| $f(x) = -4x + 7$ | Write the function. |
| $f(2) = -4(2) + 7$ | Substitute 2 for $x$. |
| $= -8 + 7$ | Multiply. |
| $= -1$ | Add. |

∴ When $x = 2$, $f(x) = -1$.

### On Your Own

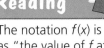
Exercises 4–9

Evaluate the function when $x = -4$, 0, and 3.

1. $f(x) = 2x - 5$
2. $g(x) = -x - 1$

### EXAMPLE 2 Solving for the Independent Variable

For $h(x) = \frac{2}{3}x - 5$, find the value of $x$ for which $h(x) = -7$.

| | |
|---|---|
| $h(x) = \frac{2}{3}x - 5$ | Write the function. |
| $-7 = \frac{2}{3}x - 5$ | Substitute $-7$ for $h(x)$. |
| $-2 = \frac{2}{3}x$ | Add 5 to each side. |
| $-3 = x$ | Multiply each side by $\frac{3}{2}$. |

∴ When $x = -3$, $h(x) = -7$.

### On Your Own

Now You're Ready
Exercises 11–16

Find the value of $x$ so that the function has the given value.

3. $f(x) = 6x + 9$; $f(x) = 21$
4. $g(x) = -\frac{1}{2}x + 3$; $g(x) = -1$

# EXAMPLE 3  Graphing a Linear Function in Function Notation

**Graph $f(x) = 2x + 5$.**

**Step 1:** Make a table of values.

| x | −2 | −1 | 0 | 1 | 2 |
|---|---|---|---|---|---|
| f(x) | 1 | 3 | 5 | 7 | 9 |

**Step 2:** Plot the ordered pairs.

**Step 3:** Draw a line through the points.

**Study Tip**
The graph of $f(x)$ consists of the points $(x, f(x))$.

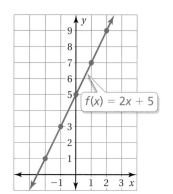

## On Your Own

**Graph the linear function.**

Now You're Ready
Exercises 22–27

**5.** $f(x) = 3x - 2$   **6.** $g(x) = -x + 4$   **7.** $h(x) = -\dfrac{3}{4}x - 1$

---

## Key Idea

**Vertical Translations**

The graph of $f(x) + k$ is a vertical translation of the graph of $f(x)$, where $k \neq 0$.

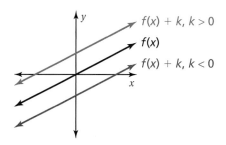

In vertical translations of graphs of linear functions, the graphs have the same slope but different y-intercepts.

Section 5.4   Function Notation

### EXAMPLE 4 — Comparing Graphs of Linear Functions

**Graph $g(x) = x - 3$. Compare the graph to the graph of $f(x) = x$.**

Use the slope and y-intercept to graph the equations.

$g(x) = x - 3$
$\phantom{g(x)} = 1x + (-3)$
↑ slope  ↑ y-intercept

$f(x) = x$
$\phantom{f(x)} = 1x + 0$
↑ slope  ↑ y-intercept

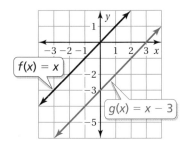

∴ The graphs have the same slope but different y-intercepts. The graph of g is a translation 3 units down of the graph of f.

### EXAMPLE 5 — Real-Life Application

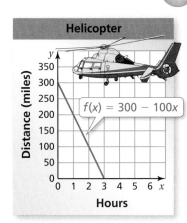

**The graph shows the number y of miles a helicopter is from its destination after x hours on its first flight. On its second flight, the helicopter travels at the same speed but 50 miles farther. Which statement is true about the graph of the function that represents the second flight compared to the graph of the function that represents the first flight?**

- **Ⓐ** The slope decreases.
- **Ⓑ** The slope increases.
- **Ⓒ** The graph is a translation 50 units down.
- **Ⓓ** The graph is a translation 50 units up.

The helicopter travels at the same speed on both flights. So, the graphs have the same slope. You can eliminate choices A and B.

Because the helicopter travels 50 miles farther on the second flight, it is 50 miles farther from its destination when $x = 0$. So, the graph of the function that represents the second flight is a vertical translation 50 units up of the graph of the function that represents the first flight.

∴ The correct answer is **Ⓓ**.

### On Your Own

*Now You're Ready*
Exercises 29–31

**Graph the function. Compare the graph to the graph of $f(x) = -2x$.**

8. $g(x) = -2x + 3$
9. $h(x) = -2x - 5$

10. **WHAT IF?** In Example 5, the helicopter travels the same distance but 50 miles per hour faster on the second flight. How does the graph of the function that represents the second flight compare to the graph of the function that represents the first flight?

## 5.4 Exercises

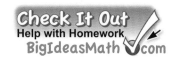

### Vocabulary and Concept Check

1. **VOCABULARY** What is function notation? Give an example.
2. **VOCABULARY** Your height can be represented by a function $h(x)$ where $x$ is your age. What does $h(13)$ represent?
3. **WRITING** What type of graph is given by $y = mx + b$? How does changing the value of $b$ affect the graph?

### Practice and Problem Solving

**Evaluate the function when $x = -2, 0,$ and $5$.**

4. $f(x) = x + 6$
5. $g(x) = 3x - 2$
6. $h(x) = -2x + 9$
7. $h(x) = -x - 7$
8. $g(x) = 6x - 3$
9. $f(x) = -5x + 2$

10. **ERROR ANALYSIS** Describe and correct the error in evaluating the function $g(x) = 4x + 6$ when $x = -2$.

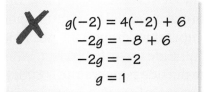

**Find the value of $x$ so that the function has the given value.**

11. $h(x) = -7x + 10;\ h(x) = 3$
12. $t(x) = -3x - 5;\ t(x) = 4$
13. $n(x) = 4x + 15;\ n(x) = 7$
14. $p(x) = 6x - 12;\ p(x) = 18$
15. $q(x) = \frac{1}{3}x - 2;\ q(x) = -4$
16. $r(x) = -\frac{4}{5}x + 7;\ r(x) = -5$

17. **SUMMER JOB** You earn $11 per hour working at a grocery store during the summer. The function $p(x) = 11x$ represents the amount you earn for working $x$ hours.

   a. You work 18 hours. How much do you earn?
   b. How many hours do you have to work to earn $275?

18. **ORCHESTRA** A group of friends are buying tickets to the orchestra. Each ticket costs $17.50 and one of the friends has a coupon for $10. The function $C(x) = 17.5x - 10$ represents the total cost of buying $x$ tickets.

   a. How much does it cost to buy 5 tickets?
   b. How many tickets can you buy with $130.00?

Section 5.4  Function Notation  229

**Match the function with its graph.**

19. $f(x) = -2x - 2$     20. $g(x) = \dfrac{1}{2}x + 2$     21. $h(x) = \dfrac{1}{2}x - 2$

A. 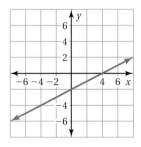     B. 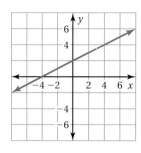     C.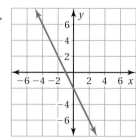

**Graph the linear function.**

22. $f(x) = 4x + 1$     23. $g(x) = -2x - 5$     24. $h(x) = -\dfrac{1}{2}x - 3$

25. $f(x) = \dfrac{3}{5}x + 2$     26. $g(x) = 7x - 4$     27. $h(x) = -6x + 3$

28. **ATMOSPHERIC TEMPERATURE** Under normal conditions, the atmospheric temperature drops 3.5°F per 1000 feet of altitude up to 40,000 feet. When the outside temperature is 80°F, the atmospheric temperature can be modeled by $t(x) = -3.5x + 80$, where $x$ is the altitude in thousands of feet.

   a. Graph the function and identify its domain and range.

   b. Find and interpret the value of $x$ so that $t(x) = -25$.

**Graph the function. Compare the graph to the graph of $f(x) = 3x$.**

29. $g(x) = 3x + 2$     30. $n(x) = 3x - 7$     31. $v(x) = 3x - \dfrac{7}{2}$

32. **DECK** The function $C(x) = 25x + 50$ represents the labor cost for Jones Remodeling to build a deck, where $x$ is the number of hours. Sample labor costs from their main competitor, Premiere Remodeling, are shown in the table.

| Hours | Cost |
|---|---|
| 2 | $130 |
| 4 | $160 |
| 6 | $190 |

   a. Which cost function has the greater rate of change? What does the rate of change represent?

   b. The graph of which cost function has the greater $y$-intercept? Interpret the $y$-intercept.

   c. The job is estimated to take 8 hours. Which company would you hire? Explain your reasoning.

230   Chapter 5   Linear Functions

**Graph the functions $f(x)$ and $g(x)$ in the same coordinate plane. Use the graph to solve $f(x) = g(x)$.**

**33.** $f(x) = x - 2$

$g(x) = 4x - 8$

**34.** $f(x) = -\dfrac{1}{5}x - 3$

$g(x) = 2x + 8$

**35.** $f(x) = \dfrac{2}{3}x - 7$

$g(x) = -x + 3$

**36. CHOOSE TOOLS** What tool would you use to solve $f(x) = g(x)$ when $f(x) = 2.5x + 17$ and $g(x) = 0.8x$? Explain. Then solve $f(x) = g(x)$.

**Given $f(x) = 2x + 1$, find the value of $k$ so that the graph is $f(x) + k$.**

**37.**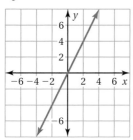

**38.**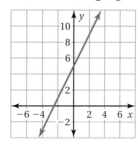

**39.**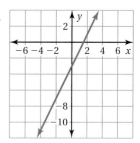

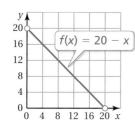

**40. PERIMETER** The graph shows the relationship between the width $y$ and length $x$ of a rectangle in inches. A second rectangle has a perimeter that is 10 inches less than the perimeter of the first rectangle. How does the graph relating the width and length of the second rectangle compare to the graph shown?

**41. CRITICAL THINKING** The graph of $y = x + 4$ is a translation 4 units up of the graph of $y = x$. How can you obtain the graph of $y = x + 4$ from the graph of $y = x$ using a horizontal translation?

**42. Structure** Given that $f(x) = 3x - 5$ and $g(x) = 4x$, write a function that represents $f(g(x))$ and a function that represents $g(f(x))$.

 **Fair Game Review** What you learned in previous grades & lessons

**Write in slope-intercept form an equation of the line that passes through the given points.** *(Section 2.6)*

**43.** (0, 0), (4, 4)

**44.** (−4, 9), (1, −1)

**45.** (−2, 1), (3, 1)

**46. MULTIPLE CHOICE** You buy a pair of gardening gloves for $2.25 and $x$ packets of seeds for $0.88 each. Which equation represents the total cost $y$?
*(Skills Review Handbook)*

**Ⓐ** $y = 0.88x - 2.25$

**Ⓑ** $y = 0.88x + 2.25$

**Ⓒ** $y = 2.25x - 0.88$

**Ⓓ** $y = 2.25x + 0.88$

# Extension 5.4 Special Functions

**Key Vocabulary**
piecewise function, p. 232
step function, p. 233
absolute value function, p. 234

**COMMON CORE**
Functions
In this extension, you will
• graph piecewise, step, and absolute value functions.
Learning Standards
F.BF.3
F.IF.1
F.IF.2
F.IF.7b

## Key Idea

**Piecewise Function**

A **piecewise function** is a function defined by two or more equations. Each "piece" of the function applies to a different part of its domain. An example is shown below.

$$y = \begin{cases} x - 2, & \text{if } x \leq 0 \\ 2x + 1, & \text{if } x > 0 \end{cases}$$

- The expression $x - 2$ gives the value of $y$ when $x$ is less than or equal to 0.
- The expression $2x + 1$ gives the value of $y$ when $x$ is greater than 0.

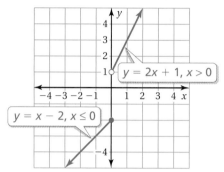

### EXAMPLE 1 — Graphing a Piecewise Function

Graph $y = \begin{cases} -x - 4, & \text{if } x < 0 \\ x, & \text{if } x \geq 0 \end{cases}$. Describe the domain and range.

**Step 1:** Graph $y = -x - 4$ for $x < 0$. Because $x$ is not equal to 0, use an open circle at $(0, -4)$.

**Step 2:** Graph $y = x$ for $x \geq 0$. Because $x$ is greater than or equal to 0, use a closed circle at $(0, 0)$.

∴ The domain is all real numbers. The range is $y > -4$.

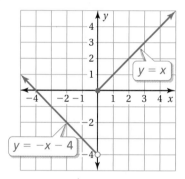

## Practice

**Graph the function. Describe the domain and range.**

1. $y = \begin{cases} x + 3, & \text{if } x \leq 0 \\ -x, & \text{if } x > 0 \end{cases}$

2. $y = \begin{cases} x - 2, & \text{if } x < 0 \\ 4x, & \text{if } x \geq 0 \end{cases}$

3. $y = \begin{cases} -3x - 2, & \text{if } x \leq 1 \\ x + 1, & \text{if } x > 1 \end{cases}$

4. $y = \begin{cases} 2x, & \text{if } x < -1 \\ -2x, & \text{if } x \geq -1 \end{cases}$

5. $y = \begin{cases} 1, & \text{if } x < -3 \\ x - 1, & \text{if } -3 \leq x \leq 3 \\ -2, & \text{if } x > 3 \end{cases}$

6. $y = \begin{cases} -x + 2, & \text{if } x \leq -2 \\ 5, & \text{if } -2 < x < 1 \\ 3x, & \text{if } x \geq 1 \end{cases}$

7. **REASONING** Does $y = \begin{cases} 1 - x, & \text{if } x \leq 0 \\ x - 1, & \text{if } x \geq -2 \end{cases}$ represent a function? Explain your reasoning.

# EXAMPLE 2  Writing a Piecewise Function

**Write a piecewise function for the graph.**

Each "piece" of the function is linear.

When $x < 0$, the graph is the line given by $y = x + 3$.

When $x \geq 0$, the graph is the line given by $y = 2x - 1$.

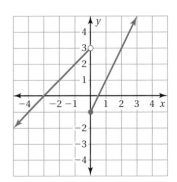

So, a piecewise function for the graph is $f(x) = \begin{cases} x + 3, & \text{if } x < 0 \\ 2x - 1, & \text{if } x \geq 0 \end{cases}$.

**Study Tip**
The graph of a step function can look like a staircase.

A **step function** is a piecewise function defined by constant values over its domain. The graph of a step function consists of a series of line segments.

# EXAMPLE 3  Graphing a Step Function

**You rent a karaoke machine for 5 days. The rental company charges $50 for the first day and $25 for each additional day. Write and graph a step function that represents the relationship between the number of days $x$ and the total cost of renting the karaoke machine.**

Use a table to organize the information.

| Time (days) | Total Cost |
|---|---|
| $0 < x \leq 1$ | 50 |
| $1 < x \leq 2$ | 75 |
| $2 < x \leq 3$ | 100 |
| $3 < x \leq 4$ | 125 |
| $4 < x \leq 5$ | 150 |

$f(x) = \begin{cases} 50, & \text{if } 0 < x \leq 1 \\ 75, & \text{if } 1 < x \leq 2 \\ 100, & \text{if } 2 < x \leq 3 \\ 125, & \text{if } 3 < x \leq 4 \\ 150, & \text{if } 4 < x \leq 5 \end{cases}$

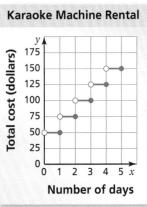

Karaoke Machine Rental

## Practice

**Write a piecewise function for the graph.**

8.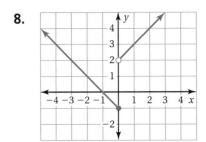

9.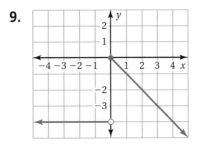

10.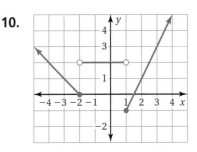

11. **LANDSCAPING** A landscaper rents a wood chipper for 4 days. The rental company charges $100 for the first day and $50 for each additional day. Write and graph a step function that represents the relationship between the number of days $x$ and the total cost of renting the chipper.

**Study Tip**

The absolute value function $f(x) = |x|$ can be written as a piecewise function.

$$f(x) = \begin{cases} -x, & \text{if } x < 0 \\ 0, & \text{if } x = 0 \\ x, & \text{if } x > 0 \end{cases}$$

**Absolute Value Function**

An **absolute value function** has a V-shaped graph that opens up or down.

The most basic absolute value function is $f(x) = |x|$.

The absolute value of a number is always nonnegative. So, the range of $f(x) = |x|$ is $y \geq 0$.

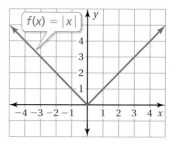

**EXAMPLE 4  Graphing Absolute Value Functions**

Graph each function. Compare the graph to the graph of $y = |x|$. Describe the domain and range.

**a.** $y = |x| + 3$

**Step 1:** Make a table of values.

| x | −2 | −1 | 0 | 1 | 2 |
|---|---|---|---|---|---|
| y | 5 | 4 | 3 | 4 | 5 |

**Step 2:** Plot the ordered pairs.

**Step 3:** Draw the V-shaped graph.

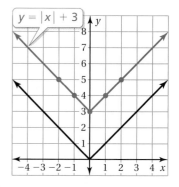

**Study Tip**

The function $y = |x| + 3$ can be written as a piecewise function.

$$f(x) = \begin{cases} -x + 3, & \text{if } x < 0 \\ x + 3, & \text{if } x \geq 0 \end{cases}$$

∴ The graph of $y = |x| + 3$ is a translation 3 units up of the graph of $y = |x|$. The domain is all real numbers. The range is $y \geq 3$.

**b.** $y = |x - 2|$

**Step 1:** Make a table of values.

| x | 0 | 1 | 2 | 3 | 4 |
|---|---|---|---|---|---|
| y | 2 | 1 | 0 | 1 | 2 |

**Step 2:** Plot the ordered pairs.

**Step 3:** Draw the V-shaped graph.

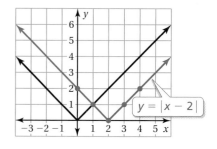

∴ The graph of $y = |x - 2|$ is a translation 2 units to the right of the graph of $y = |x|$. The domain is all real numbers. The range is $y \geq 0$.

# EXAMPLE 5  Graphing Absolute Value Functions

Graph $y = -\frac{1}{2}|x|$. Compare the graph to the graph of $y = |x|$. Describe the domain and range.

**Step 1:** Make a table of values.

| x | −2 | −1 | 0 | 1 | 2 |
|---|----|----|---|---|---|
| y | −1 | $-\frac{1}{2}$ | 0 | $-\frac{1}{2}$ | −1 |

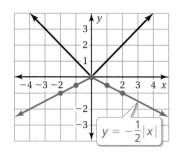

**Step 2:** Plot the ordered pairs.

**Step 3:** Draw the V-shaped graph.

∴ The graph of $y = -\frac{1}{2}|x|$ opens down and is wider than the graph of $y = |x|$. The domain is all real numbers. The range is $y \leq 0$.

## Practice

**Graph the function. Compare the graph to the graph of $y = |x|$. Describe the domain and range.**

12. $y = |x| - 1$
13. $y = |x| + 5$
14. $y = |x + 4|$
15. $y = |x - 3|$
16. $y = \frac{1}{4}|x|$
17. $y = -3|x|$
18. $y = |x + 1| - 2$
19. $y = -|x - 5| + 1$
20. $y = 4|x| - 4$

**Write an equation for the given translation of $y = |x|$.**

21. 7 units down
22. 10 units left
23. 1 unit down and 5 units right
24. 4 units up and 6 units left

25. **REASONING** Explain how the graph of each function compares to the graph of $y = |x|$ for positive and negative values of $k$, $h$, and $a$.
   a. $y = |x| + k$
   b. $y = |x - h|$
   c. $y = a|x|$

**Solve each equation using a graph. Check your solution.**

26. $|x - 1| = 3$
27. $|x + 2| - 6 = -1$
28. $2|x + 7| = 4$

29. **STRUCTURE** Rewrite the function $y = |x + 4|$ using piecewise notation.

30. **STRUCTURE** Graph $y = \begin{cases} -x + 5, & \text{if } x \leq 0 \\ |x|, & \text{if } x > 0 \end{cases}$. Describe the domain and range.

Extension 5.4 Special Functions

# 5.5 Comparing Linear and Nonlinear Functions

**Essential Question** How can you recognize when a pattern in real life is linear or nonlinear?

### 1 ACTIVITY: Finding Patterns for Similar Figures

Work with a partner. Copy and complete each table for the sequence of similar rectangles. Graph the data in each table. Decide whether each pattern is linear or nonlinear.

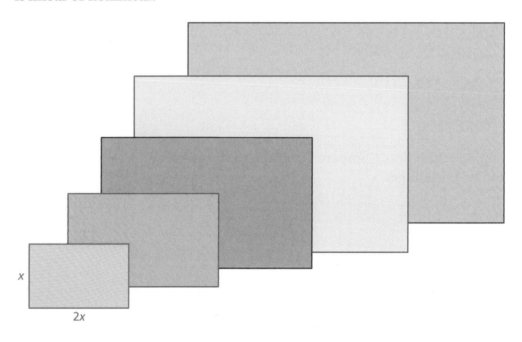

**a.** Perimeters of Similar Rectangles

| x | 1 | 2 | 3 | 4 | 5 |
|---|---|---|---|---|---|
| P |   |   |   |   |   |

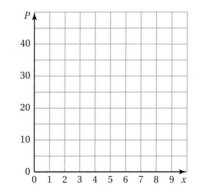

**b.** Areas of Similar Rectangles

| x | 1 | 2 | 3 | 4 | 5 |
|---|---|---|---|---|---|
| A |   |   |   |   |   |

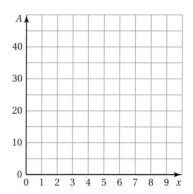

**COMMON CORE**

**Functions**
In this lesson, you will
- identify linear and nonlinear functions from tables or graphs.

Learning Standards
8.F.3
F.LE.1b

## 2 ACTIVITY: Comparing Linear and Nonlinear Functions

**Math Practice**

**Interpret Results**
How do the graphs help you to answer the question? Does your answer make sense?

Work with a partner. The table shows the height *h* (in feet) of a falling object at *t* seconds.

- Graph the data in the table.
- Decide whether the graph is linear or nonlinear.
- Compare the two falling objects. Which one has an increasing speed?

**a.** Falling parachute jumper

| *t* | 0 | 1 | 2 | 3 | 4 |
|---|---|---|---|---|---|
| *h* | 300 | 285 | 270 | 255 | 240 |

**b.** Falling bowling ball

| *t* | 0 | 1 | 2 | 3 | 4 |
|---|---|---|---|---|---|
| *h* | 300 | 284 | 236 | 156 | 44 |

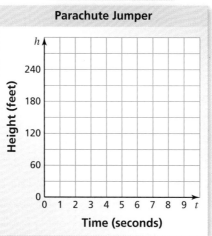

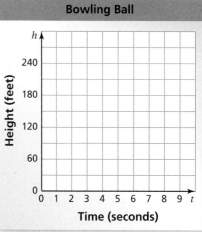

## What Is Your Answer?

3. **IN YOUR OWN WORDS** How can you recognize when a pattern in real life is linear or nonlinear? Describe two real-life patterns: one that is linear and one that is nonlinear. Use patterns that are different from those described in Activities 1 and 2.

Use what you learned about comparing linear and nonlinear functions to complete Exercises 3–6 on page 240.

# 5.5 Lesson

**Key Vocabulary**
nonlinear function, p. 238

The graph of a linear function shows a constant rate of change. A **nonlinear function** does not have a constant rate of change. So, its graph is *not* a line.

### EXAMPLE 1  Identifying Functions from Tables

**Does the table represent a *linear* or *nonlinear* function? Explain.**

**Study Tip**
A constant rate of change describes a quantity that changes by equal amounts over equal intervals.

a.
         +3  +3  +3

| x | 3  | 6  | 9  | 12 |
|---|----|----|----|----|
| y | 40 | 32 | 24 | 16 |

         −8  −8  −8

As *x* increases by 3, *y* decreases by 8. The rate of change is constant. So, the function is linear.

b.
         +2  +2  +2

| x | 1 | 3  | 5  | 7  |
|---|---|----|----|----|
| y | 2 | 11 | 33 | 88 |

         +9  +22  +55

As *x* increases by 2, *y* increases by different amounts. The rate of change is *not* constant. So, the function is nonlinear.

### EXAMPLE 2  Identifying Functions from Graphs

**Does the graph represent a *linear* or *nonlinear* function? Explain.**

a.
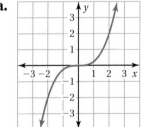

The graph is *not* a line.
So, the function is nonlinear.

b.
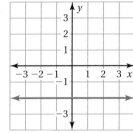

The graph is a line.
So, the function is linear.

### On Your Own

*Now You're Ready*
Exercises 3–11

**Does the table or graph represent a *linear* or *nonlinear* function? Explain.**

1.
| x  | y  |
|----|----|
| 0  | 25 |
| 7  | 20 |
| 14 | 15 |
| 21 | 10 |

2.
| x | y  |
|---|----|
| 2 | 8  |
| 4 | 4  |
| 6 | 0  |
| 8 | −4 |

3.
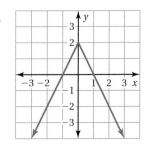

238   Chapter 5   Linear Functions

## EXAMPLE 3 — Identify a Function from an Equation

**Which equation represents a *nonlinear* function?**

- Ⓐ $y = 4.7$
- Ⓑ $y = \pi x$
- Ⓒ $y = \dfrac{4}{x}$
- Ⓓ $y = 4(x - 1)$

You can rewrite the equations $y = 4.7$, $y = \pi x$, and $y = 4(x - 1)$ in slope-intercept form. So, they are linear functions.

You cannot rewrite the equation $y = \dfrac{4}{x}$ in slope-intercept form. So, it is a nonlinear function.

∴ The correct answer is Ⓒ.

## EXAMPLE 4 — Real-Life Application

**Account A earns simple interest. Account B earns compound interest. The table shows the balances for 5 years. Graph the data and compare the graphs.**

**Study Tip**
In Example 4, the *initial value* of each function is $100.

| Year, t | Account A Balance | Account B Balance |
|---|---|---|
| 0 | $100 | $100 |
| 1 | $110 | $110 |
| 2 | $120 | $121 |
| 3 | $130 | $133.10 |
| 4 | $140 | $146.41 |
| 5 | $150 | $161.05 |

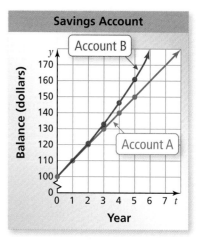

Both graphs show that the balances are positive and increasing.

The balance of Account A has a constant rate of change of $10. So, the function representing the balance of Account A is linear.

The balance of Account B increases by different amounts each year. Because the rate of change is not constant, the function representing the balance of Account B is nonlinear.

### On Your Own

**Now You're Ready**
Exercises 12–14

**Does the equation represent a *linear* or *nonlinear* function? Explain.**

4. $y = x + 5$
5. $y = \dfrac{4x}{3}$
6. $y = 1 - x^2$

Section 5.5  Comparing Linear and Nonlinear Functions

# 5.5 Exercises

## Vocabulary and Concept Check

1. **VOCABULARY** Describe how linear functions and nonlinear functions are different.

2. **WHICH ONE DOESN'T BELONG?** Which equation does *not* belong with the other three? Explain your reasoning.

   | $5y = 2x$ | $y = \dfrac{2}{5}x$ | $10y = 4x$ | $5xy = 2$ |

## Practice and Problem Solving

Graph the data in the table. Decide whether the function is *linear* or *nonlinear*.

3. 
| x | 0 | 1 | 2 | 3 |
|---|---|---|---|---|
| y | 4 | 8 | 12 | 16 |

4. 
| x | 1 | 2 | 3 | 4 |
|---|---|---|---|---|
| y | 1 | 2 | 6 | 24 |

5. 
| x | 6 | 5 | 4 | 3 |
|---|---|---|---|---|
| y | 21 | 15 | 10 | 6 |

6. 
| x | −1 | 0 | 1 | 2 |
|---|---|---|---|---|
| y | −7 | −3 | 1 | 5 |

Does the table or graph represent a *linear* or *nonlinear* function? Explain.

7.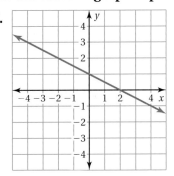

8.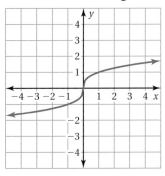

9. 
| x | 5 | 11 | 17 | 23 |
|---|---|---|---|---|
| y | 7 | 11 | 15 | 19 |

10. 
| x | −3 | −1 | 1 | 3 |
|---|---|---|---|---|
| y | 9 | 1 | 1 | 9 |

11. **VOLUME** The table shows the volume V (in cubic feet) of a cube with a side length of x feet. Does the table represent a linear or nonlinear function? Explain.

| Side Length, x | 1 | 2 | 3 | 4 | 5 | 6 | 7 | 8 |
|---|---|---|---|---|---|---|---|---|
| Volume, V | 1 | 8 | 27 | 64 | 125 | 216 | 343 | 512 |

**Does the equation represent a *linear* or *nonlinear* function? Explain.**

**12.** $2x + 3y = 7$     **13.** $y + x = 4x + 5$     **14.** $y = \dfrac{8}{x^2}$

**15. LIGHT** The frequency $y$ (in terahertz) of a light wave is a function of its wavelength $x$ (in nanometers). Does the table represent a linear or nonlinear function? Explain.

| Color | Red | Yellow | Green | Blue | Violet |
|---|---|---|---|---|---|
| Wavelength, $x$ | 660 | 595 | 530 | 465 | 400 |
| Frequency, $y$ | 454 | 504 | 566 | 645 | 749 |

**16. MODELING** The table shows the cost $y$ (in dollars) of $x$ pounds of sunflower seeds.

| Pounds, $x$ | Cost, $y$ |
|---|---|
| 2 | 2.80 |
| 3 | ? |
| 4 | 5.60 |

  **a.** What is the missing $y$-value that makes the table represent a linear function?
  **b.** Write a linear function that represents the cost $y$ of $x$ pounds of seeds.
  **c.** What is the initial value of the function?
  **d.** Does the function have a maximum value? Explain your reasoning.

**17. TREES** Tree A grows at a rate of 1.5 feet per year. The table shows the height $h$ (in feet) of Tree B after $x$ years.

| Years, $x$ | Height, $h$ |
|---|---|
| 0 | 0 |
| 2 | 3.2 |
| 5 | 8 |

  **a.** Does the table represent a linear or nonlinear function? Explain.
  **b.** Which tree is growing at a faster rate? Explain.

**18. PRECISION** The radius of the base of a cylinder is 3 feet. Is the volume of the cylinder a linear or nonlinear function of the height of the cylinder? Explain.

**19.**   The ordered pairs represent a function.

$(0, 0), (1, 1), (2, 4), (3, 9),$ and $(4, 16)$

  **a.** Graph the ordered pairs and describe the pattern. Is the function linear or nonlinear?
  **b.** Write an equation that represents the function.

**Fair Game Review** What you learned in previous grades & lessons

**Find the square root(s).** *(Skills Review Handbook)*

**20.** $\sqrt{49}$     **21.** $-\sqrt{36}$     **22.** $\pm\sqrt{9}$

**23. MULTIPLE CHOICE** Which of the following equations has a slope of $-2$ and passes through the point $(2, 3)$? *(Section 2.6)*

  **Ⓐ** $y = -2x + 6$   **Ⓑ** $y - 3 = -2(x + 2)$   **Ⓒ** $y = -2x + 7$   **Ⓓ** $y - 2 = -2(x - 3)$

## 5.6 Arithmetic Sequences

**Essential Question** How are arithmetic sequences used to describe patterns?

### 1 ACTIVITY: Describing a Pattern

Work with a partner.
- Use the figures to complete the table.
- Plot the points in your completed table.
- Describe the pattern of the *y*-values.

a.  $n = 1$     $n = 2$     $n = 3$     $n = 4$     $n = 5$

| Number of Rows, *n* | 1 | 2 | 3 | 4 | 5 |
|---|---|---|---|---|---|
| Number of Dots, *y* | | | | | |

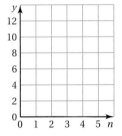

b.  $n = 1$     $n = 2$     $n = 3$     $n = 4$     $n = 5$

| Number of Stars, *n* | 1 | 2 | 3 | 4 | 5 |
|---|---|---|---|---|---|
| Number of Sides, *y* | | | | | |

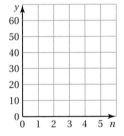

**COMMON CORE**

**Arithmetic Sequences**

In this lesson, you will
- extend and graph arithmetic sequences.
- write equations for arithmetic sequences.
- solve real-life problems.

Learning Standards
F.BF.2
F.IF.3
F.LE.2

c.  $n = 1$     $n = 2$     $n = 3$     $n = 4$     $n = 5$

| *n* | 1 | 2 | 3 | 4 | 5 |
|---|---|---|---|---|---|
| Number of Circles, *y* | | | | | |

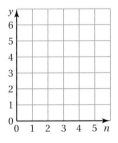

## 2 ACTIVITY: Using a Pattern in Science to Predict

Work with a partner. In chemistry, water is called $H_2O$ because each molecule of water has 2 hydrogen atoms and 1 oxygen atom.

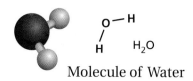
Molecule of Water

- Use the figures to complete the table.
- Describe the pattern of the $y$-values.
- Use your pattern to predict the number of atoms in 23 molecules.

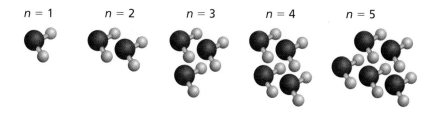

| Number of Molecules, $n$ | 1 | 2 | 3 | 4 | 5 |
|---|---|---|---|---|---|
| Number of Atoms, $y$ | | | | | |

## 3 ACTIVITY: Writing a Story

**Math Practice**

**Make a Plan**
What is your plan for writing your story?

Work with a partner.
- Describe the pattern in the table.
- Write and illustrate a story using the numbers in the table.
- Graph the data shown in the table.

| Jan | Feb | Mar | Apr | May | Jun | Jul | Aug | Sep | Oct | Nov | Dec |
|---|---|---|---|---|---|---|---|---|---|---|---|
| 12 | 20 | 28 | 36 | 44 | 52 | 60 | 68 | 76 | 84 | 92 | 100 |

### What Is Your Answer?

4. **IN YOUR OWN WORDS** How are arithmetic sequences used to describe patterns? Give an example from real life.

**Practice** — Use what you learned about arithmetic sequences to complete Exercise 3 on page 247.

## 5.6 Lesson

**Key Vocabulary**
sequence, p. 244
term, p. 244
arithmetic sequence, p. 244
common difference, p. 244

A **sequence** is an ordered list of numbers. Each number in a sequence is called a **term**. Each term $a_n$ has a specific position $n$ in the sequence.

$$5, \ 10, \ 15, \ 20, \ 25, \ldots, a_n, \ldots$$

- 1st position
- 3rd position
- $n$th position

### Key Idea

**Arithmetic Sequence**

In an **arithmetic sequence,** the difference between consecutive terms is the same. This difference is called the **common difference.** Each term is found by adding the common difference to the previous term.

$$5, \ 10, \ 15, \ 20, \ldots$$ Terms of an arithmetic sequence

$+5 \ +5 \ +5$ ← Common difference

### EXAMPLE 1  Extending an Arithmetic Sequence

**Write the next three terms of the arithmetic sequence $-7, -14, -21, -28, \ldots$.**

Use a table to organize the terms and find the pattern.

| Position | 1 | 2 | 3 | 4 |
|---|---|---|---|---|
| Term | $-7$ | $-14$ | $-21$ | $-28$ |

$+(-7) \ +(-7) \ +(-7)$

Each term is 7 less than the previous term. So, the common difference is $-7$.

Add $-7$ to a term to find the next term.

| Position | 1 | 2 | 3 | 4 | 5 | 6 | 7 |
|---|---|---|---|---|---|---|---|
| Term | $-7$ | $-14$ | $-21$ | $-28$ | $-35$ | $-42$ | $-49$ |

$+(-7) \ +(-7) \ +(-7)$

∴ The next three terms are $-35, -42,$ and $-49$.

### On Your Own

**Now You're Ready**
Exercises 13–18

**Write the next three terms of the arithmetic sequence.**

**1.** $-12, 0, 12, 24, \ldots$  **2.** $0.2, 0.6, 1, 1.4, \ldots$  **3.** $4, 3\frac{3}{4}, 3\frac{1}{2}, 3\frac{1}{4}, \ldots$

**244**   Chapter 5   Linear Functions

# EXAMPLE 2 Graphing an Arithmetic Sequence

**Graph the arithmetic sequence 4, 8, 12, 16, . . . . What do you notice?**

Make a table. Then plot the ordered pairs $(n, a_n)$.

| Position, $n$ | Term, $a_n$ |
|---|---|
| 1 | 4 |
| 2 | 8 |
| 3 | 12 |
| 4 | 16 |

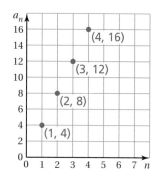

∴ The points of the graph lie on a line.

### On Your Own

**Now You're Ready**
Exercises 25–28

Write the next three terms of the arithmetic sequence. Then graph the sequence.

**4.** 3, 6, 9, 12, . . .   **5.** 4, 2, 0, −2, . . .   **6.** 1, 0.8, 0.6, 0.4, . . .

---

Because consecutive terms of an arithmetic sequence have a common difference, the sequence has a constant rate of change. So, the points of any arithmetic sequence lie on a line. You can use the first term and the common difference to write a linear function that describes an arithmetic sequence.

| Position, $n$ | Term, $a_n$ | Written using $a_1$ and $d$ | Numbers |
|---|---|---|---|
| 1 | first term, $a_1$ | $a_1$ | 4 |
| 2 | second term, $a_2$ | $a_1 + d$ | $4 + 4 = 8$ |
| 3 | third term, $a_3$ | $a_1 + 2d$ | $4 + 2(4) = 12$ |
| 4 | fourth term, $a_4$ | $a_1 + 3d$ | $4 + 3(4) = 16$ |
| ⋮ | ⋮ | ⋮ | ⋮ |
| $n$ | $n$th term, $a_n$ | $a_1 + (n-1)d$ | $4 + (n-1)(4)$ |

### Key Idea

**Equation for an Arithmetic Sequence**

Let $a_n$ be the $n$th term of an arithmetic sequence with first term $a_1$ and common difference $d$. The $n$th term is given by

$$a_n = a_1 + (n-1)d.$$

Section 5.6 Arithmetic Sequences

## EXAMPLE 3 Writing an Equation for an Arithmetic Sequence

Write an equation for the $n$th term of the arithmetic sequence 14, 11, 8, 5, . . . . Then find $a_{50}$.

The first term is 14 and the common difference is $-3$.

$a_n = a_1 + (n - 1)d$   Equation for an arithmetic sequence
$a_n = 14 + (n - 1)(-3)$   Substitute 14 for $a_1$ and $-3$ for $d$.
$a_n = -3n + 17$   Simplify.

**Study Tip**
Notice that the equation in Example 3 is of the form $y = mx + b$, where $y$ is replaced by $a_n$ and $x$ is replaced by $n$.

Use the equation to find the 50th term.

$a_n = -3n + 17$   Write the equation.
$a_{50} = -3(50) + 17$   Substitute 50 for $n$.
$= -133$   Simplify.

## EXAMPLE 4 Real-Life Application

Online bidding for a purse increases $5 for each bid after the $60 initial bid.

| Bid Number | 1 | 2 | 3 |
|---|---|---|---|
| Bid Amount | $60 | $65 | $70 |

**a.** Write an equation for the $n$th term of the arithmetic sequence.

The first term is 60 and the common difference is 5.

$a_n = a_1 + (n - 1)d$   Equation for an arithmetic sequence
$a_n = 60 + (n - 1)5$   Substitute 60 for $a_1$ and 5 for $d$.
$a_n = 5n + 55$   Simplify.

**b.** The winning bid is $90. How many bids were there?

Use the equation to find the value of $n$ for which $a_n = 90$.

$a_n = 5n + 55$   Write the equation.
$90 = 5n + 55$   Substitute 90 for $a_n$.
$35 = 5n$   Subtract 55 from each side.
$7 = n$   Divide each side by 5.

∴ There were 7 bids.

**Check**

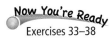

### On Your Own

Now You're Ready
Exercises 33–38

Write an equation for the $n$th term of the arithmetic sequence. Then find $a_{25}$.

**7.** 4, 5, 6, 7, . . .   **8.** 8, 16, 24, 32, . . .   **9.** $-2, -1, 0, 1, . . .$

**10. WHAT IF?** In Example 4, the winning bid is $105. How many bids were there?

## 5.6 Exercises

###  Vocabulary and Concept Check

1. **VOCABULARY** How do you find the common difference of an arithmetic sequence?

2. **WRITING** How are the graphs of arithmetic sequences and linear functions similar? How are they different?

###  Practice and Problem Solving

**Use the figures to complete the table. Then describe the pattern of the y-values.**

3.  n = 1   n = 2   n = 3   n = 4

| Number of Quarters, n | 1 | 2 | 3 | 4 |
|---|---|---|---|---|
| Number of Cents, y | | | | |

**Write the next three terms of the arithmetic sequence.**

4. First term: 2
   Common difference: 11

5. First term: 18
   Common difference: 3.5

6. First term: 0
   Common difference: $4\frac{1}{2}$

**Find the common difference of the arithmetic sequence.**

7. 5, 10, 15, 20, . . .

8. 16.1, 14.1, 12.1, 10.1, . . .

9. 100, 125, 150, 175, . . .

10. 3, $3\frac{1}{2}$, 4, $4\frac{1}{2}$, . . .

11. 6.5, 5, 3.5, 2, . . .

12. 350, 500, 650, 800, . . .

**Write the next three terms of the arithmetic sequence.**

13. 10, 13, 16, 19, . . .

14. 1, 12, 23, 34, . . .

15. 16, 21, 26, 31, . . .

16. 60, 30, 0, −30, . . .

17. 1.3, 1, 0.7, 0.4, . . .

18. $\frac{5}{6}, \frac{2}{3}, \frac{1}{2}, \frac{1}{3}, \ldots$

19. **PATTERN** Write a sequence to represent the number of smiley faces in each group. Is the sequence arithmetic? Explain.

**Determine whether the sequence is arithmetic. If so, find the common difference.**

**20.** 13, 26, 39, 52, . . .

**21.** 5, 9, 14, 20, . . .

**22.** 6, 12, 24, 48, . . .

**23.** 69, 75, 81, 87, . . .

**24. ERROR ANALYSIS** Describe and correct the error in finding the common difference of the arithmetic sequence.

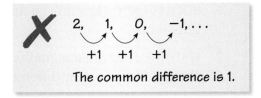

**Write the next three terms of the arithmetic sequence. Then graph the sequence.**

**25.** 7, 6.4, 5.8, 5.2, . . .

**26.** −15, 0, 15, 30, . . .

**27.** $\dfrac{1}{2}, \dfrac{5}{8}, \dfrac{3}{4}, \dfrac{7}{8}, \ldots$

**28.** −1, −3, −5, −7, . . .

**29. NUMBER SENSE** The first term of an arithmetic sequence is 3. The common difference of the sequence is 1.5 times the first term. Write the next three terms of the sequence. Then graph the sequence.

**30. DOMINOES** The first row of a dominoes display has 10 dominoes. Each row after the first has two more dominoes than the row before it. Write the first five terms of the sequence that represents the number of dominoes in each row. Then graph the sequence.

**31. ZOO** A zoo charges $8 per person for admission.

  **a.** Copy and complete the table.

  **b.** Do the costs in your table show an arithmetic sequence? If so, graph the sequence.

  **c.** What is the cost for one person to visit the zoo six times?

  **d.** An annual family pass costs $130. How many times does a family of five have to visit the zoo for the annual pass to be the better deal? Explain.

| Number of Visits in One Year | Cost |
| --- | --- |
| 1 | $8 |
| 2 | |
| 3 | |
| 4 | |

**32. REPEATED REASONING** Firewood is stacked in a pile. The bottom row has 20 logs and the top row has 14 logs. Each row has one more log than the row above it. How many logs are in the pile?

**Write an equation for the $n$th term of the arithmetic sequence. Then find $a_{10}$.**

**33.** −5, −4, −3, −2, . . .

**34.** −3, −6, −9, −12, . . .

**35.** $\dfrac{1}{2}, 1, 1\dfrac{1}{2}, 2, \ldots$

**36.** 10, 11, 12, 13, . . .

**37.** −10, −20, −30, −40, . . .

**38.** $\dfrac{1}{7}, \dfrac{2}{7}, \dfrac{3}{7}, \dfrac{4}{7}, \ldots$

**39. MOVIE REVENUE** A movie earns $100 million the first week it is released. The movie earns $20 million less each additional week. Write an equation for the $n$th term of the arithmetic sequence.

**40. REASONING** Are the terms of an arithmetic sequence independent or dependent? Explain your reasoning.

**41. SPEED** On a highway, you take 3 seconds to increase your speed from 32 to 35 miles per hour. Your speed increases the same amount each second.

   **a.** Write the first four terms of the sequence that represents your speed each second.
   **b.** Write an equation that describes the arithmetic sequence.
   **c.** The speed limit is 65 miles per hour. What is the domain of the function?

**42. OPEN-ENDED** Write the first four terms of two different arithmetic sequences with a common difference of −3. Write an equation for the $n$th term of each sequence.

**43. REASONING** Is the domain of an arithmetic sequence discrete or continuous? Describe the types of numbers in the domain.

**44. EARTH DAY** You and a group of friends take turns planting 2 trees each at a campsite. After the first person plants 2 trees, there are 12 trees at the campsite.

   **a.** Write an equation for the $n$th term of the sequence.
   **b.** What do you notice about the slope given by the equation and the common difference of the sequence?
   **c.** After 8 more people plant trees, how many trees are at the campsite?

**45.** *Critical Thinking* The number of births in a country each minute after midnight January 1st can be estimated by the sequence in the table.

   **a.** Write an equation for the $n$th term of the sequence.
   **b.** Is the domain discrete or continuous?
   **c.** Explain how to use your function to estimate the number of births in a day.

| Minutes after Midnight January 1st | 1 | 2 | 3 | 4 |
|---|---|---|---|---|
| Babies Born | 5 | 10 | 15 | 20 |

## Fair Game Review What you learned in previous grades & lessons

**Solve the system of linear equations by graphing.** *(Section 4.1)*

**46.** $y = 2x$
$y = 3x + 2$

**47.** $y = -2x + 6$
$y = \frac{1}{4}x - 3$

**48.** $y + x = 0$
$y + 2 = -\frac{1}{2}x$

**49. MULTIPLE CHOICE** What expression is equivalent to $4^5$? *(Skills Review Handbook)*

   **Ⓐ** $4 \cdot 5$  **Ⓑ** $4 \cdot 4 \cdot 4 \cdot 4$  **Ⓒ** $5^4$  **Ⓓ** $4 \cdot 4 \cdot 4 \cdot 4 \cdot 4$

# 5.4–5.6 Quiz

**Evaluate the function when $x = -4, 0,$ and $2$.** *(Section 5.4)*

1. $f(x) = x - 2$
2. $g(x) = 7x + 3$
3. $h(x) = -\dfrac{1}{4}x + 5$

**Graph the function. Compare the graph to the graph of $f(x) = 4x$.** *(Section 5.4)*

4. $g(x) = 4x + 1$
5. $h(x) = 4x - 2$
6. $n(x) = 4x - 6$

**Graph the function. Compare the graph to the graph of $y = |x|$. Describe the domain and range.** *(Section 5.4)*

7. $y = |x| + 2$
8. $y = |x - 6|$
9. $y = 2|x|$

**Does the table or graph represent a *linear* or *nonlinear* function? Explain.** *(Section 5.5)*

10.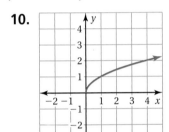
11.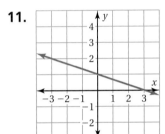
12.

| x | y |
|---|---|
| 0 | 3 |
| 3 | 0 |
| 6 | 3 |
| 9 | 6 |

**Write an equation for the $n$th term of the arithmetic sequence. Then find $a_{15}$.** *(Section 5.6)*

13. $5, 6, 7, 8, \ldots$
14. $-3, -2, -1, 0, \ldots$
15. $4, 8, 12, 16, \ldots$
16. $-1.5, -0.5, 0.5, 1.5, \ldots$

17. **HIGH-SPEED RAIL** A high-speed passenger train travels at 110 miles per hour. The function $d(x) = 1375 - 110x$ represents the distance (in miles) the train is from its destination after $x$ hours. How far is the train from its destination after 8 hours? *(Section 5.4)*

18. **CHICKEN SALAD** The equation $y = 7.9x$ represents the cost $y$ (in dollars) of buying $x$ pounds of chicken salad. Does this equation represent a linear or nonlinear function? Explain. *(Section 5.5)*

19. **PHONE BILL** The table shows your phone bill for each minute over your plan limit. *(Section 5.6)*

    a. Write an equation for the $n$th term of the arithmetic sequence.
    b. Your phone bill is $45.35. How many extra minutes were billed to your account?

| Extra Minute | 1 | 2 | 3 |
|---|---|---|---|
| Phone Bill | $40.40 | $40.85 | $41.30 |

# 5 Chapter Review

## Review Key Vocabulary

function, *p. 204*
domain, *p. 204*
range, *p. 204*
independent variable, *p. 204*
dependent variable, *p. 204*
relation, *p. 208*

Vertical Line Test, *p. 209*
discrete domain, *p. 212*
continuous domain, *p. 212*
linear function, *p. 218*
function notation, *p. 226*
piecewise function, *p. 232*

step function, *p. 233*
absolute value function, *p. 234*
nonlinear function, *p. 238*
sequence, *p. 244*
term, *p. 244*
arithmetic sequence, *p. 244*
common difference, *p. 244*

## Review Examples and Exercises

### 5.1 Domain and Range of a Function (pp. 202–209)

**Find the domain and range of the function represented by the graph.**

Write the ordered pairs. Identify the inputs and outputs.

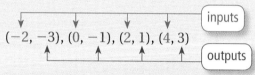

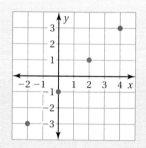

∴ The domain is $-2, 0, 2,$ and $4$.
The range is $-3, -1, 1,$ and $3$.

### Exercises

**Find the domain and range of the function represented by the graph.**

1.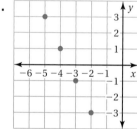

2.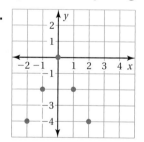

### 5.2 Discrete and Continuous Domains (pp. 210–215)

A yearbook costs $19.50. The graph shows the cost $y$ of $x$ yearbooks. Is the domain discrete or continuous?

Because you cannot buy part of a yearbook, the graph consists of individual points.

∴ So, the domain is discrete.

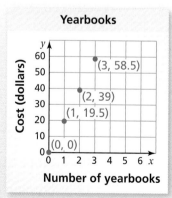

Chapter Review **251**

### Exercises

**Graph the function. Is the domain discrete or continuous?**

3. 
| Hours, x | 0 | 1 | 2 | 3 | 4 |
|---|---|---|---|---|---|
| Miles, y | 0 | 4 | 8 | 12 | 16 |

4. 
| Stamps, x | 20 | 40 | 60 | 80 | 100 |
|---|---|---|---|---|---|
| Cost, y | 8.4 | 16.8 | 25.2 | 33.6 | 42 |

## 5.3 Linear Function Patterns (pp. 216–221)

**Use the graph to write a linear function that relates y to x.**

The points lie on a line. Find the slope and y-intercept of the line.

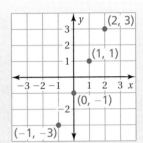

$$\text{slope} = \frac{\text{change in } y}{\text{change in } x} = \frac{3-1}{2-1} = \frac{2}{1} = 2$$

Because the line crosses the y-axis at $(0, -1)$, the y-intercept is $-1$.

∴ So, the linear function is $y = 2x - 1$.

### Exercises

**Use the graph or table to write a linear function that relates y to x.**

5.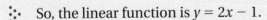

6. 
| x | −2 | 0 | 2 | 4 |
|---|---|---|---|---|
| y | −7 | −7 | −7 | −7 |

## 5.4 Function Notation (pp. 224–235)

**Evaluate $f(x) = 3x - 20$ when $x = 4$.**

$f(x) = 3x - 20$  Write the function.

$f(4) = 3(4) - 20$  Substitute 4 for x.

$= -8$  Simplify.

### Exercises

**Evaluate the function when $x = -5, 0,$ and $2$.**

7. $f(x) = 5x + 12$

8. $g(x) = -1.5x - 1$

9. $h(x) = 7 - 3x$

10. Compare the graph of $f(x) = -3x - 1$ to the graph of $g(x) = -3x$.

11. Compare the graph of $y = |x| + 1$ to the graph of $y = |x|$.

## 5.5 Comparing Linear and Nonlinear Functions (pp. 236–241)

**Does the table represent a *linear* or *nonlinear* function? Explain.**

a.
+2 +2 +2

| x | 0 | 2 | 4 | 6 |
|---|---|---|---|---|
| y | 0 | 1 | 4 | 9 |

+1 +3 +5

b.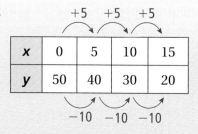

As *x* increases by 2, *y* increases by different amounts. The rate of change is *not* constant. So, the function is nonlinear.

As *x* increases by 5, *y* decreases by 10. The rate of change is constant. So, the function is linear.

### Exercises

**Does the table represent a *linear* or *nonlinear* function? Explain.**

12.
| x | 3 | 6 | 9 | 12 |
|---|---|---|---|---|
| y | 1 | 10 | 19 | 28 |

13.
| x | 1 | 3 | 5 | 7 |
|---|---|---|---|---|
| y | 3 | 1 | 1 | 3 |

## 5.6 Arithmetic Sequences (pp. 242–249)

**Write an equation for the *n*th term of the arithmetic sequence $-3, -5, -7, -9, \ldots$. Then find $a_{20}$.**

The first term is $-3$ and the common difference is $-2$.

$a_n = a_1 + (n-1)d$   Equation for an arithmetic sequence

$a_n = -3 + (n-1)(-2)$   Substitute $-3$ for $a_1$ and $-2$ for $d$.

$a_n = -2n - 1$   Simplify.

Use the equation to find the 20th term.

$a_{20} = -2(20) - 1$   Substitute 20 for *n*.

$= -41$   Simplify.

### Exercises

**Write an equation for the *n*th term of the arithmetic sequence. Then find $a_{30}$.**

14. $11, 10, 9, 8, \ldots$

15. $6, 12, 18, 24, \ldots$

16. $-9, -7, -5, -3, \ldots$

Chapter Review 253

# 5 Chapter Test

1. Find the domain and range of the function represented by the graph.

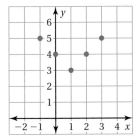

2. Graph the function. Is the domain discrete or continuous?

| Minutes, x | Gallons, y |
|---|---|
| 0 | 60 |
| 5 | 45 |
| 10 | 30 |
| 15 | 15 |

3. Use the graph to write a linear function that relates $y$ to $x$.

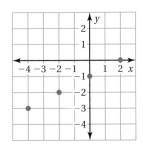

4. Does the table represent a *linear* or *nonlinear* function? Explain.

| x | 0 | 2 | 4 | 6 |
|---|---|---|---|---|
| y | 8 | 0 | −8 | −16 |

**Evaluate the function when $x = -3, 0,$ and $6$.**

5. $f(x) = 9x - 10$

6. $g(x) = 2.5x + 5$

7. $h(x) = 15 - 3x$

8. Compare the graph of $h(x) = 5x + 2$ to the graph of $f(x) = 5x$.

9. Compare the graph of $y = |x + 3| - 2$ to the graph of $y = |x|$.

10. Graph $f(x) = \begin{cases} -x, & \text{if } x \leq 0 \\ x + 5, & \text{if } x > 0 \end{cases}$. Describe the domain and range.

**Write an equation for the $n$th term of the arithmetic sequence. Then find $a_{25}$.**

11. $6, 12, 18, 24, \ldots$

12. $-6, -5, -4, -3, \ldots$

13. $3, 1, -1, -3, \ldots$

14. **FOOD DRIVE** You are putting cans of food into boxes for a food drive. One box holds 30 cans of food. Write a linear function using function notation that represents the number of cans of food that will fit in $x$ boxes. Is the domain discrete or continuous?

15. **SEATING** The first row of a theater has 20 seats. Each row after the first has two more seats than the row before it. Write an equation for the number of seats in the $n$th row. How many seats are in row 20?

16. **SURFACE AREA** A function relates the surface area $S$ (in square inches) of a cube to the side length $x$ (in inches) of the cube. Is the function linear or nonlinear? Explain.

# 5 Standards Assessment

**Test-Taking Strategy: Work Backwards**

1. The domain of the function $y = 0.2x - 5$ is 5, 10, 15, 20. What is the range of this function? *(8.F.1)*

    A. 20, 15, 10, 5

    B. 0, 5, 10, 15

    C. 4, 3, 2, 1

    D. −4, −3, −2, −1

2. A toy runs on a rechargeable battery. During use, the battery loses power at a constant rate. The percent $P$ of total power left in the battery after $x$ hours can be found using the equation shown below. When will the battery be fully discharged? *(A.REI.3)*

    $$P = -0.25x + 1$$

    F. After 4 hours of use

    G. After 1 hour of use

    H. After 0.75 hour of use

    I. After 0.25 hour of use

3. A limousine company charges a fixed cost for a limousine and an hourly rate for its driver. It costs $500 to rent the limousine for 5 hours and $800 to rent the limousine for 10 hours. What is the fixed cost, in dollars, to rent the limousine? *(8.F.4)*

4. Which graph shows a nonlinear function? *(F.LE.1b)*

    A.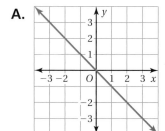

    B.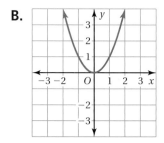

    C.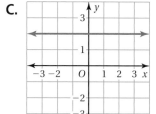

    D.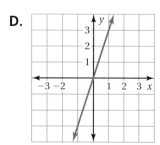

5. The equations $y = -x + 4$ and $y = \frac{1}{2}x - 8$ form a system of linear equations. The table below shows the y-value for each equation at six different values of x. (A.REI.6)

| x | 0 | 2 | 4 | 6 | 8 | 10 |
|---|---|---|---|---|---|---|
| $y = -x + 4$ | 4 | 2 | 0 | −2 | −4 | −6 |
| $y = \frac{1}{2}x - 8$ | −8 | −7 | −6 | −5 | −4 | −3 |

What can you conclude from the table?

F. The system has one solution, when $x = 0$.

G. The system has one solution, when $x = 4$.

H. The system has one solution, when $x = 8$.

I. The system has no solution.

6. The temperature fell from 54 degrees Fahrenheit to 36 degrees Fahrenheit over a six-hour period. The temperature fell by the same number of degrees each hour. How many degrees Fahrenheit did the temperature fall each hour? (A.CED.1)

7. What is the domain of the function graphed in the coordinate plane below? (8.F.1)

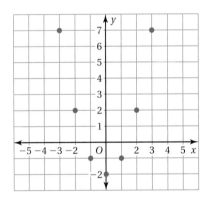

A. 0, 1, 2, 3

B. −2, −1, 2, 7

C. −3, −2, −1, 0, 1, 2, 3

D. −2, −1, 0, 1, 2, 3, 7

8. What value of w makes the equation below true? (A.REI.3)

$$\frac{w}{3} = 3(w - 1) - 1$$

F. $\frac{3}{2}$

G. $\frac{5}{4}$

H. $\frac{3}{4}$

I. $\frac{1}{2}$

9. What is the slope of the line shown in the graph below? *(F.IF.6)*

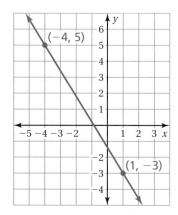

A. $-\dfrac{2}{5}$

B. $-\dfrac{2}{3}$

C. $-\dfrac{8}{5}$

D. $-\dfrac{8}{3}$

10. A line with a slope of $\dfrac{1}{3}$ passes through the point (6, 1). What is the equation of the line? *(A.CED.2)*

F. $y = \dfrac{1}{3}x$

G. $y = \dfrac{1}{3}x + 1$

H. $x - 3y = 3$

I. $x + 3y = 3$

11. The tables show how the perimeter and area of a square are related to its side length. Examine the data in the table. *(F.LE.1b)*

**Think Solve Explain**

| Side Length | 1 | 2 | 3 | 4 | 5 | 6 |
|---|---|---|---|---|---|---|
| Perimeter | 4 | 8 | 12 | 16 | 20 | 24 |

| Side Length | 1 | 2 | 3 | 4 | 5 | 6 |
|---|---|---|---|---|---|---|
| Area | 1 | 4 | 9 | 16 | 25 | 36 |

*Part A* Does the first table show a linear function? Explain your reasoning.

*Part B* Does the second table show a linear function? Explain your reasoning.

12. In many states, you must be at least 14 years old to operate a personal watercraft. Which inequality represents this situation? *(A.CED.1)*

A. $y > 14$

B. $y < 14$

C. $y \geq 14$

D. $y \leq 14$

Standards Assessment

# 6 Exponential Equations and Functions

- 6.1 Properties of Square Roots
- 6.2 Properties of Exponents
- 6.3 Radicals and Rational Exponents
- 6.4 Exponential Functions
- 6.5 Exponential Growth
- 6.6 Exponential Decay
- 6.7 Geometric Sequences

"If one flea had 100 babies, and each baby grew up and had 100 babies, ..."

"... and each of those babies grew up and had 100 babies, you would have 1,010,101 fleas."

"Here's how I remember the square root of 2."

"February is the 2nd month. It has 28 days. Split 28 into 14 and 14. Move the decimal to get 1.414."

# What You Learned Before

"It's called the Power of Negative One, Descartes!"

## Using Order of Operations (8.EE.1)

**Example 1** Evaluate $10^2 \div (30 \div 3) - 4(3-9) + 5^0$.

First: Parentheses $\quad 10^2 \div (30 \div 3) - 4(3-9) + 5^0 = 10^2 \div 10 - 4(-6) + 5^0$
Second: Exponents $\quad\quad\quad\quad\quad\quad\quad\quad\quad\quad\quad\quad\quad = 100 \div 10 - 4(-6) + 1$
Third: Multiplication and Division (from left to right) $\quad = 10 + 24 + 1$
Fourth: Addition and Subtraction (from left to right) $\quad\;\; = 35$

**Try It Yourself**
Evaluate the expression.

1. $12\left(\dfrac{14}{2}\right) - 3^3 + 15 - 2^0$
2. $5^2 \cdot 8 \div 2^2 + 20 \cdot 3 - 4$
3. $-7 + 16 \cdot 4^{-2} + (10 - 4^2)$

## Finding Square Roots (8.EE.2)

**Example 2** Find $-\sqrt{81}$.

$-\sqrt{81}$ represents the negative square root. Because $9^2 = 81$, $-\sqrt{81} = -\sqrt{9^2} = -9$.

**Try It Yourself**

4. Find $\sqrt{64}$.
5. Find $\pm\sqrt{121}$.
6. Find $-\sqrt{4}$.

## Writing an Equation for an Arithmetic Sequence (F.BF.2)

**Example 3** Write an equation for the $n$th term of the arithmetic sequence 5, 15, 25, 35, . . ..

The first term is 5 and the common difference is 10.

$a_n = a_1 + (n-1)d$ $\quad\quad$ Equation for an arithmetic sequence
$a_n = 5 + (n-1)10$ $\quad\quad$ Substitute 5 for $a_1$ and 10 for $d$.
$a_n = 10n - 5$ $\quad\quad\quad\quad$ Simplify.

**Try It Yourself**
Write an equation for the $n$th term of the arithmetic sequence.

7. 2, 4, 6, 8, . . .
8. 6, 3, 0, −3, . . .
9. 22, 15, 8, 1, . . .

# 6.1 Properties of Square Roots

**Essential Question** How can you multiply and divide square roots?

Recall that when you multiply a number by itself, you square the number.

> Symbol for squaring is 2nd power.

$4^2 = 4 \cdot 4$
$= 16$     4 squared is 16.

To "undo" this, take the square root of the number.

> Symbol for square root is a radical sign.

$\sqrt{16} = \sqrt{4^2} = 4$     The square root of 16 is 4.

## 1  ACTIVITY: Finding Square Roots

**Work with a partner.** Use a square root symbol to write the side length of the square. Then find the square root. Check your answer by multiplying.

**a. Sample:** $s = \sqrt{81} = 9$ ft

Area = 81 ft²

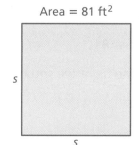

∴ The side length of the square is 9 feet.

**b.** Area = 121 yd²

**c.** Area = 324 cm²

**d.** Area = 361 mi²

**e.** Area = 2.89 in.²

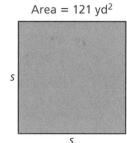

**f.** Area = 6.25 m²

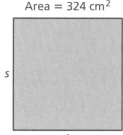

**g.** Area = $\frac{16}{25}$ ft²

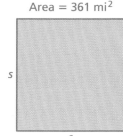

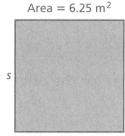
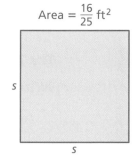

**COMMON CORE**

**Square Roots**
In this lesson, you will
- simplify and evaluate square roots.
- simplify radical expressions.

Preparing for Standard N.RN.3

260   Chapter 6   Exponential Equations and Functions

## 2 ACTIVITY: Operations with Square Roots

**Math Practice 3**

**Analyze Conjectures**
How do you know if your conclusion is accurate? How can you explain your conclusion?

Work with a partner. When you have an expression that involves two operations, you need to know whether you obtain the same result regardless of the order in which you perform the operations. In each of the following, compare the results obtained by the two orders. What can you conclude?

a. **Square Roots and Addition**

Is $\sqrt{36} + \sqrt{64}$ equal to $\sqrt{36 + 64}$?
In general, is $\sqrt{a} + \sqrt{b}$ equal to $\sqrt{a + b}$?
Explain your reasoning.

b. **Square Roots and Multiplication**

Is $\sqrt{4} \cdot \sqrt{9}$ equal to $\sqrt{4 \cdot 9}$?
In general, is $\sqrt{a} \cdot \sqrt{b}$ equal to $\sqrt{a \cdot b}$?
Explain your reasoning.

c. **Square Roots and Subtraction**

Is $\sqrt{64} - \sqrt{36}$ equal to $\sqrt{64 - 36}$?
In general, is $\sqrt{a} - \sqrt{b}$ equal to $\sqrt{a - b}$?
Explain your reasoning.

d. **Square Roots and Division**

Is $\dfrac{\sqrt{100}}{\sqrt{4}}$ equal to $\sqrt{\dfrac{100}{4}}$?
In general, is $\dfrac{\sqrt{a}}{\sqrt{b}}$ equal to $\sqrt{\dfrac{a}{b}}$?
Explain your reasoning.

## What Is Your Answer?

3. **IN YOUR OWN WORDS** How can you multiply and divide square roots? Write a rule for:

   a. The product of square roots
   b. The quotient of square roots

**Practice** — Use what you learned about square roots to complete Exercises 3–5 on page 264.

Section 6.1  Properties of Square Roots

## 6.1 Lesson

### Key Ideas

**Product Property of Square Roots**

Algebra

$\sqrt{xy} = \sqrt{x} \cdot \sqrt{y}$, where $x, y \geq 0$

Numbers

$\sqrt{9 \cdot 5} = \sqrt{9} \cdot \sqrt{5} = 3\sqrt{5}$

**Quotient Property of Square Roots**

Algebra

$\sqrt{\dfrac{x}{y}} = \dfrac{\sqrt{x}}{\sqrt{y}}$, where $x \geq 0$ and $y > 0$

Numbers

$\sqrt{\dfrac{3}{4}} = \dfrac{\sqrt{3}}{\sqrt{4}} = \dfrac{\sqrt{3}}{2}$

### EXAMPLE 1  Simplifying Square Roots

a.  $\sqrt{150} = \sqrt{25 \cdot 6}$     Factor using the greatest perfect square factor.

$\phantom{\sqrt{150}} = \sqrt{25} \cdot \sqrt{6}$     Product Property of Square Roots

$\phantom{\sqrt{150}} = 5\sqrt{6}$     Simplify.

**Remember**

A square root is simplified when the radicand has no perfect square factors other than 1.

b.  $\sqrt{\dfrac{15}{64}} = \dfrac{\sqrt{15}}{\sqrt{64}}$     Quotient Property of Square Roots

$\phantom{\sqrt{\dfrac{15}{64}}} = \dfrac{\sqrt{15}}{8}$     Simplify.

### EXAMPLE 2  Evaluating Square Roots

Evaluate $\sqrt{b^2 - 4ac}$ when $a = 2$, $b = -8$, and $c = 4$.

$\sqrt{b^2 - 4ac} = \sqrt{(-8)^2 - 4(2)(4)}$     Substitute.

$\phantom{\sqrt{b^2 - 4ac}} = \sqrt{32}$     Simplify.

$\phantom{\sqrt{b^2 - 4ac}} = \sqrt{16 \cdot 2}$     Factor.

$\phantom{\sqrt{b^2 - 4ac}} = \sqrt{16} \cdot \sqrt{2}$     Product Property of Square Roots

$\phantom{\sqrt{b^2 - 4ac}} = 4\sqrt{2}$     Simplify.

### On Your Own

**Now You're Ready**
Exercises 6–17

Simplify the expression.

1. $\sqrt{\dfrac{23}{9}}$     2. $-\sqrt{80}$     3. $\sqrt{\dfrac{27}{100}}$

4. Evaluate $\sqrt{b^2 - 4ac}$ when $a = 2$, $b = -6$, and $c = -5$.

## EXAMPLE 3 Simplifying Radical Expressions

Simplify $\dfrac{6 + \sqrt{8}}{2}$.

$$\dfrac{6 + \sqrt{8}}{2} = \dfrac{6 + \sqrt{4 \cdot 2}}{2} \qquad \text{Factor the radicand.}$$

$$= \dfrac{6 + \sqrt{4} \cdot \sqrt{2}}{2} \qquad \text{Product Property of Square Roots}$$

$$= \dfrac{6 + 2\sqrt{2}}{2} \qquad \text{Simplify.}$$

$$= 3 + \sqrt{2} \qquad \text{Divide.}$$

## EXAMPLE 4 Real-Life Application

The circumference $C$ of the art room in a mansion is given by the formula $C = 2\pi\sqrt{\dfrac{a^2 + b^2}{2}}$. Find the circumference of the room.

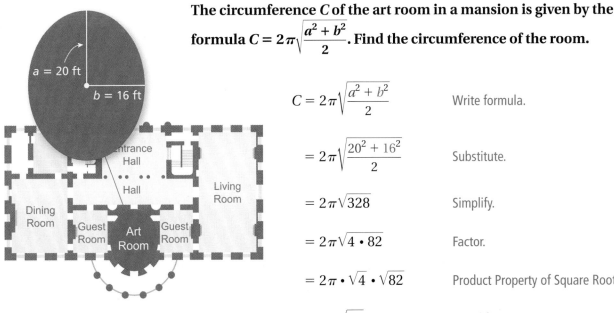

$$C = 2\pi\sqrt{\dfrac{a^2 + b^2}{2}} \qquad \text{Write formula.}$$

$$= 2\pi\sqrt{\dfrac{20^2 + 16^2}{2}} \qquad \text{Substitute.}$$

$$= 2\pi\sqrt{328} \qquad \text{Simplify.}$$

$$= 2\pi\sqrt{4 \cdot 82} \qquad \text{Factor.}$$

$$= 2\pi \cdot \sqrt{4} \cdot \sqrt{82} \qquad \text{Product Property of Square Roots}$$

$$= 4\pi\sqrt{82} \qquad \text{Simplify.}$$

∴ The circumference of the room is $4\pi\sqrt{82}$, or about 114 feet.

### On Your Own

**Now You're Ready**
Exercises 21–26

Simplify the expression.

5. $\dfrac{8 + \sqrt{32}}{2}$

6. $\dfrac{-1 - \sqrt{27}}{4}$

7. $\dfrac{2 - \sqrt{28}}{2(3)}$

8. Use the formula in Example 4 to find the circumference of an ellipse in which $a = 14$ feet and $b = 6$ feet.

Section 6.1   Properties of Square Roots

# 6.1 Exercises

## Vocabulary and Concept Check

1. **WRITING** How do you know when the square root of a positive integer is simplified?

2. **WRITING** How is the Product Property of Square Roots similar to the Quotient Property of Square Roots?

## Practice and Problem Solving

**Find the dimensions of the square. Check your answer.**

3. Area = 64 ft²

4. Area = 144 in.²

5. Area = $\frac{9}{16}$ cm²

**Simplify the expression.**

6. $\sqrt{18}$

7. $-\sqrt{200}$

8. $\sqrt{12}$

9. $\sqrt{48}$

10. $\sqrt{125}$

11. $-\sqrt{\frac{23}{64}}$

12. $-\sqrt{\frac{65}{121}}$

13. $\sqrt{\frac{18}{49}}$

14. $\sqrt{\frac{25}{36}}$

**Evaluate the expression when $x = -2$, $y = 8$, and $z = \frac{1}{2}$.**

15. $\sqrt{x^2 + yz}$

16. $\sqrt{2x^2 + y^2}$

17. $\sqrt{y - 44xz}$

18. **ERROR ANALYSIS** Describe and correct the error in simplifying the expression.

$$\sqrt{\frac{20}{9}} = \frac{\sqrt{20}}{\sqrt{9}} = \frac{\sqrt{20}}{3}$$

19. **ELECTRICITY** The electric current $I$ (in amperes) an appliance uses is given by the formula $I = \sqrt{\frac{P}{R}}$, where $P$ is the power (in watts) and $R$ is the resistance (in ohms). Find the current an appliance uses when the power is 147 watts and the resistance is 4 ohms.

20. **BASEBALL** You drop a baseball from a height of 56 feet. Use the expression $\sqrt{\dfrac{h}{16}}$, where $h$ is the height (in feet), to find the time (in seconds) it takes the baseball to hit the ground.

**Simplify the expression.**

❸ 21. $\dfrac{6 + \sqrt{44}}{2}$

22. $\dfrac{-7 - \sqrt{98}}{7}$

23. $\dfrac{10 + \sqrt{300}}{5}$

24. $\dfrac{-3 - \sqrt{80}}{6}$

25. $\dfrac{2 + \sqrt{28}}{4}$

26. $\dfrac{-4 + \sqrt{32}}{-2(5)}$

27. **VOLUME** A pet store installs a new aquarium in your teacher's classroom. What is the volume of the aquarium?

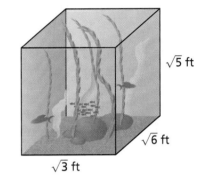

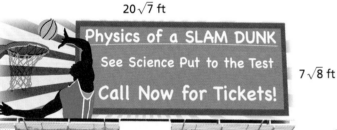

28. **BILLBOARD** What is the area of the rectangular billboard?

**Simplify the expression. Assume all variables are positive.**

29. $\sqrt{42x^2y^2}$

30. $\sqrt{25y^2z}$

31. $\sqrt{18x^3y^2z}$

32. **Modeling** Write an equation that represents the side length $s$ of a cube as a function of the surface area $A$ of the cube. Find the side length when the surface area is 72 square feet.

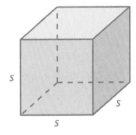

**Fair Game Review** What you learned in previous grades & lessons

**Evaluate the expression.** *(Skills Review Handbook)*

33. $3^5$

34. $2^4$

35. $5^3$

36. **MULTIPLE CHOICE** Which value is equivalent to $6(0.2)^3$? *(Skills Review Handbook)*

Ⓐ 0.008  Ⓑ 0.048  Ⓒ 1.728  Ⓓ 3.6

# Extension 6.1 Real Number Operations

**Key Vocabulary**
closed, p. 266

A set of numbers is **closed** under an operation when the operation performed on any two numbers in the set results in a number that is also in the set. For example, the set of integers is closed under addition, subtraction, and multiplication. This means that if $a$ and $b$ are two integers, then $a + b$, $a - b$, and $ab$ are also integers.

### ACTIVITY 1: Sums and Products of Rational Numbers

The table shows several sums and products of rational numbers. Complete the table.

**Remember**
A *rational number* is a number that can be written as $\frac{a}{b}$, where $a$ and $b$ are integers and $b \neq 0$. An *irrational number* cannot be written as the ratio of two integers.

| Sum or Product | Answer | Rational or Irrational? |
|---|---|---|
| $12 + 5$ | | |
| $-4 + 9$ | | |
| $\frac{4}{5} + \frac{2}{3}$ | | |
| $0.74 + 2.1$ | | |
| $3 \times 8$ | | |
| $-4 \times 6$ | | |
| $3.1 \times 0.6$ | | |
| $\frac{3}{4} \times \frac{5}{7}$ | | |

### ACTIVITY 2: Sums of Rational and Irrational Numbers

The table shows several sums of rational and irrational numbers. Complete the table.

| Sum | Answer | Rational or Irrational? |
|---|---|---|
| $1 + \sqrt{5}$ | | |
| $\sqrt{2} + \frac{5}{6}$ | | |
| $4 + \pi$ | | |
| $-8 + \sqrt{10}$ | | |

## Practice

1. Using the results in Activity 1, do you think the set of rational numbers is closed under addition? under multiplication? Explain your reasoning.

2. Using the results in Activity 2, what do you notice about the sum of a rational number and an irrational number?

## ACTIVITY 3 — Products of Rational and Irrational Numbers

**Common Core**

**Real Number Operations**
In this extension, you will
- determine whether sums or products are rational or irrational.

Learning Standard
N.RN.3

The table shows several products of rational and irrational numbers. Complete the table.

| Product | Answer | Rational or Irrational? |
|---|---|---|
| $6 \cdot \sqrt{12}$ | | |
| $-2 \cdot \pi$ | | |
| $\dfrac{2}{5} \cdot \sqrt{3}$ | | |
| $0 \times \sqrt{6}$ | | |

## ACTIVITY 4 — Sums and Products of Irrational Numbers

The table shows several sums and products of irrational numbers. Complete the table.

| Sum or Product | Answer | Rational or Irrational? |
|---|---|---|
| $3\sqrt{2} + 5\sqrt{2}$ | | |
| $\sqrt{12} + \sqrt{27}$ | | |
| $\sqrt{7} + \pi$ | | |
| $-\pi + \pi$ | | |
| $\pi \cdot \sqrt{7}$ | | |
| $\sqrt{5} \times \sqrt{2}$ | | |
| $4\pi \cdot \sqrt{3}$ | | |
| $\sqrt{3} \times \sqrt{3}$ | | |

## Practice

3. Using the results in Activity 3, is the product of a rational number and an irrational number always irrational? Explain.

4. Using the results in Activity 4, do you think the set of irrational numbers is closed under addition? under multiplication? Explain your reasoning.

5. **CRITICAL THINKING** Is the set of irrational numbers closed under division? If not, find a counterexample. (A *counterexample* is an example that shows that a statement is false.)

6. **STRUCTURE** The set of integers is closed under addition and multiplication. Use this information to show that the sum and product of two rational numbers are always rational numbers.

Extension 6.1  Real Number Operations

## 6.2 Properties of Exponents

**Essential Question** How can you use inductive reasoning to observe patterns and write general rules involving properties of exponents?

### 1 ACTIVITY: Writing a Rule for Products of Powers

Work with a partner. Write the product of the two powers as a single power. Then, write a *general rule* for finding the product of two powers with the same base.

a. **Sample:** $(3^4)(3^3) = (3 \cdot 3 \cdot 3 \cdot 3)(3 \cdot 3 \cdot 3) = 3^7$

b. $(2^2)(2^3) =$

c. $(4^1)(4^5) =$

d. $(5^3)(5^5) =$

e. $(x^2)(x^6) =$

### 2 ACTIVITY: Writing a Rule for Quotients of Powers

Work with a partner. Write the quotient of the two powers as a single power. Then, write a *general rule* for finding the quotient of two powers with the same base.

a. **Sample:** $\dfrac{3^4}{3^2} = \dfrac{3 \cdot 3 \cdot \cancel{3} \cdot \cancel{3}}{\cancel{3} \cdot \cancel{3}} = 3^2$

b. $\dfrac{4^3}{4^2} =$

c. $\dfrac{2^5}{2^2} =$

d. $\dfrac{x^6}{x^3} =$

e. $\dfrac{3^4}{3^4} =$

**COMMON CORE**

**Exponents**
In this lesson, you will
- simplify expressions using the properties of exponents.

Learning Standard
N.RN.2

### 3 ACTIVITY: Writing a Rule for Powers of Powers

Work with a partner. Write the expression as a single power. Then, write a *general rule* for finding a power of a power.

a. **Sample:** $(3^2)^3 = (3 \cdot 3)(3 \cdot 3)(3 \cdot 3) = 3^6$

b. $(2^2)^4 =$

c. $(7^3)^2 =$

d. $(y^3)^3 =$

e. $(x^4)^2 =$

268   Chapter 6   Exponential Equations and Functions

## 4  ACTIVITY: Writing a Rule for Powers of Products

**Math Practice 7**

**View as Components**
What are the different parts of the expressions? How does this help you rewrite the product?

Work with a partner. Write the expression as the product of two powers. Then, write a *general rule* for finding a power of a product.

a. **Sample:** $(2 \cdot 3)^3 = (2 \cdot 3)(2 \cdot 3)(2 \cdot 3) = (2^3)(3^3)$

b. $(2 \cdot 5)^2 = $ _____   c. $(5 \cdot 4)^3 = $ _____

d. $(6a)^4 = $ _____   e. $(3x)^2 = $ _____

## 5  ACTIVITY: Writing a Rule for Powers of Quotients

Work with a partner. Write the expression as the quotient of two powers. Then, write a *general rule* for finding a power of a quotient.

a. **Sample:** $\left(\dfrac{3}{2}\right)^4 = \dfrac{3}{2} \cdot \dfrac{3}{2} \cdot \dfrac{3}{2} \cdot \dfrac{3}{2} = \dfrac{3 \cdot 3 \cdot 3 \cdot 3}{2 \cdot 2 \cdot 2 \cdot 2} = \dfrac{3^4}{2^4}$

b. $\left(\dfrac{2}{3}\right)^2 = $ _____   c. $\left(\dfrac{4}{3}\right)^3 = $ _____

d. $\left(\dfrac{x}{2}\right)^3 = $ _____   e. $\left(\dfrac{a}{b}\right)^4 = $ _____

## What Is Your Answer?

6. **IN YOUR OWN WORDS** How can you use inductive reasoning to observe patterns and write general rules involving properties of exponents?

7. There are $3^3$ small cubes in the cube below. Write an expression for the number of small cubes in the large cube at the right.

**Practice** — Use what you learned about exponents to complete Exercises 6–11 on page 273.

## 6.2 Lesson

### Key Ideas

**Product of Powers Property**

**Words** To multiply powers with the same base, add their exponents.

**Numbers** $4^6 \cdot 4^3 = 4^{6+3} = 4^9$  **Algebra** $a^m \cdot a^n = a^{m+n}$

**Quotient of Powers Property**

**Words** To divide powers with the same base, subtract their exponents.

**Numbers** $\dfrac{4^6}{4^3} = 4^{6-3} = 4^3$  **Algebra** $\dfrac{a^m}{a^n} = a^{m-n}$, where $a \ne 0$

**Power of a Power Property**

**Words** To find a power of a power, multiply the exponents.

**Numbers** $(4^6)^3 = 4^{6 \cdot 3} = 4^{18}$  **Algebra** $(a^m)^n = a^{mn}$

**Remember**
For any integer $n$ and any nonzero integer $a$, $a^0 = 1$ and $a^{-n} = \dfrac{1}{a^n}$.

### EXAMPLE 1  Using Properties of Exponents

Simplify. Write your answer using only positive exponents.

a. $3^2 \cdot 3^6 = 3^{2+6}$    Product of Powers Property
      $= 3^8$    Simplify.

> The base is 3. Add the exponents.

b. $\dfrac{(-4)^2}{(-4)^7} = (-4)^{2-7}$    Quotient of Powers Property
      $= (-4)^{-5}$    Simplify.
      $= \dfrac{1}{(-4)^5}$    Definition of negative exponent

> The base is $-4$. Subtract the exponents.

c. $(z^4)^{-3} = z^{4 \cdot (-3)}$    Power of a Power Property
      $= z^{-12}$    Simplify.
      $= \dfrac{1}{z^{12}}$    Definition of negative exponent

> The base is $z$. Multiply the exponents.

### On Your Own

Exercises 12–17

Simplify. Write your answer using only positive exponents.

1. $10^4 \cdot 10^{-6}$
2. $x^9 \cdot x^{-9}$
3. $\dfrac{-5^8}{-5^4}$
4. $\dfrac{y^6}{y^7}$
5. $(6^{-2})^{-5}$
6. $(w^{12})^5$

## Key Ideas

**Power of a Product Property**

**Words** To find a power of a product, find the power of each factor and multiply.

**Numbers** $(3 \cdot 2)^5 = 3^5 \cdot 2^5$  **Algebra** $(ab)^m = a^m b^m$

**Power of a Quotient Property**

**Words** To find a power of a quotient, find the power of the numerator and the power of the denominator and divide.

**Numbers** $\left(\dfrac{3}{2}\right)^5 = \dfrac{3^5}{2^5}$  **Algebra** $\left(\dfrac{a}{b}\right)^m = \dfrac{a^m}{b^m}$, where $b \neq 0$

### EXAMPLE 2  Using Properties of Exponents

**Simplify. Write your answer using only positive exponents.**

a. $(-1.5y)^2 = (-1.5)^2 \cdot y^2$  Power of a Product Property
$\phantom{(-1.5y)^2} = 2.25y^2$  Simplify.

b. $\left(\dfrac{a}{-10}\right)^3 = \dfrac{a^3}{(-10)^3}$  Power of a Quotient Property
$\phantom{\left(\dfrac{a}{-10}\right)^3} = -\dfrac{a^3}{1000}$  Simplify.

c. $\left(\dfrac{2x}{3}\right)^{-5} = \dfrac{(2x)^{-5}}{3^{-5}}$  Power of a Quotient Property
$\phantom{\left(\dfrac{2x}{3}\right)^{-5}} = \dfrac{3^5}{(2x)^5}$  Definition of negative exponent
$\phantom{\left(\dfrac{2x}{3}\right)^{-5}} = \dfrac{3^5}{2^5 x^5}$  Power of a Product Property
$\phantom{\left(\dfrac{2x}{3}\right)^{-5}} = \dfrac{243}{32x^5}$  Simplify.

### On Your Own

**Now You're Ready**
Exercises 21–26

**Simplify. Write your answer using only positive exponents.**

7. $(10y)^{-3}$
8. $\left(-\dfrac{4}{n}\right)^5$
9. $\left(\dfrac{1}{2k^2}\right)^5$
10. $\left(\dfrac{6c}{7}\right)^{-2}$

**EXAMPLE 3  Simplifying an Expression**

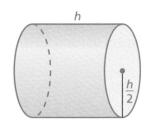

**Which expression represents the volume of the cylinder?**

Ⓐ $\dfrac{h^2}{2}$   Ⓑ $\dfrac{\pi h^2}{4}$   Ⓒ $\dfrac{\pi h^3}{2}$   Ⓓ $\dfrac{\pi h^3}{4}$

$$V = \pi r^2 h \quad \text{Formula for volume of a cylinder}$$
$$= \pi\left(\dfrac{h}{2}\right)^2 (h) \quad \text{Substitute } \dfrac{h}{2} \text{ for } r.$$
$$= \pi\left(\dfrac{h^2}{2^2}\right)(h) \quad \text{Power of a Quotient Property}$$
$$= \dfrac{\pi h^3}{4} \quad \text{Simplify.}$$

∴ The correct answer is Ⓓ.

**EXAMPLE 4  Real-Life Application**

A jellyfish emits about $1.25 \times 10^8$ particles of light, or photons, in $6.25 \times 10^{-4}$ second. How many photons does the jellyfish emit each second? Write your answer in scientific notation and in standard form.

Divide to find the unit rate.

$$\dfrac{1.25 \times 10^8}{6.25 \times 10^{-4}} \quad \begin{array}{l}\leftarrow \text{photons} \\ \leftarrow \text{seconds}\end{array} \quad \text{Write the rate.}$$

$$= \dfrac{1.25}{6.25} \times \dfrac{10^8}{10^{-4}} \quad \text{Rewrite.}$$

$$= 0.2 \times 10^{12} \quad \text{Simplify.}$$

$$= 2 \times 10^{11} \quad \text{Write in scientific notation.}$$

∴ The jellyfish emits $2 \times 10^{11}$, or 200,000,000,000 photons per second.

> **Remember**
> A number is written in scientific notation when it is of the form $a \times 10^b$, where $1 \leq a < 10$ and $b$ is an integer.

### On Your Own

**11.** In Example 3, which expression represents the area of a base of the cylinder?

**12.** It takes the Sun about $2.3 \times 10^8$ years to orbit the center of the Milky Way. It takes Pluto about $2.5 \times 10^2$ years to orbit the Sun. How many times does Pluto orbit the Sun while the Sun completes one orbit around the Milky Way? Write your answer in scientific notation.

## 6.2 Exercises

### Vocabulary and Concept Check

**MATCHING** Match the property with its example.

1. Quotient of Powers Property
2. Power of a Power Property
3. Power of a Quotient Property
4. Power of a Product Property

A. $(4^5)^2 = 4^{5 \cdot 2}$  B. $\left(\dfrac{5}{2}\right)^4 = \dfrac{5^4}{2^4}$  C. $(5 \cdot 2)^4 = 5^4 \cdot 2^4$  D. $\dfrac{4^5}{4^2} = 4^{5-2}$

5. **DIFFERENT WORDS, SAME QUESTION** Which is different? Find "both" answers.

Simplify $3^3 \cdot 3^6$.   Simplify $3^{3+6}$.   Simplify $3^{6-3}$.   Simplify $3^6 \cdot 3^3$.

### Practice and Problem Solving

Simplify the expression.

6. $(n^4)(n^3)$
7. $\dfrac{x^5}{x^3}$
8. $(c^5)^3$
9. $(4b)^3$
10. $\left(\dfrac{k}{3}\right)^5$
11. $\dfrac{(2a)^6}{a^2}$

Simplify. Write your answer using only positive exponents.

12. $8^{-2} \cdot 8^7$
13. $b^4 \cdot b^7$
14. $\dfrac{12^7}{12^2}$
15. $\dfrac{d^5}{d^8}$
16. $(5^5)^4$
17. $(x^3)^{-2}$

**ERROR ANALYSIS** Describe and correct the error in simplifying the expression.

18. 
$$x^5 \cdot x^{-2} = x^{5 \cdot (-2)}$$
$$= x^{-10}$$
$$= \dfrac{1}{x^{10}}$$

19. 
$$(m^3)^4 = m^{3+4}$$
$$= m^7$$

20. **MICROSCOPE** A microscope magnifies an object $10^5$ times. The length of an object is $10^2$ nanometers. What is its magnified length?

**Simplify. Write your answer using only positive exponents.**

21. $(6.2y)^2$

22. $\left(\dfrac{w}{4}\right)^4$

23. $\left(-\dfrac{6}{d}\right)^{-2}$

24. $(7p)^{-3}$

25. $(-5x)^5$

26. $\left(\dfrac{3n^3}{4}\right)^2$

27. **ERROR ANALYSIS** Describe and correct the error in simplifying the expression.

$$\left(\dfrac{x^3}{3}\right)^2 = \dfrac{(x^3)^2}{3} = \dfrac{x^6}{3}$$

28. **OPEN-ENDED** Use the properties of exponents to write three expressions equivalent to $x^8$.

29. **REASONING** Are the expressions $(a^4)^2$ and $a^{4^2}$ equivalent? Explain your reasoning.

30. **GEOMETRY** Consider Cube A and Cube B.

    a. Which property of exponents should you use to find the volume of each cube?

    b. How can you use the Power of a Quotient Property to find how many times greater the volume of Cube B is than the volume of Cube A?

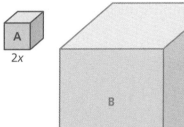

31. **SPHERE** The volume $V$ of a sphere is $V = \dfrac{4}{3}\pi r^3$, where $r$ is the radius. What is the volume of the sphere in terms of $m$ and $\pi$?

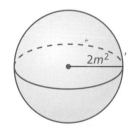

32. **PROBABILITY** The probability of rolling a 6 on a number cube is $\dfrac{1}{6}$. The probability of rolling a 6 twice in a row is $\left(\dfrac{1}{6}\right)^2 = \dfrac{1}{36}$.

    a. Write an expression that represents the probability of rolling a 6 $n$ times in a row.

    b. What is the probability of rolling a 6 five times in a row?

    c. What is the probability of flipping heads on a coin five times in a row?

**Evaluate the expression. Write your answer in scientific notation.**

**33.** $(3.4 \times 10^2)(1.5 \times 10^{-5})$     **34.** $(6.1 \times 10^{-3})(8 \times 10^9)$     **35.** $(4.8 \times 10^{-4})(7.2 \times 10^{-6})$

**36.** $\dfrac{(3 \times 10^3)}{(4 \times 10^5)}$     **37.** $\dfrac{(6.4 \times 10^{-7})}{(1.6 \times 10^{-5})}$     **38.** $\dfrac{(3.9 \times 10^{-5})}{(7.8 \times 10^{-8})}$

**Simplify. Write your answer using only positive exponents.**

**39.** $(6x^2y^{-4})^{-3}$     **40.** $\dfrac{(2m)^{-2}n^5}{-m^4n^{-3}}$     **41.** $\dfrac{15b^{-3}c^4}{(6b^{-4}c^{-5})^2}$

**42. REASONING** Write $8x^3y^3$ as the power of a product.

**43. COMPUTER CHIP** The area of a rectangular computer chip is $112a^3b^2$ square microns. The width is $8ab$ microns. What is the length?

**44. PROBLEM SOLVING** The speed of light is approximately $3 \times 10^5$ kilometers per second. The table shows the average distance each planet is from the Sun. How long does it take sunlight to reach Earth? Jupiter? Neptune?

**45. RICHTER SCALE** The Richter Scale is used to compare the intensities of earthquakes. An increase of 1 in magnitude on the Richter Scale represents a tenfold increase in intensity. An earthquake registers 7.4 on the Richter Scale and is followed by an aftershock that is 1000 times less intense. What is the magnitude of the aftershock?

| Planet | Average Distance from the Sun (km) |
|---|---|
| Mercury | $5.8 \times 10^7$ |
| Venus | $1.1 \times 10^8$ |
| Earth | $1.5 \times 10^8$ |
| Mars | $2.3 \times 10^8$ |
| Jupiter | $7.8 \times 10^8$ |
| Saturn | $1.4 \times 10^9$ |
| Uranus | $2.9 \times 10^9$ |
| Neptune | $4.5 \times 10^9$ |

**46. Precision** Find $x$ and $y$ when $\dfrac{k^{2x}}{k^y} = k^{13}$ and $(k^x k^{2y})^2 = k^{28}$. Explain how you found your answer.

## Fair Game Review   What you learned in previous grades & lessons

**Simplify the expression.** *(Section 6.1)*

**47.** $\sqrt{48}$     **48.** $\sqrt{\dfrac{70}{36}}$     **49.** $\sqrt{\dfrac{180}{121}}$

**50. MULTIPLE CHOICE** Which of the following is the solution of $\dfrac{x}{3} < -6$? *(Section 3.3)*

    **Ⓐ** $x > -2$     **Ⓑ** $x < -2$     **Ⓒ** $x > -18$     **Ⓓ** $x < -18$

# 6.3 Radicals and Rational Exponents

**Essential Question** How can you write and evaluate an *n*th root of a number?

Recall that you cube a number as follows.

Symbol for cubing is 3rd power.

$2^3 = 2 \cdot 2 \cdot 2$
$= 8$       2 cubed is 8.

To "undo" this, take the cube root of the number.

Symbol for cube root is $\sqrt[3]{\phantom{x}}$.

$\sqrt[3]{8} = \sqrt[3]{2^3} = 2$       The cube root of 8 is 2.

## 1 ACTIVITY: Finding Cube Roots

**Work with a partner.** Use a cube root symbol to write the side length of the cube. Then find the cube root. Check your answer by multiplying. Which cube is the largest? Which two are the same size? Explain your reasoning.

a. Volume = 27 ft³
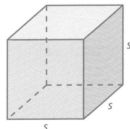

b. Volume = 125 cm³
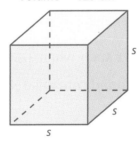

c. Volume = 3375 in.³
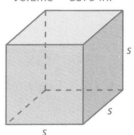

d. Volume = 3.375 m³
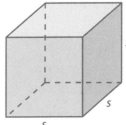

e. Volume = 1 yd³
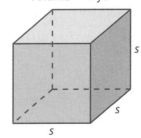

f. Volume = $\frac{125}{8}$ mm³
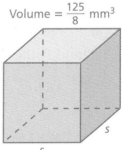

Cubes are not drawn to scale.

**COMMON CORE**

**Exponents**
In this lesson, you will
- simplify expressions with rational exponents.

Learning Standards
N.RN.1
N.RN.2

## 2 ACTIVITY: Estimating nth Roots

Work with a partner. When you raise an nth root of a number to the nth power, you get the original number.

$$(\sqrt[n]{a})^n = a$$

**Sample:** The 4th root of 16 is 2 because $2^4 = 16$.

$$\sqrt[4]{16} = 2$$

**Check:** $2^4 = 2 \cdot 2 \cdot 2 \cdot 2 = 16$ ✓

Match the nth root with the point on the number line. Justify your answer.

a. $\sqrt[4]{25}$     b. $\sqrt{0.5}$     c. $\sqrt[5]{2.5}$

d. $\sqrt[3]{65}$     e. $\sqrt[3]{55}$     f. $\sqrt[6]{20{,}000}$

## What Is Your Answer?

3. **IN YOUR OWN WORDS** How can you write and evaluate the nth root of a number?

4. The body mass m (in kilograms) of a dinosaur that walked on two feet can be modeled by

   $$m = (0.00016)C^{2.73}$$

   where C is the circumference (in millimeters) of the dinosaur's femur. The mass of a Tyrannosaurus rex was 4000 kilograms. What was the circumference of its femur?

**Math Practice 3**

Justify Conclusions

What information can you use to justify your answer?

Femur

**Practice** — Use what you learned about cube roots to complete Exercises 3–5 on page 280.

Section 6.3   Radicals and Rational Exponents   277

# 6.3 Lesson

**Key Vocabulary**
nth root, p. 278

When $b^n = a$ for an integer $n$ greater than 1, $b$ is an **nth root** of $a$.

$$\sqrt[n]{a} \quad \text{nth root of } a$$

The nth roots of a number may be real numbers or *imaginary numbers*. You will study imaginary numbers in a future course.

### EXAMPLE 1 — Finding nth Roots

Simplify each expression.

a. $\sqrt[3]{64}$

$\sqrt[3]{64} = \sqrt[3]{4 \cdot 4 \cdot 4}$
$= 4$

b. $\sqrt[4]{81}$

$\sqrt[4]{81} = \sqrt[4]{3 \cdot 3 \cdot 3 \cdot 3}$
$= 3$

**Study Tip**

In Example 1b, although $3^4 = 81$ and $(-3)^4 = 81$, $\sqrt[4]{81} = 3$ because the radical symbol indicates the positive root.

 **Key Idea**

**Rational Exponents**

**Words** The nth root of a positive number $a$ can be written as a power with base $a$ and an exponent of $1/n$.

**Numbers** $\sqrt[4]{81} = 81^{1/4}$ **Algebra** $\sqrt[n]{a} = a^{1/n}$

### EXAMPLE 2 — Simplifying Expressions with Rational Exponents

Simplify each expression.

a. $400^{1/2}$

$400^{1/2} = \sqrt{400}$    Write the expression in radical form.
$= \sqrt{20 \cdot 20}$    Rewrite.
$= 20$    Simplify.

**Reading**

When $n = 2$, the 2 is typically not written with the radical sign.

b. $243^{1/5}$

$243^{1/5} = \sqrt[5]{243}$    Write the expression in radical form.
$= \sqrt[5]{3 \cdot 3 \cdot 3 \cdot 3 \cdot 3}$    Rewrite.
$= 3$    Simplify.

● **On Your Own**

Now You're Ready
Exercises 13–18

Simplify the expression.

1. $\sqrt[3]{216}$    2. $\sqrt[5]{32}$    3. $\sqrt[4]{625}$

4. $49^{1/2}$    5. $343^{1/3}$    6. $64^{1/6}$

You can use properties of exponents to simplify expressions involving rational exponents.

### EXAMPLE 3  Using Properties of Exponents

a. $16^{3/4} = 16^{(1/4) \cdot 3}$    Rewrite the exponent.
        $= (16^{1/4})^3$    Power of a Power Property
        $= 2^3$    Evaluate the fourth root of 16.
        $= 8$    Evaluate power.

b. $27^{4/3} = 27^{1/3 \cdot 4}$    Rewrite the exponent.
        $= (27^{1/3})^4$    Power of a Power Property
        $= 3^4$    Evaluate the third root of 27.
        $= 81$    Evaluate power.

### On Your Own

Now You're Ready
Exercises 20–25

Simplify the expression.

7. $64^{2/3}$      8. $9^{5/2}$      9. $256^{3/4}$

### EXAMPLE 4  Real-Life Application

Volume = 113 cubic feet

The radius $r$ of a sphere is given by the equation $r = \left(\dfrac{3V}{4\pi}\right)^{1/3}$, where $V$ is the volume of the sphere. Find the radius of the beach ball to the nearest foot. Use 3.14 for $\pi$.

$r = \left(\dfrac{3V}{4\pi}\right)^{1/3}$    Write the equation.

$= \left[\dfrac{3(113)}{4(3.14)}\right]^{1/3}$    Substitute 113 for $V$ and 3.14 for $\pi$.

$= \left(\dfrac{339}{12.56}\right)^{1/3}$    Multiply.

$\approx 3$    Use a calculator.

∴ The radius of the beach ball is about 3 feet.

### On Your Own

10. **WHAT IF?** In Example 4, the volume of the beach ball is 17,000 cubic inches. Find the radius to the nearest inch. Use 3.14 for $\pi$.

# 6.3 Exercises

## Vocabulary and Concept Check

1. **WRITING** Explain how to simplify $81^{1/4}$.

2. **WHICH ONE DOESN'T BELONG?** Which expression does *not* belong with the other three? Explain your reasoning.

$(\sqrt[3]{27})^2$    $27^{2/3}$    $3^2$    $27^{3/2}$

## Practice and Problem Solving

**Find the dimensions of the cube. Check your answer.**

3. Volume = 64 in.³
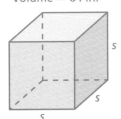

4. Volume = 216 cm³
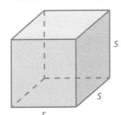

5. Volume = $\frac{343}{512}$ ft³
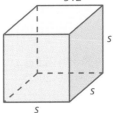

**Write the expression in rational exponent form.**

6. $\sqrt[7]{5}$

7. $(\sqrt[3]{4})^2$

8. $(\sqrt[5]{8})^4$

**Write the expression in radical form.**

9. $15^{1/3}$

10. $140^{1/7}$

11. $78^{2/5}$

12. **ERROR ANALYSIS** Describe and correct the error in writing the expression in rational exponent form.

✗ $(\sqrt[3]{2})^4 = 2^{3/4}$

**Simplify the expression.**

① 13. $\sqrt[4]{256}$

14. $\sqrt[3]{125}$

15. $\sqrt[5]{1024}$

② 16. $128^{1/7}$

17. $1000^{1/3}$

18. $81^{1/2}$

19. **BAKE SALE** A math club is having a bake sale. Find the length and width of the bake sale sign.

$4^{1/2}$ ft

$\sqrt[6]{729}$ ft

280  Chapter 6  Exponential Equations and Functions

**Simplify the expression.**

20. $32^{3/5}$
21. $125^{2/3}$
22. $36^{3/2}$
23. $243^{2/5}$
24. $128^{5/7}$
25. $343^{4/3}$

26. **PAPER CUPS** The radius $r$ of the base of a cone is given by the equation $r = \left(\dfrac{3V}{\pi h}\right)^{1/2}$, where $V$ is the volume of the cone and $h$ is the height of the cone. Find the radius of the paper cup to the nearest inch. Use 3.14 for $\pi$.

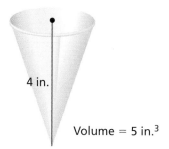

4 in.

Volume = 5 in.$^3$

27. **WRITING** Explain how to write $(\sqrt[n]{a})^m$ in rational exponent form.

28. **PROBLEM SOLVING** The formula for the volume of a regular dodecahedron is $V \approx 7.66\,\ell^3$, where $\ell$ is the length of an edge. The volume of the dodecahedron is 20 cubic feet. Estimate the edge length.

 Determine whether the statement is *always*, *sometimes*, or *never* true. Let $x$ be a nonnegative real number. Justify your answer.

29. $(x^{1/3})^3 = x$
30. $x^{1/3} = x^{-3}$
31. $x^{1/3} = \sqrt[3]{x}$
32. $x^{1/3} = x^3$
33. $\dfrac{x^{2/3}}{x^{1/3}} = \sqrt[3]{x}$
34. $x = x^{1/3} \cdot x^3$

 What you learned in previous grades & lessons

**Graph the linear equation.** *(Section 2.3 and Section 2.4)*

35. $y = -2x + 1$
36. $4x - 2y = 6$
37. $y = -\dfrac{1}{3}x - 5$

38. **MULTIPLE CHOICE** Which equation is shown in the graph? *(Section 2.1)*

Ⓐ $y = -\dfrac{1}{2}x + 1$
Ⓑ $y = -\dfrac{1}{2}x - 1$
Ⓒ $y = \dfrac{1}{2}x - 1$
Ⓓ $y = \dfrac{1}{2}x + 1$

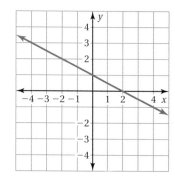

Section 6.3   Radicals and Rational Exponents   281

# 6 Study Help

You can use an **information frame** to help you organize and remember concepts. Here is an example of an information frame for the Product of Powers Property.

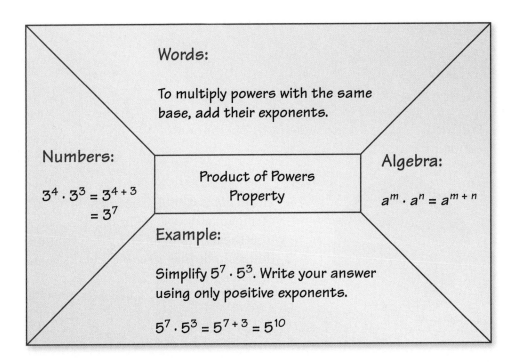

## On Your Own

**Make information frames to help you study these topics.**

1. Product Property of Square Roots
2. Quotient Property of Square Roots
3. Quotient of Powers Property
4. Power of a Power Property
5. Power of a Product Property
6. Power of a Quotient Property
7. rational exponents

**After you complete this chapter, make information frames for the following topics.**

8. exponential growth functions
9. exponential decay functions
10. geometric sequences

"Dear Mom, I am sending you an information frame card for Mother's Day!"

# 6.1–6.3 Quiz

**Simplify the expression.** *(Section 6.1)*

1. $\sqrt{20}$

2. $\sqrt{\dfrac{11}{81}}$

3. $\dfrac{4 - \sqrt{12}}{2}$

4. $\dfrac{-6 + \sqrt{45}}{3}$

**Evaluate the expression when $x = 2$, $y = -3$, and $z = 6$.** *(Section 6.1)*

5. $\sqrt{x + y^2 z}$

6. $\sqrt{3xz - y^2}$

**Simplify. Write your answer using only positive exponents.** *(Section 6.2)*

7. $3^2 \cdot 3^4$

8. $(k^4)^3$

9. $(4y)^{-2}$

10. $\left(\dfrac{r}{2}\right)^3$

**Simplify.** *(Section 6.3)*

11. $\sqrt[3]{27}$

12. $16^{1/4}$

13. $512^{2/3}$

14. $4^{5/2}$

15. **CEDAR CHEST** You store blankets in a cedar chest. What is the volume of the cedar chest? *(Section 6.1)*

16. **CRITICAL THINKING** Is the set of irrational numbers closed under subtraction? If not, find a counterexample. *(Section 6.1)*

| Unit of Mass | Mass |
|---|---|
| gigagram | $10^9$ grams |
| megagram | $10^6$ grams |
| kilogram | $10^3$ grams |
| hectogram | $10^2$ grams |
| dekagram | $10^1$ grams |
| decigram | $10^{-1}$ gram |
| centigram | $10^{-2}$ gram |
| milligram | $10^{-3}$ gram |
| microgram | $10^{-6}$ gram |
| nanogram | $10^{-9}$ gram |

17. **METRIC UNITS** The table shows several units of mass. *(Section 6.2)*

   a. How many times larger is a kilogram than a nanogram? Write your answer using only positive exponents.

   b. How many times smaller is a milligram than a hectogram? Write your answer using only positive exponents.

   c. Which is greater, 10,000 milligrams or 1000 decigrams? Explain your reasoning.

# 6.4 Exponential Functions

**Essential Question** What are the characteristics of an exponential function?

### 1 ACTIVITY: Describing an Exponential Function

Work with a partner. The graph below shows estimates of the population of Earth from 5000 B.C. through 1500 A.D. at 500-year intervals.

a. Describe the pattern.

b. Did Earth's population increase by the same *amount* or the same *percent* for each 500-year period? Explain.

c. Assume the pattern continued. Estimate Earth's population in 2000.

d. Use the Internet to find Earth's population in 2000. Did the pattern continue? If not, why did the pattern change?

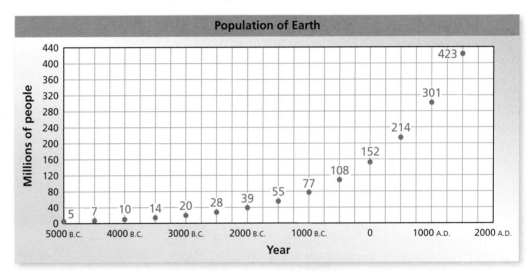

**Exponential Functions**

In this lesson, you will
- identify, evaluate, and graph exponential functions.

Learning Standards
A.REI.3
A.REI.11
F.BF.3
F.IF.7e
F.LE.1a
F.LE.2

4000 B.C.
Civilization begins to develop in Mesopotamia.

3000 B.C.
Stonehenge is built in England.

2000 B.C.
Middle Kingdom in Egypt

### ② ACTIVITY: Modeling an Exponential Function

**Math Practice 6**

**Calculate Accurately**

How can you check the accuracy of your answers?

**Work with a partner.** Use the following exponential function to complete the table. Compare the results with the data in Activity 1.

$$P = 152(1.406)^{t/500}$$

| Year | t | Population from Activity 1 | P |
|---|---|---|---|
| 5000 B.C. | −5000 | | |
| 4500 B.C. | −4500 | | |
| 4000 B.C. | −4000 | | |
| 3500 B.C. | −3500 | | |
| 3000 B.C. | −3000 | | |
| 2500 B.C. | −2500 | | |
| 2000 B.C. | −2000 | | |
| 1500 B.C. | −1500 | | |
| 1000 B.C. | −1000 | | |
| 500 B.C. | −500 | | |
| 1 B.C. | 0 | | |
| 500 A.D. | 500 | | |
| 1000 A.D. | 1000 | | |
| 1500 A.D. | 1500 | | |

1 B.C.
Augustus Caesar controls most of the Mediterranean world. (Use $t = 0$ to approximate 1 B.C.)

1000 A.D.
Song Dynasty has about one-fifth of Earth's population.

### What Is Your Answer?

3. **IN YOUR OWN WORDS** What are the characteristics of an exponential function?

4. Sketch the graph of each exponential function. Does the function match the characteristics you described in Question 3? Explain.

   a. $y = 2^x$    b. $y = 2(3)^x$    c. $y = 3(1.5)^x$

**Practice** — Use what you learned about exponential functions to complete Exercises 4 and 5 on page 289.

## 6.4 Lesson

**Key Vocabulary**
exponential function, p. 286

A function of the form $y = ab^x$, where $a \neq 0$, $b \neq 1$, and $b > 0$ is an **exponential function**. The exponential function $y = ab^x$ is a nonlinear function that changes by equal factors over equal intervals.

### EXAMPLE 1 — Identifying Functions

**Does each table represent a *linear* or an *exponential* function? Explain.**

a.

+1  +1  +1

| x | 0 | 1 | 2 | 3 |
|---|---|---|---|---|
| y | 2 | 4 | 6 | 8 |

+2  +2  +2

∴ As x increases by 1, y increases by 2. The rate of change is constant. So, the function is linear.

b.

| x | y |
|---|---|
| 0 | 4 |
| 1 | 8 |
| 2 | 16 |
| 3 | 32 |

+1 ×2, +1 ×2, +1 ×2

∴ As x increases by 1, y is multiplied by 2. So, the function is exponential.

### EXAMPLE 2 — Evaluating Exponential Functions

**Evaluate each function for the given value of x.**

a. $y = -2(5)^x$; $x = 3$

$y = -2(5)^x$    Write the function.

$= -2(5)^3$    Substitute for x.

$= -2(125)$    Evaluate the power.

$= -250$    Multiply.

b. $y = 3(0.5)^x$; $x = -2$

$y = 3(0.5)^x$

$= 3(0.5)^{-2}$

$= 3(4)$

$= 12$

### On Your Own

*Now You're Ready*
Exercises 6–15

**Does the table represent a *linear* or an *exponential* function? Explain.**

1.

| x | 0 | 1 | 2 | 3 |
|---|---|---|---|---|
| y | 8 | 4 | 2 | 1 |

2.

| x | y |
|---|---|
| −4 | 1 |
| 0 | 0 |
| 4 | −1 |
| 8 | −2 |

**Evaluate the function when $x = -2$, $0$, and $\dfrac{1}{2}$.**

3. $y = 2(9)^x$

4. $y = 1.5(2)^x$

## EXAMPLE 3  Graphing an Exponential Function

Graph $y = 2^x$. Describe the domain and range.

Step 1: Make a table of values.

| x | −2 | −1 | 0 | 1 | 2 | 3 |
|---|---|---|---|---|---|---|
| y | $\frac{1}{4}$ | $\frac{1}{2}$ | 1 | 2 | 4 | 8 |

**Study Tip**
In Example 3, you can substitute any value for x. So, the domain is all real numbers.

Step 2: Plot the ordered pairs.
Step 3: Draw a smooth curve through the points.

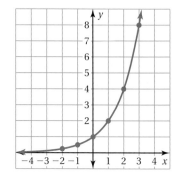

∴ From the graph, you can see that the domain is all real numbers and the range is all positive real numbers.

## EXAMPLE 4  Graphing a Vertical Translation

Graph $y = 2^x + 3$. Describe the domain and range. Compare the graph to the graph of $y = 2^x$.

Step 1: Make a table of values.

| x | −2 | −1 | 0 | 1 | 2 | 3 |
|---|---|---|---|---|---|---|
| y | $\frac{13}{4}$ | $\frac{7}{2}$ | 4 | 5 | 7 | 11 |

**Remember**
In Section 5.4, you learned that the graph of $f(x) + k$ is a vertical translation of the graph of $f(x)$.

Step 2: Plot the ordered pairs.
Step 3: Draw a smooth curve through the points.

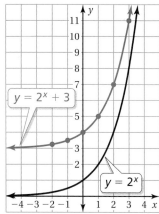

∴ From the graph, you can see that the domain is all real numbers and the range is all real numbers greater than 3. The graph of $y = 2^x + 3$ is a translation 3 units up of the graph of $y = 2^x$.

### On Your Own

**Now You're Ready**
Exercises 21–23 and 27–29

Graph the function. Describe the domain and range.

5. $y = 3^x$
6. $y = \left(\frac{1}{2}\right)^x$
7. $y = -2\left(\frac{1}{4}\right)^x$

8. Graph $y = \left(\frac{1}{2}\right)^x - 2$. Describe the domain and range. Compare the graph to the graph of $y = \left(\frac{1}{2}\right)^x$.

**Study Tip**

To find the y-intercept of the graph of $y = ab^x$, substitute 0 for x.
$y = ab^0$
$y = a(1)$
$y = a$
So, the y-intercept is a.

For an exponential function of the form $y = ab^x$, the y-values change by a factor of b as x increases by 1. Also notice that a is the y-intercept.

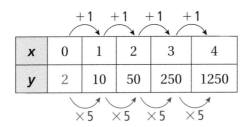

$y = 2(5)^x$

### EXAMPLE 5  Real-Life Application

The graph represents a bacteria population y after x days.

a. **Write an exponential function that represents the population.**

Use the graph to make a table of values.

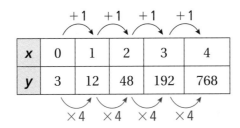

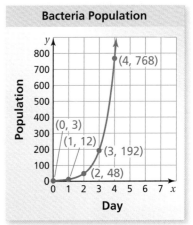

**Study Tip**

For help with rational exponents, see Section 6.3.

The y-intercept is 3 and the y-values increase by a factor of 4 as x increases by 1.

∴ So, the population can be modeled by $y = 3(4)^x$.

b. **Find the population after 12 hours and after 5 days.**

| *Population after 12 hours* | | *Population after 5 days* |
|---|---|---|
| $y = 3(4)^x$ | Write the function. | $y = 3(4)^x$ |
| $= 3(4)^{1/2}$ | Substitute for x. | $= 3(4)^5$ |
| $= 3(2)$ | Evaluate the power. | $= 3(1024)$ |
| $= 6$ | Multiply. | $= 3072$ |

12 hours = $\frac{1}{2}$ day

∴ There are 6 bacteria after 12 hours and 3072 bacteria after 5 days.

### On Your Own

Exercises 36–39

9. A bacteria population y after x days can be represented by an exponential function whose graph passes through (0, 100) and (1, 200).

   a. Write a function that represents the population.
   b. Find the population after 6 days. Does this bacteria population grow faster than the bacteria population in Example 5? Explain.

# 6.4 Exercises

## Vocabulary and Concept Check

1. **VOCABULARY** Describe how linear and exponential functions change over equal intervals.

2. **OPEN-ENDED** Sketch an increasing exponential function whose graph has a y-intercept of 2.

3. **WHICH ONE DOESN'T BELONG?** Which equation does *not* belong with the other three? Explain your reasoning.

   $y = 3^x$   $f(x) = 2(4)^x$   $f(x) = (-3)^x$   $y = 5(3)^x$

## Practice and Problem Solving

**Sketch the graph of the exponential function.**

4. $y = 4^x$

5. $y = 2(2)^x$

**Does the table represent a *linear* or an *exponential* function? Explain.**

 6.

| x | y |
|---|---|
| 0 | −2 |
| 1 | 0 |
| 2 | 2 |
| 3 | 4 |

7.

| x | y |
|---|---|
| 1 | 6 |
| 2 | 12 |
| 3 | 24 |
| 4 | 48 |

8.

| x | −1 | 0 | 1 | 2 |
|---|---|---|---|---|
| y | 0.25 | 1 | 4 | 16 |

9.

| x | −3 | 0 | 3 | 6 |
|---|---|---|---|---|
| y | 10 | 1 | −8 | −17 |

**Evaluate the function for the given value of x.**

 10. $y = 3^x$; $x = 2$

11. $f(x) = 3(2)^x$; $x = -1$

12. $y = -4(5)^x$; $x = 2$

13. $f(x) = 0.5^x$; $x = -3$

14. $f(x) = \frac{1}{3}(6)^x$; $x = 3$

15. $y = \frac{1}{4}(4)^x$; $x = \frac{3}{2}$

16. **ERROR ANALYSIS** Describe and correct the error in evaluating the function.

   $g(x) = 6(0.5)^x$; $x = -2$
   $g(-2) = 6(0.5)^{-2}$
   $= 3^{-2}$
   $= \frac{1}{9}$

17. **CALCULATOR** You graph an exponential function on a calculator. You zoom in repeatedly at 25% of the screen size. The function $y = 0.25^x$ represents the percent (in decimal form) of the original screen display that you see, where x is the number of times you zoom in. You zoom in twice. What percent of the original screen do you see?

Section 6.4   Exponential Functions   289

**Match the function with its graph.**

**18.** $f(x) = -3(4)^x$     **19.** $y = 2(0.5)^x$     **20.** $y = 4(1.5)^x$

A. 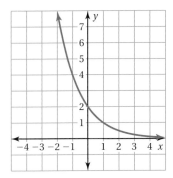    B. 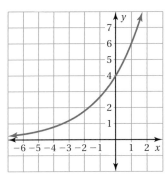    C.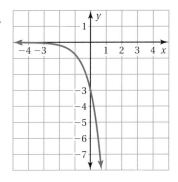

**Graph the function. Describe the domain and range.**

**21.** $y = 9^x$     **22.** $f(x) = -7^x$     **23.** $f(x) = 4\left(\dfrac{1}{4}\right)^x$

**24. LOGIC** Describe the graph of $y = a(2)^x$ when $a$ is (a) positive and (b) negative. (c) How does the graph change as $a$ changes?

**25. NUMBER SENSE** Consider the graph of $f(x) = 2(b)^x$. How do the graphs differ when $b > 1$ and $0 < b < 1$?

**26. COYOTES** A population $y$ of coyotes in a national park triples every 20 years. The function $y = 15(3)^x$ represents the population, where $x$ is the number of 20-year periods.

  a. Graph the function. Describe the domain and range.
  b. Find and interpret the $y$-intercept.
  c. How many coyotes are in the national park after 20 years?

**Graph the function. Describe the domain and range. Compare the graph to the graph of $y = 3^x$.**

**27.** $y = 3^x - 1$     **28.** $y = 3^x + 3$     **29.** $y = 3^x - \dfrac{1}{2}$

**30. REASONING** Graph the function $f(x) = -2^x$. Then graph $g(x) = -2^x - 3$.

  a. Describe the domain and range of each function.
  b. Find the $y$-intercept of the graph of each function.
  c. How are the $y$-intercept, domain, and range affected by the translation?

**31. REASONING** When does an exponential function intersect the $x$-axis? Give an example to justify your answer.

**Given $g(x) = 0.25^x - 1$, find the value of $k$ so that the graph is $g(x) + k$.**

32.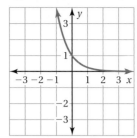

33.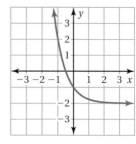

34.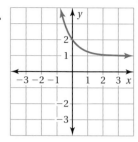

**35. REASONING** Graph $g(x) = 4^{x+2}$. Compare the graph to the graph of $f(x) = 4^x$.

**Write an exponential function represented by the graph or table.**

36.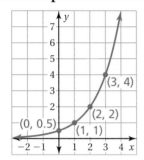

37.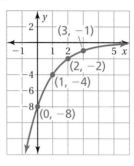

38.
| x | 0 | 1 | 2 | 3 |
|---|---|---|---|---|
| y | 2 | 8 | 32 | 128 |

39.
| x | 0 | 1 | 2 | 3 |
|---|---|---|---|---|
| y | −3 | −15 | −75 | −375 |

**40. ART GALLERY** The graph represents the number $y$ of visitors to a new art gallery after $x$ months.

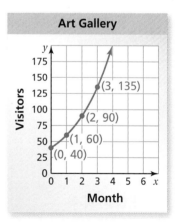

a. Write an exponential function that represents this situation.

b. Approximate the number of visitors after 5 months.

**41. SALES** A sales report shows that 3300 gas grills were purchased from a chain of hardware stores last year. The store expects grill sales to increase 6% each year. About how many grills does the store expect to sell in year 6? Use an equation to justify your answer.

**42. Structure** The graph of $g$ is a translation 4 units up and 3 units right of the graph of $f(x) = 2^x$. Write an equation for $g$.

## Fair Game Review  *What you learned in previous grades & lessons*

**Write the percent as a decimal.** *(Skills Review Handbook)*

43. 23%

44. 3%

45. 150%

**46. MULTIPLE CHOICE** Which of the following is equivalent to $100(0.95)$? *(Skills Review Handbook)*

Ⓐ 0.95    Ⓑ 9.5    Ⓒ 95    Ⓓ 950

# Extension 6.4 Solving Exponential Equations

To solve an exponential equation of the form $b^x = b^y$ when $b > 0$ and $b \neq 1$, solve the equation $x = y$.

### EXAMPLE 1 Solving Exponential Equations

a. Solve $5^x = 125$.

| | |
|---|---|
| $5^x = 125$ | Write the equation. |
| $5^x = 5^3$ | Rewrite 125 as $5^3$. |
| $x = 3$ | Equate the exponents. |

b. Solve $4^x = 2^{x-3}$.

| | |
|---|---|
| $4^x = 2^{x-3}$ | Write the equation. |
| $(2^2)^x = 2^{x-3}$ | Rewrite 4 as $2^2$. |
| $2^{2x} = 2^{x-3}$ | Power of a Power Property |
| $2x = x - 3$ | Equate the exponents. |
| $x = -3$ | Solve for $x$. |

**Check**
$4^x = 2^{x-3}$
$4^{-3} \stackrel{?}{=} 2^{-3-3}$
$\dfrac{1}{4^3} \stackrel{?}{=} \dfrac{1}{2^6}$
$\dfrac{1}{64} = \dfrac{1}{64}$ ✓

c. Solve $9^{x+2} = 27^x$.

| | |
|---|---|
| $9^{x+2} = 27^x$ | Write the equation. |
| $(3^2)^{x+2} = (3^3)^x$ | Rewrite 9 as $3^2$ and 27 as $3^3$. |
| $3^{2x+4} = 3^{3x}$ | Power of a Power Property |
| $2x + 4 = 3x$ | Equate the exponents. |
| $4 = x$ | Solve for $x$. |

**Check**
$9^{x+2} = 27^x$
$9^{4+2} \stackrel{?}{=} 27^4$
$531{,}441 = 531{,}441$ ✓

## Practice

**Solve the equation. Check your solution, if possible.**

1. $3^x = 81$
2. $2^x = 32$
3. $\dfrac{1}{16} = 4^x$
4. $10^x = 10^{x+1}$
5. $\left(\dfrac{1}{5}\right)^x = \left(\dfrac{1}{5}\right)^{3x}$
6. $6^{x-5} = 36^x$
7. $100^{5x+2} = 1000^{4x-1}$
8. $32^{1-x} = 8^{2x-2}$
9. $\left(\dfrac{1}{8}\right)^{x-5} = 4^x$

10. **NUMBER SENSE** Explain how you can use mental math to solve the equation $8^{x-4} = 1$.

11. **REASONING** Why does this method for solving $b^x = b^y$ not work when $b = 1$? Give an example to justify your answer.

# EXAMPLE 2  Solving an Equation by Graphing

**Use a graphing calculator to solve** $\left(\frac{1}{2}\right)^{x-1} = 7$.

**Exponential Functions**
In this extension, you will
- solve exponential equations algebraically and graphically.

**Learning Standards**
A.REI.3
A.REI.11
F.BF.3
F.IF.7e
F.LE.1a
F.LE.2

**Step 1:** Write a system of equations using each side of the equation.

$$y = \left(\frac{1}{2}\right)^{x-1} \quad \text{Equation 1}$$

$$y = 7 \quad \text{Equation 2}$$

**Step 2:** Enter the equations into your calculator. Then graph the equations in a standard viewing window.

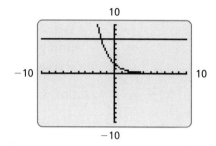

**Step 3:** Use the *intersect* feature to find the point of intersection. It is at about $(-1.81, 7)$.

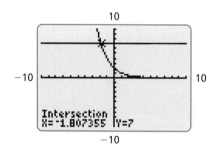

∴ So, the solution is $x \approx -1.81$.

**Check:** Check the solution algebraically.

$\left(\frac{1}{2}\right)^{x-1} = 7$     Write the equation.

$\left(\frac{1}{2}\right)^{-1.81-1} \stackrel{?}{=} 7$     Substitute −1.81 for x.

$7.01 \approx 7$ ✓     Use a calculator.

## Practice

**Use a graphing calculator to solve the equation.**

**12.** $4^{x+3} = 6$      **13.** $2^x = 1.8$      **14.** $4 = 8^x$

**15.** $\left(\frac{3}{4}\right)^{x+2} = 10$      **16.** $2^{-x-3} = 3^{x+1}$      **17.** $5^x = -4^{x+4}$

# 6.5 Exponential Growth

**Essential Question** What are the characteristics of exponential growth?

### 1 ACTIVITY: Comparing Types of Growth

Work with a partner. Describe the pattern of growth for each sequence and graph. How many of the patterns represent exponential growth? Explain your reasoning.

a. 1, 4, 7, 10, 13, 16, 19, 22, 25, 28, 31

b. 1.0, 1.4, 2.0, 2.7, 3.8, 5.4, 7.5, 10.5, 14.8, 20.7, 28.9

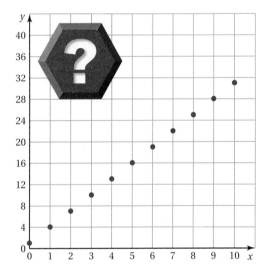
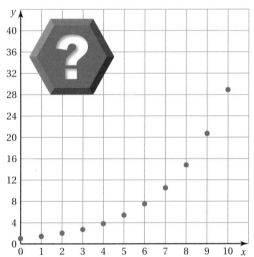

c. 1.0, 1.3, 2.3, 4.0, 6.3, 9.3, 13.0, 17.3, 22.3, 28.0, 34.3

d. 1.0, 1.6, 2.4, 3.4, 4.7, 6.4, 8.7, 11.5, 15.3, 20.2, 26.6

**Common Core**

**Exponential Functions**

In this lesson, you will
- write, interpret, and graph exponential growth functions.

Learning Standards
A.SSE.1a
A.SSE.1b
F.IF.7e

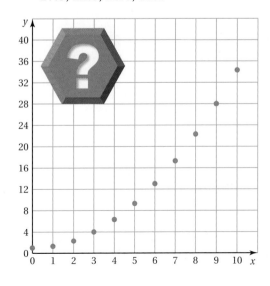
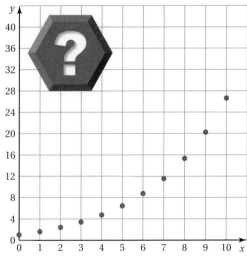

294  Chapter 6  Exponential Equations and Functions

## 2 ACTIVITY: Predicting a Future Event

**Math Practice**

**Consider Similar Problems**

How can you use the results from the previous activity to help you solve this problem?

Work with a partner. It is estimated that in 1782 there were about 100,000 nesting pairs of bald eagles in the United States. By the 1960s, this number had dropped to about 500 nesting pairs. This decline was attributed to loss of habitat, loss of prey, hunting, and the use of the pesticide DDT.

The 1940 Bald Eagle Protection Act prohibited the trapping and killing of the birds. In 1967, the bald eagle was declared an endangered species in the United States. With protection, the nesting pair population began to increase, as shown in the graph. Finally, in 2007, the bald eagle was removed from the list of endangered and threatened species.

Describe the growth pattern shown in the graph. Is it exponential growth? Assume the pattern continues. When will the population return to the levels of the late 1700s? Explain your reasoning.

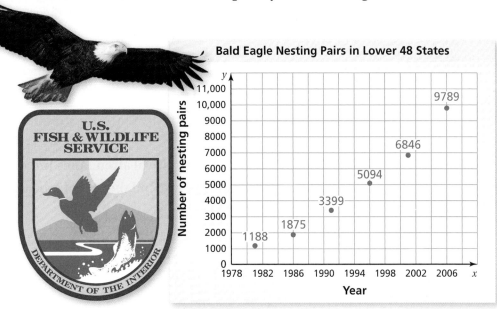

## What Is Your Answer?

3. **IN YOUR OWN WORDS** What are the characteristics of exponential growth? How can you distinguish exponential growth from other growth patterns?

4. Which of the following are examples of exponential growth? Explain.
   a. Growth of the balance of a savings account
   b. Speed of the moon in orbit around Earth
   c. Height of a ball that is dropped from a height of 100 feet

**Practice**

Use what you learned about exponential growth to complete Exercises 3 and 4 on page 298.

# 6.5 Lesson

**Key Vocabulary**
exponential growth, *p. 296*
exponential growth function, *p. 296*
compound interest, *p. 297*

**Exponential growth** occurs when a quantity increases by the same factor over equal intervals of time.

## 🔑 Key Idea

**Exponential Growth Functions**

A function of the form $y = a(1 + r)^t$, where $a > 0$ and $r > 0$, is an **exponential growth function**.

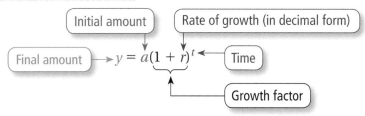

**Study Tip**
Notice that an exponential growth function is of the form $y = ab^x$, where $b$ is replaced by $1 + r$ and $x$ is replaced by $t$.

### EXAMPLE 1 Using an Exponential Growth Function

The function $y = 150,000(1.1)^t$ represents the attendance $y$ at a music festival $t$ years after 2010.

**a. By what percent does the festival attendance increase each year?**

Use the growth factor $1 + r$ to find the rate of growth.

$1 + r = 1.1$     Write an equation.
$r = 0.1$        Subtract 1 from each side.

∴ So, the festival attendance increases by 10% each year.

**b. How many people will attend the festival in 2014? Round your answer to the nearest ten thousand.**

The value $t = 4$ represents 2014.

$y = 150,000(1.1)^t$     Write exponential growth function.
$= 150,000(1.1)^4$     Substitute 4 for $t$.
$= 219,615$             Use a calculator.

∴ About 220,000 people will attend the festival in 2014.

### On Your Own

Exercises 5–10

**1.** The function $y = 500,000(1.15)^t$ represents the number $y$ of members of a website $t$ years after 2010.

    **a.** By what percent does the website membership increase each year?

    **b.** How many members will there be in 2016? Round your answer to the nearest hundred thousand.

## Key Idea

**Compound Interest**

**Compound interest** is interest earned on the principal *and* on previously earned interest. The balance $y$ of an account earning compound interest is

$$y = P\left(1 + \frac{r}{n}\right)^{nt}.$$

$P$ = principal (initial amount)
$r$ = annual interest rate (in decimal form)
$t$ = time (in years)
$n$ = number of times interest is compounded per year

### EXAMPLE 2 — Writing a Function

You deposit $100 in a savings account that earns 5% annual interest compounded yearly. Write a function for the balance after $t$ years.

**Study Tip**
For interest compounded yearly, you can substitute 1 for $n$ in the formula to get $y = P(1 + r)^t$.

$y = P\left(1 + \dfrac{r}{n}\right)^{nt}$   Write compound interest formula.

$y = 100\left(1 + \dfrac{0.05}{1}\right)^{(1)(t)}$   Substitute 100 for $P$, 0.05 for $r$, and 1 for $n$.

$y = 100(1.05)^t$   Simplify.

### EXAMPLE 3 — Real-Life Application

The table shows the balance of a money market account over time.

| Year, $t$ | Balance |
|---|---|
| 0 | $100 |
| 1 | $110 |
| 2 | $121 |
| 3 | $133.10 |
| 4 | $146.41 |
| 5 | $161.05 |

**a.** Write a function for the balance after $t$ years.

From the table, you know the balance increases 10% each year.

$y = a(1 + r)^t$   Write exponential growth function.

$y = 100(1 + 0.1)^t$   Substitute 100 for $a$ and 0.1 for $r$.

$y = 100(1.1)^t$   Simplify.

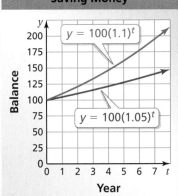

**b.** Graph the functions from part (a) and Example 2 in the same coordinate plane. Compare the account balances.

The money market account earns 10% interest each year and the savings account earns 5% interest each year. So, the balance of the money market account increases faster.

### On Your Own

Now You're Ready
Exercises 11–14

**2.** You deposit $500 in a savings account that earns 4% annual interest compounded yearly. Write and graph a function that represents the balance $y$ (in dollars) after $t$ years.

# 6.5 Exercises

## Vocabulary and Concept Check

1. **VOCABULARY** When does the exponential function $y = a(1 + r)^t$ represent an exponential growth function?

2. **VOCABULARY** The population of a city grows by 3% each year. What is the growth factor?

## Practice and Problem Solving

**Describe the pattern of growth for the sequence.**

3. 1.0, 1.2, 1.4, 1.7, 2.1, 2.5, 3.0, 3.6, 4.3, 5.2, 6.2

4. 1, 7, 13, 19, 25, 31, 37, 43, 49, 55, 61

**Identify the initial amount $a$ and the rate of growth $r$ (as a percent) of the exponential function. Evaluate the function when $t = 5$. Round your answer to the nearest tenth.**

5. $y = 25(1.2)^t$

6. $f(t) = 12(1.05)^t$

7. $d(t) = 1500(1.074)^t$

8. $y = 175(1.028)^t$

9. $g(t) = 6.7(2)^t$

10. $h(t) = 1.8^t$

**Write and graph a function that represents the situation.**

11. You deposit $800 in an account that earns 7% annual interest compounded yearly.

12. Your $35,000 annual salary increases by 4% each year.

13. A population of 210,000 increases by 12.5% each year.

14. Sales of $10,000 increase by 70% each year.

15. **ERROR ANALYSIS** The growth rate of a bacteria culture is 150% each hour. Initially, there are 10 bacteria. Describe and correct the error in finding the number of bacteria in the culture after 8 hours.

>  $b(t) = 10(1.5)^t$
> $b(8) = 10(1.5)^8 \approx 256.3$
> After 8 hours, there are about 256 bacteria in the culture.

16. **INVESTMENT** The function $y = 7500(1.08)^t$ represents the value $y$ of an investment after $t$ years.

  a. What is the initial investment?

  b. What is the value of the investment after 6 years?

17. **POPULATION** The population of a city has been increasing by 2% annually. In 2000, the population was 315,000. Predict the population of the city in 2020. Round your answer to the nearest thousand.

**Write a function that represents the situation. Find the balance in the account after the given time period.**

18. $2000 deposit that earns 5% annual interest compounded quarterly; 5 years

19. $6200 deposit that earns 8.4% annual interest compounded monthly; 18 months

20. **NUMBER SENSE** During a flu epidemic, the number of sick people triples every week. What is the growth rate as a percent? Explain your reasoning.

21. **SAVINGS** You deposit $9000 in a savings account that earns 3.6% annual interest compounded monthly. You also save $40 per month in a safe at home. Write a function $C(t) = b(t) + h(t)$, where $b(t)$ represents the balance of your savings account and $h(t)$ represents the amount in your safe after $t$ years. What does $C(t)$ represent?

22. **REASONING** The number of concert tickets sold doubles every hour. After 12 hours, all of the tickets are sold. After how many hours are about one-fourth of the tickets sold? Explain your reasoning.

23. **YOU BE THE TEACHER** The balance of a savings account can be modeled by the function $b(t) = 5000(1.024)^t$, where $t$ is the time in years. To model the monthly balance, a student writes

$$b(t) = 5000(1.024)^t = 5000(1.024)^{\left(\frac{1}{12} \cdot 12\right)t} = 5000\left(1.024^{\frac{1}{12}}\right)^{12t} \approx 5000(1.002)^{12t}.$$

Is the student correct? Explain your reasoning.

24. **Critical Thinking** Gordon Moore stated that the number of transistors that can be placed on an integrated circuit will double every 2 years. This trend is known as Moore's Law. In 1978, the Intel®8086 held 29,000 transistors on an integrated circuit.

   a. Write a function that represents Moore's Law, where $t$ is the number of years since 1978.

   b. How many transistors could be placed on an integrated circuit in 2015?

**Fair Game Review** *What you learned in previous grades & lessons*

**Simplify the expression.** *(Section 6.2)*

25. $\left(\dfrac{2}{3}\right)^2$     26. $\left(\dfrac{1}{4}\right)^3$     27. $\left(\dfrac{3}{5}\right)^4$

28. **MULTIPLE CHOICE** The domain of the function $y = 4x - 3$ is 1, 4, 7, 10, and 13. Which number is *not* in the range of the function? *(Section 5.1)*

    Ⓐ 1     Ⓑ 10     Ⓒ 13     Ⓓ 25

# 6.6 Exponential Decay

**Essential Question** What are the characteristics of exponential decay?

## 1 ACTIVITY: Comparing Types of Decay

Work with a partner. Describe the pattern of decay for each sequence and graph. Which of the patterns represent exponential decay? Explain your reasoning.

**a.** 30.0, 24.3, 19.2, 14.7, 10.8, 7.5, 4.8, 2.7, 1.2, 0.3, 0.0

**b.** 30, 27, 24, 21, 18, 15, 12, 9, 6, 3, 0

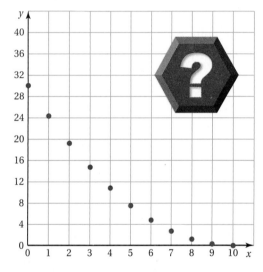
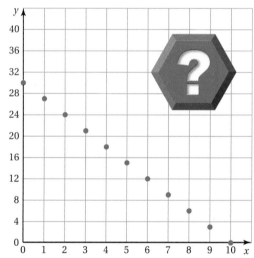

**c.** 30.0, 24.0, 19.2, 15.4, 12.3, 9.8, 7.9, 6.3, 5.0, 4.0, 3.2

**d.** 30.0, 29.7, 28.8, 27.3, 25.2, 22.5, 19.2, 15.3, 10.8, 5.7, 0.0

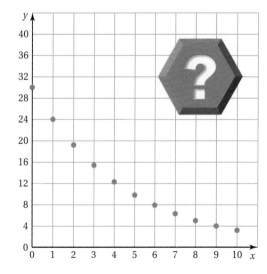
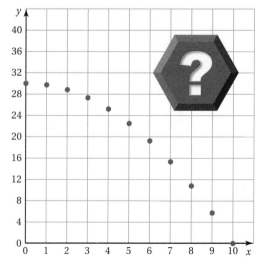

**COMMON CORE**

**Exponential Functions**

In this lesson, you will
- identify exponential growth and decay.
- write, interpret, and graph exponential decay functions.

Learning Standards
A.SSE.1a
A.SSE.1b
F.IF.7e

## 2 ACTIVITY: Describing a Decay Pattern

**Math Practice 4**

**Simplify a Situation**
How can you organize the given information to simplify this problem? How is the answer affected?

Work with a partner. Newton's Law of Cooling states that when an object at one temperature is exposed to air of another temperature, the difference in the two temperatures drops by the same percent each hour.

A forensic pathologist was called to estimate the time of death of a person. At midnight, the body temperature was 80.5°F and the room temperature was 60°F. One hour later, the body temperature was 78.5°F.

**a.** By what percent did the difference between the body temperature and the room temperature drop during the hour?

**b.** Assume that the original body temperature was 98.6°F. Use the percent decrease found in part (a) to make a table showing the body temperatures. Use the table to estimate the time of death.

| Hour | Temperature (°F) |
|---|---|
| 0 | 98.6 |
| 1 |  |
| 2 |  |
| 3 |  |
| 4 |  |
| 5 |  |
| 6 |  |
| 7 |  |
| 8 |  |
| 9 |  |
| 10 |  |

### What Is Your Answer?

**3. IN YOUR OWN WORDS** What are the characteristics of exponential decay? How can you distinguish exponential decay from other decay patterns?

**4.** Sketch a graph of the data from the table in Activity 2. Do the data represent exponential decay? Explain your reasoning.

**5.** Suppose the pathologist had arrived at 6:00 A.M. What would have been the body temperature at that time?

**Practice** — Use what you learned about exponential decay to complete Exercises 3 and 4 on page 304.

Section 6.6   Exponential Decay

## 6.6 Lesson

Check It Out
Lesson Tutorials
BigIdeasMath.com

**Key Vocabulary**
exponential decay, p. 302
exponential decay function, p. 302

**Exponential decay** occurs when a quantity decreases by the same factor over equal intervals of time.

### Key Idea

**Exponential Decay Functions**

A function of the form $y = a(1 - r)^t$, where $a > 0$ and $0 < r < 1$, is an **exponential decay function**.

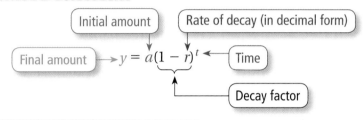

**Study Tip**
Notice that an exponential decay function is of the form $y = ab^x$, where $b$ is replaced by $1 - r$ and $x$ is replaced by $t$.

For exponential growth, the value inside the parentheses is greater than 1 because $r$ is added to 1. For exponential decay, the value inside the parentheses is less than 1 because $r$ is subtracted from 1.

### EXAMPLE 1  Identifying Exponential Growth and Decay

Determine whether each table represents an *exponential growth function*, an *exponential decay function*, or *neither*.

a.

| x | y |
|---|---|
| 0 | 270 |
| 1 | 90 |
| 2 | 30 |
| 3 | 10 |

+1 between each x; $\times \frac{1}{3}$ between each y.

As $x$ increases by 1, $y$ is multiplied by $\frac{1}{3}$.

So, the table represents an exponential decay function.

b.

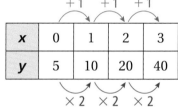

| x | 0 | 1 | 2 | 3 |
|---|---|---|---|---|
| y | 5 | 10 | 20 | 40 |

As $x$ increases by 1, $y$ is multiplied by 2.

So, the table represents an exponential growth function.

### On Your Own

Now You're Ready
Exercises 8–13

Determine whether the table represents an *exponential growth function*, an *exponential decay function*, or *neither*.

1.
| x | 0 | 1 | 2 | 3 |
|---|---|---|---|---|
| y | 64 | 16 | 4 | 1 |

2.
| x | 1 | 3 | 5 | 7 |
|---|---|---|---|---|
| y | 4 | 11 | 18 | 25 |

# EXAMPLE 2  Interpreting an Exponential Decay Function

The function $P = 4870(0.94)^t$ represents the population $P$ of a town after $t$ years. By what percent does the population decrease each year?

Use the decay factor $1 - r$ to find the rate of decay.

$1 - r = 0.94$   Write an equation.

$r = 0.06$   Solve for $r$.

∴ So, the population of the town decreases by 6% each year.

### On Your Own

**Now You're Ready**
Exercises 15–17

3. The function $A = 275\left(\dfrac{9}{10}\right)^t$ represents the area $A$ (in square miles) of a coral reef after $t$ years. By what percent does the area of the coral reef decrease each year?

# EXAMPLE 3  Real-Life Application

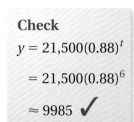

The value of a car is $21,500. It loses 12% of its value every year.

a. **Write a function that represents the value $y$ (in dollars) of the car after $t$ years.**

$y = a(1 - r)^t$   Write exponential decay function.

$y = 21{,}500(1 - 0.12)^t$   Substitute 21,500 for $a$ and 0.12 for $r$.

$y = 21{,}500(0.88)^t$   Simplify.

b. **Graph the function from part (a). Use the graph to estimate the value of the car after 6 years.**

From the graph, you can see that the $y$-value is about 10,000 when $t = 6$.

∴ So, the value of the car is about $10,000 after 6 years.

**Check**

$y = 21{,}500(0.88)^t$

$= 21{,}500(0.88)^6$

$\approx 9985$ ✓

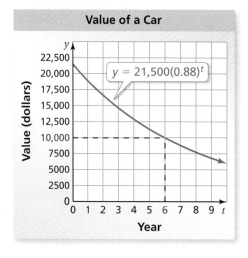

Value of a Car

$y = 21{,}500(0.88)^t$

### On Your Own

**Now You're Ready**
Exercise 22

4. **WHAT IF?** The car loses 9% of its value every year.

a. Write a function that represents the value $y$ (in dollars) of the car after $t$ years.

b. Graph the function from part (a). Estimate the value of the car after 12 years. Round your answer to the nearest thousand.

# 6.6 Exercises

## Vocabulary and Concept Check

1. **WRITING** When does the function $y = ab^x$ represent exponential growth? exponential decay?

2. **VOCABULARY** What is the decay factor in the function $y = a(1 - r)^t$?

## Practice and Problem Solving

**Describe the pattern of decay for the sequence.**

3. 28, 26, 24, 22, 20, 18, 16, 14, 12, 10, 8

4. 256, 192, 144, 108, 81, 60.8, 45.6, 34.2, 25.6, 19.2, 14.4

**Determine whether the graph represents an *exponential growth function*, an *exponential decay function*, or *neither*.**

5.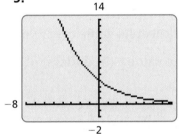
6.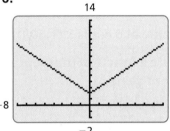
7.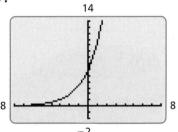

**Determine whether the table represents an *exponential growth function*, an *exponential decay function*, or *neither*.**

8. 
| x | 0 | 1 | 2 | 3 |
|---|---|---|---|---|
| y | 17 | 51 | 153 | 459 |

9. 
| x | 1 | 2 | 3 | 4 |
|---|---|---|---|---|
| y | 32 | 28 | 24 | 20 |

10. 
| x | 1 | 2 | 3 | 4 |
|---|---|---|---|---|
| y | 625 | 125 | 25 | 5 |

11. 
| x | 2 | 4 | 6 | 8 |
|---|---|---|---|---|
| y | 256 | 64 | 16 | 4 |

12. 
| x | 2 | 4 | 6 | 8 |
|---|---|---|---|---|
| y | 35 | 42 | 49 | 42 |

13. 
| x | 3 | 5 | 7 | 9 |
|---|---|---|---|---|
| y | 6 | 216 | 7776 | 279,936 |

14. **CAMPER** The table shows the value of a camper $t$ years after it is purchased.

   a. Determine whether the table represents an *exponential growth function*, an *exponential decay function*, or *neither*.

   b. What is the value of the camper after 5 years?

| t | Value |
|---|---|
| 1 | $24,000 |
| 2 | $19,200 |
| 3 | $15,360 |
| 4 | $12,288 |

**Write the rate of decay of the function as a percent.**

**15.** $y = 4(0.8)^t$

**16.** $f(t) = 30(0.95)^t$

**17.** $g(t) = \left(\dfrac{3}{4}\right)^t$

**Match the exponential function with its graph.**

**18.** $y = 10(1.3)^t$

**19.** $h(t) = 6\left(\dfrac{7}{8}\right)^t$

**20.** $y = 2(0.6)^t$

A.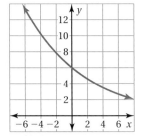

B.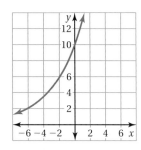

C.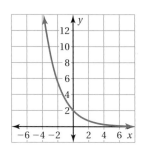

**21. CHOOSE TOOLS** When would you graph an exponential decay function by hand? When would you use a graphing calculator? Explain your reasoning.

**22. POPULATION** A city has a population of 250,000. The population is expected to decrease by 1.5% annually for the next decade. Write a function that represents this situation. Then predict the population in 10 years.

**23. TIRE PRESSURE** At noon on Monday, the air pressure of a tire is 32 pounds per square inch (psi). The tire loses 8% of its air every day. The tire pressure monitoring system (TPMS) will alert the driver when the tire pressure is less than or equal to 24 psi. On what day of the week will the TPMS alert the driver? Use the *trace* feature of a graphing calculator to help find the answer.

**24. Structure** The graph of an exponential function passes through $\left(2, \dfrac{3}{2}\right)$ and $\left(4, \dfrac{3}{8}\right)$.

  **a.** Do the *y*-values increase or decrease as *x* increases? How do you know?

  **b.** Find the *y*-intercept of the graph.

  **c.** Write an exponential function that represents the graph.

### Fair Game Review  *What you learned in previous grades & lessons*

**Write an equation for the *n*th term of the arithmetic sequence. Then find $a_{15}$.** *(Section 5.6)*

**25.** 9, 12, 15, 18, . . .

**26.** 3, 1, −1, −3, . . .

**27.** −7, −11, −15, −19, . . .

**28. MULTIPLE CHOICE** What is the solution of the linear system? *(Section 4.3)*

  **Ⓐ** (−2, −3)  **Ⓑ** (−2, 3)
  **Ⓒ** (2, −3)  **Ⓓ** (2, 3)

  $2x - 5y = 11$
  $5x - 3y = -1$

# 6.7 Geometric Sequences

**Essential Question** How are geometric sequences used to describe patterns?

### 1 ACTIVITY: Describing Calculator Patterns

Work with a partner.
- Enter the keystrokes on a calculator and record the results in the table.
- Describe the pattern.

a. Step 1   [2] [=]
   Step 2   [×] [2] [=]
   Step 3   [×] [2] [=]
   Step 4   [×] [2] [=]
   Step 5   [×] [2] [=]

| Step | 1 | 2 | 3 | 4 | 5 |
|---|---|---|---|---|---|
| Calculator Display | | | | | |

b. Step 1   [6] [4] [=]
   Step 2   [×] [.] [5] [=]
   Step 3   [×] [.] [5] [=]
   Step 4   [×] [.] [5] [=]
   Step 5   [×] [.] [5] [=]

| Step | 1 | 2 | 3 | 4 | 5 |
|---|---|---|---|---|---|
| Calculator Display | | | | | |

c. Use a calculator to make your own sequence. Start with any number and multiply by 3 each time. Record your results in the table.

| Step | 1 | 2 | 3 | 4 | 5 |
|---|---|---|---|---|---|
| Calculator Display | | | | | |

**COMMON CORE**

**Geometric Sequences**
In this lesson, you will
- extend and graph geometric sequences.
- write equations for geometric sequences.
- solve real-life problems.

Learning Standards
F.BF.2
F.IF.3
F.LE.2

306   Chapter 6   Exponential Equations and Functions

## 2 ACTIVITY: Folding a Sheet of Paper

**Math Practice 8**

**Repeat Calculations**
What calculations are repeated? How does this help you answer the question?

Work with a partner. A sheet of paper is about 0.1 mm thick.

a. How thick would it be if you folded it in half once?

b. How thick would it be if you folded it in half a second time?

c. How thick would it be if you folded it in half 6 times?

d. What is the greatest number of times you can fold a sheet of paper in half? How thick is the result?

e. Do you agree with the statement below? Explain your reasoning.

*"If it were possible to fold the paper 15 times, it would be taller than you."*

## 3 ACTIVITY: Writing a Story

**The King and the Beggar**

A king offered a beggar fabulous meals for one week. Instead, the beggar asked for a single grain of rice the first day, 2 grains the second day, and double the amount each day after for one month. The king agreed. But, as the month progressed, he realized that he would lose his entire kingdom.

**Work with a partner.**

- Why does the king think he will lose his entire kingdom?
- Write your own story about doubling or tripling a small object many times.
- Draw pictures for your story.
- Include a table to organize the amounts.
- Write your story so that one of the characters is surprised by the size of the final number.

### What Is Your Answer?

4. **IN YOUR OWN WORDS** How are geometric sequences used to describe patterns? Give an example from real life.

**Practice**

Use what you learned about geometric sequences to complete Exercise 4 on page 310.

Section 6.7 Geometric Sequences 307

## 6.7 Lesson

**Check It Out**
Lesson Tutorials
BigIdeasMath.com

**Key Vocabulary**
geometric sequence, p. 308
common ratio, p. 308

### Key Idea

**Geometric Sequence**

In a **geometric sequence,** the ratio between consecutive terms is the same. This ratio is called the **common ratio.** Each term is found by multiplying the previous term by the common ratio.

1, 5, 25, 125, . . .   Terms of a geometric sequence
×5  ×5  ×5 ← Common ratio

### EXAMPLE 1  Extending a Geometric Sequence

**Write the next three terms of the geometric sequence 3, 6, 12, 24, . . . .**

Use a table to organize the terms and extend the pattern.

| Position | 1 | 2 | 3 | 4 | 5 | 6 | 7 |
|---|---|---|---|---|---|---|---|
| Term | 3 | 6 | 12 | 24 | 48 | 96 | 192 |

×2  ×2  ×2  ×2  ×2  ×2

Each term is twice the previous term. So, the common ratio is 2.

Multiply a term by 2 to find the next term.

∴ The next three terms are 48, 96, and 192.

### EXAMPLE 2  Graphing a Geometric Sequence

**Graph the geometric sequence 32, 16, 8, 4, 2, . . . . What do you notice?**

Make a table. Then plot the ordered pairs $(n, a_n)$.

| Position, $n$ | 1 | 2 | 3 | 4 | 5 |
|---|---|---|---|---|---|
| Term, $a_n$ | 32 | 16 | 8 | 4 | 2 |

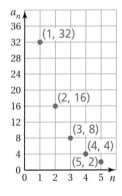

∴ The points of the graph appear to lie on an exponential curve.

### On Your Own

**Now You're Ready**
Exercises 11–16

Write the next three terms of the geometric sequence. Then graph the sequence.

1. 1, 3, 9, 27, . . .
2. 64, 16, 4, 1, . . .
3. 80, −40, 20, −10, . . .

---

308   Chapter 6   Exponential Equations and Functions   Multi-Language Glossary at BigIdeasMath.com

Because consecutive terms of a geometric sequence change by equal factors, the points of any geometric sequence with a positive common ratio lie on an exponential curve. You can use the first term and the common ratio to write an exponential function that describes a geometric sequence.

| Position, $n$ | Term, $a_n$ | Written using $a_1$ and $r$ | Numbers |
|---|---|---|---|
| 1 | first term, $a_1$ | $a_1$ | 1 |
| 2 | second term, $a_2$ | $a_1 r$ | $1 \cdot 5 = 5$ |
| 3 | third term, $a_3$ | $a_1 r^2$ | $1 \cdot 5^2 = 25$ |
| 4 | fourth term, $a_4$ | $a_1 r^3$ | $1 \cdot 5^3 = 125$ |
| ⋮ | ⋮ | ⋮ | ⋮ |
| $n$ | $n$th term, $a_n$ | $a_1 r^{n-1}$ | $1 \cdot 5^{n-1}$ |

## Key Idea

**Equation for a Geometric Sequence**

Let $a_n$ be the $n$th term of a geometric sequence with first term $a_1$ and common ratio $r$. The $n$th term is given by

$$a_n = a_1 r^{n-1}.$$

**Study Tip**

Notice that $a_n = a_1 r^{n-1}$ is of the form $y = ab^x$.

### EXAMPLE 3 — Real-Life Application

Clicking the *zoom-out* button on a mapping website doubles the side length of the square map.

| Zoom-out Clicks | 1 | 2 | 3 |
|---|---|---|---|
| Map Side Length (miles) | 5 | 10 | 20 |

**a.** Write an equation for the $n$th term of the geometric sequence.

The first term is 5 and the common ratio is 2.

$a_n = a_1 r^{n-1}$     Equation for a geometric sequence

$a_n = 5(2)^{n-1}$     Substitute 5 for $a_1$ and 2 for $r$.

**b.** Find and interpret $a_8$.

Use the equation to find the 8th term.

$a_n = 5(2)^{n-1}$     Write the equation.

$\quad = 5(2)^{8-1}$     Substitute 8 for $n$.

$\quad = 640$     Simplify.

∴ The side length of the square map after 8 clicks is 640 miles.

### On Your Own

**Now You're Ready**
Exercises 25–28

**4. WHAT IF?** After how many clicks on the *zoom-out* button is the side length of the map 2560 miles?

## 6.7 Exercises

### Vocabulary and Concept Check

1. **WRITING** How are arithmetic sequences and geometric sequences different?
2. **REASONING** Compare and contrast the two sequences.

   2, 4, 6, 8, 10, . . .       2, 4, 8, 16, 32, . . .

3. **CRITICAL THINKING** Why do the points of a geometric sequence lie on an exponential curve only when the common ratio is positive?

### Practice and Problem Solving

4. Enter 4 on a calculator. Multiply by 6 four times. Record your results in the table. Describe the pattern.

| Step | 1 | 2 | 3 | 4 | 5 |
|---|---|---|---|---|---|
| Calculator Display | | | | | |

**Find the common ratio of the geometric sequence.**

5. 3, −12, 48, −192, . . .
6. 200, 100, 50, 25, . . .
7. 7640, 764, 76.4, 7.64, . . .
8. 9, −18, 36, −72, . . .
9. 0.1, 0.9, 8.1, 72.9, . . .
10. 5, 1, $\frac{1}{5}$, $\frac{1}{25}$, . . .

**Write the next three terms of the geometric sequence. Then graph the sequence.**

11. 2, 10, 50, 250, . . .
12. −7, 14, −28, 56, . . .
13. 81, −27, 9, −3, . . .
14. −375, −75, −15, −3, . . .
15. 36, 6, 1, $\frac{1}{6}$, . . .
16. $\frac{1}{49}$, $\frac{1}{7}$, 1, 7, . . .

17. **ERROR ANALYSIS** Describe and correct the error in writing the next three terms of the geometric sequence.

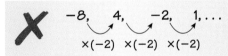

The next three terms are −2, 4, and −8.

18. **BADMINTON** A badminton tournament begins with 128 teams. After the first round, 64 teams remain. After the second round, 32 teams remain. How many teams remain after the third, fourth, and fifth rounds?

**Tell whether the sequence is *geometric*, *arithmetic*, or *neither*.**

19. −8, 0, 8, 16, . . .
20. −1, 3, −5, 7, . . .
21. 1, 4, 9, 16, . . .
22. $\frac{3}{49}$, $\frac{3}{7}$, 3, 21, . . .
23. 192, 24, 3, $\frac{3}{8}$, . . .
24. −25, −18, −12, −7, . . .

310   Chapter 6   Exponential Equations and Functions

**Write an equation for the nth term of the geometric sequence. Then find $a_7$.**

**25.** 1, −5, 25, −125, . . .

**26.** 2, 8, 32, 128, . . .

**27.**

| n | 1 | 2 | 3 | 4 |
|---|---|---|---|---|
| $a_n$ | 5 | 15 | 45 | 135 |

**28.**

| n | 1 | 2 | 3 | 4 |
|---|---|---|---|---|
| $a_n$ | 2 | 14 | 98 | 686 |

**29. CHAIN EMAIL** You start a chain email and send it to 6 friends. The process continues and each of your friends forwards the email to 6 people.

  a. Write an equation for the nth term of the geometric sequence.

  b. Describe the domain. Is the domain discrete or continuous?

**30. REASONING** What is the 9th term of a geometric sequence where $a_3 = 81$ and $r = 3$?

**31. PRECISION** Are the terms of a geometric sequence independent or dependent? Explain your reasoning.

**32. ROOM AND BOARD** A college student makes a deal with her parents to live at home instead of living on campus. She will pay her parents $0.01 for the first day of the month, $0.02 for the second day, $0.04 for the third day, and so on.

  a. Write an equation for the nth term of the geometric sequence.

  b. What will she pay on the 25th day?

  c. Did the student make a good choice or should she have chosen to live on campus? Explain.

**33. Repeated Reasoning** A soup kitchen makes 16 gallons of soup. Each day, a quarter of the soup is served and the rest is saved for the next day.

  a. Write the first five terms of the sequence of the number of fluid ounces of soup left each day.

  b. Write an equation to represent the sequence.

  c. When is all the soup gone? Explain.

**Fair Game Review** What you learned in previous grades & lessons

**Simplify the expression.** *(Skills Review Handbook)*

**34.** $2n - 6n$

**35.** $2(4x - 5) + x$

**36.** $4(y - 1) - (y + 2)$

**37. MULTIPLE CHOICE** What is the solution of $6(3 - x) = -4x + 12$? *(Section 1.3)*

  Ⓐ $x = -3$   Ⓑ $x = -2$   Ⓒ $x = 3$   Ⓓ $x = 6$

# Extension 6.7 Recursively Defined Sequences

**Key Vocabulary**
recursive rule, p. 312

In Sections 5.6 and 6.7, you wrote *explicit* equations for sequences. Now, you will write *recursive* equations for sequences. A **recursive rule** gives the beginning term(s) of a sequence and an equation that indicates how any term $a_n$ in the sequence relates to the previous term.

## Key Idea

**Recursive Equation for an Arithmetic Sequence**
$a_n = a_{n-1} + d$, where $d$ is the common difference.

**Recursive Equation for a Geometric Sequence**
$a_n = r \cdot a_{n-1}$, where $r$ is the common ratio.

### EXAMPLE 1 Writing Terms of Recursively Defined Sequences

Write the first six terms of each sequence. Then graph each sequence.

a. $a_1 = 2, a_n = a_{n-1} + 3$

$a_1 = 2$
$a_2 = a_1 + 3 = 2 + 3 = 5$
$a_3 = a_2 + 3 = 5 + 3 = 8$
$a_4 = a_3 + 3 = 8 + 3 = 11$
$a_5 = a_4 + 3 = 11 + 3 = 14$
$a_6 = a_5 + 3 = 14 + 3 = 17$

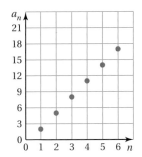

b. $a_1 = 1, a_n = 3a_{n-1}$

$a_1 = 1$
$a_2 = 3a_1 = 3(1) = 3$
$a_3 = 3a_2 = 3(3) = 9$
$a_4 = 3a_3 = 3(9) = 27$
$a_5 = 3a_4 = 3(27) = 81$
$a_6 = 3a_5 = 3(81) = 243$

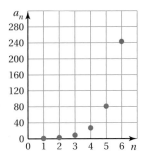

## Practice

Write the first six terms of the sequence. Then graph the sequence.

1. $a_1 = 0, a_n = a_{n-1} - 8$

2. $a_1 = -7.5, a_n = a_{n-1} + 2.5$

3. $a_1 = -36, a_n = \frac{1}{2}a_{n-1}$

4. $a_1 = 0.7, a_n = 10a_{n-1}$

# EXAMPLE 2  Writing Recursive Rules

**Recursive Sequences**

In this extension, you will
- write the terms of recursively defined sequences.
- write recursive equations for sequences.

Learning Standards
F.BF.2
F.IF.3
F.LE.2

**Write a recursive rule for each sequence.**

**a.** $-30, -18, -6, 6, 18, \ldots$

Use a table to organize the terms and find the pattern.

| Position | 1 | 2 | 3 | 4 | 5 |
|----------|---|---|---|---|---|
| Term | $-30$ | $-18$ | $-6$ | 6 | 18 |

$+12 \quad +12 \quad +12 \quad +12$

The sequence is arithmetic with first term $-30$ and common difference 12.

$a_n = a_{n-1} + d$    Recursive equation (arithmetic)

$a_n = a_{n-1} + 12$    Substitute 12 for $d$.

So, a recursive rule for the sequence is $a_1 = -30$, $a_n = a_{n-1} + 12$.

**b.** $500, 100, 20, 4, 0.8, \ldots$

Use a table to organize the terms and find the pattern.

| Position | 1 | 2 | 3 | 4 | 5 |
|----------|---|---|---|---|---|
| Term | 500 | 100 | 20 | 4 | 0.8 |

$\times \frac{1}{5} \quad \times \frac{1}{5} \quad \times \frac{1}{5} \quad \times \frac{1}{5}$

The sequence is geometric with first term 500 and common ratio $\frac{1}{5}$.

$a_n = r \cdot a_{n-1}$    Recursive equation (geometric)

$a_n = \frac{1}{5} a_{n-1}$    Substitute $\frac{1}{5}$ for $r$.

So, a recursive rule for the sequence is $a_1 = 500$, $a_n = \frac{1}{5} a_{n-1}$.

## Practice

**Write a recursive rule for the sequence.**

**5.** $8, 3, -2, -7, -12, \ldots$

**6.** $1.3, 2.6, 3.9, 5.2, 6.5, \ldots$

**7.** $4, 20, 100, 500, 2500, \ldots$

**8.** $1600, -400, 100, -25, 6.25, \ldots$

**9. SUNFLOWERS** Write a recursive rule for the height of the sunflower over time.

1 month: 2 feet    2 months: 3.5 feet    3 months: 5 feet    4 months: 6.5 feet

## EXAMPLE 3  Translating Recursive Rules into Explicit Equations

**Write an explicit equation for each recursive rule.**

**a.** $a_1 = 25, a_n = a_{n-1} - 10$

The recursive rule represents an arithmetic sequence with first term 25 and common difference $-10$.

$a_n = a_1 + (n-1)d$     Equation for an arithmetic sequence

$a_n = 25 + (n-1)(-10)$     Substitute 25 for $a_1$ and $-10$ for $d$.

$a_n = -10n + 35$     Simplify.

**b.** $a_1 = 19.6, a_n = -0.5a_{n-1}$

The recursive rule represents a geometric sequence with first term 19.6 and common ratio $-0.5$.

$a_n = a_1 r^{n-1}$     Equation for a geometric sequence

$a_n = 19.6(-0.5)^{n-1}$     Substitute 19.6 for $a_1$ and $-0.5$ for $r$.

## EXAMPLE 4  Translating Explicit Equations into Recursive Rules

**Write a recursive rule for each explicit equation.**

**a.** $a_n = -2n + 3$

The explicit equation represents an arithmetic sequence with first term $-2(1) + 3 = 1$ and common difference $-2$.

$a_n = a_{n-1} + d$     Recursive equation (arithmetic)

$a_n = a_{n-1} + (-2)$     Substitute $-2$ for $d$.

So, a recursive rule for the sequence is $a_1 = 1, a_n = a_{n-1} - 2$.

**b.** $a_n = -3(2)^{n-1}$

The explicit equation represents a geometric sequence with first term $-3$ and common ratio 2.

$a_n = r \cdot a_{n-1}$     Recursive equation (geometric)

$a_n = 2a_{n-1}$     Substitute 2 for $r$.

So, a recursive rule for the sequence is $a_1 = -3, a_n = 2a_{n-1}$.

## Practice

**Write an explicit equation for the recursive rule.**

**10.** $a_1 = -45, a_n = a_{n-1} + 20$

**11.** $a_1 = 13, a_n = -3a_{n-1}$

**Write a recursive rule for the explicit equation.**

**12.** $a_n = -n + 1$

**13.** $a_n = -2.5(2)^{n-1}$

You can write recursive rules for sequences that are neither arithmetic nor geometric. One way is to look for patterns in the sums of consecutive terms.

**EXAMPLE 5** **Writing Recursive Rules for Other Sequences**

Write a recursive rule for the sequence 1, 1, 2, 3, 5, 8, …. Then write the next 3 terms of the sequence.

The sequence does not have a common difference or a common ratio. Find the sums of consecutive terms.

$a_1 + a_2 = 1 + 1 = 2$   2 is the third term.
$a_2 + a_3 = 1 + 2 = 3$   3 is the fourth term.
$a_3 + a_4 = 2 + 3 = 5$   5 is the fifth term.
$a_4 + a_5 = 3 + 5 = 8$   8 is the sixth term.

So, a recursive equation for the sequence is $a_n = a_{n-2} + a_{n-1}$. Use the equation to find the next three terms.

$a_7 = a_5 + a_6$     $a_8 = a_6 + a_7$     $a_9 = a_7 + a_8$
$= 5 + 8$           $= 8 + 13$        $= 13 + 21$
$= 13$              $= 21$              $= 34$

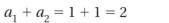

∴ A recursive rule for the sequence is $a_1 = 1$, $a_2 = 1$, $a_n = a_{n-2} + a_{n-1}$. The next three terms are 13, 21, and 34.

The sequence in Example 5 is called the *Fibonacci sequence*. This pattern is naturally occurring in many objects, such as flowers.

**Write a recursive rule for the sequence. Then write the next 3 terms of the sequence.**

**14.** 5, 6, 11, 17, 28, …

**15.** −3, −4, −7, −11, −18, …

**16.** 1, 1, 0, −1, −1, 0, 1, 1, …

**17.** 4, 3, 1, 2, −1, 3, −4, …

**Use a pattern in the products of consecutive terms to write a recursive rule for the sequence. Then write the next 2 terms of the sequence.**

**18.** 2, 3, 6, 18, 108, …

**19.** −2, 2.5, −5, −12.5, 62.5, …

**20. GEOMETRY** Consider squares 1–6 in the diagram.

  **a.** Write a sequence in which each term $a_n$ is the side length of square $n$.

  **b.** What is the name of this sequence? What is the next term of this sequence?

  **c.** Use the term in part (b) to add another square to the diagram and extend the spiral.

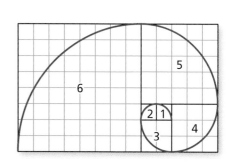

# 6.4–6.7 Quiz

**Does the table represent a *linear* or an *exponential* function? Explain.** *(Section 6.4)*

1. 
| x | 1 | 2 | 3 | 4 |
|---|---|---|---|---|
| y | 5 | 10 | 15 | 20 |

2. 
| x | 2 | 4 | 6 | 8 |
|---|---|---|---|---|
| y | 5 | 10 | 20 | 40 |

**Graph the function. Describe the domain and range.** *(Section 6.4)*

3. $y = 5^x$

4. $y = -2\left(\dfrac{1}{6}\right)^x$

**Solve the equation. Check your solution, if possible.** *(Section 6.4)*

5. $8^{x+2} = 64^{4x+1}$

6. $7^{2x-6} = 49^{3x-11}$

**Determine whether the table represents an *exponential growth function*, an *exponential decay function*, or *neither*.** *(Section 6.6)*

7. 
| x | 0 | 1 | 2 | 3 |
|---|---|---|---|---|
| y | 7 | 21 | 63 | 189 |

8. 
| x | 1 | 2 | 3 | 4 |
|---|---|---|---|---|
| y | 14,641 | 1331 | 121 | 11 |

**Write the next three terms of the geometric sequence. Then graph the sequence.** *(Section 6.7)*

9. $15, -45, 135, -405, \ldots$

10. $768, 192, 48, 12, \ldots$

**Write a recursive rule for the sequence.** *(Section 6.7)*

11. $5, 11, 17, 23, \ldots$

12. $-14, 28, -56, 112, \ldots$

13. **SAVINGS ACCOUNT** You deposit $2500 in a savings account that earns 6% annual interest compounded yearly. *(Section 6.5)*

   a. Write and graph a function that represents the balance $y$ (in dollars) after $t$ years.

   b. What is the balance after 5 years?

14. **CURRENCY** A country's base unit of currency is valued at US$2. The country's base unit of currency loses about 3.9% of its value every month. *(Section 6.6)*

   a. Write a function that represents the value $y$ (in U.S. dollars) of the base unit of currency after $t$ months.

   b. What is the value of the country's base unit of currency after 1.5 years?

# 6 Chapter Review

## Review Key Vocabulary

closed, *p. 266*
*n*th root, *p. 278*
exponential function, *p. 286*
exponential growth, *p. 296*
exponential growth function, *p. 296*
compound interest, *p. 297*
exponential decay, *p. 302*
exponential decay function, *p. 302*
geometric sequence, *p. 308*
common ratio, *p. 308*
recursive rule, *p. 312*

## Review Examples and Exercises

### 6.1 Properties of Square Roots (pp. 260–267)

Evaluate $\sqrt{b^2 - 4ac}$ when $a = -2$, $b = 2$, and $c = 5$.

$\sqrt{b^2 - 4ac} = \sqrt{2^2 - 4(-2)(5)}$     Substitute.

$\qquad\qquad\;\; = \sqrt{44}$     Simplify.

$\qquad\qquad\;\; = \sqrt{4 \cdot 11}$     Factor.

$\qquad\qquad\;\; = \sqrt{4} \cdot \sqrt{11}$     Product Property of Square Roots

$\qquad\qquad\;\; = 2\sqrt{11}$     Simplify.

### Exercises

Evaluate the expression when $x = 3$, $y = 4$, and $z = 2$.

1. $\sqrt{xy^2 z}$
2. $\sqrt{2z + y}$
3. $\dfrac{8 + \sqrt{xy}}{z}$

### 6.2 Properties of Exponents (pp. 268–275)

Simplify $\left(\dfrac{3x}{4}\right)^{-4}$. Write your answer using only positive exponents.

$\left(\dfrac{3x}{4}\right)^{-4} = \dfrac{(3x)^{-4}}{4^{-4}}$     Power of a Quotient Property

$\qquad\qquad = \dfrac{4^4}{(3x)^4}$     Definition of negative exponent

$\qquad\qquad = \dfrac{4^4}{3^4 x^4}$     Power of a Product Property

$\qquad\qquad = \dfrac{256}{81x^4}$     Simplify.

Chapter Review   317

## Exercises

**Simplify. Write your answer using only positive exponents.**

4. $y^3 \cdot y^{-3}$
5. $\dfrac{x^4}{x^7}$
6. $(xy^2)^3$
7. $\left(\dfrac{2x}{5y}\right)^{-2}$

### 6.3 Radicals and Rational Exponents (pp. 276–281)

**Simplify each expression.**

a. $\sqrt[3]{512} = \sqrt[3]{8 \cdot 8 \cdot 8} = 8$  Rewrite and simplify.

b. $900^{1/2} = \sqrt{900}$  Write the expression in radical form.
   $= \sqrt{30 \cdot 30}$  Rewrite.
   $= 30$  Simplify.

## Exercises

**Simplify the expression.**

8. $\sqrt[3]{8}$
9. $64^{1/2}$
10. $625^{3/4}$

### 6.4 Exponential Functions (pp. 284–293)

a. **Graph $y = 4^x$.**

   **Step 1:** Make a table of values.

   | x | −1 | 0 | 1 | 2 | 3 |
   |---|---|---|---|---|---|
   | y | 0.25 | 1 | 4 | 16 | 64 |

   **Step 2:** Plot the ordered pairs.

   **Step 3:** Draw a smooth curve through the points.

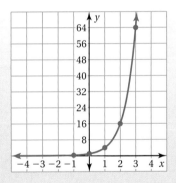

b. **Write an exponential function represented by the graph.**

   Use the graph to make a table of values.

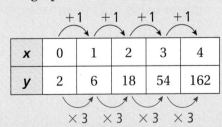

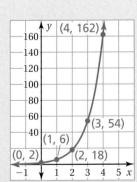

   The y-intercept is 2 and the y-values increase by a factor of 3 as x increases by 1.

   ∴ So, the exponential function is $y = 2(3)^x$.

### Exercises

**11.** Graph $y = -2(4)^x + 3$. Describe the domain and range. Compare the graph to the graph of $y = -2(4)^x$.

**Write an exponential function represented by the graph or table.**

**12.**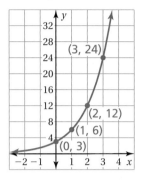

**13.**

| x | 0 | 1 | 2 | 3 |
|---|---|---|---|---|
| y | 2 | 1 | 0.5 | 0.25 |

**Solve the equation. Check your solution, if possible.**

**14.** $3^x = 27$

**15.** $5^x = 5^{x-2}$

**16.** $2^{5x} = 8^{2x-4}$

## 6.5 Exponential Growth (pp. 294–299)

The enrollment at a high school increases by 4% each year. In 2010, there were 800 students enrolled at the school.

**a.** Write a function that represents the enrollment $y$ of the high school after $t$ years.

$y = a(1 + r)^t$      Write exponential growth function.

$y = 800(1 + 0.04)^t$      Substitute 800 for $a$ and 0.04 for $r$.

$y = 800(1.04)^t$      Simplify.

**b.** How many students will be enrolled at the high school in 2020?

The value $t = 10$ represents 2020.

$y = 800(1.04)^t$      Write exponential growth function.

$y = 800(1.04)^{10}$      Substitute 10 for $t$.

$\approx 1184$      Use a calculator.

### Exercises

**17. PLUMBER** A plumber charges $22 per hour. The hourly rate increases by 3% each year.

**a.** Write a function that represents the plumber's hourly rate $y$ (in dollars) after $t$ years.

**b.** What is the plumber's hourly rate after 8 years?

## 6.6 Exponential Decay (pp. 300–305)

The table shows the value of a boat over time.

| Year, t | 0 | 1 | 2 | 3 |
|---|---|---|---|---|
| Value, y | $6000 | $4800 | $3840 | $3072 |

**a.** Determine whether the table represents an *exponential growth function*, an *exponential decay function*, or *neither*.

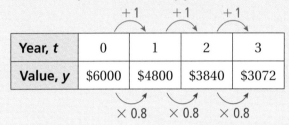

As $x$ increases by 1, $y$ is multiplied by 0.8. So, the table represents an exponential decay function.

**b.** The boat loses 20% of its value every year. Write a function that represents the value $y$ (in dollars) of the boat after $t$ years.

$y = a(1 - r)^t$   Write exponential decay function.

$y = 6000(1 - 0.2)^t$   Substitute 6000 for $a$ and 0.2 for $r$.

$y = 6000(0.8)^t$   Simplify.

**c.** Graph the function from part (b). Use the graph to estimate the value of the boat after 8 years.

From the graph, you can see that the $y$-value is about 1000 when $t = 8$.

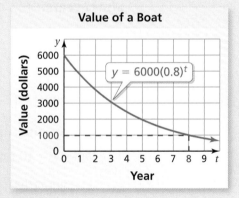

So, the value of the boat is about $1000 after 8 years.

### Exercises

Determine whether the table represents an *exponential growth function*, an *exponential decay function*, or *neither*.

**18.**

| x | 0 | 1 | 2 | 3 |
|---|---|---|---|---|
| y | 3 | 6 | 12 | 24 |

**19.**

| x | 1 | 2 | 3 | 4 |
|---|---|---|---|---|
| y | 162 | 108 | 72 | 48 |

**20. DISCOUNT** The price of a TV is $1500. The price decreases by 6% each month. Write and graph a function that represents the price $y$ (in dollars) of the TV after $t$ months. Use the graph to estimate the price of the TV after 1 year.

## 6.7 Geometric Sequences (pp. 306–315)

**a. Write the next three terms of the geometric sequence 2, 6, 18, 54, . . . .**

Use a table to organize the terms and extend the pattern.

| Position | 1 | 2 | 3 | 4 | 5 | 6 | 7 |
|---|---|---|---|---|---|---|---|
| Term | 2 | 6 | 18 | 54 | 162 | 486 | 1458 |

× 3, × 3, × 3, × 3, × 3, × 3

Each term is 3 times the previous term. So, the common ratio is 3.

Multiply a term by 3 to find the next term.

∴ The next three terms are 162, 486, and 1458.

**b. Graph the geometric sequence 24, 12, 6, 3, 1.5, . . . . What do you notice?**

Make a table. Then plot the ordered pairs $(n, a_n)$.

| Position, $n$ | 1 | 2 | 3 | 4 | 5 |
|---|---|---|---|---|---|
| Term, $a_n$ | 24 | 12 | 6 | 3 | 1.5 |

∴ The points of the graph appear to lie on an exponential curve.

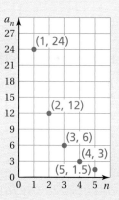

### Exercises

**Write the next three terms of the geometric sequence. Then graph the sequence.**

**21.** −3, 9, −27, 81, . . .

**22.** 48, 12, 3, $\frac{3}{4}$, . . .

**Write an equation for the $n$th term of the geometric sequence.**

**23.**

| $n$ | 1 | 2 | 3 | 4 |
|---|---|---|---|---|
| $a_n$ | 1 | 4 | 16 | 64 |

**24.**

| $n$ | 1 | 2 | 3 | 4 |
|---|---|---|---|---|
| $a_n$ | 5 | −10 | 20 | −40 |

**Write a recursive rule for the sequence.**

**25.** 3, 8, 13, 18, 23, . . .

**26.** 3, 6, 12, 24, 48, . . .

# 6 Chapter Test

**Simplify the expression.**

1. $\sqrt{98}$
2. $\sqrt{\dfrac{19}{25}}$
3. $\dfrac{6 - \sqrt{48}}{2}$

**Simplify. Write your answer using only positive exponents.**

4. $z^{-2} \cdot z^4$
5. $\dfrac{b^{-5}}{b^{-8}}$
6. $\left(\dfrac{2c^4}{5}\right)^{-3}$

**Simplify the expression.**

7. $\sqrt[4]{16}$
8. $729^{1/6}$
9. $32^{7/5}$

10. Graph $y = 7^x + 1$. Describe the domain and range. Compare the graph to the graph of $y = 7^x$.

**Write an exponential function represented by the table.**

11. 
| x | 0 | 1 | 2 | 3 |
|---|---|---|---|---|
| y | −1 | −2 | −4 | −8 |

12. 
| x | 0 | 1 | 2 | 3 |
|---|---|---|---|---|
| y | 3 | −12 | 48 | −192 |

**Solve the equation. Check your solution, if possible.**

13. $2^x = 128$
14. $256^{x+2} = 16^{3x-1}$

**Write and graph a function that represents the situation.**

15. Your $42,500 annual salary increases by 3% each year.

16. You deposit $500 in an account that earns 6.5% annual interest compounded yearly.

**Determine whether the table represents an *exponential growth function*, an *exponential decay function*, or *neither*.**

17. 
| x | 0 | 1 | 2 | 3 |
|---|---|---|---|---|
| y | 15 | 30 | 60 | 120 |

18. 
| x | 0 | 1 | 2 | 3 |
|---|---|---|---|---|
| y | 400 | 100 | 25 | 6.25 |

19. **TRAINING** You follow the training schedule from your coach.

   a. Write an equation for the *n*th term of the geometric sequence.

   b. Write a recursive rule for the explicit equation in part (a).

   c. On what day do you run approximately 3 kilometers?

   *Training On Your Own*
   Day 1: Run 1 km.
   Each day after Day 1: Run 20% farther than the previous day.

# 6 Standards Assessment

1. Which point is a solution of the system of inequalities shown below? *(A.REI.12)*

   $$y \geq 4x - 3$$
   $$3x - 2y < 4$$

   **A.** $(-2, -7)$     **C.** $(-4, -8)$

   **B.** $(1, 1)$     **D.** $(4, 5)$

2. What is the value of the function $y = -10(5)^x$ when $x = -3$? *(F.IF.7e)*

   **F.** $-\dfrac{2}{25}$     **H.** $\dfrac{2}{25}$

   **G.** $-\dfrac{2}{125}$     **I.** 30

3. Which graph shows the solution of $x - 1.9 \geq 0.3$? *(A.CED.1)*

   **A.**

   **C.**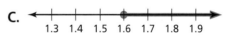

   **B.**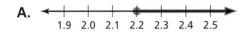

   **D.**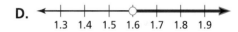

4. What is the value of $27^{4/3}$? *(N.RN.2)*

5. Which graph represents the equation $-5x - 5y = 25$? *(A.REI.10)*

   **F.**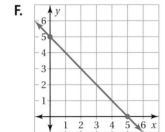

   **H.**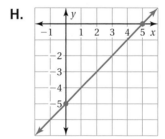

   **G.**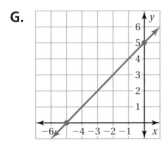

   **I.**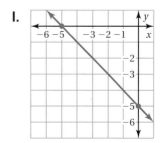

6. A system of two linear equations has infinitely many solutions. What can you conclude about the graphs of the two equations? *(8.EE.8b)*

   A. The lines have the same slope and the same y-intercept.

   B. The lines have the same slope and different y-intercepts.

   C. The lines have different slopes and the same y-intercept.

   D. The lines have different slopes and different y-intercepts.

7. The domain of the function $y = -5x + 19$ is 0, 2, 4, and 6. What is the range of the function? *(F.IF.1)*

   F. −19, −9, 1, 11

   G. −11, −1, 9, 19

   H. 6, 4, 2, 0

   I. −19, −11, −9, 1

8. What is the 50th term of the sequence 20, 9, −2, −13, . . . ? *(F.LE.2)*

9. Which graph shows an exponential decay function? *(F.IF.7E)*

   A.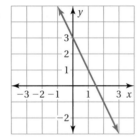

   B.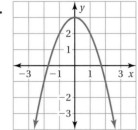

   C.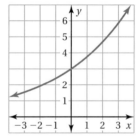

   D.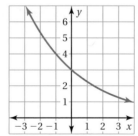

10. The lowest temperature ever recorded on Earth is −129° Fahrenheit. The highest temperature ever recorded on Earth is 136° Fahrenheit. Let $t$ represent the temperature, in degrees Fahrenheit. Which inequality represents all temperatures ever recorded on Earth? *(A.CED.1)*

    F. $-129 < t < 136$

    G. $-129 \leq t < 136$

    H. $-129 \leq t \leq 136$

    I. $-129 < t \leq 136$

11. The graph of which equation is perpendicular to the line that passes through the points $(-3, -6)$ and $(5, -2)$?  *(F.IF.6)*

   A. $y = \dfrac{1}{2}x + 3$

   B. $y = -2x + 7$

   C. $y = -\dfrac{1}{2}x - 3$

   D. $y = 2x + 1$

12. Which of the following is true about the graph of the linear equation $y = -7x + 5$?  *(F.IF.4)*

   F. The slope is 5 and the $y$-intercept is $-7$.

   G. The slope is $-5$ and the $y$-intercept is $-7$.

   H. The slope is $-7$ and the $y$-intercept is $-5$.

   I. The slope is $-7$ and the $y$-intercept is 5.

13. At the beginning of a tennis tournament, there are 256 players. After each round, one-half of the remaining players are eliminated.  *(F.IF.7e)*

   *Part A* Write a function that represents the number of players left in the tournament after each round.

   *Part B* Does the function in part (a) represent exponential growth or exponential decay? Explain your reasoning.

   *Part C* Graph the function in part (a).

   *Part D* How many tennis matches does a player have to win to win the tournament? Explain your reasoning.

14. Which sequence is neither arithmetic nor geometric?  *(F.IF.3)*

   A. 1, 50, 2500, 125,000, . . .

   B. 10, 0, $-10$, $-20$, . . .

   C. 4, $-4$, 4, $-4$, . . .

   D. 0, 1, 3, 6, . . .

15. For $f(x) = -3x - 10$, what value of $x$ makes $f(x) = -7$?  *(F.IF.2)*

   F. $-7$    G. $-1$    H. 1    I. 11

16. Which expression is equivalent to $20\sqrt{200}$?  *(N.RN.3)*

   A. $40\sqrt{2}$    B. $40\sqrt{10}$    C. $200\sqrt{2}$    D. 400

# 7 Polynomial Equations and Factoring

- 7.1 Polynomials
- 7.2 Adding and Subtracting Polynomials
- 7.3 Multiplying Polynomials
- 7.4 Special Products of Polynomials
- 7.5 Solving Polynomial Equations in Factored Form
- 7.6 Factoring Polynomials Using the GCF
- 7.7 Factoring $x^2 + bx + c$
- 7.8 Factoring $ax^2 + bx + c$
- 7.9 Factoring Special Products

"Here's how it goes, Descartes."

"The friends of my friends are my friends. The friends of my enemies are my enemies."

"The enemies of my friends are my enemies. The enemies of my enemies are my friends."

"Descartes, which one is the monomial and which one is the polynomial?"

"Remember that poly means many and mono means one."

# What You Learned Before

"Dear Editor, I disagree with your claim that the sum of two binomials is always a binomial."

## ● Simplifying Algebraic Expressions (7.EE.1)

**Example 1** Simplify $5x + 7 - 2x - 3$.

$5x + 7 - 2x - 3 = 5x - 2x + 7 - 3$     Commutative Property of Addition
$\phantom{5x + 7 - 2x - 3} = (5 - 2)x + 7 - 3$     Distributive Property
$\phantom{5x + 7 - 2x - 3} = 3x + 4$     Simplify.

**Example 2** Simplify $-7(y - 2) + 3y$.

$-7(y - 2) + 3y = -7(y) - (-7)(2) + 3y$     Distributive Property
$\phantom{-7(y - 2) + 3y} = -7y + 14 + 3y$     Multiply.
$\phantom{-7(y - 2) + 3y} = -7y + 3y + 14$     Commutative Property of Addition
$\phantom{-7(y - 2) + 3y} = (-7 + 3)y + 14$     Distributive Property
$\phantom{-7(y - 2) + 3y} = -4y + 14$     Add coefficients.

### Try It Yourself
**Simplify the expression.**

1. $3x - 8 + 4x$
2. $3t - 4 - 6t + 7$
3. $-7z + 3 + 2z + 4z + 5$
4. $3(w + 2) - 5$
5. $4g - 2(g + 6)$
6. $3(n + 1) - 4(n - 3)$

## ● Finding the Greatest Common Factor (6.NS.4)

**Example 3** Find the greatest common factor of 50 and 75.

$50 = 2 \cdot 5 \cdot 5$
$75 = 3 \cdot 5 \cdot 5$

∴ The GCF is $5 \cdot 5 = 25$.

**Example 4** Find the greatest common factor of 30 and 42.

$30 = 2 \cdot 3 \cdot 5$
$42 = 2 \cdot 3 \cdot 7$

∴ The GCF is $2 \cdot 3 = 6$.

### Try It Yourself
**Find the greatest common factor.**

7. 28, 64
8. 60, 72
9. 24, 27

## 7.1 Polynomials

**Essential Question** How can you use algebra tiles to model and classify polynomials?

### 1 ACTIVITY: Meaning of Prefixes

Work with a partner. Think of a word that uses one of the prefixes with one of the base words. Then define the word and write a sentence that uses the word.

| Prefix | Base Word |
|---|---|
| Mono | Dactyl |
| Bi | Cycle |
| Tri | Ped |
| Poly | Syllabic |

### 2 ACTIVITY: Classifying Polynomials Using Algebra Tiles

Work with a partner. Six different algebra tiles are shown at the right.

Write the polynomial that is modeled by the algebra tiles. Then classify the polynomial as a monomial, binomial, or trinomial. Explain your reasoning.

a.

b.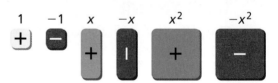

c.

d.

e.

f.

**Common Core**

**Polynomials**

In this lesson, you will
- find the degrees of monomials.
- classify polynomials.

Applying Standard A.SSE.1a

## 3  ACTIVITY: Solving an Algebra Tile Puzzle

**Math Practice 2**

**Use Expressions**
What do the shapes and colors of the tiles represent? How does this help you write a polynomial?

Work with a partner. Write the polynomial modeled by the algebra tiles, evaluate the polynomial at the given value, and write the result in the corresponding square of the Sudoku puzzle. Then solve the puzzle.

### A3, H7
Value when $x = 2$

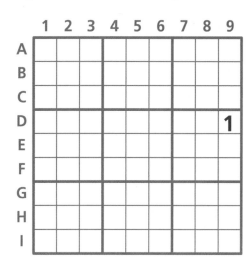

### A4, B3, E5, G6, I7
Value when $x = 2$

### A6, D7, E2, H5
Value when $x = -3$

### B5, F1, H3
Value when $x = -1$

### A7, F9, I4
Value when $x = 3$

### E8, F3, I6
Value when $x = -1$

### C4, I3
Value when $x = 3$

### B7, D1
Value when $x = -2$

## What Is Your Answer?

4. **IN YOUR OWN WORDS** How can you use algebra tiles to model and classify polynomials? Explain why algebra tiles have the dimensions, shapes, and colors that they have.

**Practice** → Use what you learned about modeling polynomials to complete Exercises 5 and 6 on page 332.

Section 7.1  Polynomials  329

# 7.1 Lesson

Check It Out
Lesson Tutorials
BigIdeasMath.com

**Key Vocabulary**
monomial, p. 330
degree of a
 monomial, p. 330
polynomial, p. 331
binomial, p. 331
trinomial, p. 331
degree of a
 polynomial, p. 331

A **monomial** is a number, a variable, or a product of a number and one or more variables with whole number exponents.

| Monomials | Not monomials | Reason |
|---|---|---|
| $-4$ | $x^{1.5}$ | Monomials must have whole number exponents. |
| $\frac{1}{2}y^2$ | $-\frac{2}{z}$ | Monomials cannot have variables in the denominator. |
| $2.5x^2y$ | $7^y$ | Monomials cannot have variable exponents. |

The **degree of a monomial** is the sum of the exponents of the variables in the monomial.

### EXAMPLE 1   Finding the Degrees of Monomials

**Find the degree of each monomial.**

**a.** $5x^2$

The exponent of $x$ is 2.

∴ So, the degree of the monomial is 2.

**b.** $-\frac{1}{2}xy^3$

The exponent of $x$ is 1 and the exponent of $y$ is 3.

The sum of the exponents is $1 + 3 = 4$.

∴ So, the degree of the monomial is 4.

**c.** $-3$

You can rewrite $-3$ as $-3x^0$.

The exponent of $x$ is 0.

∴ So, the degree of the monomial is 0.

**Remember**
For any nonzero number $a$, $a^0 = 1$.

### On Your Own

Now You're Ready
Exercises 7–14

**Find the degree of the monomial.**

**1.** $-3x^4$

**2.** $7c^3d^2$

**3.** $\frac{5}{3}y$

**4.** $-20.5$

A **polynomial** is a monomial or a sum of monomials. Each monomial is called a *term* of the polynomial.

A polynomial with two terms is a **binomial**.

$5x + 2$

A polynomial with three terms is a **trinomial**.

$x^2 + 5x + 2$

The **degree of a polynomial** is the greatest degree of its terms. A polynomial in one variable is in *standard form* when the exponents of the terms decrease from left to right.

### EXAMPLE 2  Classifying Polynomials

**Write each polynomial in standard form. Identify the degree and classify each polynomial by the number of terms.**

| | Polynomial | Standard Form | Degree | Type of Polynomial |
|---|---|---|---|---|
| a. | $-3z^4$ | $-3z^4$ | 4 | monomial |
| b. | $4 + 5x^2 - x$ | $5x^2 - x + 4$ | 2 | trinomial |
| c. | $8q + q^5$ | $q^5 + 8q$ | 5 | binomial |

### EXAMPLE 3  Real-Life Application

The polynomial $-16t^2 + v_0 t + s_0$ represents the height (in feet) of an object, where $v_0$ is the initial vertical velocity (in feet per second), $s_0$ is the initial height of the object (in feet), and $t$ is the time (in seconds).

a. **Write a polynomial that represents the height of the baseball.**

$-16t^2 + v_0 t + s_0 = -16t^2 + 30t + 4$   Substitute 30 for $v_0$ and 4 for $s_0$.

b. **What is the height of the baseball after 1 second?**

$-16t^2 + 30t + 4 = -16(1)^2 + 30(1) + 4$   Substitute 1 for $t$.

$= -16 + 30 + 4$   Simplify.

$= 18$   Add.

∴ The height of the baseball after 1 second is 18 feet.

### On Your Own

**Now You're Ready**
Exercises 15–23

**Write the polynomial in standard form. Identify the degree and classify the polynomial by the number of terms.**

**5.** $4 - 9z$   **6.** $t^2 - t^3 - 10t$   **7.** $2.8x + x^3$

**8.** In Example 3, the initial height is 5 feet. What is the height of the baseball after 2 seconds?

Section 7.1   Polynomials   331

# 7.1 Exercises

## Vocabulary and Concept Check

1. **WRITING** Is $-\dfrac{\pi}{3}$ a monomial? Explain your reasoning.

2. **VOCABULARY** When is a polynomial in one variable in standard form?

3. **OPEN-ENDED** Write a trinomial of degree 5 in standard form.

4. **WHICH ONE DOESN'T BELONG?** Which expression does *not* belong with the other three? Explain your reasoning.

   $a^3 + 4a$    $8^x$    $b - 2^{-1}$    $-6y^8z$

## Practice and Problem Solving

**Use algebra tiles to represent the polynomial.**

5. $x^2 + 2x - 4$

6. $2x^2 - x + 3$

**Find the degree of the monomial.**

7. $4g$

8. $23x^4$

9. $s^8 t$

10. $-\dfrac{4}{9}$

11. $1.75k^2$

12. $\dfrac{1}{8}m^2 n^4$

13. $2\pi$

14. $-3q^4 r s^6$

**Write the polynomial in standard form. Identify the degree and classify the polynomial by the number of terms.**

15. $7 + 3p^2$

16. $2w^6$

17. $8d - 2 - 4d^3$

18. $6.5c^2 + 1.2c^4 - c$

19. $4v^{11} - v^{12}$

20. $-\dfrac{1}{4}y - \dfrac{3}{8}y^2$

21. $7.4z^5$

22. $\sqrt{3}n^7 - 19 + \sqrt{2}n^3$

23. $\pi r^2 - \dfrac{5}{7}r^8 + 2r^5$

24. **ERROR ANALYSIS** Describe and correct the error in writing the polynomial in standard form.

    ✗ polynomial: $3m^2 - 5m^5 + m^4$
    standard form: $-5m^5 + 3m^2 + m^4$

25. **SPHERE** The expression $\dfrac{4}{3}\pi r^3$ represents the volume of a sphere with radius $r$. Why is this expression a monomial? What is its degree?

**Tell whether the expression is a polynomial. If so, identify the degree and classify the polynomial by the number of terms.**

**26.** $-g^3$

**27.** $7^x - 2x^2$

**28.** $y^{-3} + 1.5$

**29.** $8k^5 + 4k^3 - k$

**30. LOGIC** The polynomial $d^2 - \pi r^2$ represents the area of a region, where $d$ is the diameter of a circle and $r$ is the radius of the circle. How can this happen? Justify your answer with a diagram.

**Use the polynomial $-16t^2 + v_0 t + s_0$ to write a polynomial that represents the height of the object. Then find the height of the object after 1 second.**

**31. WATER BALLOON** You throw a water balloon from a building.

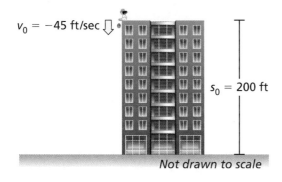

**32. TENNIS** You bounce a tennis ball on a racket.

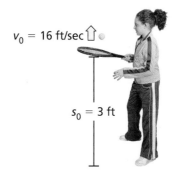

**33. Number Sense** The polynomial $-w^2 + 28w$ represents the area of a rectangular garden with a width of $w$ feet.

a. Use guess, check, and revise to find the width of the garden with the maximum area. (*Hint:* The width is between 10 feet and 18 feet.)

b. What is the perimeter of the garden?

c. How many seed packets do you need for the garden?

**Fair Game Review** *What you learned in previous grades & lessons*

**Simplify the expression.** *(Skills Review Handbook)*

**34.** $2x + 4y + 3x + 13$

**35.** $4x - x + 5y - 7y$

**36.** $-11 + 5x - 3x + x$

**37. MULTIPLE CHOICE** What is the surface area of the prism? *(Skills Review Handbook)*

Ⓐ $11x$ ft

Ⓑ $(24x + 36)$ ft

Ⓒ $(36x + 24)$ ft

Ⓓ $(60x + 12)$ ft

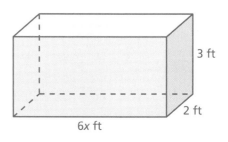

## 7.2 Adding and Subtracting Polynomials

**Essential Question** How can you add polynomials? How can you subtract polynomials?

### 1 EXAMPLE: Adding Polynomials Using Algebra Tiles

Work with a partner. Six different algebra tiles are shown at the right.

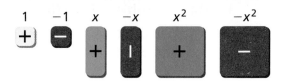

Write the polynomial addition steps shown by the algebra tiles.

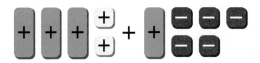

**Step 1:**  Group like tiles.

**Step 2:**  Remove zero pairs.

**Step 3:**  Simplify.

### 2 ACTIVITY: Adding Polynomials Using Algebra Tiles

**Polynomials**
In this lesson, you will
- add and subtract polynomials.

Learning Standard
A.APR.1

Use algebra tiles to find the sum of the polynomials.

**a.** $(x^2 + 2x - 1) + (2x^2 - 2x + 1)$     **b.** $(4x + 3) + (x - 2)$

**c.** $(x^2 + 2) + (3x^2 + 2x + 5)$     **d.** $(2x^2 - 3x) + (x^2 - 2x + 4)$

**e.** $(x^2 - 3x + 2) + (x^2 + 4x - 1)$     **f.** $(4x - 3) + (2x + 1) + (-3x + 2)$

**g.** $(-x^2 + 3x) + (2x^2 - 2x)$     **h.** $(x^2 + 2x - 5) + (-x^2 - 2x + 5)$

**334** Chapter 7 Polynomial Equations and Factoring

### ③ EXAMPLE: Subtracting Polynomials Using Algebra Tiles

**Write the polynomial subtraction steps shown by the algebra tiles.**

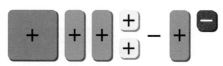

**Step 1:** 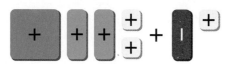  To subtract, add the opposite.

**Step 2:**   Group like tiles.

**Step 3:**   Remove zero pairs.

**Step 4:** 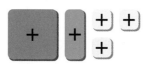  Simplify.

**Math Practice 7**

**View as Components**
How can you use algebra tiles to represent the sums and differences of polynomials?

### ④ ACTIVITY: Subtracting Polynomials Using Algebra Tiles

**Use algebra tiles to find the difference of the polynomials.**

a. $(x^2 + 2x - 1) - (2x^2 - 2x + 1)$  b. $(4x + 3) - (x - 2)$

c. $(x^2 + 2) - (3x^2 + 2x + 5)$  d. $(2x^2 - 3x) - (x^2 - 2x + 4)$

### What Is Your Answer?

5. **IN YOUR OWN WORDS** How can you add polynomials? Use the results of Activity 2 to summarize a procedure for adding polynomials without using algebra tiles.

6. **IN YOUR OWN WORDS** How can you subtract polynomials? Use the results of Activity 4 to summarize a procedure for subtracting polynomials without using algebra tiles.

**Practice**  Use what you learned about adding and subtracting polynomials to complete Exercises 3 and 4 on page 338.

Section 7.2  Adding and Subtracting Polynomials  335

## 7.2 Lesson

You can add polynomials using a vertical or horizontal method to combine like terms.

### EXAMPLE 1 Adding Polynomials

**Find each sum.**

a. $(3a^2 + 8) + (5a - 1)$

b. $(-x^2 + 5x + 4) + (3x^2 - 8x + 9)$

a. **Vertical method:** Align like terms vertically and add.

$$\begin{array}{r} 3a^2 \phantom{+5a} + 8 \\ +\phantom{3a^2+} 5a - 1 \\ \hline 3a^2 + 5a + 7 \end{array}$$

Leave a space for the missing term.

b. **Horizontal method:** Group like terms and simplify.

$(-x^2 + 5x + 4) + (3x^2 - 8x + 9) = (-x^2 + 3x^2) + [5x + (-8x)] + (4 + 9)$
$= 2x^2 - 3x + 13$

To subtract one polynomial from another polynomial, add the opposite.

### EXAMPLE 2 Subtracting Polynomials

**Find each difference.**

a. $(y^2 + 4y + 2) - (2y^2 - 5y - 3)$

b. $(5x^2 + 4x - 1) - (2x^2 - 6)$

**Study Tip**

You can think of finding the opposite of a polynomial as finding the opposite of each term's coefficient.

a. Use the vertical method.

$$\begin{array}{r} (y^2 + 4y + 2) \\ -(2y^2 - 5y - 3) \end{array}$$

Add the opposite.

$$\begin{array}{r} y^2 + 4y + 2 \\ +\; -2y^2 + 5y + 3 \\ \hline -y^2 + 9y + 5 \end{array}$$

b. Use the horizontal method.

$(5x^2 + 4x - 1) - (2x^2 - 6) = (5x^2 + 4x - 1) + (-2x^2 + 6)$
$= [5x^2 + (-2x^2)] + 4x + (-1 + 6)$
$= 3x^2 + 4x + 5$

### ● On Your Own

**Now You're Ready**
Exercises 5–10 and 12–17

**Find the sum or difference.**

1. $(b - 10) + (4b - 3)$
2. $(x^2 - x - 2) + (7x^2 - x)$
3. $(p^2 + p + 3) - (-4p^2 - p + 3)$
4. $(-k + 5) - (3k^2 - 6)$

### EXAMPLE 3 — Adding Polynomials

Which polynomial represents the sum of $x^2 - 2xy - y^2$ and $x^2 + xy + y^2$?

**A** $-3xy$    **B** $-3xy - 2y^2$    **C** $2x^2 - xy$    **D** $2x^2 + 3xy + 2y^2$

Use the horizontal method to find the sum.

$$(x^2 - 2xy - y^2) + (x^2 + xy + y^2) = (x^2 + x^2) + (-2xy + xy) + (-y^2 + y^2)$$
$$= 2x^2 - xy$$

∴ The correct answer is **C**.

### EXAMPLE 4 — Real-Life Application

A penny is thrown straight downward from a height of 200 feet. At the same time, a paintbrush falls from a height of 100 feet. The polynomials represent the heights (in feet) of the objects after $t$ seconds.

**a.** Write a polynomial that represents the distance between the penny and the paintbrush after $t$ seconds.

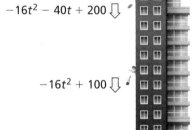

Not drawn to scale

To find the distance between the objects after $t$ seconds, subtract the polynomials.

         *Penny*        *Paintbrush*

$$(-16t^2 - 40t + 200) - (-16t^2 + 100)$$
$$= (-16t^2 - 40t + 200) + (16t^2 - 100)$$
$$= (-16t^2 + 16t^2) - 40t + [200 + (-100)]$$
$$= -40t + 100$$

∴ The polynomial $-40t + 100$ represents the distance between the objects after $t$ seconds.

**Study Tip**
To check your answer, substitute 2 into the original polynomials and verify that the difference of the heights is 20.

**b.** What is the distance between the objects after 2 seconds?

Find the value of $-40t + 100$ when $t = 2$.

$-40t + 100 = -40(2) + 100$     Substitute 2 for $t$.
                $= 20$                 Simplify.

∴ After 2 seconds, the distance between the objects is 20 feet.

### On Your Own

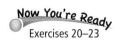
Exercises 20–23

**5.** In Example 3, which polynomial represents the difference of the two polynomials?

**6.** In Example 4, the polynomial $-16t^2 - 25t + 200$ represents the height of the penny after $t$ seconds. What is the distance between the objects after 1 second?

# 7.2 Exercises

## Vocabulary and Concept Check

1. **WRITING** How do you add $(4x^2 - 3 + 2y^3)$ and $(-6x^2 - 15)$ using a vertical method? a horizontal method?

2. **REASONING** Describe how subtracting polynomials is similar to subtracting integers.

## Practice and Problem Solving

**Use algebra tiles to find the sum or difference of the polynomials.**

3. $(x^2 - 3x + 2) + (x^2 + 4x - 1)$

4. $(x^2 + 2x - 5) - (-x^2 - 2x + 5)$

**Find the sum.**

5. $(5y + 4) + (-2y + 6)$

6. $(3g^2 - g) + (3g^2 - 8g + 4)$

7. $(2n^2 - 5n - 6) + (-n^2 - 3n + 11)$

8. $(-3p^2 + 5p - 2) + (-p^2 - 8p - 15)$

9. $(-a^3 + 4a - 3) + (5a^3 - a)$

10. $\left(-s^2 - \dfrac{2}{9}s + 1\right) + \left(-\dfrac{5}{9}s - 4\right)$

11. **ERROR ANALYSIS** Describe and correct the error in finding the sum of the polynomials.

$$\begin{array}{r} -5x^2 + 1 \\ +\phantom{-5x^2} 2x - 8 \\ \hline -3x - 7 \end{array}$$

**Find the difference.**

12. $(d^2 - 9) - (3d - 1)$

13. $(k^2 - 7k + 2) - (k^2 - 12)$

14. $(x^2 - 4x + 9) - (3x^2 - 6x - 7)$

15. $(-r - 10) - (-4r^2 + r + 7)$

16. $(t^4 - t^2 + t) - (-9t^2 + 7t - 12)$

17. $\left(\dfrac{1}{6}q^2 + \dfrac{2}{3}\right) - \left(\dfrac{1}{12}q^2 - \dfrac{1}{3}\right)$

18. **ERROR ANALYSIS** Describe and correct the error in finding the difference of the polynomials.

$$\begin{aligned}(x^2 - 5x) - (-3x^2 + 2x) &= (x^2 - 5x) + (3x^2 + 2x) \\ &= (x^2 + 3x^2) + (-5x + 2x) \\ &= 4x^2 - 3x\end{aligned}$$

19. **COST** The cost (in dollars) of making $b$ bracelets is represented by $4 + 5b$. The cost (in dollars) of making $b$ necklaces is $8b + 6$. Write a polynomial that represents how much more it costs to make $b$ necklaces than $b$ bracelets.

338   Chapter 7   Polynomial Equations and Factoring

**Find the sum or difference.**

20. $(c^2 - 6d^2) + (c^2 - 2cd + 2d^2)$

21. $(-x^2 + 9xy) - (x^2 + 6xy - 8y^2)$

22. $(2s^2 - 5st - t^2) - (s^2 + 7st - t^2)$

23. $(a^2 - 3ab + 2b^2) + (-4a^2 + 5ab - b^2)$

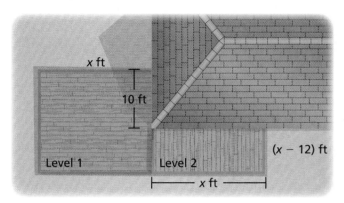

24. **MODELING** You are building a multi-level deck.
   a. Write a polynomial that represents the area of each level.
   b. Write a polynomial that represents the total area of the deck.
   c. What is the total area of the deck when $x = 20$?
   d. A gallon of deck sealant covers 400 square feet. How many gallons of sealant do you need to cover the deck once? Explain.

25. **Problem Solving** You drop a ball from a height of 98 feet. At the same time, your friend throws a ball upward. The polynomials represent the heights (in feet) of the balls after $t$ seconds.

   a. Write a polynomial that represents the distance between your ball and your friend's ball after $t$ seconds.
   b. What is the distance between the balls after 1.5 seconds?
   c. After how many seconds are the balls at the same height? How far are they from the ground? Explain your reasoning.

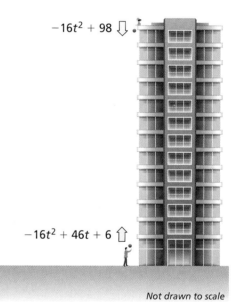

*Not drawn to scale*

## Fair Game Review  *What you learned in previous grades & lessons*

**Simplify the expression.** *(Skills Review Handbook)*

26. $2(x - 1) + 3(x + 2)$

27. $(4y - 3) - 2(y - 5)$

28. $-5(2w + 1) - 3(-4w + 2)$

29. **MULTIPLE CHOICE** Which inequality is represented by the graph? *(Section 3.1)*

Ⓐ $x < -2$    Ⓑ $x > -2$    Ⓒ $x \le -2$    Ⓓ $x \ge -2$

# 7.3 Multiplying Polynomials

**Essential Question** How can you multiply two binomials?

### 1 ACTIVITY: Multiplying Binomials Using Algebra Tiles

Work with a partner. Six different algebra tiles are shown below.

Write the product of the two binomials shown by the algebra tiles.

**a.** $(x + 3)(x - 2) = $ _____

**b.** $(2x - 1)(2x + 1) = $ _____

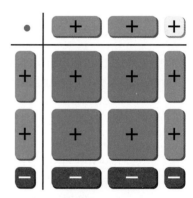

**c.** $(x + 2)(2x - 1) = $ _____

**d.** $(-x - 2)(x - 3) = $ _____

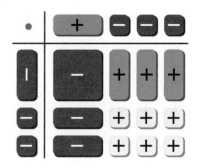

**COMMON CORE**

**Polynomials**
In this lesson, you will
- multiply binomials using the Distributive Property, a table, or the FOIL method.
- multiply binomials and trinomials.

Learning Standard
A.APR.1

## 2   ACTIVITY: Multiplying Monomials Using Algebra Tiles

**Math Practice**

**Use a Diagram**
How can you represent the product of polynomials using diagrams?

Work with a partner. Write each product. Explain your reasoning.

a. $(+) \cdot (+) =$     b. $(+) \cdot (-) =$

c. $(-) \cdot (-) =$     d. $(+) \cdot (+) =$

e. $(+) \cdot (-) =$     f. $(-) \cdot (+) =$

g. $(-) \cdot (-) =$

h. $(+) \cdot (+) =$

i. $(+) \cdot (-) =$

j. $(-) \cdot (-) =$

## 3   ACTIVITY: Multiplying Binomials Using Algebra Tiles

Use algebra tiles to find each product.

a. $(2x - 2)(2x + 1)$     b. $(4x + 3)(x - 2)$

c. $(-x + 2)(2x + 2)$     d. $(2x - 3)(x + 4)$

e. $(3x + 2)(-x - 1)$     f. $(2x + 1)(-3x + 2)$

g. $(x - 2)^2$     h. $(2x - 3)^2$

### What Is Your Answer?

4. **IN YOUR OWN WORDS** How can you multiply two binomials? Use the results of Activity 3 to summarize a procedure for multiplying binomials without using algebra tiles.

5. Find two binomials with the given product.

   a. $x^2 - 3x + 2$     b. $x^2 - 4x + 4$

**Practice**

Use what you learned about multiplying binomials to complete Exercises 3 and 4 on page 345.

Section 7.3   Multiplying Polynomials

# 7.3 Lesson

**Key Vocabulary**
FOIL Method, *p. 343*

In Section 1.2, you used the Distributive Property to multiply a binomial by a monomial. You can also use the Distributive Property to multiply two binomials.

### EXAMPLE 1  Multiplying Binomials Using the Distributive Property

**Find each product.**

**a.** $(x + 2)(x + 5)$

Use the horizontal method.

$(x + 2)(x + 5) = x(x + 5) + 2(x + 5)$    Distribute $(x + 5)$ to each term of $(x + 2)$.

$\phantom{(x + 2)(x + 5)} = x(x) + x(5) + 2(x) + 2(5)$    Distributive Property

$\phantom{(x + 2)(x + 5)} = x^2 + 5x + 2x + 10$    Multiply.

$\phantom{(x + 2)(x + 5)} = x^2 + 7x + 10$    Combine like terms.

**b.** $(x + 3)(x - 4)$

Use the vertical method.

$$\begin{array}{r} x + 3 \\ \times \quad x - 4 \\ \hline -4x - 12 \\ x^2 + 3x \phantom{-00} \\ \hline x^2 - x - 12 \end{array}$$

Multiply $-4(x + 3)$.    Align like terms vertically.
Distributive Property
Multiply $x(x + 3)$.    Distributive Property
Combine like terms.

∴ The product is $x^2 - x - 12$.

### EXAMPLE 2  Multiplying Binomials Using a Table

**Find $(2x - 3)(x + 5)$.**

**Step 1:** Write each binomial as a sum of terms.

$(2x - 3)(x + 5) = [2x + (-3)](x + 5)$

**Step 2:** Make a table of products.

|   | $2x$ | $-3$ |
|---|------|------|
| $x$ | $2x^2$ | $-3x$ |
| $5$ | $10x$ | $-15$ |

∴ The product is $2x^2 - 3x + 10x - 15$, or $2x^2 + 7x - 15$.

### On Your Own

Exercises 5–13 and 16–21

**Use the Distributive Property to find the product.**

**1.** $(y + 4)(y + 1)$      **2.** $(z - 2)(z + 6)$

**Use a table to find the product.**

**3.** $(p + 3)(p - 8)$      **4.** $(r - 5)(2r - 1)$

The **FOIL Method** is a shortcut for multiplying two binomials.

 **Key Idea**

**FOIL Method**

To multiply two binomials using the FOIL Method, find the sum of the products of the

First terms,   $(x + 1)(x + 2)$   ➡   $x(x) = x^2$

Outer terms,   $(x + 1)(x + 2)$   ➡   $x(2) = 2x$

Inner terms, and   $(x + 1)(x + 2)$   ➡   $1(x) = x$

Last terms.   $(x + 1)(x + 2)$   ➡   $1(2) = 2$

$$(x + 1)(x + 2) = x^2 + 2x + x + 2 = x^2 + 3x + 2$$

**EXAMPLE 3** **Multiplying Binomials Using the FOIL Method**

**Find each product.**

**a.** $(x - 3)(x - 6)$

$\qquad$ First $\quad$ Outer $\quad$ Inner $\quad$ Last
$(x - 3)(x - 6) = x(x) + x(-6) + (-3)(x) + (-3)(-6)$   Use the FOIL Method.
$\qquad\qquad\quad = x^2 + (-6x) + (-3x) + 18$   Multiply.
$\qquad\qquad\quad = x^2 - 9x + 18$   Combine like terms.

**b.** $(2x + 1)(3x - 5)$

$\qquad$ First $\quad$ Outer $\quad$ Inner $\quad$ Last
$(2x + 1)(3x - 5) = 2x(3x) + 2x(-5) + 1(3x) + 1(-5)$   Use the FOIL Method.
$\qquad\qquad\quad = 6x^2 + (-10x) + 3x + (-5)$   Multiply.
$\qquad\qquad\quad = 6x^2 - 7x - 5$   Combine like terms.

● **On Your Own**

Now You're Ready
Exercises 22–30

**Use the FOIL Method to find the product.**

**5.** $(m + 5)(m - 6)$ $\qquad$ **6.** $(x - 4)(x + 2)$

**7.** $(k + 5)(6k + 3)$ $\qquad$ **8.** $\left(2u + \dfrac{1}{2}\right)\left(u - \dfrac{3}{2}\right)$

Section 7.3   Multiplying Polynomials   343

### EXAMPLE 4  Multiplying a Binomial and a Trinomial

Find $(x + 5)(x^2 - 3x - 2)$.

$$
\begin{array}{r}
x^2 - 3x - 2 \\
\times \quad x + 5 \\
\hline
5x^2 - 15x - 10 \\
x^3 - 3x^2 - 2x \quad\quad\quad \\
\hline
x^3 + 2x^2 - 17x - 10
\end{array}
$$

Align like terms vertically.

Multiply $5(x^2 - 3x - 2)$. → Distributive Property

Multiply $x(x^2 - 3x - 2)$. → Distributive Property

Combine like terms.

∴ The product is $x^3 + 2x^2 - 17x - 10$.

### EXAMPLE 5  Real-Life Application

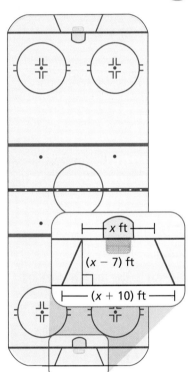

In hockey, a goalie behind the goal line can only play a puck in a trapezoidal region.

**a.** Write a polynomial that represents the area of the trapezoidal region.

$$\frac{1}{2}h(b_1 + b_2) = \frac{1}{2}(x - 7)[x + (x + 10)] \quad \text{Substitute.}$$

$$= \frac{1}{2}(x - 7)(2x + 10) \quad \text{Combine like terms.}$$

$$\overset{\text{F} \quad \text{O} \quad \text{I} \quad \text{L}}{= \frac{1}{2}[2x^2 + 10x + (-14x) + (-70)]} \quad \text{Use the FOIL Method.}$$

$$= \frac{1}{2}(2x^2 - 4x - 70) \quad \text{Combine like terms.}$$

$$= x^2 - 2x - 35 \quad \text{Distributive Property}$$

**b.** Find the area of the trapezoidal region when the shorter base is 18 feet.

Find the value of $x^2 - 2x - 35$ when $x = 18$.

$$x^2 - 2x - 35 = 18^2 - 2(18) - 35 \quad \text{Substitute 18 for } x.$$

$$= 324 - 36 - 35 \quad \text{Simplify.}$$

$$= 253 \quad \text{Subtract.}$$

∴ The area of the trapezoidal region is 253 square feet.

### On Your Own

Exercises 40–45

**Find the product.**

**9.** $(x + 1)(x^2 + 5x + 8)$  **10.** $(n - 3)(n^2 - 2n + 4)$

**11. WHAT IF?** How does the polynomial in Example 5 change if the longer base is extended by 1 foot? Explain.

# 7.3 Exercises

## Vocabulary and Concept Check

1. **VOCABULARY** Describe two ways to find the product of two binomials.
2. **WRITING** Explain how the letters of the word FOIL can help you remember how to multiply two binomials.

## Practice and Problem Solving

**Write the product of the two binomials shown by the algebra tiles.**

3. $(x - 2)(x + 2) =$ ⬚

4. $(-x + 3)(2x - 1) =$ ⬚

**Use the Distributive Property to find the product.**

5. $(x + 1)(x + 3)$
6. $(y + 6)(y + 4)$
7. $(z - 5)(z + 3)$
8. $(a + 8)(a - 3)$
9. $(g - 7)(g - 2)$
10. $(n - 6)(n - 4)$
11. $(3m + 1)(m + 9)$
12. $(2p - 4)(3p + 2)$
13. $(6 - 5s)(2 - s)$

14. **ERROR ANALYSIS** Describe and correct the error in finding the product.

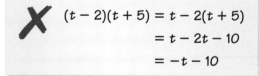

$$(t - 2)(t + 5) = t - 2(t + 5)$$
$$= t - 2t - 10$$
$$= -t - 10$$

15. **CALCULATOR** The width of a calculator can be represented by $(3x + 1)$ inches. The length of the calculator is twice the width. Write a polynomial that represents the area of the calculator.

Section 7.3   Multiplying Polynomials   345

**Use a table to find the product.**

**16.** $(x + 3)(x + 1)$   **17.** $(y + 10)(y - 5)$   **18.** $(h - 8)(h - 9)$

**19.** $(-3 + 2j)(4j - 7)$   **20.** $(5c + 6)(6c + 5)$   **21.** $(5d - 12)(-7 + 3d)$

**Use the FOIL Method to find the product.**

**22.** $(b + 3)(b + 7)$   **23.** $(w + 9)(w + 6)$   **24.** $(k + 5)(k - 1)$

**25.** $(x - 4)(x + 8)$   **26.** $(q - 3)(q - 4)$   **27.** $(z - 5)(z - 9)$

**28.** $(t + 2)(2t + 1)$   **29.** $(5v - 3)(2v + 4)$   **30.** $(9 - r)(2 - 3r)$

**31. ERROR ANALYSIS** Describe and correct the error in finding the product.

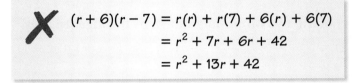

**32. OPEN-ENDED** Write two binomials whose product includes the term 12.

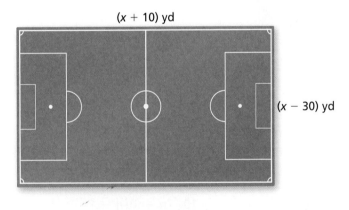

$(x + 10)$ yd

$(x - 30)$ yd

**33. SOCCER** The soccer field is rectangular.

a. Write a polynomial that represents the area of the soccer field.

b. Use the polynomial in part (a) to find the area of the field when $x = 90$.

c. A groundskeeper mows 200 square yards in 3 minutes. How long does it take the groundskeeper to mow the field?

**Write a polynomial that represents the area of the shaded region.**

**34.** 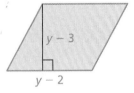   **35.** 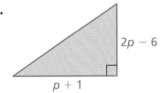   **36.**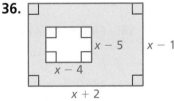

**Find the product.**

**37.** $(n + 3)(2n^2 + 1)$   **38.** $(x + y)(2x - y)$   **39.** $(2r + s)(r - 3s)$

**40.** $(x - 4)(x^2 - 3x + 2)$   **41.** $(f^2 + 4f - 8)(f - 1)$   **42.** $(3 + i)(i^2 + 8i - 2)$

**43.** $(t^2 - 5t + 1)(-3 + t)$   **44.** $(b - 4)(5b^2 - 5b + 4)$   **45.** $(3e^2 - 5e + 7)(6e + 1)$

**46. REASONING** Can you use the FOIL method to multiply a binomial by a trinomial? a trinomial by a trinomial? Explain your reasoning.

**47. AMUSEMENT PARK** You go to an amusement park $(x + 1)$ times each year and pay $(x + 40)$ dollars each time, where $x$ is the number of years after 2011.

  a. Write a polynomial that represents your yearly admission cost.

  b. What is your yearly admission cost in 2013?

**48. PRECISION** You use the Distributive Property to multiply $(x + 3)(x - 5)$. Your friend uses the FOIL Method to multiply $(x - 5)(x + 3)$. Should your answers be equivalent? Justify your answer.

**49. REASONING** The product of $(x + m)(x + n)$ is $x^2 + bx + c$.

  a. What do you know about $m$ and $n$ when $c < 0$?

  b. What do you know about $m$ and $n$ when $c > 0$?

**50. PICTURE** You design the wooden picture frame and paint the front surface.

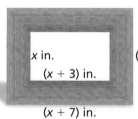

  a. Write a polynomial that represents the area of wood you paint.

  b. You design the picture frame to display a 5-inch by 8-inch photograph. How much wood do you paint?

$x$ in.
$(x + 3)$ in.
$(x + 4)$ in.
$(x + 7)$ in.

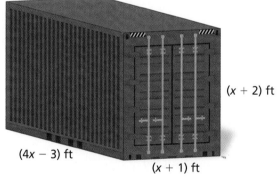

**51.**  **Number Sense** The shipping container is a rectangular prism. Write a polynomial that represents the volume of the container.

$(4x - 3)$ ft
$(x + 1)$ ft
$(x + 2)$ ft

# Fair Game Review  *What you learned in previous grades & lessons*

**Write the polynomial in standard form. Identify the degree and classify the polynomial by the number of terms.** *(Section 7.1)*

**52.** $2x - 5x^2 - x^3$

**53.** $z^2 - \dfrac{5}{7}z$

**54.** $-15y^7$

**55. MULTIPLE CHOICE** Which system of linear equations does the graph represent? *(Section 4.1)*

  Ⓐ  $y = 3x + 4$
     $y = -2x - 6$

  Ⓑ  $y = 2x + 1$
     $y = -x - 2$

  Ⓒ  $y = -x + 7$
     $y = 4x - 8$

  Ⓓ  $y = x + 10$
     $y = -3x + 2$

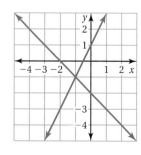

# 7.4 Special Products of Polynomials

**Essential Question** What are the patterns in the special products $(a + b)(a - b)$, $(a + b)^2$, and $(a - b)^2$?

### 1 ACTIVITY: Finding a Sum and Difference Pattern

Work with a partner. Six different algebra tiles are shown below.

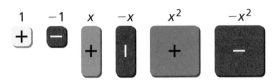

Write the product of the two binomials shown by the algebra tiles.

a. $(x + 2)(x - 2) = $ _____

b. $(2x - 1)(2x + 1) = $ _____

### 2 ACTIVITY: Describing a Sum and Difference Pattern

**COMMON CORE**

**Polynomials**
In this lesson, you will
- use patterns to multiply polynomials.

Learning Standard
A.APR.1

Work with a partner.

a. Describe the pattern for the special product: $(a + b)(a - b)$.

b. Use the pattern you described to find each product. Check your answers using algebra tiles.

  i. $(x + 3)(x - 3)$   ii. $(x - 4)(x + 4)$   iii. $(3x + 1)(3x - 1)$

  iv. $(3y + 4)(3y - 4)$   v. $(2x - 5)(2x + 5)$   vi. $(z + 1)(z - 1)$

**348** Chapter 7  Polynomial Equations and Factoring

### 3  ACTIVITY: Finding the Square of a Binomial Pattern

**Write the product of the two binomials shown by the algebra tiles.**

a. $(x + 2)^2 = $ ▭

b. $(2x - 1)^2 = $ ▭

### 4  ACTIVITY: Describing the Square of a Binomial Pattern

**Math Practice**

**Find General Methods**
What did the products of the binomials in the previous activity have in common? How does this help in describing the pattern?

**Work with a partner.**

a. Describe the pattern for the special product: $(a + b)^2$.

b. Describe the pattern for the special product: $(a - b)^2$.

c. Use the patterns you described to find each product. Check your answers using algebra tiles.

  i. $(x + 3)^2$   ii. $(x - 2)^2$   iii. $(3x + 1)^2$

  iv. $(3y + 4)^2$   v. $(2x - 5)^2$   vi. $(z + 1)^2$

## What Is Your Answer?

5. **IN YOUR OWN WORDS** What are the patterns in the special products $(a + b)(a - b)$, $(a + b)^2$, and $(a - b)^2$? Use the results of Activities 2 and 4 to write formulas for these special products.

*Practice* — Use what you learned about the patterns in special products to complete Exercises 3–5 on page 352.

Section 7.4   Special Products of Polynomials   349

# 7.4 Lesson

Check It Out
Lesson Tutorials
BigIdeasMath.com

Some pairs of binomials show patterns when multiplied. You can use these patterns to multiply other similar pairs of binomials.

## Key Idea

**Sum and Difference Pattern**

**Algebra**

$(a + b)(a - b) = a^2 - b^2$

**Example**

$(x + 3)(x - 3) = x^2 - 3^2$
$\phantom{(x + 3)(x - 3)} = x^2 - 9$

**Study Tip**

Because multiplication is commutative, the pattern also applies to $(a - b)(a + b)$.

### EXAMPLE 1  Using the Sum and Difference Pattern

Find each product.

**a.** $(x + 7)(x - 7)$

$(a + b)(a - b) = a^2 - b^2$     Sum and Difference Pattern
$(x + 7)(x - 7) = x^2 - 7^2$     Use pattern.
$\phantom{(x + 7)(x - 7)} = x^2 - 49$     Simplify.

**b.** $(3x - 1)(3x + 1)$

$(a - b)(a + b) = a^2 - b^2$     Sum and Difference Pattern
$(3x - 1)(3x + 1) = (3x)^2 - 1^2$     Use pattern.
$\phantom{(3x - 1)(3x + 1)} = 9x^2 - 1$     Simplify.

**Check**
Use the FOIL Method.
$(3x - 1)(3x + 1)$
$= 9x^2 + 3x - 3x - 1$
$= 9x^2 - 1$ ✓

### On Your Own

Now You're Ready
Exercises 6–14

Find the product.

**1.** $(x - 4)(x + 4)$     **2.** $(b + 10)(b - 10)$     **3.** $(2g + 5)(2g - 5)$

## Key Idea

**Square of a Binomial Pattern**

**Algebra**

$(a + b)^2 = a^2 + 2ab + b^2$

$(a - b)^2 = a^2 - 2ab + b^2$

**Example**

$(x + 3)^2 = x^2 + 2(x)(3) + 3^2$
$\phantom{(x + 3)^2} = x^2 + 6x + 9$

$(x - 3)^2 = x^2 - 2(x)(3) + 3^2$
$\phantom{(x - 3)^2} = x^2 - 6x + 9$

350     Chapter 7     Polynomial Equations and Factoring

### EXAMPLE 2  Using the Square of a Binomial Pattern

Find each product.

**a.** $(y + 1)^2$

$(a + b)^2 = a^2 + 2ab + b^2$  Square of a Binomial Pattern

$(y + 1)^2 = y^2 + 2(y)(1) + 1^2$  Use pattern.

$= y^2 + 2y + 1$  Simplify.

**Check**
Use the FOIL Method.
$(2z - 3)^2 = (2z - 3)(2z - 3)$
$= 4z^2 - 6z - 6z + 9$
$= 4z^2 - 12z + 9$ ✓

**b.** $(2z - 3)^2$

$(a - b)^2 = a^2 - 2ab + b^2$  Square of a Binomial Pattern

$(2z - 3)^2 = (2z)^2 - 2(2z)(3) + 3^2$  Use pattern.

$= 4z^2 - 12z + 9$  Simplify.

### EXAMPLE 3  Real-Life Application

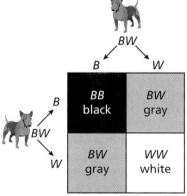

A diagram that models possible gene combinations in offspring is called a Punnett square.

Each of two dogs has one black gene (*B*) and one white gene (*W*). The diagram shows the possible gene combinations of an offspring and the resulting colors.

**a.** What percent of the possible gene combinations result in black?

Use the diagram. One of the four possible gene combinations results in black.

So, $\frac{1}{4}$ or 25% of the possible gene combinations result in black.

**b.** The genetic makeup of an offspring can be modeled by $(0.5B + 0.5W)^2$. Use the square of a binomial pattern to model the possible gene combinations of an offspring.

$(a + b)^2 = a^2 + 2ab + b^2$  Square of a Binomial Pattern

$(0.5B + 0.5W)^2 = (0.5B)^2 + 2(0.5B)(0.5W) + (0.5W)^2$  Use pattern.

$= 0.25B^2 + 0.5BW + 0.25W^2$  Simplify.

25% BB (black)   50% BW (gray)   25% WW (white)

### On Your Own

Exercises 16–24

Find the product.

**4.** $(w + 2)^2$   **5.** $(x - 7)^2$   **6.** $(3y - 1)^2$   **7.** $(5z + 4)^2$

# 7.4 Exercises

## Vocabulary and Concept Check

1. **OPEN-ENDED** Write two binomials whose product can be found using the sum and difference pattern.

2. **WHICH ONE DOESN'T BELONG?** Which expression does *not* belong with the other three? Explain your reasoning.

$(x + 1)(x - 1)$   $(3x + 2)(3x - 2)$   $(x + 2)(x - 3)$   $(2x + 5)(2x - 5)$

## Practice and Problem Solving

**Use algebra tiles to find the product.**

3. $(x + 6)(x - 6)$
4. $(3y - 2)(3y + 2)$
5. $(2z + 2)^2$

**Find the product.**

6. $(x + 2)(x - 2)$
7. $(g - 5)(g + 5)$
8. $(z - 8)(z + 8)$
9. $(b + 12)(b - 12)$
10. $(2x + 1)(2x - 1)$
11. $(3x - 4)(3x + 4)$
12. $(6x + 7)(6x - 7)$
13. $(9 - c)(9 + c)$
14. $(8 - 3m)(8 + 3m)$

15. **REASONING** Write two binomials whose product is $x^2 - 16$. Explain how you found your answer.

**Find the product.**

16. $(b - 2)^2$
17. $(y + 8)^2$
18. $(n + 6)^2$
19. $(d - 10)^2$
20. $(2f - 1)^2$
21. $(5p + 2)^2$
22. $(4b - 5)^2$
23. $(12 - x)^2$
24. $(4 + 7t)^2$

**ERROR ANALYSIS** Describe and correct the error in finding the product.

25.
$$(k + 4)^2 = k^2 + 4^2$$
$$= k^2 + 16$$

26.
$$(s + 5)(s - 5)$$
$$= s^2 + 2(s)(5) - 5^2$$
$$= s^2 + 10s - 25$$

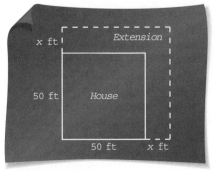

27. **CONSTRUCTION** A contractor extends a house on two sides.

   a. The area of the first level of the house after the renovation is represented by $(x + 50)^2$. Find this product.

   b. Use the polynomial in part (a) to find the area of the first level when $x = 15$. What is the area of the extension?

**Write a polynomial that represents the area of the figure.**

28.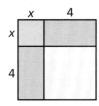

29.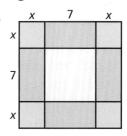

**Find the product.**

30. $(x^2 + 1)(x^2 - 1)$ 

31. $(x + y)(x - y)$ 

32. $(2x - y)^2$

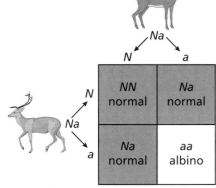

33. **GENETICS** In deer, the gene $N$ is for normal coloring and the gene $a$ is for no coloring, or albino. Any gene combination with an $N$ results in normal coloring. The diagram shows the possible gene combinations of an offspring and the resulting colors from parents that both have the gene combination $Na$.

   a. What percent of the possible gene combinations result in normal coloring?

   b. The genetic makeup of an offspring can be modeled by $(0.5N + 0.5a)^2$. Use the square of a binomial pattern to model the possible gene combinations of an offspring.

34. **VISION** Your iris controls the amount of light that enters your eye by changing the size of your pupil.

   a. Write a polynomial that represents the area of your pupil. Write your answer in terms of $\pi$.

   b. The width $x$ of your iris decreases from 4 millimeters to 2 millimeters when you enter a dark room. How many times greater is the area of your pupil after entering the room than before entering the room? Explain.

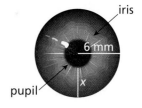

35. **Repeated Reasoning** Find $(x + 1)^3$ and $(x + 2)^3$. Find a pattern in the terms and use it to write a pattern for the cube of a binomial $(a + b)^3$.

## Fair Game Review  *What you learned in previous grades & lessons*

**Find the product.** *(Section 7.3)*

36. $(x + 4)(x + 9)$ 

37. $(y - 7)(y + 3)$ 

38. $(z - 10)(z - 1)$

39. **MULTIPLE CHOICE** What is the solution of the linear system? *(Section 4.2)*

   $y = 2x - 5$
   $3x - 8y = 1$

   Ⓐ $(-3, -1)$   Ⓑ $(-3, 1)$   Ⓒ $(3, -1)$   Ⓓ $(3, 1)$

Section 7.4   Special Products of Polynomials   353

# 7 Study Help

You can use an **idea and examples chart** to organize information about a concept. Here is an example of an idea and examples chart for using the FOIL Method to multiply binomials.

> **FOIL Method:** To multiply two binomials using the FOIL Method, find the sum of the products of the **F**irst terms, **O**uter terms, **I**nner terms, and **L**ast terms.

Example

$$(x - 2)(x + 3) = \overset{\text{First}}{x(x)} + \overset{\text{Outer}}{x(3)} + \overset{\text{Inner}}{(-2)(x)} + \overset{\text{Last}}{(-2)(3)}$$ Use the FOIL Method.
$$= x^2 + (3x) + (-2x) + (-6)$$ Multiply.
$$= x^2 + x - 6$$ Combine like terms.

Example

$$(3x - 1)(2x - 2) = \overset{\text{First}}{3x(2x)} + \overset{\text{Outer}}{3x(-2)} + \overset{\text{Inner}}{(-1)(2x)} + \overset{\text{Last}}{(-1)(-2)}$$ Use the FOIL Method.
$$= 6x^2 + (-6x) + (-2x) + 2$$ Multiply.
$$= 6x^2 - 8x + 2$$ Combine like terms.

## On Your Own

**Make idea and examples charts to help you study these topics.**

1. degree of a polynomial
2. adding and subtracting polynomials
3. special products of polynomials

**After you complete this chapter, make idea and examples charts for the following topics.**

4. factored form of a polynomial
5. factoring polynomials using the GCF
6. factoring polynomials of the form $x^2 + bx + c$
7. factoring polynomials of the form $ax^2 + bx + c$

"I made an **idea and examples chart** to give my owner ideas for my birthday next week."

# 7.1–7.4 Quiz

**Write the polynomial in standard form. Identify the degree and classify the polynomial by the number of terms.** *(Section 7.1)*

1. $-8q^3$
2. $-9 + d^2 - 3d$
3. $\frac{2}{3}m^4 - \frac{5}{6}m^6$
4. $-1.3z + 2z^4 + 7.4z^2$

**Find the sum or difference.** *(Section 7.2)*

5. $(2x^2 + 5) + (-x^2 + 4)$
6. $(-3n^2 + n) - (2n^2 + 7)$
7. $(-p^2 + 4p) - (p^2 - 3p + 15)$
8. $(a^2 - 3ab + b^2) + (-a^2 + ab + b^2)$

**Find the product.** *(Section 7.3 and Section 7.4)*

9. $(w + 6)(w + 7)$
10. $(y + 9)(y - 3)$
11. $(d - 2)(d - 5)$
12. $(2z - 3)(3z + 5)$
13. $(h - 1)(h + 1)$
14. $(p + 9)(p - 9)$
15. $(t + 5)^2$
16. $(q - 2)^2$

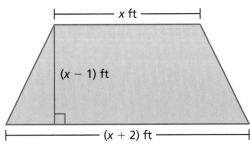

17. **WINDOW SEAT** A window seat is in the shape of a trapezoid. *(Section 7.3)*

   a. Write a polynomial that represents the area of the window seat.

   b. What is the area of the window seat when $x = 3$?

18. **COMPOUND INTEREST** You are saving for a guitar. You deposit $100 in an account that earns interest compounded annually. The expression $100(1 + r)^2$ represents the balance after 2 years, where $r$ is the annual interest rate in decimal form. *(Section 7.4)*

   a. Write a polynomial that represents the balance of your account.

   b. What is the balance of your account when the interest rate is 12%?

   c. How much more money do you need to save to buy the guitar?

# 7.5 Solving Polynomial Equations in Factored Form

**Essential Question** How can you solve a polynomial equation that is written in factored form?

Two polynomial equations are equivalent when they have the same solutions. For instance, the following equations are equivalent because the only solutions of each equation are $x = 1$ and $x = 2$.

| Factored Form | Standard Form | Nonstandard Form |
|---|---|---|
| $(x - 1)(x - 2) = 0$ | $x^2 - 3x + 2 = 0$ | $x^2 - 3x = -2$ |

✓ Check this by substituting 1 and 2 for $x$ in each equation.

## 1 ACTIVITY: Matching Equivalent Forms of an Equation

Work with a partner. Match each factored form of the equation with two other forms of equivalent equations. Notice that an equation is considered to be in factored form only when the product of the factors is equal to 0.

**Factored Form**
a. $(x - 1)(x - 3) = 0$
b. $(x - 2)(x - 3) = 0$
c. $(x + 1)(x - 2) = 0$
d. $(x - 1)(x + 2) = 0$
e. $(x + 1)(x - 3) = 0$

**Standard Form**
A. $x^2 - x - 2 = 0$
B. $x^2 + x - 2 = 0$
C. $x^2 - 4x + 3 = 0$
D. $x^2 - 5x + 6 = 0$
E. $x^2 - 2x - 3 = 0$

**Nonstandard Form**
1. $x^2 - 5x = -6$
2. $(x - 1)^2 = 4$
3. $x^2 - x = 2$
4. $x(x + 1) = 2$
5. $x^2 - 4x = -3$

## 2 ACTIVITY: Writing a Conjecture

Work with a partner. Substitute

1, 2, 3, 4, 5, and 6 for $x$

in each equation. Write a conjecture describing what you discovered.

a. $(x - 1)(x - 2) = 0$  b. $(x - 2)(x - 3) = 0$  c. $(x - 3)(x - 4) = 0$

d. $(x - 4)(x - 5) = 0$  e. $(x - 5)(x - 6) = 0$  f. $(x - 6)(x - 1) = 0$

**COMMON CORE**

**Polynomial Equations**
In this lesson, you will
- solve polynomial equations in factored form.

Learning Standard
A.REI.4b

### 3 ACTIVITY: Special Properties of 0 and 1

**Work with a partner.** The numbers 0 and 1 have special properties that are shared by no other numbers. For each of the following, decide whether the property is true for 0, 1, both, or neither. Explain your reasoning.

**a.** If you add ▢ to a number $n$, you get $n$.

**b.** If the product of two numbers is ▢, then one or both numbers are 0.

**c.** The square of ▢ is equal to itself.

**d.** If you multiply a number $n$ by ▢, you get $n$.

**e.** If you multiply a number $n$ by ▢, you get 0.

**f.** The opposite of ▢ is equal to itself.

### 4 ACTIVITY: Writing About Solving Equations

**Math Practice**

**Use Definitions**
What previous examples, information, and definitions can you use to reply to the student's comment?

**Work with a partner.** Imagine that you are part of a study group in your algebra class. One of the students in the group makes the following comment.

"I don't see why we spend so much time solving equations that are equal to zero. Why don't we spend more time solving equations that are equal to other numbers?"

Write an answer for this student.

### What Is Your Answer?

**5.** One of the properties in Activity 3 is called the Zero-Product Property. It is one of the most important properties in all of algebra. Which property is it? Explain how it is used in algebra and why it is so important.

**6. IN YOUR OWN WORDS** How can you solve a polynomial equation that is written in factored form?

**Practice** — Use what you learned about solving polynomial equations to complete Exercises 4–6 on page 360.

Section 7.5 Solving Polynomial Equations in Factored Form

## 7.5 Lesson

**Key Vocabulary**
factored form, *p. 358*
Zero-Product Property, *p. 358*
root, *p. 358*

A polynomial is in **factored form** when it is written as a product of factors.

| Standard form | Factored form |
|---|---|
| $x^2 + 2x$ | $x(x + 2)$ |
| $x^2 + 5x - 24$ | $(x - 3)(x + 8)$ |

When one side of an equation is a polynomial in factored form and the other side is 0, use the **Zero-Product Property** to solve the polynomial equation. The solutions of a polynomial equation are also called **roots**.

### Key Idea

**Zero-Product Property**

**Words** If the product of two real numbers is 0, then at least one of the numbers is 0.

**Algebra** If $a$ and $b$ are real numbers and $ab = 0$, then $a = 0$ or $b = 0$.

### EXAMPLE 1  Solving Polynomial Equations

Solve each equation.

**a.** $x(x + 8) = 0$

**Check**

Substitute each solution in the original equation.

$0(0 + 8) \stackrel{?}{=} 0$
$0(8) \stackrel{?}{=} 0$
$0 = 0$ ✓

$-8(-8 + 8) \stackrel{?}{=} 0$
$-8(0) \stackrel{?}{=} 0$
$0 = 0$ ✓

$x(x + 8) = 0$   Write equation.
$x = 0$ or $x + 8 = 0$   Use Zero-Product Property.
      $x = -8$   Solve for $x$.

∴ The roots are $x = 0$ and $x = -8$.

**b.** $(x + 6)(x - 5) = 0$

$(x + 6)(x - 5) = 0$   Write equation.
$x + 6 = 0$ or $x - 5 = 0$   Use Zero-Product Property.
$x = -6$ or $x = 5$   Solve for $x$.

∴ The roots are $x = -6$ and $x = 5$.

### On Your Own

Exercises 4–9

Solve the equation.

1. $x(x - 1) = 0$
2. $3t(t + 2) = 0$
3. $(z - 4)(z - 6) = 0$
4. $(b + 7)^2 = 0$

### EXAMPLE 2 Solving a Polynomial Equation

What are the solutions of $(2a + 7)(2a - 7) = 0$?

- **(A)** $-7$ and $7$
- **(B)** $-\dfrac{7}{2}$ and $\dfrac{7}{2}$
- **(C)** $-2$ and $2$
- **(D)** $-\dfrac{2}{7}$ and $\dfrac{2}{7}$

$(2a + 7)(2a - 7) = 0$     Write equation.

$2a + 7 = 0$   or   $2a - 7 = 0$     Use Zero-Product Property.

$a = -\dfrac{7}{2}$   or   $a = \dfrac{7}{2}$     Solve for $a$.

∴ The correct answer is **(B)**.

### EXAMPLE 3 Real-Life Application

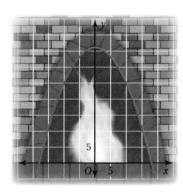

The arch of a fireplace can be modeled by $y = -\dfrac{1}{9}(x + 18)(x - 18)$, where $x$ and $y$ are measured in inches. The $x$-axis represents the floor. Find the width of the arch at floor level.

Use the $x$-coordinates at floor level to find the width. At floor level, $y = 0$. So, substitute 0 for $y$ and solve for $x$.

$y = -\dfrac{1}{9}(x + 18)(x - 18)$     Write equation.

$0 = -\dfrac{1}{9}(x + 18)(x - 18)$     Substitute 0 for $y$.

$0 = (x + 18)(x - 18)$     Multiply each side by $-9$.

$x + 18 = 0$   or   $x - 18 = 0$     Use Zero-Product Property.

$x = -18$   or   $x = 18$     Solve for $x$.

The width is the distance between the $x$-coordinates, $-18$ and $18$.

∴ So, the width of the arch at floor level is $18 - (-18) = 36$ inches.

### On Your Own

*Now You're Ready*
Exercises 10–15

Solve the equation.

**5.** $(3p + 5)(3p - 5) = 0$     **6.** $(12 - 6x)^2 = 0$

**7.** The entrance to a mine shaft can be modeled by $y = -\dfrac{1}{2}(x + 4)(x - 4)$, where $x$ and $y$ are measured in feet. The $x$-axis represents the ground. Find the width of the entrance at ground level.

# 7.5 Exercises

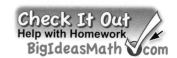

## Vocabulary and Concept Check

1. **REASONING** Is $x = 3$ a solution of $(x - 3)(x + 6) = 0$? Explain.

2. **WRITING** Describe how to solve $(x - 2)(x + 1) = 0$ using the Zero-Product Property.

3. **WHICH ONE DOESN'T BELONG?** Which statement does *not* belong with the other three? Explain your reasoning.

   $(n - 9)(n + 3)$      $(2k + 5)(k - 3)$

   $(g + 2)^2$      $x^2 + 4x$

## Practice and Problem Solving

**Solve the equation.**

4. $x(x + 7) = 0$
5. $12t(t - 5) = 0$
6. $(s - 9)(s - 1) = 0$
7. $(q + 3)(q - 2) = 0$
8. $(h - 8)^2 = 0$
9. $(m + 4)^2 = 0$
10. $(5 - k)(5 + k) = 0$
11. $(3 - g)(7 - g) = 0$
12. $(3p + 6)^2 = 0$
13. $(4z - 12)^2 = 0$
14. $\left(\frac{1}{2}y + 4\right)(y - 8) = 0$
15. $\left(\frac{1}{3}d - 2\right)\left(\frac{1}{3}d + 2\right) = 0$

16. **ERROR ANALYSIS** Describe and correct the error in solving the equation.

    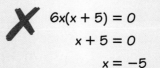

    $6x(x + 5) = 0$
    $x + 5 = 0$
    $x = -5$
    The root is $x = -5$.

**Find the $x$-coordinates of the points where the graph crosses the $x$-axis.**

17.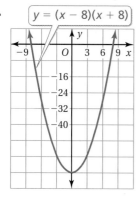
    $y = (x - 8)(x + 8)$

18.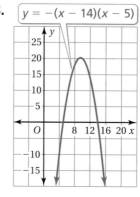
    $y = -(x - 14)(x - 5)$

19.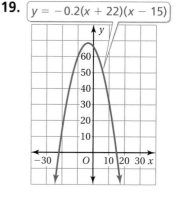
    $y = -0.2(x + 22)(x - 15)$

20. **CHOOSE TOOLS** The entrance of a tunnel can be modeled by $y = -\frac{11}{50}(x - 4)(x - 24)$, where $x$ and $y$ are measured in feet. The $x$-axis represents the ground. Find the width of the tunnel at ground level.

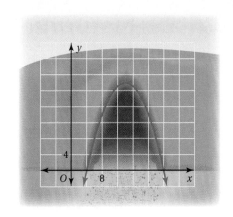

**Solve the equation.**

21. $5z(z + 2)(z - 1) = 0$

22. $w(w - 6)^2 = 0$

23. $(r - 4)(r + 4)(r + 8) = 0$

24. $(2p + 3)(2p - 3)(p + 7) = 0$

25. **GATEWAY ARCH** The Gateway Arch in St. Louis can be modeled by $y = -\frac{2}{315}(x + 315)(x - 315)$, where $x$ and $y$ are measured in feet. The $x$-axis represents the ground.

    a. Find the width of the arch at ground level.
    b. How tall is the arch?

26.  Find the values of $x$ in terms of $y$ that are solutions of the equation.

    a. $(x + y)(2x - y) = 0$
    b. $(x^2 - y^2)(4x + 16y) = 0$

## Fair Game Review  *What you learned in previous grades & lessons*

**Find the greatest common factor of the numbers.** *(Skills Review Handbook)*

27. 21 and 63

28. 12 and 27

29. 30, 75, and 90

30. **MULTIPLE CHOICE** What is the slope of the line? *(Section 2.2)*

    Ⓐ $-3$   Ⓑ $-\frac{1}{3}$

    Ⓒ $\frac{1}{3}$   Ⓓ $3$

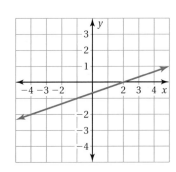

# 7.6 Factoring Polynomials Using the GCF

**Essential Question** How can you use common factors to write a polynomial in factored form?

### 1  ACTIVITY: Finding Monomial Factors

Work with a partner. Six different algebra tiles are shown below.

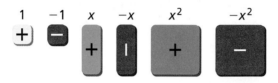

**Sample:**

**Step 1:** Look at the rectangular array for $x^2 + 3x$.

**Step 2:** Use algebra tiles to label the dimensions of the rectangle.

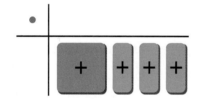  

**Step 3:** Write the polynomial in factored form by finding the dimensions of the rectangle.

$$\text{Area} = x^2 + 3x = x(x + 3)$$

width → $x$, length → $(x+3)$

**Use algebra tiles to write each polynomial in factored form.**

a.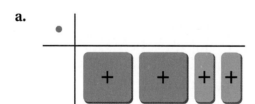

b.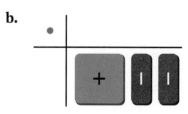

c.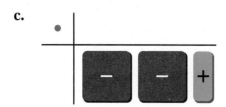

d.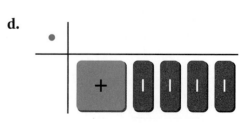

**COMMON CORE**

**Polynomial Equations**

In this lesson, you will
- factor polynomials using the greatest common factor.
- solve polynomial equations by factoring.

Learning Standards
A.REI.4b
A.SSE.3a

362   Chapter 7   Polynomial Equations and Factoring

## 2 ACTIVITY: Finding Monomial Factors

**Math Practice**

**Interpret Results**
What does your answer represent? How can you make sure your answer makes sense?

Work with a partner. Use algebra tiles to write each polynomial in factored form.

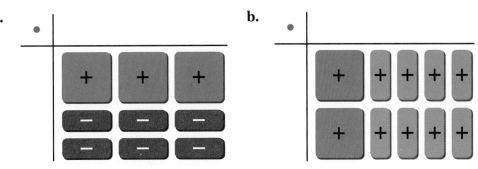

## 3 ACTIVITY: Finding Monomial Factors

Work with a partner. Use algebra tiles to model each polynomial as a rectangular array. Then write the polynomial in factored form by finding the dimensions of the rectangle.

a. $3x^2 - 9x$      b. $7x + 14x^2$      c. $-2x^2 + 6x$

## What Is Your Answer?

4. Consider the polynomial $4x^2 + 8x$.

   a. What are the terms of the polynomial?
   b. List all the factors that are common to both terms.
   c. Of the common factors, which is the greatest? Explain your reasoning.

5. **IN YOUR OWN WORDS** How can you use common factors to write a polynomial in factored form?

**Practice**

Use what you learned about factoring polynomials to complete Exercises 3–5 on page 366.

Section 7.6  Factoring Polynomials Using the GCF

## 7.6 Lesson

Writing a polynomial as a product of factors is called *factoring*. When the terms of a polynomial have a common factor, you can factor the polynomial as shown below.

**Factoring Polynomials Using the GCF**

**Step 1:** Find the greatest common factor (GCF) of the terms.

**Step 2:** Use the Distributive Property to write the polynomial as a product of the GCF and its remaining factors.

### EXAMPLE 1  Factoring Polynomials

**Factor each polynomial.**

**a.** $2x^2 + 18$

**Step 1:** Find the GCF of the terms.

$2x^2 = \boxed{2} \cdot x \cdot x$
$18 = \boxed{2} \cdot 3 \cdot 3$

The GCF is 2.

> **Study Tip**
> When you factor a polynomial, you *undo* the multiplication of its factors.

**Step 2:** Write the polynomial as a product of the GCF and its remaining factors.

$2x^2 + 18 = 2(x^2) + 2(9)$    Factor out GCF.
$= 2(x^2 + 9)$    Distributive Property

**b.** $15y^3 + 10y^2$

**Step 1:** Find the GCF of the terms.

$15y^3 = 3 \cdot \boxed{5} \cdot \boxed{y} \cdot \boxed{y} \cdot y$
$10y^2 = 2 \cdot \boxed{5} \cdot \boxed{y} \cdot \boxed{y}$

The GCF is $5 \cdot y \cdot y = 5y^2$.

**Step 2:** Write the polynomial as a product of the GCF and its remaining factors.

$15y^3 + 10y^2 = 5y^2(3y) + 5y^2(2)$    Factor out GCF.
$= 5y^2(3y + 2)$    Distributive Property

● **On Your Own**

Exercises 6–11

**Factor the polynomial.**

**1.** $5z^2 + 30$    **2.** $3x^2 + 14x$    **3.** $8y^2 - 24y$

364    Chapter 7    Polynomial Equations and Factoring

To solve an equation using the Zero-Product Property, you may need to first collect the terms on one side of the equation and then factor.

**EXAMPLE 2    Solving an Equation by Factoring**

Solve $4g^2 = -6g$.

| | |
|---|---|
| $4g^2 = -6g$ | Write equation. |
| $4g^2 + 6g = 0$ | Add $6g$ to each side. |
| $2g(2g + 3) = 0$ | Factor the polynomial. |
| $2g = 0$ or $2g + 3 = 0$ | Use Zero-Product Property. |
| $g = 0$ or $g = -\dfrac{3}{2}$ | Solve for $g$. |

∴ The solutions are $g = 0$ and $g = -\dfrac{3}{2}$.

### On Your Own

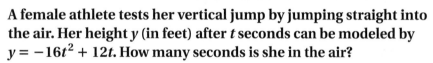

Solve the equation.

4. $3x^2 + 21x = 0$    5. $5z^2 = 5z$    6. $18y = 6y^2$

**EXAMPLE 3    Real-Life Application**

A female athlete tests her vertical jump by jumping straight into the air. Her height $y$ (in feet) after $t$ seconds can be modeled by $y = -16t^2 + 12t$. How many seconds is she in the air?

She is on the ground when $y = 0$. So, substitute 0 for $y$ and solve for $t$.

| | |
|---|---|
| $y = -16t^2 + 12t$ | Write equation. |
| $0 = -16t^2 + 12t$ | Substitute 0 for $y$. |
| $0 = 4t(-4t + 3)$ | Factor the polynomial. |
| $4t = 0$ or $-4t + 3 = 0$ | Use Zero-Product Property. |
| $t = 0$ or $t = 0.75$ | Solve for $t$. |

She starts the jump at $t = 0$ and lands when $t = 0.75$.

∴ So, she is in the air for 0.75 second.

### On Your Own

7. **WHAT IF?** The height of a male athlete testing his vertical jump can be modeled by $y = -16t^2 + 14t$. How many seconds is he in the air?

# 7.6 Exercises

## Vocabulary and Concept Check

1. **REASONING** What is the greatest common factor of $12y$ and $30y^2$?
2. **WRITING** Describe how to factor a polynomial using the greatest common factor.

## Practice and Problem Solving

**Use algebra tiles to factor the polynomial.**

3. $4x + 8$
4. $2x^2 + 4x$
5. $x^2 - 4x$

**Factor the polynomial.**

 6. $5z^2 + 45z$
7. $8m^2 + 4m$
8. $3y^3 - 9y^2$
9. $20x^3 + 30x^2$
10. $4w^3 - 8w + 12$
11. $5t^2 + 20t + 50$

12. **ERROR ANALYSIS** Describe and correct the error in factoring the polynomial.

   $2x^2 + 2x = 2(x^2) + 2(x)$
   $= 2(x^2 + x)$

13. **INTEREST** You deposit $100 in a savings account that earns simple interest. The balance of the account can be represented by $100 + 100rt$, where $r$ is the annual interest rate and $t$ is the time in years. Factor the polynomial.

**Solve the equation.**

 14. $2q + 10 = 0$
15. $10x + 15 = 0$
16. $4p^2 - p = 0$
17. $6m^2 + 12m = 0$
18. $3n^2 = 9n$
19. $4r^2 = -28r$
20. $4a^3 = 44a^2$
21. $6k^3 + 39k^2 = 0$
22. $2y^2 = 2\pi y$

23. **ERROR ANALYSIS** Describe and correct the error in solving the equation.

   $3x^2 = 15x$
   $3x^2 - 15x = 0$
   $3x(x - 15) = 0$
   $3x = 0$ or $x - 15 = 0$
   $x = 0$ or $x = 15$
   The roots are $x = 0$ and $x = 15$.

24. **AGES** Your brother is $y$ years old. Your older cousins are $2y^2$ and $6y$ years old. The difference between your cousins' ages is zero. Your brother is older than 1 year old. How old is he?

**Solve the equation.**

**25.** $5b^2 - 20b = b^2$

**26.** $5n^2 + 40n = 5n$

**27.** $2s^3 + 15s^2 = 3s^2$

**28.** $8g^3 - 2g^2 = 2g^3 - 5g^2$

**29. OPEN-ENDED** Write a binomial whose terms have a GCF of $3x$.

**30. SCHOOL SIGN** The area (in square feet) of the school sign can be represented by $15x^2 - 6x$.

   **a.** Write an expression that represents the length of the sign.

   **b.** Describe two ways to find the area of the sign when $x = 2$.

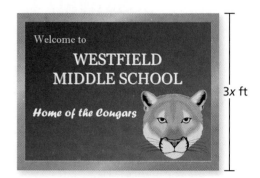

**31. DOLPHIN** A dolphin jumps straight into the air during a performance. The dolphin's height $y$ (in feet) after $t$ seconds can be modeled by $y = -16t^2 + 24t$.

   **a.** How many seconds is the dolphin in the air?

   **b.** The dolphin reaches its maximum height after 0.75 second. What is the maximum height of the jump?

**32. Modeling** Your teacher's work station is made up of two identical desks arranged as shown.

   **a.** Write an equation in terms of $x$ that relates the area of Desk 1 to the area of Desk 2.

   **b.** What is the value of $x$?

   **c.** Find the area of the top of your teacher's work station.

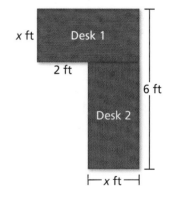

# Fair Game Review  *What you learned in previous grades & lessons*

**Find the product.** *(Section 7.3)*

**33.** $(y + 4)(y + 6)$

**34.** $(m - 2)(m - 9)$

**35.** $(2k + 1)(2k - 3)$

**36. MULTIPLE CHOICE** An African elephant weighs 5,200,000 grams. Write this number in scientific notation. *(Skills Review Handbook)*

   **Ⓐ** $0.52 \times 10^{-7}$ g

   **Ⓑ** $5.2 \times 10^{-6}$ g

   **Ⓒ** $52 \times 10^5$ g

   **Ⓓ** $5.2 \times 10^6$ g

# 7.7 Factoring $x^2 + bx + c$

**Essential Question** How can you factor the trinomial $x^2 + bx + c$ into the product of two binomials?

### 1  ACTIVITY: Finding Binomial Factors

Work with a partner. Six different algebra tiles are shown below.

**Sample:**

**Step 1:** Arrange the algebra tiles into a rectangular array to model $x^2 + 5x + 6$.

**Step 2:** Use algebra tiles to label the dimensions of the rectangle.

**Step 3:** Write the polynomial in factored form by finding the dimensions of the rectangle.

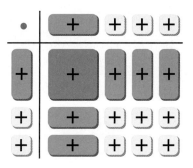

$$\text{Area} = x^2 + 5x + 6 = (x + 2)(x + 3)$$

with width and length labeled.

**Common Core**

**Polynomial Equations**
In this lesson, you will
- factor trinomials of the form $x^2 + bx + c$.

Learning Standards
A.REI.4b
A.SSE.3a

Use algebra tiles to write each polynomial as the product of two binomials. Check your answer by multiplying.

a.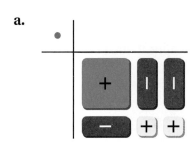

b.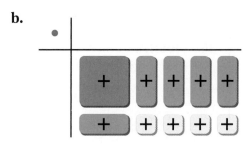

## 2 ACTIVITY: Finding Binomial Factors

Work with a partner. Use algebra tiles to write each polynomial as the product of two binomials. Check your answer by multiplying.

a.
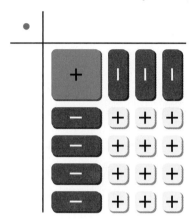

b.
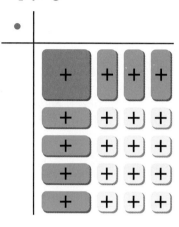

## 3 ACTIVITY: Finding Binomial Factors

**Math Practice**

**Make Sense of Quantities**
What is the relationship between a polynomial and its binomial factors?

Work with a partner. Write each polynomial as the product of two binomials. Check your answer by multiplying.

a. $x^2 + 6x + 9$   b. $x^2 - 6x + 9$   c. $x^2 + 6x + 8$

d. $x^2 - 6x + 8$   e. $x^2 + 6x + 5$   f. $x^2 - 6x + 5$

### What Is Your Answer?

4. **IN YOUR OWN WORDS** How can you factor the trinomial $x^2 + bx + c$ into the product of two binomials?

   a. Describe a strategy that uses algebra tiles.
   b. Describe a strategy that does not use algebra tiles.

5. Use one of your strategies to factor each trinomial.

   a. $x^2 + 6x - 16$   b. $x^2 - 6x - 16$   c. $x^2 + 6x - 27$

**Practice**

Use what you learned about factoring trinomials to complete Exercises 3–5 on page 373.

## 7.7 Lesson

Consider the polynomial $x^2 + bx + c$, where $b$ and $c$ are integers. To factor this polynomial as $(x + p)(x + q)$, you need to find integers $p$ and $q$ such that $p + q = b$ and $pq = c$.

$$(x + p)(x + q) = x^2 + px + qx + pq$$
$$= x^2 + (p + q)x + pq$$

### Key Idea

**Factoring $x^2 + bx + c$ When $c$ Is Positive**

**Algebra**   $x^2 + bx + c = (x + p)(x + q)$ when $p + q = b$ and $pq = c$.

When $c$ is positive, $p$ and $q$ have the same sign as $b$.

**Examples**   $x^2 + 6x + 5 = (x + 1)(x + 5)$
$x^2 - 6x + 5 = (x - 1)(x - 5)$

### EXAMPLE 1  Factoring $x^2 + bx + c$ When $b$ and $c$ Are Positive

**Factor $x^2 + 10x + 16$.**

Notice that $b = 10$ and $c = 16$.

- Because $c$ is positive, the factors $p$ and $q$ must have the same sign so that $pq$ is positive.
- Because $b$ is also positive, $p$ and $q$ must each be positive so that $p + q$ is positive.

Find two positive integer factors of 16 whose sum is 10.

**Check**
Use the FOIL Method.
$(x + 2)(x + 8)$
$= x^2 + 8x + 2x + 16$
$= x^2 + 10x + 16$ ✓

| Factors of 16 | Sum of Factors |
|---|---|
| 1, 16 | 17 |
| 2, 8 | 10 |
| 4, 4 | 8 |

The values of $p$ and $q$ are 2 and 8.

∴ So, $x^2 + 10x + 16 = (x + 2)(x + 8)$.

### On Your Own

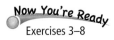
Exercises 3–8

**Factor the polynomial.**

1. $x^2 + 2x + 1$
2. $x^2 + 9x + 8$
3. $y^2 + 6y + 8$
4. $z^2 + 11z + 24$

370   Chapter 7   Polynomial Equations and Factoring

**EXAMPLE 2** **Factoring $x^2 + bx + c$ When $b$ Is Negative and $c$ Is Positive**

Factor $x^2 - 8x + 12$.

Notice that $b = -8$ and $c = 12$.

- Because $c$ is positive, the factors $p$ and $q$ must have the same sign so that $pq$ is positive.
- Because $b$ is negative, $p$ and $q$ must each be negative so that $p + q$ is negative.

**Check**
Use the FOIL Method.
$(x - 2)(x - 6)$
$= x^2 - 6x - 2x + 12$
$= x^2 - 8x + 12$ ✓

Find two negative integer factors of 12 whose sum is $-8$.

| Factors of 12 | $-1, -12$ | $-2, -6$ | $-3, -4$ |
|---|---|---|---|
| Sum of Factors | $-13$ | $-8$ | $-7$ |

The values of $p$ and $q$ are $-2$ and $-6$.

∴ So, $x^2 - 8x + 12 = (x - 2)(x - 6)$.

**On Your Own**

Now You're Ready
Exercises 10–15

Factor the polynomial.

5. $w^2 - 4w + 3$   6. $n^2 - 12n + 35$   7. $x^2 - 14x + 24$

## Key Idea

**Factoring $x^2 + bx + c$ When $c$ Is Negative**

**Algebra** $x^2 + bx + c = (x + p)(x + q)$ when $p + q = b$ and $pq = c$.
When $c$ is negative, $p$ and $q$ have different signs.

**Example** $x^2 - 4x - 5 = (x + 1)(x - 5)$

**EXAMPLE 3** **Factoring $x^2 + bx + c$ When $c$ Is Negative**

Factor $x^2 + 4x - 21$.

Notice that $b = 4$ and $c = -21$. Because $c$ is negative, the factors $p$ and $q$ must have different signs so that $pq$ is negative.

Find two integer factors of $-21$ whose sum is 4.

| Factors of $-21$ | $-21, 1$ | $-1, 21$ | $-7, 3$ | $-3, 7$ |
|---|---|---|---|---|
| Sum of Factors | $-20$ | $20$ | $-4$ | $4$ |

The values of $p$ and $q$ are $-3$ and $7$.

∴ So, $x^2 + 4x - 21 = (x - 3)(x + 7)$.

# EXAMPLE 4  Real-Life Application

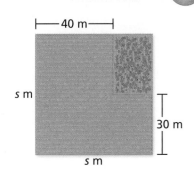

**A farmer plants a rectangular pumpkin patch in the northeast corner of the square plot of land. The area of the pumpkin patch is 600 square meters. What is the area of the square plot of land?**

The length of the pumpkin patch is $(s - 30)$ meters and the width is $(s - 40)$ meters. Write and solve an equation for its area.

| | |
|---|---|
| $600 = (s - 30)(s - 40)$ | Write an equation. |
| $600 = s^2 - 70s + 1200$ | Multiply. |
| $0 = s^2 - 70s + 600$ | Subtract 600 from each side. |
| $0 = (s - 10)(s - 60)$ | Factor the polynomial. |
| $s - 10 = 0$ or $s - 60 = 0$ | Use Zero-Product Property. |
| $s = 10$ or $s = 60$ | Solve for $s$. |

The diagram shows that the side length is at least 30 meters, so 10 meters does not make sense in this situation. The width is 60 meters.

∴ So, the area of the square plot of land is $60(60) = 3600$ square meters.

## On Your Own

**Now You're Ready**
Exercises 21–29

**Factor the polynomial.**

8. $x^2 + 2x - 15$    9. $y^2 + 13y - 30$    10. $v^2 + v - 20$

11. $z^2 - z - 12$    12. $m^2 - 11m - 26$    13. $x^2 - 3x - 40$

14. **WHAT IF?** In Example 4, the area of the pumpkin patch is 200 square meters. What is the area of the square plot of land?

# Summary

### Factoring $x^2 + bx + c$ as $(x + p)(x + q)$

The diagram shows the relationships between the signs of $b$ and $c$ and the signs of $p$ and $q$.

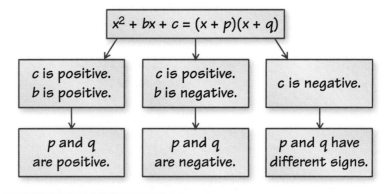

## 7.7 Exercises

### Vocabulary and Concept Check

1. **WRITING** You are factoring $x^2 + 11x - 26$. What do the signs of the terms tell you about the factors? Explain.

2. **OPEN-ENDED** Write a trinomial that can be factored as $(x + p)(x + q)$ where $p$ and $q$ are positive.

### Practice and Problem Solving

**Factor the polynomial.**

3. $x^2 + 8x + 7$
4. $z^2 + 7z + 12$
5. $n^2 + 8n + 12$
6. $s^2 + 11s + 30$
7. $h^2 + 11h + 18$
8. $y^2 + 13y + 40$

9. **ERROR ANALYSIS** Describe and correct the error in factoring the polynomial.

$$t^2 + 14t + 48 = (t + 4)(t + 12)$$

**Factor the polynomial.**

10. $v^2 - 5v + 4$
11. $x^2 - 9x + 20$
12. $d^2 - 5d + 6$
13. $k^2 - 10k + 24$
14. $w^2 - 17w + 72$
15. $j^2 - 13j + 42$

**Solve the equation.**

16. $m^2 + 3m + 2 = 0$
17. $x^2 + 11x + 28 = 0$
18. $n^2 - 9n + 18 = 0$

19. **PROFIT** A company's profit (in millions of dollars) can be represented by $x^2 - 6x + 8$, where $x$ is the number of years since the company started. When did the company have a profit of $3 million?

20. **PROJECTION** A projector displays an image on a wall. The area (in square feet) of the rectangular projection can be represented by $x^2 - 8x + 15$.

   a. Write a binomial that represents the height of the projection.

   b. Find the perimeter of the projection when the height of the wall is 8 feet.

**Factor the polynomial.**

**21.** $x^2 + 3x - 4$   **22.** $z^2 + 7z - 18$   **23.** $n^2 + 4n - 12$

**24.** $s^2 + 3s - 40$   **25.** $h^2 + 6h - 27$   **26.** $y^2 + 2y - 48$

**27.** $m^2 - 6m - 7$   **28.** $x^2 - x - 20$   **29.** $t^2 - 6t - 16$

**Solve the equation.**

**30.** $v^2 + 3v - 4 = 0$   **31.** $x^2 + 5x - 14 = 0$   **32.** $n^2 - 5n = 24$

**33. ERROR ANALYSIS** Describe and correct the error in solving the equation.

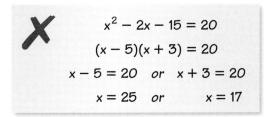

**34. DENTIST** A dentist's office and parking lot are on a rectangular piece of land. The area (in square meters) of the land can be represented by $x^2 + x - 30$.

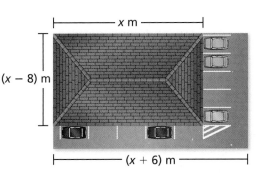

  a. Write a binomial that represents the width of the land.
  b. Write an expression that represents the area of the parking lot.
  c. Evaluate the expressions in parts (a) and (b) when $x = 20$.

**Find the dimensions of the polygon with the given area.**

**35.** Area = 44 square feet

**36.** Area = 35 square centimeters

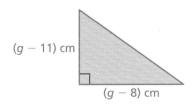

**37.** Area = 120 square feet

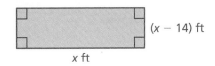

**38.** Area = 75 square centimeters

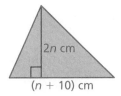

**39. COMPUTER** A web browser is open on your computer screen.

   **a.** The area of the browser is 24 square inches. Find the value of $x$.

   **b.** The browser covers $\frac{3}{13}$ of the screen. What are the dimensions of the screen?

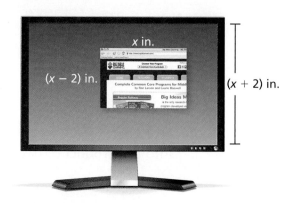

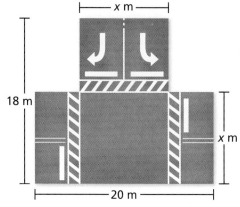

**40. LOGIC** Road construction workers are paving the area shown.

   **a.** Write an expression that represents the area being paved.

   **b.** The area being paved is 280 square meters. Write and solve an equation to find $x$.

   **c.** The equation in part (b) has two solutions. Explain why one of the solutions is not reasonable.

**41. PHOTOGRAPHY** You enlarge a photograph on a computer. The area (in square inches) of the enlarged photograph can be represented by $x^2 + 17x + 70$.

   **a.** Write binomials that represent the length and width of the enlarged photograph.

   **b.** How many inches greater is the length of the enlarged photograph than the width? Explain.

   **c.** The area of the enlarged photograph is 154 square inches. Find the dimensions of each photograph.

**42.**  **Number Sense** Find all of the integer values of $b$ for which the trinomial $x^2 + bx - 12$ has two binomial factors of the form $(x + p)$ and $(x + q)$.

# Fair Game Review  What you learned in previous grades & lessons

**Factor the polynomial.** *(Section 7.6)*

**43.** $2y - 18$  **44.** $7n^2 + 23n$  **45.** $8z^3 + 28z^2$

**46. MULTIPLE CHOICE** Which expression is *not* equivalent to $\sqrt{\frac{9}{4}}$? *(Section 6.1)*

  **Ⓐ** $\frac{3}{2}$    **Ⓑ** $\sqrt{2.25}$    **Ⓒ** $2\sqrt{3}$    **Ⓓ** $3\sqrt{\frac{1}{4}}$

# 7.8 Factoring $ax^2 + bx + c$

**Essential Question** How can you factor the trinomial $ax^2 + bx + c$ into the product of two binomials?

### 1  ACTIVITY: Finding Binomial Factors

**Work with a partner. Six different algebra tiles are shown below.**

**Sample:**

**Step 1:** Arrange the algebra tiles into a rectangular array to model $2x^2 + 5x + 2$.

**Step 2:** Use algebra tiles to label the dimensions of the rectangle.

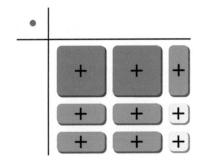 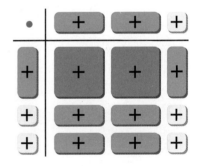

**Step 3:** Write the polynomial in factored form by finding the dimensions of the rectangle.

$$\text{Area} = 2x^2 + 5x + 2 = (2x + 1)(x + 2)$$

with length and width labeled.

**COMMON CORE**

**Polynomial Equations**

In this lesson, you will
- factor trinomials of the form $ax^2 + bx + c$.

Learning Standards
A.REI.4b
A.SSE.3a

Use algebra tiles to write the polynomial as the product of two binomials. Check your answer by multiplying.

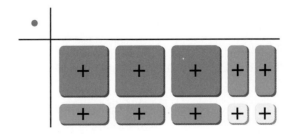

376  Chapter 7  Polynomial Equations and Factoring

## 2 ACTIVITY: Finding Binomial Factors

Work with a partner. Use algebra tiles to write each polynomial as the product of two binomials. Check your answer by multiplying.

a.
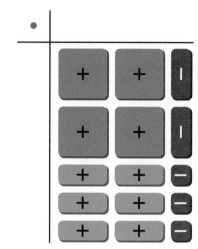

b.
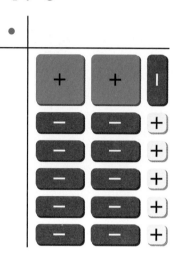

## 3 ACTIVITY: Finding Binomial Factors

**Math Practice**

**Find Entry Points**
What should you do first when factoring a polynomial using algebra tiles?

Work with a partner. Write each polynomial as the product of two binomials. Check your answer by multiplying.

a. $2x^2 + 5x - 3$
b. $3x^2 + 10x - 8$
c. $4x^2 + 4x - 3$

d. $2x^2 + 11x + 15$
e. $9x^2 - 6x + 1$
f. $4x^2 + 11x - 3$

### What Is Your Answer?

4. **IN YOUR OWN WORDS** How can you factor the trinomial $ax^2 + bx + c$ into the product of two binomials?

5. Use your strategy to factor each trinomial.

   a. $4x^2 + 4x + 1$
   b. $3x^2 + 5x - 2$
   c. $2x^2 - 13x + 15$

**Practice** Use what you learned about factoring trinomials to complete Exercises 3–5 on page 380.

Section 7.8  Factoring $ax^2 + bx + c$  **377**

## 7.8 Lesson

In Section 7.7, you factored polynomials of the form $ax^2 + bx + c$, where $a = 1$. To factor polynomials of the form $ax^2 + bx + c$, where $a \neq 1$, first look for the GCF of the terms of the polynomial.

### EXAMPLE 1  Factoring Out the GCF

**Factor $5x^2 + 15x + 10$.**

Notice that the GCF of the terms $5x^2$, $15x$, and 10 is 5.

$5x^2 + 15x + 10 = 5(x^2 + 3x + 2)$    Factor out GCF.
$\qquad\qquad\qquad\quad = 5(x + 1)(x + 2)$    Factor $x^2 + 3x + 2$.

So, $5x^2 + 15x + 10 = 5(x + 1)(x + 2)$.

When there is no GCF, consider the possible factors of $a$ and $c$.

### EXAMPLE 2  Factoring $ax^2 + bx + c$ When $ac$ Is Positive

**a. Factor $4x^2 + 13x + 3$.**

Consider the possible factors of $a = 4$ and $c = 3$.

Factors are 1, 2, and 4. ⟶ $4x^2 + 13x + 3$ ⟵ Factors are 1 and 3.

These factors lead to the following possible products.

$(1x + 1)(4x + 3) \qquad (1x + 3)(4x + 1) \qquad (2x + 1)(2x + 3)$

Multiply to find the product that is equal to the original polynomial.

$(x + 1)(4x + 3) = 4x^2 + 7x + 3$ ✗   $(2x + 1)(2x + 3) = 4x^2 + 8x + 3$ ✗
$(x + 3)(4x + 1) = 4x^2 + 13x + 3$ ✓

So, $4x^2 + 13x + 3 = (x + 3)(4x + 1)$.

**b. Factor $3x^2 - 7x + 2$.**

Consider the possible factors of $a = 3$ and $c = 2$. Because $b$ is negative and $c$ is positive, both factors of $c$ must be negative.

Factors are 1 and 3. ⟶ $3x^2 - 7x + 2$ ⟵ Factors are $-2$ and $-1$.

**Study Tip**
When $ac$ is positive, the sign of $b$ determines whether the factors of $c$ are positive or negative.

These factors lead to the following possible products.

$(1x - 1)(3x - 2) \qquad (1x - 2)(3x - 1)$

Multiply to find the product that is equal to the original polynomial.

$(x - 1)(3x - 2) = 3x^2 - 5x + 2$ ✗   $(x - 2)(3x - 1) = 3x^2 - 7x + 2$ ✓

So, $3x^2 - 7x + 2 = (x - 2)(3x - 1)$.

**Now You're Ready**
Exercises 6–11
and 13–15

### On Your Own
**Factor the polynomial.**

1. $8x^2 - 56x + 48$
2. $2x^2 + 11x + 5$
3. $2x^2 - 7x + 5$
4. $3x^2 - 14x + 8$

### EXAMPLE 3 — Factoring $ax^2 + bx + c$ When $ac$ Is Negative

**Factor $2x^2 - 5x - 7$.**

Consider the possible factors of $a = 2$ and $c = -7$. Because $b$ and $c$ are both negative, the factors of $c$ must have different signs.

Factors are 1 and 2. → $2x^2 - 5x - 7$ ← Factors are $\pm 1$ and $\pm 7$.

**Study Tip**
For polynomials of the form $ax^2 + bx + c$, where $a$ is negative, factor out $-1$ first to make factoring easier. Just be sure to put $-1$ back in your final answer.

These factors lead to the following possible products.

$(x + 1)(2x - 7)$   $(x + 7)(2x - 1)$   $(x - 1)(2x + 7)$   $(x - 7)(2x + 1)$

Multiply to find the product that is equal to the original polynomial.

$(x + 1)(2x - 7) = 2x^2 - 5x - 7$ ✓   $(x - 1)(2x + 7) = 2x^2 + 5x - 7$ ✗

$(x + 7)(2x - 1) = 2x^2 + 13x - 7$ ✗   $(x - 7)(2x + 1) = 2x^2 - 13x - 7$ ✗

∴ So, $2x^2 - 5x - 7 = (x + 1)(2x - 7)$.

### EXAMPLE 4 — Real-Life Application

The length of a rectangular game reserve is 1 mile longer than twice the width. The area of the reserve is 55 square miles. How wide is the reserve?

(A) 2 mi   (B) 2.5 mi   (C) 5 mi   (D) 5.5 mi

Write an equation that represents the area of the reserve. Then solve by factoring. Let $w$ represent the width. Then $2w + 1$ represents the length.

| | |
|---|---|
| $w(2w + 1) = 55$ | Area of the reserve |
| $2w^2 + w - 55 = 0$ | Multiply. Then subtract 55 from each side. |
| $(w - 5)(2w + 11) = 0$ | Factor left side of the equation. |
| $w - 5 = 0$ or $2w + 11 = 0$ | Use Zero-Product Property. |
| $w = 5$ or $w = -\dfrac{11}{2}$ | Solve for $w$. Use the positive solution. |

∴ The correct answer is (C).

**Now You're Ready**
Exercises 16–21

### On Your Own
**Factor the polynomial.**

5. $6x^2 + x - 12$
6. $4x^2 - 19x - 5$

7. **WHAT IF?** In Example 4, the area of the reserve is 136 square miles. How wide is the reserve?

# 7.8 Exercises

## Vocabulary and Concept Check

1. **WRITING** Describe how to factor polynomials of the form $ax^2 + bx + c$.

2. **WHICH ONE DOESN'T BELONG?** Which factored polynomial does *not* belong with the other three? Explain your reasoning.

   $(2x - 3)(x + 2)$    $x(2x - 3) + 2(2x - 3)$    $(2x + 3)(x - 2)$    $2x(x + 2) - 3(x + 2)$

## Practice and Problem Solving

**Use algebra tiles to write the polynomial as the product of two binomials.**

3. $2x^2 - 3x + 1$
4. $3x^2 + x - 2$
5. $4x^2 + 11x + 6$

**Factor the polynomial.**

6. $3x^2 + 3x - 6$
7. $8v^2 + 8v - 48$
8. $4k^2 + 28k + 48$
9. $6y^2 - 24y + 18$
10. $9r^2 - 36r - 45$
11. $7d^2 - 63d + 140$

12. **ERROR ANALYSIS** Describe and correct the error in factoring the polynomial.

    $$2x^2 + 2x - 4 = 2x(x + 1 - 2)$$
    $$= 2x(x - 1)$$

**Factor the polynomial.**

13. $3h^2 + 11h + 6$
14. $6x^2 - 5x + 1$
15. $8m^2 + 30m + 7$
16. $18v^2 - 15v - 18$
17. $2n^2 - 5n - 3$
18. $4z^2 - 4z - 3$
19. $8g^2 - 10g - 12$
20. $10w^2 + 19w - 15$
21. $14d^2 + 3d - 2$

22. **ERROR ANALYSIS** Describe and correct the error in factoring the polynomial.

    $$6x^2 - 7x - 3 = (3x - 3)(2x + 1)$$

23. **DANCE FLOOR** The area (in square feet) of a rectangular lighted dance floor can be represented by $8x^2 + 22x + 5$. Write the expressions that represent the dimensions of the dance floor.

**Solve the equation.**

**24.** $5x^2 - 5x - 30 = 0$

**25.** $2k^2 - 5k - 18 = 0$

**26.** $12m^2 + 11m = 15$

**Factor the polynomial.**

**27.** $-3w^2 - 2w + 8$

**28.** $-12x^2 + 48x + 27$

**29.** $-40n^2 + 70n - 15$

**30. CLIFF DIVING** The height $h$ (in feet) above the water of a cliff diver is modeled by $h = -16t^2 + 8t + 80$, where $t$ is the time (in seconds). How long is the diver in the air?

**31. REASONING** For what values of $t$ can $2x^2 + tx + 10$ be written as the product of two binomials?

**32. INVITATION** The length of a rectangular birthday party invitation is 1 inch less than twice its width. The area of the invitation is 15 square inches. Will the invitation fit in a $3\frac{5}{8}$-inch by $5\frac{1}{8}$-inch envelope without being folded? Explain your reasoning.

**33. SWIMMING POOL** A rectangular swimming pool is bordered by a concrete patio. The width of the patio is the same on every side. The surface area of the pool is equal to the area of the patio border. What is the width of the patio border?

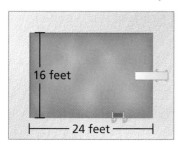

**34. REASONING** When is it *not* possible to factor $ax^2 + bx + c$, where $a \neq 1$? Give an example.

**35. CHOOSE TOOLS** A vendor can sell 50 bobbleheads per day when the price is $40 each. For every $2 decrease in price, 5 more bobbleheads are sold each day.

   **a.** The revenue from yesterday was $2160. What was the price per bobblehead? (*Note:* revenue = units sold × unit price)

   **b.** How much should the vendor charge per bobblehead to maximize the daily revenue? Explain how you found your answer.

**Structure Factor the polynomial.**

**36.** $40k^3 + 6k^2 - 4k$

**37.** $6x^2 + 5xy - 4y^2$

**38.** $18m^3 + 39m^2n - 15mn^2$

## Fair Game Review  *What you learned in previous grades & lessons*

**Find the product.** *(Section 7.4)*

**39.** $(2x - 7)(2x + 7)$

**40.** $(k + 5)^2$

**41.** $(3b - 4)^2$

**42. MULTIPLE CHOICE** Two angles are supplementary. The measure of one of the angles is 58°. What is the measure of the other angle? *(Skills Review Handbook)*

   **A** 22°    **B** 32°    **C** 58°    **D** 122°

# 7.9 Factoring Special Products

**Essential Question** How can you recognize and factor special products?

### 1 ACTIVITY: Factoring Special Products

Work with a partner. Six different algebra tiles are shown below.

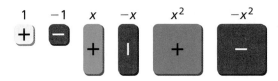

Use algebra tiles to write each polynomial as the product of two binomials. Check your answer by multiplying. State whether the product is a "special product" that you studied in Lesson 7.4.

a.
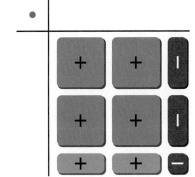

b.
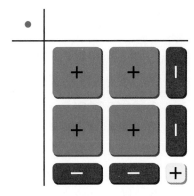

c.
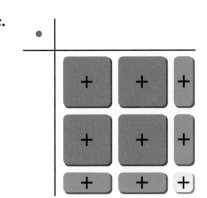

d.
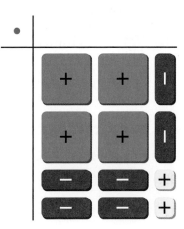

**COMMON CORE**

**Polynomial Equations**

In this lesson, you will
- factor differences of two squares.
- factor perfect square trinomials.

Learning Standards
A.REI.4b
A.SSE.2
A.SSE.3a

**382** Chapter 7 Polynomial Equations and Factoring

## 2 ACTIVITY: Factoring Special Products

**Work with a partner. Use algebra tiles to complete the rectangular array in three different ways, so that each way represents a different special product. Write each special product in polynomial form and also in factored form.**

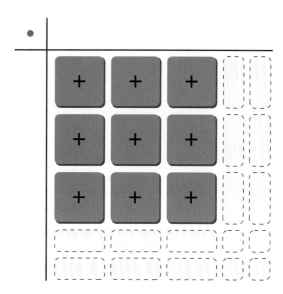

## 3 ACTIVITY: Finding Binomial Factors

**Work with a partner. Write each polynomial as the product of two binomials. Check your answer by multiplying.**

a. $4x^2 - 12x + 9$    b. $4x^2 - 9$    c. $4x^2 + 12x + 9$

**Math Practice**

**Maintain Oversight**
How is factoring a special product similar to factoring a polynomial?

### What Is Your Answer?

4. **IN YOUR OWN WORDS** How can you recognize and factor special products? Describe a strategy for recognizing which polynomials can be factored as special products.

5. Use your strategy to factor each polynomial.

   a. $25x^2 + 10x + 1$    b. $25x^2 - 10x + 1$    c. $25x^2 - 1$

**Practice** → Use what you learned about factoring polynomials as special products to complete Exercises 4–6 on page 386.

## 7.9 Lesson

You can use special product patterns to factor polynomials.

### 🔑 Key Idea

**Difference of Two Squares Pattern**

**Algebra**
$a^2 - b^2 = (a + b)(a - b)$

**Example**
$x^2 - 9 = x^2 - 3^2$
$= (x + 3)(x - 3)$

### EXAMPLE 1 — Factoring the Difference of Two Squares

**Factor each polynomial.**

**a.** $x^2 - 25$

$x^2 - 25 = x^2 - 5^2$     Write as $a^2 - b^2$.
$\quad\quad\quad\,\, = (x + 5)(x - 5)$     Difference of Two Squares Pattern

**b.** $64 - y^2$

$64 - y^2 = 8^2 - y^2$     Write as $a^2 - b^2$.
$\quad\quad\quad\, = (8 + y)(8 - y)$     Difference of Two Squares Pattern

**c.** $4z^2 - 1$

$4z^2 - 1 = (2z)^2 - 1^2$     Write as $a^2 - b^2$.
$\quad\quad\quad\, = (2z + 1)(2z - 1)$     Difference of Two Squares Pattern

**Remember**
You can check your answers using the FOIL Method.

### 🟠 On Your Own

**Now You're Ready** Exercises 4–8

**Factor the polynomial.**

1. $x^2 - 36$     2. $100 - m^2$    3. $9n^2 - 16$    4. $16h^2 - 49$

### 🔑 Key Idea

**Perfect Square Trinomial Pattern**

**Algebra**
$a^2 + 2ab + b^2 = (a + b)^2$

**Example**
$x^2 + 6x + 9 = x^2 + 2(x)(3) + 3^2$
$= (x + 3)^2$

$a^2 - 2ab + b^2 = (a - b)^2$

$x^2 - 6x + 9 = x^2 - 2(x)(3) + 3^2$
$= (x - 3)^2$

### EXAMPLE 2  Factoring Perfect Square Trinomials

Factor each polynomial.

**a.** $n^2 + 8n + 16$

$n^2 + 8n + 16 = n^2 + 2(n)(4) + 4^2$  Write as $a^2 + 2ab + b^2$.
$= (n + 4)^2$  Perfect Square Trinomial Pattern

**b.** $x^2 - 18x + 81$

$x^2 - 18x + 81 = x^2 - 2(x)(9) + 9^2$  Write as $a^2 - 2ab + b^2$.
$= (x - 9)^2$  Perfect Square Trinomial Pattern

#### On Your Own

*Now You're Ready*
*Exercises 9–12*

Factor the polynomial.

**5.** $m^2 - 2m + 1$    **6.** $d^2 - 10d + 25$    **7.** $z^2 + 20z + 100$

### EXAMPLE 3  Real-Life Application

$y = 81 - 16t^2$

A bird picks up a golf ball and drops it while flying. The function represents the height $y$ (in feet) of the golf ball $t$ seconds after it is dropped. The ball hits the top of a 32-foot tall pine tree. After how many seconds does the ball hit the tree?

Substitute 32 for $y$ and solve for $t$.

$y = 81 - 16t^2$  Write equation.
$32 = 81 - 16t^2$  Substitute 32 for $y$.
$0 = 49 - 16t^2$  Subtract 32 from each side.
$0 = 7^2 - (4t)^2$  Write as $a^2 - b^2$.
$0 = (7 + 4t)(7 - 4t)$  Difference of Two Squares Pattern
$7 + 4t = 0$  or  $7 - 4t = 0$  Use Zero-Product Property.
$t = -\dfrac{7}{4}$  or  $t = \dfrac{7}{4}$  Solve for $t$.

A negative time does not make sense in this situation.

∴ So, the golf ball hits the tree after $\dfrac{7}{4}$, or 1.75 seconds.

#### On Your Own

**8. WHAT IF?** The golf ball does not hit the pine tree. After how many seconds does the ball hit the ground?

## 7.9 Exercises

### Vocabulary and Concept Check

1. **WRITING** Describe two ways to show that $x^2 - 16$ is equal to $(x + 4)(x - 4)$.

2. **REASONING** Can you use the perfect square trinomial pattern to factor $y^2 + 16y + 64$? Explain.

3. **WHICH ONE DOESN'T BELONG?** Which polynomial does *not* belong with the other three? Explain your reasoning.

$n^2 - 4$   $g^2 - 6g + 9$   $r^2 + 12r + 36$   $k^2 + 25$

### Practice and Problem Solving

**Factor the polynomial.**

4. $m^2 - 49$
5. $9 - r^2$
6. $4x^2 - 25$
7. $81d^2 - 64$
8. $121 - 16t^2$
9. $h^2 + 12h + 36$
10. $x^2 - 4x + 4$
11. $w^2 - 14w + 49$
12. $g^2 + 24g + 144$

13. **ERROR ANALYSIS** Describe and correct the error in factoring the polynomial.

$$n^2 - 16n + 64 = n^2 - 2(n)(8) + 8^2$$
$$= (n + 8)^2$$

**Solve the equation.**

14. $z^2 - 4 = 0$
15. $s^2 + 20s + 100 = 0$
16. $k^2 - 16k + 64 = 0$
17. $4x^2 = 49$
18. $n^2 + 9 = -6n$
19. $y^2 = 12y - 36$

20. **REASONING** Tell whether the polynomial can be factored. If not, change the constant term so that the polynomial can be factored using the perfect square trinomial pattern.

   a. $w^2 + 18w + 84$
   b. $y^2 - 10y + 23$
   c. $x^2 - 14x + 50$

21. **COASTER** The area (in square centimeters) of a square coaster can be represented by $d^2 + 8d + 16$. Write an expression that represents the side length of the coaster.

**Factor the polynomial.**

22. $3z^2 - 27$

23. $2m^3 - 50m$

24. $x^4 + 8x^3 + 16x^2$

25. $5f^3 - 20f^2 + 20f$

26. **PROBLEM SOLVING** The polynomial represents the area (in square feet) of the square playground.

   a. Write a polynomial that represents the side length of the playground.

   b. Write an expression for the perimeter of the playground.

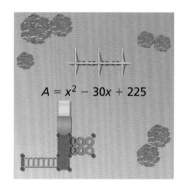

27. **NUMBER SENSE** Solve $28 = 64 - 9x^2$ in two ways.

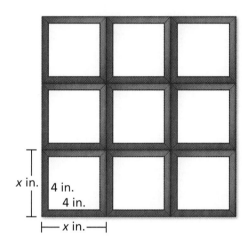

28. **INTERIOR DESIGN** You hang 9 identical square picture frames on a wall.

   a. Write a polynomial that represents the area of the picture frames, not including the pictures.

   b. The area in part (a) is 81 square inches. What is the side length of one of the picture frames? Explain your reasoning.

**Factor the polynomial.**

29. $4y^2 + 4y + 1$

30. $16v^2 - 24v + 9$

31. $9m^2 + 36m + 36$

32. **Geometry** A composite solid is made up of a cube and a rectangular prism.

   a. Write a polynomial that represents the volume of the composite solid.

   b. The volume of the composite solid is equal to $25x$. What is the value of $x$? Explain your reasoning.

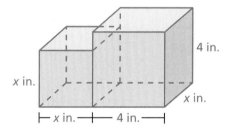

**Fair Game Review** What you learned in previous grades & lessons

**Factor the polynomial.** *(Section 7.7)*

33. $w^2 + w - 12$

34. $x^2 - 5x - 36$

35. $d^2 - 4d - 60$

36. **MULTIPLE CHOICE** You deposit $3000 in a savings account. The account earns 4% simple interest per year. What is the balance after 2 years? *(Skills Review Handbook)*

   Ⓐ $240    Ⓑ $3000    Ⓒ $3240    Ⓓ $5400

# Extension 7.9 Factoring Polynomials Completely

**Key Vocabulary**
factoring by grouping, p. 388
prime polynomial, p. 389
factored completely, p. 389

To factor polynomials with four terms, group the terms into pairs, factor the GCF out of each pair of terms, and look for a common binomial factor. This process is called **factoring by grouping.**

### EXAMPLE 1 Factoring by Grouping

**Factor each polynomial.**

**a.** $x^3 + 3x^2 + 2x + 6$

$x^3 + 3x^2 + 2x + 6 = (x^3 + 3x^2) + (2x + 6)$    Group terms with common factors.

Common binomial factor is $x + 3$. → $= x^2(x + 3) + 2(x + 3)$    Factor out GCF of each pair of terms.

$= (x + 3)(x^2 + 2)$    Factor out $(x + 3)$.

**b.** $x^3 - 7 - x^2 + 7x$

The terms $x^3$ and $-7$ do not have a common factor. Rearrange the terms of the polynomial so you can group terms with common factors.

$x^3 - 7 - x^2 + 7x = x^3 - x^2 + 7x - 7$    Rewrite polynomial.

$= (x^3 - x^2) + (7x - 7)$    Group terms with common factors.

Common binomial factor is $x - 1$. → $= x^2(x - 1) + 7(x - 1)$    Factor out GCF of each pair of terms.

$= (x - 1)(x^2 + 7)$    Factor out $(x - 1)$.

**c.** $x^2 + y + x + xy$

$x^2 + y + x + xy = x^2 + x + xy + y$    Rewrite polynomial.

$= (x^2 + x) + (xy + y)$    Group terms with common factors.

$= x(x + 1) + y(x + 1)$    Factor out GCF of each pair of terms.

$= (x + 1)(x + y)$    Factor out $(x + 1)$.

## Practice

**Factor the polynomial by grouping.**

1. $n^3 + 2n^2 + 5n + 10$
2. $p^3 - 7p^2 + 3p - 21$
3. $2y^3 + 8y^2 + 3y + 12$
4. $6s^3 - 16s^2 + 21s - 56$
5. $8v^3 + 48v - 5v^2 - 30$
6. $2w^3 - w^2 - 18w + 9$
7. $x^2 + xy + 3x + 3y$
8. $a - ab + a^2 - b$
9. $4xy + 20y + 3x + 15$

A **prime polynomial** is a polynomial that cannot be factored as a product of polynomials with integer coefficients. A factorable polynomial with integer coefficients is said to be **factored completely** when no more factors can be found and it is written as the product of prime factors.

### EXAMPLE 2  Factoring Completely

**COMMON CORE**

**Polynomial Equations**
In this extension, you will
- factor polynomials by grouping.
- factor polynomials completely.

Learning Standards
A.REI.4b
A.SSE.2
A.SSE.3a

Factor each polynomial completely.

**a.** $3x^3 - 18x^2 + 24x$

$3x^3 - 18x^2 + 24x = 3x(x^2 - 6x + 8)$     Factor out $3x$.

$= 3x(x - 2)(x - 4)$     Factor $x^2 - 6x + 8$.

**b.** $7x^4 - 28x^2$

$7x^4 - 28x^2 = 7x^2(x^2 - 4)$     Factor out $7x^2$.

$= 7x^2(x^2 - 2^2)$     Write as $a^2 - b^2$.

$= 7x^2(x + 2)(x - 2)$     Difference of Two Squares Pattern

**c.** $p^2 + 4p - 2$

The terms of $p^2 + 4p - 2$ have no common factors. There are no integer factors of $-2$ whose sum is 4. So, this polynomial is already factored completely.

### EXAMPLE 3  Solving an Equation by Factoring Completely

$2x^3 + 8x^2 = 10x$     Original equation

$2x^3 + 8x^2 - 10x = 0$     Subtract $10x$ from each side.

$2x(x^2 + 4x - 5) = 0$     Factor out $2x$.

$2x(x + 5)(x - 1) = 0$     Factor $x^2 + 4x - 5$.

$2x = 0$   or   $x + 5 = 0$   or   $x - 1 = 0$     Use Zero-Product Property.

$x = 0$   or   $x = -5$   or   $x = 1$     Solve for $x$.

∴ The solutions are $x = -5$, $x = 0$, and $x = 1$.

## Practice

**Factor the polynomial completely, if possible.**

**10.** $2x^3 + 10x^2 - 48x$

**11.** $5z^4 - 5z^2$

**12.** $20c + 4c^3 - 24c^2$

**13.** $y^2 + 6y - 5$

**14.** $q^2 - q + 7$

**15.** $3n^4 - 48n^2$

**Solve the equation.**

**16.** $k^3 - 6k^2 + 9k = 0$

**17.** $3x^3 + 6x^2 = 72x$

**18.** $4y^3 - 12y^2 - 40y = 0$

# 7.5–7.9 Quiz

**Factor the polynomial.** *(Sections 7.6–7.9)*

1. $3d^2 + 11d$
2. $9z^2 - 18z$
3. $x^2 + 9x + 20$
4. $r^2 - 3r - 18$
5. $2x^2 - 3x + 1$
6. $3b^2 - 13b + 4$
7. $x^2 - 9$
8. $z^2 + 22z + 121$

**Solve the equation.** *(Sections 7.5–7.9)*

9. $m^2 - 11m + 18 = 0$
10. $w^3 - 9w^2 = 0$
11. $6m^2 - 5m + 1 = 0$
12. $h^2 - 8 = -3h + 10$
13. $4s^2 = 144$
14. $k^2 + 100 = 20k$

15. **STORAGE** The front of a storage bunker can be modeled by $y = -\dfrac{5}{216}(x-72)(x+72)$, where $x$ and $y$ are measured in inches. The $x$-axis represents the ground. Find the width of the bunker at ground level. *(Section 7.5)*

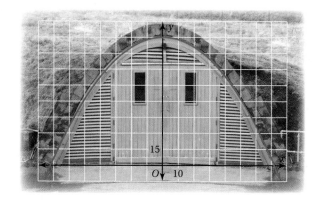

16. **DISASTER RELIEF** A helicopter drops a box of supplies after a disaster. The function represents the height $y$ (in feet) of the box $t$ seconds after it is dropped. After how many seconds does the box hit the ground? *(Section 7.9)*

17. **MAGIC SHOW** A magician's stage has a trap door. *(Section 7.7)*

   a. The total area of the stage can be represented by $x^2 + 27x + 176$. Write an expression for the width of the stage.

   b. The area of the trap door is 12 square feet. Find the value of $x$.

   c. What fraction of the area of the stage is the area of the trap door?

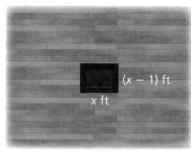

# 7 Chapter Review

## Review Key Vocabulary

monomial, *p. 330*
degree of a monomial, *p. 330*
polynomial, *p. 331*
binomial, *p. 331*
trinomial, *p. 331*

degree of a polynomial, *p. 331*
FOIL Method, *p. 343*
factored form, *p. 358*
Zero-Product Property, *p. 358*
root, *p. 358*

factoring by grouping, *p. 388*
prime polynomial, *p. 389*
factored completely, *p. 389*

## Review Examples and Exercises

### 7.1 Polynomials *(pp. 328–333)*

**a. Find the degree of $4x^2y$.**

The exponent of $x$ is 2 and the exponent of $y$ is 1.
The sum of the exponents is $2 + 1 = 3$.

∴ So, the degree of the monomial is 3.

**b. Write $x + 1 + 2x^3$ in standard form. Identify the degree and classify the polynomial by the number of terms.**

| Polynomial | Standard Form | Degree | Type of Polynomial |
|---|---|---|---|
| $x + 1 + 2x^3$ | $2x^3 + x + 1$ | 3 | trinomial |

#### Exercises

**Write the polynomial in standard form. Identify the degree and classify the polynomial by the number of terms.**

1. $2w^3 + 3 - 4w$
2. $-6y^2$
3. $-6.2 + 3t^5$

### 7.2 Adding and Subtracting Polynomials *(pp. 334–339)*

**a. $(2d^2 - 3) + (4d^2 + 2)$**

$(2d^2 - 3) + (4d^2 + 2) = (2d^2 + 4d^2) + (-3 + 2)$
$= 6d^2 - 1$

**b. $(c^2 + 5c + 1) - (c^2 - 2)$**

$(c^2 + 5c + 1) - (c^2 - 2) = (c^2 + 5c + 1) + (-c^2 + 2)$
$= [c^2 + (-c^2)] + 5c + (1 + 2) = 5c + 3$

#### Exercises

**Find the sum or difference.**

4. $(3a + 7) + (a - 1)$
5. $(x^2 + 4x - 2) + (6x^2 + 6)$
6. $(-y^2 + y + 2) - (y^2 - 5y - 2)$
7. $(p - 9) - (-8p^2 + 7)$

## 7.3 Multiplying Polynomials (pp. 340–347)

Find $(x + 1)(x - 4)$.

$$(x + 1)(x - 4) = x(x) + x(-4) + (1)(x) + (1)(-4)$$ Use the FOIL Method.
$$= x^2 + (-4x) + (x) + (-4)$$ Multiply.
$$= x^2 - 3x - 4$$ Combine like terms.

### Exercises

**Find the product.**

8. $(y + 4)(y - 2)$
9. $(q - 3)(2q + 7)$
10. $(-3v + 1)(v^2 - v - 2)$

## 7.4 Special Products of Polynomials (pp. 348–353)

Find each product.

**a.** $(x + 3)(x - 3)$

$$(a + b)(a - b) = a^2 - b^2$$ Sum and Difference Pattern
$$(x + 3)(x - 3) = x^2 - 3^2$$ Use pattern.
$$= x^2 - 9$$ Simplify.

**b.** $(y + 2)^2$

$$(a + b)^2 = a^2 + 2ab + b^2$$ Square of a Binomial Pattern
$$(y + 2)^2 = y^2 + 2(y)(2) + 2^2$$ Use pattern.
$$= y^2 + 4y + 4$$ Simplify.

### Exercises

**Find the product.**

11. $(y + 9)(y - 9)$
12. $(2x + 4)(2x - 4)$
13. $(h + 4)^2$
14. $(-1 + 2d)^2$

## 7.5 Solving Polynomial Equations in Factored Form (pp. 356–361)

Solve $(x + 4)(x - 3) = 0$.

$(x + 4)(x - 3) = 0$   Write equation.
$x + 4 = 0$  or  $x - 3 = 0$   Use Zero-Product Property.
$x = -4$  or  $x = 3$   Solve for $x$.

∴ The roots are $x = -4$ and $x = 3$.

### Exercises

**Solve the equation.**

**15.** $x(x + 2) = 0$

**16.** $(t - 3)(t - 8) = 0$

**17.** $(a + 10)^2 = 0$

**18.** $2s(s + 1)(s - 4) = 0$

## 7.6 Factoring Polynomials Using the GCF (pp. 362–367)

**Factor $4z^2 + 32$.**

**Step 1:** Find the GCF of the terms.

$$4z^2 = \boxed{2} \cdot \boxed{2} \cdot z \cdot z$$
$$32 = \boxed{2} \cdot \boxed{2} \cdot 2 \cdot 2 \cdot 2$$

The GCF is $2 \cdot 2 = 4$.

**Step 2:** Write the polynomial as a product of the GCF and its remaining factors.

$4z^2 + 32 = 4(z^2) + 4(8)$   Factor out GCF.

$\phantom{4z^2 + 32} = 4(z^2 + 8)$   Distributive Property

### Exercises

**Factor the polynomial.**

**19.** $6t^2 + 36$

**20.** $2x^2 - 20x$

**21.** $15y^3 + 3y^2$

## 7.7 Factoring $x^2 + bx + c$ (pp. 368–375)

**Factor $x^2 + 12x + 27$.**

Notice that $b = 12$ and $c = 27$.

- Because $c$ is positive, the factors $p$ and $q$ must have the same sign so that $pq$ is positive.

- Because $b$ is also positive, $p$ and $q$ must each be positive so that $p + q$ is positive.

Find two positive integer factors of 27 whose sum is 12.

| Factors of 27 | Sum of Factors |
|---|---|
| 1, 27 | 28 |
| 3, 9 | 12 |

The values of $p$ and $q$ are 3 and 9.

So, $x^2 + 12x + 27 = (x + 3)(x + 9)$.

### Exercises

**Factor the polynomial.**

22. $p^2 + 2p - 35$
23. $b^2 + 9b + 20$
24. $z^2 - 4z - 21$

## 7.8 Factoring $ax^2 + bx + c$  (pp. 376–381)

**a. Factor $2x^2 + 13x + 15$.**

Consider the possible factors of $a = 2$ and $c = 15$.

Factors are 1 and 2. ⟶ $2x^2 + 13x + 15$ ⟵ Factors are 1, 3, 5, and 15.

These factors lead to the following possible products.

$(1x + 1)(2x + 15)$     $(1x + 3)(2x + 5)$

$(1x + 15)(2x + 1)$     $(1x + 5)(2x + 3)$

Multiply to find the product that is equal to the original polynomial.

$(x + 1)(2x + 15) = 2x^2 + 17x + 15$ ✗

$(x + 15)(2x + 1) = 2x^2 + 31x + 15$ ✗

$(x + 3)(2x + 5) = 2x^2 + 11x + 15$ ✗

$(x + 5)(2x + 3) = 2x^2 + 13x + 15$ ✓

So, $2x^2 + 13x + 15 = (x + 5)(2x + 3)$.

**b. Factor $5x^2 + 4x - 9$.**

Consider the possible factors of $a = 5$ and $c = -9$. Because $b$ is positive and $c$ is negative, the factors of $c$ must have different signs.

Factors are 1 and 5. ⟶ $5x^2 + 4x - 9$ ⟵ Factors are $\pm 1$, $\pm 3$, and $\pm 9$.

These factors lead to the following possible products.

$(1x + 1)(5x - 9)$     $(1x - 1)(5x + 9)$     $(1x - 3)(5x + 3)$

$(1x + 9)(5x - 1)$     $(1x - 9)(5x + 1)$     $(1x + 3)(5x - 3)$

Multiply to find the product that is equal to the original polynomial.

$(x + 1)(5x - 9) = 5x^2 - 4x - 9$ ✗

$(x + 9)(5x - 1) = 5x^2 + 44x - 9$ ✗

$(x - 1)(5x + 9) = 5x^2 + 4x - 9$ ✓

$(x - 9)(5x + 1) = 5x^2 - 44x - 9$ ✗

$(x - 3)(5x + 3) = 5x^2 - 12x - 9$ ✗

$(x + 3)(5x - 3) = 5x^2 + 12x - 9$ ✗

So, $5x^2 + 4x - 9 = (x - 1)(5x + 9)$.

### Exercises

**Factor the polynomial.**

25. $10a^2 + 11a + 3$
26. $4z^2 + 11z + 6$
27. $2x^2 - 27x - 14$
28. $-2p^2 + 2p + 4$

29. **OUTSIDE PATIO** You are installing new tile on an outside patio. The area (in square feet) of the rectangular patio can be represented by $8x^2 + 33x + 4$. Write the expressions that represent the dimensions of the patio.

## 7.9 Factoring Special Products (pp. 382–389)

**Factor each polynomial.**

**a.** $x^2 - 16$

$$x^2 - 16 = x^2 - 4^2 \quad \text{Write as } a^2 - b^2.$$
$$= (x + 4)(x - 4) \quad \text{Difference of Two Squares Pattern}$$

**b.** $x^2 - 2x + 1$

$$x^2 - 2x + 1 = x^2 - 2(x)(1) + 1^2 \quad \text{Write as } a^2 - 2ab + b^2.$$
$$= (x - 1)^2 \quad \text{Perfect Square Trinomial Pattern}$$

**c.** $x^3 + 4x^2 + 3x + 12$

$$x^3 + 4x^2 + 3x + 12 = (x^3 + 4x^2) + (3x + 12) \quad \text{Group terms with common factors.}$$
$$= x^2(x + 4) + 3(x + 4) \quad \text{Factor out GCF of each pair of terms.}$$
$$= (x + 4)(x^2 + 3) \quad \text{Factor out } (x + 4).$$

Common binomial factor is $x + 4$.

**d.** $2x^4 - 8x^2$

$$2x^4 - 8x^2 = 2x^2(x^2 - 4) \quad \text{Factor out } 2x^2.$$
$$= 2x^2(x^2 - 2^2) \quad \text{Write as } a^2 - b^2.$$
$$= 2x^2(x + 2)(x - 2) \quad \text{Difference of Two Squares Pattern}$$

### Exercises

**Factor the polynomial.**

30. $x^2 - 9$
31. $y^2 - 100$
32. $z^2 + 6z + 9$
33. $m^2 + 16m + 64$
34. $x^2 - 3x + 4ax - 12a$
35. $n^3 - 9n$

# 7 Chapter Test

**Write the polynomial in standard form. Identify the degree and classify the polynomial by the number of terms.**

1. $-2.1w^3$
2. $7k + 4 - 3k^2$
3. $-c^8 + 9c^{12}$

**Find the sum or difference.**

4. $(-2p + 4) - (p^2 - 6p + 8)$
5. $(4s^2 + 2st + t) + (-3s^2 + 5st - 4t)$

**Find the product.**

6. $(h - 5)(h - 8)$
7. $(2w - 3)(2w + 5)$
8. $(z + 11)(z - 11)$

**Factor the polynomial.**

9. $7x^2 - 21x$
10. $n^2 + 7n + 10$
11. $m^2 - 2m - 24$
12. $6g^2 + 23g + 7$
13. $y^2 - 100$
14. $b^3 - 2b^2 + 3b - 6$

**Solve the equation.**

15. $(n - 1)(n + 6) = 0$
16. $3h^2 = -12h$
17. $s^2 - 15s + 50 = 0$
18. $5k^2 + 22k - 15 = 0$
19. $d^2 + 14d + 49 = 0$
20. $6x^4 + 8x^2 = 26x^3$

21. **TIME** The expression $\pi(r - 3)^2$ represents the area covered by the hour hand on a clock in one rotation, where $r$ is the radius of the entire clock. Write a polynomial that represents the area covered by the hour hand in one rotation.

22. **TRAMPOLINE** You are jumping on a trampoline. Your height $y$ (in feet) above the trampoline after $t$ seconds can be represented by $y = -16t^2 + 24t$. How many seconds are you in the air?

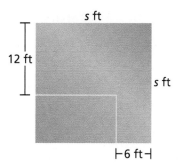

23. **CEMENT** You pour cement in a rectangular region of a square garage. The area of the rectangular region is 112 square feet.

    a. What is the area of the garage floor?

    b. You place caution tape along the two sides of the newly cemented region that are not on the wall. How many feet of caution tape do you use?

24. **ARCHERY** The area (in square inches) of the target can be represented by $\pi(x^2 + 6x + 9)$.

    a. Find the areas of the red bull's eye and the gray ring when the area of the target is $25\pi$ square inches. Write your answer in terms of $\pi$.

    b. Write a binomial that represents the radius of the target.

    c. What is the width of the gray ring? Does it change as $x$ changes? Does its area change as $x$ changes? Explain.

# 7 Standards Assessment

1. Which inequality is shown in the coordinate plane? *(A.REI.12)*

   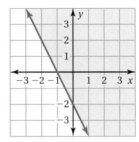

   A. $y > -2x - 2$
   
   B. $y < -2x - 2$
   
   C. $y \geq -2x - 2$
   
   D. $y \leq -2x - 2$

   **Test-Taking Strategy**
   **Solve Directly or Eliminate Choices**

   You are having $x$ cat treats for dinner where $x^2 - x - 6 = 0$. How many is that?
   Ⓐ $-3$  Ⓑ $-2$  Ⓒ $2$  Ⓓ $3$
   Dinnertime!

   "You can eliminate A and B. Then, solve directly to determine that the correct answer is D."

2. Which expression is equivalent to $\left(\dfrac{a^3}{a^{-2}}\right)^{-3}$? *(N.RN.2)*

   F. $a^2$
   
   G. $\dfrac{1}{a^3}$
   
   H. $\dfrac{1}{a^2}$
   
   I. $\dfrac{1}{a^{15}}$

3. What is the degree of the polynomial shown below? *(A.SSE.1a)*

   $$p^3 + 2p - 5p^4$$

4. Which of the following is the equation of the line that passes through the points $(-1, -6)$ and $(2, 6)$? *(A.CED.2)*

   A. $y = -\dfrac{1}{4}x - \dfrac{25}{4}$
   
   B. $y = -4x - 10$
   
   C. $y = \dfrac{1}{4}x - \dfrac{23}{4}$
   
   D. $y = 4x - 2$

5. What are the roots of $(5b + 3)(5b - 3) = 0$? *(A.REI.4b)*

   F. $-\dfrac{3}{5}$ and $\dfrac{3}{5}$
   
   G. $-3$ and $3$
   
   H. $-5$ and $5$
   
   I. $-\dfrac{5}{3}$ and $\dfrac{5}{3}$

Standards Assessment 397

6. What is the range of the function graphed in the coordinate plane below? *(F.IF.1)*

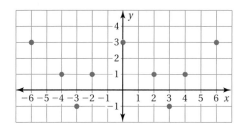

 A. 1, 3

 B. −1, 1, 3

 C. 0, 2, 3, 4, 6

 D. −6, −4, −3, −2, 0, 2, 3, 4, 6

7. Which polynomial represents the product of $2x - 4$ and $x^2 + 6x - 2$? *(A.APR.1)*

 F. $2x^3 + 8x^2 - 4x + 8$

 G. $2x^3 + 8$

 H. $2x^3 + 8x^2 - 28x + 8$

 I. $2x^3 - 24x - 2$

8. For what value of $b$ does the system of linear equations shown below have no solution? *(8.EE.8a)*

$$y = 6x + 3$$
$$bx - 2y = -10$$

9. The graph of which equation is shown in the coordinate plane? *(F.IF.7e)*

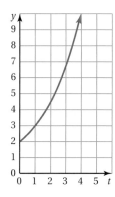

 A. $y = 2(1.5)^t$

 B. $y = 2^t$

 C. $y = 2(0.5)^t$

 D. $y = (0.5)^t$

10. The playing area of a hole on a miniature golf course is 216 square feet. What is the perimeter of the playing area? Explain. *(A.REI.4b)*

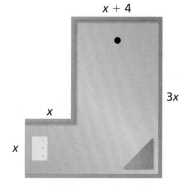

11. Which graph shows the solution of the system of linear inequalities shown below?  (A.REI.12)

$$2x + 3y > 6$$
$$2x - y \leq -2$$

F.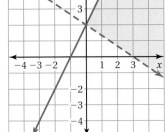

H.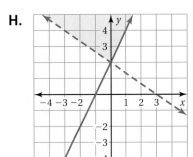

G.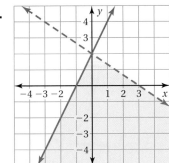

I.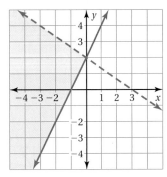

12. Andy was factoring the polynomial in the box below.  (A.SSE.2)

$$16t^2 - 49 = 4t^2 - 7^2$$
$$= (2t + 7)(2t - 7)$$

What should Andy do to correct the error that he made?

A. Rewrite $16t^2$ as $(16t)^2$.

B. Rewrite $4t^2 - 7^2$ as $(2t - 7)^2$.

C. Rewrite $16t^2$ as $(4t)^2$.

D. Rewrite $16t^2 - 49$ as $(4t - 7)^2$.

13. What is the common ratio of the sequence $243, -81, 27, -9, \ldots$?  (F.LE.2)

F. $\dfrac{1}{3}$

G. $-\dfrac{1}{3}$

H. 3

I. $-3$

# 8 Graphing Quadratic Functions

8.1  Graphing $y = ax^2$
8.2  Focus of a Parabola
8.3  Graphing $y = ax^2 + c$
8.4  Graphing $y = ax^2 + bx + c$
8.5  Comparing Linear, Exponential, and Quadratic Functions

# What You Learned Before

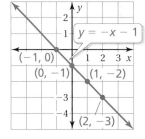

"Okay, Descartes, this will test the theory of parabolic flight paths."

## ● Graphing a Linear Equation (A.CED.2)

**Example 1** Graph $y = -x - 1$.

**Step 1:** Make a table of values.

| x | y = −x − 1 | y | (x, y) |
|---|---|---|---|
| −1 | y = −(−1) − 1 | 0 | (−1, 0) |
| 0 | y = −(0) − 1 | −1 | (0, −1) |
| 1 | y = −(1) − 1 | −2 | (1, −2) |
| 2 | y = −(2) − 1 | −3 | (2, −3) |

**Step 2:** Plot the ordered pairs.

**Step 3:** Draw a line through the points.

### Try It Yourself
**Graph the linear equation.**

1. $y = 2x - 3$
2. $y = -3x + 4$
3. $y = x + 5$

## ● Evaluating an Expression (6.EE.2c)

**Example 2** Evaluate $2x^2 + 3x - 5$ when $x = -1$.

$2x^2 + 3x - 5 = 2(-1)^2 + 3(-1) - 5$   Substitute −1 for x.

$\qquad = 2(1) + 3(-1) - 5$   Evaluate the power.

$\qquad = 2 - 3 - 5$   Multiply.

$\qquad = -6$   Subtract.

### Try It Yourself
**Evaluate the expression when $x = -2$.**

4. $-x^2 - 4x + 1$
5. $3x^2 + x - 2$
6. $-2x^2 - 4x + 3$

# 8.1 Graphing $y = ax^2$

**Essential Question** What are the characteristics of the graph of the quadratic function $y = ax^2$? How does the value of $a$ affect the graph of $y = ax^2$?

### 1 ACTIVITY: Graphing a Quadratic Function

**Work with a partner.**

- Complete the input-output table.
- Plot the points in the table.
- Sketch the graph by connecting the points with a smooth curve.
- What do you notice about the graphs?

a.

| x | $y = x^2$ |
|---|---|
| −3 | |
| −2 | |
| −1 | |
| 0 | |
| 1 | |
| 2 | |
| 3 | |

b.

| x | $y = -x^2$ |
|---|---|
| −3 | |
| −2 | |
| −1 | |
| 0 | |
| 1 | |
| 2 | |
| 3 | |

**COMMON CORE**

**Graphing Quadratic Functions**

In this lesson, you will
- identify characteristics of quadratic functions.
- graph quadratic functions.

Learning Standard
F.BF.3

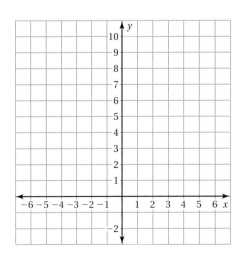
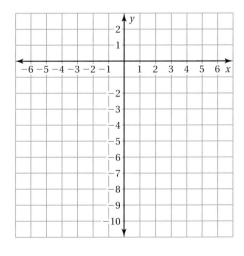

402 Chapter 8 Graphing Quadratic Functions

## 2 ACTIVITY: Graphing a Quadratic Function

**Math Practice 7**

**Look for Patterns**
What pattern do you notice when comparing each equation with its graph?

Work with a partner. Graph each function. How does the value of $a$ affect the graph of $y = ax^2$?

a. $y = 3x^2$

b. $y = -5x^2$

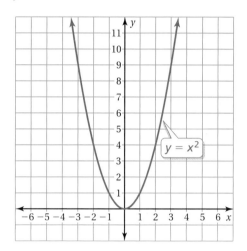 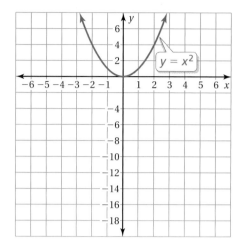

c. $y = -0.2x^2$

d. $y = \dfrac{1}{10}x^2$

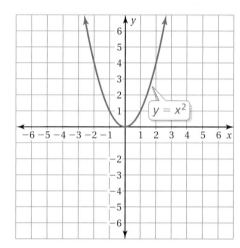 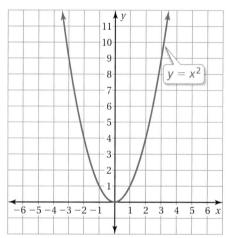

## What Is Your Answer?

3. **IN YOUR OWN WORDS** What are the characteristics of the graph of the quadratic function $y = ax^2$? How does the value of $a$ affect the graph of $y = ax^2$? Consider $a < 0$, $|a| > 1$, and $0 < |a| < 1$ in your answer.

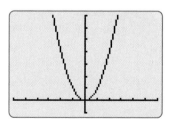

**Practice**  Use what you learned about the graphs of quadratic functions to complete Exercises 5–7 on page 407.

Section 8.1  Graphing $y = ax^2$  **403**

## 8.1 Lesson

**Key Vocabulary**
quadratic function, p. 404
parabola, p. 404
vertex, p. 404
axis of symmetry, p. 404

A **quadratic function** is a nonlinear function that can be written in the standard form $y = ax^2 + bx + c$, where $a \neq 0$. The U-shaped graph of a quadratic function is called a **parabola**.

### Key Idea

**Characteristics of Quadratic Functions**

The most basic quadratic function is $y = x^2$.

The lowest or highest point on a parabola is the **vertex**.

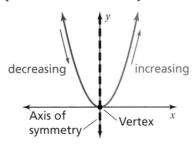

The vertical line that divides the parabola into two symmetric parts is the **axis of symmetry**. The axis of symmetry passes through the vertex.

### EXAMPLE 1  Identifying Characteristics of a Quadratic Function

**Consider the graph of the quadratic function.**

Using the graph, you can identify the vertex, axis of symmetry, and the behavior of the graph as shown.

You can also determine the following:

- The domain is all real numbers.
- The range is all real numbers greater than or equal to $-2$.
- When $x < -1$, $y$ increases as $x$ decreases.
- When $x > -1$, $y$ increases as $x$ increases.

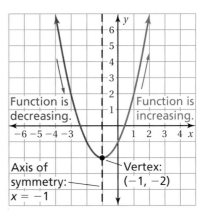

### On Your Own

**Now You're Ready**
Exercises 8–10

Identify characteristics of the graph of the quadratic function.

1.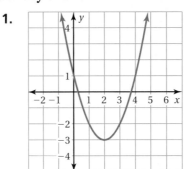

2.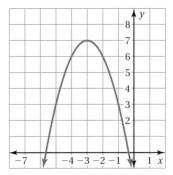

### Key Ideas

**Graphing $y = ax^2$ When $a > 0$**

- When $0 < a < 1$, the graph of $y = ax^2$ opens up and is wider than the graph of $y = x^2$.
- When $a > 1$, the graph of $y = ax^2$ opens up and is narrower than the graph of $y = x^2$.

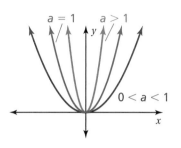

**Graphing $y = ax^2$ When $a < 0$**

- When $-1 < a < 0$, the graph of $y = ax^2$ opens down and is wider than the graph of $y = x^2$.
- When $a < -1$, the graph of $y = ax^2$ opens down and is narrower than the graph of $y = x^2$.

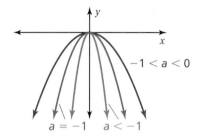

### EXAMPLE 2  Graphing $y = ax^2$ When $a > 0$

**Graph $y = 2x^2$. Compare the graph to the graph of $y = x^2$.**

**Step 1:** Make a table of values.

| x | −2 | −1 | 0 | 1 | 2 |
|---|----|----|---|---|---|
| y | 8  | 2  | 0 | 2 | 8 |

**Step 2:** Plot the ordered pairs.

**Step 3:** Draw a smooth curve through the points.

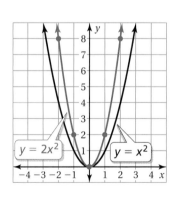

∴ Both graphs open up and have the same vertex, (0, 0), and the same axis of symmetry, $x = 0$. The graph of $y = 2x^2$ is narrower than the graph of $y = x^2$.

### On Your Own

Now You're Ready
Exercises 11–16

Graph the function. Compare the graph to the graph of $y = x^2$.

3. $y = 5x^2$
4. $y = \dfrac{1}{3}x^2$
5. $y = \dfrac{3}{2}x^2$

Section 8.1  Graphing $y = ax^2$  405

### EXAMPLE 3 — Graphing $y = ax^2$ When $a < 0$

Graph $y = -\dfrac{1}{3}x^2$. Compare the graph to the graph of $y = x^2$.

**Step 1:** Make a table of values. Choose $x$-values that make the calculations simple.

| x | −6  | −3 | 0 | 3  | 6   |
|---|-----|----|---|----|-----|
| y | −12 | −3 | 0 | −3 | −12 |

**Step 2:** Plot the ordered pairs.

**Step 3:** Draw a smooth curve through the points.

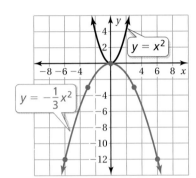

The graphs have the same vertex, (0, 0), and the same axis of symmetry, $x = 0$, but the graph of $y = -\dfrac{1}{3}x^2$ opens down. The graph of $y = -\dfrac{1}{3}x^2$ is wider than the graph of $y = x^2$.

### EXAMPLE 4 — Real-Life Application

The diagram shows the cross section of a satellite dish, where $x$ and $y$ are measured in meters. Find the width and depth of the dish.

Use the domain of the function to find the width of the dish. Use the range to find the depth.

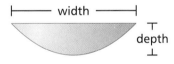

The leftmost point on the graph is (−2, 1) and the rightmost point is (2, 1). So, the domain is $-2 \leq x \leq 2$, which represents 4 meters.

The lowest point on the graph is (0, 0) and the highest points on the graph are (−2, 1) and (2, 1). So, the range is $0 \leq y \leq 1$, which represents 1 meter.

So, the satellite dish is 4 meters wide and 1 meter deep.

### On Your Own

**Now You're Ready**
Exercises 18–23 and 34

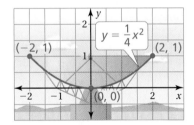

Graph the function. Compare the graph to the graph of $y = x^2$.

6. $y = -3x^2$

7. $y = -0.1x^2$

8. $y = -\dfrac{1}{4}x^2$

9. The cross section of a spotlight can be modeled by the graph of $y = 0.5x^2$, where $x$ and $y$ are measured in inches and $-2 \leq x \leq 2$. Find the width and depth of the spotlight.

# 8.1 Exercises

Check It Out
Help with Homework
BigIdeasMath.com

## Vocabulary and Concept Check

1. **VOCABULARY** Describe the vertex and axis of symmetry of the graph of $y = ax^2$.

2. **VOCABULARY** What is the U-shaped graph of a quadratic function called?

3. **WRITING** Without graphing, which graph is wider, $y = 6x^2$ or $y = \frac{1}{6}x^2$? Explain your reasoning.

4. **WRITING** When does the graph of a quadratic function open up? open down?

## Practice and Problem Solving

Use a graphing calculator to graph the function. Compare the graph to the graph of $y = -4x^2$.

5. $y = -0.4x^2$
6. $y = -0.04x^2$
7. $y = -0.004x^2$

Identify characteristics of the graph of the quadratic function.

8.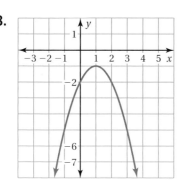
9.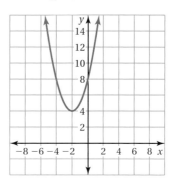
10.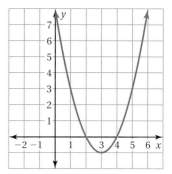

Graph the function. Compare the graph to the graph of $y = x^2$.

11. $y = 6x^2$
12. $y = 8x^2$
13. $y = \frac{1}{4}x^2$

14. $y = \frac{3}{4}x^2$
15. $y = \frac{5}{2}x^2$
16. $y = \frac{7}{5}x^2$

17. **WATERFALL** A fish swims over the waterfall. The distance $y$ (in feet) that the fish falls is given by the function $y = 16t^2$, where $t$ is the time (in seconds).

   a. Describe the domain and range of the function.
   b. Graph the function using the domain in part (a).
   c. Use the graph to determine when the fish lands in the water below.

**Graph the function. Compare the graph to the graph of $y = x^2$.**

18. $y = -2x^2$

19. $y = -7x^2$

20. $y = -\dfrac{1}{5}x^2$

21. $y = -\dfrac{5}{8}x^2$

22. $y = -\dfrac{5}{3}x^2$

23. $y = -\dfrac{9}{4}x^2$

24. **ERROR ANALYSIS** Describe and correct the error in graphing and comparing $y = x^2$ and $y = 0.5x^2$.

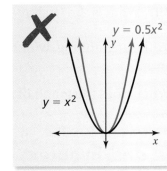

**Describe the possible values of $a$.**

25.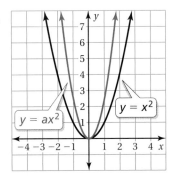

26.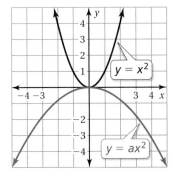

27.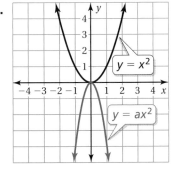

28. **REASONING** A parabola opens up and passes through $(-4, 2)$ and $(6, -3)$. How do you know that $(-4, 2)$ is not the vertex?

29. **REASONING** Describe the domain and range of the function $y = ax^2$ when (a) $a > 0$ and (b) $a < 0$.

**Determine whether the statement is *always*, *sometimes*, or *never* true. Explain your reasoning.**

30. The graph of $y = ax^2$ is narrower than the graph of $y = x^2$ when $a > 0$.

31. The graph of $y = ax^2$ is narrower than the graph of $y = x^2$ when $|a| > 1$.

32. The graph of $y = ax^2$ is wider than the graph of $y = x^2$ when $0 < |a| < 1$.

33. The graph of $y = ax^2$ is wider than the graph of $y = dx^2$ when $|a| > |d|$.

34. **BRIDGE** The arch support of a bridge can be modeled by $y = -0.0012x^2$, where $x$ and $y$ are measured in feet. Find the height and width of the arch.

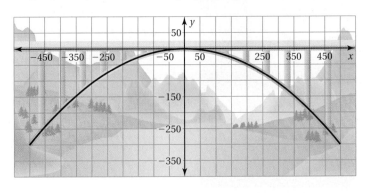

**In Exercises 35–37, use the graph of the function $y = ax^2$.**

35. When is the function increasing?

36. When is the function decreasing?

37. Find the value of $a$ when the graph passes through $(-2, 3)$.

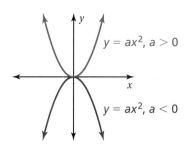

38. **MODELING** The diagram shows the cross section of a swirling glass of water, where $x$ and $y$ are measured in centimeters. The surface of the cross section of the rotating liquid is a parabola.

    a. About how wide is the mouth of the glass?
    b. Suppose the rotational speed of the liquid changes. The cross section can now be modeled by $y = 0.1x^2$. Did the rotational speed *increase* or *decrease*? Explain your reasoning.

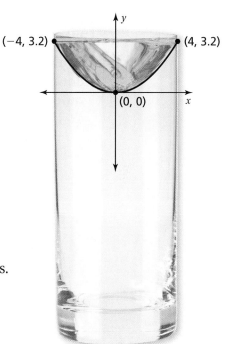

39. **ASSEMBLY LINE** The number $y$ of units an assembly line can produce in 1 hour can be modeled by the function $y = 0.5x^2$, where $x$ is the number of employees. The assembly line has a capacity of 10 employees.

    a. Describe the domain of the function.
    b. Graph the function using the domain in part (a).
    c. Is it better for the company to run one assembly line at full capacity or two assembly lines at half capacity? Explain your reasoning.

40. **Logic** Is the $x$-intercept of the graph of $y = ax^2$ always 0? Justify your answer.

## Fair Game Review  what you learned in previous grades & lessons

**Solve the proportion using the Cross Products Property.** *(Skills Review Handbook)*

41. $\dfrac{x}{6} = \dfrac{13}{2}$

42. $\dfrac{5}{3} = \dfrac{n}{9}$

43. $\dfrac{4}{b} = \dfrac{6}{21}$

44. $\dfrac{14}{9} = \dfrac{7}{y}$

45. **MULTIPLE CHOICE** What is the completely factored form of $2x^5 - 8x^3$? *(Section 7.9)*

    Ⓐ $2x^3(x-2)^2$  
    Ⓑ $2x^3(x-2)(x+2)$  
    Ⓒ $2x^3(x+2)^2$  
    Ⓓ $2x^3(x^2-4)$

Section 8.1  Graphing $y = ax^2$

## 8.2 Focus of a Parabola

**Essential Question** Why do satellite dishes and spotlight reflectors have parabolic shapes?

### 1 ACTIVITY: A Property of Satellite Dishes

Work with a partner. Rays are coming straight down. When they hit the parabola, they reflect off at the same angle at which they entered.

- Draw the outgoing part of each ray so that it intersects the *y*-axis.
- What do you notice about where the reflected rays intersect the *y*-axis?
- Where is the receiver for the satellite dish? Explain.

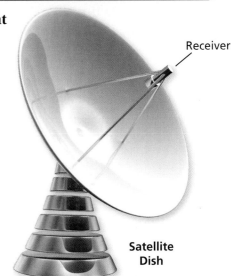

Receiver

Satellite Dish

**COMMON CORE**

**Graphing Quadratic Functions**

In this lesson, you will
- find the foci of parabolas.
- write equations of parabolas with vertices at the origin given the foci.

Learning Standard
F.IF.4

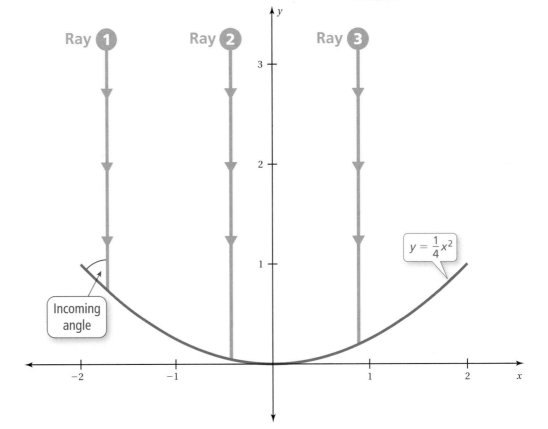

Ray 1   Ray 2   Ray 3

$y = \frac{1}{4}x^2$

Incoming angle

410  Chapter 8  Graphing Quadratic Functions

## 2 ACTIVITY: A Property of Spotlights

Work with a partner. Beams of light are coming from the bulb in a spotlight. When the beams hit the parabola, they reflect off at the same angle at which they entered.

- Draw the outgoing part of each beam. What do they have in common? Explain.

**Math Practice 3**

**Justify Conclusions**
What information can you use to justify your conclusion?

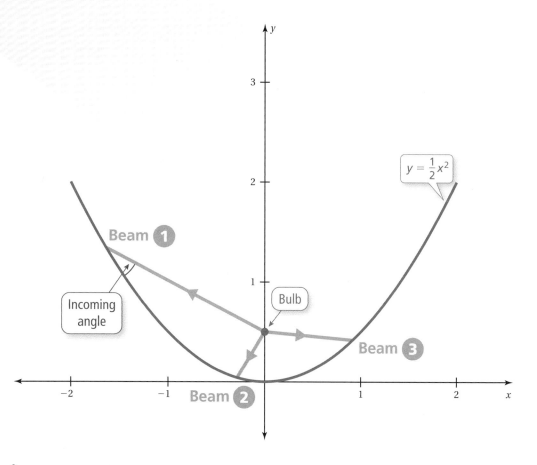

## What Is Your Answer?

3. **IN YOUR OWN WORDS** Why do satellite dishes and spotlight reflectors have parabolic shapes?

4. Design and draw a parabolic satellite dish. Label the dimensions of the dish. Label the receiver.

**Practice**

Use what you learned about parabolas to complete Exercises 4–6 on page 414.

Section 8.2  Focus of a Parabola  411

## 8.2 Lesson

**Key Vocabulary**
focus, p. 412

### Key Idea

**The Focus of a Parabola**

The **focus** of a parabola is a fixed point on the interior of a parabola that lies on the axis of symmetry. A parabola "wraps" around the focus.

For functions of the form $y = ax^2$, the focus is $\left(0, \dfrac{1}{4a}\right)$.

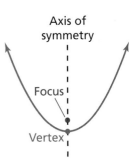

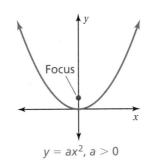

$y = ax^2, a > 0$

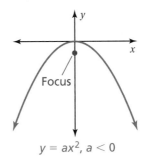

$y = ax^2, a < 0$

### EXAMPLE 1 — Finding the Focus of a Parabola

Graph $y = -\dfrac{1}{4}x^2$. Identify the focus.

**Step 1:** Make a table of values. Then graph.

| x | −4 | −2 | 0 | 2 | 4 |
|---|----|----|---|---|---|
| y | −4 | −1 | 0 | −1 | −4 |

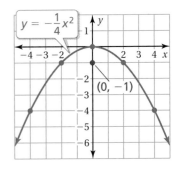

**Step 2:** Identify the focus. The function is of the form $y = ax^2$, so $a = -\dfrac{1}{4}$.

$$\dfrac{1}{4a} = \dfrac{1}{4\left(-\dfrac{1}{4}\right)} \qquad \text{Substitute } -\dfrac{1}{4} \text{ for } a.$$

$$= \dfrac{1}{-1}, \text{ or } -1 \qquad \text{Multiply.}$$

∴ So, the focus of the function $y = -\dfrac{1}{4}x^2$ is $(0, -1)$.

### On Your Own

Now You're Ready
Exercises 7–12

Graph the function. Identify the focus.

1. $y = 2x^2$
2. $y = \dfrac{1}{6}x^2$
3. $y = -3x^2$

## EXAMPLE 2 Writing an Equation of a Parabola

**Write an equation of the parabola with focus (0, 4) and vertex at the origin.**

For $y = ax^2$, the focus is $\left(0, \dfrac{1}{4a}\right)$. Use the given focus, (0, 4), to write an equation to find $a$.

$\dfrac{1}{4a} = 4$     Equate the y-coordinates.

$1 = 16a$     Multiply each side by 4a.

$\dfrac{1}{16} = a$     Divide each side by 16.

∴ An equation of the parabola is $y = \dfrac{1}{16}x^2$.

## EXAMPLE 3 Real-Life Application

A birdwatcher uses a parabolic microphone to collect and record bird sounds. The cross section of the microphone can be modeled by $y = \dfrac{1}{24}x^2$, where $x$ and $y$ are measured in inches. The focus is located at the end of the receiver arm. What is the length of the receiver arm?

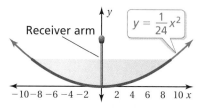

The arm length is the distance from the focus to the vertex. Identify the focus. For the function $y = \dfrac{1}{24}x^2$, $a = \dfrac{1}{24}$.

$\dfrac{1}{4a} = \dfrac{1}{4\left(\dfrac{1}{24}\right)}$     Substitute $\dfrac{1}{24}$ for a.

$= \dfrac{1}{\dfrac{1}{6}}$     Multiply.

$= 6$     Divide.

The focus is (0, 6). The vertex is (0, 0). The distance from (0, 0) to (0, 6) is 6 units.

∴ So, the length of the receiver arm is 6 inches.

### On Your Own

Now You're Ready
Exercises 14–19

**4.** Write an equation of the parabola with focus (0, −3) and vertex at the origin.

**5. WHAT IF?** In Example 3, the cross section of the microphone can be modeled by $y = \dfrac{1}{40}x^2$. What is the length of the receiver arm?

## 8.2 Exercises

### Vocabulary and Concept Check

1. **VOCABULARY** What is the relationship between the focus and the axis of symmetry of a parabola?

2. **WRITING** When the focus of a parabola lies below the vertex, does the parabola open up or down? Is $a > 0$ or $a < 0$? Explain.

3. **OPEN-ENDED** Write an equation of a parabola whose focus is below the $x$-axis.

### Practice and Problem Solving

**Determine whether the shape is parabolic.**

4.

5.

6.

**Graph the function. Identify the focus.**

7. $y = x^2$

8. $y = 4x^2$

9. $y = -12x^2$

10. $y = \frac{1}{4}x^2$

11. $y = \frac{1}{2}x^2$

12. $y = -0.75x^2$

13. **ERROR ANALYSIS** Describe and correct the error in identifying the focus.

**Write an equation of the parabola with a vertex at the origin and the given focus.**

14. $(0, 1)$

15. $(0, -2)$

16. $(0, 5)$

17. $\left(0, -\frac{1}{4}\right)$

18. $(0, -1)$

19. $(0, 0.5)$

20. **COMET** A comet travels along a parabolic path around the Sun. The Sun is the focus of the path. When the comet is at the vertex of the path, it is 60,000,000 kilometers from the Sun. Write an equation that represents the path of the comet. Assume the focus is on the positive $y$-axis and the vertex is $(0, 0)$.

**Match the equation with its graph.**

**21.** $y = 2x^2$    **22.** $y = 3x^2$    **23.** $y = 2.5x^2$

**A.** 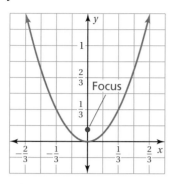    **B.** 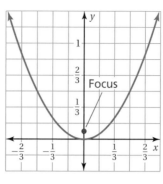    **C.**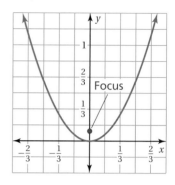

**Determine whether the statement is *sometimes*, *always*, or *never* true for the function $y = ax^2$. Explain your reasoning.**

**24.** The vertex and focus of a parabola can occur at the same point.

**25.** If a parabola opens down, then the focus lies below the *x*-axis.

**26.** The *y*-coordinate of the focus of a parabola is greater than the *y*-coordinate of the vertex.

**27. REASONING** Describe how the graph of $y = ax^2$ changes as the distance between the vertex and focus increases.

**28. WHISPER DISH** Whisper dishes are parabolic sound reflectors that transmit and receive sound waves from opposite ends of a room. For the best sound reception, you place your ear at the focus of the dish. Write an equation for the cross section of a dish when your ear is 3 feet from the vertex.

**29. SOLAR COOKING** You make a solar cooker using a parabolic reflective surface. You suspend a sausage with wire through the focus of each end piece of the cooker. How far from the bottom should you place the wire?

**30. Structure** For what values of *a* will the distance between the focus and the vertex of the graph of $y = ax^2$ be less than the distance between the focus and the vertex of the graph of $y = (ax)^2$?

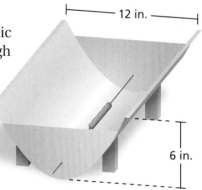

 **Fair Game Review** What you learned in previous grades & lessons

**Evaluate the expression when $x = -3$ and $n = 2$.** *(Skills Review Handbook)*

**31.** $x^2 + 7$    **32.** $3n^2 - 2$    **33.** $x + n^2$

**34. MULTIPLE CHOICE** Which number is equivalent to $32^{3/5}$? *(Section 6.3)*

　Ⓐ 2    Ⓑ 4    Ⓒ 8    Ⓓ 10

Section 8.2    Focus of a Parabola    **415**

## 8.3 Graphing $y = ax^2 + c$

**Essential Question** How does the value of $c$ affect the graph of $y = ax^2 + c$?

### 1 ACTIVITY: Graphing $y = ax^2 + c$

Work with a partner. Sketch the graphs of both functions in the same coordinate plane. How does the value of $c$ affect the graph of $y = ax^2 + c$?

a. $y = x^2$ and $y = x^2 + 2$

b. $y = 2x^2$ and $y = 2x^2 - 2$

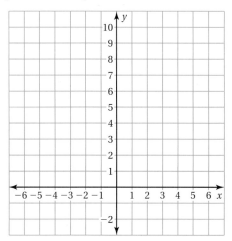
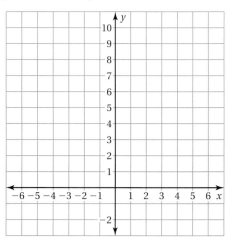

c. $y = -x^2 + 4$ and $y = -x^2 + 9$

d. $y = \frac{1}{2}x^2$ and $y = \frac{1}{2}x^2 - 8$

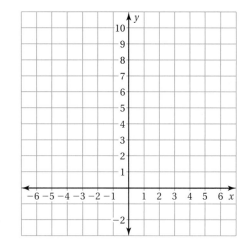
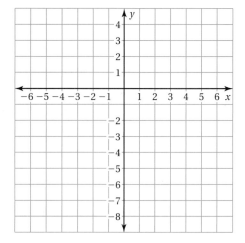

**COMMON CORE**

**Graphing Quadratic Functions**

In this lesson, you will
- graph quadratic functions of the form $y = ax^2 + c$ and compare to the graph of $y = x^2$.

Learning Standard
F.BF.3

416 Chapter 8 Graphing Quadratic Functions

## 2 ACTIVITY: Finding x-Intercepts of Graphs

**Math Practice 6**

**Communicate Precisely**
What have you included in your answer to make sure your explanation is precise?

Work with a partner. Graph each function. Find the x-intercepts of the graph. Explain how you found the x-intercepts.

a. $y = x^2 - 4$

b. $y = 2x^2 - 8$

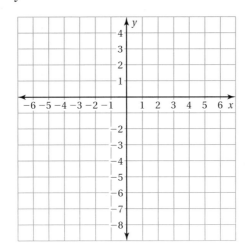
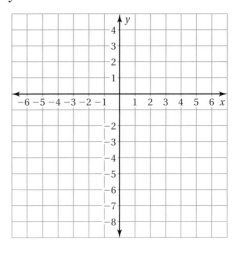

c. $y = -x^2 + 1$

d. $y = \frac{1}{3}x^2 - 3$

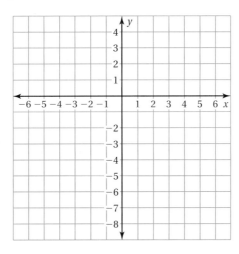
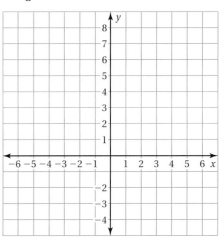

## What Is Your Answer?

3. **IN YOUR OWN WORDS** How does the value of $c$ affect the graph of $y = ax^2 + c$? Use a graphing calculator to verify your conclusions.

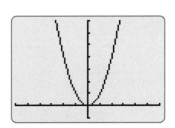

**Practice** — Use what you learned about the graphs of quadratic functions to complete Exercises 7–9 on page 420.

Section 8.3  Graphing $y = ax^2 + c$  **417**

## 8.3 Lesson

**Key Vocabulary**
zero, p. 419

### Key Idea

**Graphing $y = x^2 + c$**

- When $c > 0$, the graph of $y = x^2 + c$ is a vertical translation $c$ units up of the graph of $y = x^2$.
- When $c < 0$, the graph of $y = x^2 + c$ is a vertical translation $|c|$ units down of the graph of $y = x^2$.

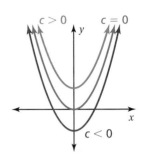

### EXAMPLE 1  Graphing $y = x^2 + c$

Graph $y = x^2 - 2$. Compare the graph to the graph of $y = x^2$.

**Step 1:** Make a table of values.

| x | −2 | −1 | 0 | 1 | 2 |
|---|----|----|---|---|---|
| y | 2  | −1 | −2 | −1 | 2 |

**Step 2:** Plot the ordered pairs.

**Step 3:** Draw a smooth curve through the points.

∴ Both graphs open up and have the same axis of symmetry, $x = 0$. The graph of $y = x^2 - 2$ is a translation 2 units down of the graph of $y = x^2$.

### EXAMPLE 2  Graphing $y = ax^2 + c$

Graph $y = 4x^2 + 1$. Compare the graph to the graph of $y = x^2$.

**Step 1:** Make a table of values.

| x | −2 | −1 | 0 | 1 | 2 |
|---|----|----|---|---|---|
| y | 17 | 5  | 1 | 5 | 17 |

**Step 2:** Plot the ordered pairs.

**Step 3:** Draw a smooth curve through the points.

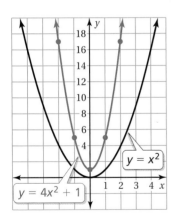

∴ Both graphs open up and have the same axis of symmetry, $x = 0$. The graph of $y = 4x^2 + 1$ is narrower than the graph of $y = x^2$. The vertex of the graph of $y = 4x^2 + 1$ is a translation 1 unit up of the vertex of the graph of $y = x^2$.

### On Your Own

**Now You're Ready**
Exercises 7–15

Graph the function. Compare the graph to the graph of $y = x^2$.

1. $y = x^2 + 3$
2. $y = 2x^2 - 5$
3. $y = -\dfrac{1}{2}x^2 + 4$

### EXAMPLE 3  Translating the Graph of $y = x^2 + c$

**Which of the following is true when you translate the graph of $y = x^2 - 5$ to the graph of $y = x^2 + 2$?**

**Ⓐ** The graph shifts 3 units up.  **Ⓑ** The graph shifts 7 units up.

**Ⓒ** The graph shifts 7 units down.  **Ⓓ** The graph shifts 3 units down.

Both graphs open up and have the same axis of symmetry, $x = 0$. The vertex of $y = x^2 - 5$ is $(0, -5)$. The vertex of $y = x^2 + 2$ is $(0, 2)$. To move the vertex from $(0, -5)$ to $(0, 2)$, you must translate the graph 7 units up.

∴ The correct answer is **Ⓑ**.

A **zero** of a function $f(x)$ is an $x$-value for which $f(x) = 0$. A zero is located at the $x$-intercept of the graph of the function.

### EXAMPLE 4  Real-Life Application

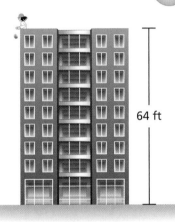

The function $f(t) = -16t^2 + s_0$ gives the approximate height (in feet) of a falling object $t$ seconds after it is dropped from an initial height $s_0$ (in feet). An egg is dropped from a height of 64 feet. When does the egg hit the ground?

The initial height is 64 feet. So, the function $f(t) = -16t^2 + 64$ gives the height of the egg after $t$ seconds. It hits the ground when $f(t) = 0$.

**Step 1:** Make a table of values and sketch the graph.

| t | 0 | 1 | 2 |
|---|---|---|---|
| f(t) | 64 | 48 | 0 |

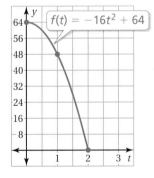

**Step 2:** Find the zero of the function. When $t = 2$, $f(t) = 0$. So, the zero is 2.

∴ The egg hits the ground 2 seconds after it is dropped.

**Common Error**
The graph in Example 4 shows the height of the object over time, not the path of the object.

### On Your Own

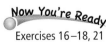
Exercises 16–18, 21

**4.** The graph of $y = 2x^2 + 1$ is shifted to $y = 2x^2 - 1$. Describe the translation.

**5. REASONING** Explain why only nonnegative values of $t$ are used in Example 4.

**6. WHAT IF?** In Example 4, the egg is dropped from a height of 100 feet. When does the egg hit the ground?

# 8.3 Exercises

## Vocabulary and Concept Check

1. **VOCABULARY** Describe the vertex and axis of symmetry of the graph of $y = ax^2 + c$. How is the value of $c$ related to the vertex of the graph?

2. **NUMBER SENSE** Without graphing, which graph has the greater $y$-intercept, $y = x^2 + 4$ or $y = x^2 - 4$? Explain your reasoning.

3. **WRITING** How does the graph of $y = ax^2 + c$ compare to the graph of $y = ax^2$?

## Practice and Problem Solving

Match each function with its graph.

4. $y = 2x^2 + 3$
5. $y = -2x^2 + 3$
6. $y = 2x^2 - 3$

A. 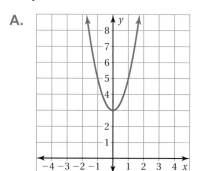   B. 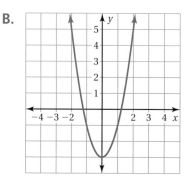   C.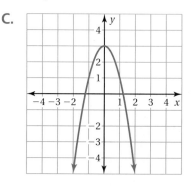

Graph the function. Compare the graph to the graph of $y = x^2$.

7. $y = x^2 + 4$
8. $y = x^2 - 3$
9. $y = -x^2 + 5$

10. $y = -x^2 - 9$
11. $y = 2x^2 - 4$
12. $y = \frac{1}{2}x^2 + 2$

Use a graphing calculator to graph the function. Compare the graph to the graph of $y = x^2$.

13. $y = 3x^2 + 4$
14. $y = -2x^2 - 1$
15. $y = -\frac{1}{4}x^2 - \frac{1}{2}$

Describe how to translate the graph of $y = x^2 + 2$ to the graph of the given function.

16. $y = x^2 + 4$
17. $y = x^2 - 1$
18. $y = x^2 - 4.5$

19. **ERROR ANALYSIS** Describe and correct the error in comparing the graphs.

20. **REASONING** The domain of $y = ax^2 + c$ is all real numbers. Describe the range when (a) $a > 0$ and (b) $a < 0$.

21. **WATER BALLOON** A water balloon is dropped from a height of 16 feet. The function $h = -16x^2 + 16$ gives the height $h$ of the balloon after $x$ seconds. When does it hit the ground?

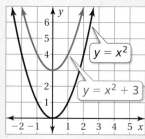

The graph of $y = x^2 + 3$ is a translation 3 units down of the graph of $y = x^2$.

**22. APPLE** The function $y = -16x^2 + 36$ gives the height $y$ (in feet) of an apple after falling $x$ seconds. Find and interpret the $x$- and $y$-intercepts.

**Find the zeros of the function.**

**23.** $y = x^2 - 1$     **24.** $y = x^2 - 4$     **25.** $y = -x^2 + 9$

**Sketch a quadratic function with the given characteristics.**

**26.** The parabola opens up and the vertex is $(0, 3)$.

**27.** The parabola opens down, the vertex is $(0, 4)$, and one of the $x$-intercepts is 2.

**28.** The function is increasing when $x < 0$ and the $x$-intercepts of the parabola are $-1$ and $1$.

**29.** The graph is below the $x$-axis and the highest point on the parabola is $(0, -5)$.

**30. REASONING** Describe two algebraic methods you can use to find the zeros of the function $f(t) = -16t^2 + 64$.

**31. REASONING** Can the focus and the vertex of a parabola lie on opposite sides of the $x$-axis? Explain your reasoning.

**32. PROBLEM SOLVING** The paths of water from three different garden waterfalls are given below. Each function gives the height $h$ (in feet) and the horizontal distance $d$ (in feet) of the water.

**Waterfall 1:** $h = -3.1d^2 + 4.8$
**Waterfall 2:** $h = -3.5d^2 + 1.9$
**Waterfall 3:** $h = -1.1d^2 + 1.6$

**a.** Which waterfall drops water from the highest point?
**b.** Which waterfall follows the narrowest path?
**c.** Which waterfall sends water the farthest?

**33. Logic** Let $f(x)$ be a quadratic function of the form $f(x) = ax^2 + c$.

**a.** How does the graph of $f(x) + k$ compare to the graph of $f(x)$ when $k < 0$? when $k > 0$?

**b.** Let $k$ be a real number not equal to 0 or 1. How does the graph of $k \cdot f(x)$ compare to the graph of $f(x)$?

### Fair Game Review What you learned in previous grades & lessons

**Factor the polynomial.** *(Section 7.7 and Section 7.8)*

**34.** $x^2 - 2x - 8$     **35.** $2x^2 - 7x + 3$     **36.** $x^2 + 2x - 35$

**37. MULTIPLE CHOICE** What is the product of $(x - 2)$ and $(x - 4)$? *(Section 7.3)*

   **Ⓐ** $x^2 - 2$     **Ⓑ** $x^2 - 6x + 8$     **Ⓒ** $x^2 - 2x - 6$     **Ⓓ** $x^2 - 4x - 8$

# 8 Study Help

You can use a **summary triangle** to explain a topic. Here is an example of a summary triangle for graphing a quadratic function of the form $y = ax^2$ when $a > 0$.

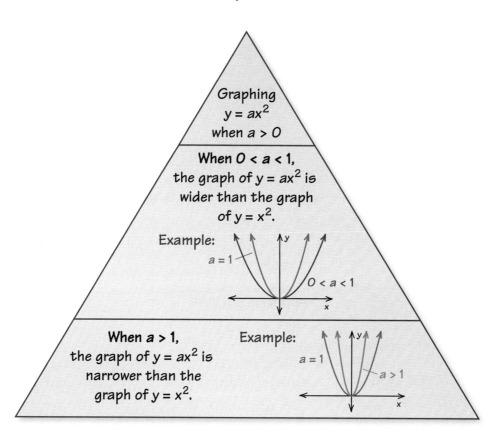

## On Your Own

**Make summary triangles to help you study these topics.**

1. graphing $y = ax^2$ when $a < 0$
2. identifying the focus of a parabola
3. graphing $y = ax^2 + c$

**After you complete this chapter, make summary triangles for the following topics.**

4. graphing $y = ax^2 + bx + c$
5. graphing $y = a(x - h)^2 + k$

"What do you call a cheese summary triangle that isn't yours?"

# 8.1–8.3 Quiz

**Identify characteristics of the graph of the quadratic function.** *(Section 8.1)*

1.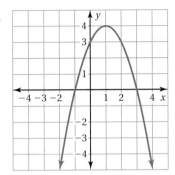

2.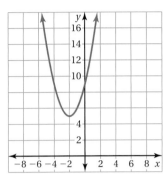

3.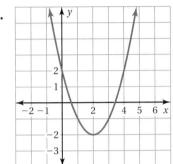

4.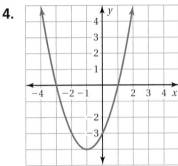

**Graph the function. Compare the graph to the graph of $y = x^2$.** *(Section 8.1)*

5. $y = -x^2$

6. $y = 4x^2$

7. $y = \dfrac{2}{5}x^2$

**Graph the function. Identify the focus.** *(Section 8.2)*

8. $y = 5x^2$

9. $y = -6x^2$

10. $y = \dfrac{1}{3}x^2$

**Write an equation of the parabola with a vertex at the origin and the given focus.** *(Section 8.2)*

11. $(0, -4)$

12. $(0, 2)$

13. $\left(0, \dfrac{1}{5}\right)$

**Graph the function. Compare the graph to the graph of $y = x^2$.** *(Section 8.3)*

14. $y = x^2 + 5$

15. $y = 2x^2 - 2$

16. $y = -x^2 + 3$

17. **PINEAPPLE** The distance $y$ (in feet) that a pineapple falls is given by the function $y = 16t^2$, where $t$ is the time (in seconds). Use a graph to determine how many seconds it takes for the pineapple to fall 32 feet. *(Section 8.1)*

18. **SMARTPHONE** A new smartphone application is available for download. The number $y$ of downloads can be modeled by the function $y = 6.3x^2 + 3000$, where $x$ is the number of hours since the new application was released. How many hours does it take for the number of downloads to reach 3630? *(Section 8.3)*

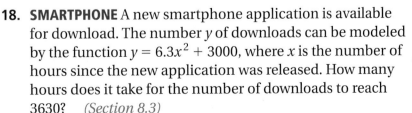

# 8.4 Graphing $y = ax^2 + bx + c$

**Essential Question** How can you find the vertex of the graph of $y = ax^2 + bx + c$?

## 1 ACTIVITY: Comparing Two Graphs

**Work with a partner.**

- Sketch the graphs of $y = 2x^2 - 8x$ and $y = 2x^2 - 8x + 6$.
- What do you notice about the *x*-value of the vertex of each graph?

$y = 2x^2 - 8x$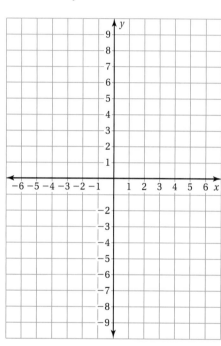

$y = 2x^2 - 8x + 6$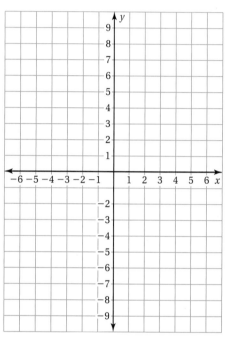

### Common Core

**Graphing Quadratic Functions**

In this lesson, you will
- find the axes of symmetry and the vertices of graphs.
- graph quadratic functions of the form $y = ax^2 + bx + c$.
- find maximum and minimum values.

Learning Standards
F.BF.3
F.IF.4
F.IF.7a

## 2 ACTIVITY: Comparing x-Intercepts with the Vertex

**Work with a partner.**

- Use the graph in Activity 1 to find the *x*-intercepts of the graph of $y = 2x^2 - 8x$. Verify your answer by solving $0 = 2x^2 - 8x$.
- Compare the location of the vertex to the location of the *x*-intercepts.

424   Chapter 8   Graphing Quadratic Functions

### 3  ACTIVITY: Finding Intercepts

**Math Practice**

**Use Prior Results**

How can you use results from the previous activities to complete the table?

**Work with a partner.**

- Solve $0 = ax^2 + bx$ by factoring.
- What are the x-intercepts of the graph of $y = ax^2 + bx$?
- Copy and complete the table to verify your answer.

| x | $y = ax^2 + bx$ |
|---|---|
| 0 | |
| $-\dfrac{b}{a}$ | |

### 4  ACTIVITY: Deductive Reasoning

**Work with a partner. Complete the following logical argument.**

The x-intercepts of the graph of $y = ax^2 + bx$ are 0 and $-\dfrac{b}{a}$.

The vertex of the graph of $y = ax^2 + bx$ occurs when $x = $ _____ .

The vertices of the graphs of $y = ax^2 + bx$ and $y = ax^2 + bx + c$ have the same x-value.

The vertex of $y = ax^2 + bx + c$ occurs when $x = $ _____ .

## What Is Your Answer?

5. **IN YOUR OWN WORDS** How can you find the vertex of the graph of $y = ax^2 + bx + c$?

6. Without graphing, find the vertex of the graph of $y = x^2 - 4x + 3$. Check your result by graphing.

**Practice**

Use what you learned about the vertices of the graphs of quadratic functions to complete Exercises 6–8 on page 429.

# 8.4 Lesson

**Key Vocabulary**
maximum value, p. 427
minimum value, p. 427

## Key Idea

**Properties of the Graph of $y = ax^2 + bx + c$**

- The graph opens up when $a > 0$ and the graph opens down when $a < 0$.
- The $y$-intercept is $c$.
- The $x$-coordinate of the vertex is $-\dfrac{b}{2a}$.
- The axis of symmetry is $x = -\dfrac{b}{2a}$.

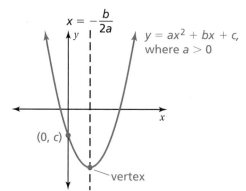

### EXAMPLE 1  Finding the Axis of Symmetry and the Vertex of a Graph

Find (a) the axis of symmetry and (b) the vertex of the graph of $y = 2x^2 + 8x - 1$.

**a.** Find the axis of symmetry when $a = 2$ and $b = 8$.

$x = -\dfrac{b}{2a}$     Write the equation for the axis of symmetry.

$x = -\dfrac{8}{2(2)}$     Substitute 2 for $a$ and 8 for $b$.

$x = -2$     Simplify.

∴ The axis of symmetry is $x = -2$.

**b.** The axis of symmetry is $x = -2$, so the $x$-coordinate of the vertex is $-2$. Use the function to find the $y$-coordinate of the vertex.

$y = 2x^2 + 8x - 1$     Write the function.

$= 2(-2)^2 + 8(-2) - 1$     Substitute $-2$ for $x$.

$= -9$     Simplify.

∴ The vertex is $(-2, -9)$.

### On Your Own

**Now You're Ready**
Exercises 6–11

Find (a) the axis of symmetry and (b) the vertex of the graph of the function.

**1.** $y = 3x^2 - 2x$     **2.** $y = x^2 + 6x + 5$     **3.** $y = -\dfrac{1}{2}x^2 + 7x - 4$

### EXAMPLE 2  Graphing $y = ax^2 + bx + c$

Graph $y = 3x^2 - 6x + 5$. Describe the domain and range.

**Step 1:** Find and graph the axis of symmetry.

$$x = -\frac{b}{2a} = -\frac{(-6)}{2(3)} = 1$$

**Step 2:** Find and plot the vertex.

The axis of symmetry is $x = 1$, so the x-coordinate of the vertex is 1. Use the function to find the y-coordinate of the vertex.

$y = 3x^2 - 6x + 5$     Write the function.

$\phantom{y} = 3(1)^2 - 6(1) + 5$     Substitute 1 for x.

$\phantom{y} = 2$     Simplify.

So, the vertex is (1, 2).

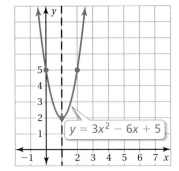

**Step 3:** Use the y-intercept to find two more points on the graph.

The y-intercept is 5. So, (0, 5) lies on the graph. Because the axis of symmetry is $x = 1$, the point (2, 5) also lies on the graph.

**Step 4:** Draw a smooth curve through the points.

∴ The domain is all real numbers. The range is $y \geq 2$.

### On Your Own

**Now You're Ready**
Exercises 13–18

Graph the function. Describe the domain and range.

**4.** $y = 2x^2 + 4x + 1$    **5.** $y = x^2 - 8x + 7$    **6.** $y = -5x^2 - 10x - 2$

### Key Ideas

**Maximum and Minimum Values**

The y-coordinate of the vertex of the graph of $y = ax^2 + bx + c$ is the **maximum value** of the function when $a < 0$ or the **minimum value** of the function when $a > 0$.

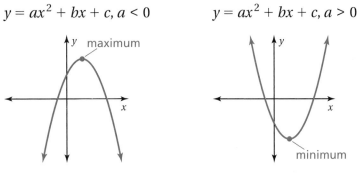

Section 8.4    Graphing $y = ax^2 + bx + c$

**EXAMPLE 3** **Finding Maximum and Minimum Values**

Tell whether the function $f(x) = -4x^2 - 24x - 19$ has a minimum value or a maximum value. Then find the value.

For $f(x) = -4x^2 - 24x - 19$, $a = -4$ and $-4 < 0$. So, the parabola opens down and the function has a maximum value. To find the maximum value, find the y-coordinate of the vertex.

$$x = -\frac{b}{2a} = -\frac{(-24)}{2(-4)} = -3 \qquad \text{The x-coordinate of the vertex is } -\frac{b}{2a}.$$

$$f(-3) = -4(-3)^2 - 24(-3) - 19 \qquad \text{Substitute } -3 \text{ for } x.$$

$$= 17 \qquad \text{Simplify.}$$

∴ The maximum value is 17.

### On Your Own

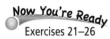
Exercises 21–26

Tell whether the function has a minimum value or a maximum value. Then find the value.

7. $g(x) = 8x^2 - 8x + 6$

8. $h(x) = -\frac{1}{4}x^2 + 3x + 1$

**EXAMPLE 4** **Real-Life Application**

The function $f(t) = -16t^2 + 80t + 5$ gives the height (in feet) of a water balloon $t$ seconds after it is launched. Use a graphing calculator to find the maximum height of the water balloon.

**Step 1:** Enter the function $f(t) = -16t^2 + 80t + 5$ into your calculator and graph it. Because time cannot be negative, use only nonnegative values of $t$.

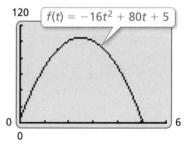

**Step 2:** Use the *maximum* feature to find the maximum value of the function.

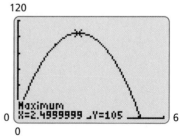

**Study Tip**

The *minimum* feature of a graphing calculator can be used for parabolas that open up.

∴ The maximum height of the water balloon is 105 feet.

### On Your Own

9. When does the water balloon reach its maximum height?

# 8.4 Exercises

## Vocabulary and Concept Check

1. **VOCABULARY** Explain how you can tell whether a quadratic function has a maximum value or a minimum value without graphing the function.

2. **DIFFERENT WORDS, SAME QUESTION** Consider the quadratic function $y = -2x^2 + 8x + 24$. Which is different? Find "both" answers.

   What is the maximum value of the function?

   What is the $y$-coordinate of the vertex of the graph?

   What is the greatest number in the range of the function?

   What is the axis of symmetry of the graph of the function?

## Practice and Problem Solving

Find the vertex, the axis of symmetry, and the $y$-intercept of the graph.

3.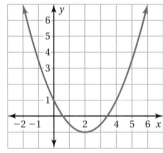
4.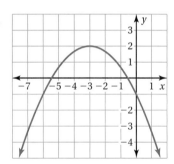
5.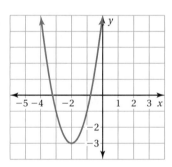

Find (a) the axis of symmetry and (b) the vertex of the graph of the function.

6. $y = 2x^2 - 4x$
7. $y = 3x^2 + 12x$
8. $y = -8x^2 - 16x - 1$
9. $y = -6x^2 + 24x - 20$
10. $y = \dfrac{2}{5}x^2 - 4x + 14$
11. $y = -\dfrac{3}{4}x^2 + 6x - 18$

12. **ERROR ANALYSIS** Describe and correct the error in finding the axis of symmetry of the graph of $y = 3x^2 - 12x + 11$.

    $$x = -\dfrac{b}{2a} = \dfrac{-12}{2(3)} = -2$$

    The axis of symmetry is $x = -2$.

Graph the function. Describe the domain and range.

13. $y = 2x^2 + 12x + 14$
14. $y = 4x^2 + 24x + 31$
15. $y = -8x^2 - 16x - 9$
16. $y = -5x^2 + 30x - 47$
17. $y = \dfrac{2}{3}x^2 - 8x + 19$
18. $y = -\dfrac{1}{2}x^2 - 8x - 25$

Section 8.4  Graphing $y = ax^2 + bx + c$  **429**

19. **FIREWORK** The function shown represents the height $h$ (in feet) of a firework $t$ seconds after it is launched. The firework explodes at its highest point.

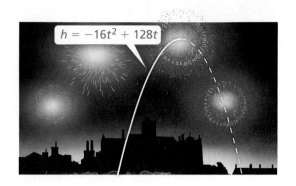

a. When does the firework explode?

b. Describe the domain and range of $h$.

20. **REASONING** Given the quadratic equation $y = ax^2 + bx + c$, find the axis of symmetry when $b = 0$.

**Tell whether the function has a minimum value or a maximum value. Then find the value.**

21. $y = 3x^2 - 18x + 15$

22. $y = -5x^2 + 10x + 7$

23. $y = -4x^2 + 4x - 2$

24. $y = 2x^2 - 10x + 13$

25. $y = -\frac{1}{2}x^2 + 8x + 20$

26. $y = \frac{1}{5}x^2 - 12x + 27$

27. **PRECISION** The vertex of a graph of a quadratic function is $(3, -1)$. One point on the graph is $(6, 8)$. Find another point on the graph. Justify your answer.

28. **SUSPENSION BRIDGE** The cables between the two towers of a suspension bridge can be modeled by $y = \frac{1}{400}x^2 - x + 150$, where $x$ and $y$ are measured in feet. The cables are at road level midway between the towers. How high is the road above the water?

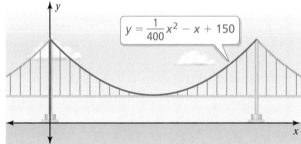

29. **STEEPLECHASE** The function $h(t) = -16t^2 + 16t$ gives the height $h$ (in feet) of a horse $t$ seconds after it jumps during a steeplechase.

a. When does the horse reach its maximum height?

b. Can the horse clear a fence that is 3.5 feet tall? If so, by how much?

c. How long is the horse in the air?

Use the *minimum* or *maximum* feature of a graphing calculator to approximate the vertex of the graph of the function.

**30.** $y = -6.2x^2 + 4.8x - 1$   **31.** $y = 0.5x^2 + \sqrt{2}x - 3$   **32.** $y = \pi x^2 + 3x$

**33. CHOOSE TOOLS** The graph of a quadratic function passes through (4, 0), (5, 3), and (6, 4). Does the graph open up or down? Explain your reasoning.

**34. REASONING** For a quadratic function $f$, what does $f\left(-\dfrac{b}{2a}\right)$ represent? Explain your reasoning.

**35. CALCULATORS** An office supply store sells about 60 graphing calculators per month for $100 each. For each $5 decrease in price, the store expects to sell 5 more calculators.

  **a.** Write a quadratic function that represents the revenue from calculator sales. (*Note:* revenue = units sold × unit price)

  **b.** How much should the store charge per calculator to maximize monthly revenue?

**36. AIR CANNON** At a basketball game, an air cannon is used to launch T-shirts into the crowd. The function $y = -\dfrac{1}{8}x^2 + 4x$ gives the path of a T-shirt. The function $3y = 2x - 14$ gives the height of the bleachers. In both functions, $y$ represents height (in feet) and $x$ represents horizontal distance (in feet). At what height does the T-shirt land in the bleachers?

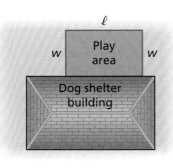

**37. DOG SHELTER** The owners of a dog shelter want to enclose a rectangular play area on the side of their building. They have $k$ feet of fencing. What is the maximum area of the outside enclosure in terms of $k$? (*Hint:* Find the $y$-coordinate of the vertex of the graph of the area function.)

## Fair Game Review  What you learned in previous grades & lessons

**Graph the function.** *(Section 2.4 and Section 6.5)*

**38.** $-4x + y = 3$   **39.** $y = 20(1.2)^t$   **40.** $r(t) = 400(1.05)^t$

**41. MULTIPLE CHOICE** What is the value of $3(4)^x$ when $x = 2$? *(Section 6.4)*

  Ⓐ 6     Ⓑ 24     Ⓒ 48     Ⓓ 144

# Extension 8.4 Graphing $y = a(x - h)^2 + k$

Check It Out
Lesson Tutorials
BigIdeasMath.com

**Key Vocabulary**
vertex form, p. 432

The **vertex form** of a quadratic function is $y = a(x - h)^2 + k$, where $a \neq 0$. The vertex of the parabola is $(h, k)$.

## Key Idea

**Graphing $y = (x - h)^2$**

- When $h > 0$, the graph of $y = (x - h)^2$ is a horizontal translation $h$ units to the right of the graph of $y = x^2$.
- When $h < 0$, the graph of $y = (x - h)^2$ is a horizontal translation $h$ units to the left of the graph of $y = x^2$.

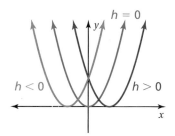

### EXAMPLE 1 Graphing $y = (x - h)^2$

**COMMON CORE**

**Graphing Quadratic Functions**

In this extension, you will
- graph quadratic functions of the form $y = a(x - h)^2 + k$ and compare to the graph of $y = x^2$.

Learning Standards
F.BF.3
F.IF.4
F.IF.7a

Graph $y = (x - 4)^2$. Compare the graph to the graph of $y = x^2$.

**Step 1:** Make a table of values.

| x | 0 | 2 | 4 | 6 | 8 |
|---|---|---|---|---|---|
| y | 16 | 4 | 0 | 4 | 16 |

**Step 2:** Plot the ordered pairs.

**Step 3:** Draw a smooth curve through the points.

The graph of $y = (x - 4)^2$ is a translation 4 units to the right of the graph of $y = x^2$.

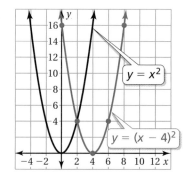

## Practice

**Graph the function. Compare the graph to the graph of $y = x^2$. Use a graphing calculator to check your answer.**

1. $y = (x + 3)^2$
2. $y = (x - 1)^2$
3. $y = (x - 6)^2$
4. $y = (x + 10)^2$
5. $y = (x - 1.5)^2$
6. $y = \left(x + \dfrac{5}{2}\right)^2$

7. **REASONING** Compare the graphs of $y = x^2 + 6x + 9$ and $y = x^2$ without graphing the functions. How can factoring help you compare the parabolas? Explain.

8. **STRUCTURE** Write the function in Example 1 in the form $y = ax^2 + bx + c$. Describe advantages and disadvantages of writing the function in each form.

# EXAMPLE 2  Graphing $y = (x - h)^2 + k$

Graph $y = (x + 5)^2 - 1$. Compare the graph to the graph of $y = x^2$.

**Step 1:** Make a table of values.

| x | −7 | −6 | −5 | −4 | −3 |
|---|----|----|----|----|----|
| y | 3  | 0  | −1 | 0  | 3  |

**Step 2:** Plot the ordered pairs.

**Step 3:** Draw a smooth curve through the points.

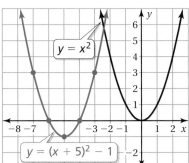

∴ The graph of $y = (x + 5)^2 - 1$ is a translation 5 units to the left and 1 unit down of the graph of $y = x^2$.

# EXAMPLE 3  Graphing $y = a(x - h)^2 + k$

Graph $y = -2(x + 2)^2 + 3$. Compare the graph to the graph of $y = x^2$.

**Step 1:** Make a table of values.

| x | −4 | −3 | −2 | −1 | 0  |
|---|----|----|----|----|----|
| y | −5 | 1  | 3  | 1  | −5 |

**Step 2:** Plot the ordered pairs.

**Step 3:** Draw a smooth curve through the points.

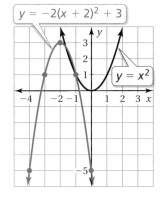

**Study Tip**

Notice what the values in vertex form represent:
- $a$: opens up or down, and is wider or narrower
- $h$: horizontal translation
- $k$: vertical translation
- $(h, k)$: vertex

∴ The graph of $y = -2(x + 2)^2 + 3$ opens down and is narrower than the graph of $y = x^2$. The vertex of the graph of $y = -2(x + 2)^2 + 3$ is a translation 2 units to the left and 3 units up of the vertex of the graph of $y = x^2$.

## Practice

**Graph the function. Compare the graph to the graph of $y = x^2$. Use a graphing calculator to check your answer.**

9. $y = (x - 2)^2 + 4$
10. $y = (x + 1)^2 - 7$
11. $y = (x - 8)^2 - 8$
12. $y = 3(x - 1)^2 + 6$
13. $y = -(x - 3)^2 - 5$
14. $y = \frac{1}{2}(x + 4)^2 - 2$

**Describe how the graph of $g(x)$ compares to the graph of $f(x)$.**

15. $g(x) = f(x) - 7$
16. $g(x) = f(x + 10)$
17. $g(x) = 5f(x)$
18. $g(x) = f(2x)$

19. **REASONING** The graph of $y = x^2$ is translated 2 units right and 5 units down. Write an equation for the function in vertex form and in standard form.

20. **REASONING** Does $k$ represent the $y$-intercept of the graph of $y = a(x - h)^2 + k$? Explain.

# 8.5 Comparing Linear, Exponential, and Quadratic Functions

**Essential Question** How can you compare the growth rates of linear, exponential, and quadratic functions?

### 1 ACTIVITY: Comparing Speeds

Work with a partner. Three cars start traveling at the same time. The distance traveled in $t$ minutes is $y$ miles. Complete each table and sketch all three graphs in the same coordinate plane. Compare the speeds of the three cars. Which car has a constant speed? Which car is accelerating the most? Explain your reasoning.

| $t$ | $y = t$ |
|-----|---------|
| 0   |         |
| 0.2 |         |
| 0.4 |         |
| 0.6 |         |
| 0.8 |         |
| 1.0 |         |

| $t$ | $y = 2^t - 1$ |
|-----|---------------|
| 0   |               |
| 0.2 |               |
| 0.4 |               |
| 0.6 |               |
| 0.8 |               |
| 1.0 |               |

| $t$ | $y = t^2$ |
|-----|-----------|
| 0   |           |
| 0.2 |           |
| 0.4 |           |
| 0.6 |           |
| 0.8 |           |
| 1.0 |           |

**COMMON CORE**

**Linear, Quadratic, and Exponential Functions**

In this lesson, you will
- identify linear, quadratic, and exponential functions using graphs or tables.

Learning Standards
F.IF.4
F.IF.6
F.IF.7a
F.LE.3

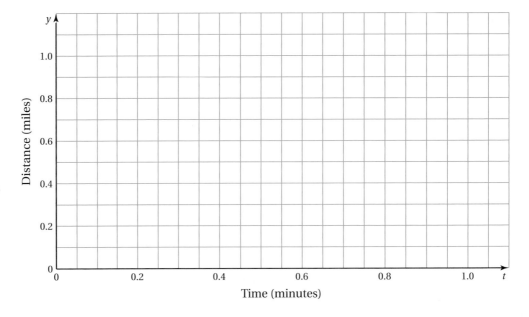

**434** Chapter 8  Graphing Quadratic Functions

## 2 ACTIVITY: Comparing Speeds

**Math Practice 4**

**Analyze Relationships**
What is the relationship between the speeds of the cars?

Work with a partner. Analyze the speeds of the three cars over the given time periods. The distance traveled in $t$ minutes is $y$ miles. Which car eventually overtakes the others?

a.

| $t$ | $y = t$ |
|---|---|
| 1 | |
| 2 | |
| 3 | |
| 4 | |

| $t$ | $y = 2^t - 1$ |
|---|---|
| 1 | |
| 2 | |
| 3 | |
| 4 | |

| $t$ | $y = t^2$ |
|---|---|
| 1 | |
| 2 | |
| 3 | |
| 4 | |

b.

| $t$ | $y = t$ |
|---|---|
| 4 | |
| 5 | |
| 6 | |
| 7 | |
| 8 | |
| 9 | |

| $t$ | $y = 2^t - 1$ |
|---|---|
| 4 | |
| 5 | |
| 6 | |
| 7 | |
| 8 | |
| 9 | |

| $t$ | $y = t^2$ |
|---|---|
| 4 | |
| 5 | |
| 6 | |
| 7 | |
| 8 | |
| 9 | |

## What Is Your Answer?

3. **IN YOUR OWN WORDS** How can you compare the growth rates of linear, exponential, and quadratic functions? Which type of growth eventually leaves the other two in the dust? Explain your reasoning.

**Practice**

Use what you learned about comparing functions to complete Exercises 3–5 on page 439.

## 8.5 Lesson

Check It Out
Lesson Tutorials
BigIdeasMath com

### Key Idea

| Linear Function | Exponential Function | Quadratic Function |
|---|---|---|
|  | | |
| Line | Curve | Parabola |
| $y = mx + b$ | $y = ab^x$ | $y = ax^2 + bx + c$ |

### EXAMPLE 1 — Identifying Functions Using Graphs

Plot the points. Tell whether the points represent a *linear*, an *exponential*, or a *quadratic* function.

**a.** $(4, 4), (2, 0), (0, 0),$
$\left(1, -\dfrac{1}{2}\right), (-2, 4)$

**b.** $(0, 1), (2, 4), (4, 7),$
$(-2, -2), (-4, -5)$

**c.** $(0, 2), (2, 8), (1, 4),$
$(-1, 1), \left(-2, \dfrac{1}{2}\right)$

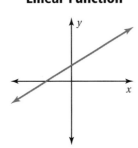
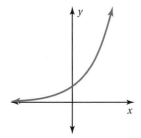
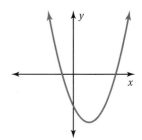

∴ Quadratic function  ∴ Linear function  ∴ Exponential function

### On Your Own

**Now You're Ready**
Exercises 9–12

Plot the points. Tell whether the points represent a *linear*, an *exponential*, or a *quadratic* function.

**1.** $(-1, 5), (2, -1), (0, -1), (3, 5), (1, -3)$

**2.** $(-1, 2), (-2, 8), (-3, 32), \left(0, \dfrac{1}{2}\right), \left(1, \dfrac{1}{8}\right)$

**3.** $(-3, 5), (0, -1), (2, -5), (-4, 7), (1, -3)$

# Key Idea

**Differences and Ratios of Functions**

**Linear Function:** $y = 2x + 5$

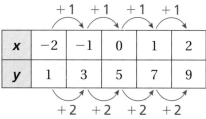

The $y$-values have a common *difference* of 2.

**Exponential Function:** $y = 4(2)^x$

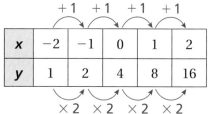

The $y$-values have a common *ratio* of 2.

**Quadratic Function:** $y = x^2 + 2x - 1$

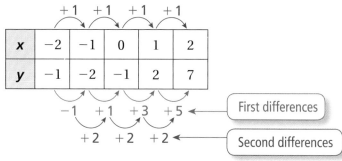

For quadratic functions, the second differences are constant.

**Study Tip**

For a linear function, the first differences are constant.

## EXAMPLE 2 — Identifying Functions Using Differences or Ratios

Tell whether the table of values represents a *linear*, an *exponential*, or a *quadratic* function.

a.

|  | +1 | +1 | +1 | +1 |  |
|---|---|---|---|---|---|
| x | −3 | −2 | −1 | 0 | 1 |
| y | 11 | 8 | 5 | 2 | −1 |

−3  −3  −3  −3

∴ The $y$-values have a common difference of −3. So, the table represents a linear function.

b.

| x | −1 | 0 | 1 | 2 | 3 |
|---|---|---|---|---|---|
| y | 0 | −1 | 2 | 9 | 20 |

−1  +3  +7  +11
   +4  +4  +4

∴ The second differences are constant. So, the table represents a quadratic function.

**Study Tip**

For a quadratic function, the $y$-values will increase, then decrease, or the $y$-values will decrease, then increase.

### On Your Own

**Now You're Ready**
Exercises 14–17

4. Tell whether the table of values represents a *linear*, an *exponential*, or a *quadratic* function.

| x | −1 | 0 | 1 | 2 | 3 |
|---|---|---|---|---|---|
| y | 1 | 3 | 9 | 27 | 81 |

Section 8.5 Comparing Linear, Exponential, and Quadratic Functions

# EXAMPLE 3  Identifying and Writing a Function

| x | y |
|---|---|
| 0 | 0 |
| 2 | 1 |
| 4 | 4 |
| 6 | 9 |
| 8 | 16 |

Tell whether the table of values represents a *linear*, an *exponential*, or a *quadratic* function. Then write an equation for the function using the form $y = mx + b$, $y = ab^x$, or $y = ax^2$.

**Step 1:** Graph the data. The function appears to be exponential or quadratic.

**Step 2:** Check the *y*-values. If there is no common difference or ratio, check the second differences.

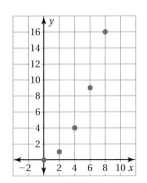

Second differences are constant.

The function is quadratic.

**Step 3:** Use the form $y = ax^2$.

$1 = a(2)^2$   Use the point (2, 1). Substitute 2 for *x* and 1 for *y*.

$\dfrac{1}{4} = a$   Solve for *a*.

∴ So, an equation for the quadratic function is $y = \dfrac{1}{4}x^2$.

**Study Tip**
To check your function in Example 3, substitute the other points from the table to see if they satisfy the function.

### On Your Own

Now You're Ready
Exercises 19–24

5. Tell whether the table of values represents a *linear*, an *exponential*, or a *quadratic* function. Then write an equation for the function using the form $y = mx + b$, $y = ab^x$, or $y = ax^2$.

| x | −1 | 0 | 1 | 2 | 3 |
|---|---|---|---|---|---|
| y | 16 | 8 | 4 | 2 | 1 |

## Summary

**Linear Function**
$y = mx + b$

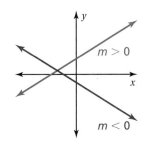

**Exponential Function**
$y = ab^x$, $a \neq 0$, $b \neq 1$, and $b > 0$

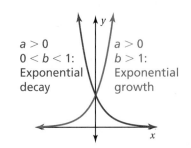

$a > 0$
$0 < b < 1$:
Exponential decay

$a > 0$
$b > 1$:
Exponential growth

**Quadratic Function**
$y = ax^2 + bx + c$, $a \neq 0$

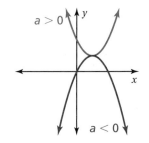

# 8.5 Exercises

## Vocabulary and Concept Check

1. **VOCABULARY** How can you decide whether to use a linear, a quadratic, or an exponential function to model a data set?

2. **WHICH ONE DOESN'T BELONG?** Which graph does *not* belong with the other three? Explain your reasoning.

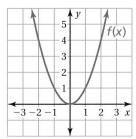 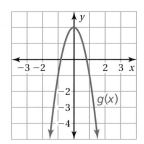 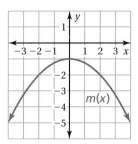 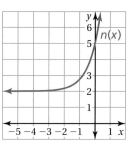

## Practice and Problem Solving

**Find the values of *x* when *f(x)* is greater than *g(x)*.**

3. $f(x) = x^2$
   $g(x) = x$

4. $f(x) = 3^x$
   $g(x) = 2x$

5. $f(x) = 4^x$
   $g(x) = 2x^2$

**Match the function type with its graph.**

6. Linear function
7. Exponential function
8. Quadratic function

A.

B.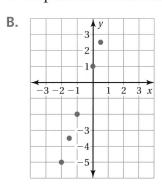

C.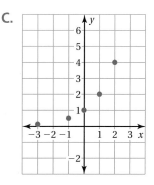

**Plot the points. Tell whether the points represent a *linear*, an *exponential*, or a *quadratic* function.**

9. $(-2, -1), (-1, 0), (1, 2), (2, 3), (0, 1)$

10. $(1, 8), \left(-4, \frac{1}{4}\right), \left(-3, \frac{1}{2}\right), (-2, 1), (-1, 2)$

11. $(0, -3), (1, 0), (2, 9), (-2, 9), (-1, 0)$

12. $(-1, -3), (-3, 5), (0, -1), (1, 5), (2, 15)$

13. **SUBWAY** A student takes a subway to a public library. The table shows the distance *d* (in miles) the student travels in *t* minutes. Tell whether the data can be modeled by a *linear*, an *exponential*, or a *quadratic* function.

| Time, *t* | 0.5 | 1 | 3 | 5 |
|---|---|---|---|---|
| Distance, *d* | 0.335 | 0.67 | 2.01 | 3.35 |

Section 8.5  Comparing Linear, Exponential, and Quadratic Functions  439

Tell whether the table of values represents a *linear*, an *exponential*, or a *quadratic* function.

**14.**

| x | −2 | −1 | 0 | 1 | 2 |
|---|---|---|---|---|---|
| y | 0 | 0.5 | 1 | 1.5 | 2 |

**15.**

| x | −1 | 0 | 1 | 2 | 3 |
|---|---|---|---|---|---|
| y | 0.2 | 1 | 5 | 25 | 125 |

**16.**

| x | −2 | −1 | 0 | 1 | 2 |
|---|---|---|---|---|---|
| y | 0.75 | 1.5 | 3 | 6 | 12 |

**17.**

| x | 2 | 3 | 4 | 5 | 6 |
|---|---|---|---|---|---|
| y | 2 | 4.5 | 8 | 12.5 | 18 |

**18. REASONING** Can the *y*-values of a data set have both a common difference and a common ratio? Explain your reasoning.

Tell whether the data values represent a *linear*, an *exponential*, or a *quadratic* function. Then write an equation for the function using the form $y = mx + b$, $y = ab^x$, or $y = ax^2$.

**19.** (−2, 8), (−1, 2), (0, 0), (1, 2), (2, 8)

**20.** (−3, 8), (−2, 4), (−1, 2), (0, 1), (1, 0.5)

**21.**

| x | −2 | −1 | 0 | 1 | 2 |
|---|---|---|---|---|---|
| y | 4 | 1 | −2 | −5 | −8 |

**22.**

| x | −1 | 0 | 1 | 2 | 3 |
|---|---|---|---|---|---|
| y | 2.5 | 5 | 10 | 20 | 40 |

**23.**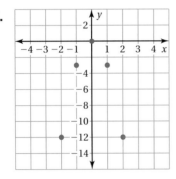

**24.**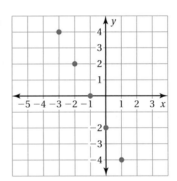

**25. ERROR ANALYSIS** Describe and correct the error in writing an equation for the function represented by the ordered pairs.

(−1, 4), (0, 0), (1, 4), (2, 16), (3, 36)

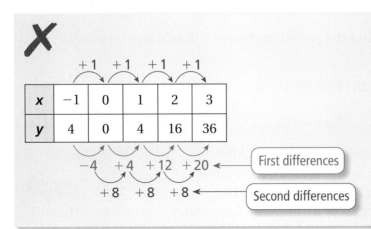

**26. HIGH SCHOOL FOOTBALL** The table shows the number of people attending the first five football games at a high school.

a. Plot the points.

b. Does a *linear*, an *exponential*, or a *quadratic* function represent this situation? Explain.

| Game, g | 1 | 2 | 3 | 4 | 5 |
|---|---|---|---|---|---|
| Number of People, p | 252 | 325 | 270 | 249 | 310 |

**27. CRITICAL THINKING** Is the graph of a set of points enough to determine whether the points represent a *linear*, an *exponential*, or a *quadratic* function? Justify your answer.

**28. RECORDING STUDIO** The table shows the amount of money (in dollars) that a musician pays for using a recording studio.

| Number of Hours, h | 1 | 2 | 3 | 4 |
|---|---|---|---|---|
| Amount, m (dollars) | 110 | 145 | 180 | 215 |

a. Plot the points. Then determine the type of function that best represents this situation.

b. Write a function that models the data.

c. How much does it cost to use the studio for 10 hours?

**29. TOURNAMENT** At the beginning of a basketball tournament, there are 64 teams. After each round, one-half of the remaining teams are eliminated.

a. Make a table showing the number of teams remaining after each round.

b. Determine the type of function that best represents this situation.

c. Write a function that models the data.

d. After which round do you know the team that won the tournament?

**30. Repeated Reasoning** Write a function that has constant second differences of 3.

## Fair Game Review  What you learned in previous grades & lessons

**Find the *x*-intercept(s) of the graph.** *(Section 2.3 and Section 8.3)*

31.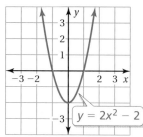

32.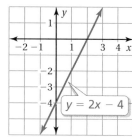

33.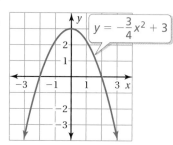

**34. MULTIPLE CHOICE** What is the factored form of $8x^3 - 18x$? *(Section 7.9)*

Ⓐ $2x(2x + 3)^2$   Ⓑ $(2x + 3)(2x - 3)$   Ⓒ $(2x - 3)^2$   Ⓓ $2x(2x + 3)(2x - 3)$

# Extension 8.5 Comparing Graphs of Functions

You have already learned that the average rate of change (or slope) between any two points on a line is the change in y divided by the change in x. You can find the average rate of change between two points of a nonlinear function using the same method.

## ACTIVITY 1 Rates of Change of a Quadratic Function

In Example 4 on page 428, the function $f(t) = -16t^2 + 80t + 5$ gives the height (in feet) of a water balloon $t$ seconds after it is launched.

**COMMON CORE**

Linear, Quadratic, and Exponential Functions

In this extension, you will
- compare graphs of linear, quadratic, and exponential functions.
- solve real-life problems.

Learning Standards
F.IF.4
F.IF.6
F.IF.7a
F.LE.3

a. Copy and complete the table for $f(t)$.

| t | 0 | 0.5 | 1 | 1.5 | 2 | 2.5 | 3 | 3.5 | 4 | 4.5 | 5 |
|---|---|-----|---|-----|---|-----|---|-----|---|-----|---|
| f(t) | | | | | | | | | | | |

b. Graph the ordered pairs from part (a). Then draw a smooth curve through the points.

c. For what values is the function increasing? decreasing?

d. Copy and complete the tables to find the average rate of change for each interval.

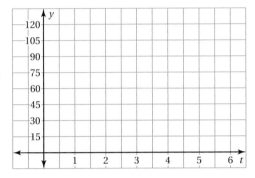

| Time Interval | 0 to 0.5 sec | 0.5 to 1 sec | 1 to 1.5 sec | 1.5 to 2 sec | 2 to 2.5 sec |
|---|---|---|---|---|---|
| Average Rate of Change (ft/sec) | | | | | |

| Time Interval | 2.5 to 3 sec | 3 to 3.5 sec | 3.5 to 4 sec | 4 to 4.5 sec | 4.5 to 5 sec |
|---|---|---|---|---|---|
| Average Rate of Change (ft/sec) | | | | | |

## Practice

1. Compared to the average rate of change of a linear function, what do you notice about the average rate of change in part (d) of Activity 1?

2. Is the average rate of change increasing or decreasing from 0 to 2.5 seconds? How can you use the graph to justify your answer?

3. What do you notice about the average rate of change when the function is increasing and when the function is decreasing?

4. In Example 4 on page 419, the function $f(t) = -16t^2 + 64$ gives the height of an egg $t$ seconds after it is dropped. (a) Make a table of values. Use the domain $0 \leq t \leq 2$ with intervals of 0.5 second. (b) Graph the ordered pairs and draw a smooth curve through the points. (c) Describe where the function is increasing and decreasing. (d) Find the average rate of change for each interval in the table. What do you notice?

## ACTIVITY 2  Rates of Change of Different Functions

The graphs show the numbers *y* of videos on three video-sharing websites *x* hours after the websites are launched.

*Linear*  *Quadratic*  *Exponential*

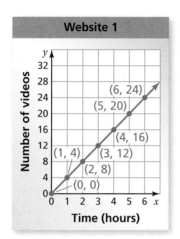
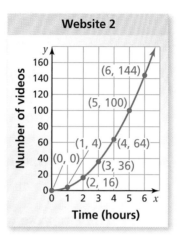
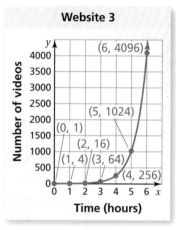

a. Do the three websites ever have the same number of videos?

b. Copy and complete the table for each function.

| Time Interval | 0 to 1 h | 1 to 2 h | 2 to 3 h | 3 to 4 h | 4 to 5 h | 5 to 6 h |
|---|---|---|---|---|---|---|
| Average Rate of Change (videos/hour) | | | | | | |

c. What do you notice about the average rate of change of the linear function?

d. What do you notice about the average rate of change of the quadratic function?

e. What do you notice about the average rate of change of the exponential function?

f. Which average rate of change increases more quickly, the quadratic function or the exponential function?

## Practice

5. **REASONING** How does a quantity that is increasing exponentially compare to a quantity that is increasing linearly or quadratically?

6. **REASONING** Explain why the average rate of change of a linear function is constant and the average rate of change of a quadratic or exponential function is not constant.

Extension 8.5  Comparing Graphs of Functions

# 8.4–8.5 Quiz

**Find (a) the axis of symmetry and (b) the vertex of the graph of the function.** *(Section 8.4)*

1. $y = x^2 - 2x - 3$
2. $y = -2x^2 + 12x + 5$

**Graph the function. Describe the domain and range.** *(Section 8.4)*

3. $y = -4x^2 - 4x + 7$
4. $y = 2x^2 + 8x - 5$
5. $y = -4x^2 - 8x + 12$

**Tell whether the function has a minimum value or a maximum value. Then find the value.** *(Section 8.4)*

6. $y = 5x^2 + 10x - 3$
7. $y = -\frac{1}{2}x^2 + 2x + 16$
8. $y = -2x^2 + 8x + 3$

**Graph the function. Compare the graph to the graph of $y = x^2$.** *(Section 8.4)*

9. $y = (x - 5)^2$
10. $y = (x + 6)^2 - 2$

**Plot the points. Tell whether the points represent a *linear*, an *exponential*, or a *quadratic* function.** *(Section 8.5)*

11. (3, 6), (4, 16), (5, 30), (0, 0), (−1, 6)
12. (1, 7.5), (3, 6.5), (4, 6), (2, 7), (5, 5.5)

**Tell whether the table of values represents a *linear*, an *exponential*, or a *quadratic* function. Then write an equation for the function using the form $y = mx + b$, $y = ab^x$, or $y = ax^2$.** *(Section 8.5)*

13.

| x | y |
|---|---|
| −1 | 1 |
| 0 | 3 |
| 1 | 9 |
| 2 | 27 |
| 3 | 81 |

14.

| x | y |
|---|---|
| −3 | −3 |
| −2 | −1 |
| −1 | 1 |
| 0 | 3 |
| 1 | 5 |

15.

| x | y |
|---|---|
| 1 | −5 |
| 2 | −20 |
| 3 | −45 |
| 4 | −80 |
| 5 | −125 |

16. **FOOTBALL** The function $h(t) = -16t^2 + 20t + 6$ gives the height (in feet) of a football $t$ seconds after it is thrown. Describe the domain and range. Find the maximum height of the football. *(Section 8.4)*

17. **DOWNLOADING MUSIC** The table shows the amounts of money (in dollars) that you pay to download songs from a website. *(Section 8.5)*

    a. Plot the points. Tell whether the points represent a *linear*, an *exponential*, or a *quadratic* function.

    | Number of Songs, s | 2 | 3 | 4 | 5 |
    |---|---|---|---|---|
    | Amount, a (dollars) | 2.58 | 3.87 | 5.16 | 6.45 |

    b. Write a function that models the data.
    c. How much does it cost to download 15 songs?

# 8 Chapter Review

## Review Key Vocabulary

quadratic function, *p. 404*
parabola, *p. 404*
vertex, *p. 404*
axis of symmetry, *p. 404*
focus, *p. 412*
zero, *p. 419*
maximum value, *p. 427*
minimum value, *p. 427*
vertex form, *p. 432*

## Review Examples and Exercises

### 8.1 Graphing $y = ax^2$ (pp. 402–409)

**Graph $y = -4x^2$. Compare the graph to the graph of $y = x^2$.**

**Step 1:** Make a table of values.

| x | −2  | −1 | 0 | 1  | 2   |
|---|-----|----|---|----|-----|
| y | −16 | −4 | 0 | −4 | −16 |

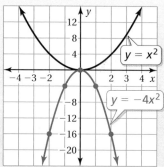

**Step 2:** Plot the ordered pairs.
**Step 3:** Draw a smooth curve through the points.

• The graphs have the same vertex, (0, 0), and the same axis of symmetry, $x = 0$, but the graph of $y = -4x^2$ opens down. The graph of $y = -4x^2$ is narrower than the graph of $y = x^2$.

### Exercises

**Graph the function. Compare the graph to the graph of $y = x^2$.**

1. $y = 7x^2$
2. $y = \frac{1}{2}x^2$
3. $y = -\frac{3}{4}x^2$

### 8.2 Focus of a Parabola (pp. 410–415)

**Write an equation of the parabola with focus (0, 2) and vertex at the origin.**

For $y = ax^2$, the focus is $\left(0, \frac{1}{4a}\right)$. Use the given focus, (0, 2), to write an equation to find $a$.

$\frac{1}{4a} = 2$   Equate the y-coordinates.

$\frac{1}{8} = a$   Solve for $a$.

• An equation of the parabola is $y = \frac{1}{8}x^2$.

### Exercises

**4.** Graph $y = -\frac{1}{2}x^2$. Identify the focus.

**5.** Write an equation of the parabola with focus (0, 10) and vertex at the origin.

## 8.3 Graphing $y = ax^2 + c$ (pp. 416–421)

**Graph $y = 2x^2 + 3$. Compare the graph to the graph of $y = x^2$.**

**Step 1:** Make a table of values.

| x | −2 | −1 | 0 | 1 | 2 |
|---|---|---|---|---|---|
| y | 11 | 5 | 3 | 5 | 11 |

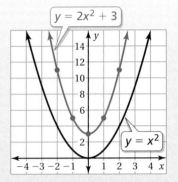

**Step 2:** Plot the ordered pairs.

**Step 3:** Draw a smooth curve through the points.

∴ Both graphs open up and have the same axis of symmetry, $x = 0$. The graph of $y = 2x^2 + 3$ is narrower than the graph of $y = x^2$. The vertex of the graph of $y = 2x^2 + 3$ is a translation 3 units up of the vertex of the graph of $y = x^2$.

### Exercises

Graph the function. Compare the graph to the graph of $y = x^2$.

**6.** $y = x^2 + 6$  **7.** $y = -x^2 - 4$  **8.** $y = 3x^2 - 5$

## 8.4 Graphing $y = ax^2 + bx + c$ (pp. 424–433)

**Graph $y = 4x^2 + 8x - 1$. Describe the domain and range.**

**Step 1:** Find and graph the axis of symmetry: $x = -\frac{b}{2a} = -\frac{8}{2(4)} = -1$.

**Step 2:** Find and plot the vertex. The $x$-coordinate of the vertex is $-1$. The $y$-coordinate is: $y = 4(-1)^2 + 8(-1) - 1 = -5$. So, the vertex is $(-1, -5)$.

**Step 3:** Use the $y$-intercept to find two more points on the graph. The $y$-intercept is $-1$. So, $(0, -1)$ lies on the graph. Because the axis of symmetry is $x = -1$, the point $(-2, -1)$ also lies on the graph.

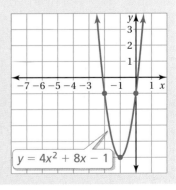

**Step 4:** Draw a smooth curve through the points.

∴ The domain is all real numbers. The range is $y \geq -5$.

### Exercises

**Graph the function. Describe the domain and range.**

9. $y = x^2 - 2x + 7$    10. $y = -3x^2 + 3x - 4$    11. $y = \frac{1}{2}x^2 - 6x + 10$

12. The function $f(t) = -16t^2 + 75t + 12$ gives the height (in feet) of a pumpkin $t$ seconds after it is launched from a catapult. Use a graphing calculator to find the maximum height of the pumpkin. When does the pumpkin reach its maximum height?

**Graph the function. Compare the graph to the graph of $y = x^2$. Use a graphing calculator to check your answer.**

13. $y = (x + 5)^2$    14. $y = (x + 3)^2 - 2$    15. $y = -(x - 1)^2 + 1$

## 8.5 Comparing Linear, Exponential, and Quadratic Functions
(pp. 434–443)

**Tell whether the data values represent a *linear*, an *exponential*, or a *quadratic* function.**

a. $(-4, 1), (-3, -2), (-2, -3), (-1, -2), (0, 1)$

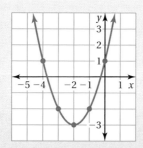

:: The points represent a quadratic function.

b.

| x | -1 | 0 | 1 | 2 | 3 |
|---|---|---|---|---|---|
| y | 15 | 8 | 1 | -6 | -13 |

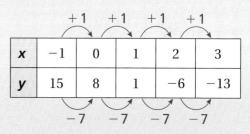

:: The $y$-values have a common difference of $-7$. So, the table represents a linear function.

### Exercises

16. Tell whether the table of values represents a *linear*, an *exponential*, or a *quadratic* function. Then write an equation for the function using the form $y = mx + b$, $y = ab^x$, or $y = ax^2$.

| x | -1 | 0 | 1 | 2 | 3 |
|---|---|---|---|---|---|
| y | 16 | 8 | 4 | 2 | 1 |

17. The function $f(t) = -16t^2 + 75t + 4$ gives the height (in feet) of a baseball $t$ seconds after it is thrown. Sketch a graph of the function. Find the average rate of change from 0 to 1 second and from 1 to 2 seconds. Is the average rate of change increasing or decreasing? How does the graph justify your answer?

# 8 Chapter Test

**Graph the function. Compare the graph to the graph of $y = x^2$.**

1. $y = 3x^2$
2. $y = 2x^2 + 2$
3. $y = -\frac{1}{2}x^2 - 1$
4. $y = (x - 3)^2$
5. $y = (x + 1)^2 - 1$
6. $y = -2(x - 5)^2$

**Graph the function. Identify the focus.**

7. $y = 6x^2$
8. $y = \frac{1}{5}x^2$
9. $y = -1.5x^2$

**Graph the function. Describe the domain and range.**

10. $y = x^2 + 2x - 1$
11. $y = -x^2 - 3x + 3$
12. $y = 2x^2 + 4x - 4$

13. Describe how the graph of $g(x) = f(x + 6)$ compares to the graph of $f(x)$.

**Tell whether the table of values represents a *linear*, an *exponential*, or a *quadratic* function. Then write an equation for the function using the form $y = mx + b$, $y = ab^x$, or $y = ax^2$.**

14.
| x | −1 | 0 | 1 | 2 | 3 |
|---|---|---|---|---|---|
| y | 4 | 8 | 16 | 32 | 64 |

15.
| x | −2 | −1 | 0 | 1 | 2 |
|---|---|---|---|---|---|
| y | −8 | −2 | 0 | −2 | −8 |

16. **EARTH'S ORBIT** The table shows the distance $d$ (in miles) that the Earth moves in its orbit around the Sun after $t$ seconds. Tell whether the data can be modeled by a *linear*, an *exponential*, or a *quadratic* function.

| Time, t | 1 | 2 | 3 | 4 | 5 |
|---|---|---|---|---|---|
| Distance, d | 19 | 38 | 57 | 76 | 95 |

17. **RADIO TELESCOPE** An observatory uses a radio telescope to collect data from another galaxy. The cross section of the telescope's dish can be modeled by $y = \frac{1}{120}x^2$, where $x$ and $y$ are measured in meters. The telescope's receiver is located at the focus of the parabola. What is the distance from the vertex of the parabola to the receiver?

18. **REASONING** Consider the function $f(x) = x^2 + 3$. Is the average rate of change increasing or decreasing from $x = 0$ to $x = 4$? Explain.

# 8 Standards Assessment

1. What is the equation of the parabola shown in the graph? *(F.IF.4)*

   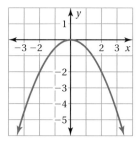

   **A.** $y = \dfrac{1}{2}x^2$

   **B.** $y = 2x^2$

   **C.** $y = -\dfrac{1}{2}x^2$

   **D.** $y = -2x^2$

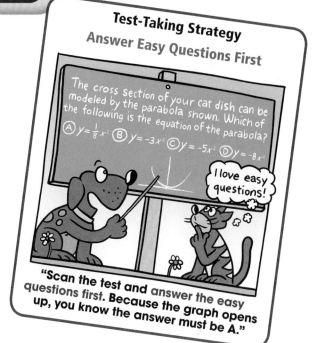

**Test-Taking Strategy**
**Answer Easy Questions First**

"Scan the test and answer the easy questions first. Because the graph opens up, you know the answer must be A."

2. Which expression is equivalent to $(b^{-5})^{-4}$? *(N.RN.2)*

   **F.** $b^{-20}$         **H.** $b^9$

   **G.** $b^{-9}$        **I.** $b^{20}$

3. What is the axis of symmetry of the graph of $y = 3x^2 - 6x - 14$? *(F.IF.4)*

   **A.** $x = -3$      **C.** $x = 1$

   **B.** $x = -1$      **D.** $x = 3$

4. What is the minimum value of the function $y = 3x^2 + 12x + 6$? *(F.IF.7a)*

5. What are the solutions of $16a^2 - 49 = 0$? *(A.REI.4b)*

   **F.** $a = -\dfrac{7}{4}$ and $a = \dfrac{7}{4}$      **H.** $a = -4$ and $a = 4$

   **G.** $a = -7$ and $a = 7$      **I.** $a = -\dfrac{4}{7}$ and $a = \dfrac{4}{7}$

6. What is an equation of the parabola with focus $\left(0, \frac{1}{8}\right)$ and vertex at the origin?   (F.IF.4)

   A. $y = 2x^2$
   
   B. $y = \frac{1}{2}x^2$
   
   C. $y = \frac{1}{8}x^2$
   
   D. $y = \frac{1}{32}x^2$

7. Which expression is equivalent to $(t - 4)^2$?   (A.APR.1)

   F. $t^2 + 16$
   
   G. $t^2 - 16$
   
   H. $t^2 - 8t + 16$
   
   I. $t^2 - 8t - 16$

8. What is the value of $8^{2/3}$?   (N.RN.2)

9. Which of the following is the graph of $y = 4(0.5)^t$?   (F.IF.7e)

   A.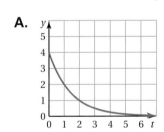
   
   C.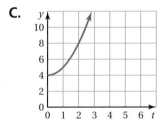
   
   B.

   D.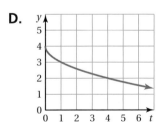

10. The function $f(t) = -16t^2 + 32t + 4$ gives the height (in feet) of a softball $t$ seconds after it is thrown.   (F.IF.4)

   *Part A*  Does the function have a minimum value or a maximum value? Explain your reasoning.
   
   *Part B*  Find the value from Part A.

**11.** Which of the following is not a solution of the system of linear equations when $a \neq 0$, $b \neq 0$, and $c \neq 0$?  *(8.EE.8a)*

$$ax + by = c$$
$$2ax + 2by = 2c$$

**F.** $\left(\dfrac{c}{2a}, \dfrac{c}{2b}\right)$

**H.** $\left(c, \dfrac{c - ac}{b}\right)$

**G.** $\left(\dfrac{c}{a}, 0\right)$

**I.** $\left(\dfrac{c}{b}, 0\right)$

**12.** Amanda was graphing $y = \dfrac{1}{4}x^2 + 2$. Her work is shown below. *(F.BF.3)*

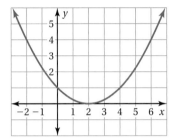

What should Amanda do to correct the error that she made?

**A.** Shift the graph 2 units left.

**B.** Shift the graph 2 units up.

**C.** Shift the graph 2 units left and 2 units up.

**D.** Shift the graph 2 units left and 2 units down.

**13.** Which of the following is an equation of the line that passes through the points (4, 4) and (5, 7)? *(A.CED.2)*

**F.** $y = 3x - 8$

**H.** $y = \dfrac{1}{3}x + \dfrac{8}{3}$

**G.** $y = \dfrac{1}{3}x + \dfrac{16}{3}$

**I.** $y = -3x + 16$

# 9 Solving Quadratic Equations

- 9.1 Solving Quadratic Equations by Graphing
- 9.2 Solving Quadratic Equations Using Square Roots
- 9.3 Solving Quadratic Equations by Completing the Square
- 9.4 Solving Quadratic Equations Using the Quadratic Formula
- 9.5 Solving Systems of Linear and Quadratic Equations

"Do you know why the quadratic equation $x^2 + 1 = 0$ has no real solutions?"

"It's because the graph of $y = x^2 + 1$ doesn't cross the x-axis!"

"Imaginary that!"

"Okay, you hold your tail straight so that there are exactly two points of intersection."

"That's perfect Descartes!"

"I need a firefighter to rescue me."

# What You Learned Before

"Descartes, I'm going to teach you to sing the Quadratic Formula."

## ● Finding Square Roots (8.EE.2)

**Example 1** Find $\sqrt{144}$.

Because $12^2 = 144$, $\sqrt{144} = \sqrt{12^2} = 12$.
↑ *Positive square root*

**Example 2** Find $-\sqrt{225}$.

Because $15^2 = 225$, $-\sqrt{225} = -\sqrt{15^2} = -15$.
↑ *Negative square root*

### Try It Yourself
**Find the square root(s).**

1. $\sqrt{81}$
2. $-\sqrt{169}$
3. $\pm\sqrt{\dfrac{9}{25}}$
4. $-\sqrt{6.25}$

## ● Simplifying Square Roots (N.RN.2)

**Example 3** Simplify $\sqrt{75}$.

$\sqrt{75} = \sqrt{25 \cdot 3}$   Factor using the greatest perfect square factor.

$\phantom{\sqrt{75}} = \sqrt{25} \cdot \sqrt{3}$   Use the Product Property of Square Roots.

$\phantom{\sqrt{75}} = 5\sqrt{3}$   Simplify.

### Try It Yourself
5. Simplify $\sqrt{54}$.
6. Simplify $\sqrt{80}$.
7. Simplify $\sqrt{200}$.

## ● Factoring Perfect Square Trinomials (A.SSE.2)

**Example 4** Factor $x^2 + 14x + 49$.

$x^2 + 14x + 49 = x^2 + 2(x)(7) + 7^2$   Write as $a^2 + 2ab + b^2$.

$\phantom{x^2 + 14x + 49} = (x + 7)^2$   Perfect Square Trinomial Pattern

**Example 5** Factor $y^2 - 10y + 25$.

$y^2 - 10y + 25 = y^2 - 2(y)(5) + 5^2$   Write as $a^2 - 2ab + b^2$.

$\phantom{y^2 - 10y + 25} = (y - 5)^2$   Perfect Square Trinomial Pattern

### Try It Yourself
**Factor the trinomial.**

8. $x^2 + 10x + 25$
9. $m^2 - 20m + 100$
10. $p^2 + 12p + 36$

# 9.1 Solving Quadratic Equations by Graphing

**Essential Question** How can you use a graph to solve a quadratic equation in one variable?

Earlier in the book, you learned that the *x*-intercept of the graph of

$$y = ax + b \quad \text{2 variables}$$

is the same as the solution of

$$ax + b = 0. \quad \text{1 variable}$$

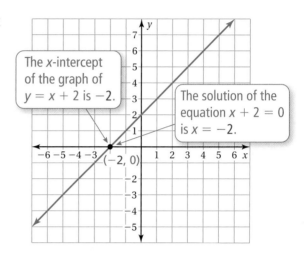

The *x*-intercept of the graph of $y = x + 2$ is $-2$.

The solution of the equation $x + 2 = 0$ is $x = -2$.

## 1 ACTIVITY: Solving a Quadratic Equation by Graphing

**Work with a partner.**

a. Sketch the graph of $y = x^2 - 2x$.

b. What is the definition of an *x*-intercept of a graph? How many *x*-intercepts does this graph have? What are they?

c. What is the definition of a solution of an equation in *x*? How many solutions does the equation $x^2 - 2x = 0$ have? What are they?

d. Explain how you can verify that the *x*-values found in part (c) are solutions of $x^2 - 2x = 0$.

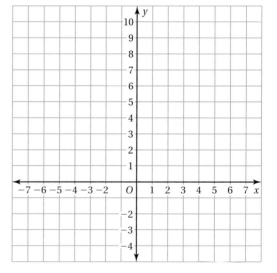

**COMMON CORE**

**Solving Quadratic Equations**

In this lesson, you will
- solve quadratic equations by graphing.

Learning Standards
A.REI.4b
A.REI.11

454 Chapter 9 Solving Quadratic Equations

## 2 ACTIVITY: Solving Quadratic Equations by Graphing

**Math Practice**

**Use Clear Definitions**
How is the solution of the equation represented by the graph of the equation?

Work with a partner. Solve each equation by graphing.

a. $x^2 - 4 = 0$

b. $x^2 + 3x = 0$

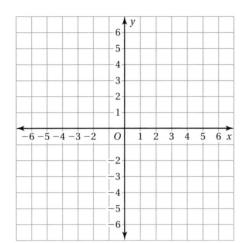
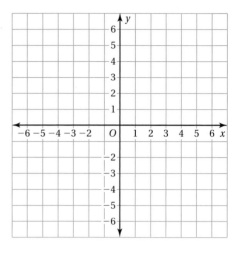

c. $-x^2 + 2x = 0$

d. $x^2 - 2x + 1 = 0$

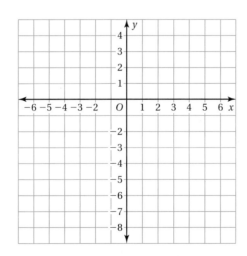
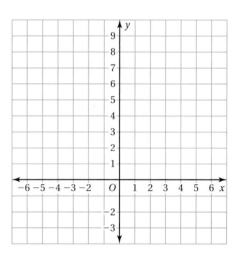

## What Is Your Answer?

3. **IN YOUR OWN WORDS** How can you use a graph to solve a quadratic equation in one variable?

4. After you find a solution graphically, how can you check your result algebraically? Use your solutions in Activity 2 as examples.

**Practice** — Use what you learned about solving quadratic equations to complete Exercises 5–7 on page 459.

## 9.1 Lesson

**Key Vocabulary**
quadratic equation, p. 456

A **quadratic equation** is a nonlinear equation that can be written in the standard form $ax^2 + bx + c = 0$, where $a \neq 0$.

In Chapter 7, you solved quadratic equations by factoring. You can also solve quadratic equations in standard form by finding the $x$-intercept(s) of the graph of the related function $y = ax^2 + bx + c$.

### EXAMPLE 1 — Solving a Quadratic Equation: Two Real Solutions

Solve $x^2 + 2x - 3 = 0$ by graphing.

**Step 1:** Graph the related function $y = x^2 + 2x - 3$.

**Step 2:** Find the $x$-intercepts. They are $-3$ and $1$.

∴ So, the solutions are $x = -3$ and $x = 1$.

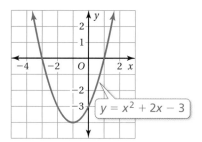

**Remember**
The solutions of a quadratic equation are also called roots.

**Check** Check each solution in the original equation.

| | | |
|---|---|---|
| $x^2 + 2x - 3 = 0$ | Original equation | $x^2 + 2x - 3 = 0$ |
| $(-3)^2 + 2(-3) - 3 \stackrel{?}{=} 0$ | Substitute. | $1^2 + 2(1) - 3 \stackrel{?}{=} 0$ |
| $0 = 0$ ✓ | Simplify. | $0 = 0$ ✓ |

### EXAMPLE 2 — Solving a Quadratic Equation: One Real Solution

Solve $x^2 - 8x = -16$ by graphing.

**Step 1:** Rewrite the equation in standard form.

$x^2 - 8x = -16$   Write the equation.

$x^2 - 8x + 16 = 0$   Add 16 to each side.

**Step 2:** Graph the related function $y = x^2 - 8x + 16$.

**Step 3:** Find the $x$-intercept. The only $x$-intercept is at the vertex $(4, 0)$.

∴ So, the solution is $x = 4$.

**Study Tip**
You can also solve the equation in Example 2 by factoring.
$x^2 - 8x + 16 = 0$
$(x - 4)(x - 4) = 0$
So, $x = 4$.

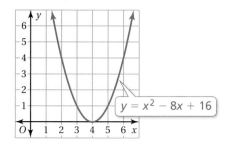

### On Your Own

**Now You're Ready**
Exercises 8–10

Solve the equation by graphing. Check your solution(s).

1. $x^2 - x - 2 = 0$
2. $x^2 + 7x + 10 = 0$
3. $x^2 + x = 12$
4. $x^2 + 1 = 2x$
5. $x^2 + 4x = 0$
6. $x^2 + 10x = -25$

**EXAMPLE 3** **Solving a Quadratic Equation: No Real Solutions**

Solve $-x^2 = 4x + 5$ by graphing.

**Method 1:** Rewrite the equation in standard form and graph the related function $y = x^2 + 4x + 5$.

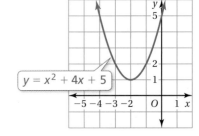

∴ There are no $x$-intercepts. So, $-x^2 = 4x + 5$ has no real solutions.

**Method 2:** Graph each side of the equation.

$y = -x^2$      Left side
$y = 4x + 5$      Right side

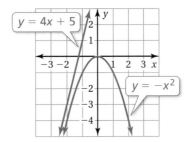

∴ The graphs do not intersect. So, $-x^2 = 4x + 5$ has no real solutions.

### On Your Own

Now You're Ready
Exercises 11–16

**Solve the equation by graphing. Check your solution(s).**

**7.** $x^2 = 3x - 3$      **8.** $x^2 + 7x = -6$      **9.** $2x + 5 = -x^2$

### GO Summary

Quadratic equations may have two real solutions, one real solution, or no real solutions.

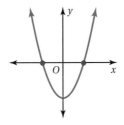
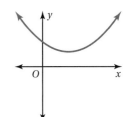

- two real solutions
- two $x$-intercepts

- one real solution
- one $x$-intercept

- no real solutions
- no $x$-intercepts

Section 9.1    Solving Quadratic Equations by Graphing    457

## EXAMPLE 4  Real-Life Application

A football player kicks a football 2 feet above the ground with an upward velocity of 75 feet per second. The function $h = -16t^2 + 75t + 2$ gives the height $h$ (in feet) of the football after $t$ seconds. After how many seconds is the football 50 feet above the ground?

To determine when the football is 50 feet above the ground, find the $t$-values for which $h = 50$. So, solve the equation $-16t^2 + 75t + 2 = 50$.

**Step 1:** Rewrite the equation in standard form.

$$-16t^2 + 75t + 2 = 50 \quad \text{Write the equation.}$$
$$-16t^2 + 75t - 48 = 0 \quad \text{Subtract 50 from each side.}$$

**Step 2:** Use a graphing calculator to graph the related function $h = -16t^2 + 75t - 48$.

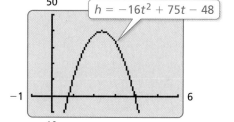

**Step 3:** Use the *zero* feature to find the zeros of the function.

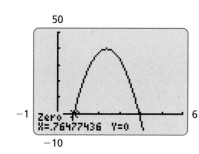
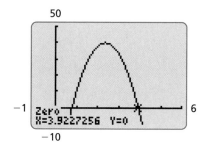

∴ The football is 50 feet above the ground after about 0.8 second and about 3.9 seconds.

### Remember
A zero of a function $y = f(x)$ is an $x$-value for which the value of the function is zero.

### On Your Own

Now You're Ready
Exercise 18

**10. WHAT IF?** After how many seconds is the football 65 feet above the ground?

---

## Summary

- The *solutions*, or *roots*, of $x^2 - 6x + 5 = 0$ are $x = 1$ and $x = 5$.

- The *x-intercepts* of the graph of $y = x^2 - 6x + 5$ are 1 and 5.

- The *zeros* of the function $f(x) = x^2 - 6x + 5$ are 1 and 5.

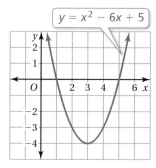

# 9.1 Exercises

## Vocabulary and Concept Check

1. **VOCABULARY** What is a quadratic equation?

2. **WHICH ONE DOESN'T BELONG?** Which equation does *not* belong with the other three? Explain your reasoning.

   $x^2 + 5x = 20$     $x^2 + x - 4 = 0$     $x^2 - 6 = 4x$     $7x + 12 = x^2$

3. **WRITING** How can you use a graph to find the number of solutions of a quadratic equation?

4. **WRITING** How are solutions, roots, $x$-intercepts, and zeros related?

## Practice and Problem Solving

**Determine the solution(s) of the equation. Check your solution(s).**

5. $x^2 - 10x + 24 = 0$

6. $-x^2 - 4x - 6 = 0$

7. $x^2 + 12x + 36 = 0$

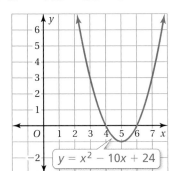 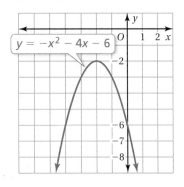 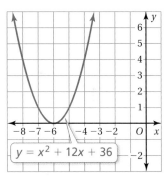

**Solve the equation by graphing. Check your solution(s).**

8. $x^2 - 4x = 0$

9. $x^2 - 6x + 9 = 0$

10. $x^2 - 6x - 7 = 0$

11. $x^2 - 2x + 5 = 0$

12. $x^2 + x - 2 = 0$

13. $x^2 + 4x + 4 = 0$

14. $-x^2 - 2x + 15 = 0$

15. $-x^2 + 14x - 49 = 0$

16. $-x^2 + 4x - 7 = 0$

17. **FLOP SHOT** The height $y$ (in yards) of a flop shot in golf can be modeled by $y = -x^2 + 5x$, where $x$ is the horizontal distance (in yards).

   a. Interpret the $x$-intercepts of the graph of the equation.

   b. How far away does the golf ball land?

18. **VOLLEYBALL** The height $h$ (in feet) of an underhand volleyball serve can be modeled by $h = -16t^2 + 30t + 4$, where $t$ is the time in seconds. After how many seconds is the ball 16 feet above the ground?

Section 9.1    Solving Quadratic Equations by Graphing    459

**Rewrite the equation in standard form. Then solve the equation by graphing. Check your solution(s) with a graphing calculator.**

19. $x^2 = 6x - 8$
20. $x^2 = -1 - 2x$
21. $x^2 = -x - 3$
22. $x^2 = 2x - 4$
23. $5x - 6 = x^2$
24. $3x - 18 = -x^2$

**Solve the equation by using Method 2 from Example 3. Check your solution(s).**

25. $x^2 = 10 - 3x$
26. $4 - 4x = -x^2$
27. $5x - 7 = x^2$
28. $x^2 = 6x - 10$
29. $x^2 = -2x - 1$
30. $x^2 - 8x = 9$

31. **REASONING** Example 3 shows two methods for solving a quadratic equation. Which method do you prefer? Explain your reasoning.

32. **ERROR ANALYSIS** Describe and correct the error in solving the equation.

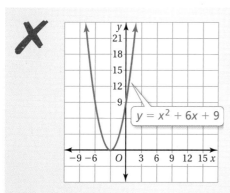

The only solution of the equation $x^2 + 6x + 9 = 0$ is $x = 9$.

33. **BASEBALL** A baseball player throws a baseball with an upward velocity of 24 feet per second. The release point is 6 feet above the ground. The function $h = -16t^2 + 24t + 6$ gives the height $h$ (in feet) of the baseball after $t$ seconds.

   a. How long is the ball in the air if no one catches it?

   b. How long does the ball remain above 6 feet?

34. **SOFTBALL** You throw a softball straight up into the air with an upward velocity of 40 feet per second. The release point is 5 feet above the ground. The function $h = -16t^2 + 40t + 5$ gives the height $h$ (in feet) of the softball after $t$ seconds.

   a. How long is the ball in the air if you miss it?

   b. How long is the ball in the air if you catch it at a height of 5 feet?

Use a graphing calculator to approximate the zeros of the function to the nearest tenth.

**35.** $f(x) = x^2 + 6x + 1$  **36.** $f(x) = x^2 - 3x - 2$  **37.** $f(x) = x^2 + 5x - 4$

**38.** $f(x) = -x^2 - 2x + 5$  **39.** $f(x) = -x^2 + 4x - 2$  **40.** $f(x) = -x^2 + 9x - 6$

**41. MODELING** A dirt bike launches off a ramp that is 8 feet tall. The upward velocity of the dirt bike is 20 feet per second.

  **a.** Write a function that models the height $h$ (in feet) of the dirt bike after $t$ seconds.

  **b.** After how many seconds does the dirt bike land?

**42. WORLD'S STRONGEST MAN** One of the events in the World's Strongest Man competition is the keg toss. In this event, competitors try to throw kegs of various weights over a wall that is 16 feet 6 inches high.

  **a.** A competitor releases a keg 5 feet above the ground with an upward velocity of 27 feet per second. Is this throw high enough to clear the wall? Explain your reasoning.

  **b.** Do the heights of the competitors factor into their success at this event? Explain your reasoning.

**Reasoning** Determine whether the statement is *sometimes*, *always*, or *never* true. Justify your answer.

**43.** The graph of $y = ax^2 + c$ has two $x$-intercepts when $a = -2$.

**44.** The graph of $y = ax^2 + c$ has no $x$-intercepts when $a$ and $c$ have the same sign.

**45.** The graph of $y = ax^2 + bx + c$ has more than two zeros when $a \neq 0$.

 **Fair Game Review** *What you learned in previous grades & lessons*

**Simplify the expression.** *(Section 6.1)*

**46.** $4\sqrt{36}$  **47.** $9\sqrt{81}$  **48.** $2\sqrt{27}$  **49.** $5\sqrt{50}$

**50. MULTIPLE CHOICE** Which expression is equivalent to $\left(\dfrac{2x^3}{3m^5}\right)^2$? *(Section 6.2)*

  **Ⓐ** $\dfrac{2x^5}{3m^7}$  **Ⓑ** $\dfrac{2x^6}{3m^{10}}$  **Ⓒ** $\dfrac{4x^5}{9m^7}$  **Ⓓ** $\dfrac{4x^6}{9m^{10}}$

# 9.2 Solving Quadratic Equations Using Square Roots

**Essential Question** How can you determine the number of solutions of a quadratic equation of the form $ax^2 + c = 0$?

### 1 ACTIVITY: The Number of Solutions of $ax^2 + c = 0$

Work with a partner. Solve each equation by graphing. Explain how the number of solutions of

$$ax^2 + c = 0 \quad \text{Quadratic equation}$$

**relates to the graph of**

$$y = ax^2 + c. \quad \text{Quadratic function}$$

**a.** $x^2 - 4 = 0$

**b.** $2x^2 + 5 = 0$

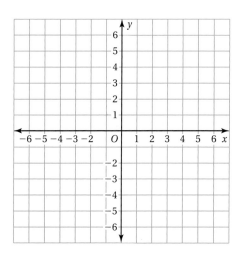
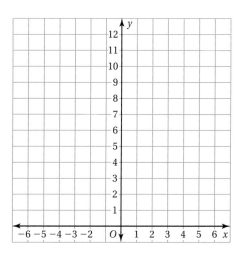

**c.** $x^2 = 0$

**d.** $x^2 - 5 = 0$

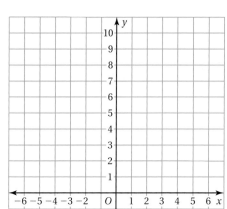
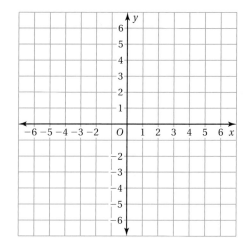

**Common Core**

**Solving Quadratic Equations**

In this lesson, you will
- solve quadratic equations by taking square roots.

Learning Standard
A.REI.4b

462  Chapter 9  Solving Quadratic Equations

## 2  ACTIVITY: Estimating Solutions

Work with a partner. Complete each table. Use the completed tables to estimate the solutions of $x^2 - 5 = 0$. Explain your reasoning.

a.

| $x$ | $x^2 - 5$ |
|---|---|
| 2.21 | |
| 2.22 | |
| 2.23 | |
| 2.24 | |
| 2.25 | |
| 2.26 | |

b.

| $x$ | $x^2 - 5$ |
|---|---|
| −2.21 | |
| −2.22 | |
| −2.23 | |
| −2.24 | |
| −2.25 | |
| −2.26 | |

## 3  ACTIVITY: Using Technology to Estimate Solutions

**Math Practice 5**

**Choose Appropriate Tools**
What different types of technology can be used to answer the questions? Which tool would be the most appropriate and why?

Work with a partner. Two equations are equivalent when they have the same solutions.

a. Are the equations
$$x^2 - 5 = 0 \quad \text{and} \quad x^2 = 5$$
equivalent? Explain your reasoning.

b. Use the square root key on a calculator to estimate the solutions of $x^2 - 5 = 0$. Describe the accuracy of your estimates.

c. Write the *exact* solutions of $x^2 - 5 = 0$.

### What Is Your Answer?

4. **IN YOUR OWN WORDS** How can you determine the number of solutions of a quadratic equation of the form $ax^2 + c = 0$?

5. Write the exact solutions of each equation. Then use a calculator to estimate the solutions.

   a. $x^2 - 2 = 0$     b. $3x^2 - 15 = 0$     c. $x^2 = 8$

**Practice** — Use what you learned about quadratic equations to complete Exercises 3–5 on page 466.

## 9.2 Lesson

In Section 6.1, you studied properties of square roots. Here you will use square roots to solve quadratic equations of the form $ax^2 + c = 0$.

### Key Idea

**Solving Quadratic Equations Using Square Roots**

You can solve $x^2 = d$ by taking the square root of each side.

- When $d > 0$, $x^2 = d$ has two real solutions, $x = \pm\sqrt{d}$.
- When $d = 0$, $x^2 = d$ has one real solution, $x = 0$.
- When $d < 0$, $x^2 = d$ has no real solutions.

**EXAMPLE 1** **Solving Quadratic Equations Using Square Roots**

a. Solve $3x^2 - 27 = 0$ using square roots.

$3x^2 - 27 = 0$     Write the equation.
$3x^2 = 27$     Add 27 to each side.
$x^2 = 9$     Divide each side by 3.
$x = \pm\sqrt{9}$     Take the square root of each side.
$x = \pm 3$     Simplify.

∴ The solutions are $x = 3$ and $x = -3$.

b. Solve $x^2 - 10 = -10$ using square roots.

$x^2 - 10 = -10$     Write the equation.
$x^2 = 0$     Add 10 to each side.
$x = 0$     Take the square root of each side.

∴ The only solution is $x = 0$.

c. Solve $-5x^2 + 11 = 16$ using square roots.

$-5x^2 + 11 = 16$     Write the equation.
$-5x^2 = 5$     Subtract 11 from each side.
$x^2 = -1$     Divide each side by $-5$.

∴ The equation has no real solutions.

**Remember**

The square of a real number cannot be negative. That is why the equation in part (c) has no real solutions.

### On Your Own

**Now You're Ready** Exercises 12–20

Solve the equation using square roots.

1. $-3x^2 = -75$
2. $x^2 + 12 = 10$
3. $4x^2 - 15 = -15$

### EXAMPLE 2  Solving a Quadratic Equation Using Square Roots

Solve $(x - 1)^2 = 25$ using square roots.

$(x - 1)^2 = 25$   Write the equation.
$x - 1 = \pm 5$   Take the square root of each side.
$x = 1 \pm 5$   Add 1 to each side.

So, the solutions are $x = 1 + 5 = 6$ and $x = 1 - 5 = -4$.

**Check**

Use a graphing calculator to check your answer. Rewrite the equation as $(x - 1)^2 - 25 = 0$. Graph the related function $y = (x - 1)^2 - 25$ and find the x-intercepts, or zeros. The zeros are $-4$ and $6$, so the solution checks.

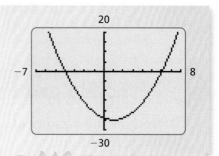

### EXAMPLE 3  Real-Life Application

A touch tank has a height of 3 feet. Its length is 3 times its width. The volume of the tank is 270 cubic feet. Find the length and width of the tank.

3 feet

The length $\ell$ is 3 times the width $w$, so $\ell = 3w$. Write an equation using the formula for the volume of a rectangular prism.

$V = \ell w h$   Write the formula.
$270 = 3w(w)(3)$   Substitute 270 for $V$, $3w$ for $\ell$, and 3 for $h$.
$270 = 9w^2$   Multiply.
$30 = w^2$   Divide each side by 9.
$\pm\sqrt{30} = w$   Take the square root of each side.

**Study Tip**

Use the positive square root because negative solutions do not make sense in this context. Length and width cannot be negative.

The solutions are $\sqrt{30}$ and $-\sqrt{30}$. Use the positive solution.

So, the width is $\sqrt{30} \approx 5.5$ feet and the length is $3\sqrt{30} \approx 16.4$ feet.

#### On Your Own

Exercises 23–28 and 32

**Solve the equation using square roots.**

4. $(x + 7)^2 = 0$
5. $4(x - 3)^2 = 9$
6. $(2x + 1)^2 = 35$

7. **WHAT IF?** In Example 3, the volume of the tank is 315 cubic feet. Find the length and width of the tank.

Section 9.2  Solving Quadratic Equations Using Square Roots  465

## 9.2 Exercises

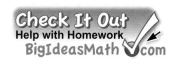

### Vocabulary and Concept Check

1. **REASONING** How many real solutions does the equation $x^2 = d$ have when $d$ is positive? 0? negative?

2. **WHICH ONE DOESN'T BELONG?** Which equation does *not* belong with the other three? Explain your reasoning.

   $x^2 = 9$   $x^2 = 2$   $x^2 = -7$   $x^2 = 21$

### Practice and Problem Solving

**Determine the number of solutions of the equation. Then use a calculator to estimate the solutions.**

3. $x^2 - 11 = 0$
4. $x^2 + 10 = 0$
5. $2x^2 - 3 = 0$

**Determine the number of solutions of the equation. Then solve the equation using square roots.**

6. $x^2 = 25$
7. $x^2 = -36$
8. $x^2 = 8$
9. $x^2 = 21$
10. $x^2 = 0$
11. $x^2 = 169$

**Solve the equation using square roots.**

12. $x^2 - 16 = 0$
13. $x^2 + 12 = 0$
14. $x^2 + 6 = 0$
15. $x^2 - 61 = 0$
16. $2x^2 - 98 = 0$
17. $-x^2 + 9 = 9$
18. $x^2 + 13 = 7$
19. $-4x^2 - 5 = -5$
20. $-3x^2 + 8 = 8$

21. **ERROR ANALYSIS** Describe and correct the error in solving the equation.

    ✗ Solve $2x^2 - 33 = 39$.
    
    $2x^2 = 72$    Add 33 to each side.
    $x^2 = 36$    Divide each side by 2.
    $x = 6$    Take the square root of each side.
    
    ∴ The solution is $x = 6$.

22. **WAREHOUSE** A box falls off a warehouse shelf from a height of 16 feet. The function $h = -16x^2 + 16$ gives the height $h$ (in feet) of the box after $x$ seconds. When does it hit the floor?

**Solve the equation using square roots. Use a graphing calculator to check your solution(s).**

② 23. $(x + 3)^2 = 0$   24. $(x - 1)^2 = 4$   25. $(2x - 1)^2 = 81$

26. $(4x - 5)^2 = 9$   27. $9(x + 1)^2 = 16$   28. $4(x - 2)^2 = 25$

**Use the given area $A$ to find the dimensions of the figure.**

29. $A = 64$ in.$^2$   30. $A = 78$ cm$^2$   31. $A = 144\pi$ ft$^2$

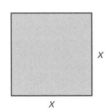  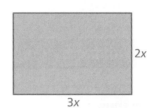  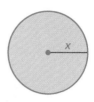

③ 32. **POND** An in-ground pond has the shape of a rectangular prism. The pond has a height of 24 inches and a volume of 33,000 cubic inches. The pond's length is 2 times its width. Find the length and width of the pond.

33. **AREA RUG** The design of a square area rug for your living room is shown. You want the area of the inner square to be 25% of the total area of the rug. Find the side length $x$ of the inner square.

34. **WRITING** How can you approximate the roots of a quadratic equation when the roots are not integers?

35. **LOGIC** Given the equation $ax^2 + c = 0$, describe the values of $a$ and $c$ so the equation has the following number of solutions.

   a. two solutions   b. one solution   c. no solutions

36. **Reasoning** Without graphing, where do the graphs of $y = x^2$ and $y = 9$ intersect? Explain.

## Fair Game Review  *What you learned in previous grades & lessons*

**Find the product.** *(Section 7.4)*

37. $(x + 5)^2$   38. $(w - 7)^2$   39. $(2y - 3)^2$

40. **MULTIPLE CHOICE** What is an explicit equation for $a_1 = -3$, $a_n = a_{n-1} + 2$? *(Section 6.7)*

   Ⓐ $a_n = 2n - 3$   Ⓑ $a_n = 2n - 5$   Ⓒ $a_n = n + 2$   Ⓓ $a_n = -3n + 2$

# 9.3 Solving Quadratic Equations by Completing the Square

**Essential Question** How can you use "completing the square" to solve a quadratic equation?

### 1 EXAMPLE: Solving by Completing the Square

Work with a partner. Five different algebra tiles are shown at the right.

Solve $x^2 + 4x = -2$ by completing the square.

**Step 1:** Use algebra tiles to model the equation
$$x^2 + 4x = -2.$$

**Step 2:** Add four yellow tiles to the left side of the equation so that it is a perfect square. Balance the equation by also adding four yellow tiles to the right side.
$$x^2 + 4x + 4 = -2 + 4$$
$$(x + 2)^2 = 2$$

Add 4 tiles to each side.

**Step 3:** Take the square root of each side of the equation and simplify.
$$x + 2 = \pm\sqrt{2}$$
$$x = -2 \pm \sqrt{2}$$

**Check** Check each solution in the original equation.

$$x^2 + 4x = -2$$
$$(-2 + \sqrt{2})^2 + 4(-2 + \sqrt{2}) \stackrel{?}{=} -2$$
$$4 - 4\sqrt{2} + 2 - 8 + 4\sqrt{2} \stackrel{?}{=} -2$$
$$4 + 2 - 8 \stackrel{?}{=} -2$$
$$-2 = -2 \checkmark$$

Now you check the other solution.

**COMMON CORE**

**Solving Quadratic Equations**

In this lesson, you will
- solve quadratic equations by completing the square.
- solve real-life problems.

Learning Standards
A.REI.4a
A.REI.4b
A.SSE.3b
F.IF.8a

**468** Chapter 9 Solving Quadratic Equations

## 2 ACTIVITY: Solving by Completing the Square

**Work with a partner.**

- Write the equation modeled by the algebra tiles.
- Use algebra tiles to complete the square.
- Write the solutions of the equation.
- Check each solution.

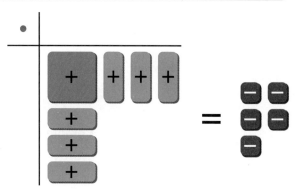

## 3 ACTIVITY: Writing a Rule

**Math Practice**

**State the Meaning of Symbols**
Which algebra tiles do you need to add to complete the square? How can you represent the tiles in the equation?

**Work with a partner.**

- What does this group of tiles represent?
- How is the coefficient of $x$ for this group of tiles related to the coefficient of $x$ in the equation from Activity 2? How is it related to the number of tiles you add to each side when completing the square?

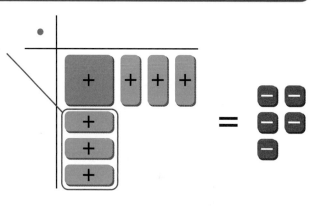

- **WRITE A RULE** Fill in the blanks.

  To complete the square, take _____ of the coefficient of the $x$-term and _____ it. _____ this number to each side of the equation.

## What Is Your Answer?

4. **IN YOUR OWN WORDS** How can you use "completing the square" to solve a quadratic equation?

5. Solve each quadratic equation by completing the square.

   **a.** $x^2 - 2x = 1$   **b.** $x^2 - 4x = -1$   **c.** $x^2 + 4x = -3$

**Practice** — Use what you learned about quadratic equations to complete Exercises 3–5 on page 472.

Section 9.3  Solving Quadratic Equations by Completing the Square

## 9.3 Lesson

**Key Vocabulary**
completing the square, p. 470

Another method for solving quadratic equations is **completing the square**. In this method, a constant $c$ is added to the expression $x^2 + bx$ so that $x^2 + bx + c$ is a perfect square trinomial.

### Key Idea

**Completing the Square**

**Words** To complete the square for an expression of the form $x^2 + bx$, follow these steps.

**Step 1:** Find one-half of $b$, the coefficient of $x$.
**Step 2:** Square the result from Step 1.
**Step 3:** Add the result from Step 2 to $x^2 + bx$.

Factor the resulting expression as the square of a binomial.

**Algebra** $x^2 + bx + \left(\dfrac{b}{2}\right)^2 = \left(x + \dfrac{b}{2}\right)^2$

### EXAMPLE 1 — Completing the Square

**Complete the square for each expression. Then factor the trinomial.**

a. $x^2 + 6x$

**Step 1:** Find one-half of $b$.  $\dfrac{b}{2} = \dfrac{6}{2} = 3$

**Step 2:** Square the result from Step 1.  $3^2 = 9$

**Step 3:** Add the result from Step 2 to $x^2 + bx$.  $x^2 + 6x + 9$

∴ $x^2 + 6x + 9 = (x + 3)^2$

b. $x^2 - 9x$

**Step 1:** Find one-half of $b$.  $\dfrac{b}{2} = \dfrac{-9}{2}$

**Step 2:** Square the result from Step 1.  $\left(\dfrac{-9}{2}\right)^2 = \dfrac{81}{4}$

**Step 3:** Add the result from Step 2 to $x^2 + bx$.  $x^2 - 9x + \dfrac{81}{4}$

∴ $x^2 - 9x + \dfrac{81}{4} = \left(x - \dfrac{9}{2}\right)^2$

### On Your Own

**Now You're Ready**
Exercises 12–17

Complete the square for each expression. Then factor the trinomial.

1. $x^2 + 10x$
2. $x^2 - 4x$
3. $x^2 + 7x$

To solve a quadratic equation by completing the square, write the equation in the form $x^2 + bx = d$.

### EXAMPLE 2  Solving a Quadratic Equation by Completing the Square

**Solve $x^2 + 8x - 3 = 17$ by completing the square.**

| | |
|---|---|
| $x^2 + 8x - 3 = 17$ | Write the equation. |
| $x^2 + 8x = 20$ | Add 3 to each side. |
| Complete the square. → $x^2 + 8x + 16 = 20 + 16$ | Add $\left(\frac{8}{2}\right)^2$, or 16, to each side. |
| $(x + 4)^2 = 36$ | Factor $x^2 + 8x + 16$. |
| $x + 4 = \pm 6$ | Take the square root of each side. |
| $x = -4 \pm 6$ | Subtract 4 from each side. |

**Common Error**
When completing the square, be sure to add to *both* sides of the equation.

∴ The solutions are $x = -4 + 6 = 2$ and $x = -4 - 6 = -10$.

### EXAMPLE 3  Real-Life Application

**You throw a stone from a height of 16 feet with an upward velocity of 32 feet per second. The function $h = -16t^2 + 32t + 16$ gives the height $h$ of the stone after $t$ seconds. When does the stone land in the water?**

Find the $t$-values for which $h = 0$. So, solve $-16t^2 + 32t + 16 = 0$.

| | |
|---|---|
| $-16t^2 + 32t + 16 = 0$ | Write the equation. |
| $t^2 - 2t - 1 = 0$ | Divide each side by $-16$. |
| $t^2 - 2t = 1$ | Add 1 to each side. |
| Complete the square. → $t^2 - 2t + 1 = 1 + 1$ | Add $\left(\frac{-2}{2}\right)^2$, or 1, to each side. |
| $(t - 1)^2 = 2$ | Factor $t^2 - 2t + 1$. |
| $t - 1 = \pm\sqrt{2}$ | Take the square root of each side. |
| $t = 1 \pm \sqrt{2}$ | Add 1 to each side. |

The solutions are $x = 1 + \sqrt{2} \approx 2.4$ and $x = 1 - \sqrt{2} \approx -0.4$. Use the positive solution.

**Study Tip**
Before completing the square, make sure the leading coefficient is 1.

∴ The stone lands in the water after about 2.4 seconds.

### On Your Own

**Now You're Ready**
Exercises 18–23

**Solve the equation by completing the square.**

4. $x^2 - 6x = 27$   5. $x^2 + 12x + 3 = 1$   6. $2x^2 + 4x + 10 = 58$

7. **WHAT IF?** In Example 3, the function $h = -16t^2 + 64t + 16$ gives the height $h$ (in feet) of the stone after $t$ seconds. When does the stone land in the water?

## 9.3 Exercises

### Vocabulary and Concept Check

1. **VOCABULARY** Explain how to complete the square for an expression of the form $x^2 + bx$.

2. **WRITING** For what values of $b$ is it easier to complete the square for $x^2 + bx$? Explain.

### Practice and Problem Solving

Use algebra tiles to complete the square. Then write the perfect square trinomial.

3.    4. 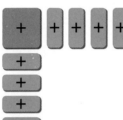   5.

Find the value of $c$ that completes the square.

6. $x^2 - 8x + c$
7. $x^2 + 4x + c$
8. $x^2 - 2x + c$
9. $x^2 - 14x + c$
10. $x^2 + 12x + c$
11. $x^2 + 18x + c$

Complete the square for the expression. Then factor the trinomial.

12. $x^2 - 10x$
13. $x^2 + 16x$
14. $x^2 + 22x$
15. $x^2 - 40x$
16. $x^2 - 3x$
17. $x^2 + 5x$

Solve the equation by completing the square.

18. $x^2 + 2x = 3$
19. $x^2 - 6x = 16$
20. $x^2 + 4x + 7 = -6$
21. $x^2 + 5x - 7 = -14$
22. $2x^2 - 8x = 10$
23. $2x^2 - 3x + 1 = 0$

24. **ERROR ANALYSIS** Describe and correct the error in solving the equation.

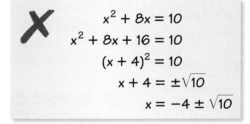

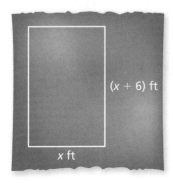

25. **PATIO** The area of the new patio is 216 square feet.
   a. Write an equation for the area of the patio.
   b. Find the dimensions of the patio by completing the square.

472   Chapter 9   Solving Quadratic Equations

26. **NUMBER SENSE** Find the value of $b$ that makes $x^2 + bx + 25$ a perfect square trinomial.

27. **REASONING** You are completing the square to solve $3x^2 + 6x = 12$. What is the first step?

28. **REASONING** Consider the equation $x^2 + 4x - 12 = 0$.

    a. Solve the equation by completing the square.

    b. Explain how to use the solutions to find the minimum value of $y = x^2 + 4x - 12$.

29. **TOY ROCKET** The function $h = -16t^2 + 64t + 32$ gives the height $h$ (in feet) of a toy rocket after $t$ seconds.

    a. When does the rocket hit the ground?

    b. What is the maximum height of the rocket? Justify your answer.

30. **ROSE GARDEN** You plant a rectangular rose garden along the side of your garage. You enclose 3 sides of the garden with 40 feet of fencing. The total area of the garden is 100 square feet. Find the possible dimensions of the garden. Round to the nearest tenth. Which size garden would you choose?

31. **PRECISION** The product of two consecutive positive integers is 42. Write and solve an equation to find the integers.

32. **Structure** Begin solving $x^2 + 4x + 3 = 0$ by completing the square. Stop when you obtain an equation of the form $(x + p)^2 = q$.

    a. Write the related function in vertex form. Without graphing, determine the maximum or minimum value of the function.

    b. Find the minimum value of $y = x^2 + bx + c$.

## Fair Game Review  *What you learned in previous grades & lessons*

**Simplify $\sqrt{b^2 - 4ac}$ for the given values.** *(Section 6.1)*

33. $a = 3, b = -6, c = 2$      34. $a = -2, b = 4, c = 7$      35. $a = 1, b = 6, c = 4$

36. **MULTIPLE CHOICE** What are the solutions of $x^2 - 49 = 0$? *(Section 9.2)*

    Ⓐ $x = 7$      Ⓑ $x = -7, x = 7$      Ⓒ $x = 0, x = 7$      Ⓓ no solution

# 9 Study Help

You can use an **information wheel** to organize information about a topic. Here is an example of an information wheel for quadratic equations.

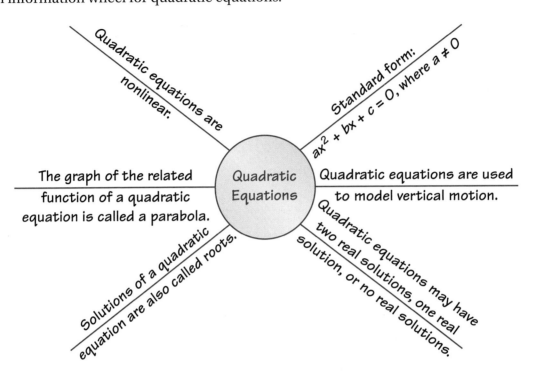

## On Your Own

**Make information wheels to help you study these topics.**

1. solving quadratic equations by graphing
2. solving quadratic equations using square roots
3. solving quadratic equations by completing the square

**After you complete this chapter, make information wheels for the following topics.**

4. solving quadratic equations using the quadratic formula
5. choosing a solution method for solving quadratic equations
6. solving systems of linear and quadratic equations

"My information wheel for Fluffy has matching adjectives and nouns."

## 9.1–9.3 Quiz

**Determine the solution(s) of the equation. Check your solution(s).** *(Section 9.1)*

1. $x^2 - 2x - 3 = 0$
2. $x^2 - 2x + 3 = 0$
3. $x^2 + 10x + 25 = 0$

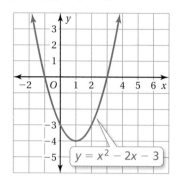 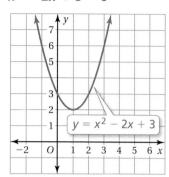 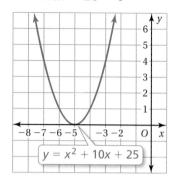

**Solve the equation by graphing. Check your solution(s).** *(Section 9.1)*

4. $x^2 + 9x + 14 = 0$
5. $x^2 - 7x = 8$
6. $x + 1 = -x^2$

**Solve the equation using square roots.** *(Section 9.2)*

7. $4x^2 = 64$
8. $-3x^2 + 6 = 10$
9. $(x - 8)^2 = 1$

**Solve the equation by completing the square.** *(Section 9.3)*

10. $x^2 + 4x = 45$
11. $x^2 - 2x - 1 = 8$
12. $2x^2 + 12x + 20 = 34$
13. $-4x^2 + 8x + 44 = 16$

14. **REASONING** Explain how to determine the number of real solutions of $x^2 = 100$ without solving. *(Section 9.2)*

15. **VOLUME** The length of a rectangular prism is 4 times its width. The volume of the prism is 380 cubic meters. Find the length and width of the prism. *(Section 9.2)*

16. **PROBLEM SOLVING** A cannon launches a cannonball from a height of 3 feet with an upward velocity of 40 feet per second. The function $h = -16t^2 + 40t + 3$ gives the height $h$ (in feet) of the cannonball after $t$ seconds. *(Section 9.1 and Section 9.3)*

   a. After how many seconds is the cannonball 10 feet above the ground?

   b. What is the maximum height of the cannonball?

# 9.4 Solving Quadratic Equations Using the Quadratic Formula

**Essential Question** How can you use the discriminant to determine the number of solutions of a quadratic equation?

### 1 ACTIVITY: Deriving the Quadratic Formula

Work with a partner. The following steps show one method of solving $ax^2 + bx + c = 0$. Explain what was done in each step.

$$ax^2 + bx + c = 0 \quad \leftarrow \text{1. Write the equation.}$$

$$4a^2x^2 + 4abx + 4ac = 0 \quad \leftarrow \text{2. What was done?}$$

$$4a^2x^2 + 4abx + 4ac + b^2 = b^2 \quad \leftarrow \text{3. What was done?}$$

$$4a^2x^2 + 4abx + b^2 = b^2 - 4ac \quad \leftarrow \text{4. What was done?}$$

$$(2ax + b)^2 = b^2 - 4ac \quad \leftarrow \text{5. What was done?}$$

$$2ax + b = \pm\sqrt{b^2 - 4ac} \quad \leftarrow \text{6. What was done?}$$

$$2ax = -b \pm \sqrt{b^2 - 4ac} \quad \leftarrow \text{7. What was done?}$$

**Quadratic Formula:** $x = \dfrac{-b \pm \sqrt{b^2 - 4ac}}{2a} \quad \leftarrow \text{8. What was done?}$

**Common Core**

**Solving Quadratic Equations**
In this lesson, you will
- solve quadratic equations by the quadratic formula.
- use discriminants to determine the number of real solutions of quadratic equations.

Learning Standards
A.REI.4a
A.REI.4b

### 2 ACTIVITY: Deriving the Quadratic Formula by Completing the Square

- Solve $ax^2 + bx + c = 0$ by completing the square. (*Hint:* Subtract $c$ from each side, divide each side by $a$, and then proceed by completing the square.)

- Compare this method with the method in Activity 1. Explain why you think $4a$ and $b^2$ were chosen in Steps 2 and 3 of Activity 1.

476   Chapter 9   Solving Quadratic Equations

## 3 ACTIVITY: Analyzing the Solutions of an Equation

**Math Practice**

**Explain the Meaning**
What does it mean for an equation to have a solution? How does this compare to the graph of the equation?

Work with a partner. In the quadratic formula in Activity 1, the expression under the radical sign, $b^2 - 4ac$, is called the **discriminant**. For each graph, decide whether the corresponding discriminant is equal to 0, is greater than 0, or is less than 0. Explain your reasoning.

**a.** 1 rational solution

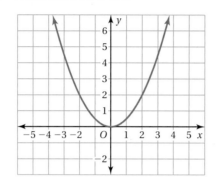

**b.** 2 rational solutions

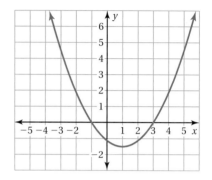

**c.** 2 irrational solutions

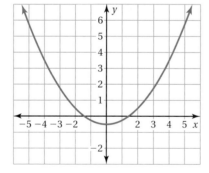

**d.** no real solutions

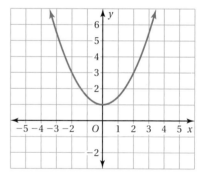

## What Is Your Answer?

**4. IN YOUR OWN WORDS** How can you use the discriminant to determine the number of solutions of a quadratic equation?

**5.** Use the quadratic formula to solve each quadratic equation.

   **a.** $x^2 + 2x - 3 = 0$    **b.** $x^2 - 4x + 4 = 0$    **c.** $x^2 + 4x + 5 = 0$

**6.** Use the Internet to research *imaginary numbers*. How are they related to quadratic equations?

**Practice**

Use what you learned about quadratic equations to complete Exercises 9–11 on page 481.

Section 9.4   Solving Quadratic Equations Using the Quadratic Formula

## 9.4 Lesson

**Key Vocabulary**
quadratic formula, *p. 478*
discriminant, *p. 480*

Another way to solve quadratic equations is to use the *quadratic formula*.

### Key Idea

**Quadratic Formula**

The real solutions of the quadratic equation $ax^2 + bx + c = 0$ are

$$x = \frac{-b \pm \sqrt{b^2 - 4ac}}{2a}$$

where $a \neq 0$ and $b^2 - 4ac \geq 0$. This is called the **quadratic formula.**

### EXAMPLE 1  Solving a Quadratic Equation Using the Quadratic Formula

Solve $2x^2 - 5x + 3 = 0$ using the quadratic formula.

$$x = \frac{-b \pm \sqrt{b^2 - 4ac}}{2a} \quad \text{Quadratic Formula}$$

$$= \frac{-(-5) \pm \sqrt{(-5)^2 - 4(2)(3)}}{2(2)} \quad \text{Substitute 2 for } a, -5 \text{ for } b, \text{ and } 3 \text{ for } c.$$

$$= \frac{5 \pm \sqrt{1}}{4} \quad \text{Simplify.}$$

$$= \frac{5 \pm 1}{4} \quad \text{Evaluate the square root.}$$

**Study Tip**
You can use the roots of a quadratic equation to factor the related expression. In Example 1, you can use 1 and $\frac{3}{2}$ to factor $2x^2 - 5x + 3$ as $(x - 1)(2x - 3)$.

∴ So, the solutions are $x = \frac{5 + 1}{4} = \frac{3}{2}$ and $x = \frac{5 - 1}{4} = 1$.

**Check** Check each solution in the original equation.

$$2x^2 - 5x + 3 = 0 \quad \text{Original equation} \quad 2x^2 - 5x + 3 = 0$$

$$2\left(\frac{3}{2}\right)^2 - 5\left(\frac{3}{2}\right) + 3 \stackrel{?}{=} 0 \quad \text{Substitute.} \quad 2(1)^2 - 5(1) + 3 \stackrel{?}{=} 0$$

$$\frac{9}{2} - \frac{15}{2} + 3 \stackrel{?}{=} 0 \quad \text{Simplify.} \quad 2 - 5 + 3 \stackrel{?}{=} 0$$

$$0 = 0 \checkmark \quad \text{Simplify.} \quad 0 = 0 \checkmark$$

### On Your Own

Now You're Ready
Exercises 12–14

Solve the equation using the quadratic formula.

1. $x^2 - 6x + 5 = 0$
2. $4x^2 + x - 3 = 0$
3. $-6x^2 + 7x - 2 = 0$

### EXAMPLE 2 Solving a Quadratic Equation Using the Quadratic Formula

**Solve $4x^2 + 20x + 25 = 0$ using the quadratic formula.**

$$x = \frac{-b \pm \sqrt{b^2 - 4ac}}{2a}$$ Quadratic Formula

$$= \frac{-20 \pm \sqrt{20^2 - 4(4)(25)}}{2(4)}$$ Substitute 4 for $a$, 20 for $b$, and 25 for $c$.

$$= \frac{-20 \pm \sqrt{0}}{8} = -\frac{5}{2}$$ Simplify.

∴ The solution is $x = -\dfrac{5}{2}$.

### EXAMPLE 3 Real-Life Application

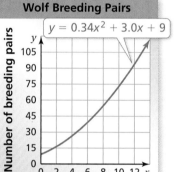

**Wolf Breeding Pairs**
$y = 0.34x^2 + 3.0x + 9$
Number of breeding pairs
Years since 1995

The number $y$ of Northern Rocky Mountain wolf breeding pairs $x$ years since 1995 can be modeled by $y = 0.34x^2 + 3.0x + 9$. When were there about 30 breeding pairs?

To determine when there were 30 breeding pairs, find the $x$-values for which $y = 30$. So, solve the equation $30 = 0.34x^2 + 3.0x + 9$.

$30 = 0.34x^2 + 3.0x + 9$  Write the equation.

$0 = 0.34x^2 + 3.0x - 21$  Write in standard form.

$$x = \frac{-b \pm \sqrt{b^2 - 4ac}}{2a}$$ Quadratic Formula

$$= \frac{-3.0 \pm \sqrt{3.0^2 - 4(0.34)(-21)}}{2(0.34)}$$ Substitute 0.34 for $a$, 3.0 for $b$, and $-21$ for $c$.

$$= \frac{-3.0 \pm \sqrt{37.56}}{0.68}$$ Simplify.

The solutions are $x = \dfrac{-3.0 + \sqrt{37.56}}{0.68} \approx 5$ and $x = \dfrac{-3.0 - \sqrt{37.56}}{0.68} \approx -13$.

∴ Because $x$ represents the number of years since 1995, $x$ is greater than or equal to zero. So, there were about 30 breeding pairs 5 years after 1995, in 2000.

### On Your Own

Now You're Ready
Exercises 15–23

**Solve the equation using the quadratic formula.**

**4.** $4x^2 - 4x + 1 = 0$  **5.** $-5x^2 + x = -4$  **6.** $3x^2 + 2x = 5$

**7. WHAT IF?** In Example 3, when were there about 85 breeding pairs?

Section 9.4  Solving Quadratic Equations Using the Quadratic Formula  479

The expression $b^2 - 4ac$ in the quadratic formula is the **discriminant**.

$$x = \frac{-b \pm \sqrt{b^2 - 4ac}}{2a} \leftarrow \text{discriminant}$$

You can use the discriminant to determine the number of real solutions of a quadratic equation.

## Key Idea

**Interpreting the Discriminant**

| $b^2 - 4ac > 0$ | $b^2 - 4ac = 0$ | $b^2 - 4ac < 0$ |
|---|---|---|
| 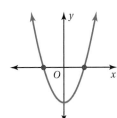 | 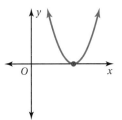 | 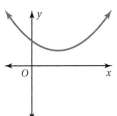 |
| • two real solutions<br>• two $x$-intercepts | • one real solution<br>• one $x$-intercept | • no real solutions<br>• no $x$-intercepts |

**Study Tip**

The solutions of a quadratic equation may be real numbers or *imaginary numbers*. You will study imaginary numbers in a future course.

### EXAMPLE 4 — Determining the Number of Real Solutions

**a.** Determine the number of real solutions of $x^2 + 8x - 3 = 0$.

$b^2 - 4ac = 8^2 - 4(1)(-3)$    Substitute 1 for $a$, 8 for $b$, and $-3$ for $c$.

$= 64 + 12$    Simplify.

$= 76$    Add.

∴ The discriminant is greater than 0, so the equation has two real solutions.

**b.** Determine the number of real solutions of $2x^2 + 7 = 6x$.

Write the equation in standard form: $2x^2 - 6x + 7 = 0$.

$b^2 - 4ac = (-6)^2 - 4(2)(7)$    Substitute 2 for $a$, $-6$ for $b$, and 7 for $c$.

$= 36 - 56$    Simplify.

$= -20$    Subtract.

∴ The discriminant is less than 0, so the equation has no real solutions.

### On Your Own

Now You're Ready
Exercises 27–32

Determine the number of real solutions of the equation.

**8.** $-x^2 + 4x - 4 = 0$    **9.** $6x^2 + 2x = -1$    **10.** $\frac{1}{2}x^2 = 7x - 1$

## 9.4 Exercises

### Vocabulary and Concept Check

1. **VOCABULARY** Write the formula that can be used to solve any quadratic equation.
2. **VOCABULARY** What does the discriminant tell you about the number of solutions of a quadratic equation?

### Practice and Problem Solving

**Write the equation in standard form. Then identify the values of *a*, *b*, and *c* that you would use to solve the equation using the quadratic formula.**

3. $x^2 = 7x$
4. $x^2 - 4x = -12$
5. $-2x^2 + 1 = 5x$
6. $3x + 2 = 4x^2$
7. $4 - 6x = -x^2$
8. $-8x = 3x^2 + 3$

**Solve the equation using the quadratic formula. Round to the nearest tenth, if necessary.**

9. $x^2 - 12x + 36 = 0$
10. $x^2 + 7x + 16 = 0$
11. $x^2 - 10x - 11 = 0$
12. $2x^2 - x - 1 = 0$
13. $2x^2 - 6x + 5 = 0$
14. $9x^2 - 6x + 1 = 0$
15. $6x^2 - 13x = -6$
16. $-3x^2 + 6x = 4$
17. $1 - 8x = -16x^2$
18. $x^2 - 5x + 3 = 0$
19. $x^2 + 2x = 9$
20. $5x^2 - 2 = 4x$

**ERROR ANALYSIS** Describe and correct the error in solving the equation.

21. $3x^2 - 7x - 6 = 0$

$$x = \frac{-7 \pm \sqrt{(-7)^2 - 4(3)(-6)}}{2(3)}$$
$$= \frac{-7 \pm \sqrt{121}}{6}$$
$$x = \frac{2}{3} \text{ and } x = -3$$

22. $-2x^2 + 9x = 4$

$$x = \frac{-9 \pm \sqrt{9^2 - 4(-2)(4)}}{2(-2)}$$
$$= \frac{-9 \pm \sqrt{113}}{-4}$$
$$x \approx -0.41 \text{ and } x \approx 4.91$$

23. **PIER** A swimmer takes a running jump off a pier. The path of the swimmer can be modeled by the equation $h = -0.1d^2 + 0.1d + 3$, where $h$ is the height (in feet) and $d$ is the horizontal distance (in feet). How far from the pier does the swimmer enter the water?

**Match the discriminant with the corresponding graph.**

**24.** $b^2 - 4ac > 0$  **25.** $b^2 - 4ac = 0$  **26.** $b^2 - 4ac < 0$

A. 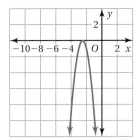  B. 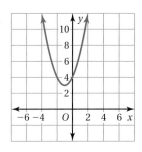  C.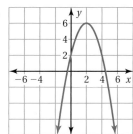

**Use the discriminant to determine the number of real solutions of the equation.**

**④ 27.** $x^2 - 6x + 10 = 0$  **28.** $x^2 - 5x - 3 = 0$  **29.** $2x^2 - 12x = -18$

**30.** $4x^2 = 4x - 1$  **31.** $-\frac{1}{4}x^2 + 4x = -2$  **32.** $-5x^2 + 8x = 9$

**33. REPEATED REASONING** You use the quadratic formula to solve an equation.

  a. You obtain solutions that are integers. Could you have used factoring to solve the equation? Explain your reasoning.

  b. You obtain solutions that are fractions. Could you have used factoring to solve the equation? Explain your reasoning.

  c. Make a generalization about quadratic equations with rational solutions.

**34. STOPPING A CAR** The distance $d$ (in feet) it takes to stop a car traveling $v$ miles per hour can be modeled by $d = 0.05v^2 + 2.2v$. It takes a car 235 feet to stop. How fast was the car going when the brakes were applied?

**35. FISHING** The amount $y$ of trout (in tons) caught in a lake from 1990 to 2009 can be modeled by $y = -0.08x^2 + 1.6x + 10$, where $x$ is the number of years since 1990.

  a. When were about 15 tons of trout caught in the lake?

  b. Do you think this model can be used for future years? Explain your reasoning.

**36. ERROR ANALYSIS** Describe and correct the error in finding the number of solutions of the equation $2x^2 - 5x - 2 = -11$.

$$b^2 - 4ac = (-5)^2 - 4(2)(-2)$$
$$= 25 - (-16)$$
$$= 41$$
The equation has two solutions.

**482** Chapter 9 Solving Quadratic Equations

Use the discriminant to determine how many times the graph of the related function intersects the $x$-axis.

**37.** $x^2 + 5x - 1 = 0$  **38.** $4x^2 + 4x = -1$  **39.** $4 - 3x = -6x^2$

Give a value for $c$ where (a) you can factor to solve the equation and (b) you must use the quadratic formula to solve the equation.

**40.** $x^2 + 3x + c = 0$  **41.** $x^2 - 6x + c = 0$  **42.** $x^2 - 8x + c = 0$

**43. REASONING** How many solutions does $ax^2 + bx + c = 0$ have when $a$ and $c$ have different signs? Explain your reasoning.

**44. REASONING** When the discriminant is a perfect square, are the solutions of $ax^2 + bx + c = 0$ rational or irrational? Assume $a$, $b$, and $c$ are integers. Explain your reasoning.

**45. PROBLEM SOLVING** A rancher constructs two rectangular horse pastures that share a side, as shown. The pastures are enclosed by 1050 feet of fencing. Each pasture has an area of 15,000 square feet.

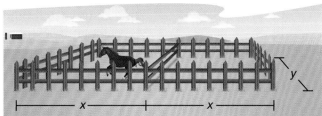

  **a.** Show that $y = 350 - \frac{4}{3}x$.

  **b.** Find the possible lengths and widths of each pasture.

**46. Critical Thinking** You are trying to hang a tire swing. To get the rope over a tree branch that is 15 feet high, you tie the rope to a weight and throw it over the branch. You release the weight at a height of 5.5 feet. What is the minimum upward velocity needed to reach the branch?

## Fair Game Review  *What you learned in previous grades & lessons*

**Solve the system of linear equations.**  *(Section 4.4)*

**47.** $x + y = 0$
      $3x + 2y = 1$

**48.** $2x - 2y = 4$
      $-x + y = -2$

**49.** $2x - 4y = -1$
      $-3x + 6y = -5$

**50. MULTIPLE CHOICE** What is the solution of the equation $7x + 3x = 5x - 10$? *(Section 1.3)*

  **Ⓐ** $x = -2$   **Ⓑ** $x = -\frac{2}{3}$   **Ⓒ** $x = 2$   **Ⓓ** $x = 4$

# Extension 9.4 Choosing a Solution Method

The table shows five methods for solving quadratic equations. While there is no one correct method, some methods may be easier to use than others. Some advantages and disadvantages of each method are shown.

**Solving Quadratic Equations**

In this extension, you will
- choose a method to solve quadratic equations.

Learning Standards
A.REI.4a
A.REI.4b

## Key Ideas

**Methods for Solving Quadratic Equations**

| Method | Advantages | Disadvantages |
|---|---|---|
| Factoring (Lessons 7.6–7.9) | • Straightforward when equation can be factored easily | • Some equations are not factorable. |
| Graphing (Lesson 9.1) | • Can easily see the number of solutions<br>• Use when approximate solutions are sufficient.<br>• Can use a graphing calculator | • May not give exact solutions |
| Using Square Roots (Lesson 9.2) | • Use to solve equations of the form $x^2 = d$. | • Can only be used for certain equations |
| Completing the Square (Lesson 9.3) | • Best used when $a = 1$ and $b$ is even | • May involve difficult calculations |
| Quadratic Formula (Lesson 9.4) | • Can be used for *any* quadratic equation<br>• Gives exact solutions | • Takes time to do calculations |

### EXAMPLE 1 Solving a Quadratic Equation Using Different Methods

**Solve $x^2 + 8x + 12 = 0$ using two different methods.**

**Method 1:** Solve by graphing. Graph the related function $y = x^2 + 8x + 12$.

The $x$-intercepts are $-6$ and $-2$.

∴ So, the solutions are $x = -6$ and $x = -2$.

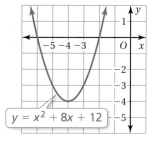

**Method 2:** Solve by factoring.

$$x^2 + 8x + 12 = 0$$    Write the equation.
$$(x + 2)(x + 6) = 0$$    Factor left side.
$$x + 2 = 0 \quad \text{or} \quad x + 6 = 0$$    Use Zero-Product Property.
$$x = -2 \quad \text{or} \quad x = -6$$    Solve for $x$.

∴ The solutions are $x = -2$ and $x = -6$.

**Study Tip**

Notice that each method produces the same solutions, $x = -6$ and $x = -2$.

## EXAMPLE 2  Choosing a Method

Solve $x^2 - 10x = 1$ using any method. Explain your choice of method.

The coefficient of the $x^2$-term is 1 and the coefficient of the $x$-term is an even number. So, solve by completing the square.

| | |
|---|---|
| $x^2 - 10x = 1$ | Write the equation. |
| Complete the square. → $x^2 - 10x + 25 = 1 + 25$ | Add $\left(\dfrac{-10}{2}\right)^2$, or 25, to each side. |
| $(x - 5)^2 = 26$ | Factor $x^2 - 10x + 25$. |
| $x - 5 = \pm\sqrt{26}$ | Take the square root of each side. |
| $x = 5 \pm \sqrt{26}$ | Add 5 to each side. |

∴ The solutions are $x = 5 + \sqrt{26} \approx 10.1$ and $x = 5 - \sqrt{26} \approx -0.1$.

## EXAMPLE 3  Choosing a Method

Solve $2x^2 - 13x - 24 = 0$ using any method. Explain your choice of method.

The equation is not easily factorable and the numbers are somewhat large. So, solve using the quadratic formula.

| | |
|---|---|
| $x = \dfrac{-b \pm \sqrt{b^2 - 4ac}}{2a}$ | Quadratic Formula |
| $= \dfrac{-(-13) \pm \sqrt{(-13)^2 - 4(2)(-24)}}{2(2)}$ | Substitute 2 for $a$, $-13$ for $b$, and $-24$ for $c$. |
| $= \dfrac{13 \pm \sqrt{361}}{4}$ | Simplify. |
| $= \dfrac{13 \pm 19}{4}$ | Evaluate the square root. |

∴ The solutions are $x = \dfrac{13 + 19}{4} = 8$ and $x = \dfrac{13 - 19}{4} = -\dfrac{3}{2}$.

## Practice

**Solve the equation using two different methods.**

1. $x^2 + 14x = -8$
2. $x^2 - 10x + 9 = 0$
3. $-4x^2 + 144 = 0$

**Solve the equation using any method. Explain your choice of method.**

4. $x^2 + 11x - 12 = 0$
5. $9x^2 - 5 = 4$
6. $5x^2 - x - 1 = 0$
7. $x^2 - 3x - 40 = 0$
8. $x^2 + 12x + 5 = -15$
9. $x^2 = 2x - 5$
10. $-8x^2 - 2 = 14$
11. $x^2 + x - 12 = 0$
12. $x^2 + 6x + 9 = 16$

# 9.5 Solving Systems of Linear and Quadratic Equations

**Essential Question** How can you solve a system of two equations when one is linear and the other is quadratic?

### 1 ACTIVITY: Solving a System of Equations

Work with a partner. Solve the system of equations using the given strategy. Which strategy do you prefer? Why?

**System of Equations:**

$y = x + 2$   Linear

$y = x^2 + 2x$   Quadratic

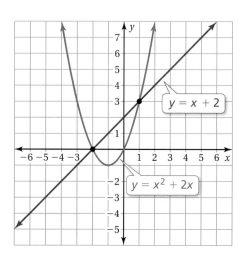

**a. Solve by Graphing**

*Graph* each equation and find the points of intersection of the line and the parabola.

**b. Solve by Substitution**

*Substitute* the expression for $y$ from the quadratic equation into the linear equation to obtain

$$x^2 + 2x = x + 2.$$

Solve this equation and substitute each $x$-value into the linear equation $y = x + 2$ to find the corresponding $y$-value.

**COMMON CORE**

**Solving Systems of Equations**

In this lesson, you will
- solve systems of linear and quadratic equations.

Learning Standard
A.REI.7

**c. Solve by Elimination**

*Eliminate* $y$ by subtracting the linear equation from the quadratic equation to obtain

$$\begin{aligned} y &= x^2 + 2x \\ y &= \phantom{x^2 +{}} x + 2 \\ \hline 0 &= x^2 + \phantom{x}x - 2. \end{aligned}$$

Solve this equation and substitute each $x$-value into the linear equation $y = x + 2$ to find the corresponding $y$-value.

486   Chapter 9   Solving Quadratic Equations

## 2  ACTIVITY: Analyzing Systems of Equations

**Math Practice 8**

**Evaluate Results**
How can you check the solution of the system of equations to verify that your answer is reasonable?

Work with a partner. Match each system of equations with its graph. Then solve the system of equations.

**a.** $y = x^2 - 4$
$y = -x - 2$

**b.** $y = x^2 - 2x + 2$
$y = 2x - 2$

**c.** $y = x^2 + 1$
$y = x - 1$

**d.** $y = x^2 - x - 6$
$y = 2x - 2$

**A.**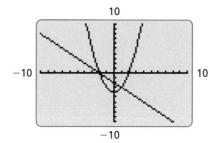

**B.**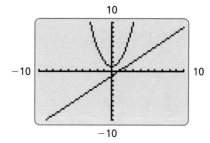

**C.**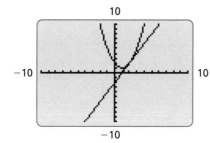

**D.**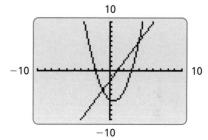

### What Is Your Answer?

3. **IN YOUR OWN WORDS** How can you solve a system of two equations when one is linear and the other is quadratic?

4. Summarize your favorite strategy for solving a system of two equations when one is linear and the other is quadratic.

5. Write a system of equations (one linear and one quadratic) that has the following number of solutions.

   **a.** no solutions   **b.** one solution   **c.** two solutions

   Your systems should be different from those in the activities.

**Practice** — Use what you learned about systems of equations to complete Exercises 3–5 on page 490.

Section 9.5  Solving Systems of Linear and Quadratic Equations

## 9.5 Lesson

You learned methods for solving systems of linear equations in Chapter 4. You can use similar methods to solve systems of linear and quadratic equations.

- Solving by Graphing (Section 4.1 and Section 9.1)
- Solving by Substitution (Section 4.2)
- Solving by Elimination (Section 4.3)

### EXAMPLE 1  Solving a System of Linear and Quadratic Equations

**Solve the system by substitution.**

$y = x^2 + x - 1$     Equation 1
$y = -2x + 3$     Equation 2

**Step 1:** The equations are already solved for $y$.

**Step 2:** Substitute $-2x + 3$ for $y$ in Equation 1 and solve for $x$.

| | |
|---|---|
| $y = x^2 + x - 1$ | Equation 1 |
| $-2x + 3 = x^2 + x - 1$ | Substitute $-2x + 3$ for $y$. |
| $3 = x^2 + 3x - 1$ | Add $2x$ to each side. |
| $0 = x^2 + 3x - 4$ | Subtract 3 from each side. |
| $0 = (x + 4)(x - 1)$ | Factor right side. |
| $x + 4 = 0$   or   $x - 1 = 0$ | Use Zero-Product Property. |
| $x = -4$   or   $x = 1$ | Solve for $x$. |

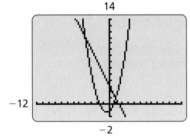

**Step 3:** Substitute $-4$ and $1$ for $x$ in Equation 2 and solve for $y$.

| | | | |
|---|---|---|---|
| $y = -2x + 3$ | Equation 2 | $y = -2x + 3$ | |
| $= -2(-4) + 3$ | Substitute. | $= -2(1) + 3$ | |
| $= 8 + 3$ | Multiply. | $= -2 + 3$ | |
| $= 11$ | Add. | $= 1$ | |

∴ So, the solutions are $(-4, 11)$ and $(1, 1)$.

### On Your Own

**Now You're Ready**
Exercises 6–11

Solve the system by substitution. Check your solution(s).

1. $y = x^2 + 9$
$y = 9$

2. $y = -5x$
$y = x^2 - 3x - 3$

3. $y = -3x^2 + 2x + 1$
$y = 5 - 3x$

## EXAMPLE 2  Solving a System of Linear and Quadratic Equations

**Solve the system by elimination.**

$y = x^2 - 3x - 2$   Equation 1
$y = -3x - 8$   Equation 2

**Step 1:** Subtract.

$y = x^2 - 3x - 2$   Equation 1
$y = \phantom{x^2}\; -3x - 8$   Equation 2
$\overline{0 = x^2 \phantom{-3x}\; + 6}$   Subtract the equations.

**Step 2:** Solve for $x$.

$0 = x^2 + 6$   Equation from Step 1
$-6 = x^2$   Subtract 6 from each side.

∴ The square of a real number cannot be negative. So, the system has no real solutions.

## EXAMPLE 3  Analyze a System of Equations

**Which statement about the system is valid?**

$y = 2x^2 + 5x - 1$   Equation 1
$y = x - 3$   Equation 2

**(A)** There is one solution because the graph of $y = x - 3$ has one $y$-intercept.

**(B)** There is one solution because $y = x - 3$ has one zero.

**(C)** There is one solution because the graphs of $y = 2x^2 + 5x - 1$ and $y = x - 3$ intersect at one point.

**(D)** There are two solutions because the graph of $y = 2x^2 + 5x - 1$ has two $x$-intercepts.

Use a graphing calculator to graph the system. The graphs of $y = 2x^2 + 5x - 1$ and $y = x - 3$ intersect at only one point, $(-1, -4)$.

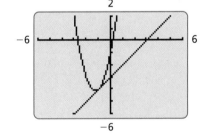

∴ So, the correct answer is **(C)**.

### On Your Own

**Now You're Ready**
Exercises 12–17

**Solve the system by elimination. Check your solution(s).**

4. $y = x^2 + x$
   $y = x + 5$

5. $y = 9x^2 + 8x - 6$
   $y = 5x - 4$

6. $y = 2x + 5$
   $y = -3x^2 + x - 4$

7. **WHAT IF?** In Example 3, does the system still have one solution when Equation 2 is changed to $y = x - 2$? Explain.

Section 9.5  Solving Systems of Linear and Quadratic Equations

# 9.5 Exercises

## Vocabulary and Concept Check

1. **VOCABULARY** What is a solution of a system of linear and quadratic equations?

2. **WRITING** How is solving a system of linear and quadratic equations similar to solving a system of linear equations? How is it different?

## Practice and Problem Solving

**Match the system of equations with its graph. Then solve the system.**

3. $y = x^2 - 2x + 1$
   $y = x + 1$

4. $y = x^2 + 3x + 2$
   $y = -x - 3$

5. $y = x - 1$
   $y = -x^2 + x - 1$

A.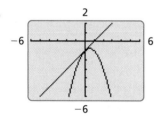
B.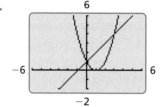
C.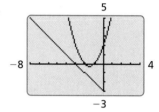

**Solve the system by substitution. Check your solution(s).**

6. $y = x - 5$
   $y = x^2 + 4x - 5$

7. $y = -2x^2$
   $y = 4x + 2$

8. $y = -x + 7$
   $y = -x^2 - 2x - 1$

9. $y = -x^2 + 7$
   $y - 2x = 4$

10. $y - 5 = -x^2$
    $y = 5$

11. $y = 2x^2 + 3x - 4$
    $y - 4x = 2$

**Solve the system by elimination. Check your solution(s).**

12. $y = -x^2 - 2x + 2$
    $y = 4x + 2$

13. $y = -2x^2 + x - 3$
    $y = 2x - 2$

14. $y = 2x - 1$
    $y = x^2$

15. $y = -2x$
    $y - x^2 = 3x$

16. $y - 1 = x^2 + x$
    $y = -x - 2$

17. $y = \frac{1}{2}x - 7$
    $y + 4x = x^2 - 2$

18. **MOVIES** The attendances $y$ for two movies can be modeled by the following equations, where $x$ is the number of days since the movies opened.

    $y = -x^2 + 35x + 100$    Movie A
    $y = -5x + 275$    Movie B

    When is the attendance for each movie the same?

**Solve the system using a graphing calculator.**

**19.** $y = x^2 + 6x - 1$
$y = 3x - 3$

**20.** $y = \frac{1}{4}x - 12$
$y = x^2 - 6x$

**21.** $y = \frac{1}{2}x^2$
$y = 2x - 2$

**22. CHOOSE TOOLS** Do you prefer to solve systems of equations by hand or using a graphing calculator? Explain your reasoning.

**23. ERROR ANALYSIS** Describe and correct the error in solving the system of equations.

$y = x^2 - 3x + 4$
$y = 2x + 4$

The only solution of the system of equations is (0, 4).

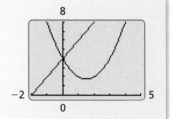

**24. WEBSITES** The function $y = -x^2 + 65x + 256$ models the number $y$ of subscribers to a website, where $x$ is the number of days since the website was launched. The number of subscribers to a competitor's website can be modeled by a linear function. The websites have the same number of subscribers on days 1 and 34.

   **a.** Write a linear function that models the number of subscribers to the competitor's website.

   **b.** Solve the system to verify the function from part (a).

**25. REASONING** The graph shows a quadratic function and the linear function $y = c$.

   **a.** How many solutions will the system have when you change the linear equation to $y = c + 2$?

   **b.** How many solutions will the system have when you change the linear equation to $y = c - 2$?

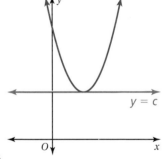

**26. Writing** Can a system of linear and quadratic equations have an infinite number of solutions? Explain your reasoning.

## Fair Game Review  *What you learned in previous grades & lessons*

**Solve the equation by graphing. Check your solution(s).** *(Section 9.1)*

**27.** $x^2 = 1$

**28.** $x^2 - 4x - 5 = 0$

**29.** $-x^2 = 2x + 7$

**30. MULTIPLE CHOICE** What is the factored form of the polynomial $x^2 - 36$? *(Section 7.9)*

   **Ⓐ** $(x + 6)^2$  **Ⓑ** $(x - 6)^2$  **Ⓒ** $(x + 6)(x - 6)$  **Ⓓ** $x + 6$

# 9.4–9.5 Quiz

**Solve the equation using the quadratic formula.** *(Section 9.4)*

1. $x^2 + 8x - 20 = 0$
2. $13x = 2x^2 + 6$
3. $9 - 24x = -16x^2$

**Use the discriminant to determine the number of real solutions of the equation.** *(Section 9.4)*

4. $x^2 + 6x - 13 = 0$
5. $-8x^2 - x = 5$
6. $\frac{3}{4}x^2 = 3x - 3$

7. Solve $x^2 + 10x + 21 = 0$ using two different methods. *(Section 9.4)*

**Solve the equation using any method. Explain your choice of method.** *(Section 9.4)*

8. $x^2 + 4x - 11 = 0$
9. $-4x^2 + 1 = 0$
10. $52 = x^2 - 2x$

**Solve the system.** *(Section 9.5)*

11. $y = x^2 - 16$
    $y = -7$
12. $y = x^2 + 2x + 1$
    $y = 2x + 2$
13. $y = x^2 - 5x + 8$
    $y = -3x - 4$

14. **BACTERIA** The numbers $y$ of two types of bacteria after $t$ hours are given by the models below. *(Section 9.5)*

    $y = 3t^2 + 8t + 20$     Type 1
    $y = 27t + 60$     Type 2

    a. As $t$ increases, which type grows more quickly? Explain.

    b. When are the numbers of Type 1 and Type 2 bacteria the same?

    c. When are there more Type 1 bacteria than Type 2? When are there more Type 2 bacteria than Type 1? Use a graph to support your answer.

15. **CELLULAR PHONE CALLS** The average monthly bill $y$ (in dollars) for a customer's cell phone $x$ years after 2000 can be modeled by $y = -0.2x^2 + 2x + 45$. When was the average monthly bill about $50? *(Section 9.4)*

16. **REASONING** Do you think the model in Exercise 15 can be used for future years? Explain using a graphing calculator to support your answer. *(Section 9.4)*

# 9 Chapter Review

## Review Key Vocabulary

quadratic equation, *p. 456*  
completing the square, *p. 470*  
quadratic formula, *p. 478*  
discriminant, *p. 480*

## Review Examples and Exercises

### 9.1 Solving Quadratic Equations by Graphing (pp. 454–461)

Solve $x^2 + 3x - 4 = 0$ by graphing.

**Step 1:** Graph the related function $y = x^2 + 3x - 4$.

**Step 2:** Find the *x*-intercepts. They are $-4$ and $1$.

So, the solutions are $x = -4$ and $x = 1$.

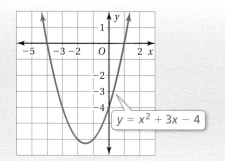

#### Exercises

Solve the equation by graphing. Check your solution(s).

1. $x^2 - 9x + 18 = 0$  
2. $x^2 - 2x = -4$  
3. $-8x - 16 = x^2$

### 9.2 Solving Quadratic Equations Using Square Roots (pp. 462–467)

A sprinkler sprays water that covers a circular region of $90\pi$ square feet. Find the diameter of the circle.

Write an equation using the formula for the area of a circle.

$A = \pi r^2$   Write the formula.

$90\pi = \pi r^2$   Substitute $90\pi$ for $A$.

$90 = r^2$   Divide each side by $\pi$.

$\pm\sqrt{90} = r$   Take the square root of each side.

A diameter cannot be negative, so use the positive square root. The diameter is twice the radius. So, the diameter is $2\sqrt{90}$.

The diameter of the circle is $2\sqrt{90} \approx 19$ feet.

#### Exercises

Solve the equation using square roots.

4. $x^2 - 10 = -10$  
5. $4x^2 = -100$  
6. $(x + 2)^2 = 64$

## 9.3 Solving Quadratic Equations by Completing the Square (pp. 468–473)

**Solve $x^2 - 6x + 4 = 11$ by completing the square.**

| | |
|---|---|
| $x^2 - 6x + 4 = 11$ | Write the equation. |
| $x^2 - 6x = 7$ | Subtract 4 from each side. |
| Complete the square. → $x^2 - 6x + 9 = 7 + 9$ | Add $\left(\dfrac{-6}{2}\right)^2$, or 9, to each side. |
| $(x - 3)^2 = 16$ | Factor $x^2 - 6x + 9$. |
| $x - 3 = \pm 4$ | Take the square root of each side. |
| $x = 3 \pm 4$ | Add 3 to each side. |

∴ The solutions are $x = 3 + 4 = 7$ and $x = 3 - 4 = -1$.

### Exercises

**Solve the equation by completing the square.**

7. $x^2 + x + 10 = 0$
8. $x^2 + 2x + 5 = 4$
9. $2x^2 - 4x = 10$

10. **CREDIT CARD** The width $w$ of a credit card is 3 centimeters shorter than the length $\ell$. The area is 46.75 square centimeters. Find the perimeter.

## 9.4 Solving Quadratic Equations Using the Quadratic Formula (pp. 476–485)

**Solve $-3x^2 + x = -8$ using the quadratic formula.**

| | |
|---|---|
| $-3x^2 + x = -8$ | Write original equation. |
| $-3x^2 + x + 8 = 0$ | Write in standard form. |
| $x = \dfrac{-b \pm \sqrt{b^2 - 4ac}}{2a}$ | Quadratic Formula |
| $= \dfrac{-1 \pm \sqrt{1^2 - 4(-3)(8)}}{2(-3)}$ | Substitute $-3$ for $a$, 1 for $b$, and 8 for $c$. |
| $= \dfrac{-1 \pm \sqrt{97}}{-6}$ | Simplify. |

∴ The solutions are $x = \dfrac{-1 + \sqrt{97}}{-6} \approx -1.5$ and $x = \dfrac{-1 - \sqrt{97}}{-6} \approx 1.8$.

### Exercises

**Solve the equation using the quadratic formula.**

11. $x^2 + 2x - 15 = 0$
12. $2x^2 - x + 8 = 3$
13. $-5x^2 + 10x = 5$

**Solve the equation using any method. Explain your choice of method.**

14. $x^2 - 121 = 0$
15. $x^2 - 4x + 4 = 0$
16. $x^2 - 4x = -1$

## 9.5 Solving Systems of Linear and Quadratic Equations  (pp. 486–491)

Solve the system by substitution.

$y = 2x^2 - 5$  Equation 1
$y = -x + 1$  Equation 2

**Step 1:** The equations are already solved for $y$.

**Step 2:** Substitute $-x + 1$ for $y$ in Equation 1 and solve for $x$.

| | |
|---|---|
| $y = 2x^2 - 5$ | Equation 1 |
| $-x + 1 = 2x^2 - 5$ | Substitute $-x + 1$ for $y$. |
| $1 = 2x^2 + x - 5$ | Add $x$ to each side. |
| $0 = 2x^2 + x - 6$ | Subtract 1 from each side. |
| $0 = (2x - 3)(x + 2)$ | Factor right side. |
| $2x - 3 = 0$  or  $x + 2 = 0$ | Use Zero-Product Property. |
| $x = \dfrac{3}{2}$  or  $x = -2$ | Solve for $x$. |

**Step 3:** Substitute $\dfrac{3}{2}$ and $-2$ for $x$ in Equation 2 and solve for $y$.

$y = -x + 1$     Equation 2         $y = -x + 1$
$\phantom{y} = -\dfrac{3}{2} + 1$     Substitute.         $\phantom{y} = -(-2) + 1$
$\phantom{y} = -\dfrac{1}{2}$     Simplify.         $\phantom{y} = 3$

∴ So, the solutions are $\left(\dfrac{3}{2}, -\dfrac{1}{2}\right)$ and $(-2, 3)$.

**Check**

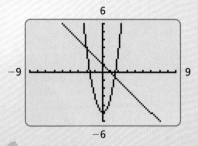

### Exercises

Solve the system. Check your solution(s).

**17.** $y = x^2 - 2x - 4$
     $y = -5$

**18.** $y = x^2 - 9$
     $y = 2x + 5$

**19.** $y = 2 - 3x$
     $y = -x^2 - 5x - 4$

Chapter Review   495

# 9 Chapter Test

**Solve the equation by graphing.**

1. $x^2 - 7x + 12 = 0$
2. $x^2 + 12x = -36$
3. $x + 1 = -x^2$

**Solve the equation using square roots.**

4. $14 = 2x^2$
5. $x^2 + 9 = 5$
6. $(4x + 3)^2 = 16$

**Solve the equation by completing the square.**

7. $x^2 - 8x + 15 = 0$
8. $x^2 - 6x = 10$
9. $x^2 - 8x = -9$
10. $16 = x^2 - 16x - 20$

**Solve the equation using the quadratic formula.**

11. $5x^2 + x - 4 = 0$
12. $9x^2 + 6x + 1 = 0$
13. $-2x^2 + 3x + 7 = 0$

14. **REASONING** Use the discriminant to determine how many times the graph of $y = 4x^2 - 4x + 1$ intersects the $x$-axis.

15. **CHOOSING A METHOD** Solve $x^2 - 9x - 10 = 0$ using any method. Explain your choice of method.

**Solve the system.**

16. $y = x^2 - 4x - 2$
    $y = -4x + 2$

17. $y = -5x^2 + x - 1$
    $y = -7$

18. **GEOMETRY** The area of the triangle is 35 square feet. Use a quadratic equation to find the length of the base. Round your answer to the nearest tenth.

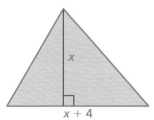

19. **SNOWBOARDING** A snowboarder leaves an 8-foot-tall ramp with an upward velocity of 28 feet per second. The function $h = -16t^2 + 28t + 8$ gives the height $h$ (in feet) of the snowboarder after $t$ seconds. How many points does the snowboarder earn with a perfect landing?

| Criteria | Scoring |
|---|---|
| Maximum height | 1 point per foot |
| Time in air | 5 points per second |
| Perfect landing | 25 points |

# 9 Standards Assessment

1. Which expression represents the area of the square? *(A.APR.1)*

   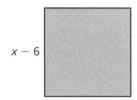

   **A.** $x^2 + 12x + 36$  **C.** $x^2 + 36$
   
   **B.** $x^2 - 12x + 36$  **D.** $x^2 - 12x - 36$

2. Which of the following equations has *no* real solutions? *(A.REI.4b)*

   **F.** $(x - 25)^2 = 5$  **H.** $(3x + 1)^2 = 9$
   
   **G.** $-4x^2 = 0$  **I.** $2x^2 + 1 = -1$

---

**Test-Taking Strategy**
**Answer Easy Questions First**

The function $h = -16t^2 + 45t + 20$ gives the height $h$ (in feet) of a cannonball after $t$ seconds. At what height was it launched?
Ⓐ -20  Ⓑ 0  Ⓒ 20  Ⓓ 45

"Shiver me whiskers!"

"Scan the test and answer the easy questions first. You know that the constant term, 20, is the initial height. So, the answer is C."

---

3. Use the solution below to determine which statement about the function $y = x^2 + 4x + 3$ is false. *(A.SSE.3a)*

   $$x^2 + 4x + 3 = 0$$
   $$(x + 3)(x + 1) = 0$$
   $$x + 3 = 0 \text{ or } x + 1 = 0$$
   $$x = -3 \quad\quad x = -1$$

   **A.** The zeros are $-1$ and $-3$.
   
   **B.** The graph crosses the $x$-axis at $(-1, 0)$ and $(-3, 0)$.
   
   **C.** The axis of symmetry of the graph is $x = -2$.
   
   **D.** The maximum value occurs when $x = -2$.

4. For $f(x) = 5x + 8$, what value of $x$ makes $f(x) = -12$? *(F.IF.2)*

5. Which line represents the axis of symmetry of the graph of the quadratic function? *(F.IF.4)*

   **F.** $x = -2$  **H.** $x = -\dfrac{3}{2}$
   
   **G.** $x = -\dfrac{5}{3}$  **I.** $x = -1$

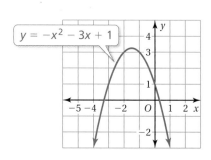

$y = -x^2 - 3x + 1$

**6.** What are the *exact* roots of the quadratic equation $3x^2 + x - 1 = 0$?   (A.REI.4b)

**A.** $-0.8, 0.4$

**B.** $-0.77, 0.43$

**C.** $\dfrac{-1 - \sqrt{13}}{6}, \dfrac{-1 + \sqrt{13}}{6}$

**D.** $-\dfrac{3}{4}, \dfrac{1}{2}$

**7.** The function $h = -16t^2 + 60t + 2$ gives the height $h$ (in feet) of a soccer ball after $t$ seconds. Which of the following statements is true?   (A.REI.4b)

**F.** The soccer ball reaches a height of 60 feet.

**G.** It takes the soccer ball 2.5 seconds to reach its maximum height.

**H.** The soccer ball hits the ground after about 5 seconds.

**I.** The soccer ball is kicked from a height of 2 feet.

**8.** Which *best* describes the solutions of the system of equations below?   (A.REI.7)

$$y = x^2 + 2x - 8 \quad \text{Equation 1}$$
$$y = 5x + 2 \quad \text{Equation 2}$$

**A.** Their graphs intersect at one point, $(-2, -8)$. So, there is one solution.

**B.** Their graphs intersect at two points, $(-2, -8)$ and $(5, 27)$. So, there are two solutions.

**C.** Their graphs do not intersect. So, there is no solution.

**D.** The graph of $y = x^2 + 2x - 8$ has two $x$-intercepts. So, there are two solutions.

**9.** Which graph shows exponential growth?   (F.LE.1c)

**F.**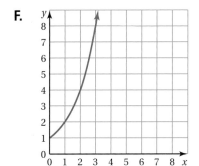

**H.**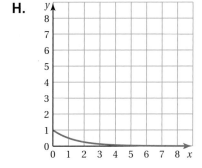

**G.**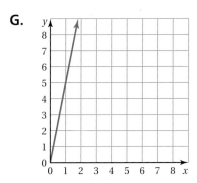

**I.**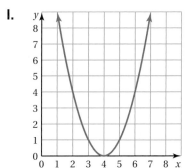

498   Chapter 9   Solving Quadratic Equations

10. For Parts A–D, use the function $y = 3x^2 + 3x + 4$ to find each characteristic without using a graph. Show your work and explain your reasoning. *(F.IF.4)*

   *Part A* direction the graph of the function opens

   *Part B* y-intercept of the graph of the function

   *Part C* axis of symmetry of the graph of the function

   *Part D* vertex of the graph of the function

11. Jamie is solving the equation $x^2 - 14x + 7 = 18$ by completing the square.

$$x^2 - 14x + 7 = 18$$
$$x^2 - 14x = 11$$
$$x^2 - 14x + 49 = 11$$
$$(x - 7)^2 = 11$$
$$x - 7 = \pm\sqrt{11}$$
$$x = 7 \pm \sqrt{11}$$

   What should Jamie do to correct the error that he made? *(A.REI.4b)*

   **A.** Add 49 to each side of the equation.

   **B.** Factor $x^2 - 14x + 49$ as $(x + 7)^2$.

   **C.** Subtract 49 from each side of the equation instead of adding 49.

   **D.** Only use the positive square root of 11.

12. What is the value of the discriminant for the quadratic equation $1.5x^2 - 6x = 13$? *(A.REI.4b)*

13. Which of the following statements is true about the quadratic function shown in the graph? *(A.REI.4b)*

   **F.** The range is all real numbers.

   **G.** The domain is all real numbers.

   **H.** The zeros are $(-1, 0)$ and $(5, 0)$.

   **I.** A minimum occurs at the vertex.

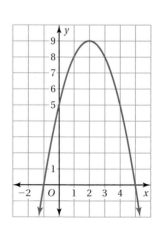

# 10 Square Root Functions and Geometry

**10.1** Graphing Square Root Functions

**10.2** Solving Square Root Equations

**10.3** The Pythagorean Theorem

**10.4** Using the Pythagorean Theorem

"I'm pretty sure that Pythagoras was a Greek."

"I said 'Greek', not 'Geek'."

"Let's figure out how we can measure the height of the giant hyena standing right behind you."

# What You Learned Before

## ● Evaluating an Expression Involving a Square Root (8.EE.2)

**Example 1** Evaluate $-4(\sqrt{121} - 16)$.

$$-4(\sqrt{121} - 16) = -4(11 - 16) \quad \text{Evaluate the square root.}$$
$$= -4(-5) \quad \text{Subtract.}$$
$$= 20 \quad \text{Multiply.}$$

### Try It Yourself
**Evaluate the expression.**

1. $7\sqrt{25} + 10$
2. $-8 - \sqrt{\dfrac{64}{16}}$
3. $-2(3\sqrt{4} + 13)$

## ● Factoring $x^2 + bx + c$ (A.SSE.3a)

**Example 2** Factor $x^2 - 3x - 28$.

Notice that $b = -3$ and $c = -28$. Because $c$ is negative, the factors $p$ and $q$ must have different signs so that $pq$ is negative.

Find two integer factors of $-28$ whose sum is $-3$.

| Factors of −28 | −28, 1 | −1, 28 | −14, 2 | −2, 14 | −7, 4 | −4, 7 |
|---|---|---|---|---|---|---|
| Sum of Factors | −27 | 27 | −12 | 12 | −3 | 3 |

The values of $p$ and $q$ are $-7$ and $4$.

∴ So, $x^2 - 3x - 28 = (x - 7)(x + 4)$.

### Try It Yourself
**Factor the polynomial.**

4. $y^2 + 12y + 27$
5. $n^2 - 11n + 10$
6. $w^2 - 2w - 48$
7. $z^2 + 25z + 100$

# 10.1 Graphing Square Root Functions

**Essential Question** How can you sketch the graph of a square root function?

## 1 ACTIVITY: Graphing Square Root Functions

**Work with a partner.**

- Make a table of values for the function.
- Use the table to sketch the graph of the function.
- Describe the domain of the function.
- Describe the range of the function.

**a.** $y = \sqrt{x}$

**b.** $y = \sqrt{x} + 2$

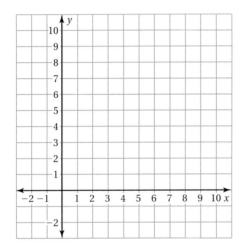
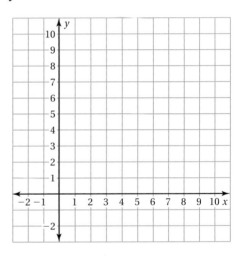

**c.** $y = \sqrt{x + 1}$

**d.** $y = -\sqrt{x}$

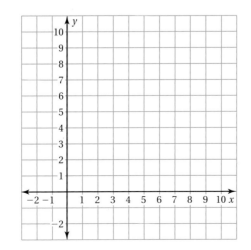
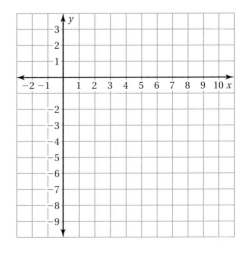

**COMMON CORE**

**Square Root Functions**

In this lesson, you will
- graph square root functions.
- compare graphs of square root functions.

Learning Standards
F.IF.4
F.IF.7b

## 2 ACTIVITY: Writing Square Root Functions

**Math Practice 4**

**Use a Table**
How can you use values of the domain and range to help write a function?

Work with a partner. Write a square root function, $y = f(x)$, that has the given values. Then use the function to complete the table.

a.
| x | f(x) |
|---|------|
| −4 | 0 |
| −3 | 1 |
| −2 | |
| −1 | |
| 0 | 2 |
| 1 | |

b.
| x | f(x) |
|---|------|
| −4 | 1 |
| −3 | 2 |
| −2 | |
| −1 | |
| 0 | 3 |
| 1 | |

## 3 ACTIVITY: Writing a Square Root Function

Work with a partner. Write a square root function, $y = f(x)$, that has the given points on its graph. Explain how you found your function.

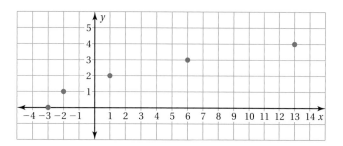

## What Is Your Answer?

4. **IN YOUR OWN WORDS** How can you sketch the graph of a square root function? Summarize a procedure for sketching the graph. Then use your procedure to sketch the graph of each function.

   a. $y = 2\sqrt{x}$
   
   b. $y = \sqrt{x} - 1$
   
   c. $y = \sqrt{x-1}$
   
   d. $y = -2\sqrt{x}$

**Practice** → Use what you learned about the graphs of square root functions to complete Exercises 3–8 on page 506.

Section 10.1  Graphing Square Root Functions  503

## 10.1 Lesson

**Key Vocabulary**
square root function, p. 504

### Key Idea

**Square Root Function**

A **square root function** is a function that contains a square root with the independent variable in the radicand. The most basic square root function is $y = \sqrt{x}$.

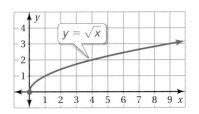

The value of the radicand in the square root function cannot be negative. So, the domain of a square root function includes $x$-values for which the radicand is greater than or equal to 0.

### EXAMPLE 1  Finding the Domain of a Square Root Function

Find the domain of $y = 3\sqrt{x-5}$.

The radicand cannot be negative. So, $x - 5$ is greater than or equal to 0.

$x - 5 \geq 0$    Write an inequality for the domain.

$x \geq 5$    Add 5 to each side.

∴ The domain is the set of real numbers greater than or equal to 5.

### On Your Own

**Now You're Ready**
Exercises 9–14

Find the domain of the function.

1. $y = 10\sqrt{x}$
2. $y = \sqrt{x} + 7$
3. $y = \sqrt{-x+1}$

### EXAMPLE 2  Comparing Graphs of Square Root Functions

Graph $y = \sqrt{x} + 3$. Describe the domain and range. Compare the graph to the graph of $y = \sqrt{x}$.

**Step 1:** Make a table of values.

| x | 0 | 1 | 4 | 9 | 16 |
|---|---|---|---|---|----|
| y | 3 | 4 | 5 | 6 | 7  |

**Step 2:** Plot the ordered pairs.

**Step 3:** Draw a smooth curve through the points.

**Remember**
When graphing, remember $f(x) + k$ is a vertical translation of $f(x)$.

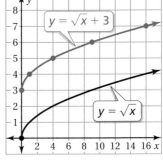

∴ From the graph, you can see that the domain is $x \geq 0$ and the range is $y \geq 3$. The graph of $y = \sqrt{x} + 3$ is a translation 3 units up of the graph of $y = \sqrt{x}$.

### EXAMPLE 3 Comparing Graphs of Square Root Functions

Graph $y = -\sqrt{x-2}$. Describe the domain and range. Compare the graph to the graph of $y = \sqrt{x}$.

**Step 1:** Make a table of values.

| x | 2 | 3 | 4 | 5 | 6 |
|---|---|---|---|---|---|
| y | 0 | −1 | −1.4 | −1.7 | −2 |

**Study Tip**
The graph of $f(x − h)$ is a horizontal translation of $f(x)$.

**Step 2:** Plot the ordered pairs.

**Step 3:** Draw a smooth curve through the points.

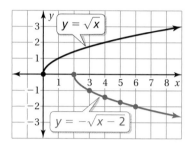

From the graph, you can see that the domain is $x \geq 2$ and the range is $y \leq 0$. The graph of $y = -\sqrt{x-2}$ is a reflection of the graph of $y = \sqrt{x}$ in the x-axis and then a translation 2 units to the right.

#### On Your Own

*Now You're Ready*
Exercises 16–21

Graph the function. Describe the domain and range. Compare the graph to the graph of $y = \sqrt{x}$.

**4.** $y = \sqrt{x} - 4$   **5.** $y = \sqrt{x+5}$   **6.** $y = -\sqrt{x+1} + 2$

---

### EXAMPLE 4 Real-Life Application

The velocity y (in meters per second) of a tsunami can be modeled by the function $y = \sqrt{9.8x}$, where x is the water depth (in meters). Use a graphing calculator to graph the function. At what depth does the velocity of the tsunami exceed 200 meters per second?

**Step 1:** Enter the function $y = \sqrt{9.8x}$ into your calculator and graph it. Because the radicand cannot be negative, use only nonnegative values of x.

**Step 2:** Use the *trace* feature to find where the value of y is about 200.

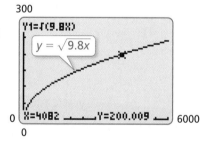

The velocity exceeds 200 meters per second at a depth of about 4100 meters.

#### On Your Own

**7.** Find the domain and range of the function in Example 4.

**8.** **WHAT IF?** In Example 4, at what depth does the velocity of the tsunami exceed 100 meters per second?

# 10.1 Exercises

## Vocabulary and Concept Check

1. **VOCABULARY** Is $y = 2x\sqrt{5}$ a square root function? Explain.
2. **REASONING** How do you find the domain of a square root function?

## Practice and Problem Solving

**Match the function with its graph.**

3. $y = 8\sqrt{x}$
4. $y = \dfrac{5}{4}\sqrt{x}$
5. $y = -4\sqrt{x}$

A.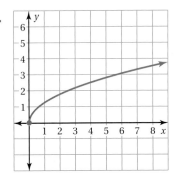

B.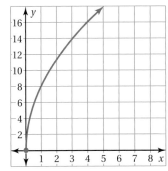

C.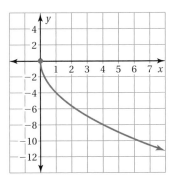

**Graph the function. Describe the domain.**

6. $y = 3\sqrt{x}$
7. $y = 7\sqrt{x}$
8. $y = -0.5\sqrt{x}$

**Find the domain of the function.**

9. $y = 5\sqrt{x}$
10. $y = \sqrt{x} + 1$
11. $y = \sqrt{x - 2}$
12. $y = \sqrt{-x - 1}$
13. $y = 2\sqrt{x + 4}$
14. $y = \dfrac{1}{2}\sqrt{-x + 2}$

15. **FIRE** The nozzle pressure of a fire hose allows firefighters to control the amount of water they spray on a fire. The flow rate $f$ (in gallons per minute) can be modeled by the function $f = 120\sqrt{p}$, where $p$ is the nozzle pressure (in pounds per square inch).

   a. Use a graphing calculator to graph the function.
   
   b. Use the *trace* feature to approximate the nozzle pressure that results in a flow rate of 300 gallons per minute.

506 Chapter 10 Square Root Functions and Geometry

**Graph the function. Describe the domain and range. Compare the graph to the graph of $y = \sqrt{x}$.**

**16.** $y = \sqrt{x} - 2$      **17.** $y = \sqrt{x} + 4$      **18.** $y = \sqrt{x+4}$

**19.** $y = \sqrt{x+2} - 2$      **20.** $y = -\sqrt{x-3}$      **21.** $y = -\sqrt{x-1} + 3$

**22. ERROR ANALYSIS** Describe and correct the error in graphing the function $y = \sqrt{x+1}$.

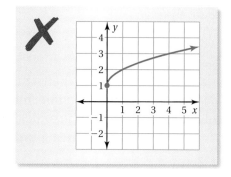

**23. OPEN-ENDED** Consider the graph of $y = \sqrt{x}$.

    **a.** Write a function that is a vertical translation of the graph of $y = \sqrt{x}$.

    **b.** Write a function that is a reflection of the graph of $y = \sqrt{x}$.

**24. REASONING** Can the domain of a square root function include negative numbers? Can the range include negative numbers? Explain your reasoning.

**25. GEOMETRY** The radius of a circle is given by $r = \sqrt{\dfrac{A}{\pi}}$, where $A$ is the area of the circle.

    **a.** Find the domain of the function. Use a graphing calculator to graph the function.

    **b.** Use the *trace* feature to approximate the area of a circle with a radius of 3 inches.

**26. PROBLEM SOLVING** The speed $S$ (in miles per hour) of a van before it skids to a stop can be modeled by the equation $S = \sqrt{30df}$, where $d$ is the length (in feet) of the skid marks and $f$ is the drag factor of the road surface. Suppose the drag factor is 0.75 and the speed of the van was 40 miles per hour. Is the length of the skid marks more than 65 feet long? Explain your reasoning.

**27. Precision** Compare the graphs of the functions $f(x) = \sqrt{x}$ and $g(x) = \sqrt[3]{x}$.

## Fair Game Review    *What you learned in previous grades & lessons*

**Solve the equation.** *(Section 7.5)*

**28.** $x(x-8) = 0$      **29.** $(x+3)^2 = 0$      **30.** $(x+2)(x-3) = 0$

**31. MULTIPLE CHOICE** What are the next three terms of the geometric sequence 240, 120, 60, 30, . . . ? *(Section 6.7)*

    Ⓐ 20, 10, 5      Ⓑ 15, 7.5, 3.75      Ⓒ 20, 10, 0      Ⓓ 15, 10, 5

# Extension 10.1 Rationalizing the Denominator

**Key Vocabulary**
simplest form of a radical expression, *p. 508*
rationalizing the denominator, *p. 508*
conjugates, *p. 509*

In Section 6.1, you used properties to simplify radical expressions. A radical expression is in **simplest form** when the following are true.

- No radicands have perfect square factors other than 1.
- No radicands contain fractions.
- No radicals appear in the denominator of a fraction.

When a radicand in the denominator of a fraction is not a perfect square, multiply the fraction by an appropriate form of 1 to eliminate the radical from the denominator. This process is called **rationalizing the denominator**.

### EXAMPLE 1 Simplifying a Radical Expression

Simplify $\sqrt{\frac{1}{3}}$.

**Study Tip**
Rationalizing the denominator works because you multiply the numerator and denominator by the same nonzero number $a$, which is the same as multiplying by $\frac{a}{a}$, or 1.

$$\sqrt{\frac{1}{3}} = \frac{\sqrt{1}}{\sqrt{3}} \quad \text{Quotient Property of Square Roots}$$

$$= \frac{\sqrt{1}}{\sqrt{3}} \cdot \frac{\sqrt{3}}{\sqrt{3}} \quad \text{Multiply by } \frac{\sqrt{3}}{\sqrt{3}}.$$

$$= \frac{\sqrt{1 \cdot 3}}{\sqrt{3 \cdot 3}} \quad \text{Product Property of Square Roots}$$

$$= \frac{\sqrt{3}}{\sqrt{9}} \quad \text{Simplify.}$$

$$= \frac{\sqrt{3}}{3} \quad \text{Evaluate the square root.}$$

## Practice

**Simplify the expression.**

1. $\dfrac{1}{\sqrt{10}}$

2. $\dfrac{\sqrt{2}}{\sqrt{7}}$

3. $\sqrt{\dfrac{9}{2}}$

4. $\sqrt{\dfrac{10}{21}}$

5. $\sqrt{\dfrac{5}{18}}$

6. $\sqrt{\dfrac{40}{48}}$

7. $\dfrac{4}{\sqrt{5}} + \dfrac{1}{\sqrt{5}}$

8. $\sqrt{3} - \dfrac{2}{\sqrt{12}}$

9. $\sqrt{\dfrac{16}{15}} - \dfrac{1}{3}$

10. **REASONING** Explain why for any number $a$, $\sqrt{a^2} = |a|$. Use this rule to simplify the expression $\sqrt{\dfrac{x^2}{2}}$.

**Square Root Functions**

In this extension, you will
- simplify radical expressions.

Learning Standards
F.IF.4
F.IF.7b

The binomials $a\sqrt{b} + c\sqrt{d}$ and $a\sqrt{b} - c\sqrt{d}$ are called **conjugates.** You can use conjugates to simplify radical expressions that involve a sum or difference of radicals in the denominator.

### EXAMPLE 2  Simplifying a Radical Expression

Simplify $\dfrac{1}{3 + \sqrt{5}}$.

$\dfrac{1}{3 + \sqrt{5}} = \dfrac{1}{3 + \sqrt{5}} \cdot \dfrac{3 - \sqrt{5}}{3 - \sqrt{5}}$    The conjugate of $3 + \sqrt{5}$ is $3 - \sqrt{5}$.

$= \dfrac{1(3 - \sqrt{5})}{3^2 - (\sqrt{5})^2}$    Sum and Difference Pattern

$= \dfrac{3 - \sqrt{5}}{4}$    Simplify.

**Study Tip**
Notice that the product of conjugates is a rational number.

### EXAMPLE 3  Real-Life Application

The distance $d$ (in miles) that you can see to the horizon with your eye level $h$ feet above the water is given by $d = \sqrt{\dfrac{3h}{2}}$. How far can you see when your eye level is 5 feet above the water?

$d = \sqrt{\dfrac{3(5)}{2}}$    Substitute 5 for $h$.

$= \dfrac{\sqrt{15}}{\sqrt{2}} \cdot \dfrac{\sqrt{2}}{\sqrt{2}}$    Multiply by $\dfrac{\sqrt{2}}{\sqrt{2}}$.

$= \dfrac{\sqrt{30}}{2}$    Simplify.

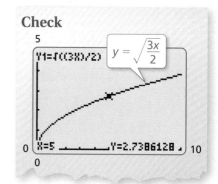

Check

∴ You can see $\dfrac{\sqrt{30}}{2}$, or about 2.74 miles.

### Practice

**Simplify the expression.**

11. $\dfrac{6}{1 + \sqrt{3}}$

12. $\dfrac{5}{\sqrt{3} - 2}$

13. $\dfrac{10}{\sqrt{2} + \sqrt{7}}$

14. **WHAT IF?** In Example 3, how far can you see when your eye level is 35 feet above the water?

Extension 10.1   Rationalizing the Denominator

# 10.2 Solving Square Root Equations

**Essential Question** How can you solve an equation that contains square roots?

### 1  ACTIVITY: Analyzing a Free-Falling Object

**Work with a partner.** The table shows the time *t* (in seconds) that it takes a free-falling object (with no air resistance) to fall *d* feet.

a. Sketch the graph of *t* as a function of *d*.

b. Use your graph to estimate the time it takes for a free-falling object to fall 240 feet.

c. The relationship between *d* and *t* is given by the function
$$t = \sqrt{\frac{d}{16}}.$$
Use this function to check the estimate you obtained from the graph.

d. Consider a free-falling object that takes 5 seconds to hit the ground. How far did it fall? Explain your reasoning.

| d feet | t seconds |
|---|---|
| 0 | 0.00 |
| 32 | 1.41 |
| 64 | 2.00 |
| 96 | 2.45 |
| 128 | 2.83 |
| 160 | 3.16 |
| 192 | 3.46 |
| 224 | 3.74 |
| 256 | 4.00 |
| 288 | 4.24 |
| 320 | 4.47 |

**COMMON CORE**

**Radical Functions**

In this lesson, you will
- solve square root equations, including those with square roots on both sides.
- identify extraneous solutions.

Applying Standard
N.RN.2

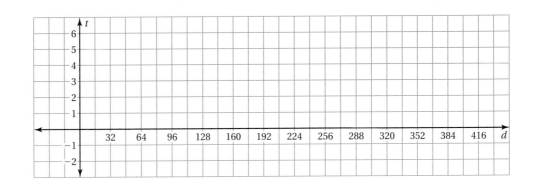

**510** Chapter 10  Square Root Functions and Geometry

## 2 ACTIVITY: Solving a Square Root Equation

Work with a partner. Sketch the graph of each function. Then find the value of $x$ such that $f(x) = 2$. Explain your reasoning.

**Math Practice**

**Interpret a Solution**
How can you use a graph to find the value of $x$ given the value of the function?

a. $f(x) = \sqrt{x} - 2$

b. $f(x) = \sqrt{x-1}$

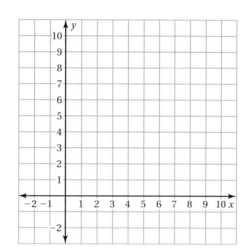
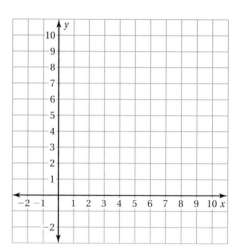

## 3 ACTIVITY: Solving a Square Root Equation

Work with a partner. The speed $s$ (in feet per second) of the free-falling object in Activity 1 is given by the function

$$s = \sqrt{64d}.$$

Find the distance traveled for each speed.

a. $s = 8$ ft/sec

b. $s = 16$ ft/sec

c. $s = 24$ ft/sec

## What Is Your Answer?

4. **IN YOUR OWN WORDS** How can you solve an equation that contains square roots? Summarize a procedure for solving a square root equation. Then use your procedure to solve each equation.

a. $\sqrt{x} + 2 = 3$

b. $4 - \sqrt{x} = 1$

c. $5 = \sqrt{x+20}$

d. $-3 = -2\sqrt{x}$

**Practice**

Use what you learned about solving square root equations to complete Exercises 3–5 on page 515.

# 10.2 Lesson

**Key Vocabulary**
square root equation, *p. 512*
extraneous solution, *p. 513*

A **square root equation** is an equation that contains a square root with a variable in the radicand. To solve a square root equation, use properties of equality to isolate the square root by itself on one side of the equation, then use the following property.

## Key Idea

**Squaring Each Side of an Equation**

**Words** If two expressions are equal, then their squares are also equal.

**Algebra** If $a = b$, then $a^2 = b^2$.

### EXAMPLE 1 Solving Square Root Equations

a. Solve $\sqrt{x} + 5 = 13$.

| | |
|---|---|
| $\sqrt{x} + 5 = 13$ | Write the equation. |
| $\sqrt{x} = 8$ | Subtract 5 from each side. |
| $(\sqrt{x})^2 = 8^2$ | Square each side of the equation. |
| $x = 64$ | Simplify. |

∴ The solution is $x = 64$.

**Check**
$\sqrt{x} + 5 = 13$
$\sqrt{64} + 5 \stackrel{?}{=} 13$
$8 + 5 \stackrel{?}{=} 13$
$13 = 13$ ✓

b. Solve $3 - \sqrt{x} = 0$.

| | |
|---|---|
| $3 - \sqrt{x} = 0$ | Write the equation. |
| $3 = \sqrt{x}$ | Add $\sqrt{x}$ to each side. |
| $3^2 = (\sqrt{x})^2$ | Square each side of the equation. |
| $9 = x$ | Simplify. |

∴ The solution is $x = 9$.

**Check**
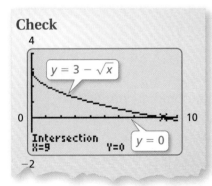

### On Your Own

**Now You're Ready**
Exercises 6–11

Solve the equation. Check your solution.

1. $\sqrt{x} = 6$
2. $\sqrt{x} - 7 = 3$
3. $\sqrt{x} + 15 = 22$
4. $1 - \sqrt{x} = -2$

## EXAMPLE 2 Solving a Square Root Equation

Solve $4\sqrt{x+2} + 3 = 19$.

| | |
|---|---|
| $4\sqrt{x+2} + 3 = 19$ | Write the equation. |
| $4\sqrt{x+2} = 16$ | Subtract 3 from each side. |
| $\sqrt{x+2} = 4$ | Divide each side by 4. |
| $(\sqrt{x+2})^2 = 4^2$ | Square each side of the equation. |
| $x + 2 = 16$ | Simplify. |
| $x = 14$ | Subtract 2 from each side. |

∴ The solution is $x = 14$.

**Check**
$4\sqrt{x+2} + 3 = 19$
$4\sqrt{14+2} + 3 \stackrel{?}{=} 19$
$4\sqrt{16} + 3 \stackrel{?}{=} 19$
$4(4) + 3 \stackrel{?}{=} 19$
$16 + 3 \stackrel{?}{=} 19$
$19 = 19$ ✓

### On Your Own

Now You're Ready
Exercises 13–18

Solve the equation. Check your solution.

**5.** $\sqrt{x+4} + 7 = 11$    **6.** $8\sqrt{x-1} = 24$    **7.** $15 = 6 + \sqrt{3x-9}$

## EXAMPLE 3 Solving an Equation with Square Roots on Both Sides

Solve $\sqrt{2x-1} = \sqrt{x+4}$.

| | |
|---|---|
| $\sqrt{2x-1} = \sqrt{x+4}$ | Write the equation. |
| $(\sqrt{2x-1})^2 = (\sqrt{x+4})^2$ | Square each side of the equation. |
| $2x - 1 = x + 4$ | Simplify. |
| $x - 1 = 4$ | Subtract x from each side. |
| $x = 5$ | Add 1 to each side. |

∴ The solution is $x = 5$.

**Check**

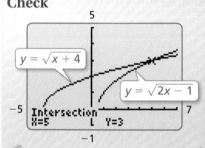

### On Your Own

Now You're Ready
Exercises 22–27

Solve the equation. Check your solution.

**8.** $\sqrt{3x+1} = \sqrt{4x-7}$    **9.** $\sqrt{x} = \sqrt{5x-1}$

Squaring each side of an equation can sometimes introduce a solution that is *not* a solution of the original equation. This solution is called an **extraneous solution**. Be sure to always substitute your solutions into the original equation to check for extraneous solutions.

**EXAMPLE 4** **Identifying an Extraneous Solution**

| | | |
|---|---|---|
| $x = \sqrt{x+6}$ | | Original equation |
| $x^2 = (\sqrt{x+6})^2$ | | Square each side of the equation. |
| $x^2 = x + 6$ | | Simplify. |
| $x^2 - x - 6 = 0$ | | Subtract $x$ and 6 from each side. |
| $(x-3)(x+2) = 0$ | | Factor. |
| $(x-3) = 0$ or $(x+2) = 0$ | | Use Zero-Product Property. |
| $x = 3$ or $x = -2$ | | Solve for $x$. |

**Check**

$3 \stackrel{?}{=} \sqrt{3+6}$   Substitute for $x$.   $-2 \stackrel{?}{=} \sqrt{-2+6}$

$3 \stackrel{?}{=} \sqrt{9}$   Simplify.   $-2 \stackrel{?}{=} \sqrt{4}$

$3 = 3$ ✓   $-2 \ne 2$ ✗

∴ Because $x = -2$ does not check in the original equation, it is an extraneous solution. The only solution is $x = 3$.

**EXAMPLE 5** **Real-Life Application**

The period $P$ (in seconds) of a pendulum is given by the function $P = 2\pi\sqrt{\dfrac{L}{32}}$, where $L$ is the pendulum length (in feet). What is the length of a pendulum that has a period of 2 seconds?

**Study Tip**
The period of a pendulum is the amount of time it takes for the pendulum to swing back and forth.

| | |
|---|---|
| $2 = 2\pi\sqrt{\dfrac{L}{32}}$ | Substitute 2 for $P$ in the function. |
| $\dfrac{1}{\pi} = \sqrt{\dfrac{L}{32}}$ | Divide each side by $2\pi$ and simplify. |
| $\dfrac{1}{\pi^2} = \dfrac{L}{32}$ | Square each side and simplify. |
| $\dfrac{32}{\pi^2} = L$ | Multiply both sides by 32. |
| $3.2 \approx L$ | Use a calculator. |

∴ The length of the pendulum is about 3.2 feet.

**On Your Own**

Now You're Ready
Exercises 31–36

10. Solve $\sqrt{x-1} = x - 3$. Check your solution.

11. **WHAT IF?** In Example 5, what is the length of a pendulum that has a period of 4 seconds? Is your result twice the length in Example 5? Explain.

# 10.2 Exercises

## Vocabulary and Concept Check

1. **VOCABULARY** Is $x\sqrt{3} = 4$ a square root equation? Explain your reasoning.

2. **WRITING** Why should you check every solution of a square root equation?

## Practice and Problem Solving

**Sketch the graph of the function. Then find the value of $x$ such that $f(x) = 3$.**

3. $f(x) = \sqrt{x} + 1$
4. $f(x) = \sqrt{x-3}$
5. $f(x) = \sqrt{x+1} - 2$

**Solve the equation. Check your solution.**

6. $\sqrt{x} = 9$
7. $7 = \sqrt{x} - 5$
8. $\sqrt{x} + 6 = 10$
9. $\sqrt{x} + 12 = 23$
10. $4 - \sqrt{x} = 4$
11. $-8 = 7 - \sqrt{x}$

12. **ERROR ANALYSIS** Describe and correct the error in solving the equation.

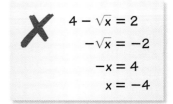

**Solve the equation. Check your solution.**

13. $\sqrt{x-3} + 5 = 9$
14. $2\sqrt{x+4} = 16$
15. $25 = 7 + 3\sqrt{x-9}$
16. $\sqrt{\dfrac{x}{2} - 1} + 14 = 18$
17. $-1 = \sqrt{5x+1} - 7$
18. $12 = 19 - \sqrt{3x-11}$

19. **CUBE** The formula $s = \sqrt{\dfrac{A}{6}}$ gives the edge length $s$ of a cube with a surface area of $A$. What is the surface area of a cube with an edge length of 4 inches?

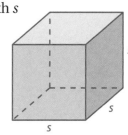

20. **BASE JUMPING** The Cave of Swallows is a natural open-air pit cave in the state of San Luis Potosí, Mexico. The 1220-foot deep cave is a popular destination for BASE jumpers. The formula $t = \sqrt{\dfrac{d}{16}}$ gives the distance $d$ (in feet) a BASE jumper free falls in $t$ seconds. How far does the BASE jumper fall in 3 seconds?

21. **WRITING** Explain how you would solve $\sqrt{m+4} - \sqrt{3m} = 0$.

**Solve the equation. Check your solution.**

22. $\sqrt{2x-9} = \sqrt{x}$
23. $\sqrt{x+1} = \sqrt{4x-8}$
24. $\sqrt{3x+1} = \sqrt{7x-19}$
25. $\sqrt{8x-7} = \sqrt{6x+7}$
26. $\sqrt{2x+1} - \sqrt{4x} = 0$
27. $\sqrt{5x} - \sqrt{8x-2} = 0$

**Find the value of x.**

28. Perimeter = 28 cm

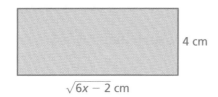

29. Area = $\sqrt{5x-4}$ ft²

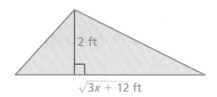

30. **OPEN-ENDED** Write a square root equation of the form $\sqrt{ax} + b = c$ that has a solution of 9.

**Solve the equation. Check your solution.**

31. $x = \sqrt{5x-4}$
32. $\sqrt{9x-14} = x$
33. $\sqrt{3x+10} = x$
34. $2x = \sqrt{6-10x}$
35. $x - 1 = \sqrt{3-x}$
36. $\sqrt{-4x-19} = x+4$

37. **ERROR ANALYSIS** Describe and correct the error in solving the equation.

$$x = \sqrt{12-4x}$$
$$x^2 = 12 - 4x$$
$$x^2 + 4x - 12 = 0$$
$$(x-2)(x+6) = 0$$
$$x = 2 \text{ or } x = -6$$

38. **REASONING** Explain how to use mental math to find the solution of $\sqrt{2x} + 5 = 1$.

**Determine whether the statement is *true* or *false*.**

39. If $\sqrt{a} = b$, then $(\sqrt{a})^2 = b^2$.
40. If $\sqrt{a} = \sqrt{b}$, then $a = b$.
41. If $a^2 = b^2$, then $a = b$.
42. If $a^2 = \sqrt{b}$, then $a^4 = (\sqrt{b})^2$.

43. **ELECTRICITY** The formula $V = \sqrt{PR}$ relates the voltage $V$ (in volts), power $P$ (in watts), and resistance $R$ (in ohms) of an electrical circuit. What is the resistance of a 1000-watt hair dryer on a 120-volt circuit?

**44. CHOOSE TOOLS** Consider the equation $x + 2 = \sqrt{2x - 3}$.

   a. Graph each side of the equation in the same coordinate plane. Solve the equation by finding points of intersection.
   b. Solve the equation algebraically. How does your solution compare to the solution in part (a)?
   c. Which method do you prefer? Explain your reasoning.

**45. TRAPEZE** The time $t$ (in seconds) it takes a trapeze artist to swing back and forth is given by the function $t = 2\pi\sqrt{\dfrac{r}{32}}$, where $r$ is the rope length (in feet). It takes 6 seconds to swing back and forth. How long is the rope? Use 3.14 for $\pi$.

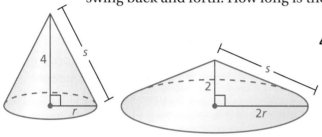

**46. GEOMETRY** The formula $s = \sqrt{r^2 + h^2}$ gives the slant height $s$ of a cone, where $r$ is the radius of the base, and $h$ is the height. The slant heights of the two cones are equal. Find the radius of each cone.

**47. CRITICAL THINKING** How is squaring $\sqrt{x + 2}$ different than squaring $\sqrt{x} + 2$?

**Solve the equation. Check your solution.**

**48.** $\sqrt{x + 15} = \sqrt{x} + \sqrt{5}$

**49.** $2 - \sqrt{x + 1} = \sqrt{x + 2}$

**50. Modeling** The formula $h = \sqrt{2A - b_2 h}$ gives the height $h$ of the speaker box, where $A$ is the area of one trapezoidal side, and $b_2$ is the length of base 2.

   a. Given that $A = 168$ square inches and $b_2 = 16$ inches, find $h$.
   b. What is the length of $b_1$ (base 1)?
   c. Speakers work best when the volume of the speaker box is ±10% of the manufacturer's recommendation. Find the range of the widths $w$ when the manufacturer recommends a volume of 1.5 cubic feet.

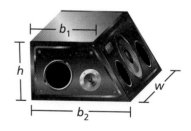

## Fair Game Review  *What you learned in previous grades & lessons*

**Two angle measures of a triangle are given. Find the measure of the missing angle.** *(Skills Review Handbook)*

**51.** 40°, 48°  **52.** 45°, 55°  **53.** 36°, 54°

**54. MULTIPLE CHOICE** Which function is represented by the ordered pairs $(-1, 0.5)$, $(0, 1)$, $(1, 2)$, $(2, 4)$, and $(3, 8)$? *(Section 8.5)*

   Ⓐ $y = 0.5x^2$   Ⓑ $y = 2^x$   Ⓒ $y = 2x^2$   Ⓓ $y = 2x$

Section 10.2  Solving Square Root Equations

# 10 Study Help

You can use a **word magnet** to organize information associated with a vocabulary word. Here is an example of a word magnet for square root functions.

**Square Root Function**

**Definition:** A function that contains a square root with the independent variable in the radicand.

**Examples:**
$y = \sqrt{x} + 3$
$y = \sqrt{x-1}$
$y = \sqrt{x+5} - 4$

**Sample Graph:**

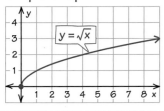

**Domain:** The value of the radicand cannot be negative. So, the domain is limited to x-values for which the radicand is greater than or equal to 0.

**Graph:** Make a table of values. Plot the ordered pairs. Draw a smooth curve through the points. Find the domain and range.

**Compare:** When graphing a square root function f(x):
- f(x) + k is a vertical translation of f(x).
- f(x + h) is a horizontal translation of f(x).
- −f(x) is a reflection of f(x) in the x-axis.

## On Your Own

**Make word magnets to help you study these topics.**

1. rationalizing the denominator
2. solving a square root equation
3. extraneous solution

**After you complete this chapter, make word magnets for the following topics.**

4. Pythagorean Theorem
5. converse of the Pythagorean Theorem
6. distance formula

"How do you like the word magnet I made for 'Beagle'?"

# 10.1–10.2 Quiz

**Find the domain of the function.** *(Section 10.1)*

1. $y = 15\sqrt{x}$
2. $y = \sqrt{x-3}$
3. $y = \sqrt{3-x}$

**Graph the function. Describe the domain and range. Compare the graph to the graph of $y = \sqrt{x}$.** *(Section 10.1)*

4. $y = \sqrt{x} + 5$
5. $y = \sqrt{x-4}$
6. $y = -\sqrt{x-2} + 1$

**Simplify the expression.** *(Section 10.1)*

7. $\sqrt{\dfrac{6}{42}}$
8. $\dfrac{2}{\sqrt{3}} - \dfrac{1}{\sqrt{3}}$
9. $\dfrac{7}{\sqrt{5}+2}$

**Solve the equation.** *(Section 10.2)*

10. $\sqrt{x-1} + 7 = 15$
11. $\sqrt{x} = \sqrt{6x-20}$
12. $x = \sqrt{21-4x}$

**Find the value of $x$.** *(Section 10.2)*

13. Perimeter = 24 mi

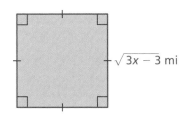

14. Area = $2\sqrt{4x-7}$ m²

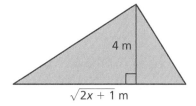

15. **BRIDGE** The time $t$ (in seconds) it takes an object to drop $h$ feet is given by $t = \dfrac{1}{4}\sqrt{h}$. *(Section 10.1)*

    a. Graph the function. Describe the domain and range.

    b. It takes about 7.4 seconds for a stone dropped from the New River Gorge Bridge in West Virginia, to reach the water below. About how high is the bridge above the New River?

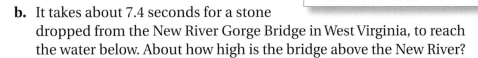

16. **SPEED OF SOUND** The speed of sound $s$ (in meters per second) through air is given by $s = 20\sqrt{T+273}$, where $T$ is the temperature in degrees Celsius. *(Section 10.2)*

    a. What is the temperature when the speed of sound is 340 meters per second?

    b. How long does it take you to hear the wolf howl when the temperature is $-17°C$?

# 10.3 The Pythagorean Theorem

**Essential Question** How are the lengths of the sides of a right triangle related?

Pythagoras was a Greek mathematician and philosopher who discovered one of the most famous rules in mathematics. In mathematics, a rule is called a **theorem**. So, the rule that Pythagoras discovered is called the Pythagorean Theorem.

Pythagoras
(c. 570 B.C.–c. 490 B.C.)

## 1   ACTIVITY: Discovering the Pythagorean Theorem

**Work with a partner.**

a. On grid paper, draw any right triangle. Label the lengths of the two shorter sides (the **legs**) $a$ and $b$.

b. Label the length of the longest side (the **hypotenuse**) $c$.

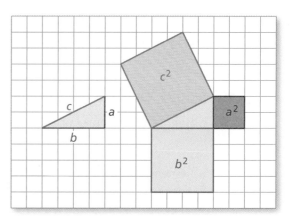

c. Draw squares along each of the three sides. Label the areas of the three squares $a^2$, $b^2$, and $c^2$.

**COMMON CORE**

**Pythagorean Theorem**
In this lesson, you will
- discover the Pythagorean Theorem.
- find missing side lengths of right triangles.
- solve real-life problems.

Learning Standards
8.G.6
8.G.7

d. Cut out the three squares. Make eight copies of the right triangle and cut them out. Arrange the figures to form two identical larger squares.

e. What does this tell you about the relationship among $a^2$, $b^2$, and $c^2$?

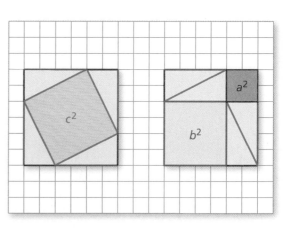

520  Chapter 10  Square Root Functions and Geometry

### 2  ACTIVITY: Finding the Length of the Hypotenuse

**Math Practice 8**

**Evaluate Results**
How can you verify that your answer is reasonable?

Work with a partner. Use the result of Activity 1 to find the length of the hypotenuse of each right triangle.

a.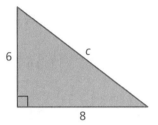

b.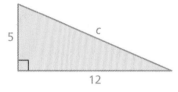

c.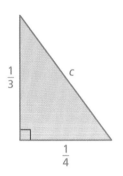

d.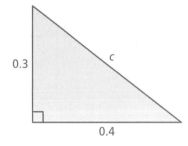

### 3  ACTIVITY: Finding the Length of a Leg

Work with a partner. Use the result of Activity 1 to find the length of the leg of each right triangle.

a.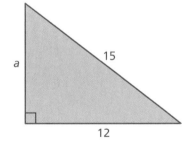

b.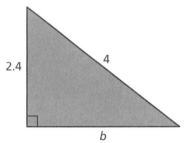

## What Is Your Answer?

4. **IN YOUR OWN WORDS** How are the lengths of the sides of a right triangle related? Give an example using whole numbers.

**Practice** — Use what you learned about the Pythagorean Theorem to complete Exercises 3–5 on page 524.

# 10.3 Lesson

**Key Vocabulary**
theorem, *p. 520*
legs, *p. 522*
hypotenuse, *p. 522*
Pythagorean Theorem, *p. 522*

## Key Ideas

### Sides of a Right Triangle

The sides of a right triangle have special names.

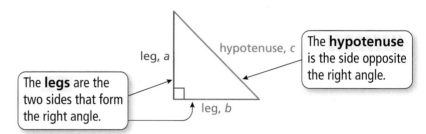

The **legs** are the two sides that form the right angle.

The **hypotenuse** is the side opposite the right angle.

**Study Tip**
In a right triangle, the legs are the shorter sides and the hypotenuse is always the longest side.

### The Pythagorean Theorem

**Words** In any right triangle, the sum of the squares of the lengths of the legs is equal to the square of the length of the hypotenuse.

**Algebra** $a^2 + b^2 = c^2$

### EXAMPLE 1 Finding the Length of a Hypotenuse

**Find the length of the hypotenuse of the triangle.**

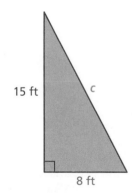

| | |
|---|---|
| $a^2 + b^2 = c^2$ | Write the Pythagorean Theorem. |
| $15^2 + 8^2 = c^2$ | Substitute 15 for $a$ and 8 for $b$. |
| $225 + 64 = c^2$ | Evaluate powers. |
| $289 = c^2$ | Add. |
| $\sqrt{289} = \sqrt{c^2}$ | Take positive square root of each side. |
| $17 = c$ | Simplify. |

∴ The length of the hypotenuse is 17 feet.

### On Your Own

**Find the length of the hypotenuse of the triangle.**

1. 
2. 

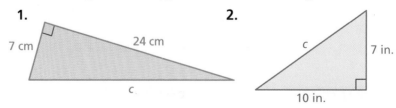

### EXAMPLE 2 — Finding the Length of a Leg

**Find the missing length of the triangle.**

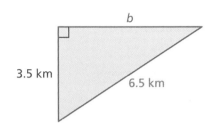

$a^2 + b^2 = c^2$    Write the Pythagorean Theorem.
$3.5^2 + b^2 = 6.5^2$    Substitute 3.5 for *a* and 6.5 for *c*.
$12.25 + b^2 = 42.25$    Evaluate powers.
$b^2 = 30$    Subtract 12.25 from each side.
$b = \sqrt{30}$    Take positive square root of each side.

∴ The length of the leg is $\sqrt{30} \approx 5.5$ kilometers.

### EXAMPLE 3 — Real-Life Application

**Paintball Team A is located 70 feet north and 60 feet east of the base. Team B is located 30 feet north and 30 feet east of the base. How far apart are the teams?**

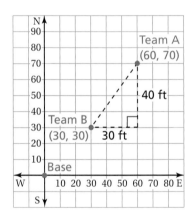

**Step 1:** Draw the situation in a coordinate plane. Let the base be at the origin. From the descriptions, you can plot Team A at (60, 70) and Team B at (30, 30).

**Step 2:** Draw a right triangle with a hypotenuse that represents the distance between the teams. The lengths of the legs are 30 feet and 40 feet.

**Step 3:** Use the Pythagorean Theorem to find the length of the hypotenuse.

$a^2 + b^2 = c^2$    Write the Pythagorean Theorem.
$30^2 + 40^2 = c^2$    Substitute 30 for *a* and 40 for *b*.
$900 + 1600 = c^2$    Evaluate powers.
$2500 = c^2$    Add.
$50 = c$    Take positive square root of each side.

∴ The teams are 50 feet apart.

### On Your Own

**Now You're Ready** — Exercises 3–8

**Find the missing length of the triangle.**

3.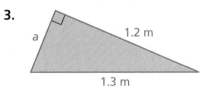

4.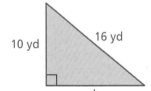

5. **WHAT IF?** In Example 3, Team B moves 10 feet to the west. How far apart are the teams to the nearest foot?

# 10.3 Exercises

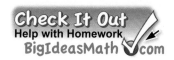

## Vocabulary and Concept Check

1. **VOCABULARY** You are given the lengths of the hypotenuse and one leg of a right triangle. Describe how you can find the length of the other leg.

2. **DIFFERENT WORDS, SAME QUESTION** Which is different? Find "both" answers.

   Which side is a leg?

   Which side is shortest?

   Which side is longest?

   Which side is part of a right angle?

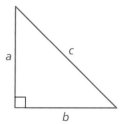

## Practice and Problem Solving

**Find the missing length of the triangle.**

3.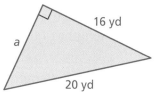

4.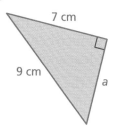

5.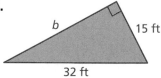

6.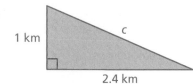

7.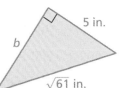

8.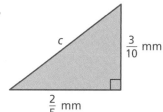

9. **ERROR ANALYSIS** Describe and correct the error in finding the missing length of the triangle.

$$a^2 + b^2 = c^2$$
$$6^2 + 20^2 = c^2$$
$$436 = c^2$$
$$\sqrt{436} = c$$

10. **TRIPOD** The center of the tripod forms a 90° angle with the ground. Find the length of the support leg to the nearest tenth of an inch.

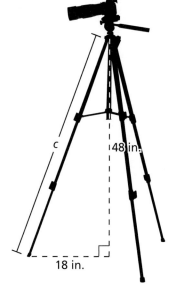

524 Chapter 10 Square Root Functions and Geometry

11. **TELEVISIONS** Televisions are advertised by the lengths of their diagonals. Approximate the length of the diagonal of the television to the nearest inch.

12. **TENNIS** A tennis player asks the referee a question. The sound of the player's voice only travels 50 feet. Can the referee hear the question? Explain.

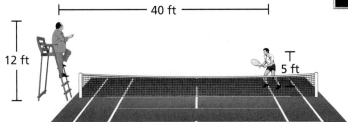

13. **GOLF** The figure shows the location of a golf ball after a tee shot. How many feet from the hole is the ball?

14. **SNOWBALLS** You and a friend throw snowballs at each other. You are 20 feet north and 15 feet east of your house. Your friend is 25 feet east and 10 feet north of your house. How far apart are you and your friend?

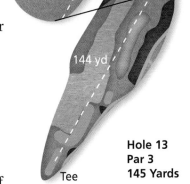

Hole 13
Par 3
145 Yards

15. **PRECISION** The legs of a right triangle have lengths of 28 meters and 21 meters. The hypotenuse has a length of $5x$ meters.
    a. Write an equation to solve for $x$.
    b. Describe how to solve the equation by factoring and by taking a square root. Which method do you prefer? Explain.
    c. What is the value of $x$?

16. **Structure** The side lengths of a right triangle are three consecutive integers.
    a. Write an expression that represents each side length. Which side length represents the hypotenuse?
    b. Write and solve an equation to find the three integers.

 **Fair Game Review** *What you learned in previous grades & lessons*

**Graph the function. Compare the graph to the graph of $y = x^2$.** *(Section 8.3)*

17. $y = -2x^2 + 4$
18. $y = -x^2 - 6$
19. $y = 3x^2 + 8$

20. **MULTIPLE CHOICE** Which polynomial is equivalent to $(x^2 - 3x + 1) - (-2x^2 + x - 4)$? *(Section 7.2)*

Ⓐ $3x^2 - 4x - 3$   Ⓑ $-x^2 - 2x - 5$   Ⓒ $3x^2 - 4x + 5$   Ⓓ $-x^2 + 4x + 3$

# 10.4 Using the Pythagorean Theorem

**Essential Question** In what other ways can you use the Pythagorean Theorem?

The *converse* of a statement switches the hypothesis and the conclusion.

| Statement: | Converse of the statement: |
|---|---|
| If $p$, then $q$. | If $q$, then $p$. |

## 1 ACTIVITY: Analyzing Converses of Statements

**Work with a partner. Write the converse of the true statement. Determine whether the converse is true or false. If it is false, give a counterexample.**

a. **Sample:** If $a = b$, then $a^2 = b^2$.
   Converse: If $a^2 = b^2$, then $a = b$.
   The converse is false. A counterexample is $a = -2$ and $b = 2$.

b. If two nonvertical lines have the same slope, then the lines are parallel.

c. If a sequence has a common difference, then it is an arithmetic sequence.

d. If $a$ and $b$ are rational numbers, then $a + b$ is a rational number.

**Is the converse of a true statement always true? always false? Explain.**

## 2 ACTIVITY: The Converse of the Pythagorean Theorem

**Work with a partner.** The converse of the Pythagorean Theorem states: "If the equation $a^2 + b^2 = c^2$ is true for the side lengths of a triangle, then the triangle is a right triangle."

a. Do you think the converse of the Pythagorean Theorem is true or false? How could you use deductive reasoning to support your answer?

b. Consider $\triangle DEF$ with side lengths $a$, $b$, and $c$, such that $a^2 + b^2 = c^2$. Also consider $\triangle JKL$ with leg lengths $a$ and $b$, where $\angle K = 90°$.

- What does the Pythagorean Theorem tell you about $\triangle JKL$?
- What does this tell you about $c$ and $x$?
- What does this tell you about $\triangle DEF$ and $\triangle JKL$?
- What does this tell you about $\angle E$?
- What can you conclude?

**COMMON CORE**

**Pythagorean Theorem**
In this lesson, you will
- identify right triangles.
- find distances between two points.
- solve real-life problems.

Learning Standards
8.G.6
8.G.7
8.G.8

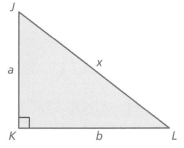

### 3 ACTIVITY: Developing the Distance Formula

Work with a partner. Follow the steps below to write a formula that you can use to find the distance between any two points in a coordinate plane.

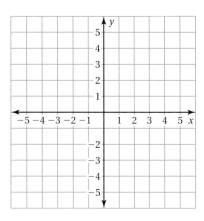

**Math Practice 6**

**Communicate Precisely**

What steps can you take to make sure that you have written the distance formula accurately?

**Step 1:** Choose two points in the coordinate plane that do not lie on the same horizontal or vertical line. Label the points $(x_1, y_1)$ and $(x_2, y_2)$.

**Step 2:** Draw a line segment connecting the points. This will be the hypotenuse of a right triangle.

**Step 3:** Draw horizontal and vertical line segments from the points to form the legs of the right triangle.

**Step 4:** Use the $x$-coordinates to write an expression for the length of the horizontal leg.

**Step 5:** Use the $y$-coordinates to write an expression for the length of the vertical leg.

**Step 6:** Substitute the expressions for the lengths of the legs into the Pythagorean Theorem.

**Step 7:** Solve the equation in Step 6 for the hypotenuse $c$.

**What does the length of the hypotenuse tell you about the two points?**

## What Is Your Answer?

4. **IN YOUR OWN WORDS** In what other ways can you use the Pythagorean Theorem?

5. What kind of real-life problems do you think the converse of the Pythagorean Theorem can help you solve?

 Use what you learned about the converse of a true statement to complete Exercises 3 and 4 on page 530.

Section 10.4 Using the Pythagorean Theorem

# 10.4 Lesson

**Key Vocabulary**
distance formula, p. 528

## Key Ideas

**Converse of the Pythagorean Theorem**

If the equation $a^2 + b^2 = c^2$ is true for the side lengths of a triangle, then the triangle is a right triangle.

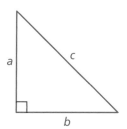

### EXAMPLE 1  Identifying a Right Triangle

**Study Tip**
A *Pythagorean triple* is a set of three positive integers $a$, $b$, and $c$, where $a^2 + b^2 = c^2$.

Tell whether each triangle is a right triangle.

a.
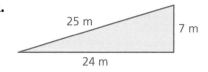

b.
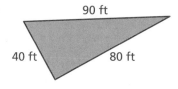

**Common Error**
When using the converse of the Pythagorean Theorem, always substitute the length of the longest side for $c$.

a.
$a^2 + b^2 = c^2$
$7^2 + 24^2 \stackrel{?}{=} 25^2$
$49 + 576 \stackrel{?}{=} 625$
$625 = 625$ ✓

∴ It *is* a right triangle.

b.
$a^2 + b^2 = c^2$
$40^2 + 80^2 \stackrel{?}{=} 90^2$
$1600 + 6400 \stackrel{?}{=} 8100$
$8000 \neq 8100$ ✗

∴ It *is not* a right triangle.

### On Your Own

**Now You're Ready**
Exercises 5–10

Tell whether the triangle with the given side lengths is a right triangle.

1. 15 cm, 10 cm, 18 cm
2. 50 yd, 40 yd, 30 yd

---

On page 527, you used the Pythagorean Theorem to develop the *distance formula*. You can use the **distance formula** to find the distance between any two points in a coordinate plane.

## Key Idea

**Distance Formula**

The distance $d$ between any two points $(x_1, y_1)$ and $(x_2, y_2)$ is given by the formula
$d = \sqrt{(x_2 - x_1)^2 + (y_2 - y_1)^2}$.

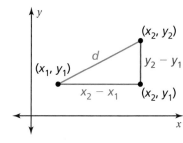

### EXAMPLE 2  Finding the Distance Between Two Points

Find the distance between $(-3, 5)$ and $(2, -1)$.

Let $(x_1, y_1) = (-3, 5)$ and $(x_2, y_2) = (2, -1)$.

$d = \sqrt{(x_2 - x_1)^2 + (y_2 - y_1)^2}$   Write the distance formula.

$\phantom{d} = \sqrt{[2 - (-3)]^2 + (-1 - 5)^2}$   Substitute.

$\phantom{d} = \sqrt{5^2 + (-6)^2}$   Simplify.

$\phantom{d} = \sqrt{25 + 36}$   Evaluate powers.

$\phantom{d} = \sqrt{61}$   Add.

### EXAMPLE 3  Real-Life Application

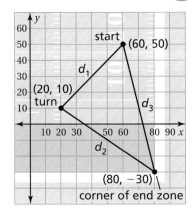

A football coach designs a passing play in which a receiver runs down the field, makes a 90° turn, and runs to the corner of the end zone. A receiver runs the play as shown. Did the receiver run the play as designed? Each unit of the grid represents 10 feet.

Use the distance formula to find the lengths of the three sides.

$d_1 = \sqrt{(60 - 20)^2 + (50 - 10)^2} = \sqrt{40^2 + 40^2} = \sqrt{3200}$ feet

$d_2 = \sqrt{(80 - 20)^2 + (-30 - 10)^2} = \sqrt{60^2 + (-40)^2} = \sqrt{5200}$ feet

$d_3 = \sqrt{(80 - 60)^2 + (-30 - 50)^2} = \sqrt{20^2 + (-80)^2} = \sqrt{6800}$ feet

Use the converse of the Pythagorean Theorem to determine if the side lengths form a right triangle.

$(\sqrt{3200})^2 + (\sqrt{5200})^2 \stackrel{?}{=} (\sqrt{6800})^2$

$3200 + 5200 \stackrel{?}{=} 6800$

$8400 \neq 6800$ ✗

It is not a right triangle. So, the receiver did not make a 90° turn.

∴ The receiver did not run the play as designed.

### On Your Own

**Now You're Ready**
Exercises 11–16

Find the distance between the two points.

**3.** $(0, 4), (5, 2)$   **4.** $(-1, 3), (4, -8)$   **5.** $(-10, -6), (6, 2)$

**6. WHAT IF?** In Example 3, the receiver made the turn at $(30, 20)$. Did the receiver run the play as designed? Explain.

Section 10.4  Using the Pythagorean Theorem

# 10.4 Exercises

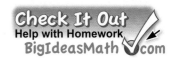

## Vocabulary and Concept Check

1. **WRITING** Describe two ways to find the distance between two points in a coordinate plane.

2. **WHICH ONE DOESN'T BELONG?** Which set of numbers does not belong with the other three? Explain your reasoning.

   | 3, 4, 5 | 8, 15, 17 | 18, 22, 29 | 9, 40, 41 |

## Practice and Problem Solving

**Write the converse of the true statement. Determine whether the converse is true or false.**

3. If $a$ is an even number, then $a^2$ is even.
4. If $a$ is positive, then $|a| = a$.

**Tell whether the triangle with the given side lengths is a right triangle.**

5.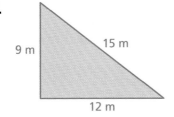
6.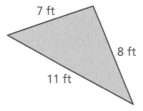
7.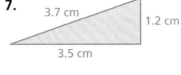

8. 16 in., 18 in., 24 in.
9. 30 yd, 22 yd, 15 yd
10. 8 mm, 15 mm, 17 mm

**Find the distance between the two points.**

11. $(4, -3), (-2, -5)$
12. $(1, 1), (4, 5)$
13. $(-7, -1), (-4, 8)$
14. $(1, 3), (7, 7)$
15. $(2, -8), (4, -1)$
16. $(-7, 4), (-3, 2)$

17. **ERROR ANALYSIS** Describe and correct the error in finding the distance between the points $(-4, -3)$ and $(2, -1)$.

    $$d = \sqrt{[2 - (-4)]^2 - [-1 - (-3)]^2}$$
    $$= \sqrt{36 - 4}$$
    $$= 4\sqrt{2}$$

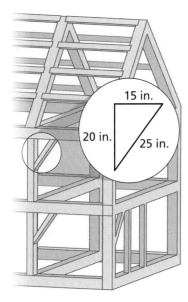

18. **CONSTRUCTION** A post and beam frame for a shed is shown in the diagram. Does the brace form a right triangle with the post and beam? Explain.

**Tell whether the set of measurements can be the side lengths of a right triangle.**

**19.** $5\sqrt{5}$, 10, 15

**20.** 7, $3\sqrt{10}$, 6

**21.** 21, 72, 75

**22. REASONING** Plot the points $(-1, -2)$, $(2, 1)$, and $(-3, 6)$ in a coordinate plane. Are the points the vertices of a right triangle? Explain.

**23. GEOCACHING** You spend the day looking for hidden containers in a wooded area using a global positioning system (GPS). You park your car on the side of the road and then locate Container 1 and Container 2 before going back to the car. Does your path form a right triangle? Explain. Each unit of the grid represents 10 yards.

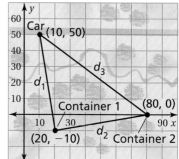

**24. REASONING** Your teacher wants the class to find the distance between the two points $(3, 2)$ and $(8, 6)$. You choose $(3, 2)$ for $(x_1, y_1)$ and your friend chooses $(8, 6)$ for $(x_1, y_1)$. Do you and your friend obtain the same answer? Justify your answer.

**25. AIRPORT** Which plane is closer to the base of the airport tower? Explain.

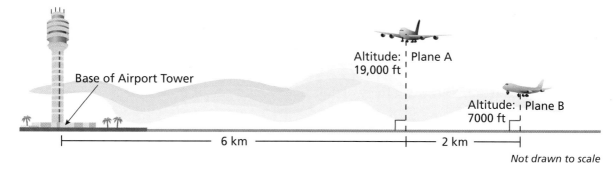

**26. Structure** Consider the two points $(x_1, y_1)$ and $(x_2, y_2)$ in the coordinate plane. How can you find the point $(x_m, y_m)$ located in the middle of the two given points? Justify your answer using the distance formula.

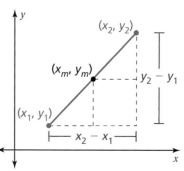

## Fair Game Review  *What you learned in previous grades & lessons*

**Solve the equation using the quadratic formula.** *(Section 9.4)*

**27.** $2x^2 - 5x + 3 = 0$

**28.** $2x^2 - x - 1 = 0$

**29.** $x^2 + 3x + 5 = 0$

**30. MULTIPLE CHOICE** Which point is the focus of the graph of $y = 2x^2$? *(Section 8.2)*

Ⓐ $\left(0, \dfrac{1}{8}\right)$  Ⓑ $\left(0, \dfrac{1}{4}\right)$  Ⓒ $\left(0, \dfrac{1}{2}\right)$  Ⓓ $\left(0, -\dfrac{1}{8}\right)$

# 10.3–10.4 Quiz

**Find the missing length of the triangle.** *(Section 10.3)*

1.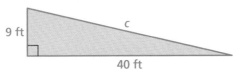

2.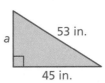

3.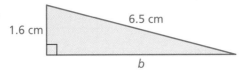

4.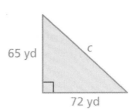

**Tell whether the triangle with the given side lengths is a right triangle.** *(Section 10.4)*

5.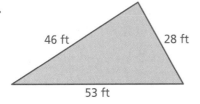

6.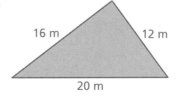

**Find the distance between the two points.** *(Section 10.4)*

7. $(-3, -1), (-1, -5)$

8. $(-4, 2), (5, 1)$

9. $(1, -2), (4, -5)$

10. $(-1, 1), (7, 4)$

11. $(-6, 5), (-4, -6)$

12. $(-1, 4), (1, 3)$

13. **FABRIC** You cut a rectangular piece of fabric in half along the diagonal. The fabric measures 28 inches wide and $1\frac{1}{4}$ yards long. What is the length (in inches) of the diagonal? *(Section 10.3)*

**Use the figure to answer Exercises 14–17.** *(Section 10.4)*

14. How far is the cabin from the peak?

15. How far is the fire tower from the lake?

16. How far is the lake from the peak?

17. You are standing at $(-5, -6)$. How far are you from the lake?

# 10 Chapter Review

## Review Key Vocabulary

square root function, *p. 504*
simplest form of a radical expression, *p. 508*
rationalizing the denominator, *p. 508*
conjugates, *p. 509*
square root equation, *p. 512*

extraneous solution, *p. 513*
theorem, *p. 520*
legs, *p. 522*
hypotenuse, *p. 522*
Pythagorean Theorem, *p. 522*
distance formula, *p. 528*

## Review Examples and Exercises

### 10.1 Graphing Square Root Functions *(pp. 502–509)*

a. **Graph $y = \sqrt{x} - 1$. Describe the domain and range. Compare the graph to the graph of $y = \sqrt{x}$.**

   **Step 1:** Make a table of values.

   | x | 0 | 1 | 4 | 9 | 16 |
   |---|---|---|---|---|---|
   | y | −1 | 0 | 1 | 2 | 3 |

   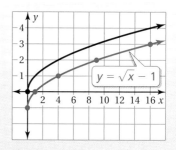

   **Step 2:** Plot the ordered pairs.
   **Step 3:** Draw a smooth curve through the points.

   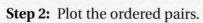 The domain is $x \geq 0$. The range is $y \geq -1$. The graph of $y = \sqrt{x} - 1$ is a translation 1 unit down of the graph of $y = \sqrt{x}$.

b. **Graph $y = \sqrt{x + 2}$. Describe the domain and range. Compare the graph to the graph of $y = \sqrt{x}$.**

   **Step 1:** Make a table of values.

   | x | −2 | −1 | 0 | 1 | 2 |
   |---|---|---|---|---|---|
   | y | 0 | 1 | 1.4 | 1.7 | 2 |

   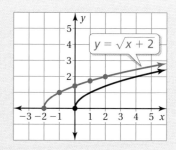

   **Step 2:** Plot the ordered pairs.
   **Step 3:** Draw a smooth curve through the points.

   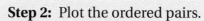 The domain is $x \geq -2$. The range is $y \geq 0$. The graph of $y = \sqrt{x + 2}$ is a translation 2 units to the left of the graph of $y = \sqrt{x}$.

### Exercises

**Graph the function. Describe the domain and range. Compare the graph to the graph of $y = \sqrt{x}$.**

1. $y = \sqrt{x} + 7$
2. $y = \sqrt{x - 6}$
3. $y = -\sqrt{x + 3} - 1$

Chapter Review  533

## 10.2 Solving Square Root Equations (pp. 510–517)

Solve $\sqrt{12-x} = x$.

| | |
|---|---|
| $\sqrt{12-x} = x$ | Write the equation. |
| $(\sqrt{12-x})^2 = x^2$ | Square each side of the equation. |
| $12 - x = x^2$ | Simplify. |
| $0 = x^2 + x - 12$ | Rewrite equation. |
| $0 = (x-3)(x+4)$ | Factor. |
| $(x-3) = 0$ or $(x+4) = 0$ | Use Zero-Product Property. |
| $x = 3$ or $x = -4$ | Solve for $x$. |

**Check**

| | | | |
|---|---|---|---|
| $\sqrt{12-3} \stackrel{?}{=} 3$ | Substitute for $x$. | $\sqrt{12-(-4)} \stackrel{?}{=} -4$ | |
| $\sqrt{9} \stackrel{?}{=} 3$ | Simplify. | $\sqrt{16} \stackrel{?}{=} -4$ | |
| $3 = 3$ ✓ | | $4 \neq -4$ ✗ | |

∴ Because $x = -4$ does not check in the original equation, it is an extraneous solution. The only solution is $x = 3$.

### Exercises

**Solve the equation. Check your solution.**

4. $8 + \sqrt{x} = 18$
5. $\sqrt{x-1} + 9 = 15$
6. $\sqrt{5x-9} = \sqrt{4x}$
7. $x = \sqrt{3x+4}$

## 10.3 The Pythagorean Theorem (pp. 520–525)

**Find the length of the hypotenuse of the triangle.**

| | |
|---|---|
| $a^2 + b^2 = c^2$ | Write the Pythagorean Theorem. |
| $10^2 + 24^2 = c^2$ | Substitute. |
| $100 + 576 = c^2$ | Evaluate powers. |
| $676 = c^2$ | Add. |
| $\sqrt{676} = \sqrt{c^2}$ | Take positive square root of each side. |
| $26 = c$ | Simplify. |

∴ The length of the hypotenuse is 26 meters.

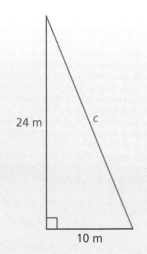

## Exercises

**Find the missing length of the triangle.**

8.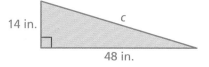

9.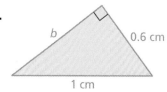

### 10.4 Using the Pythagorean Theorem (pp. 526–531)

**a. Is the triangle formed by the rope and the tent a right triangle?**

$$a^2 + b^2 = c^2$$
$$64^2 + 48^2 \stackrel{?}{=} 80^2$$
$$4096 + 2304 \stackrel{?}{=} 6400$$
$$6400 = 6400 \checkmark$$

∴ It *is* a right triangle.

**b. Find the distance between (−3, 1) and (4, 7).**

Let $(x_1, y_1) = (-3, 1)$ and $(x_2, y_2) = (4, 7)$.

$d = \sqrt{(x_2 - x_1)^2 + (y_2 - y_1)^2}$   Write the distance formula.

$\phantom{d} = \sqrt{[4 - (-3)]^2 + (7 - 1)^2}$   Substitute.

$\phantom{d} = \sqrt{7^2 + 6^2}$   Simplify.

$\phantom{d} = \sqrt{49 + 36}$   Evaluate powers.

$\phantom{d} = \sqrt{85}$   Add.

## Exercises

**Tell whether the triangle is a right triangle.**

10.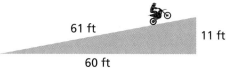

11.

**Find the distance between the two points.**

12. $(-2, -5), (3, 5)$

13. $(-4, 7), (4, 0)$

# 10 Chapter Test

**Graph the function. Describe the domain and range. Compare the graph to the graph of $y = \sqrt{x}$.**

1. $y = \sqrt{x} - 6$
2. $y = \sqrt{x + 10}$
3. $y = -\sqrt{x - 2} + 3$

**Solve the equation.**

4. $9 - \sqrt{x} = 3$
5. $\sqrt{2x - 7} - 3 = 6$
6. $\sqrt{8x - 21} = \sqrt{18 - 5x}$
7. $x + 5 = \sqrt{7x + 53}$

**Find the missing length of the triangle.**

8.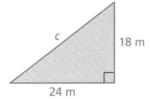

9.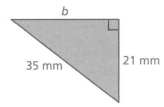

**Tell whether the triangle is a right triangle.**

10.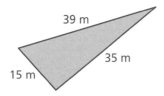

11.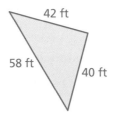

**Find the distance between the two points.**

12. $(-2, 3), (6, 9)$
13. $(0, -5), (4, 1)$
14. $(-3, -4), (2, -7)$

15. **ROLLER COASTER** The velocity $v$ (in meters per second) of a roller coaster at the bottom of a hill is given by $v = \sqrt{19.6h}$, where $h$ is the height of the hill (in meters). (a) Graph the function. Describe the domain and range. (b) How tall must the hill be for the velocity of the roller coaster at the bottom of the hill to be at least 28 meters per second?

16. **FINANCE** The average annual interest rate $r$ (in decimal form) that an investment earns over 2 years is given by $r = \sqrt{\dfrac{V_2}{V_0}} - 1$, where $V_0$ is the initial investment and $V_2$ is the value of the investment after 2 years. You initially invest $800 which earns an average annual interest of 6% over 2 years. What is the value of $V_2$?

17. **BUTTERFLY** Approximate the wingspan of the butterfly.

536   Chapter 10   Square Root Functions and Geometry

# 10 Standards Assessment

1. Which function is shown in the graph? *(F.IF.7b)*

   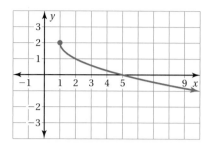

   A. $y = -\sqrt{x+1} + 2$

   B. $y = -\sqrt{x-1} + 2$

   C. $y = -\sqrt{x+2} + 1$

   D. $y = -\sqrt{x-2} + 1$

**Test-Taking Strategy**
**Solve Problem Before Looking at Choices**

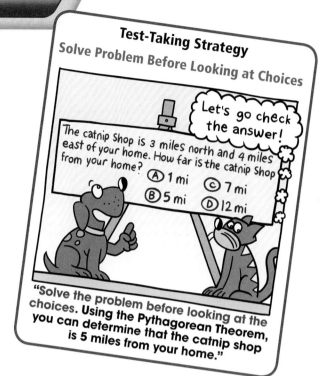

"Solve the problem before looking at the choices. Using the Pythagorean Theorem, you can determine that the catnip shop is 5 miles from your home."

2. Which number is equivalent to $\dfrac{8^{10} \cdot 2^{-5}}{16^4}$? *(N.RN.2)*

   F. 1

   G. $2^9$

   H. $2^{24}$

   I. $2^{-11}$

3.  What value of $x$ makes the equation below true? *(N.RN.2)*

   $$x = \sqrt{2x+8}$$

4. A line with a slope of $-2$ passes through the point $(1, -6)$. Which of the following is not a point on the line? *(A.CED.2)*

   A. $(-8, 12)$

   B. $(-4, 4)$

   C. $(4, -4)$

   D. $(8, -20)$

5. What value(s) of $x$ make the equation below true? *(N.RN.2)*

   $$\sqrt{2x-4} = x - 2$$

   F. Only $-2$

   G. 2 and 4

   H. Only 4

   I. $-2$ and $-4$

6. What is the focus of the parabola? *(F.IF.4)*

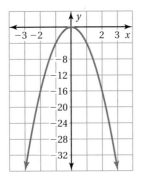

A. $\left(0, \dfrac{1}{16}\right)$

B. $(0, 16)$

C. $\left(0, -\dfrac{1}{16}\right)$

D. $(0, -16)$

7. The range of the function $y = 6x - 8$ is all real numbers from 1 to 10. What is the domain of the function? *(F.IF.1)*

F. all real numbers from 1.5 to 3

G. all real numbers from $-2$ to 52

H. all integers from $-2$ to 52

I. all real numbers

8. What is the distance between the two points in the coordinate plane? *(8.G.8)*

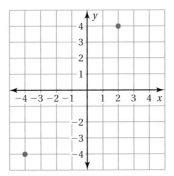

9. Which ordered pair is a solution of the system of inequalities shown in the graph? *(A.REI.12)*

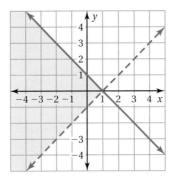

A. $(1, 0)$

B. $(0, 1)$

C. $(0, -1)$

D. $(2, 0)$

**10.** Two nature trails are shown below. Which trail is longer? By how much? Explain your reasoning. *(8.G.7)*

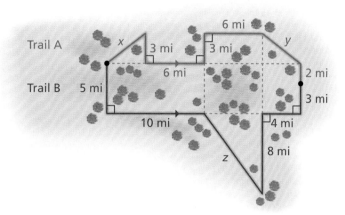

**11.** The solution of which inequality is shown in the graph below? *(A.REI.3)*

**F.** $5x - 7 \geq 3$

**G.** $4x + 3 \leq 11$

**H.** $12 - 3x < 6$

**I.** $10 - 2x > 6$

**12.** Tom was graphing $y = \sqrt{x + 2} - 1$. His work is shown below. *(F.IF.7b)*

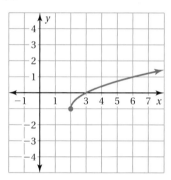

What should Tom do to correct the error that he made?

**A.** Shift the graph 1 unit down and 1 unit left.

**B.** Shift the graph 4 units left.

**C.** Shift the graph 3 units up and 3 units left.

**D.** Shift the graph 1 unit down and 3 units left.

**13.** What is the vertex of the graph of $y = 2x^2 - 4x + 6$? *(F.IF.4)*

**F.** $(-1, 12)$

**G.** $(-2, 22)$

**H.** $(1, 4)$

**I.** $(2, 6)$

# 11 Rational Equations and Functions

- **11.1** Direct and Inverse Variation
- **11.2** Graphing Rational Functions
- **11.3** Simplifying Rational Expressions
- **11.4** Multiplying and Dividing Rational Expressions
- **11.5** Dividing Polynomials
- **11.6** Adding and Subtracting Rational Expressions
- **11.7** Solving Rational Equations

"Descartes, in your homework problem, you are being chased by a cat-eating hyena."

"Both of you must stay on the graph of $y = 1/x$. The safe zone is the $x$-axis."

"Can you ever reach the safe zone? Explain your reasoning."

"Descartes, I am keeping track of how many doggy treats my owner gives me each day."

"I am finding that my happiness is directly proportional to the day of the week."

# What You Learned Before

"Your namesake, René Descartes, believed in rational thinking."

*I get it. He thought. Therefore, I am.*

### ● Evaluating Expressions with Fractions (5.NF.1, 6.NS.1)

**Example 1** Find $\dfrac{1}{10} + \dfrac{3}{5}$.

$$\dfrac{1}{10} + \dfrac{3}{5} = \dfrac{1}{10} + \dfrac{6}{10}$$
$$= \dfrac{1+6}{10}$$
$$= \dfrac{7}{10}$$

**Example 2** Find $\dfrac{4}{3} \div \dfrac{5}{3}$.

$$\dfrac{4}{3} \div \dfrac{5}{3} = \dfrac{4}{3} \cdot \dfrac{3}{5}$$
$$= \dfrac{4 \cdot \cancel{3}}{\cancel{3} \cdot 5}$$
$$= \dfrac{4}{5}$$

**Try It Yourself**
Evaluate the expression.

1. $\dfrac{1}{6} + \dfrac{5}{6}$
2. $\dfrac{2}{3} - \dfrac{1}{6}$
3. $\dfrac{5}{2} \cdot \dfrac{2}{3}$
4. $\dfrac{5}{6} \div \dfrac{8}{3}$

### ● Solving Proportions (7.RP.3)

**Example 3** Solve $\dfrac{4}{x} = \dfrac{5}{12}$.

$\dfrac{4}{x} = \dfrac{5}{12}$   Write the proportion.

$4 \cdot 12 = x \cdot 5$   Use the Cross Products Property.

$48 = 5x$   Multiply.

$9.6 = x$   Divide each side by 5.

**Try It Yourself**
Solve the proportion.

5. $\dfrac{4}{6} = \dfrac{2}{x}$
6. $\dfrac{3}{12} = \dfrac{w}{8}$
7. $\dfrac{15}{y} = \dfrac{25}{13}$

# 11.1 Direct and Inverse Variation

**Essential Question** How can you recognize when two variables vary directly? How can you recognize when they vary inversely?

### 1  ACTIVITY: Recognizing Direct Variation

**Work with a partner. You hang different weights from the same spring.**

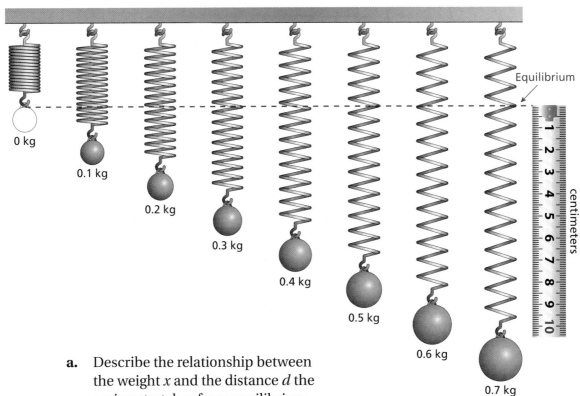

**Common Core**

**Direct and Inverse Variation**

In this lesson, you will
- identify direct and inverse variation.
- write and graph direct and inverse variation equations.

Learning Standard
A.REI.10

a. Describe the relationship between the weight $x$ and the distance $d$ the spring stretches from equilibrium. Explain why the distance is said to vary *directly* with the weight.

b. Graph the relationship between $x$ and $d$. What are the characteristics of the graph?

c. Write an equation that represents $d$ as a function of $x$.

d. In physics, the relationship between $d$ and $x$ is described by Hooke's Law. How would you describe Hooke's Law?

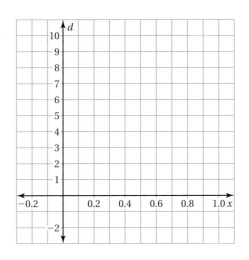

542   Chapter 11   Rational Equations and Functions

## 2  ACTIVITY: Recognizing Inverse Variation

**Math Practice 6**

**Calculate Accurately**

How can you verify that your graph and equation represent the relationship between $x$ and $y$?

Work with a partner. The area of each rectangle is 64 square inches.

$x = 64$ in.   $y = 1$ in.

$x = 4$ in.
$y = 16$ in.

$x = 2$ in.
$y = 32$ in.

$x = 32$ in.   $y = 2$ in.

$x = 16$ in.   $y = 4$ in.

$x = 8$ in.   $y = 8$ in.

$x = 1$ in.
$y = 64$ in.

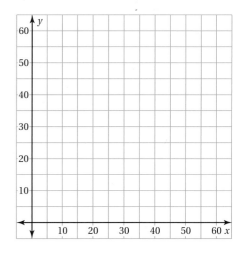

a. Describe the relationship between $x$ and $y$. Explain why $y$ is said to vary inversely with $x$.

b. Graph the relationship between $x$ and $y$. What are the characteristics of the graph?

c. Write an equation that represents $y$ as a function of $x$.

## What Is Your Answer?

3. **IN YOUR OWN WORDS** How can you recognize when two variables vary directly? How can you recognize when they vary inversely?

4. Does the flapping rate of a bird's wings vary directly or inversely with the length of its wings? Explain your reasoning.

**Practice** — Use what you learned about direct and inverse variation to complete Exercises 3 and 4 on page 547.

Section 11.1   Direct and Inverse Variation

## 11.1 Lesson

**Key Vocabulary**
direct variation, p. 544
inverse variation, p. 544

### Key Ideas

**Direct Variation**

Two quantities $x$ and $y$ show **direct variation** when $y = kx$, where $k$ is a nonzero constant.

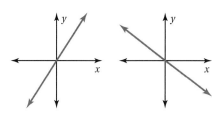

$y = kx, k > 0$  $y = kx, k < 0$

The ratio $\dfrac{y}{x}$ is constant.

**Inverse Variation**

Two quantities $x$ and $y$ show **inverse variation** when $y = \dfrac{k}{x}$, where $k$ is a nonzero constant.

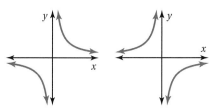

$y = \dfrac{k}{x}, k > 0$  $y = \dfrac{k}{x}, k < 0$

The product $xy$ is constant.

**Study Tip**
The constant $k$ is called the *constant of proportionality* or the *constant of variation*.

### EXAMPLE 1 — Identifying Direct and Inverse Variation

**Tell whether $x$ and $y$ show *direct variation*, *inverse variation*, or *neither*. Explain your reasoning.**

a.

| x | 1 | 2 | 3 | 4 |
|---|---|---|---|---|
| y | 5 | 10 | 15 | 20 |

The products $xy$ are not constant. So, the table does not show inverse variation.

Check each ratio $\dfrac{y}{x}$:  $\dfrac{5}{1} = 5$,  $\dfrac{10}{2} = 5$,  $\dfrac{15}{3} = 5$,  $\dfrac{20}{4} = 5$

∴ The ratios are constant. So, $x$ and $y$ show direct variation.

b. $4xy = -4$

$y = -\dfrac{1}{x}$   Divide each side by $4x$.

∴ The equation is of the form $y = \dfrac{k}{x}$. So, $x$ and $y$ show inverse variation.

### On Your Own

**Now You're Ready** Exercises 3–6

**Tell whether $x$ and $y$ show *direct variation*, *inverse variation*, or *neither*. Explain your reasoning.**

1.

| x | 1 | 2 | 3 | 4 |
|---|---|---|---|---|
| y | 24 | 12 | 8 | 6 |

2. $y = 3x + 1$

### EXAMPLE 2  Writing and Graphing a Direct Variation Equation

**The variable $y$ varies directly with $x$. When $x = 12$, $y = -6$. Write and graph a direct variation equation that relates $x$ and $y$.**

Find the value of $k$.

$y = kx$   Write the direct variation equation.

$-6 = k(12)$   Substitute 12 for $x$ and $-6$ for $y$.

$-\dfrac{1}{2} = k$   Divide each side by 12.

So, an equation that relates $x$ and $y$ is $y = -\dfrac{1}{2}x$.

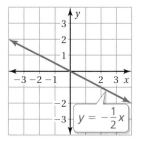

**Reading**

For direct variation equations, you can say "$y$ varies directly with $x$" or "$y$ is directly proportional to $x$." For inverse variation equations, you can say "$y$ varies inversely with $x$" or "$y$ is inversely proportional to $x$."

### EXAMPLE 3  Writing and Graphing an Inverse Variation Equation

**The variable $y$ varies inversely with $x$. When $x = 2$, $y = 5$.**

**a. Write an inverse variation equation that relates $x$ and $y$.**

Find the value of $k$.

$y = \dfrac{k}{x}$   Write the inverse variation equation.

$5 = \dfrac{k}{2}$   Substitute 2 for $x$ and 5 for $y$.

$10 = k$   Multiply each side by 2.

So, an equation that relates $x$ and $y$ is $y = \dfrac{10}{x}$.

**b. Graph the inverse variation equation. Describe the domain and range.**

Make a table of values.

| x | −10 | −5 | −2 | 0 | 2 | 5 | 10 |
|---|-----|----|----|----|----|----|----|
| y | −1 | −2 | −5 | undef. | 5 | 2 | 1 |

Plot the ordered pairs. Draw a smooth curve through the points in each quadrant. Both the domain and range are all real numbers except 0.

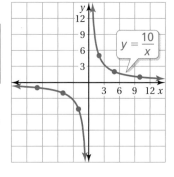

**Study Tip**

Notice that the equation $y = \dfrac{10}{x}$ is undefined when $x = 0$. So, there is no point on the graph for $x = 0$.

### On Your Own

**Now You're Ready**
Exercises 7–12

**3.** The variable $y$ varies directly with $x$. When $x = 3$, $y = 15$. Write and graph a direct variation equation that relates $x$ and $y$.

**4.** The variable $y$ varies inversely with $x$. When $x = 5$, $y = 4$. Write and graph an inverse variation equation that relates $x$ and $y$.

# EXAMPLE 4  Identifying Inverse Variation

**Which situation represents inverse variation?**

- **A** You buy several movie tickets for $7.50 each.
- **B** You earn $0.50 for each pound of aluminum cans you recycle.
- **C** The cost of a $600 cabin rental is shared equally by a group of friends.
- **D** You download several songs for $0.99 each.

Make a table of values for each situation.

**A**

| Number of tickets, x | 1 | 2 | 3 |
|---|---|---|---|
| Total cost, y | 7.50 | 15 | 22.50 |

The ratio $\frac{y}{x}$ is constant.

**B**

| Number of pounds, x | 1 | 2 | 3 |
|---|---|---|---|
| Total earned, y | 0.50 | 1 | 1.50 |

The ratio $\frac{y}{x}$ is constant.

**C**

| Number of people, x | 1 | 2 | 3 |
|---|---|---|---|
| Cost per person, y | 600 | 300 | 200 |

The product $xy$ is constant.

**D**

| Number of songs, x | 1 | 2 | 3 |
|---|---|---|---|
| Total cost, y | 0.99 | 1.98 | 2.97 |

The ratio $\frac{y}{x}$ is constant.

∴ The correct answer is **C**.

# EXAMPLE 5  Real-Life Application

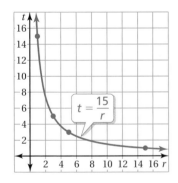

You bike 15 miles each morning. Your time $t$ (in hours) to bike 15 miles is given by $t = \frac{15}{r}$, where $r$ is your average speed (in miles per hour). **Graph the function. Make a conclusion from the graph.**

Because average speed cannot be negative, use only nonnegative values of $r$.

| r | 0 | 1 | 3 | 5 | 15 |
|---|---|---|---|---|---|
| t | undef. | 15 | 5 | 3 | 1 |

From the graph, you can see that as your average speed increases, the time it takes you to bike 15 miles decreases.

## On Your Own

Exercises 23–25

**5.** The cost of a taxi ride is shared equally by several friends. Does this situation represent direct variation or inverse variation? Explain.

**6. WHAT IF?** In Example 5, you bike 12 miles each morning. Write and graph a function that represents your time. Then make a conclusion from the graph.

# 11.1 Exercises

## Vocabulary and Concept Check

1. **VOCABULARY** Explain how direct variation equations and inverse variation equations are different.

2. **WHICH ONE DOESN'T BELONG?** Which graph does *not* belong with the other three? Explain your reasoning.

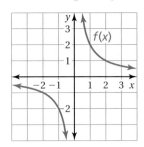 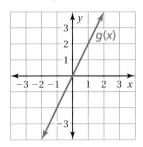 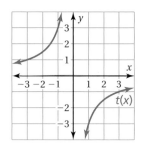 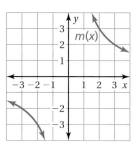

## Practice and Problem Solving

**Tell whether $x$ and $y$ show *direct variation*, *inverse variation*, or *neither*. Explain your reasoning.**

3. 
| x | 1 | 2 | 3 | 4 |
|---|---|---|---|---|
| y | 2 | 4 | 6 | 8 |

4. 
| x | 1 | 2 | 3 | 4 |
|---|---|---|---|---|
| y | 12 | 6 | 4 | 3 |

5. $2y = x$

6. $-3xy = 6$

**The variable $y$ varies directly with $x$. Write and graph a direct variation equation that relates $x$ and $y$.**

7. When $x = 2$, $y = 6$.

8. When $x = 3$, $y = 12$.

9. When $x = 30$, $y = 5$.

**The variable $y$ varies inversely with $x$. Write and graph an inverse variation equation that relates $x$ and $y$.**

10. When $x = 3$, $y = 5$.

11. When $x = 5$, $y = 9$.

12. When $x = 5$, $y = 6$.

13. **VOLUNTEERS** You want to raise $500 for a charity. You volunteer $h$ hours and raise $r$ dollars each hour. The equation $hr = 500$ represents this situation. Does this represent direct variation, inverse variation, or neither? Explain your reasoning.

Section 11.1  Direct and Inverse Variation  547

14. **ERROR ANALYSIS** The variable $y$ varies inversely with $x$. When $x = 8$, $y = 5$. Describe and correct the error in writing an inverse variation equation that relates $x$ and $y$.

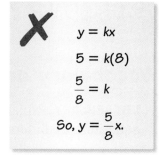

**Graph the equation. Describe the domain and range.**

15. $y = \dfrac{1}{x}$

16. $\dfrac{y}{x} = -\dfrac{1}{2}$

17. $xy = 9$

18. **REASONING** When $y$ varies directly with $x$, does $x$ vary directly with $y$? If so, describe the relationship between the constants of proportionality. Explain your reasoning.

**The variable $y$ varies inversely with $x$. Write an inverse variation equation that relates $x$ and $y$. Then find the missing value of $x$ or $y$.**

19. When $x = 6$, $y = 2$. Find $x$ when $y = 1$.

20. When $x = 4$, $y = 2$. Find $x$ when $y = \dfrac{1}{2}$.

21. When $x = -2$, $y = -5$. Find $y$ when $x = 4$.

22. When $x = 20$, $y = \dfrac{4}{5}$. Find $y$ when $x = 8$.

**Determine whether the situation represents *direct variation* or *inverse variation*. Justify your answer.**

23. You have enough money to buy 5 hats for $10 each or 10 hats for $5 each.

24. Your cousin earns $50 for mowing 2 lawns or $75 for mowing 3 lawns.

25. The money the swim team earns from a car wash is divided evenly among the members.

26. **RUNNING** You race in a 200-meter dash. Your average speed $r$ (in meters per second) is given by $r = \dfrac{200}{t}$, where $t$ is the time (in seconds) it takes you to finish the race. Graph the function. Make a conclusion from the graph.

27. **VACATION** The amount $v$ of vacation time (in hours) that an employee earns varies directly with the amount $t$ of time (in months) she works. An employee who works 2 months earns 36 hours of vacation time.

   a. Write and graph a direct variation equation that relates $v$ and $t$.

   b. How many hours of vacation time does the employee earn after working 5 months?

28. **REASONING** Make a table using positive $x$-values for the inverse variation equation $v = \dfrac{6}{x}$ and the direct variation equation $d = 6x$. How does the rate of change of $v$ differ from the rate of change of $d$?

29. **THEATER** A performing arts company is hiring actors as extras for a theater performance. The amount $t$ of performance time (in hours per person) varies inversely with the number $p$ of extras hired. The director estimates that he will need 20 extras performing 250 hours each.

    a. Write an inverse variation equation that relates $t$ and $p$.
    b. The director decides to hire 25 extras. How much performance time will each extra receive?

30. **STRUCTURE** To balance the board in the diagram, the distance (in feet) of each animal from the center of the board must vary inversely with its weight (in pounds). What is the distance of each animal from the fulcrum?

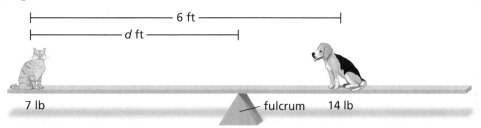

A function $f$ is odd if $f(-x) = -f(x)$. A function $f$ is even if $f(-x) = f(x)$. Determine whether the function is *odd*, *even*, or *neither*.

31. $f(x) = x$
32. $f(x) = \dfrac{1}{x}$
33. $f(x) = x^2$
34. $f(x) = \sqrt{x}$
35. $f(x) = |x|$
36. $f(x) = 2^x$

37. **REASONING** Describe the symmetry shown in the graph of (a) an even function and (b) an odd function. Justify your answers.

38.  Are all direct variation and inverse variation equations odd functions? Explain.

## Fair Game Review *What you learned in previous grades & lessons*

**Graph the function. Compare the graph to the graph of $y = x^2$.**
*(Section 8.1 and Section 8.2)*

39. $y = 3x^2$
40. $y = x^2 + 2$
41. $y = x^2 - 1$

42. **MULTIPLE CHOICE** What is the solution of the equation $\sqrt{x} - 5 = 4$? *(Section 10.2)*

    Ⓐ 1    Ⓑ 3    Ⓒ 9    Ⓓ 81

# 11.2 Graphing Rational Functions

**Essential Question** What are the characteristics of the graph of a rational function?

### 1 ACTIVITY: Graphing a Rational Function

Work with a partner. As a fundraising project, your math club is publishing an optical illusion calendar. The cost of the art, typesetting, and paper is $850. In addition to this one-time cost, the unit cost of printing each calendar is $3.25.

a. Let $A$ represent the average cost of each calendar. Write a rational function that gives the average cost of printing $x$ calendars.

$$A = \frac{\phantom{xxxxxxxx}}{x}$$

b. Make a table showing the average costs for several different production amounts. Then use the table to graph the average cost function.

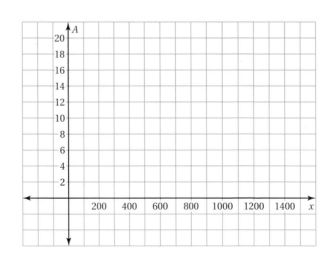

**COMMON CORE**

**Graphing Rational Functions**

In this lesson, you will
- graph rational functions.
- identify asymptotes.
- compare graphs of rational functions.

Learning Standards
A.REI.10
F.BF.4a

550  Chapter 11  Rational Equations and Functions

### 2 ACTIVITY: Analyzing the Graph of a Rational Function

**Work with a partner. Use the graph in Activity 1.**

**Math Practice**

**Justify Conclusions**
What information can you use to justify your conclusion?

a. What is the *greatest* average cost of a calendar? Explain your reasoning.

b. What is the *least* average cost of a calendar? Explain your reasoning. What characteristic of the graph is associated with the least average cost?

### 3 ACTIVITY: Analyzing Profit and Revenue

**Work with a partner. Consider the calendar project in Activity 1. Suppose your club sells 1400 calendars for $10 each.**

a. Find the revenue your club earns from the calendars.

b. How much profit does your club earn? Explain your reasoning.

## What Is Your Answer?

4. **IN YOUR OWN WORDS** What are the characteristics of the graph of a rational function? Illustrate your answer with the graphs of the following rational functions.

   a. $y = \dfrac{x+1}{x}$    b. $y = \dfrac{x+2}{x}$    c. $y = \dfrac{x+3}{x}$

**Practice**  Use what you learned about the graphs of rational functions to complete Exercises 4 and 5 on page 555.

Section 11.2   Graphing Rational Functions   551

# 11.2 Lesson

Key Vocabulary
rational function, p. 552
excluded value, p. 552
asymptote, p. 553

The inverse variation equations in Section 11.1 are *rational functions*.

## Key Idea

**Rational Function**

A **rational function** is a function of the form $y = \dfrac{\text{polynomial}}{\text{polynomial}}$, where the denominator does not equal 0. The most basic rational function is $y = \dfrac{1}{x}$.

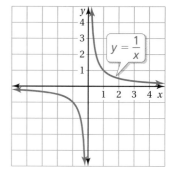

Because division by 0 is undefined, the value of the denominator of a rational function cannot be 0. So, the domain of a rational function *excludes* values that make the denominator 0. These values are called **excluded values** of the rational function.

### EXAMPLE 1  Finding the Excluded Value of a Rational Function

Find the excluded value of $y = \dfrac{2}{x + 5}$.

Find the value of $x$ that makes the denominator 0.

$x + 5 = 0$      Use the denominator to write an equation.

$x = -5$      Subtract 5 from each side.

∴ The excluded value is $x = -5$.

### EXAMPLE 2  Graphing a Rational Function

Graph $y = \dfrac{1}{x - 1}$. Describe the domain and range.

The excluded value is $x = 1$, so choose $x$-values on either side of 1.

**Step 1:** Make a table of values.

| x | −2 | −1 | 0 | 0.5 | 1 | 1.5 | 2 | 3 | 4 |
|---|---|---|---|---|---|---|---|---|---|
| y | $-\dfrac{1}{3}$ | $-\dfrac{1}{2}$ | −1 | −2 | undef. | 2 | 1 | $\dfrac{1}{2}$ | $\dfrac{1}{3}$ |

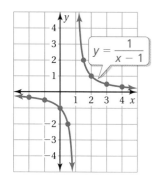

**Step 2:** Plot the ordered pairs.

**Step 3:** Draw a smooth curve through the points on each side of $x = 1$.

∴ The domain is all real numbers except 1 and the range is all real numbers except 0.

**Find the excluded value of the function.**

Exercises 6–17

1. $y = \dfrac{3}{2x}$
2. $y = \dfrac{1}{x-4}$
3. $y = \dfrac{8}{3x+1}$

**Graph the function. Describe the domain and range.**

4. $y = -\dfrac{8}{x}$
5. $y = \dfrac{1}{x+2}$
6. $y = \dfrac{1}{x} - 1$

The excluded value in Example 2 is $x = 1$. Notice that the graph approaches the vertical line $x = 1$, but never intersects it. The graph also approaches the horizontal line $y = 0$, but never intersects it. These lines are called *asymptotes*. An **asymptote** is a line that a graph approaches, but never intersects.

### Key Idea

**Asymptotes**

The graph of a rational function of the form $y = \dfrac{a}{x-h} + k$, where $a \ne 0$, has a vertical asymptote $x = h$ and a horizontal asymptote $y = k$.

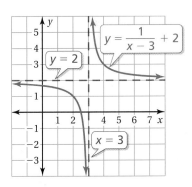

**EXAMPLE 3**  **Identifying Asymptotes**

Identify the asymptotes of the graph of $y = \dfrac{1}{x-2} - 4$. Then describe the domain and range.

Rewrite the function to find the asymptotes.

$y = \dfrac{1}{x-2} + (-4)$

Horizontal Asymptote: $y = -4$

Vertical Asymptote: $x = 2$

∴ The vertical asymptote is $x = 2$ and the horizontal asymptote is $y = -4$. So, the domain of the function is all real numbers except 2 and the range is all real numbers except $-4$.

**Check**

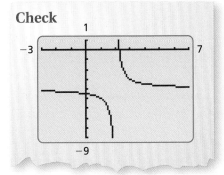

### On Your Own

Exercises 19–24

**Identify the asymptotes of the graph of the function. Then describe the domain and range.**

7. $y = \dfrac{2}{x} + 1$
8. $y = \dfrac{1}{x+5}$
9. $y = \dfrac{8}{x-3} - 2$

Section 11.2  Graphing Rational Functions  553

### EXAMPLE 4 — Comparing Graphs of Rational Functions

Graph $y = \dfrac{1}{x+2} + 3$. Compare the graph to the graph of $y = \dfrac{1}{x}$.

**Step 1:** Make a table of values. The vertical asymptote is $x = -2$, so choose $x$-values on either side of $-2$.

| x | −4  | −3 | −2.5 | −2     | −1.5 | −1 | 0   |
|---|-----|----|------|--------|------|----|-----|
| y | 2.5 | 2  | 1    | undef. | 5    | 4  | 3.5 |

**Study Tip:** Use the asymptotes to help you draw the ends of the graph.

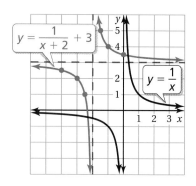

**Step 2:** Use dashed lines to graph the asymptotes $x = -2$ and $y = 3$. Then plot the ordered pairs.

**Step 3:** Draw a smooth curve through the points on each side of the vertical asymptote.

∴ The graph of $y = \dfrac{1}{x+2} + 3$ is a translation 3 units up and 2 units left of the graph of $y = \dfrac{1}{x}$.

### EXAMPLE 5 — Real-Life Application

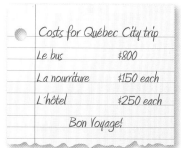

Costs for Québec City trip
Le bus — $800
La nourriture — $150 each
L'hôtel — $250 each
Bon Voyage!

The French club is planning a trip to Québec City. The function $y = \dfrac{800}{x+2} + 400$ represents the cost $y$ (in dollars) per student when $x$ students and 2 chaperones go on the trip. Use a graphing calculator to graph the function. How many students must go on the trip for the cost per student to be about $450?

**Step 1:** Use a graphing calculator to graph the function. Because the number of students cannot be negative, use only nonnegative values of $x$.

**Step 2:** Use the *trace* feature to find where the value of $y$ is about 450.

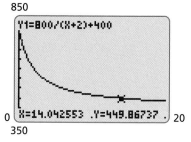

∴ About 14 students must go on the trip for the cost per student to be about $450.

### On Your Own

**Now You're Ready** — Exercises 28–33

Graph the function. Compare the graph to the graph of $y = \dfrac{1}{x}$.

**10.** $y = \dfrac{1}{x-4}$

**11.** $y = \dfrac{1}{x} - 6$

**12.** $y = -\dfrac{1}{x+3} - 3$

**13. WHAT IF?** In Example 5, how many students must go on the trip for the cost per student to be about $480?

## 11.2 Exercises

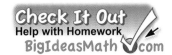

### Vocabulary and Concept Check

1. **VOCABULARY** Is $y = \dfrac{1}{\sqrt{x} + 1}$ a rational function? Explain.

2. **VOCABULARY** How is an excluded value related to a vertical asymptote?

3. **WRITING** How can you use asymptotes to help graph a rational function?

### Practice and Problem Solving

**Describe the characteristics of the graph.**

4.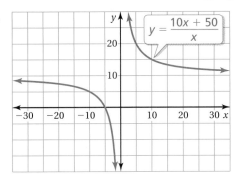

5.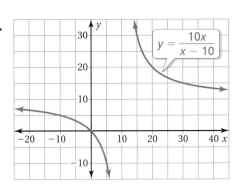

**Find the excluded value of the function.**

6. $y = \dfrac{3}{4x}$

7. $y = \dfrac{2}{x + 4}$

8. $y = \dfrac{1}{x + 3}$

9. $y = \dfrac{5}{x - 9}$

10. $y = \dfrac{7}{8 - 2x}$

11. $y = \dfrac{4}{3 + 6x}$

**Graph the function. Describe the domain and range.**

12. $y = \dfrac{5}{x}$

13. $y = \dfrac{2}{5x}$

14. $y = -\dfrac{3}{8x}$

15. $y = \dfrac{1}{x - 3}$

16. $y = \dfrac{4}{x + 1}$

17. $y = \dfrac{1}{4 - 2x}$

18. **HIKING** You hike 12 miles through a national forest to a famous landmark. Your average speed $y$ (in miles per hour) is represented by $y = \dfrac{12}{x}$, where $x$ is the total time (in hours) of the hike.

   a. Find the excluded value of the function.

   b. Graph the function. Describe the domain and range.

**Identify the asymptotes of the graph of the function. Then describe the domain and range.**

**19.** $y = -\dfrac{6}{x}$

**20.** $y = \dfrac{4}{x} + 8$

**21.** $y = \dfrac{1}{x-2} + 7$

**22.** $y = \dfrac{3}{x+4} - 4$

**23.** $y = \dfrac{-2}{x-5} - 2$

**24.** $y = 10 - \dfrac{7}{x+9}$

**25. ERROR ANALYSIS** Describe and correct the error in identifying the asymptotes of the graph of the function.

> ✗ $y = \dfrac{3}{x+4} + 5$
> The horizontal asymptote is y = 5.
> The vertical asymptote is x = 4.

**26. REASONING** Describe the domain and range of a rational function of the form $y = \dfrac{a}{x-h} + k$.

**27. OPEN-ENDED** Write a rational function whose graph has the vertical asymptote $x = 6$ and the horizontal asymptote $y = -9$.

**Graph the function. Compare the graph to the graph of $y = \dfrac{1}{x}$.**

**28.** $y = \dfrac{1}{x} + 2$

**29.** $y = \dfrac{1}{x-6}$

**30.** $y = \dfrac{1}{x+4} - 2$

**31.** $y = \dfrac{1}{x+7} + 3$

**32.** $y = \dfrac{-1}{x-1} - 5$

**33.** $y = 4 - \dfrac{1}{x+8}$

**34. SOFTBALL** A softball team buys a new $250 bat for a softball tournament. The cost of the bat is shared equally by the players on the team. Each player must also pay a $10 registration fee. The amount $y$ (in dollars) each player pays is represented by $y = \dfrac{250}{p} + 10$, where $p$ is the number of players on the team. Graph the function. How many players must be on the team for the cost per player to be about $28?

**35. GEOMETRY** The formula $h = \dfrac{2A}{b_1 + b_2}$ gives the height $h$ of a trapezoid, where $A$ is the area and $b_1$ and $b_2$ are the base lengths. Suppose $A = 60$ and $b_1 = 8$.

   **a.** Graph the function. Describe the domain and range.
   **b.** Use the graph to find $b_2$ when $h = 6$.

**36. ROAD TRIP** The function $t = \dfrac{280}{r} + 1$ models the total time $t$ (in hours) it takes to drive 280 miles at $r$ miles per hour. The model allows for two half-hour breaks. Graph the function. What does your average speed need to be for the total travel time to be 6 hours?

**Write a function for the graph.**

37.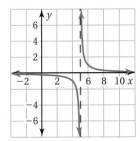

38.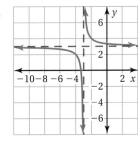

39.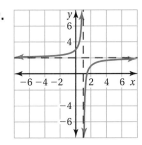

40. **REPEATED REASONING** Use a graphing calculator to graph the function $y = \dfrac{a}{x-1} + 2$ for several values of $a$. How does the value of $a$ affect the graph? Consider $a < 0$, $|a| > 1$, and $0 < |a| < 1$ in your answer.

41. **THUNDERSTORM** The time $t$ (in seconds) it takes for sound to travel 1 kilometer can be represented by $t = \dfrac{1000}{0.6T + 331}$, where $T$ is the temperature in degrees Celsius. Use a graphing calculator to graph the function for $0 \le T \le 100$. During a thunderstorm, lightning strikes 1 kilometer away. You hear the thunder 2.9 seconds later. What is the temperature?

**Graph the function. Identify the asymptotes.**

42. $y = \dfrac{x}{x+1}$

43. $y = \dfrac{1}{x^2 - 4}$

44. $y = \dfrac{x+1}{x^2 - 1}$

45. **Modeling** To qualify for a mortgage, the ratio $r$ of your expected monthly housing expenses to your gross monthly income cannot be greater than 0.28. Suppose your gross monthly income is $3500 and you expect to pay $1050 per month in housing expenses. You also expect to get a raise of $m$ dollars this month.

   a. Write and graph an equation that gives $r$ as a function of $m$.
   b. How much must the raise be in order for you to qualify for a mortgage?

 **Fair Game Review** What you learned in previous grades & lessons

**Does the equation represent a *linear* or *nonlinear* function? Explain.** *(Section 5.5)*

46. $3xy = 12$

47. $4x + 2y = 5$

48. $2y - x^2 = 8$

49. **MULTIPLE CHOICE** Which function models exponential decay? *(Section 6.6)*

   Ⓐ $y = -3\left(\dfrac{1}{2}\right)^x$ 
   Ⓑ $y = -\dfrac{1}{2}(3)^x$ 
   Ⓒ $y = \dfrac{1}{2}(3)^x$ 
   Ⓓ $y = 3\left(\dfrac{1}{2}\right)^x$

# Extension 11.2 Inverse of a Function

**Key Vocabulary**
inverse relation, p. 558
inverse function, p. 559

Recall that a *relation* pairs inputs with outputs. An **inverse relation** switches the input and output values of the original relation. For example, if a relation contains $(a, b)$, then the inverse relation contains $(b, a)$.

### EXAMPLE 1 — Finding Inverse Relations

**Find the inverse of each relation.**

**a.** $(-4, 7), (-2, 4), (0, 1), (2, -2), (4, -5)$

Switch the coordinates of each ordered pair.

$(7, -4), (4, -2), (1, 0), (-2, 2), (-5, 4)$

**COMMON CORE**

**Graphing Rational Functions**

In this extension, you will
• find inverse functions.

Learning Standards
A.REI.10
F.BF.4a

**b.**

| Input  | −1 | 0  | 1  | 2  | 3  | 4  |
|--------|----|----|----|----|----|----|
| Output | 5  | 10 | 15 | 20 | 25 | 30 |

Inverse relation:

| Input  | 5  | 10 | 15 | 20 | 25 | 30 |
|--------|----|----|----|----|----|----|
| Output | −1 | 0  | 1  | 2  | 3  | 4  |

Switch the inputs and outputs.

## Practice

**Find the inverse of the relation.**

1. $(-5, 8), (-5, 6), (0, 0), (5, 6), (10, 8)$

2. $(-3, -4), (-2, 0), (-1, 4), (0, 8), (1, 12), (2, 16), (3, 20)$

3.

| Input  | −2 | −1 | 0 | 1 | 2 |
|--------|----|----|---|---|---|
| Output | 4  | 1  | 0 | 1 | 4 |

4.

| Input  | −2 | −1 | 0 | 0 | 1 | 2 |
|--------|----|----|---|---|---|---|
| Output | 3  | 4  | 5 | 6 | 7 | 8 |

5. **WRITING** How do the domain and range of a relation compare to the domain and range of its inverse relation? Explain.

6. **CRITICAL THINKING** Recall that you can use the Vertical Line Test to determine whether a graph represents a function. What kind of similar test do you think you could use to determine whether a function has an inverse that is also a function? Explain.

**Reading**

The −1 in $f^{-1}(x)$ is not an exponent.

When a relation and its inverse are functions, they are called **inverse functions**. The inverse of a function $f$ is written as $f^{-1}(x)$. To find the inverse of a function represented by an equation, switch $x$ and $y$ and then solve for $y$.

### EXAMPLE 2  Finding Inverse Functions

Find the inverse of each function. Graph the inverse function.

a. $f(x) = 2x - 5$

$y = 2x - 5$  Replace $f(x)$ with $y$.

$x = 2y - 5$  Switch $x$ and $y$.

$x + 5 = 2y$  Add 5 to each side.

$\frac{1}{2}x + \frac{5}{2} = y$  Divide each side by 2.

$\frac{1}{2}x + \frac{5}{2} = f^{-1}(x)$  Replace $y$ with $f^{-1}(x)$.

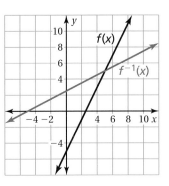

**Study Tip**

The domain is nonnegative in Example 2b, so the range of the inverse must be nonnegative. This is why you take only the positive square root of each side.

b. $f(x) = x^2$, where $x \geq 0$

$y = x^2$  Replace $f(x)$ with $y$.

$x = y^2$  Switch $x$ and $y$.

$\sqrt{x} = y$  Take the positive square root of each side.

$\sqrt{x} = f^{-1}(x)$  Replace $y$ with $f^{-1}(x)$.

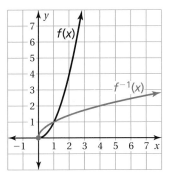

### Practice

Find the inverse of the function. Graph the inverse function.

7. $f(x) = 3x - 1$

8. $f(x) = -\frac{1}{2}x + 3$

9. $f(x) = 2x^2$, where $x \geq 0$

10. $f(x) = x^2 - 3$, where $x \geq 0$

11. $f(x) = \frac{1}{x}$

12. $f(x) = \frac{1}{x - 2}$

13. **REASONING** Suppose $f$ and $f^{-1}$ are inverse functions and $f(-2) = 5$. What is the value of $f^{-1}(5)$?

14. **REASONING** Draw the line $y = x$ on the graph in each part of Example 2. What do you notice?

15. **LOGIC** Suppose $f$ and $g$ are inverse functions. What do you know about $f(g(x))$ and $g(f(x))$? Explain.

# 11.3 Simplifying Rational Expressions

**Essential Question** How can you simplify a rational expression? What are the excluded values of a rational expression?

### 1 ACTIVITY: Simplifying a Rational Expression

**Work with a partner.**

**Sample:** You can see that the rational expressions

$$\frac{x^2 + 3x}{x^2} \quad \text{and} \quad \frac{x+3}{x}$$

are equivalent by graphing the related functions

$$y = \frac{x^2 + 3x}{x^2} \quad \text{and} \quad y = \frac{x+3}{x}.$$

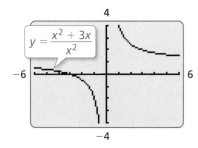
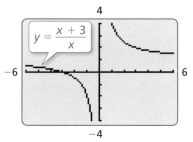

Both functions have the same graph.

**Match each rational expression with its equivalent rational expression. Use a graphing calculator to check your answers.**

a. $\dfrac{x^2 + x}{x^2}$    b. $\dfrac{x^2}{x^2 + x}$    c. $\dfrac{x+1}{x^2 - 1}$    d. $\dfrac{x+1}{x^2 + 2x + 1}$    e. $\dfrac{x^2 + 2x + 1}{x+1}$

A. $\dfrac{1}{x+1}$    B. $x+1$    C. $\dfrac{x+1}{x}$    D. $\dfrac{1}{x-1}$    E. $\dfrac{x}{x+1}$

**COMMON CORE**

**Rational Expressions**
In this lesson, you will
- simplify rational expressions.

Learning Standard
A.SSE.2

## 2  ACTIVITY: Finding Excluded Values

Work with a partner. Are the graphs of

$$y = \frac{x^2 + x}{x} \quad \text{and} \quad y = x + 1$$

*exactly* the same? Explain your reasoning.

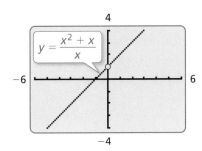
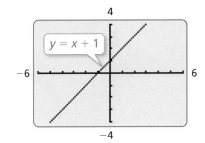

## 3  ACTIVITY: Simplifying and Finding Excluded Values

**Math Practice**

**Explain the Meaning**
What does it mean for a simplified expression to have an excluded value?

Work with a partner. Simplify each rational expression, if possible. Then compare the excluded value(s) of the original expression with the excluded value(s) of the simplified expression.

a. $\dfrac{x^2 + 2x}{x^2}$   b. $\dfrac{x^2}{x^2 + 2x}$   c. $\dfrac{x^2}{x}$

d. $\dfrac{x^2 + 4x + 4}{x + 2}$   e. $\dfrac{x - 2}{x^2 - 4}$   f. $\dfrac{1}{x^2 + 1}$

### What Is Your Answer?

4. **IN YOUR OWN WORDS** How can you simplify a rational expression? What are the excluded values of a rational expression? Include the following rational expressions in your answer.

a. $\dfrac{x(x + 1)}{x}$   b. $\dfrac{x^2 + 3x + 2}{x + 2}$   c. $\dfrac{x + 3}{x^2 - 9}$

**Practice**

Use what you learned about simplifying rational expressions to complete Exercises 3–5 on page 564.

Section 11.3  Simplifying Rational Expressions  561

## 11.3 Lesson

**Key Vocabulary**
rational expression, p. 562
simplest form of a rational expression, p. 562

A **rational expression** is an expression that can be written as a fraction whose numerator and denominator are polynomials. Values that make the denominator of the expression zero are *excluded values*.

### Key Idea

**Simplifying Rational Expressions**

**Words** A rational expression is in **simplest form** when the numerator and denominator have no common factors except 1. To simplify a rational expression, factor the numerator and denominator and *divide out* any common factors.

**Algebra** Let $a$, $b$, and $c$ be polynomials, where $b, c \neq 0$.

$$\frac{ac}{bc} = \frac{a \cdot \cancel{c}}{b \cdot \cancel{c}} = \frac{a}{b}$$

**Example**
$$\frac{2(x+1)}{5(x+1)} = \frac{2}{5}; x \neq -1$$

**Study Tip**
You can see why you can *divide out* common factors by rewriting the expression.
$$\frac{ac}{bc} = \frac{a}{b} \cdot \frac{c}{c} = \frac{a}{b} \cdot 1 = \frac{a}{b}$$

### EXAMPLE 1 Simplifying Rational Expressions

**Simplify each rational expression, if possible. State the excluded value(s).**

a. $\dfrac{12}{2x^2} = \dfrac{\cancel{2} \cdot 2 \cdot 3}{\cancel{2} \cdot x \cdot x}$   Divide out the common factor.

$= \dfrac{6}{x^2}$   Simplify.

∴ The excluded value is $x = 0$.

b. $\dfrac{n}{n+8}$

∴ The expression is in simplest form. The excluded value is $n = -8$.

**Study Tip**
Make sure you find excluded values using the *original* expression.

c. $\dfrac{3y^2}{6y(y-7)} = \dfrac{\cancel{3} \cdot \cancel{y} \cdot y}{2 \cdot \cancel{3} \cdot \cancel{y} \cdot (y-7)}$   Divide out the common factors.

$= \dfrac{y}{2(y-7)}$   Simplify.

∴ The excluded values are $y = 0$ and $y = 7$.

### On Your Own

**Now You're Ready**
Exercises 3–8

Simplify the rational expression, if possible. State the excluded value(s).

1. $\dfrac{5y^3}{2y^2}$

2. $\dfrac{8x(x+1)}{12x^2}$

3. $\dfrac{m+1}{m(m+3)}$

## EXAMPLE 2  Simplifying Rational Expressions

Simplify each rational expression, if possible. State the excluded value(s).

a. $\dfrac{1-z^2}{z-1} = \dfrac{(1-z)(1+z)}{z-1}$   Difference of Two Squares Pattern

$= \dfrac{-(z-1)(1+z)}{z-1}$   Rewrite $1-z$ as $-(z-1)$.

$= \dfrac{-\cancel{(z-1)}(1+z)}{\cancel{z-1}}$   Divide out the common factor.

$= -z - 1$   Simplify.

∴ The excluded value is $z = 1$.

b. $\dfrac{c^2 + c - 12}{c^2 - c - 20} = \dfrac{\cancel{(c+4)}(c-3)}{\cancel{(c+4)}(c-5)}$   Factor. Divide out the common factor.

$= \dfrac{c-3}{c-5}$   Simplify.

∴ The excluded values are $c = -4$ and $c = 5$.

## EXAMPLE 3  Real-Life Application

In general, as the surface area to volume ratio of a substance increases, it reacts faster with other substances. Write and simplify this ratio for a block of ice that has the shape shown.

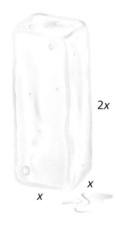

$\dfrac{\text{Surface area}}{\text{Volume}} = \dfrac{2(x^2) + 4(2x^2)}{x(x)(2x)}$   Write an expression.

$= \dfrac{\overset{5}{\cancel{10}}x^2}{\cancel{2}x^{\cancel{3}}}$   Simplify. Divide out the common factors.

$= \dfrac{5}{x}$   Simplify.

### On Your Own

**Now You're Ready**
Exercises 10–15

Simplify the rational expression, if possible. State the excluded value(s).

4. $\dfrac{2b+8}{7b+28}$

5. $\dfrac{2a-6}{4a^2-12a}$

6. $\dfrac{z^2-6z-16}{8-z}$

7. What is the surface area to volume ratio of a cube-shaped substance with edge length $x$?

Section 11.3   Simplifying Rational Expressions   563

# 11.3 Exercises

## Vocabulary and Concept Check

1. **VOCABULARY** Is $\dfrac{\sqrt{x}-1}{x+3}$ a rational expression? Explain.

2. **REASONING** Why is it necessary to state excluded values of a rational expression?

## Practice and Problem Solving

**Simplify the rational expression, if possible. State the excluded value(s).**

3. $\dfrac{6}{18x}$

4. $\dfrac{15y^3}{5y^2}$

5. $\dfrac{n-1}{n+1}$

6. $\dfrac{9w^3}{12w^4}$

7. $\dfrac{4t^2}{2t(t+11)}$

8. $\dfrac{16x^2 y}{24xy^3}$

9. **ERROR ANALYSIS** Describe and correct the error in stating the excluded value(s).

$$\dfrac{x^3}{x^2(x-3)} = \dfrac{x}{x-3}$$

The excluded value is $x = 3$.

**Simplify the rational expression. State the excluded value(s).**

10. $\dfrac{3b+9}{8b+24}$

11. $\dfrac{5-2z}{2z-5}$

12. $\dfrac{6a^2+12a}{9a^3+18a^2}$

13. $\dfrac{4-y^2}{y^2-3y-10}$

14. $\dfrac{n^2+5n+6}{n^2+8n+15}$

15. $\dfrac{3x^3-12x}{6x^3-24x^2+24x}$

16. **WRITING** Is $\dfrac{(x+2)(x-5)}{(x-2)(5-x)}$ in simplest form? Explain.

17. **RECYCLING** You hang recycling posters on bulletin boards at your school. Simplify the dimensions of the poster.

$\dfrac{(x+3)^3}{(x+3)^2}$

$\dfrac{x^2-3x}{2x-6}$

Write and simplify a rational expression for the ratio of the perimeter of the figure to its area.

18. 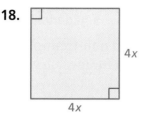   19.    20.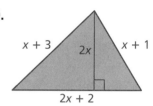

21. **OPEN-ENDED** Write a rational expression whose excluded values are $-3$ and $-5$.

22. **WRITING** Is $\dfrac{x^2 - 4}{x + 2}$ equivalent to $x - 2$? Justify your answer.

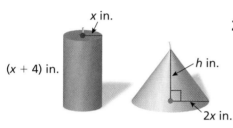

23. **PROBLEM SOLVING** The candles shown have the same volume. Write and simplify an expression for the height of the cone-shaped candle.

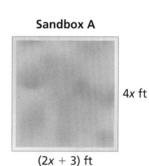 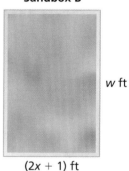

Sandbox A

Sandbox B

24. **SANDBOX** The area of Sandbox B is 4 square feet greater than the area of Sandbox A. Write and simplify an expression for the width $w$ of Sandbox B.

25.  Find two polynomials whose simplified ratio is $\dfrac{4x + 1}{2x - 1}$ and whose sum is $6x^2 + 12x$. Explain your reasoning.

## Fair Game Review  What you learned in previous grades & lessons

**Graph the function. Is the domain discrete or continuous?** *(Section 5.2)*

26.
| Input Boxes, x | Output Number of Shoes, y |
|---|---|
| 1 | 2 |
| 2 | 4 |
| 3 | 6 |

27.
| Input Months, x | Output Height of Plant, y (inches) |
|---|---|
| 1 | 1.3 |
| 2 | 2.1 |
| 3 | 2.9 |

28. **MULTIPLE CHOICE** Consider $f(x) = 2x - 4$. What is the value of $x$ so that $f(x) = 8$? *(Section 5.4)*

  Ⓐ 2   Ⓑ 4   Ⓒ 6   Ⓓ 7

# 11 Study Help

You can use an **example and non-example chart** to list examples and non-examples of a vocabulary word or term. Here is an example and non-example chart for inverse variation equations.

### Inverse Variation Equations

| Examples | Non-Examples |
|---|---|
| $y = \dfrac{2}{x}$ | $y = 2x$ |
| $2 = xy$ | $2 = \dfrac{y}{x}$ |
| $x = \dfrac{2}{y}$ | $y = \dfrac{x}{2}$ |
| $3xy = 6$ | $y = 2x + 1$ |

## On Your Own

**Make example and non-example charts to help you study these topics.**

1. direct variation equations
2. rational functions
3. excluded values
4. asymptotes
5. rational expressions
6. simplest form of a rational expression

**After you complete this chapter, make example and non-example charts for the following topics.**

7. multiplying and dividing rational expressions
8. least common denominator of rational expressions
9. adding and subtracting rational expressions
10. rational equations

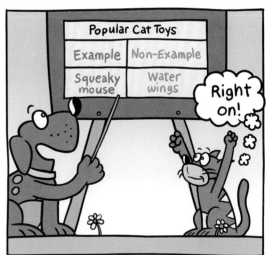

"What do you think of my example & non-example chart for popular cat toys?"

# 11.1–11.3 Quiz

**Tell whether $x$ and $y$ show *direct variation*, *inverse variation*, or *neither*. Explain your reasoning.** *(Section 11.1)*

1.
| x | y |
|---|---|
| 1 | −60 |
| 2 | −30 |
| 3 | −20 |
| 4 | −15 |

2.
| x | y |
|---|---|
| 1 | 6 |
| 2 | 12 |
| 3 | 18 |
| 4 | 24 |

3.
| x | y |
|---|---|
| 1 | −2 |
| 2 | 1 |
| 3 | 5 |
| 4 | 10 |

4. The variable $y$ varies directly with $x$. When $x = 3$, $y = 15$. Write and graph a direct variation equation that relates $x$ and $y$. *(Section 11.1)*

5. The variable $y$ varies inversely with $x$. When $x = 2$, $y = 7$. Write and graph an inverse variation equation that relates $x$ and $y$. *(Section 11.1)*

**Find the excluded value of the function.** *(Section 11.2)*

6. $y = \dfrac{2}{5x}$

7. $y = \dfrac{1}{4x - 5}$

**Identify the asymptotes of the graph of the function. Then describe the domain and range.** *(Section 11.2)*

8. $y = \dfrac{2}{x} - 5$

9. $y = -\dfrac{10}{x}$

10. $y = \dfrac{3}{x+6} + 9$

**Find the inverse of the function. Graph the inverse function.** *(Section 11.2)*

11. $f(x) = 2x + 3$

12. $f(x) = x^2 + 1$, where $x \geq 0$

**Simplify the rational expression, if possible. State the excluded value(s).** *(Section 11.3)*

13. $\dfrac{12y^4}{24y^5}$

14. $\dfrac{2z - 1}{z + 2}$

15. $\dfrac{x^2 - 2x - 8}{2x^2 + 11x + 14}$

16. **DIMENSIONS** Simplify the dimensions of the computer monitor. *(Section 11.3)*

$\dfrac{6x^4}{3x^2}$

$\dfrac{2x^2 + 5x - 3}{x + 3}$

17. **FISHING BOAT** The cost $c$ per person to charter a fishing boat varies inversely with the number $n$ of people fishing. The cost to charter a boat for an entire day is $400. *(Section 11.1)*

   a. Write an inverse variation equation that relates $c$ and $n$.
   b. How much does each person pay when 8 people fish?

## 11.4 Multiplying and Dividing Rational Expressions

**Essential Question** How can you multiply and divide rational expressions?

### 1 ACTIVITY: Matching Quotients and Products

Work with a partner. Match each quotient with a product and then with a simplified expression. Explain your reasoning.

*Quotient of Two Rational Expressions*

a. $\dfrac{2x^2}{5} \div \dfrac{14x}{10}$

b. $\dfrac{2x^2}{5} \div \dfrac{10}{14x}$

c. $\dfrac{5x^2}{2} \div \dfrac{14x}{10}$

d. $\dfrac{5x^2}{2} \div \dfrac{10}{14x}$

*Product of Two Rational Expressions*

A. $\dfrac{5x^2}{2} \cdot \dfrac{10}{14x}$

B. $\dfrac{2x^2}{5} \cdot \dfrac{10}{14x}$

C. $\dfrac{5x^2}{2} \cdot \dfrac{14x}{10}$

D. $\dfrac{2x^2}{5} \cdot \dfrac{14x}{10}$

*Simplified Expression*

1. $\dfrac{25x}{14}$

2. $\dfrac{7x^3}{2}$

3. $\dfrac{14x^3}{25}$

4. $\dfrac{2x}{7}$

e. $\dfrac{x^2-1}{x+2} \div \dfrac{x+1}{x^2-4}$

f. $\dfrac{x^2-1}{x-2} \div \dfrac{x+1}{x^2-4}$

g. $\dfrac{x^2-1}{x-2} \div \dfrac{x-1}{x^2-4}$

h. $\dfrac{x^2-1}{x+2} \div \dfrac{x-1}{x^2-4}$

E. $\dfrac{x^2-1}{x-2} \cdot \dfrac{x^2-4}{x+1}$

F. $\dfrac{x^2-1}{x-2} \cdot \dfrac{x^2-4}{x-1}$

G. $\dfrac{x^2-1}{x+2} \cdot \dfrac{x^2-4}{x-1}$

H. $\dfrac{x^2-1}{x+2} \cdot \dfrac{x^2-4}{x+1}$

5. $x^2 - x - 2$

6. $x^2 - 3x + 2$

7. $x^2 + x - 2$

8. $x^2 + 3x + 2$

i. $\dfrac{x-1}{2} \div (x-1)$

j. $\dfrac{x^2-1}{2} \div (x-1)$

k. $\dfrac{x-1}{2} \div \dfrac{1}{x-1}$

l. $\dfrac{2}{x^2-1} \div \dfrac{1}{x-1}$

I. $\dfrac{2}{x^2-1} \cdot (x-1)$

J. $\dfrac{x^2-1}{2} \cdot \dfrac{1}{x-1}$

K. $\dfrac{x-1}{2} \cdot \dfrac{1}{x-1}$

L. $\dfrac{x-1}{2} \cdot (x-1)$

9. $\dfrac{x+1}{2}$

10. $\dfrac{1}{2}$

11. $\dfrac{2}{x+1}$

12. $\dfrac{(x-1)^2}{2}$

**COMMON CORE**

**Rational Expressions**

In this lesson, you will
- multiply and divide rational expressions.

Learning Standard
A.SSE.2

## 2 ACTIVITY: Solving a Math Crossword Puzzle

**Math Practice 3**

**Use Definitions**
How do previously established definitions and results help you to solve this puzzle?

Work with a partner. Solve the crossword puzzle.

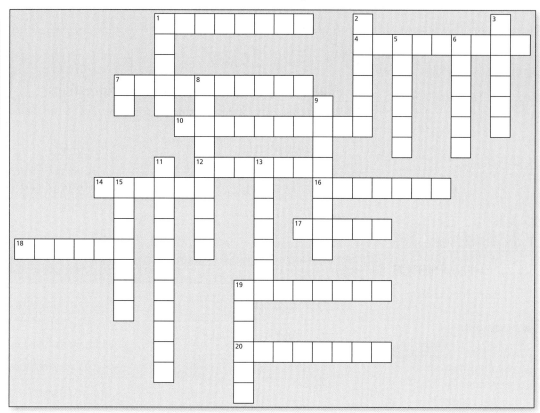

### Across
1. Inverse of subtraction
4. △ or △
7. Greek mathematician
10. Longest side of a right triangle
12. $x(x+1)$
14. $y = kx$
16. $y = \dfrac{k}{x}$
17. ∠
18. 3 ft$^3$
19. $\dfrac{2}{3}$ or $\dfrac{4}{5}$
20. △

### Down
1. 30°
2. $C$ of this is $2\pi r$
3. Dimension of □
5. 120°
6. Is the same as
7. About 3.14
8. Graph approaches $x = h$
9. $\dfrac{x}{x+1}$
11. Two numbers whose product is 1
13. $x$ in $\dfrac{1}{x}$
15. $-1, 0, 1,$ etc.
19. $x$ in $x(x+1)$

## What Is Your Answer?

3. **IN YOUR OWN WORDS** How can you multiply and divide rational expressions? Include the following in your answer.

   a. $\dfrac{x+3}{x} \cdot \dfrac{1}{x+3}$
   
   b. $\dfrac{x+3}{x} \div \dfrac{1}{x}$

**Practice**

Use what you learned about multiplying and dividing rational expressions to complete Exercises 4 and 10 on page 572.

Section 11.4  Multiplying and Dividing Rational Expressions

## 11.4 Lesson

You can use the same rules that you used for multiplying and dividing fractions to multiply and divide rational expressions.

### Key Idea

**Multiplying and Dividing Rational Expressions**

Let $a$, $b$, $c$, and $d$ be polynomials.

Multiplying: $\dfrac{a}{b} \cdot \dfrac{c}{d} = \dfrac{ac}{bd}$, where $b, d \neq 0$

Dividing: $\dfrac{a}{b} \div \dfrac{c}{d} = \dfrac{a}{b} \cdot \dfrac{d}{c} = \dfrac{ad}{bc}$, where $b, c, d \neq 0$

**EXAMPLE 1** **Multiplying Rational Expressions**

Find each product.

**Remember**
Remember that expressions may have excluded values. In Example 1a, the excluded values are $x = -1$ and $x = 0$.

a. $\dfrac{5}{2x^3} \cdot \dfrac{4x^3}{x+1} = \dfrac{5 \cdot 4x^3}{2x^3(x+1)}$   Multiply numerators and denominators.

$= \dfrac{5 \cdot \overset{2}{\cancel{4}}\cancel{x^3}}{\cancel{2}\cancel{x^3}(x+1)}$   Divide out the common factors.

$= \dfrac{10}{x+1}$   Simplify.

b. $\dfrac{h}{h+2} \cdot \dfrac{h^2 + 5h + 6}{h^2}$

$= \dfrac{h}{h+2} \cdot \dfrac{(h+3)(h+2)}{h^2}$   Factor $h^2 + 5h + 6$.

$= \dfrac{h(h+3)(h+2)}{h^2(h+2)}$   Multiply numerators and denominators.

$= \dfrac{\cancel{h}(h+3)\cancel{(h+2)}}{h^{\cancel{2}}\cancel{(h+2)}}$   Divide out the common factors.

$= \dfrac{h+3}{h}$   Simplify.

### On Your Own

**Now You're Ready**
Exercises 4–9

Find the product.

1. $\dfrac{8y^2}{y-5} \cdot \dfrac{3}{4y}$

2. $\dfrac{16}{8-c} \cdot (c-8)$

3. $\dfrac{2z-4}{6} \cdot \dfrac{3}{z^2 - 7z + 10}$

**EXAMPLE 2** **Dividing Rational Expressions**

Which expression is equivalent to $\dfrac{8}{w-4} \div \dfrac{w}{w-4}$ when $w \neq 4$?

  **Ⓐ** $\dfrac{8}{w}$    **Ⓑ** $\dfrac{w}{8}$    **Ⓒ** $\dfrac{8w}{(w-4)^2}$    **Ⓓ** $\dfrac{8w}{w^2 - 8w + 16}$

$\dfrac{8}{w-4} \div \dfrac{w}{w-4} = \dfrac{8}{w-4} \cdot \dfrac{w-4}{w}$   Multiply by the reciprocal.

$= \dfrac{8(w-4)}{w(w-4)}$   Multiply numerators and denominators.

$= \dfrac{8\cancel{(w-4)}}{w\cancel{(w-4)}}$   Divide out the common factor.

$= \dfrac{8}{w}$   Simplify.

∴ The correct answer is **Ⓐ**.


**EXAMPLE 3** **Dividing Rational Expressions**

Find the quotient $\dfrac{p^2 - p - 6}{p + 1} \div (p^2 - 4)$.

$\dfrac{p^2 - p - 6}{p+1} \div \dfrac{p^2 - 4}{1}$   Write $p^2 - 4$ as a fraction.

$= \dfrac{p^2 - p - 6}{p+1} \cdot \dfrac{1}{p^2 - 4}$   Multiply by the reciprocal.

$= \dfrac{(p-3)(p+2)}{p+1} \cdot \dfrac{1}{(p-2)(p+2)}$   Factor.

$= \dfrac{(p-3)(p+2)}{(p+1)(p-2)(p+2)}$   Multiply numerators and denominators.

$= \dfrac{(p-3)\cancel{(p+2)}}{(p+1)(p-2)\cancel{(p+2)}}$   Divide out the common factor.

$= \dfrac{p-3}{(p+1)(p-2)}$   Simplify.

**On Your Own**

*Now You're Ready*
Exercises 10–15

Find the quotient.

**4.** $\dfrac{t-2}{2t} \div \dfrac{t-2}{4t^2}$    **5.** $(g+1) \div \dfrac{g^2 + g}{g-1}$    **6.** $\dfrac{d+5}{d-1} \div (d^2 + 4d - 5)$

Section 11.4   Multiplying and Dividing Rational Expressions

# 11.4 Exercises

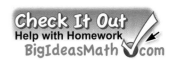

## Vocabulary and Concept Check

1. **WRITING** Describe how to multiply rational expressions.
2. **WRITING** Describe how to divide rational expressions.
3. **NUMBER SENSE** Consider the expressions $\dfrac{x}{x-2}$ and $\dfrac{x+1}{x}$. For what value(s) is the product of the expressions undefined? For what value(s) is the quotient of the expressions undefined?

## Practice and Problem Solving

**Find the product.**

4. $\dfrac{5}{3c^2} \cdot \dfrac{c^5}{15(c-2)}$

5. $\dfrac{n+3}{8n^6} \cdot \dfrac{4n^2}{7}$

6. $(d^2 - d) \cdot \dfrac{14}{1-d}$

7. $\dfrac{x+4}{6x} \cdot \dfrac{x^2}{x^2 - x - 20}$

8. $\dfrac{k^2 - 8k + 15}{5k^3} \cdot \dfrac{3k}{k-5}$

9. $\dfrac{-r-6}{2r^2 + 8r} \cdot \dfrac{4r^2 + 16r}{r^2 - 36}$

**Find the quotient.**

10. $\dfrac{2h}{h+8} \div \dfrac{16}{h+8}$

11. $\dfrac{t-5}{9t} \div \dfrac{t-5}{6t^2}$

12. $\dfrac{y+7}{7y} \div \dfrac{3y^2 + 21y}{14y - 5}$

13. $\dfrac{p^2 - 16}{p-3} \div (p-4)$

14. $\dfrac{g^2 - 4g - 21}{4g^2 + 12g} \div (g-7)$

15. $\dfrac{3z - 27}{z-6} \div (z^2 - 15z + 54)$

**ERROR ANALYSIS** Describe and correct the error in finding the quotient.

16.

$$\dfrac{6w}{w+1} \div \dfrac{3w}{w+1} = \dfrac{6w}{w+1} \cdot \dfrac{3w}{w+1}$$
$$= \dfrac{6w \cdot 3w}{(w+1)(w+1)}$$
$$= \dfrac{18w^2}{(w+1)^2}$$

17.

$$\dfrac{v-2}{4v} \div \dfrac{v-2}{6v^2} = \dfrac{4v}{v-2} \cdot \dfrac{v-2}{6v^2}$$
$$= \dfrac{4v(v-2)}{6v^2(v-2)}$$
$$= \dfrac{2}{3v}$$

**Find the total area of the red rectangle in terms of $w$.**

18.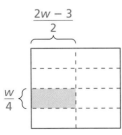

19.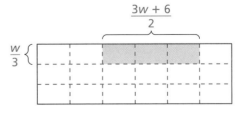

**Find the product or quotient.**

20. $\dfrac{2b^2 - b - 3}{b^2 - 6b - 7} \cdot \dfrac{b^2 - 3b - 28}{4b^2 - 4b - 3}$

21. $\dfrac{8y^2 + 6y - 5}{1 - 4y^2} \div \dfrac{12y^2 - y - 20}{6y^2 - 5y - 4}$

22. **REASONING** What are the excluded values of $\dfrac{x^2 + x - 2}{x - 3} \div \dfrac{x + 2}{x + 1}$?

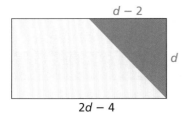

23. **CAMPSITE** A campsite is in the shape of a rectangle. The green region represents campsites with shade. The yellow represents campsites without shade. Your campsite is randomly assigned. What is the probability that your campsite has shade?

 24. **TECHNOLOGY** You can use a graphing calculator to check your answers when multiplying or dividing rational expressions. For instance, graph $y = \dfrac{5}{2x^3} \cdot \dfrac{4x^3}{x + 1}$ and $y = \dfrac{10}{x + 1}$ from Example 1a in the same viewing window.

   a. What do you notice about the graphs?
   b. How can you use the *table* feature to find the excluded values?

25. **CHARITY** The revenue $R$ (in thousands of dollars) and the average ticket price $P$ (in dollars) for a charity event can be modeled by $R = \dfrac{50 - x}{1 - 0.05x}$ and $P = 0.05x^2 + 5$, where $x$ is the number of years since 2000. (*Note:* revenue = tickets sold × ticket price)

   a. Write an equation that models the number $T$ of tickets sold as a function of $x$.
   b. In what year will this model become invalid? Explain your reasoning.

26. **Structure** Write $\dfrac{8x^3}{x + 1} \div \dfrac{2x - 2}{3x} \div \dfrac{16x^2}{x + 1}$ in simplest form.

 **Fair Game Review** What you learned in previous grades & lessons

**Graph the function. Describe the domain and range.** *(Section 8.4)*

27. $y = 2x^2 - 4x - 3$    28. $y = \dfrac{1}{4}x^2 - 5x + 2$    29. $y = -4x^2 + 8x + 5$

30. **MULTIPLE CHOICE** What is the distance between (2, 3) and (6, 5)? *(Section 10.4)*

   Ⓐ $\sqrt{6}$    Ⓑ 4    Ⓒ $3\sqrt{2}$    Ⓓ $2\sqrt{5}$

# 11.5 Dividing Polynomials

**Essential Question** How can you divide one polynomial by another polynomial?

### 1 ACTIVITY: Dividing Polynomials

**Work with a partner. Six different algebra tiles are shown below.**

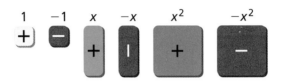

**Sample:**

**Step 1:** Arrange tiles to model
$(x^2 + 5x + 4) \div (x + 1)$
in a rectangular pattern.

**Step 2:** Complete the pattern.

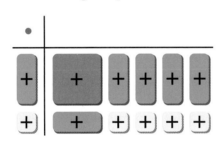
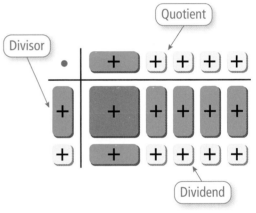

**Step 3:** Use the completed pattern to write

$(x^2 + 5x + 4) \div (x + 1) = x + 4.$

Dividend ÷ Divisor = Quotient

**Common Core**

**Rational Expressions**
In this lesson, you will
- divide polynomials by monomials.
- divide polynomials by binomials.

Learning Standard
A.SSE.2

**Complete the pattern and write the division problem.**

a.

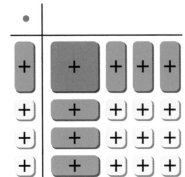

b.

574    Chapter 11    Rational Equations and Functions

### 2 ACTIVITY: Dividing Polynomials

**Math Practice**

**Make Sense of Quantities**
What do the algebra tiles represent? How can you use the tiles to divide one polynomial by another polynomial?

Work with a partner. Write two different polynomial division problems that can be associated with the given algebra tile pattern. Check your answers by multiplying.

a.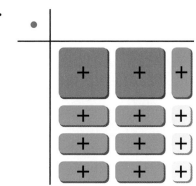

b.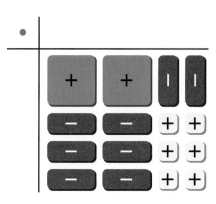

c.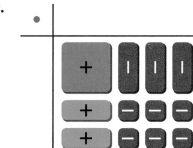

d.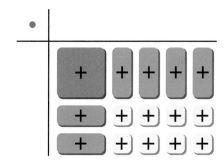

### 3 ACTIVITY: Dividing Polynomials

Work with a partner. Solve each polynomial division problem.

a. $(3x^2 - 8x - 3) \div (x - 3)$

b. $(8x^2 - 2x - 3) \div (4x - 3)$

## What Is Your Answer?

4. **IN YOUR OWN WORDS** How can you divide one polynomial by another polynomial? Include the following in your answer.

   a. $(3x^2 + 20x - 7) \div (x + 7)$

   b. $(4x^2 - 4x - 3) \div (2x - 3)$

**Practice**  Use what you learned about dividing polynomials to complete Exercises 4 and 5 on page 578.

## 11.5 Lesson

To divide a polynomial by a monomial, divide each term of the polynomial by the monomial.

### EXAMPLE 1  Dividing a Polynomial by a Monomial

**Find $(3x^2 + x - 6) \div 3x$.**

$(3x^2 + x - 6) \div 3x = \dfrac{3x^2 + x - 6}{3x}$   Write as a fraction.

$= \dfrac{3x^2}{3x} + \dfrac{x}{3x} - \dfrac{6}{3x}$   Divide each term by $3x$.

$= \dfrac{\cancel{3}x^{\cancel{2}}}{\cancel{3}\cancel{x}} + \dfrac{\cancel{x}}{3\cancel{x}} - \dfrac{\overset{2}{\cancel{6}}}{\cancel{3}x}$   Divide out the common factors.

$= x + \dfrac{1}{3} - \dfrac{2}{x}$   Simplify.

**On Your Own**

*Now You're Ready* Exercises 6 and 7

**Find the quotient.**

1. $(4z^2 - 18z) \div 2z$
2. $(n^2 - 4n + 8) \div n$
3. $(y^3 - 4y^2 + 9y) \div 4y$

You can use long division to divide a polynomial by a binomial.

### EXAMPLE 2  Dividing a Polynomial by a Binomial: No Remainder

**Find $(m^2 + 4m + 3) \div (m + 1)$.**

**Step 1:** Divide the first term of the dividend by the first term of the divisor.

Align like terms in the quotient and dividend.

$$\begin{array}{r} m\phantom{+000} \\ m+1\overline{)m^2 + 4m + 3} \\ \underline{m^2 + \phantom{0}m\phantom{+0}} \\ 3m + 3 \end{array}$$

Divide: $m^2 \div m = m$.

Multiply: $m(m + 1)$.
Subtract. Bring down the 3.

**Study Tip**

There is no remainder in Example 2, so you could have factored the dividend and divided out a common factor.

$\dfrac{m^2 + 4m + 3}{m + 1}$

$= \dfrac{(m + 3)(\cancel{m + 1})}{\cancel{m + 1}}$

$= m + 3$

**Step 2:** Divide the first term of $3m + 3$ by the first term of the divisor.

$$\begin{array}{r} m + 3 \\ m+1\overline{)m^2 + 4m + 3} \\ \underline{m^2 + \phantom{0}m\phantom{+0}} \\ 3m + 3 \\ \underline{3m + 3} \\ 0 \end{array}$$

Divide: $3m \div m = 3$.

Multiply: $3(m + 1)$.
Subtract.

So, $(m^2 + 4m + 3) \div (m + 1) = m + 3$.

When you use long division to divide polynomials and you obtain a nonzero remainder, use the following rule.

$$\text{Dividend} \div \text{Divisor} = \text{Quotient} + \frac{\text{Remainder}}{\text{Divisor}}$$

### EXAMPLE 3 — Dividing a Polynomial by a Binomial: Remainder

Find $(2 - 7y + y^2) \div (y - 3)$.   *Write the dividend in standard form.*

$$\begin{array}{r} y - 4 \phantom{xxxx}\\ y-3 \overline{\smash{\big)}\, y^2 - 7y + 2} \\ \underline{y^2 - 3y} \phantom{xxxx} \\ -4y + 2 \\ \underline{-4y + 12} \\ -10 \end{array}$$

Multiply: $y(y - 3)$.
Subtract. Bring down the 2.
Multiply: $-4(y - 3)$.
Subtract.

So, $(2 - 7y + y^2) \div (y - 3) = y - 4 - \dfrac{10}{y - 3}$.

#### On Your Own

**Now You're Ready** Exercises 8–15

**Find the quotient.**

**4.** $(s^2 - 3s - 28) \div (s - 7)$   **5.** $(x^2 + 4x - 5) \div (2 + x)$

### EXAMPLE 4 — Inserting a Missing Term

Find $(3q^2 - 8) \div (q - 2)$.   *Include a q-term with a coefficient of 0.*

$$\begin{array}{r} 3q + 6 \phantom{xxx}\\ q-2 \overline{\smash{\big)}\, 3q^2 + 0q - 8} \\ \underline{3q^2 - 6q} \phantom{xxxx} \\ 6q - 8 \\ \underline{6q - 12} \\ 4 \end{array}$$

Multiply: $3q(q - 2)$.
Subtract. Bring down the $-8$.
Multiply: $6(q - 2)$.
Subtract.

So, $(3q^2 - 8) \div (q - 2) = 3q + 6 + \dfrac{4}{q - 2}$.

**Study Tip**

When dividing polynomials using long division, first write the polynomials in standard form and insert any missing terms.

#### On Your Own

**Now You're Ready** Exercises 19–22

**Find the quotient.**

**6.** $(z^2 + 6) \div (z + 9)$   **7.** $(9y^2 - 4) \div (3y + 2)$

# 11.5 Exercises

## Vocabulary and Concept Check

1. **WRITING** How do you divide a polynomial by a monomial? by a binomial?

2. **REASONING** How can you check your answer when dividing polynomials?

3. **NUMBER SENSE** How do you know whether a binomial is a factor of a polynomial?

## Practice and Problem Solving

**Use algebra tiles to find the quotient.**

4. $(2x^2 + 6x - 8) \div (x + 4)$

5. $(4x^2 - 5x - 6) \div (4x + 3)$

**Find the quotient.**

6. $(8c^2 + 6c - 7) \div 8c$

7. $(3n^3 - 4n^2 + 12) \div 6n$

8. $(m^2 - 6m - 16) \div (m + 2)$

9. $(z^2 + 10z + 21) \div (z + 3)$

10. $(5y + 8) \div (y - 4)$

11. $(3h^2 + 2h - 1) \div (1 + h)$

12. $(3 - a + 2a^2) \div (a + 5)$

13. $(2 + 8k^2 - 9k) \div (k - 1)$

14. $(6x^2 + 5 + 17x) \div (1 + 3x)$

15. $(g - 7 + 6g^2) \div (2g - 3)$

**ERROR ANALYSIS** Describe and correct the error in finding the quotient.

16.
$$\begin{array}{r} 4 \\ x+1\overline{\smash{\big)}\,4x+3} \\ \underline{4x+4} \\ -1 \end{array}$$

$(4x + 3) \div (x + 1) = 4 - \dfrac{1}{4x+3}$

17.
$$\begin{array}{r} x - 4 \\ x-3\overline{\smash{\big)}\,x^2-x-6} \\ \underline{x^2-3x} \\ -4x - 6 \\ \underline{-4x+12} \\ 6 \end{array}$$

$(x^2 - x - 6) \div (x - 3) = x - 4 + \dfrac{6}{x-3}$

18. **AMUSEMENT PARK** The cost of a field trip to an amusement park is represented by $35x + 300$, where $x$ is the number of students going on the trip. The cost is shared equally by all the students except for three students whose parents are acting as chaperones. Find $(35x + 300) \div (x - 3)$ to find an expression for how much each student pays.

**Find the quotient.**

 19. $(d^2 - 9) \div (d + 3)$

20. $(r^2 + 10) \div (r + 5)$

21. $(8n^2 + 3) \div (2n - 1)$

22. $(10y^2 - 9y) \div (5y - 2)$

23. **ERROR ANALYSIS** Describe and correct the error in finding the quotient.

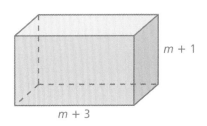

24. **REASONING** Find $k$ when $(x - 4)$ is a factor of $2x^2 - 3x + k$.

25. **CRITICAL THINKING** When dividing polynomials, how are the degrees of the dividend, divisor, and quotient related?

26. **TECHNOLOGY** Rewrite the rational function $y = \dfrac{3x - 8}{x - 3}$ in the form $y = \dfrac{a}{x - h} + k$. Graph both functions in the same viewing window of a graphing calculator.

   a. What do you notice about the graphs?
   b. What are the asymptotes of the graph of $y = \dfrac{3x - 8}{x - 3}$?

27. **GEOMETRY** The volume of the rectangular prism is $m^3 - 13m - 12$. Write an expression for the width of the prism.

   $m + 1$

   $m + 3$

28. **CHOOSE TOOLS** Would you use factoring or long division to simplify $\dfrac{x^8 - 1}{x - 1}$? Explain your reasoning.

29. **Repeated Reasoning** Find each quotient in the table and identify the pattern. Then predict the quotient $(x^5 - x^4 + x^3 - x^2 + x - 1) \div (x + 1)$ without calculating. Verify your prediction.

| Quotient |
|---|
| $(x^2 - x + 1) \div (x + 1)$ |
| $(x^3 - x^2 + x - 1) \div (x + 1)$ |
| $(x^4 - x^3 + x^2 - x + 1) \div (x + 1)$ |

 **Fair Game Review** *What you learned in previous grades & lessons*

**Solve the equation by completing the square.** *(Section 9.3)*

30. $x^2 - 4x = 5$

31. $x^2 + 8x - 7 = 0$

32. $2x^2 - 12x - 8 = 10$

33. **MULTIPLE CHOICE** What is the solution of $4^{3x} = 2^{x+1}$? *(Section 6.4)*

   Ⓐ $\dfrac{1}{5}$   Ⓑ $\dfrac{1}{2}$   Ⓒ 2   Ⓓ 3

Section 11.5  Dividing Polynomials  579

# 11.6 Adding and Subtracting Rational Expressions

**Essential Question** How can you add and subtract rational expressions?

### 1 ACTIVITY: Adding Rational Expressions

Work with a partner. You and a friend have a summer job mowing lawns. Working alone it takes you 40 hours to mow all of the lawns. Working alone it takes your friend 60 hours to mow all of the lawns.

**a.** Write a rational expression that represents the portion of the lawns you can mow in $t$ hours.

Portion you mow in $t$ hours = ☐ · ▭  (Time · Rate)

**b.** Write a rational expression that represents the portion of the lawns your friend can mow in $t$ hours.

Portion your friend mows in $t$ hours = ☐ · ▭  (Time · Rate)

**COMMON CORE**

**Rational Expressions**
In this lesson, you will
- add and subtract rational expressions.
- find least common denominators of two rational expressions.

Learning Standard
A.SSE.2

**c.** Add the two expressions to write a rational expression for the portion of the lawns that the two of you working together can mow in $t$ hours.

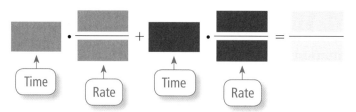

**d.** Use the expression in part (c) to find the total time it takes both of you working together to mow all of the lawns. Explain your reasoning.

## 2 ACTIVITY: Adding Rational Expressions

**Math Practice 4**

**Apply Mathematics**
How do the units of measure in a problem help you choose a formula? How does the formula help you write an expression?

Work with a partner. You are hang gliding. For the first 10,000 feet, you travel $x$ feet per minute. You then enter a valley in which the wind is greater, and for the next 6000 feet, you travel $2x$ feet per minute.

**a.** Use the formula $d = rt$ to write a rational expression that represents the time it takes you to travel the first 10,000 feet.

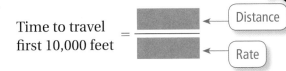

Time to travel first 10,000 feet = Distance / Rate

**b.** Use the formula $d = rt$ to write a rational expression that represents the time it takes you to travel the next 6000 feet.

Time to travel next 6000 feet = Distance / Rate

**c.** Add the two expressions to write a rational expression that represents the total time it takes you to travel 16,000 feet.

**d.** Use the expression in part (c) to find the total time it takes you to travel 16,000 feet when your rate during the first 10,000 feet is 2000 feet per minute.

## What Is Your Answer?

**3. IN YOUR OWN WORDS** How can you add and subtract rational expressions? Include the following in your answer.

**a.** $\dfrac{x}{5} + \dfrac{x}{10}$   **b.** $\dfrac{3}{x} + \dfrac{4}{x}$   **c.** $\dfrac{9}{x} + \dfrac{2}{3x}$

**d.** $\dfrac{x}{2} - \dfrac{x}{4}$   **e.** $\dfrac{x+1}{3} - \dfrac{1}{3}$   **f.** $\dfrac{1}{x} - \dfrac{1}{x^2}$

**Practice**

Use what you learned about adding and subtracting rational expressions to complete Exercises 3–5 on page 585.

Section 11.6  Adding and Subtracting Rational Expressions

# 11.6 Lesson

**Check It Out**
Lesson Tutorials
BigIdeasMath.com

**Key Vocabulary**
least common denominator of rational expressions, p. 583

You can use the same rules that you used for adding and subtracting fractions to add and subtract rational expressions.

## Key Idea

**Adding and Subtracting Rational Expressions with Like Denominators**

Let $a$, $b$, and $c$ be polynomials, where $c \neq 0$.

Adding: $\dfrac{a}{c} + \dfrac{b}{c} = \dfrac{a+b}{c}$  Subtracting: $\dfrac{a}{c} - \dfrac{b}{c} = \dfrac{a-b}{c}$

### EXAMPLE 1  Adding and Subtracting with Like Denominators

Find the sum or difference.

a. $\dfrac{5}{2x} + \dfrac{7}{2x} = \dfrac{5+7}{2x}$  Add the numerators.

$= \dfrac{12}{2x}$  Simplify.

$= \dfrac{\cancel{12}^{\,6}}{\cancel{2}x}$  Divide out the common factor.

$= \dfrac{6}{x}$  Simplify.

b. $\dfrac{3y}{y+4} - \dfrac{y-8}{y+4} = \dfrac{3y-(y-8)}{y+4}$  Subtract the numerators.

$= \dfrac{3y - y + 8}{y+4}$  Use the Distributive Property.

$= \dfrac{2y + 8}{y+4}$  Combine like terms.

$= \dfrac{2(\cancel{y+4})}{\cancel{y+4}}$  Factor. Divide out the common factor.

$= 2$  Simplify.

**Common Error**
When subtracting rational expressions, remember to distribute the negative to each term of the numerator of the expression being subtracted.

### On Your Own

**Now You're Ready**
Exercises 6–11

Find the sum or difference.

1. $\dfrac{4}{9z} - \dfrac{8}{9z}$

2. $\dfrac{3w+1}{w-1} + \dfrac{w}{w-1}$

3. $\dfrac{x+3}{x^2+x-2} - \dfrac{1}{x^2+x-2}$

To add or subtract rational expressions with unlike denominators, rewrite the expressions so they have like denominators. You can do this by finding the least common multiple of the denominators, called the **least common denominator (LCD)**.

### EXAMPLE 2  Finding the LCD of Two Rational Expressions

Find the LCD of $\dfrac{3}{10g^2}$ and $\dfrac{5}{12g}$.

First write the prime factorization of each denominator.

$$10g^2 = 2 \cdot 5 \cdot g^2 \qquad\qquad 12g = 2^2 \cdot 3 \cdot g$$

Use the greatest power of each factor that appears in either denominator to find the LCM of the denominators.

$$\text{LCM} = 2^2 \cdot 3 \cdot 5 \cdot g^2 = 60g^2$$

So, the LCD of $\dfrac{3}{10g^2}$ and $\dfrac{5}{12g}$ is $60g^2$.

#### On Your Own

**Now You're Ready**
Exercises 13–18

Find the LCD of the rational expressions.

4. $\dfrac{2}{7g}, -\dfrac{15}{4g^3}$

5. $\dfrac{8}{n}, \dfrac{n}{n+1}$

6. $\dfrac{t}{t^2-4}, \dfrac{9}{t-2}$

7. $\dfrac{x+1}{x^2-x-6}, \dfrac{5}{x(x-3)}$

### EXAMPLE 3  Adding with Unlike Denominators

Find the sum $\dfrac{1}{8x} + \dfrac{x-2}{6x^2}$.

Because the expressions have unlike denominators, find the LCD.

$$8x = 2^3 \cdot x \qquad\qquad 6x^2 = 2 \cdot 3 \cdot x^2$$

The LCD is $2^3 \cdot 3 \cdot x^2 = 24x^2$.

$\dfrac{1}{8x} + \dfrac{x-2}{6x^2} = \dfrac{1(3x)}{8x(3x)} + \dfrac{(x-2)(4)}{6x^2(4)}$   Rewrite using the LCD, $24x^2$.

$= \dfrac{3x}{24x^2} + \dfrac{4x-8}{24x^2}$   Simplify.

$= \dfrac{3x + 4x - 8}{24x^2}$   Add the numerators.

$= \dfrac{7x - 8}{24x^2}$   Simplify.

**Study Tip**
To rewrite each expression using the LCD, multiply the numerator and denominator of each expression by the factor that makes its denominator the LCD.

Section 11.6  Adding and Subtracting Rational Expressions

### EXAMPLE 4 Subtracting with Unlike Denominators

Find the difference $\dfrac{x+3}{x^2-8x+12} - \dfrac{2}{x-6}$.

$\dfrac{x+3}{x^2-8x+12} - \dfrac{2}{x-6} = \dfrac{x+3}{(x-6)(x-2)} - \dfrac{2}{x-6}$    Factor $x^2 - 8x + 12$.

$= \dfrac{x+3}{(x-6)(x-2)} - \dfrac{2(x-2)}{(x-6)(x-2)}$    Rewrite using the LCD, $(x-6)(x-2)$.

$= \dfrac{(x+3) - 2(x-2)}{(x-6)(x-2)}$    Subtract the numerators.

$= \dfrac{-x+7}{(x-6)(x-2)}$    Simplify.

### EXAMPLE 5 Real-Life Application

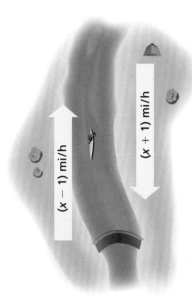

You row your kayak 5 miles downstream from your campsite to a dam, and then you row back to your campsite. You row $x$ miles per hour during the entire trip, and the river current is 1 mile per hour. Write an expression for the total time of the trip.

Solving the formula $d = rt$ for time $t$ gives $t = \dfrac{d}{r}$. Use this to write an expression for the total time of the trip.

$\underbrace{\text{Distance downstream} \div \text{Speed downstream}}_{\text{Time downstream}} + \underbrace{\text{Distance upstream} \div \text{Speed upstream}}_{\text{Time upstream}}$

$\dfrac{5}{x+1} + \dfrac{5}{x-1} = \dfrac{5(x-1)}{(x+1)(x-1)} + \dfrac{5(x+1)}{(x-1)(x+1)}$    Write an expression. Rewrite using the LCD.

$= \dfrac{5(x-1) + 5(x+1)}{(x+1)(x-1)}$    Add the numerators.

$= \dfrac{10x}{(x+1)(x-1)}$    Simplify.

### On Your Own

Now You're Ready
Exercises 20–25

Find the sum or difference.

8. $\dfrac{x+5}{3x^3} - \dfrac{2}{9x^2}$    9. $\dfrac{2k}{k+1} + \dfrac{k}{k-2}$    10. $\dfrac{3y-1}{y^2-64} - \dfrac{3}{y+8}$

11. **WHAT IF?** In Example 5, the river current is 2 miles per hour. Write an expression for the total time of the trip.

**584**    Chapter 11    Rational Equations and Functions

# 11.6 Exercises

## Vocabulary and Concept Check

1. **WRITING** Explain how finding the least common denominator of two rational expressions is similar to finding the least common denominator of two numeric fractions.

2. **REASONING** Describe how to rewrite the expressions $\dfrac{1}{x+4}$ and $\dfrac{1}{x^2-16}$ so that they have the same denominator.

## Practice and Problem Solving

**Find the sum or difference.**

3. $\dfrac{4s}{9} + \dfrac{s}{9}$

4. $\dfrac{r}{8} - \dfrac{r}{16}$

5. $\dfrac{2}{w} + \dfrac{5}{w}$

6. $\dfrac{7}{3y} + \dfrac{1}{3y}$

7. $\dfrac{5}{x+2} - \dfrac{2}{x+2}$

8. $\dfrac{2z}{4(z-1)} + \dfrac{6z}{4(z-1)}$

9. $\dfrac{3t^2}{t^2-1} - \dfrac{t+4}{t^2-1}$

10. $\dfrac{2n+3}{n^2-n-2} + \dfrac{-n-2}{n^2-n-2}$

11. $\dfrac{p-2}{p^2-5p+4} - \dfrac{2}{p^2-5p+4}$

12. **ERROR ANALYSIS** Describe and correct the error in adding the rational expressions.

$$\dfrac{1}{x-3} + \dfrac{4}{x-3} = \dfrac{5}{2x-6} \quad \text{✗}$$

**Find the LCD of the rational expressions.**

13. $\dfrac{9}{2x}, \dfrac{7}{x}$

14. $\dfrac{1}{12y}, \dfrac{5}{18y}$

15. $\dfrac{m}{m+5}, \dfrac{9}{m-4}$

16. $2g, \dfrac{1}{g}$

17. $\dfrac{h}{h+3}, \dfrac{1}{h^2+2h-3}$

18. $\dfrac{s-7}{s^2-2s-8}, \dfrac{3}{s-4}$

19. **CEREAL** The height of a cereal box is given by $\dfrac{S}{2(\ell+w)} - \dfrac{2\ell w}{2(\ell+w)}$, where $S$ is the surface area, $\ell$ is the length, and $w$ is the width. Find the difference.

**Find the sum or difference.**

**20.** $\dfrac{x+1}{2x} + \dfrac{2x-1}{5x}$

**21.** $\dfrac{y-3}{6y} + \dfrac{y+4}{8y}$

**22.** $3 - \dfrac{c-2}{c+2}$

**23.** $\dfrac{2m}{m-7} + \dfrac{4}{7-m}$

**24.** $\dfrac{x+2}{x^2+3x-10} + \dfrac{3}{2-x}$

**25.** $\dfrac{2p+3}{p^2-7p+12} - \dfrac{2}{p-3}$

**26. ERROR ANALYSIS** Describe and correct the error in adding the rational expressions.

$$\dfrac{x}{x-1} + \dfrac{2}{x+2} = \dfrac{x(x-1) + 2(x+2)}{(x-1)(x+2)}$$
$$= \dfrac{x^2 - x + 2x + 4}{(x-1)(x+2)}$$
$$= \dfrac{x^2 + x + 4}{(x-1)(x+2)}$$

**27. REASONING** Can you find a common denominator of two rational expressions by finding the product of the denominators? Is this product always going to be the least common denominator? Justify your answers.

**28. RUNNING** You run 3 miles up a hill and 3 miles down the hill. You run 25% faster going down the hill than going up the hill. Let $r$ be your speed (in miles per hour) while running up the hill. Write an expression that represents the amount of time you spend running on the hill.

**29. OPEN-ENDED** Write two rational expressions with unlike denominators.

  **a.** Find the least common denominator of the two expressions.
  **b.** Add the two expressions.

**Write an expression for the perimeter of the figure.**

**30.**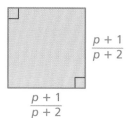

**31.**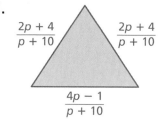

**32.**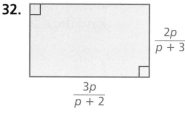

586  Chapter 11  Rational Equations and Functions

**Simplify the expression.**

33. $\dfrac{3c+1}{c-1} + \dfrac{c+1}{c^2-4c+3} - \dfrac{c-1}{c-3}$

34. $-\dfrac{11d-8}{d^2+d-20} + \dfrac{d}{d-4} + \dfrac{5}{d+5}$

35. $\dfrac{x}{x+4} + \dfrac{x^2}{x^2-x-20} \div \dfrac{x}{x+4}$

36. $\dfrac{1}{y^2+7y-8} - \dfrac{2}{2y-2} \cdot \dfrac{y+8}{2}$

37. **HOMEWORK** You have 20 more math exercises for homework than biology exercises. You finish 15 exercises before dinner and 18 exercises after dinner. Write an expression that represents the portion of exercises that are complete.

38. **WAKEBOARDING** You are wakeboarding on a river. You travel 2 miles downstream to a marina for supplies, and then you travel 3 miles upstream to a dock. The boat travels $x$ miles per hour during the entire trip, and the river current is 3 miles per hour.

    a. Write an expression that represents the total time of the trip.

    b. How long will the trip take when the speed of the boat is 18 miles per hour?

39. **Logic** Let $a$, $b$, $c$, and $d$ be polynomials. Find two rational expressions $\dfrac{a}{b}$ and $\dfrac{c}{d}$ so that $\dfrac{a}{b} - \dfrac{c}{d} = \dfrac{2x+11}{(x+4)(x+3)}$.

## Fair Game Review  *What you learned in previous grades & lessons*

**Graph the system of linear inequalities.** *(Section 4.5)*

40. $y > 2x + 1$
    $y < -3x + 4$

41. $y \geq -\dfrac{1}{2}x$
    $y < -x - 7$

42. $y \leq 4x - 3$
    $y \geq -3x + 1$

43. **MULTIPLE CHOICE** The graph of which function is shown at the right? *(Section 10.1)*

    Ⓐ $y = 2\sqrt{x}$   Ⓑ $y = -2\sqrt{x}$

    Ⓒ $y = \dfrac{1}{2}\sqrt{x}$   Ⓓ $y = 5\sqrt{x}$

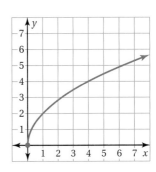

Section 11.6  Adding and Subtracting Rational Expressions

# 11.7 Solving Rational Equations

**Essential Question** How can you solve a rational equation?

### 1 ACTIVITY: Solving Rational Equations

Work with a partner. A hockey goalie faces 799 shots and saves 707 of them.

**a.** What is his save percentage?

$$\text{Save Percentage} = \frac{\text{Shots saved}}{\text{Shots faced}}$$

National Hockey League goalies typically have a save percentage above .900.

**b.** Suppose the goalie has $x$ additional consecutive saves. Write an expression for his new save percentage.

$$\text{Save Percentage} = \frac{707 \text{ plus } x \text{ additional saves}}{799 \text{ plus } x \text{ additional shots faced}}$$

**Rational Functions**

In this lesson, you will
- solve rational equations using cross products.
- solve rational equations using least common denominators.
- solve real-life problems.

Applying Standard
A.CED.1

**c.** Complete the table showing the goalie's save percentage as $x$ increases.

| Additional Saves, $x$ | 0 | 20 | 40 | 60 | 80 | 100 | 120 | 140 |
|---|---|---|---|---|---|---|---|---|
| Save Percentage | | | | | | | | |

**d.** The goalie wants to end the season with a save percentage of .900. How many additional consecutive saves must he have to achieve this? Justify your answer by solving an equation.

## 2 ACTIVITY: Solving Rational Equations

**Math Practice**

**Find General Methods**
What method did you use to complete the table? How can you use this information to write and solve an equation in part (d)?

Work with a partner. A baseball player has been at bat 47 times and has 8 hits.

a. What is his batting average?

$$\text{Batting Average} = \frac{\rule{2cm}{0.4pt}}{\rule{2cm}{0.4pt}}$$

The league batting average in Major League Baseball is usually between .250 and .270.

b. Suppose the player has $x$ additional consecutive hits. Write an expression for his new batting average.

$$\text{Batting Average} = \frac{\rule{2cm}{0.4pt}}{\rule{2cm}{0.4pt}}$$

c. Complete the table showing the player's batting average as $x$ increases.

| Additional Hits, $x$ | 0 | 1 | 2 | 3 | 4 | 5 | 6 | 7 |
|---|---|---|---|---|---|---|---|---|
| Batting Average | | | | | | | | |

d. The player wants to end the season with a batting average of .250. How many additional consecutive hits must he have to achieve this? Justify your answer by solving an equation.

## What Is Your Answer?

3. **IN YOUR OWN WORDS** How can you solve a rational equation? Include the following in your answer.

   a. $\dfrac{x-6}{6} = \dfrac{2}{3}$
   
   b. $\dfrac{x+56}{6} = \dfrac{1}{2}$
   
   c. $\dfrac{x}{4} + \dfrac{x}{2} = \dfrac{2x}{3}$

**Practice** — Use what you learned about solving rational equations to complete Exercise 4 on page 592.

Section 11.7  Solving Rational Equations

## 11.7 Lesson

**Check It Out**
Lesson Tutorials
BigIdeasMath com

**Key Vocabulary**
rational equation, p. 590

A **rational equation** is an equation that contains rational expressions. One way to solve rational equations is to use the Cross Products Property. You can use this method when each side of a rational equation consists of one rational expression.

### EXAMPLE 1  Solving Rational Equations Using Cross Products

Solve each equation.

a. $\dfrac{5}{x+4} = \dfrac{4}{x-4}$

**Check**
$\dfrac{5}{x+4} = \dfrac{4}{x-4}$
$\dfrac{5}{36+4} \stackrel{?}{=} \dfrac{4}{36-4}$
$\dfrac{1}{8} = \dfrac{1}{8}$ ✓

$\dfrac{5}{x+4} = \dfrac{4}{x-4}$   Write the equation.
$5(x-4) = 4(x+4)$   Cross Products Property
$5x - 20 = 4x + 16$   Distributive Property
$5x = 4x + 36$   Add 20 to each side.
$x = 36$   Subtract 4x from each side.

b. $\dfrac{5}{y} = \dfrac{y-2}{7}$

$\dfrac{5}{y} = \dfrac{y-2}{7}$   Write the equation.
$5(7) = y(y-2)$   Cross Products Property
$35 = y^2 - 2y$   Simplify.
$0 = y^2 - 2y - 35$   Subtract 35 from each side.
$0 = (y-7)(y+5)$   Factor.
$y - 7 = 0$  or  $y + 5 = 0$   Zero-Product Property
$y = 7$  or  $y = -5$   Solve for y.

**Check**
$\dfrac{5}{7} \stackrel{?}{=} \dfrac{7-2}{7}$   Substitute for y.   $\dfrac{5}{-5} \stackrel{?}{=} \dfrac{-5-2}{7}$
$\dfrac{5}{7} = \dfrac{5}{7}$ ✓   Simplify.   $-1 = -1$ ✓

### ● On Your Own

**Now You're Ready**
Exercises 5–10

Solve the equation. Check your solution(s).

1. $\dfrac{2}{x-3} = \dfrac{4}{x-7}$

2. $\dfrac{4}{z+4} = \dfrac{z}{z+1}$

3. $\dfrac{3y}{4} = \dfrac{6}{y+7}$

When there is more than one rational expression on one or both sides of a rational equation, multiply each side by the LCD and then solve.

**EXAMPLE 2** **Solving a Rational Equation Using the LCD**

Solve $\dfrac{z}{z-2} - \dfrac{2}{3} = \dfrac{2}{z-2}$.

$$3(z-2) \cdot \left(\dfrac{z}{z-2} - \dfrac{2}{3}\right) = 3(z-2) \cdot \dfrac{2}{z-2} \quad \text{Multiply each side by the LCD, } 3(z-2).$$

$$\dfrac{z \cdot 3(z-2)}{z-2} - \dfrac{2 \cdot 3(z-2)}{3} = \dfrac{2 \cdot 3(z-2)}{z-2} \quad \text{Multiply. Then divide out common factors.}$$

$$3z - 2z + 4 = 6 \quad \text{Simplify.}$$

$$z = 2 \quad \text{Solve for } z.$$

Because each side of the equation is undefined when $z = 2$, it is an extraneous solution.

∴ The equation has no solution.

**EXAMPLE 3** **Real-Life Application**

Your starter deck for a collectible card game has 50 cards. The deck contains 17 creature cards. You add creature cards to the deck until it contains 50% creature cards. How many do you add?

Write an equation for the ratio of creature cards to total cards after adding $x$ creature cards.

Creature cards → $\dfrac{x+17}{x+50} = 0.5$ ← Desired percent of creature cards
Total cards →

$$0.5(x + 50) = x + 17 \quad \text{Cross Products Property}$$
$$0.5x + 25 = x + 17 \quad \text{Distributive Property}$$
$$8 = 0.5x \quad \text{Simplify.}$$
$$16 = x \quad \text{Divide each side by 0.5.}$$

∴ You add 16 creature cards to the deck.

**On Your Own**

Now You're Ready
Exercises 13–18

Solve the equation. Check your solution(s).

4. $\dfrac{1}{p} - \dfrac{2}{3} = \dfrac{7}{p}$

5. $\dfrac{2}{n} + \dfrac{1}{n+3} = \dfrac{5}{n+3}$

6. $\dfrac{4}{a-6} + 1 = \dfrac{9}{a^2-36}$

7. **WHAT IF?** In Example 3, you add creature cards until the deck contains 40% creature cards. How many do you add?

# 11.7 Exercises

## Vocabulary and Concept Check

1. **VOCABULARY** Describe two methods for solving rational equations.

2. **OPEN-ENDED** Write a rational equation that can be solved by multiplying each side by $2x(x + 1)$.

3. **WRITING** Why should you check the solutions of a rational equation?

## Practice and Problem Solving

4. A basketball player attempts 64 free throws and makes 50 of them.

   a. What is her free throw percentage?

   b. Suppose the player makes $x$ additional consecutive free throws. Write an expression for her new free throw percentage.

   c. The player wants to end the season with a free throw percentage of .800. How many additional consecutive free throws must she make to achieve this?

**Solve the equation. Check your solution(s).**

5. $\dfrac{2}{b} = \dfrac{6}{b+2}$

6. $\dfrac{2}{x-1} = \dfrac{3}{x+1}$

7. $\dfrac{4}{m-4} = \dfrac{m}{3}$

8. $\dfrac{z-1}{8} = \dfrac{z}{z+9}$

9. $\dfrac{k}{2k+5} = \dfrac{1}{k-2}$

10. $\dfrac{3w}{w+1} = \dfrac{w}{3-w}$

11. **ERROR ANALYSIS** Describe and correct the error in solving the equation.

$$\dfrac{x}{x+1} = \dfrac{2}{x+1}$$
$$x(x+1) = 2(x+1)$$
$$x^2 + x = 2x + 2$$
$$x^2 - x - 2 = 0$$
$$(x-2)(x+1) = 0$$
$$x = 2 \text{ or } x = -1$$

So, the solutions are $x = 2$ and $x = -1$.

12. **WATER RESCUE** The table shows information about a water rescue team.

    a. Solve the rational equation $\dfrac{4}{x} = \dfrac{7}{x+6}$ to find the upstream speed of the rescue team.

    b. What is the downstream speed of the rescue team?

| Water Rescue | | | |
|---|---|---|---|
| Direction | Distance | Rate | Time |
| Upstream | 4 miles | $x$ mi/h | $t$ hours |
| Downstream | 7 miles | $(x+6)$ mi/h | $t$ hours |

**Solve the equation. Check your solution(s).**

**13.** $\dfrac{4}{5} - \dfrac{1}{c} = \dfrac{3}{c}$

**14.** $\dfrac{2}{y+3} - \dfrac{5}{y} = \dfrac{12}{y+3}$

**15.** $\dfrac{10}{d(d-2)} + \dfrac{4}{d} = \dfrac{5}{d-2}$

**16.** $\dfrac{n}{n-2} + \dfrac{2}{5} = \dfrac{1}{n+4}$

**17.** $\dfrac{6}{a+5} + 2 = \dfrac{28}{a^2 - 25}$

**18.** $\dfrac{x}{x+7} + \dfrac{3}{x-6} = \dfrac{2x+27}{x^2+x-42}$

**19. REASONING** Explain how you can use the Cross Products Property to solve $\dfrac{3}{x} + 1 = \dfrac{8}{x-3}$.

**20. PAINT** A department store paint mixer contains 4 pints of equal amounts of yellow and red paint. The shade of red that you want requires a paint mixture that is 75% red and 25% yellow. How many pints of red paint need to be added to the paint mixer?

**21. RAPPELLING** A rappelling club charters a bus for a trip to the mountains for $540. To lower the bus fare per person, the club invites some hikers on the trip. After 7 hikers join the trip, the bus fare per person decreases by $7. How many members of the rappelling club are going on the trip?

**To solve *work problems*, find the portion of the job each person completes in 1 unit of time. The sum of these portions is the portion of the job completed in 1 unit of time.**

**22.** You can mop a floor in 8 minutes. Your friend can mop the same floor in 12 minutes. Working together, how much time does it take to mop the floor?

**23.** You can mow a lawn in 3 hours. Your friend can mow the same lawn in 2 hours. Working together, how much time does it take to mow the lawn?

**24. Reasoning** A roofing contractor can shingle a roof in half the time it takes his assistant. Working together, they can shingle the roof in 8 hours. How much time does it take the roofing contractor to finish the job alone?

### Fair Game Review  *What you learned in previous grades & lessons*

**Solve the equation. Check your solutions.** *(Section 1.3)*

**25.** $|x - 2| = 6$

**26.** $3|2x - 7| = 9$

**27.** $2|5x + 4| - 1 = 13$

**28. MULTIPLE CHOICE** What is the solution of $-2 < -x + 5 \leq 8$? *(Section 3.4)*

**Ⓐ** $-7 < x \leq 3$   **Ⓑ** $7 > x \geq -3$   **Ⓒ** $x \leq -3$ and $x > 7$   **Ⓓ** $x < 7$ or $x \geq -3$

# 11.4–11.7 Quiz

**Find the product or quotient.** *(Section 11.4)*

1. $\dfrac{c+2}{5c^3} \cdot \dfrac{4c^4}{6}$

2. $\dfrac{4ab^3 - 2b^3}{2a^3 + 4a^2} \cdot \dfrac{2a^2 + 4a}{4ab - 2b}$

3. $\dfrac{3k}{k+3} \div \dfrac{15}{k+3}$

4. $\dfrac{m^2 - 36}{m^3} \div \dfrac{m^2 + 12m + 36}{m^3}$

**Find the quotient.** *(Section 11.5)*

5. $(6j^3 + 12j^2 + 18j) \div 6j$

6. $(m^2 - 14m + 49) \div (m - 7)$

7. $(d^2 - 5d + 8) \div (d - 3)$

8. $(5n^2 + 7) \div (n - 1)$

**Find the sum or difference.** *(Section 11.6)*

9. $\dfrac{5}{v+1} - \dfrac{10}{v+1}$

10. $\dfrac{4r}{2r-3} + \dfrac{5r-1}{3-2r}$

11. $\dfrac{t^2 - 8}{6t} + \dfrac{-t^2 + 7}{4t}$

12. $\dfrac{3p + 10}{p^2 + p - 20} - \dfrac{2}{p - 4}$

**Solve the equation. Check your solution.** *(Section 11.7)*

13. $\dfrac{3}{s-2} = \dfrac{4}{s}$

14. $2 = \dfrac{6}{2w + 1}$

15. $-5 + \dfrac{2h}{h+2} = \dfrac{7h}{h+2}$

16. $\dfrac{2}{g} + \dfrac{5}{g(g+1)} = \dfrac{6}{g+1}$

17. **PIGPEN** You are installing a fence around a pigpen. Write an expression that represents the amount of fencing you need. *(Section 11.6)*

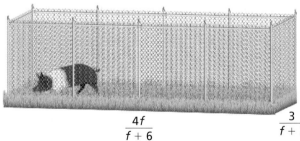

$\dfrac{4f}{f+6}$  $\dfrac{3}{f+2}$

18. **RAKING** You can rake your front yard in 30 minutes. Your friend can rake the same yard in 50 minutes. Working together, how much time does it take to rake the yard? *(Section 11.7)*

# 11 Chapter Review

## Review Key Vocabulary

direct variation, *p. 544*
inverse variation, *p. 544*
rational function, *p. 552*
excluded value, *p. 552*
asymptote, *p. 553*
inverse relation, *p. 558*
inverse function, *p. 559*
rational expression, *p. 562*
simplest form of a rational expression, *p. 562*
least common denominator of rational expressions, *p. 583*
rational equation, *p. 590*

## Review Examples and Exercises

### 11.1 Direct and Inverse Variation (pp. 542–549)

The variable $y$ varies inversely with $x$. When $x = 3$, $y = 2$.

**a.** Write an inverse variation equation that relates $x$ and $y$.

Find the value of $k$.

$y = \dfrac{k}{x}$     Write the inverse variation equation.

$2 = \dfrac{k}{3}$     Substitute 3 for $x$ and 2 for $y$.

$6 = k$     Multiply each side by 3.

So, an equation that relates $x$ and $y$ is $y = \dfrac{6}{x}$.

**b.** Graph the inverse variation equation. Describe the domain and range.

Make a table of values.

| x | −3 | −2 | −1 | 0 | 1 | 2 | 3 |
|---|----|----|----|---|---|---|---|
| y | −2 | −3 | −6 | undef. | 6 | 3 | 2 |

Plot the ordered pairs. Draw a smooth curve through the points in each quadrant. Both the domain and range are all real numbers except 0.

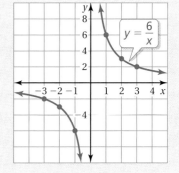

### Exercises

1. The variable $y$ varies directly with $x$. When $x = 6$, $y = 12$. Write and graph a direct variation equation that relates $x$ and $y$.

2. The variable $y$ varies inversely with $x$. When $x = 3$, $y = 8$. Write and graph an inverse variation equation that relates $x$ and $y$.

## 11.2 Graphing Rational Functions (pp. 550–559)

Graph $y = \dfrac{1}{x-2} - 1$. Compare the graph to the graph of $y = \dfrac{1}{x}$.

**Step 1:** Make a table of values. The vertical asymptote is $x = 2$, so choose $x$-values on either side of 2.

| x | 0 | 1 | 1.5 | 2 | 2.5 | 3 | 4 |
|---|---|---|---|---|---|---|---|
| y | −1.5 | −2 | −3 | undef. | 1 | 0 | −0.5 |

**Step 2:** Use dashed lines to graph the asymptotes $x = 2$ and $y = -1$. Then plot the ordered pairs.

**Step 3:** Draw a smooth curve through the points on each side of the vertical asymptote.

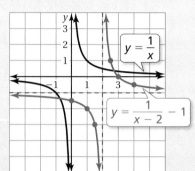

The graph of $y = \dfrac{1}{x-2} - 1$ is a translation 1 unit down and 2 units right of the graph of $y = \dfrac{1}{x}$.

Find the inverse of $f(x) = \dfrac{2}{x}$. Graph the inverse function.

$y = \dfrac{2}{x}$    Replace $f(x)$ with $y$.

$x = \dfrac{2}{y}$    Switch $x$ and $y$.

$xy = 2$    Multiply each side by $y$.

$y = \dfrac{2}{x}$    Divide each side by $x$.

$f^{-1}(x) = \dfrac{2}{x}$    Replace $y$ with $f^{-1}(x)$.

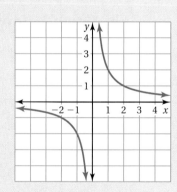

### Exercises

Graph the function. Compare the graph to the graph of $y = \dfrac{1}{x}$.

**3.** $y = \dfrac{1}{x+5}$      **4.** $y = \dfrac{1}{x} - 4$      **5.** $y = \dfrac{1}{x-7} + 1$

Find the inverse of the function. Graph the inverse function.

**6.** $f(x) = x + 2$      **7.** $f(x) = \dfrac{1}{2}x - 5$      **8.** $f(x) = \dfrac{1}{x} + 7$

## 11.3 Simplifying Rational Expressions (pp. 560–565)

Simplify $\dfrac{v^2 - 9}{v^2 - 3v}$, if possible. State the excluded value(s).

$$\dfrac{v^2 - 9}{v^2 - 3v} = \dfrac{(v-3)(v+3)}{v(v-3)} \quad \text{Factor.}$$

$$= \dfrac{\cancel{(v-3)}(v+3)}{v\cancel{(v-3)}} \quad \text{Divide out the common factor.}$$

$$= \dfrac{v+3}{v} \quad \text{Simplify.}$$

∴ The excluded values are $v = 0$ and $v = 3$.

### Exercises

Simplify the rational expression, if possible. State the excluded value(s).

9. $\dfrac{18z^2}{4z^4}$

10. $\dfrac{n^2 + 1}{n - 1}$

11. $\dfrac{b^2 + 9b + 18}{b^2 + b - 30}$

## 11.4 Multiplying and Dividing Rational Expressions (pp. 568–573)

Find the product or quotient.

a. $\dfrac{7x}{x+4} \cdot \dfrac{x+4}{x^2} = \dfrac{7x(x+4)}{x^2(x+4)}$    Multiply numerators and denominators.

$= \dfrac{7\cancel{x}\cancel{(x+4)}}{x^{\cancel{2}}\cancel{(x+4)}}$    Divide out the common factors.

$= \dfrac{7}{x}$    Simplify.

b. $\dfrac{t-6}{10} \div \dfrac{6-t}{12} = \dfrac{t-6}{10} \cdot \dfrac{12}{6-t}$    Multiply by the reciprocal.

$= \dfrac{t-6}{10} \cdot \dfrac{12}{-(t-6)}$    Rewrite $6 - t$ as $-(t-6)$.

$= \dfrac{12(t-6)}{-10(t-6)}$    Multiply numerators and denominators.

$= \dfrac{\overset{6}{\cancel{12}}\cancel{(t-6)}}{-\underset{5}{\cancel{10}}\cancel{(t-6)}}$    Divide out the common factors.

$= -\dfrac{6}{5}$    Simplify.

### Exercises

**Find the product or quotient.**

12. $\dfrac{9}{10r} \cdot \dfrac{5r^3}{6}$

13. $\dfrac{k+5}{6k^2} \div \dfrac{5+k}{12k}$

14. $\dfrac{h^2+8h}{h} \div (h^2+7h-8)$

## 11.5 Dividing Polynomials (pp. 574–579)

**Find $(-2x^2 + 8x + 1) \div 2x$.**

$$(-2x^2 + 8x + 1) \div 2x = \dfrac{-2x^2 + 8x + 1}{2x} \qquad \text{Write as a fraction.}$$

$$= \dfrac{-2x^2}{2x} + \dfrac{8x}{2x} + \dfrac{1}{2x} \qquad \text{Divide each term by } 2x.$$

$$= \dfrac{-\cancel{2}x^2}{\cancel{2}\cancel{x}} + \dfrac{\overset{4}{\cancel{8}}\cancel{x}}{\cancel{2}\cancel{x}} + \dfrac{1}{2x} \qquad \text{Divide out the common factors.}$$

$$= -x + 4 + \dfrac{1}{2x} \qquad \text{Simplify.}$$

**Find $(z^2 - 2z - 5) \div (z + 3)$.**

**Step 1:** Divide the first term of the dividend by the first term of the divisor.

Align like terms in the quotient and dividend.

$$\begin{array}{r} z\phantom{{}-2z-5} \\ z+3\overline{\smash{\big)}\,z^2 - 2z - 5\phantom{)}} \\ \underline{z^2 + 3z\phantom{-5)}} \\ -5z - 5\phantom{)} \end{array}$$

Divide: $z^2 \div z = z$.

Multiply: $z(z+3)$.
Subtract. Bring down the $-5$.

**Step 2:** Divide the first term of $-5z - 5$ by the first term of the divisor.

$$\begin{array}{r} z - 5\phantom{)} \\ z+3\overline{\smash{\big)}\,z^2 - 2z - 5\phantom{)}} \\ \underline{z^2 + 3z\phantom{-5)}} \\ -5z - 5\phantom{)} \\ \underline{-5z - 15\phantom{)}} \\ 10\phantom{)} \end{array}$$

Divide: $-5z \div z = -5$.

Multiply: $-5(z+3)$.
Subtract.

So, $(z^2 - 2z - 5) \div (z + 3) = z - 5 + \dfrac{10}{z+3}$.

### Exercises

**Find the quotient.**

15. $(8n^3 + 3n) \div 2n^2$

16. $(b^2 - 36) \div (b - 6)$

17. $(x^2 + 6x + 3) \div (x + 2)$

18. $(4c - 1) \div (c + 5)$

## 11.6 Adding and Subtracting Rational Expressions (pp. 580–587)

Find the difference $\dfrac{y+3}{7y^2} - \dfrac{4}{5y}$.

$$\dfrac{y+3}{7y^2} - \dfrac{4}{5y} = \dfrac{(y+3)(5)}{7y^2(5)} - \dfrac{4(7y)}{5y(7y)}$$ Rewrite using the LCD, $35y^2$.

$$= \dfrac{5y+15}{35y^2} - \dfrac{28y}{35y^2}$$ Simplify.

$$= \dfrac{-23y+15}{35y^2}$$ Subtract the numerators.

### Exercises

**Find the sum or difference.**

19. $\dfrac{5h-2}{h-10} - \dfrac{3h+7}{h-10}$

20. $\dfrac{5}{x} + \dfrac{8}{x+1}$

21. $\dfrac{4-x}{x^2+x-56} + \dfrac{1}{x-7}$

## 11.7 Solving Rational Equations (pp. 588–593)

You own a farm in a computer game. Twenty of the 120 animals on your farm are cows. You buy cows and increase the ratio of cows to total animals to 1 : 5. How many cows do you buy?

Write an equation for the ratio of cows to total animals after buying $x$ cows.

$5(20 + x) = 120 + x$    Cross Products Property

$100 + 5x = 120 + x$    Distributive Property

$4x = 20$    Simplify.

$x = 5$    Divide each side by 4.

∴ You buy 5 cows.

### Exercises

**Solve the equation. Check your solution(s).**

22. $\dfrac{1}{x+2} = \dfrac{12}{3x+6}$

23. $\dfrac{9}{y+6} = \dfrac{y}{y+2}$

24. $\dfrac{5}{t} - \dfrac{3}{4} = \dfrac{6}{t}$

25. **TENNIS** A tennis player lands 25 out of 40 first serves in bounds for a success rate of 62.5%. How many more consecutive first serves must she land in bounds to increase her success rate to 70%?

# 11 Chapter Test

1. The variable $y$ varies directly with $x$. When $x = 3$, $y = 18$. Write and graph a direct variation equation that relates $x$ and $y$.

2. The variable $y$ varies inversely with $x$. When $x = 6$, $y = 4$. Write and graph an inverse variation equation that relates $x$ and $y$.

**Graph the function. Compare the graph to the graph of $y = \dfrac{1}{x}$.**

3. $y = \dfrac{1}{x-6}$

4. $y = \dfrac{1}{x} + 3$

5. $y = \dfrac{1}{x+4} - 5$

**Find the inverse of the function. Graph the inverse function.**

6. $f(x) = x - 7$

7. $f(x) = \dfrac{1}{5}x - 7$

8. $f(x) = \dfrac{3}{x+4}$

**Simplify.**

9. $\dfrac{y^2 + 5y - 24}{y^2 + 10y + 16}$

10. $\dfrac{8r^4}{5} \cdot \dfrac{15r}{6r^3}$

11. $\dfrac{x^2 - 25}{x + 5} \div (x^2 - 3x - 10)$

12. $\dfrac{6k + 1}{2k - 4} + \dfrac{2k + 3}{2k - 4}$

13. $\dfrac{4}{p + 6} - \dfrac{3}{p}$

14. $\dfrac{18z + 27}{z^2 + 3z - 54} + \dfrac{z}{z + 9}$

15. Find $(12d^3 + 8d - 6) \div 3d^2$.

16. Find $(b^2 - 4b + 10) \div (b + 3)$.

**Solve the equation. Check your solution(s).**

17. $\dfrac{1}{x - 5} = \dfrac{3}{2x + 7}$

18. $\dfrac{a}{a + 3} = \dfrac{4}{a + 5}$

19. $\dfrac{6}{n} - \dfrac{2}{n - 3} = \dfrac{5}{n - 3}$

20. **BALANCE** To balance the board in the diagram, the distance (in feet) of each object from the center of the board must vary inversely with its weight (in pounds). What is the distance of the suitcase from the fulcrum?

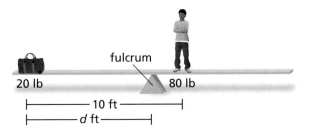

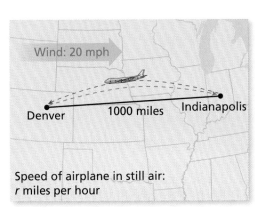

Wind: 20 mph
Denver — 1000 miles — Indianapolis
Speed of airplane in still air: $r$ miles per hour

21. **AIRPLANE** An airplane makes a round trip between two cities. The airplane flies with the wind when heading east and against the wind when heading west. Write an expression for the total time of the trip.

22. **DELIVERY TRUCK** Working alone, it takes you 30 minutes, your friend 30 minutes, and your supervisor 15 minutes to unload a delivery truck. Working together, how much time does it take all three of you to unload the truck?

# 11 Standards Assessment

**1.** Which function is shown by the graph? *(A.REI.10)*

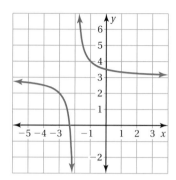

**A.** $y = \dfrac{1}{x+2} - 3$

**B.** $y = \dfrac{1}{x+3} + 2$

**C.** $y = \dfrac{1}{x+2} + 3$

**D.** $y = \dfrac{1}{x-2} - 3$

**2.** What is the value of $c$ in the triangle shown? *(8.G.7)*

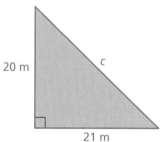

**F.** $\sqrt{41}$ m

**G.** 21 m

**H.** 29 m

**I.** 30 m

**3.** What is the $y$-coordinate of the focus of the graph of $y = \dfrac{1}{8}x^2$? *(F.IF.4)*

**4.** What is the simplest form of the rational expression $\dfrac{4x-3}{3-4x}$? *(A.SSE.2)*

**A.** $-\dfrac{3-4x}{4x-3}$

**B.** $-1$

**C.** $\dfrac{4x-3}{3-4x}$

**D.** $1$

5. What is the solution of the system of equations? (A.REI.7)
$$y = x^2 + 2x - 7$$
$$y = 2x - 7$$

F. $(-3, -4)$

G. $(0, -7)$

H. $(-7, 0)$

I. no real solutions

6. What are the solutions of the equation? (A.SSE.3a)
$$x^4 - 2x^3 - 3x^2 = 0$$

A. $x = 0$

B. $x = 1, x = -3$

C. $x = 0, x = -1, x = 3$

D. $x = 0, x = 3$

7. What is the difference of the rational expressions? (A.SSE.2)
$$\frac{5}{3x} - \frac{3}{4x}$$

F. $-\frac{2}{x}$

G. $\frac{2}{3x}$

H. $\frac{2}{x}$

I. $\frac{11}{12x}$

8. What is the slope of the line shown in the graph? (F.IF.6)

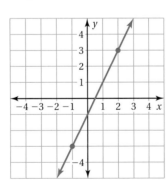

9. What is an equation for the $n$th term of the geometric sequence? (F.BF.2)

| n | 1 | 2 | 3 | 4 |
|---|---|---|---|---|
| $a_n$ | 4 | 8 | 16 | 32 |

A. $a_n = 4(2)^{n-1}$

B. $a_n = 4^{n-1}$

C. $a_n = 2(4)^{n-1}$

D. $a_n = 2^{n-1}$

10. One tablespoon of peanut butter contains 100 calories. The number $c$ of calories consumed is a function of the number $t$ of tablespoons of peanut butter eaten. *(F.IF.1)*

   *Part A* Identify the independent and dependent variables.

   *Part B* Make an input-output table.

   *Part C* Graph the function.

   *Part D* Is the domain discrete or continuous? Explain.

11. What is the solution of the inequality shown below? *(A.REI.3)*

$$\frac{y}{-3} - 4 > -12$$

   **F.** $y > 24$  **H.** $y < 24$

   **G.** $y > -24$  **I.** $y < -24$

12. John was finding the quotient of the rational expressions in the box below. *(A.SSE.2)*

$$\frac{3x}{x-4} \div \frac{2x}{4-x} = \frac{3x}{x-4} \cdot \frac{4-x}{2x}$$
$$= \frac{3x(4-x)}{2x(x-4)}$$
$$= \frac{3x}{2x}$$
$$= \frac{3}{2}$$

   What should John do to correct the error that he made?

   **A.** Do not multiply by the reciprocal.

   **B.** Factor out $-1$ from $(4 - x)$ before dividing out common factors.

   **C.** Divide $3x$ by $2x$ to get an answer of $x$.

   **D.** Use long division to find the quotient.

13. What is the solution of the system of linear equations shown below? *(A.REI.6)*

$$y = 2x - 1$$
$$y = 3x + 5$$

   **F.** $(-13, -6)$  **H.** $(-13, 6)$

   **G.** $(-6, -13)$  **I.** $(-6, 13)$

Standards Assessment

# 12 Data Analysis and Displays

- 12.1 **Measures of Central Tendency**
- 12.2 **Measures of Dispersion**
- 12.3 **Box-and-Whisker Plots**
- 12.4 **Shapes of Distributions**
- 12.5 **Scatter Plots and Lines of Fit**
- 12.6 **Analyzing Lines of Fit**
- 12.7 **Two-Way Tables**
- 12.8 **Choosing a Data Display**

"Wow. The number of minutes I can dog paddle is growing like crazy!"

"Please hold still. I am trying to find the mean of 6, 8, and 10 by dividing their sum into three equal piles."

# What You Learned Before

## Displaying Data (6.SP.5a, 6.SP.4)

**Example 1** The table shows the results of a survey. Display the data in a circle graph.

| Class Trip Location | Water park | Museum | Zoo | Other |
|---|---|---|---|---|
| Students | 25 | 11 | 5 | 4 |

A total of 45 students took the survey.

Water park:

$$\frac{25}{45} \cdot 360° = 200°$$

Museum:

$$\frac{11}{45} \cdot 360° = 88°$$

Zoo:

$$\frac{5}{45} \cdot 360° = 40°$$

Other:

$$\frac{4}{45} \cdot 360° = 32°$$

**Example 2** The frequency table shows the numbers of books that 12 people read last month. Display the data in a histogram.

| Books Read Last Month | Frequency |
|---|---|
| 0–1 | 6 |
| 2–3 | 4 |
| 4–5 | 0 |
| 6–7 | 2 |

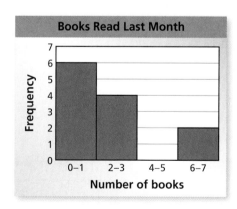

### Try It Yourself

1. Conduct a survey to determine the after-school activities of students in your class. Display the results in a circle graph.

2. Conduct a survey to determine the numbers of pets owned by students in your class. Display the results in a histogram.

# 12.1 Measures of Central Tendency

**Essential Question** How can you use measures of central tendency to distribute an amount evenly among a group of people?

### 1 ACTIVITY: Exploring Mean, Median, and Mode

**Work with a partner. Forty-five coins are arranged in nine stacks.**

a. Record the number of coins in each stack in a table.

| Stack | 1 | 2 | 3 | 4 | 5 | 6 | 7 | 8 | 9 |
|---|---|---|---|---|---|---|---|---|---|
| Coins |   |   |   |   |   |   |   |   |   |

b. Find the mean, median, and mode of the data.

c. By moving coins from one stack to another, can you change the mean? the median? the mode? Explain.

d. Is it possible to arrange the coins in stacks so that the median is 6? 8? Explain.

### 2 EXAMPLE: Drawing a Dot Plot

**Work with a partner.**

a. Draw a number line. Label the tick marks from 1 to 10.

b. Place each stack of coins in Activity 1 above the number of coins in the stack.

c. Draw a ● to represent each stack. This data display is called a *dot plot*.

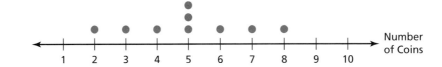

**COMMON CORE**

**Measures of Central Tendency**

In this lesson, you will
- find the mean, median, and mode of a data set.
- identify and remove outliers.
- explain the effects of changing values in data sets.

Learning Standards
S.ID.2
S.ID.3

## 3  ACTIVITY: Fair and Unfair Distributions

**Math Practice 8**

**Maintain Oversight**

What process was used to distribute the coins equally?

**Work with a partner.**

**A distribution of coins to nine people is considered *fair* if each person has the same number of coins.**

- Distribute the 45 coins into 9 stacks using a fair distribution. How is this distribution related to the mean?
- Draw a dot plot for each distribution. Which distributions seem most fair? Which distributions seem least fair? Explain your reasoning.

a.

b.

c.

d.

e.

f.

## What Is Your Answer?

4. **IN YOUR OWN WORDS** How can you use measures of central tendency to distribute an amount evenly among a group of people?

5. Use the Internet or some other reference to find examples of mean or median incomes of groups of people. Describe possible distributions that could produce the given means or medians.

**Practice** Use what you learned about measures of central tendency to complete Exercise 4 on page 610.

# 12.1 Lesson

**Key Vocabulary**
measure of central tendency, p. 608

A **measure of central tendency** is a measure that represents the center of a data set. The *mean*, *median*, and *mode* are measures of central tendency.

## Key Ideas

**Mean**

The *mean* of a data set is the sum of the data divided by the number of data values.

**Median**

Order the data. For a set with an odd number of values, the *median* is the middle value. For a set with an even number of values, the *median* is the mean of the two middle values.

**Mode**

The *mode* of a data set is the value or values that occur most often.

**Remember**

Data can have one mode, more than one mode, or no mode. When each value occurs only once, there is no mode.

### EXAMPLE 1 Finding the Mean, Median, and Mode

| Students' Hourly Wages | |
|---|---|
| $3.87 | $7.25 |
| $8.75 | $8.45 |
| $8.25 | $7.25 |
| $6.99 | $7.99 |

An amusement park hires students for the summer. The students' hourly wages are given in the table. Find the mean, median, and mode of the hourly wages.

**Mean:** (sum of the data) / (number of values) → $\frac{58.8}{8} = 7.35$

**Median:** 3.87, 6.99, 7.25, 7.25, 7.99, 8.25, 8.45, 8.75    Order the data.

$\frac{15.24}{2} = 7.62$    Mean of two middle values

**Mode:** 3.87, 6.99, 7.25, 7.25, 7.99, 8.25, 8.45, 8.75

The value 7.25 occurs most often.

∴ The mean is $7.35, the median is $7.62, and the mode is $7.25.

### On Your Own

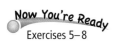
Now You're Ready
Exercises 5–8

1. **WHAT IF?** In Example 1, the park hires another student at an hourly wage of $6.99. How does this additional value affect the mean, median, and mode? Explain.

### EXAMPLE 2 Removing an Outlier

**Remember**
An *outlier* is a data value that is much greater than or much less than the other values.

**Identify the outlier in Example 1. How does the outlier affect the mean, median, and mode?**

The value $3.87 is low compared to the other wages. It is the outlier.

Find the mean, median, and mode without the outlier.

**Mean:** $\frac{54.93}{7} \approx 7.85$

**Median:** 6.99, 7.25, 7.25, 7.99, 8.25, 8.45, 8.75   The middle value, 7.99, is the median.

**Mode:** 6.99, 7.25, 7.25, 7.99, 8.25, 8.45, 8.75   The mode is 7.25.

∴ When you remove the outlier, the mean increases $7.85 − $7.35 = $0.50, the median increases $7.99 − $7.62 = $0.37, and the mode is the same.

### EXAMPLE 3 Changing the Values of a Data Set

**In Example 1, each hourly wage increases by $0.40. How does this increase affect the mean, median, and mode?**

| Students' Hourly Wages | |
|---|---|
| $4.27 | $7.65 |
| $9.15 | $8.85 |
| $8.65 | $7.65 |
| $7.39 | $8.39 |

Make a new table by adding $0.40 to each hourly wage.

**Mean:** $\frac{62}{8} = 7.75$

**Median:** 4.27, 7.39, 7.65, 7.65, 8.39, 8.65, 8.85, 9.15   Order the data.

$\frac{16.04}{2} = 8.02$   Mean of two middle values

**Mode:** 4.27, 7.39, 7.65, 7.65, 8.39, 8.65, 8.85, 9.15   The mode is 7.65.

∴ When each hourly wage increases by $0.40, the mean, median, and mode all increase by $0.40.

#### On Your Own

**Now You're Ready**
Exercises 14–19

The figure shows the altitudes of several airplanes.

2. Identify the outlier. How does the outlier affect the mean, median, and mode? Explain.

3. Each airplane increases its altitude by $1\frac{1}{2}$ miles. How does this affect the mean, median, and mode? Explain.

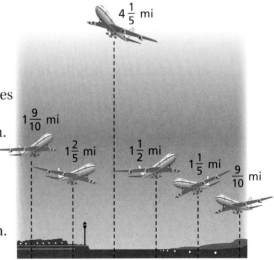

Section 12.1  Measures of Central Tendency

## 12.1 Exercises

### Vocabulary and Concept Check

1. **VOCABULARY** Describe the measures of central tendency of a data set.
2. **OPEN-ENDED** Create a data set that has more than one mode.
3. **WRITING** Describe how removing an outlier from a data set affects the mean of the data set.

### Practice and Problem Solving

4. Draw a dot plot of the data. Then find the mean, median, and mode of the data.

| Bag | 1 | 2 | 3 | 4 | 5 | 6 | 7 | 8 | 9 |
|---|---|---|---|---|---|---|---|---|---|
| Strawberries | 10 | 13 | 11 | 15 | 8 | 14 | 7 | 11 | 12 |

**Find the mean, median, and mode of the data.**

5. **Golf Scores**

| | | |
|---|---|---|
| 3 | −2 | 1 |
| 6 | 4 | −1 |
| −3 | −1 | 2 |

6. **Changes in Stock Value (dollars)**

| | | | |
|---|---|---|---|
| 1.05 | 2.03 | −1.78 | −2.41 |
| −2.64 | 0.67 | 4.02 | 1.39 |
| 0.66 | −0.38 | −3.01 | 2.20 |

7.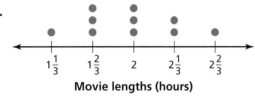
Movie lengths (hours)

8. **Available Memory (megabytes)**

| Stem | Leaf |
|---|---|
| 6 | 5 |
| 7 | 0 5 5 |
| 8 | 0 4 5 |
| 9 | 4 |

Key: 7 | 5 = 75 megabytes

**Find the value of $x$.**

9. Mean is 6; 2, 8, 9, 7, 6, $x$

10. Mean is 0; 11.5, 12.5, −10, −7.5, $x$

11. Median is 14; 9, 10, 12, $x$, 20, 25

12. Median is 51; 30, 45, $x$, 100

13. **POLAR BEARS** The table shows the masses of polar bears. Find the value of $x$ when the mean is 410 kilograms.

**Masses (kilograms)**

| | | | |
|---|---|---|---|
| 455 | 262 | 471 | 358 |
| 364 | 553 | 352 | $x$ |

610 Chapter 12 Data Analysis and Displays

14. **BASEBALL** The graph shows a player's monthly home run totals in two seasons.

    a. Identify the outlier in each season.
    b. Which measure of central tendency is most affected by removing the outlier in each season?
    c. Compare the means, medians, and modes of the home run totals in the two seasons.

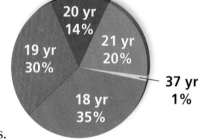

15. **TRAIL** The map shows the locations of 11 shelters along the Appalachian Trail. The distances (in miles) between these shelters are 0.1, 14.3, 5.3, 1.8, 14, 8.8, 8.8, 16.7, 6.3, and 3.3.

    a. Find the mean, median, and mode of the distances.
    b. A hiker starts at Shelter 2 and hikes to Shelter 11. How does this affect the mean, median, and mode? Explain.

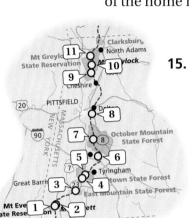

In Exercises 16–19, explain how the change affects the mean, median, and mode.

16. In Exercise 4, you add 3 strawberries to each bag.

17. You add a number $k$ to each value in a data set.

18. In Exercise 6, the value of each stock decreases by $0.05.

19. You subtract a number $k$ from each value in a data set.

20. **COLLEGE** The circle graph shows the distribution of the ages of 200 students in a college psychology class.

    a. Find the mean, median, and mode of the students' ages.
    b. Identify the outliers. How do the outliers affect the mean, median, and mode?

21. **Reasoning** The mean and median hourly wage at a bagel shop is $7.20. Hourly wages at the bagel shop increase by 10%. Where are you likely to have a greater hourly wage, at the bagel shop or at the amusement park in Example 1? Explain.

## Fair Game Review What you learned in previous grades & lessons

**Order the values from least to greatest.** *(Skills Review Handbook)*

22. $1, -3, -8, 4, 7, -5$

23. $1.2, -2.8, \frac{3}{2}, 5.4, -4.7, -\frac{2}{3}$

24. **MULTIPLE CHOICE** Which equation represents a linear function? *(Section 5.5)*

    Ⓐ $y = x^2$    Ⓑ $y = 2x$    Ⓒ $y = \frac{2}{x}$    Ⓓ $xy = 2$

# 12.2 Measures of Dispersion

**Essential Question** How can you measure the dispersion of a data set?

### 1 ACTIVITY: Measuring the Dispersion of Data

Work with a partner. The diagram shows the weights of 53 players on the Chicago Bears football team in 2011.

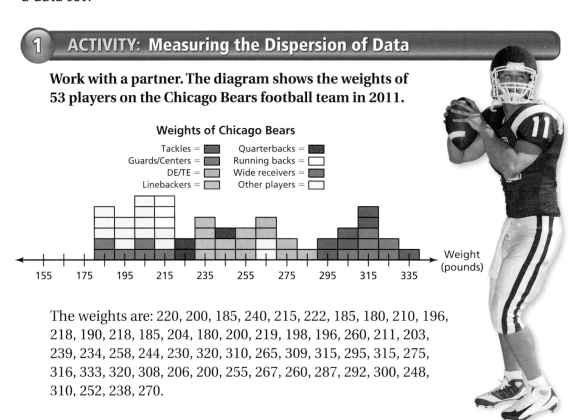

**Weights of Chicago Bears**

Tackles =
Guards/Centers =
DE/TE =
Linebackers =
Quarterbacks =
Running backs =
Wide receivers =
Other players =

The weights are: 220, 200, 185, 240, 215, 222, 185, 180, 210, 196, 218, 190, 218, 185, 204, 180, 200, 219, 198, 196, 260, 211, 203, 239, 234, 258, 244, 230, 320, 310, 265, 309, 315, 295, 315, 275, 316, 333, 320, 308, 206, 200, 255, 267, 260, 287, 292, 300, 248, 310, 252, 238, 270.

a. Describe the data. How much are the weights dispersed from the mean weight? Explain your reasoning.

> **Definition of Dispersed:** To disperse objects means to spread them over an area. For instance, the population of Texas is much more dispersed than the population of Rhode Island.

b. Does it appear that the weight of a football player is correlated to the position that he plays? Explain your reasoning. Do you think your answer is valid for other types of professional sports, such as basketball, baseball, hockey, and soccer? Explain your reasoning.

**COMMON CORE**

**Measures of Dispersion**

In this lesson, you will
- find ranges of data sets.
- compare spreads of data sets.
- find standard deviations of data sets.

Learning Standards
S.ID.2
S.ID.3

**612** Chapter 12 Data Analysis and Displays

### 2 ACTIVITY: Measuring the Dispersion of Data

**Math Practice 4**

**Analyze Relationships**

How can the diagram help you analyze the data?

Work with a partner. The diagram shows the weights of 40 players on the Los Angeles Angels baseball team in 2011.

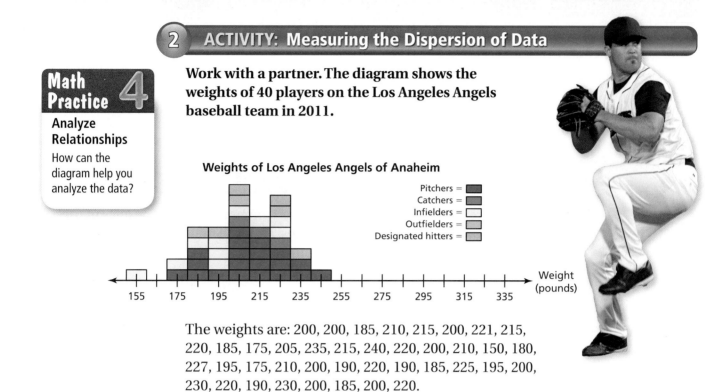

**Weights of Los Angeles Angels of Anaheim**

Pitchers = ■
Catchers = ■
Infielders = □
Outfielders = ▨
Designated hitters = ▨

The weights are: 200, 200, 185, 210, 215, 200, 221, 215, 220, 185, 175, 205, 235, 215, 240, 220, 200, 210, 150, 180, 227, 195, 175, 210, 200, 190, 220, 190, 185, 225, 195, 200, 230, 220, 190, 230, 200, 185, 200, 220.

a. Describe the data. How much are the weights dispersed from the mean weight? Explain your reasoning.

b. Compare the dispersions of the weights of players for a National Football League team and a Major League Baseball team.

c. Does it appear that the weight of a baseball player is correlated to the position that he plays? Explain your reasoning.

## What Is Your Answer?

3. **IN YOUR OWN WORDS** How can you measure the dispersion of a data set? Illustrate your answer by using the positions and weights of the 15 players on the Boston Celtics basketball team in 2011.

   Forward: 235; power forwards: 253, 295, 245; small forwards: 235, 235; centers: 255, 240, 325; point guards: 205, 186, 200; shooting guards: 205, 210, 180

   Does it appear that the weight of a basketball player is correlated to the position that he plays? Explain your reasoning.

**Practice →** Use what you learned about measuring the dispersion of data to complete Exercises 3 and 4 on page 616.

Section 12.2  Measures of Dispersion  613

# 12.2 Lesson

Check It Out
Lesson Tutorials
BigIdeasMath.com

**Key Vocabulary**
measure of dispersion, *p. 614*
range, *p. 614*
standard deviation, *p. 615*

A **measure of dispersion** is a measure that describes the spread of a data set. The simplest measure of dispersion is the range. The **range** of a data set is the difference between the greatest value and the least value.

### EXAMPLE 1 Finding the Range

Two reality cooking shows select 12 contestants each. The ages of the contestants are shown in the tables. Find the mean and range of the ages for each show. Compare your results.

| Show A | |
|---|---|
| Ages | |
| 20 | 29 |
| 19 | 22 |
| 25 | 27 |
| 27 | 29 |
| 30 | 20 |
| 21 | 31 |

| Show B | |
|---|---|
| Ages | |
| 25 | 19 |
| 20 | 27 |
| 22 | 25 |
| 27 | 22 |
| 48 | 21 |
| 32 | 24 |

**Show A:** mean = $\frac{300}{12} = 25$

Ordering the data can help you find the least and greatest ages.

19, 20, 20, 21, 22, 25, 27, 27, 29, 29, 30, 31    Order the data.

The least value is 19. The greatest value is 31.

So, the range is 31 − 19, or 12 years.

**Show B:** mean = $\frac{312}{12} = 26$

19, 20, 21, 22, 22, 24, 25, 25, 27, 27, 32, 48    Order the data.

The least value is 19. The greatest value is 48.

So, the range is 48 − 19, or 29 years.

∴ The mean ages for the shows, 25 and 26, are about the same. The range of the ages for Show A is 12 years and the range for Show B is 29 years. So, the ages for Show B are more spread out.

### On Your Own

Now You're Ready
Exercises 5 and 6

**1.** After the first week, the 25-year-old is voted off Show A. The 48-year-old is voted off Show B. How does this affect the mean and range of the remaining contestants on each show? Explain.

A disadvantage of using the range to describe the spread of a data set is that its calculation uses only two data values. A measure of dispersion that uses all the values of a data set is the *standard deviation*.

### Key Idea

**Standard Deviation**

The **standard deviation** of a data set is a measure of how much a typical value in the data set differs from the mean. It is given by

$$\text{standard deviation} = \sqrt{\frac{(x_1 - \bar{x})^2 + (x_2 - \bar{x})^2 + \cdots + (x_n - \bar{x})^2}{n}}$$

where $n$ is the number of values in the data set. The symbol $\bar{x}$ represents the mean. It is read as "$x$-bar."

A small standard deviation means that the data are clustered around the mean. A large standard deviation means that the data are more spread out.

**Remember**

The notation consisting of three dots (⋯) indicates that a pattern continues.

### EXAMPLE 2 — Finding the Standard Deviation

Find the standard deviation of the ages for Show A in Example 1. Use a table to organize your work. Interpret your result.

| $x$ | $\bar{x}$ | $x - \bar{x}$ | $(x - \bar{x})^2$ |
|---|---|---|---|
| 20 | 25 | −5 | 25 |
| 29 | 25 | 4 | 16 |
| 19 | 25 | −6 | 36 |
| 22 | 25 | −3 | 9 |
| 25 | 25 | 0 | 0 |
| 27 | 25 | 2 | 4 |
| 27 | 25 | 2 | 4 |
| 29 | 25 | 4 | 16 |
| 30 | 25 | 5 | 25 |
| 20 | 25 | −5 | 25 |
| 21 | 25 | −4 | 16 |
| 31 | 25 | 6 | 36 |

**Step 1:** Find the mean. From Example 1, the mean is 25.

**Step 2:** Find the difference between each data value and the mean, $x - \bar{x}$.

**Step 3:** Square each difference from Step 2, $(x - \bar{x})^2$.

**Step 4:** Find the mean of the squares from Step 3.

$$\frac{(x_1 - \bar{x})^2 + (x_2 - \bar{x})^2 + \cdots + (x_n - \bar{x})^2}{n} = \frac{25 + 16 + \cdots + 36}{12} \approx 17.7$$

**Step 5:** Use a calculator to find the square root.

$$\sqrt{\frac{(x_1 - \bar{x})^2 + (x_2 - \bar{x})^2 + \cdots + (x_n - \bar{x})^2}{n}} = \sqrt{17.7} \approx 4.2$$

∴ The standard deviation is 4.2. This means that the typical age of a contestant on Show A differs from the mean by about 4.2 years.

### On Your Own

*Now You're Ready*
Exercises 7–12

**2.** Find the standard deviation of the ages for Show B in Example 1. Interpret your result.

**3.** Compare the standard deviations for Show A and Show B. What can you conclude?

Section 12.2  Measures of Dispersion

## 12.2 Exercises

### Vocabulary and Concept Check

1. **VOCABULARY** In a data set, what does a measure of central tendency represent? What does a measure of dispersion represent?

2. **REASONING** What is an advantage of using the range to describe a data set? Why do you think the standard deviation is considered a more reliable measure of dispersion than the range?

### Practice and Problem Solving

**Describe the data. How much are the data dispersed from the mean? Explain your reasoning.**

3.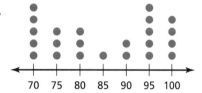

4.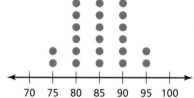

**Find the mean and range of each data set. Then compare the data sets.**

5. Heights (in inches) of two teams
   Tigers: 67, 70, 65, 72, 74, 68, 67, 69
   Centaurs: 74, 71, 68, 63, 75, 63, 65, 73

6. Numbers of fish caught during a week
   Crew A: 120, 100, 75, 112, 135, 80, 106
   Crew B: 104, 140, 159, 135, 158, 165, 140

**Find the mean and standard deviation of the data.**

7. 4, 2, 7, 3, 6, 5, 5, 8

8. 12, 4, 8, 7, 9, 13, 10

9.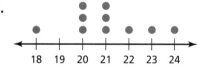

10.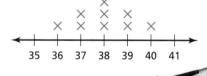

11. 
| Stem | Leaf |
|------|------|
| 4 | 0 |
| 5 | 2 |
| 6 | 1 4 5 7 |
| 7 | 3 |
| 8 | 2 |

Key: 6 | 1 = 61

12.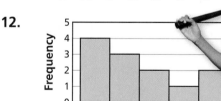

13. **GOLF** The scores for two golfers are shown.

    a. Find the mean, range, and standard deviation of the scores for each golfer. Compare your results.

    b. Which golfer do you think is more consistent? Explain.

| Kirsten | | Leah | |
|---|---|---|---|
| 83 | 88 | 89 | 87 |
| 84 | 95 | 93 | 95 |
| 91 | 89 | 92 | 94 |
| 90 | 87 | 88 | 91 |
| 98 | 95 | 89 | 92 |

**14. INCLUDING A VALUE** In Exercise 13, Kirsten's score for the next round is 90, and Leah's is 80. How does each of these scores affect the mean, range, and standard deviation of each data set? Explain.

**Find the mean, range, and standard deviation of the data.**

**15.** 4.1, 2.3, 8.7, 10.5, 6.4

**16.** −2, 0, 1, −5, 3, −4, 2, −3

**17. REASONING** Two data sets have the same range. Can you assume that the standard deviations of the two data sets are about the same? Give an example to justify your answer.

**18. ADVENTURE CLUB** The dot plots show the ages of members of three different adventure clubs. Without performing calculations, which data set has the greatest standard deviation? Which has the least standard deviation? Explain your reasoning.

a.
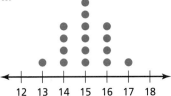

b.
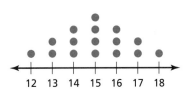

c.
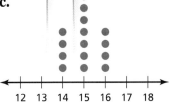

**19. PROJECT** Measure the heights (in inches) of the students in your class.

  **a.** Use a calculator to find the mean, range, and standard deviation of the heights.

  **b.** A new student who is 7 feet tall joins your class. How would you expect this person's height to affect the mean, range, and standard deviation? Verify your answer.

**20. WAITING TIMES** The waiting times at two doctors' offices are described below. At which office are you more likely to wait longer than 20 minutes? Explain. Assume the mean is at the center of each distribution and the data are evenly distributed.

Doctor's Office A: mean = 15 minutes, standard deviation = 2.5 minutes

Doctor's Office B: mean = 14 minutes, standard deviation = 5.5 minutes

**21. Critical Thinking** Can the standard deviation of a data set be 0? Can it be negative? If so, give examples to justify your answers.

## Fair Game Review  *What you learned in previous grades & lessons*

**Graph the function. Describe the domain and range.** *(Section 11.2)*

**22.** $y = -\dfrac{3}{x}$

**23.** $y = \dfrac{1}{x-6}$

**24.** $y = \dfrac{1}{x+4} - 5$

**25. MULTIPLE CHOICE** Find the quotient $(x+5) \div \dfrac{x^2+4x-5}{x+5}$. *(Section 11.4)*

Ⓐ $\dfrac{x+5}{x+1}$   Ⓑ $\dfrac{x+5}{x-1}$   Ⓒ $x^2+4x-5$   Ⓓ $\dfrac{x+5}{x-5}$

# 12.3 Box-and-Whisker Plots

**Essential Question** How can you use a box-and-whisker plot to describe a data set?

### 1 ACTIVITY: Drawing a Box-and-Whisker Plot

Work with a partner.

The numbers of first cousins of the students in an eighth-grade class are shown.

A box-and-whisker plot uses a number line to represent the data visually.

| Numbers of First Cousins | | | |
|---|---|---|---|
| 3 | 10 | 18 | 8 |
| 9 | 3 | 0 | 32 |
| 23 | 19 | 13 | 8 |
| 6 | 3 | 3 | 10 |
| 12 | 45 | 1 | 5 |
| 13 | 24 | 16 | 14 |

**a.** Order the data set and write it on a strip of grid paper with 24 equally spaced boxes.

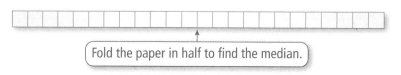

Fold the paper in half to find the median.

**b.** Fold the paper in half again to divide the data into four groups. Because there are 24 numbers in the data set, each group should have six numbers.

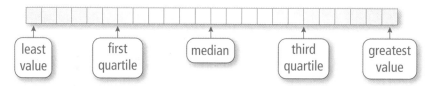

least value — first quartile — median — third quartile — greatest value

**c.** Draw a number line that includes the least value and the greatest value in the data set. Graph the five numbers that you found in part (b).

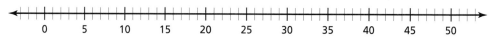

### COMMON CORE

**Box-and-Whisker Plots**

In this lesson, you will
- make and interpret box-and-whisker plots.
- find interquartile ranges of data sets.
- compare box-and-whisker plots.

Learning Standards
S.ID.1
S.ID.2
S.ID.3

**d.** Explain how the box-and-whisker plot shown below represents the data set.

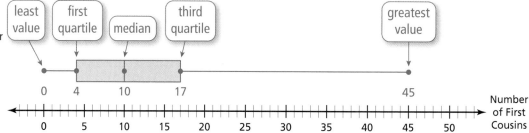

618   Chapter 12   Data Analysis and Displays

### 2 ACTIVITY: Conducting a Survey

Conduct a survey in your class. Ask each student to write the number of his or her first cousins on a piece of paper. Collect the pieces of paper and write the data on the chalkboard.

Now, work with a partner to draw a box-and-whisker plot of the data.

Two people are first cousins if they share at least one grandparent, but do not share a parent.

First Cousins

**Math Practice 7**

**View as Components**

How do the different components of a box-and-whisker plot help you interpret the values?

### 3 ACTIVITY: Reading a Box-and-Whisker Plot

Work with a partner. The box-and-whisker plots show the test score distributions of two eighth-grade standardized tests. The tests were taken by the same group of students. One test was taken in the fall and the other was taken in the spring.

a. Compare the test results.

b. Decide which box-and-whisker plot represents the results of each test. How did you make your decision?

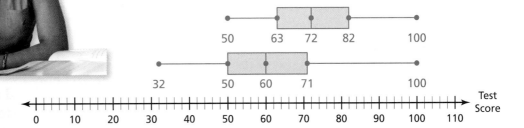

## What Is Your Answer?

4. **IN YOUR OWN WORDS** How can you use a box-and-whisker plot to describe a data set?

5. Describe who might be interested in test score distributions like those shown in Activity 3. Explain why it is important for these people to analyze test score distributions.

**Practice** — Use what you learned about box-and-whisker plots to complete Exercise 4 on page 623.

# 12.3 Lesson

**Key Vocabulary**
box-and-whisker plot, *p. 620*
quartile, *p. 620*
five-number summary, *p. 620*
interquartile range, *p. 621*

##  Key Idea

**Box-and-Whisker Plot**

A **box-and-whisker plot** displays a data set along a number line using medians. **Quartiles** divide the data set into four equal parts. The median (second quartile) divides the data set into two halves. The median of the lower half is the first quartile. The median of the upper half is the third quartile.

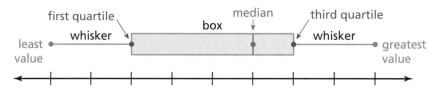

The five numbers that make up the box-and-whisker plot are called the **five-number summary** of the data set.

### EXAMPLE 1  Making a Box-and-Whisker Plot

**Make a box-and-whisker plot for the ages of the members of the U.S. women's wheelchair basketball team.**

24, 30, 30, 22, 25, 22, 18, 25, 28, 30, 25, 27

**Step 1:** Order the data. Find the median and the quartiles.

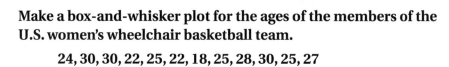

**Step 2:** Draw a number line that includes the least and greatest values. Graph points above the number line for the five-number summary.

**Step 3:** Draw a box using the quartiles. Draw a line through the median. Draw whiskers from the box to the least and greatest values.

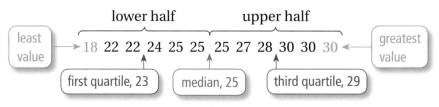

**Study Tip**
A box-and-whisker plot shows the *variability* of a data set.

### On Your Own

**Now You're Ready**
Exercises 5–7

1. A basketball player scores 14, 16, 20, 5, 22, 30, 16, and 28 points during a tournament. Make a box-and-whisker plot for the points scored by the player.

620   Chapter 12   Data Analysis and Displays

The figure shows how data are distributed in a box-and-whisker plot.

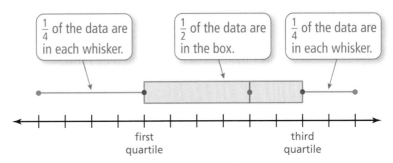

Another measure of dispersion for a data set is the **interquartile range,** which is the difference of the third quartile and the first quartile. It represents the range of the middle half of the data.

### EXAMPLE 2  Interpreting a Box-and-Whisker Plot

**The box-and-whisker plot represents the lengths of songs (in seconds) played by a rock band at a concert.**

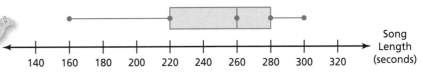

a. **Find and interpret the range of the data.**

The least value is 160. The greatest value is 300.

So, the range is 300 − 160 = 140 seconds. This means that the song lengths vary by no more than 140 seconds.

b. **Describe the distribution of the data.**

- 25% of the song lengths are between 160 and 220 seconds.
- 50% of the song lengths are between 220 and 280 seconds.
- 25% of the song lengths are between 280 and 300 seconds.

c. **Find and interpret the interquartile range of the data.**

$$\text{interquartile range} = \text{third quartile} - \text{first quartile}$$
$$= 280 - 220$$
$$= 60$$

So, the interquartile range is 60 seconds. This means that the middle half of the song lengths vary by no more than 60 seconds.

### On Your Own

Use the box-and-whisker plot in Example 1.

2. Find and interpret the range and interquartile range of the data.

3. Describe the distribution of the data.

A box-and-whisker plot shows the shape of a distribution.

## Key Ideas

**Shapes of Box-and-Whisker Plots**

*Skewed left*
- Left whisker longer than right whisker
- Most data on the right

*Symmetric*
- Whiskers about same length
- Median in the middle of the data

*Skewed right*
- Right whisker longer than left whisker
- Most data on the left

**Study Tip**

If you can draw a line through the median of a box-and-whisker plot, and each side is a mirror image of the other, then the distribution is symmetric.

### EXAMPLE 3  Comparing Box-and-Whisker Plots

**The double box-and-whisker plot represents the test scores for your class and your friend's class.**

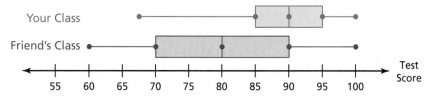

**a. Identify the shape of each distribution.**

For your class, the left whisker is longer than the right whisker, and most of the data are on the right side of the display. So, the distribution is skewed left.

For your friend's class, the whisker lengths are equal. The median is in the middle of the data. The data appear to be evenly distributed on both sides of the median. So, the distribution is symmetric.

**b. Which test scores are more spread out?**

The range of the test scores in your friend's class is greater than the range in your class. Also, because the box for your friend's class is longer than the box for your class, the interquartile range is also greater. So, the test scores in your friend's class are more spread out.

### On Your Own

**Now You're Ready**
Exercise 20

4. The double box-and-whisker plot represents the surfboard prices at Shop A and Shop B. Identify the shape of each distribution. Which shop's prices are more spread out? Explain.

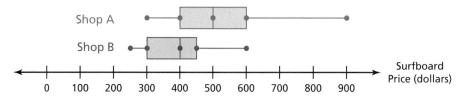

# 12.3 Exercises

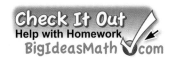

## Vocabulary and Concept Check

1. **VOCABULARY** In a box-and-whisker plot, what percent of the data is represented by each whisker? by the box?

2. **WRITING** Describe how to find the first quartile of a data set.

3. **NUMBER SENSE** What does the length of a box-and-whisker plot tell you about the data?

## Practice and Problem Solving

4. The box-and-whisker plots show the monthly car sales for a year for two sales representatives. Compare the sales for the two representatives.

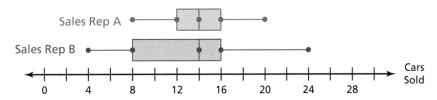

**Make a box-and-whisker plot for the data.**

5. Hours of television watched: 0, 3, 4, 5, 3, 4, 6, 5

6. Lengths (in inches) of cats: 16, 18, 20, 25, 17, 22, 23, 21

7. Elevations (in feet): −2, 0, 5, −4, 1, −3, 2, 0, 2, −3, 6, −1

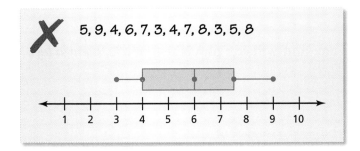

8. **ERROR ANALYSIS** Describe and correct the error in making a box-and-whisker plot for the data.

9. **FISH** The lengths (in inches) of the fish caught on a fishing trip are 9, 10, 12, 8, 13, 10, 12, 14, 7, 14, 8, and 14.

   a. What is the median of the data set?
   b. What are the first and third quartiles of the data set?
   c. Make a box-and-whisker plot for the data.

10. **INCHWORM** The table shows the lengths of 12 inchworms.

   a. Make a box-and-whisker plot for the data.
   b. Find and interpret the range of the data.
   c. Describe the distribution of the data.
   d. Find and interpret the interquartile range of the data.

   | Length (cm) | 2.5 | 2.4 | 2.3 | 2.5 | 2.7 | 2.1 | 2.8 | 2.6 | 2.1 | 2.6 | 2.9 | 2.0 |

   | Entrée Prices (dollars) | | | |
   |---|---|---|---|
   | 14.00 | 17.00 | 12.50 | 10.00 |
   | 11.00 | 18.25 | 9.00 | 8.50 |
   | 14.75 | 15.00 | 14.00 | 12.00 |

11. **ENTRÉE** The table shows the prices of entrées at a restaurant.

   a. Make a box-and-whisker plot for the data.
   b. Find and interpret the interquartile range of the data.
   c. Describe the distribution of the data.
   d. Find the standard deviation. Interpret your result.

12. **WRITING** Given the numbers 36 and 12, identify which number is the range, and which number is the interquartile range, of a set of data. Explain your reasoning.

**Determine whether the shape of the box-and-whisker plot is *symmetric*, *skewed left*, or *skewed right*. Explain.**

13.

14.

15.

16.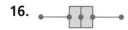

17. **ERROR ANALYSIS** Describe and correct the error in describing the box-and-whisker plot.

   ✗ The shape of the distribution is skewed right. So, there are more data values to the right of the median than to the left of the median.

18. **LOGIC** Give examples of real-life data that are symmetric and real-life data that are not symmetric. Justify your answer.

19. **CALORIES** The table shows the numbers of calories burned per hour for nine activities.

   a. Make a box-and-whisker plot for the data.
   b. Identify the outlier.
   c. Make another box-and-whisker plot without the outlier.
   d. **WRITING** Describe how the outlier affects the whiskers, the box, and the quartiles of the box-and-whisker plot.

| Calories Burned per Hour | |
|---|---|
| Fishing | 207 |
| Mowing the lawn | 325 |
| Canoeing | 236 |
| Bowling | 177 |
| Hunting | 295 |
| Fencing | 354 |
| Bike racing | 944 |
| Horseback riding | 236 |
| Dancing | 266 |

20. **CELL PHONES** The double box-and-whisker plot compares the battery lives (in hours) of two brands of cell phones.

   a. Identify the shape of each distribution.
   b. What is the range of the upper 75% of each brand?
   c. Compare the interquartile ranges of the two data sets.
   d. Which brand do you think has the greater standard deviation? Explain.

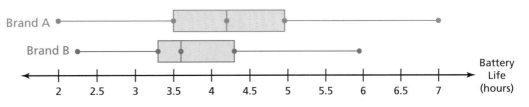

**Modeling** Create a set of data values for the box-and-whisker plot that has the given characteristic(s).

21. The least value, greatest value, quartiles, and median are all equally spaced.

22. Both whiskers are the same length as the box.

23. The box between the median and the first quartile is three times as long as the box between the median and the third quartile.

24. There is no right whisker.

**Fair Game Review** What you learned in previous grades & lessons

**Write an equation of the line that passes through the points.** *(Section 2.6)*

25. $(-4, -10), (2, 8)$

26. $(-3, 3), (0, -1)$

27. $(-4, 1), (4, -1)$

28. $(6, 7), (8, 8)$

29. **MULTIPLE CHOICE** What is the quotient of $(2z^2 - 13z + 21)$ and $(z - 3)$? *(Section 11.5)*

   (A) $2z + 7$
   (B) $2z - 7$
   (C) $z + 6$
   (D) $z - 7$

# 12.4 Shapes of Distributions

**Essential Question** How can you use a histogram to characterize the basic shape of a distribution?

### 1  ACTIVITY: Analyzing a Famous Symmetric Distribution

A famous data set was collected in Scotland in the mid-1800s. It contains the chest sizes, measured in inches, of 5738 men in the Scottish Militia.

*The Thin Red Line* is a painting by Robert Gibb. It was painted in 1881. Only the left portion of the painting is shown in the photo at the right.

| Chest Size | Number of Men |
|---|---|
| 33 | 3 |
| 34 | 18 |
| 35 | 81 |
| 36 | 185 |
| 37 | 420 |
| 38 | 749 |
| 39 | 1073 |
| 40 | 1079 |
| 41 | 934 |
| 42 | 658 |
| 43 | 370 |
| 44 | 92 |
| 45 | 50 |
| 46 | 21 |
| 47 | 4 |
| 48 | 1 |

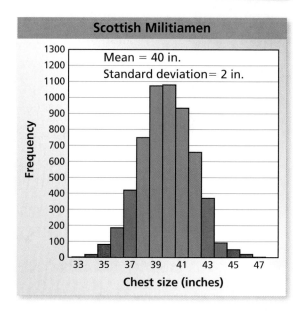

**COMMON CORE**

**Data Distributions**
In this lesson, you will
- describe shapes of distributions.
- choose appropriate measures of central tendency and dispersion to represent data sets.

Learning Standards
S.ID.2
S.ID.3

**Work with a partner.** What percent of the chest sizes lie within (a) 1 standard deviation, (b) 2 standard deviations, and (c) 3 standard deviations of the mean? Explain your reasoning.

## 2 ACTIVITY: Comparing Two Symmetric Distributions

**Math Practice**

**Make Sense of Quantities**
How can you use a histogram to understand relationships within a data set?

Work with a partner. The graphs show the distributions of the heights of 250 adult American males and 250 adult American females.

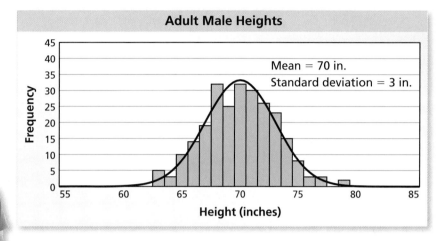

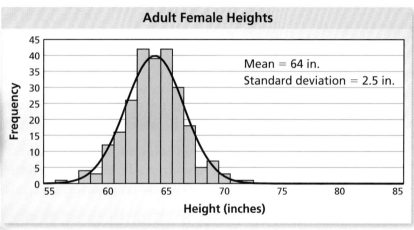

**a.** Which data set has a smaller standard deviation? Explain what this means in the real-life context.

**b.** Estimate the percent of male heights between 67 inches and 73 inches.

### What Is Your Answer?

**3. IN YOUR OWN WORDS** How can you use a histogram to characterize the basic shape of a distribution?

**4.** All three distributions in Activities 1 and 2 are roughly symmetric distributions. The histograms are called "bell-shaped."

  **a.** What are the characteristics of a symmetric distribution?

  **b.** Why is a symmetric distribution called "bell-shaped"?

  **c.** Give two other real-life examples of symmetric distributions.

**Practice**
Use what you learned about the shapes of distributions to complete Exercises 3 and 4 on page 631.

# 12.4 Lesson

Recall that a histogram is a bar graph that shows the frequency of data values in intervals of the same size. A histogram is another useful data display that shows the shape of a distribution.

## Key Ideas

**Symmetric and Skewed Distributions**

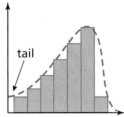 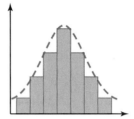 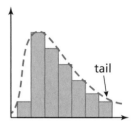

*Skewed left*      *Symmetric*      *Skewed right*

- The "tail" of the graph extends to the left.
- Most data are on the right.

- The data are evenly distributed on each side of the highest bar.

- The "tail" of the graph extends to the right.
- Most data are on the left.

**Remember**
If all the bars of a histogram are about the same height, then the distribution is a *flat*, or *uniform*, distribution. A uniform distribution is also symmetric.

### EXAMPLE 1   Describing the Shape of a Distribution

| Number of Tickets Sold | Frequency |
|---|---|
| 1–8 | 5 |
| 9–16 | 9 |
| 17–24 | 16 |
| 25–32 | 25 |
| 33–40 | 20 |
| 41–48 | 8 |
| 49–56 | 7 |

The frequency table shows the numbers of raffle tickets sold by students in your grade. Display the data in a histogram. Describe the shape of the distribution.

**Step 1:** Draw and label the axes.

**Step 2:** Draw a bar to represent the frequency of each interval.

The graph is high in the middle, and the data are about evenly distributed on each side of the highest bar.

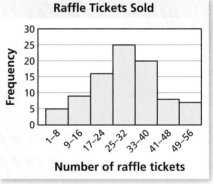

∴ So, the distribution is symmetric.

### On Your Own

*Now You're Ready*
Exercises 5 and 6

1. The frequency table shows the numbers of pounds of aluminum cans collected by students for a fundraiser. Display the data in a histogram. Describe the shape of the distribution.

| Number of Pounds | Frequency |
|---|---|
| 1–10 | 7 |
| 11–20 | 8 |
| 21–30 | 10 |
| 31–40 | 16 |
| 41–50 | 34 |
| 51–60 | 15 |

The shape of a distribution can be used to choose the most appropriate measure of central tendency that describes a data set.

For a symmetric distribution, the mean and median are about the same, although the mean should be used to describe the center.

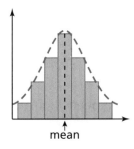

**Remember**
Outliers can affect the mean of a data set more than they affect the median.

When the distribution is skewed, the mean will be in the direction in which the distribution is skewed while the median will be less affected. So, when the data are skewed, use the median to describe the center.

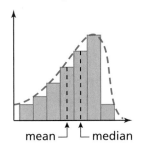 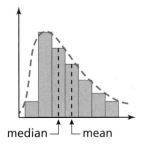

### EXAMPLE 2  Choosing an Appropriate Measure of Central Tendency

**Which measure of central tendency best represents the data? Explain your reasoning.**

a.

b.

a. Because the distribution is high on the left and the tail of the graph extends to the right, the distribution is skewed right. So, the median best represents the data.

b. Because the distribution is high in the middle and the data are about evenly distributed on both sides, the distribution is symmetric. So, the mean best represents the data.

#### On Your Own

2. Which measure of central tendency best represents the data in On Your Own Question 1? Explain your reasoning.

When a distribution is symmetric, use the standard deviation to describe the spread of the data set. When a distribution is skewed, use the five-number summary to describe the spread of the data set.

## EXAMPLE 3  Choosing Appropriate Measures

**Speeds (mi/h)**

| 32 | 44 | 39 |
|---|---|---|
| 53 | 38 | 48 |
| 56 | 41 | 42 |
| 50 | 50 | 55 |
| 55 | 45 | 49 |
| 51 | 53 | 52 |
| 54 | 60 | 55 |
| 52 | 50 | 52 |
| 55 | 40 | 60 |
| 45 | 58 | 47 |

A police officer measures the speeds (in miles per hour) of 30 motorists. The results are shown in the table at the left.

**a. Display the data in a histogram using six intervals beginning with 31–35.**

Make a frequency table using the described intervals. Then use the frequency table to make a histogram.

| Speed (mi/h) | Frequency |
|---|---|
| 31–35 | 1 |
| 36–40 | 3 |
| 41–45 | 5 |
| 46–50 | 6 |
| 51–55 | 11 |
| 56–60 | 4 |

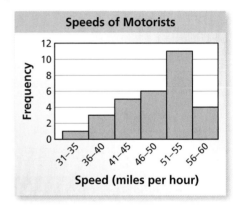

**b. Which measures of central tendency and dispersion best represent the data?**

Because the distribution is high on the right and the tail of the graph extends to the left, the distribution is skewed left. So, use the median to describe the center and the five-number summary to describe the spread.

**c. The speed limit is 45 miles per hour. How would you interpret these results?**

Because the distribution is skewed left, most of the speeds are more than 45 miles per hour. This shows that most of the motorists were speeding.

### On Your Own

Exercises 8–11

**3.** You record the numbers of email attachments sent by 30 employees of a company in one week. Your results are shown in the table.

**Email Attachments Sent**

| 74 | 105 | 98 | 68 | 64 |
|---|---|---|---|---|
| 85 | 75 | 60 | 48 | 51 |
| 65 | 55 | 58 | 45 | 38 |
| 64 | 52 | 65 | 30 | 70 |
| 72 | 5 | 45 | 77 | 83 |
| 42 | 25 | 95 | 16 | 120 |

**a.** Display the data in a histogram using six intervals beginning with 1–20.

**b.** Which measures of central tendency and dispersion best represent the data? Why?

## 12.4 Exercises

### Vocabulary and Concept Check

1. **VOCABULARY** How does the shape of a symmetric distribution differ from the shape of a skewed distribution?

2. **WRITING** How does the shape of a distribution help you decide which measure of central tendency best describes the data?

### Practice and Problem Solving

**Estimate the percent of data within 2 standard deviations of the mean.**

3.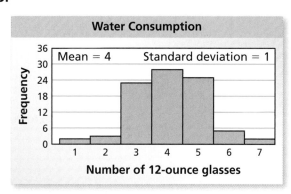

4.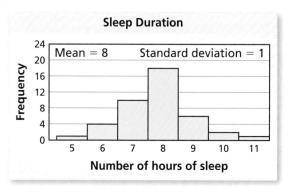

**Display the data in a histogram. Describe the shape of the distribution.**

5.

| Number of Bull's-eyes | Frequency |
|---|---|
| 1–5 | 3 |
| 6–10 | 0 |
| 11–15 | 8 |
| 16–20 | 18 |
| 21–25 | 26 |
| 26–30 | 35 |
| 31–35 | 21 |

6.

| Number of Volunteer Hours | Frequency |
|---|---|
| 1–3 | 1 |
| 4–6 | 5 |
| 7–9 | 12 |
| 10–12 | 20 |
| 13–15 | 15 |
| 16–18 | 7 |
| 19–21 | 2 |

7. **ONLINE** A survey asks people how many hours they spend online per day. The results are shown in the table. Display the data in a histogram. Describe the shape of the distribution.

| Hours Online | 0–2 | 3–5 | 6–8 | 9–11 | 12–14 |
|---|---|---|---|---|---|
| Frequency | 33 | 45 | 12 | 4 | 2 |

**Determine which measures of central tendency and dispersion best represent the data. Explain your reasoning.**

8.
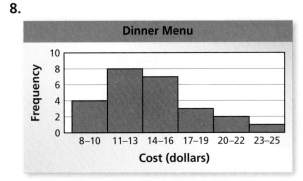

9.
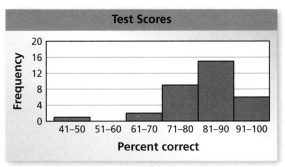

10.
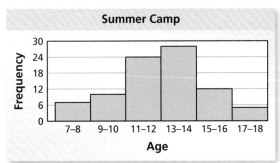

11.
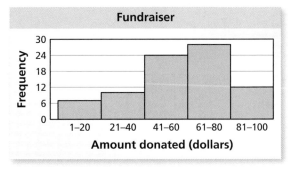

**MATCHING** Match the distribution with the corresponding box-and-whisker plot.

12.
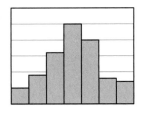

13.
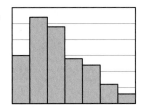

14.
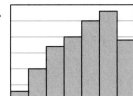

A.

B.

C.

15. **CHOOSE TOOLS** A stem-and-leaf plot is another data display that shows the distribution of data. For a large data set, would you use a stem-and-leaf plot or a histogram to show the distribution of the data? Explain.

16. **MODELING** Measuring an IQ is an inexact science. However, IQ scores have been around for years in an attempt to measure human intelligence. The greatest known IQ scores are shown in the table.

    | IQ Scores | | |
    |---|---|---|
    | 170 | 190 | 180 |
    | 160 | 180 | 210 |
    | 154 | 170 | 180 |
    | 195 | 230 | 160 |
    | 170 | 186 | 180 |
    | 225 | 190 | 170 |

    a. Display the data in a histogram using five intervals beginning with 151–166.
    b. Which measures of central tendency and dispersion best represent the data?
    c. The distribution of IQ scores for the human population is symmetric. What happens to the shape of the distribution in part (a) as you include more and more IQ scores from the population in the data set?

17. **ATM** The table shows your last 20 ATM withdrawals. What intervals would you use to display the data in a histogram? Explain your reasoning. Then display the data in a histogram.

    | ATM Withdrawals (dollars) | | | | | | | | | |
    |---|---|---|---|---|---|---|---|---|---|
    | 20 | 25 | 30 | 10 | 60 | 10 | 45 | 20 | 50 | 25 |
    | 50 | 20 | 45 | 100 | 20 | 10 | 30 | 25 | 40 | 20 |

18. **Reasoning** You record the following waiting times at a restaurant.

    | Waiting Times (minutes) | | | | | | | | | | | | | |
    |---|---|---|---|---|---|---|---|---|---|---|---|---|---|
    | 26 | 38 | 15 | 8 | 22 | 42 | 25 | 10 | 17 | 26 | 58 | 35 | 24 | 31 | 12 |
    | 29 | 25 | 0 | 34 | 44 | 32 | 20 | 18 | 7 | 40 | 42 | 19 | 32 | 13 | 21 |

    a. Display the data in a histogram using six intervals beginning with 0–9.
    b. Display the data in a histogram using twelve intervals beginning with 0–4.
    c. What happens when the number of intervals is increased?
    d. Which histogram best represents the data? Explain your reasoning.

## Fair Game Review 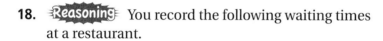 What you learned in previous grades & lessons

**Write an equation of the line that passes through the given point and is perpendicular to the given line.** *(Section 2.6)*

19. $(2, 2)$; $y = x + 3$
20. $(-3, 4)$; $y = 3x - 1$
21. $(1, -5)$; $y = -\frac{1}{2}x + 4$

22. **MULTIPLE CHOICE** Which equation represents the line that passes through $(0, 0)$ and is parallel to the line passing through $(5, -2)$ and $(1, -3)$? *(Section 2.6)*

    Ⓐ $y = -4x$   Ⓑ $y = -\frac{1}{4}x$   Ⓒ $y = \frac{1}{4}x$   Ⓓ $y = 4x$

# 12 Study Help

You can use a **concept circle** to organize information about a concept. Here is an example of a concept circle for measures of central tendency.

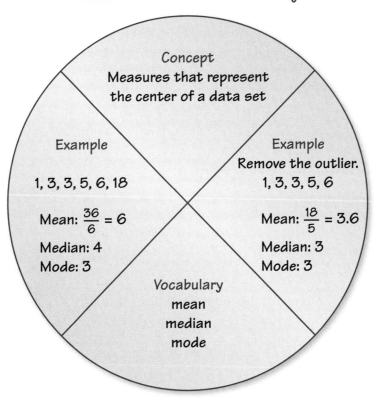

## On Your Own

**Make concept circles to help you study these topics.**

1. measures of dispersion
2. box-and-whisker plots
3. shapes of distributions

**After you complete this chapter, make concept circles for the following topics.**

4. scatter plots
5. lines of fit
6. two-way tables
7. choosing a data display

"Do you think this concept circle will help my owner understand that 'Speak' and 'Sit' need motivation?"

634  Chapter 12  Data Analysis and Displays

# 12.1–12.4 Quiz

**Find the mean, median, and mode of the data.** *(Section 12.1)*

1. 
| Checkbook Balances (dollars) | | |
|---|---|---|
| 40 | 10 | −20 |
| 0 | −10 | 40 |
| 30 | 40 | 50 |

2. 
| Hours Spent on Project | | |
|---|---|---|
| $3\frac{1}{2}$ | 5 | $2\frac{1}{2}$ |
| 3 | $3\frac{1}{2}$ | $\frac{1}{2}$ |

**Find the mean, range, and standard deviation of each data set. Then compare the data sets.** *(Section 12.2)*

3. Absent students during a week
   Girls: 6, 2, 4, 3, 4
   Boys: 5, 3, 6, 6, 9

4. Numbers of points scored
   Juniors: 19, 15, 20, 10, 14, 21, 18, 15
   Seniors: 22, 19, 29, 32, 15, 26, 30, 19

**Make a box-and-whisker plot for the data.** *(Section 12.3)*

5. Minutes of violin practice: 20, 50, 60, 40, 40, 30, 60, 40, 50, 20, 20, 35

6. Players' scores at end of first round: 200, −100, 100, 350, −50, 0, −50, 300

7. Display the data in a histogram. Describe the shape of the distribution. *(Section 12.4)*

| Bowling Scores | 51–100 | 101–150 | 151–200 | 201–250 | 251–300 |
|---|---|---|---|---|---|
| Frequency | 12 | 21 | 9 | 4 | 2 |

8. **ANOLES** The table shows the lengths of 12 green anoles. *(Section 12.1 and Section 12.3)*

   a. Find the mean, median, and mode of the data.
   b. Make a box-and-whisker plot for the data.
   c. Find and interpret the interquartile range of the data.
   d. Describe the distribution of the data.
   e. How does including 8.0 in the data set affect the mean, median, and mode?

| Length (cm) | 17.5 | 17.3 | 16.5 | 16.8 | 17.0 | 16.5 | 17.0 | 16.7 | 16.5 | 17.0 | 17.4 | 17.1 |

9. **PRESENTATIONS** The times of 20 presentations are shown in the table. *(Section 12.2 and Section 12.4)*

   a. Display the data in a histogram using five intervals beginning with 3–5.
   b. Determine and calculate the measures of central tendency and dispersion that best represent the data.
   c. The presentations are supposed to be 10 minutes long. How would you interpret these results?

| Time (minutes) | | | |
|---|---|---|---|
| 9 | 7 | 10 | 12 |
| 10 | 11 | 8 | 10 |
| 10 | 17 | 11 | 5 |
| 9 | 10 | 4 | 12 |
| 6 | 14 | 8 | 10 |

# 12.5 Scatter Plots and Lines of Fit

**Essential Question** How can you use data to predict an event?

### 1 ACTIVITY: Representing Data by a Linear Equation

Work with a partner. You have been working on a science project for 8 months. Each month, you have measured the length of a baby alligator.

My Science Project

The table shows your measurements.

September ↓                           April ↓

| Month, $x$ | 0 | 1 | 2 | 3 | 4 | 5 | 6 | 7 |
|---|---|---|---|---|---|---|---|---|
| Length (in.), $y$ | 22.0 | 22.5 | 23.5 | 25.0 | 26.0 | 27.5 | 28.5 | 29.5 |

**COMMON CORE**

**Scatter Plots**
In this lesson, you will
- interpret scatter plots.
- identify relationships from scatter plots.
- find lines of fit.
- solve real-life problems.

Learning Standards
8.SP.1
S.ID.6a
S.ID.6c
S.ID.7

Use the following steps to predict the baby alligator's length next September.

a. Graph the data in the table.

b. Draw the straight line that you think best approximates the points.

c. Write an equation of the line you drew.

d. Use the equation to predict the baby alligator's length next September.

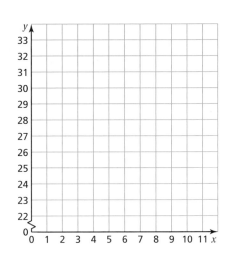

**636**    Chapter 12    Data Analysis and Displays

## 2 ACTIVITY: Representing Data by a Linear Equation

Work with a partner. You are a biologist and are studying bat populations.

You are asked to predict the number of bats that will be living in an abandoned mine after 3 years.

To start, you find the number of bats that have been living in the mine during the past 8 years.

The table shows the results of your research.

**Math Practice 6**

**Label Axes**
When labeling axes of a coordinate plane, what information do you need to label the axes correctly? Why?

7 years ago ↓                                              this year ↓

| Year, $x$ | 0 | 1 | 2 | 3 | 4 | 5 | 6 | 7 |
|---|---|---|---|---|---|---|---|---|
| Bats (thousands), $y$ | 327 | 306 | 299 | 270 | 254 | 232 | 215 | 197 |

Use the following steps to predict the number of bats that will be living in the mine after 3 years.

  a. Graph the data in the table.

  b. Draw the straight line that you think best approximates the points.

  c. Write an equation of the line you drew.

  d. Use the equation to predict the number of bats after 3 years.

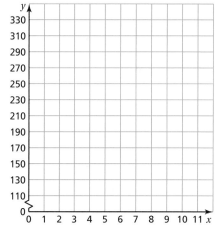

### What Is Your Answer?

3. **IN YOUR OWN WORDS** How can you use data to predict an event?

4. Use the Internet or some other reference to find data that appear to have a linear pattern. List the data in a table and graph the data. Use an equation that is based on the data to predict a future event.

**Practice**

Use what you learned about scatter plots and lines of fit to complete Exercise 3 on page 641.

# 12.5 Lesson

**Key Vocabulary**
scatter plot, p. 638
line of fit, p. 640

## Key Idea

**Scatter Plot**

A **scatter plot** is a graph that shows the relationship between two data sets. The two sets of data are graphed as ordered pairs in a coordinate plane.

### EXAMPLE 1 — Interpreting a Scatter Plot

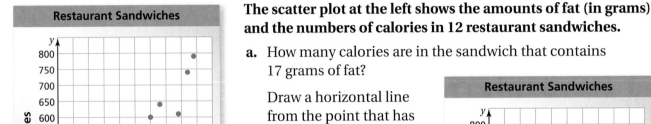

The scatter plot at the left shows the amounts of fat (in grams) and the numbers of calories in 12 restaurant sandwiches.

a. How many calories are in the sandwich that contains 17 grams of fat?

Draw a horizontal line from the point that has an $x$-value of 17. It crosses the $y$-axis at 400.

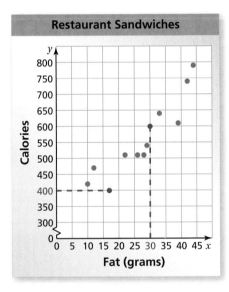

∴ So, the sandwich has 400 calories.

b. How many grams of fat are in the sandwich that contains 600 calories?

Draw a vertical line from the point that has a $y$-value of 600. It crosses the $x$-axis at 30.

∴ So, the sandwich has 30 grams of fat.

c. What tends to happen to the number of calories as the number of grams of fat increases?

Looking at the graph, the plotted points go up from left to right.

∴ So, as the number of grams of fat increases, the number of calories increases.

## On Your Own

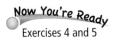

Now You're Ready
Exercises 4 and 5

1. **WHAT IF?** A sandwich has 650 calories. Based on the scatter plot in Example 1, how many grams of fat would you expect the sandwich to have? Explain your reasoning.

**Study Tip**

Scatter plots can also show unusual features of a data set, such as outliers, or gaps and clusters in the data.

A scatter plot can show that a relationship exists between two data sets.

*Positive Relationship*  *Negative Relationship*  *No Relationship*

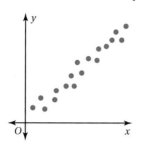 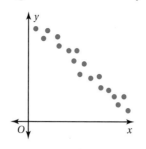 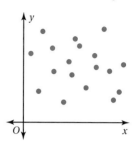

As $x$ increases, $y$ increases.

As $x$ increases, $y$ decreases.

The points show no pattern.

**EXAMPLE 2  Identifying a Relationship**

Tell whether the data show a *positive*, a *negative*, or *no* relationship.

a. Television size and price

b. Age and number of pets owned

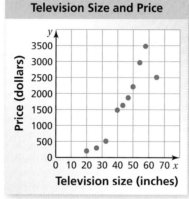 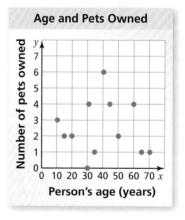

As the size of the television increases, the price increases.

The number of pets owned does not depend on a person's age.

∴ So, the scatter plot shows a positive relationship.

∴ So, the scatter plot shows no relationship.

**On Your Own**

Now You're Ready
Exercises 6–8

Make a scatter plot of the data. Tell whether the data show a *positive*, a *negative*, or *no* relationship.

**2.**

| Study Time (min), x | 30 | 20 | 60 | 90 | 45 | 10 | 30 | 75 | 120 | 80 |
|---|---|---|---|---|---|---|---|---|---|---|
| Test Score, y | 87 | 74 | 92 | 97 | 85 | 62 | 83 | 90 | 95 | 91 |

**3.**

| Age of a Car (years), x | 1 | 2 | 3 | 4 | 5 | 6 | 7 | 8 |
|---|---|---|---|---|---|---|---|---|
| Value (thousands), y | $24 | $21 | $19 | $18 | $15 | | $12 | $8 | $7 |

Section 12.5  Scatter Plots and Lines of Fit

A **line of fit** is a line drawn on a scatter plot close to most of the data points. It can be used to estimate data on a graph.

### EXAMPLE 3  Finding a Line of Fit

| Week, x | Sales (millions), y |
|---|---|
| 1 | $19 |
| 2 | $15 |
| 3 | $13 |
| 4 | $11 |
| 5 | $10 |
| 6 | $8 |
| 7 | $7 |
| 8 | $5 |

The table shows the weekly sales of a DVD and the number of weeks since its release. (a) Make a scatter plot of the data and draw a line of fit. (b) Write an equation of the line of fit. (c) Interpret the slope of the line of fit. (d) Predict the sales in week 9.

a. Plot the points in a coordinate plane. The scatter plot shows a negative relationship. Draw a line that is close to the data points. Try to have as many points above the line as below it.

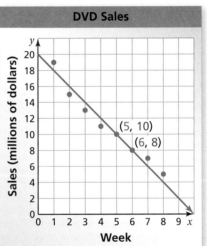

DVD Sales

b. The line passes through (5, 10) and (6, 8).

$$\text{slope} = \frac{\text{rise}}{\text{run}} = \frac{-2}{1} = -2$$

Because the line crosses the y-axis at (0, 20), the y-intercept is 20.

∴ So, an equation of the line of fit is $y = -2x + 20$.

c. The slope of the line of fit is $-2$. This means that the sales are decreasing by about $2 million each week.

d. To predict the sales in week 9, substitute 9 for $x$ in the equation of the line of fit.

$$y = -2x + 20 = -2(9) + 20 = 2$$

∴ The sales in week 9 should be about $2 million.

**Study Tip**

A line of fit does not need to pass through any of the data points.

### On Your Own

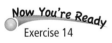
Exercise 14

4. The table shows the numbers of people who have attended a neighborhood festival over an 8-year period.

| Year, x | 1 | 2 | 3 | 4 | 5 | 6 | 7 | 8 |
|---|---|---|---|---|---|---|---|---|
| Attendance, y | 420 | 500 | 650 | 900 | 1100 | 1500 | 1750 | 2400 |

a. Make a scatter plot of the data and draw a line of fit.

b. Write an equation of the line of fit.

c. Interpret the slope of the line of fit.

d. Predict the number of people who will attend the festival in year 10.

# 12.5 Exercises

## Vocabulary and Concept Check

1. **VOCABULARY** What type of data are needed to make a scatter plot? Explain.
2. **WRITING** Explain why a line of fit is helpful when analyzing data.

## Practice and Problem Solving

3. **BLUEBERRIES** The table shows the weights $y$ of $x$ pints of blueberries.

| Number of Pints, $x$ | 0 | 1 | 2 | 3 | 4 | 5 |
|---|---|---|---|---|---|---|
| Weight (pounds), $y$ | 0 | 0.8 | 1.50 | 2.20 | 3.0 | 3.75 |

   a. Graph the data in the table.
   b. Draw the straight line that you think best approximates the points.
   c. Write an equation of the line you drew.
   d. Use the equation to predict the weight of 10 pints of blueberries.
   e. Blueberries cost $2.25 per pound. How much do 10 pints of blueberries cost?

4. **SUVS** The scatter plot shows the numbers of sport utility vehicles sold in a city from 2005 to 2010.

   a. In what year were 1000 SUVs sold?
   b. About how many SUVs were sold in 2009?
   c. Describe the relationship shown by the data.

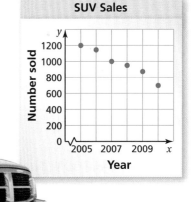

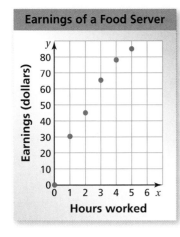

5. **EARNINGS** The scatter plot shows the total earnings (wages and tips) of a food server during 1 day.

   a. About how many hours must the server work to earn $70?
   b. About how much did the server earn for 5 hours of work?
   c. Describe the relationship shown by the data.

**Tell whether the data show a *positive*, a *negative*, or *no* relationship.**

6.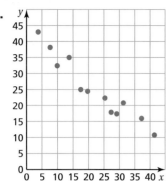

7.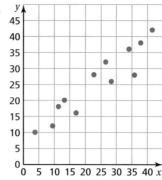

8.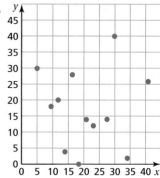

9. **HONEY** The table shows the average price per pound for honey in the United States from 2007 to 2010. What type of relationship do the data show?

| Year, x | 2007 | 2008 | 2009 | 2010 |
|---|---|---|---|---|
| Average Price per Pound, y | $1.08 | $1.42 | $1.47 | $1.60 |

10. **OPEN-ENDED** Describe a set of real-life data that has a negative relationship.

**Tell whether the line drawn on the graph is a good fit for the data. Explain your reasoning.**

11.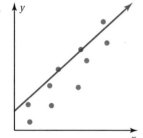

12.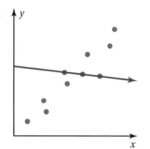

13.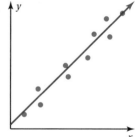

14. **VACATION** The table shows the distance you travel over a 6-hour period.

    a. Make a scatter plot of the data and draw a line of fit.
    b. Write an equation of the line of fit.
    c. Interpret the slope of the line of fit.
    d. Predict the distance you will travel in 7 hours.

| Hours, x | Distance (miles), y |
|---|---|
| 1 | 62 |
| 2 | 123 |
| 3 | 188 |
| 4 | 228 |
| 5 | 280 |
| 6 | 344 |

15. **TEST SCORES** The scatter plot shows the relationship between numbers of minutes spent studying and test scores for a science class.

    a. What type of relationship do the data show?
    b. Interpret the relationship.

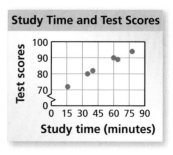

642    Chapter 12    Data Analysis and Displays

16. **REASONING** A data set has no relationship. Is it possible to find a line of fit for the data? Explain.

17. **CHOOSE TOOLS** Use a ruler or a yardstick to find the heights and arm spans of three people.

    a. Make a scatter plot using the data you collected. Then draw a line of fit for the data.
    b. Use your height and the line of fit to predict your arm span.
    c. Measure your arm span. Compare the result with your prediction in part (b).
    d. Is there a relationship between a person's height $x$ and arm span $y$? Explain.

18. **REASONING** How can an outlier be identified in a scatter plot?

**Describe the scatter plot and any relationship between the variables.**

19.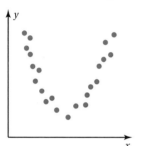

20.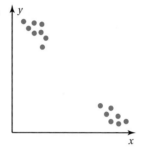

21.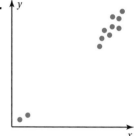

| Price of Admission (dollars), $x$ | Yearly Attendance, $y$ |
|---|---|
| 19.50 | 50,000 |
| 21.95 | 48,000 |
| 23.95 | 47,500 |
| 24.00 | 40,000 |
| 24.50 | 45,000 |
| 25.00 | 43,500 |

22. **Critical Thinking** The table shows the prices of admission to a local theater and the attendances for several years.

    a. Identify the outlier.
    b. How does the outlier affect the line of fit? Explain.
    c. Make a scatter plot of the data and draw the line of fit.
    d. Use the line of fit to predict the attendance when the admission cost is $27.

 **Fair Game Review** *What you learned in previous grades & lessons*

**Use a graph to solve the equation. Check your solution.** *(Section 4.4)*

23. $5x = 2x + 6$

24. $7x + 3 = 9x - 13$

25. $\frac{2}{3}x = -\frac{1}{3}x - 4$

26. **MULTIPLE CHOICE** The circle graph shows the super powers chosen by a class. What percent of the students want strength as their super power? *(Skills Review Handbook)*

    Ⓐ 10.5%     Ⓑ 12.5%
    Ⓒ 15%       Ⓓ 25%

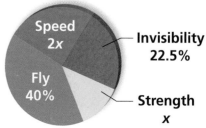

# 12.6 Analyzing Lines of Fit

**Essential Question** How can you find a line that best models a data set?

### 1  ACTIVITY: Comparing Lines of Fit

Work with a partner. You are researching the prices of liquid crystal display (LCD) televisions. The tables show the sizes and prices of several LCD televisions.

| TV Size (in.), x | 19 | 19 | 22 | 24 | 32 |
|---|---|---|---|---|---|
| Price (dollars), y | 170 | 180 | 170 | 250 | 320 |

| TV Size (in.), x | 32 | 37 | 40 | 40 | 46 |
|---|---|---|---|---|---|
| Price (dollars), y | 300 | 400 | 480 | 500 | 600 |

| TV Size (in.), x | 46 | 47 | 52 | 55 | 55 |
|---|---|---|---|---|---|
| Price (dollars), y | 850 | 800 | 950 | 1000 | 1150 |

a. Make a scatter plot of the data. Describe the pattern.

b. Draw a line of fit. Then have your partner draw a different line of fit.

c. Write an equation for each line of fit.

d. Compare your line of fit with your partner's line of fit. Are they similar? Which line of fit seems to model the data better? Why?

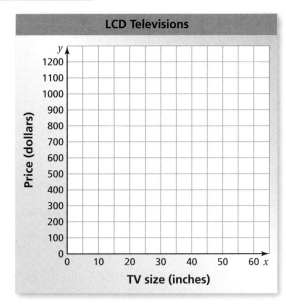

**COMMON CORE**

**Lines of Fit**

In this lesson, you will
- use residuals to determine whether models are a good fit.
- find lines of best fit using technology.
- identify correlations and causations.

Learning Standards
S.ID.6b
S.ID.7
S.ID.8
S.ID.9

### 2  ACTIVITY: Choosing a Line of Fit

Compare your line of fit with the lines of fit of the other students in your class. Which line of fit do you think best models the data? What criteria did you use when choosing the line of fit? Explain your reasoning.

### 3 ACTIVITY: Using a Graphing Calculator

**Math Practice 5**

**Recognize Usefulness of Tools**

When is it useful to use a graphing utility to find the line of best fit? What are the advantages and disadvantages?

The line of fit that models a data set most accurately is called the line of best fit. Graphing calculators use a method called linear regression to find a line of best fit. Use a graphing calculator to find an equation of the line of best fit for the data in Activity 1.

a. Enter the data from the tables into your calculator.

b. Use the *linear regression* feature of your calculator to find the equation of the line of best fit. The steps used to find the line of best fit depend on the calculator model that you have.

c. Compare the lines of fit from Activities 1 and 2 with the line of best fit. Are they similar? Explain.

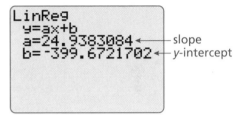

### 4 ACTIVITY: Using a Line of Best Fit

Copy and complete the table, which shows the sizes of four LCD televisions. Predict the price of each television using the line of fit from Activity 2 and the line of best fit from Activity 3. Then find the difference between the prices.

| TV Size (in.), x | Price Using Line of Fit from Activity 2 | Price Using Line of Best Fit: $y = 24.9x - 400$ | Difference Between the Prices |
|---|---|---|---|
| 26 | | | |
| 42 | | | |
| 50 | | | |
| 60 | | | |

How close are the predicted prices?

## What Is Your Answer?

5. **IN YOUR OWN WORDS** How can you find a line that best models a data set?

Use what you learned about analyzing lines of fit to complete Exercise 4 on page 649.

# 12.6 Lesson

**Key Vocabulary**
residual, *p. 646*
linear regression, *p. 647*
line of best fit, *p. 647*
correlation coefficient, *p. 647*
causation, *p. 648*

One way to determine how well a line of fit models a data set is to analyze *residuals*.

### Residuals

A **residual** is the difference between the *y*-value of a data point and the corresponding *y*-value found using the line of fit. A residual can be positive, negative, or zero.

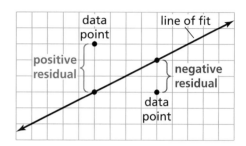

A scatter plot of the residuals shows how well a model fits a data set. If the model is a good fit, then the residual points will be randomly dispersed about the horizontal axis. If the model is not a good fit, then the residual points will form some type of pattern.

### EXAMPLE 1 — Using Residuals

| Week, $x$ | Sales (millions), $y$ |
|---|---|
| 1 | $19 |
| 2 | $15 |
| 3 | $13 |
| 4 | $11 |
| 5 | $10 |
| 6 | $8 |
| 7 | $7 |
| 8 | $5 |

In Example 3 in Section 12.5, the equation $y = -2x + 20$ models the data in the table at the left. Is the model a good fit?

**Step 1:** Calculate the residuals and organize your results in a table.

**Step 2:** Use the points (*x*, residual) to make a scatter plot.

| $x$ | $y$ | $y$-Value from Model | Residual |
|---|---|---|---|
| 1 | 19 | 18 | $19 - 18 = 1$ |
| 2 | 15 | 16 | $15 - 16 = -1$ |
| 3 | 13 | 14 | $13 - 14 = -1$ |
| 4 | 11 | 12 | $11 - 12 = -1$ |
| 5 | 10 | 10 | $10 - 10 = 0$ |
| 6 | 8 | 8 | $8 - 8 = 0$ |
| 7 | 7 | 6 | $7 - 6 = 1$ |
| 8 | 5 | 4 | $5 - 4 = 1$ |

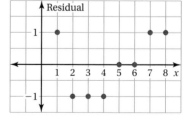

The points are randomly dispersed about the horizontal axis. So, the equation $y = -2x + 20$ is a good fit.

# EXAMPLE 2 Using Residuals

| Age, x | Salary, y |
|---|---|
| 35 | 42 |
| 37 | 44 |
| 41 | 47 |
| 43 | 50 |
| 45 | 52 |
| 47 | 51 |
| 53 | 49 |
| 55 | 45 |

The table at the left shows the ages $x$ and salaries $y$ (in thousands of dollars) of eight employees at a company. The equation $y = 0.2x + 38$ models the data. Is the model a good fit?

**Step 1:** Calculate the residuals and organize your results in a table.

**Step 2:** Use the points $(x, \text{residual})$ to make a scatter plot.

| x | y | y-Value from Model | Residual |
|---|---|---|---|
| 35 | 42 | 45.0 | $42 - 45.0 = -3.0$ |
| 37 | 44 | 45.4 | $44 - 45.4 = -1.4$ |
| 41 | 47 | 46.2 | $47 - 46.2 = 0.8$ |
| 43 | 50 | 46.6 | $50 - 46.6 = 3.4$ |
| 45 | 52 | 47.0 | $52 - 47.0 = 5.0$ |
| 47 | 51 | 47.4 | $51 - 47.4 = 3.6$ |
| 53 | 49 | 48.6 | $49 - 48.6 = 0.4$ |
| 55 | 45 | 49.0 | $45 - 49.0 = -4.0$ |

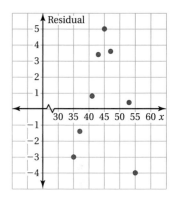

The points form a ∩-shaped pattern. So, the equation $y = 0.2x + 38$ does not model the data well.

## On Your Own

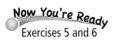
Exercises 5 and 6

1. The table shows the attendance $y$ (in thousands) at an amusement park from 2000 to 2009, where $x = 0$ represents the year 2000. The equation $y = -9.8x + 850$ models the data. Is the model a good fit?

| Year, x | 0 | 1 | 2 | 3 | 4 | 5 | 6 | 7 | 8 | 9 |
|---|---|---|---|---|---|---|---|---|---|---|
| Attendance, y | 850 | 845 | 828 | 798 | 800 | 792 | 785 | 781 | 775 | 760 |

**Study Tip**

You know how to use two points to find an equation of a line of fit. When finding an equation of the line of best fit, every point in the data set is used.

Graphing calculators use a method called **linear regression** to find a precise line of fit called a **line of best fit.** This line best models a set of data. A calculator often gives a value $r$ called the **correlation coefficient.** This value tells whether the correlation is positive or negative, and how closely the equation models the data. Values of $r$ range from $-1$ to $1$. When $r$ is close to 1 or $-1$, there is a strong correlation between the variables. As $r$ gets closer to 0, the correlation becomes weaker.

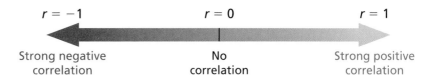

### EXAMPLE 3  Finding a Line of Best Fit Using Technology

The table shows the worldwide movie ticket sales $y$ (in billions of dollars) from 2000 to 2010, where $x = 0$ represents the year 2000. Use a graphing calculator to find an equation of the line of best fit. Identify and interpret the correlation coefficient.

| Year, $x$ | 0 | 1 | 2 | 3 | 4 | 5 | 6 | 7 | 8 | 9 | 10 |
|---|---|---|---|---|---|---|---|---|---|---|---|
| Ticket Sales, $y$ | 16 | 17 | 20 | 20 | 25 | 23 | 26 | 26 | 28 | 29 | 32 |

**Step 1:** Enter the data from the table into your calculator.

**Step 2:** Use the *linear regression* feature.

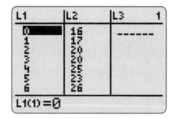

slope — $a = 1.5$
$y$-intercept — $b = 16.31818182$
correlation coefficient — $r = .9763484534$

An equation of the line of best fit is $y = 1.5x + 16$. The correlation coefficient is about 0.976. This means that the relationship between the years and ticket sales is a strong positive correlation and the equation closely models the data.

**Study Tip**

The slope of 1.5 indicates that sales are increasing by about $1.5 billion each year. The $y$-intercept of 16 represents the ticket sales of $16 billion for 2000.

When a change in one variable $x$ results in a change in another variable $y$, it is called **causation**. Causation produces a strong correlation between the two variables. The converse of the statement is not true. In other words, correlation does not imply causation.

### EXAMPLE 4  Identifying Correlation and Causation

Tell whether a correlation is likely in the situation. If so, tell whether there is a causal relationship. Explain your reasoning.

a. time spent exercising and the number of calories burned

  There is a positive correlation and a causal relationship because the more time you spend exercising, the more calories you burn.

**Reading**

A causal relationship exists when one variable causes a change in another variable.

b. the number of banks and the population of a city

  There may be a positive correlation but no causal relationship. Building more banks will not cause the population to increase.

### On Your Own

Now You're Ready
Exercises 7, 8, and 13–16

2. Use a graphing calculator to find an equation of the line of best fit for the data in On Your Own Question 1. Identify and interpret the correlation coefficient.

3. Is there a correlation between time spent playing video games and grade point average? If so, is there a causal relationship? Explain.

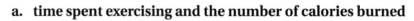

# 12.6 Exercises

## Vocabulary and Concept Check

1. **VOCABULARY** When is a residual positive? When is it negative?
2. **WRITING** Explain how you can use residuals to determine how well a line of fit models a data set.
3. **NUMBER SENSE** Which correlation coefficient indicates a stronger relationship, $-0.98$ or $0.91$? Explain.

## Practice and Problem Solving

4. **ANTLERS** The table shows the weekly growth $y$ (in inches) of an elk's antlers.

   | Week, $x$ | 1 | 2 | 3 | 4 | 5 |
   |---|---|---|---|---|---|
   | Growth, $y$ | 6.0 | 5.5 | 4.7 | 3.9 | 3.3 |

   a. Find a line of fit for the data.
   b. Use a graphing calculator to find an equation of the line of best fit.
   c. Use each model to predict the antler growth in week 6.

**Is the given model a good fit for the data in the table? Explain.**

5. $y = 4x - 5$

   | x | −4 | −3 | −2 | −1 | 0 | 1 | 2 | 3 | 4 |
   |---|---|---|---|---|---|---|---|---|---|
   | y | −18 | −13 | −10 | −7 | −2 | 0 | 6 | 10 | 15 |

6. $y = -1.3x + 1$

   | x | −8 | −6 | −4 | −2 | 0 | 2 | 4 | 6 | 8 |
   |---|---|---|---|---|---|---|---|---|---|
   | y | 9 | 10 | 5 | 8 | −1 | 1 | −4 | −12 | −7 |

**Use a graphing calculator to find an equation of the line of best fit for the data. Identify and interpret the correlation coefficient.**

7. 
   | x | 0 | 1 | 2 | 3 | 4 | 5 | 6 | 7 |
   |---|---|---|---|---|---|---|---|---|
   | y | 8 | 5 | 2 | −1 | −1 | 2 | 5 | 8 |

8. 
   | x | −8 | −6 | −4 | −2 | 0 | 2 | 4 | 6 | 8 | 10 |
   |---|---|---|---|---|---|---|---|---|---|---|
   | y | 20 | 8 | 17 | 7 | 8 | 1 | 5 | −2 | 2 | −8 |

9. **EARTHQUAKE** The table shows the total number $y$ of people reporting an earthquake $x$ minutes after it ended. Use a graphing calculator to find an equation of the line of best fit. In the same viewing window, graph the line and plot the data.

   | Minutes, $x$ | 1 | 2 | 3 | 4 | 5 | 6 | 7 | 8 |
   |---|---|---|---|---|---|---|---|---|
   | People, $y$ | 10 | 100 | 400 | 900 | 1400 | 1800 | 2100 | 2200 |

**MATCHING** Match the graph of the data with its corresponding linear regression screen.

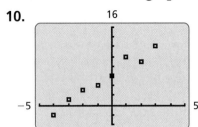

10.

11.

12.

A. LinReg
y=ax+b
a=-1.071428571
b=6.5
r²=.9361997226
r=-.9675741432

B. LinReg
y=ax+b
a=1.916666667
b=6.333333333
r²=.965076883
r=.9823832668

C. LinReg
y=ax+b
a=-.3571428571
b=1.035714286
r²=.0744047619
r=-.2727723628

**Tell whether a correlation is likely in the situation. If so, tell whether there is a causal relationship. Explain your reasoning.**

 13. the amount of time spent talking on a cell phone and the remaining battery life

14. the height of a toddler and the size of the toddler's vocabulary

15. the number of hats you own and the size of your head

16. the weight of a dog and the length of its tail

17. **FUEL MILEAGE** The table shows the prices $x$ (in thousands of dollars) and fuel economies $y$ (in miles per gallon) of several automobiles.

| Price (thousands of dollars), $x$ | 24 | 32 | 30 | 28 | 35 | 20 | 22 | 26 |
|---|---|---|---|---|---|---|---|---|
| Fuel Economy (miles per gallon), $y$ | 30 | 30 | 34 | 35 | 28 | 25 | 28 | 36 |

a. Use a graphing calculator to find an equation of the line of best fit. Identify and interpret the correlation coefficient.

b. Calculate the residuals. Then make a scatter plot of the residuals and interpret the results.

 18. **TEXTING** The table shows the numbers $y$ (in billions) of text messages sent from 2006 to 2011, where $x = 6$ represents the year 2006.

| Year, $x$ | Text Messages (billions), $y$ |
|---|---|
| 6 | 113 |
| 7 | 241 |
| 8 | 601 |
| 9 | 1360 |
| 10 | 1806 |
| 11 | 2206 |

a. Use a graphing calculator to find an equation of the line of best fit. Identify and interpret the correlation coefficient.

b. Interpret the slope of the line of best fit.

c. Calculate the residuals. Then make a scatter plot of the residuals and interpret the results.

d. Predict the number of text messages sent in 2015.

19. **GRADES** The table shows the numbers $x$ of hours spent watching television each week, and the grade point averages $y$ of several students.

    | Hours, $x$ | Grade Point Average, $y$ |
    |---|---|
    | 10 | 3.0 |
    | 5 | 3.4 |
    | 3 | 3.5 |
    | 12 | 2.7 |
    | 20 | 2.1 |
    | 15 | 2.8 |
    | 8 | 3.0 |
    | 4 | 3.7 |
    | 16 | 2.5 |

    a. Use a graphing calculator to find an equation of the line of best fit. Identify and interpret the correlation coefficient.

    b. Interpret the slope and $y$-intercept of the line of best fit.

    c. Another student watches about 14 hours of television each week. Predict the student's grade point average.

    d. Do you think watching more television each week may cause a lower grade point average? Explain.

20. **REASONING** A student spends 2 hours each week watching television and has a grade point average of 2.4. Include this information in the data set in Exercise 19. How does including this value affect the correlation coefficient? Explain.

21. **Modeling** Consider the earthquake data in Exercise 9.

    a. Copy and complete the table to show the number $y$ of people reporting the earthquake in the $x$th minute after the earthquake ended.

    | $x$ | 1 | 2 | 3 | 4 | 5 | 6 | 7 | 8 |
    |---|---|---|---|---|---|---|---|---|
    | $y$ | 10 | 90 | 300 | | | | | |

    b. Describe how the $y$-values change as $x$ increases. Do you think a linear function will fit the data well? If not, what type of function do you think will fit the data well? Explain.

    c. Use a graphing calculator to find the model in part (b).

## Fair Game Review  What you learned in previous grades & lessons

**Find the mean, range, and standard deviation of the data.** *(Section 12.2)*

22. 59, 70, 62, 68, 75, 77, 58

23. 15, 14, 11, 15, 16, 19, 14, 16, 15

24. **MULTIPLE CHOICE** What is the interquartile range of the box-and-whisker plot? *(Section 12.3)*

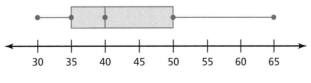

   Ⓐ 5    Ⓑ 10    Ⓒ 15    Ⓓ 35

# 12.7 Two-Way Tables

**Essential Question** How can you read and make a two-way table?

Two categories of data can be displayed in a **two-way table**.

### 1 ACTIVITY: Reading a Two-Way Table

Work with a partner. You are the manager of a sports shop. The two-way table shows the numbers of soccer T-shirts that your shop has left in stock at the end of the season.

|  |  | T-Shirt Size |  |  |  |  | Total |
|---|---|---|---|---|---|---|---|
|  |  | S | M | L | XL | XXL |  |
| Color | Blue/White | 5 | 4 | 1 | 0 | 2 |  |
|  | Blue/Gold | 3 | 6 | 5 | 2 | 0 |  |
|  | Red/White | 4 | 2 | 4 | 1 | 3 |  |
|  | Black/White | 3 | 4 | 1 | 2 | 1 |  |
|  | Black/Gold | 5 | 2 | 3 | 0 | 2 |  |
|  | Total |  |  |  |  |  | 65 |

**a.** Complete the totals for the rows and columns.

**b.** Are there any black and gold XL T-shirts in stock? Justify your answer.

**c.** The numbers of T-shirts you ordered at the beginning of the season are shown below. Complete the two-way table.

|  |  | T-Shirt Size |  |  |  |  | Total |
|---|---|---|---|---|---|---|---|
|  |  | S | M | L | XL | XXL |  |
| Color | Blue/White | 5 | 6 | 7 | 6 | 5 |  |
|  | Blue/Gold | 5 | 6 | 7 | 6 | 5 |  |
|  | Red/White | 5 | 6 | 7 | 6 | 5 |  |
|  | Black/White | 5 | 6 | 7 | 6 | 5 |  |
|  | Black/Gold | 5 | 6 | 7 | 6 | 5 |  |
|  | Total |  |  |  |  |  |  |

**COMMON CORE**

**Two-Way Tables**

In this lesson, you will
- read two-way tables.
- find marginal frequencies.
- make two-way tables.
- find relationships in two-way tables.

Learning Standards
8.SP.4
S.ID.5

**d.** How would you alter the numbers of T-shirts you order for next season? Explain your reasoning.

### 2 ACTIVITY: Analyzing Data

**Math Practice 3**

**Construct Arguments**

What are the advantages of using a table instead of a graph to analyze data?

Work with a partner. The three-dimensional two-way table shows information about the numbers of hours students at a high school work at part-time jobs during the school year.

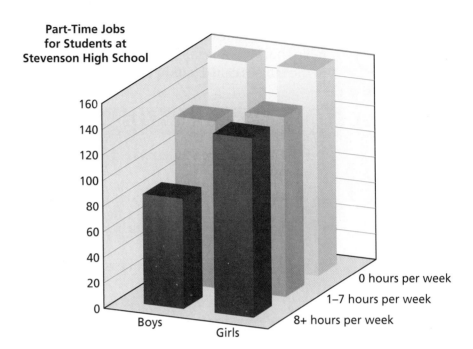

a. Make a two-way table showing the data. Use estimation to find the entries in your table.

b. Write two observations you can make that summarize the data in your table.

c. A newspaper article claims that more boys than girls drop out of high school to work full-time. Do the data support this claim? Explain your reasoning.

## What Is Your Answer?

3. **IN YOUR OWN WORDS** How can you read and make a two-way table?

4. Find a real-life data set that can be represented by a two-way table. Then make a two-way table for the data set.

**Practice** — Use what you learned about two-way tables to complete Exercises 5 and 6 on page 656.

# 12.7 Lesson

**Key Vocabulary**
two-way table, *p. 654*
joint frequency, *p. 654*
marginal frequency, *p. 654*

A **two-way table** displays two categories of data collected from the same source.

You randomly survey students in your school about their grades on the last test and whether they studied for the test. The two-way table shows your results. Each entry in the table is called a **joint frequency**.

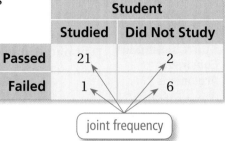

### EXAMPLE 1  Reading a Two-Way Table

**How many of the students in the survey above studied for the test and passed?**

The entry in the "Studied" column and "Passed" row is 21.

∴ So, 21 of the students in the survey studied for the test and passed.

The sums of the rows and columns in a two-way table are called **marginal frequencies**.

### EXAMPLE 2  Finding Marginal Frequencies

**Find and interpret the marginal frequencies for the survey above.**

Create a new column and row for the sums. Then add the entries.

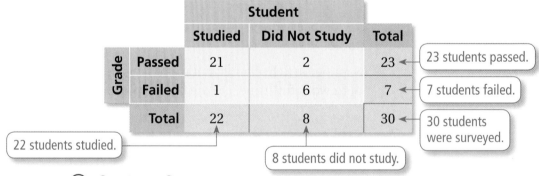

### On Your Own

*Now You're Ready*
Exercises 3–6

1. You randomly survey students in a cafeteria about their plans for a football game and a school dance. The two-way table shows your results.

    a. How many students will attend the dance but not the football game?

    b. Find and interpret the marginal frequencies for the survey.

|  |  | Football Game | |
|---|---|---|---|
|  |  | Attend | Not Attend |
| **Dance** | Attend | 35 | 5 |
|  | Not Attend | 16 | 20 |

654   Chapter 12   Data Analysis and Displays

## EXAMPLE 3  Making a Two-Way Table

**Rides bus**

| Age | Tally |
|---|---|
| 12–13 | ||||  ||||  ||||  ||||  |||| |
| 14–15 | ||||  ||||  || |
| 16–17 | ||||  ||||  |||| |

You randomly survey students between the ages of 12 and 17 about whether they ride the bus to school. The results are shown in the tally sheets. Make a two-way table that includes the marginal frequencies.

The two categories for the table are the ages and whether or not they ride the bus. Use the tally sheets to calculate each joint frequency. Then add to find each marginal frequency.

**Does not ride bus**

| Age | Tally |
|---|---|
| 12–13 | ||||  ||||  ||||  | |
| 14–15 | ||||  ||||  ||| |
| 16–17 | ||||  ||||  ||||  ||||  | |

|  |  | Age |  |  | |
|---|---|---|---|---|---|
|  |  | 12–13 | 14–15 | 16–17 | Total |
| **Student** | Rides Bus | 24 | 12 | 14 | 50 |
|  | Does Not Ride Bus | 16 | 13 | 21 | 50 |
|  | Total | 40 | 25 | 35 | 100 |

## EXAMPLE 4  Finding a Relationship in a Two-Way Table

Use the two-way table in Example 3.

**a.** For each age group, what percent of the students in the survey ride the bus to school? do not ride the bus to school? Organize the results in a two-way table. Explain what one of the entries represents.

|  |  | Age |  |  |
|---|---|---|---|---|
|  |  | 12–13 | 14–15 | 16–17 |
| **Student** | Rides Bus | 60% | 48% | 40% |
|  | Does Not Ride Bus | 40% | 52% | 60% |

$\frac{14}{35} = 0.4$

So, 40% of the 16- and 17-year-old students in the survey ride the bus to school.

**b.** Does the table in part (a) show a relationship between age and whether students ride the bus to school? Explain.

∴ The table shows that as age increases, students are less likely to ride the bus to school.

### On Your Own

*Now You're Ready*
Exercises 10 and 11

**2.** You randomly survey students in a school about whether they buy a school lunch or pack a lunch. Your results are shown.

**a.** Make a two-way table that includes the marginal frequencies.

**b.** For each grade level, what percent of the students in the survey pack a lunch? buy a school lunch? Organize the results in a two-way table. Explain what one of the entries represents.

**c.** Does the table in part (b) show a relationship between grade level and lunch choice? Explain.

**Grade 6 students**
11 pack lunch, 9 buy school lunch

**Grade 7 students**
23 pack lunch, 27 buy school lunch

**Grade 8 students**
16 pack lunch, 14 buy school lunch

# 12.7 Exercises

## Vocabulary and Concept Check

1. **VOCABULARY** Explain the relationship between joint frequencies and marginal frequencies.

2. **OPEN-ENDED** Describe how you can use a two-way table to organize data you collect from a survey.

## Practice and Problem Solving

You randomly survey students about participating in their class's yearly fundraiser. You display the two categories of data in the two-way table.

3. Find the total of each row.

4. Find the total of each column.

 5. How many female students will be participating in the fundraiser?

6. How many male students will *not* be participating in the fundraiser?

|  |  | Fundraiser | |
|---|---|---|---|
|  |  | No | Yes |
| Gender | Female | 22 | 51 |
|  | Male | 30 | 29 |

**Find and interpret the marginal frequencies.**

 7.

|  |  | School Play | |
|---|---|---|---|
|  |  | Attend | Not Attend |
| Class | Junior | 41 | 30 |
|  | Senior | 52 | 23 |

8.

|  |  | Cell Phone Minutes | |
|---|---|---|---|
|  |  | Limited | Unlimited |
| Text Plan | Limited | 78 | 0 |
|  | Unlimited | 175 | 15 |

9. **GOALS** You randomly survey students in your school. You ask whether grades, popularity, or sports is most important to them. You display your results in the two-way table.

   a. How many 10th graders chose sports? How many 11th graders chose grades?

   b. Find and interpret the marginal frequencies for the survey.

   c. What percent of students in the survey are 9th graders who chose popularity?

|  |  | Goal | | |
|---|---|---|---|---|
|  |  | Grades | Popularity | Sports |
| Grade | 9th | 31 | 18 | 23 |
|  | 10th | 39 | 16 | 19 |
|  | 11th | 42 | 6 | 17 |

10. **PETS** You randomly survey students in your school about whether they own a pet. The results are shown in the tally sheets. Make a two-way table that includes the marginal frequencies.

**Own a Pet**

| Owner | Tally |
|---|---|
| Male | ||||  ||||  ||| |
| Female | ||||  | |

**Don't Own a Pet**

| Owner | Tally |
|---|---|
| Male | ||||  ||||  ||||  ||||  |||| |
| Female | ||||  ||||  |||| |

| Eye Color of Males Surveyed | | |
|---|---|---|
| Green | Blue | Brown |
| 5 | 16 | 27 |

| Eye Color of Females Surveyed | | |
|---|---|---|
| Green | Blue | Brown |
| 3 | 19 | 18 |

11. **EYE COLOR** You randomly survey students in your school about the color of their eyes. The results are shown in the tables.

   a. Make a two-way table.

   b. Find and interpret the marginal frequencies for the survey.

   c. For each eye color, what percent of the students in the survey are male? female? Organize the results in a two-way table. Explain what two of the entries represent.

12. **REASONING** Use the information from Exercise 11. For each gender, what percent of the students in the survey have green eyes? blue eyes? brown eyes? Organize the results in a two-way table. Explain what two of the entries represent.

13. **Precision** What percent of students in the survey in Exercise 11 are either female or have green eyes? What percent of students in the survey are males that do not have green eyes? Find and explain the sum of these two percents.

## Fair Game Review  What you learned in previous grades & lessons

Simplify the rational expression, if possible. State the excluded value(s). *(Section 11.3)*

14. $\dfrac{1-x}{x-1}$

15. $\dfrac{x^2 - x - 6}{x^2 + x - 12}$

16. $\dfrac{15x^3 - 6x^2}{21x^3 + 3x^2}$

17. **MULTIPLE CHOICE** What is the solution of $\dfrac{1}{x} = \dfrac{3}{x+4}$? *(Section 11.7)*

   Ⓐ $x = -2$   Ⓑ $x = 2$   Ⓒ $x = 3$   Ⓓ $x = 4$

# 12.8 Choosing a Data Display

**Essential Question** How can you display data in a way that helps you make decisions?

### 1  ACTIVITY: Displaying Data

Work with a partner. Analyze and display each data set in a way that best describes the data. Explain your choice of display.

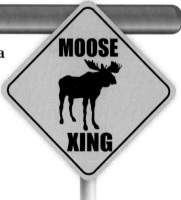

a. **ROAD KILL** A group of schools in New England participated in a 2-month study and reported 3962 dead animals.

Birds 307      Mammals 2746
Amphibians 145      Reptiles 75
Unknown 689

b. **BLACK BEAR ROAD KILL** The data below show the numbers of black bears killed on a state's roads from 1993 to 2012.

| 1993 | 30 | 2000 | 47  | 2007 | 99  |
|------|----|------|-----|------|-----|
| 1994 | 37 | 2001 | 49  | 2008 | 129 |
| 1995 | 46 | 2002 | 61  | 2009 | 111 |
| 1996 | 33 | 2003 | 74  | 2010 | 127 |
| 1997 | 43 | 2004 | 88  | 2011 | 141 |
| 1998 | 35 | 2005 | 82  | 2012 | 135 |
| 1999 | 43 | 2006 | 109 |      |     |

c. **RACCOON ROAD KILL** A 1-week study along a 4-mile section of road found the following weights (in pounds) of raccoons that had been killed by vehicles.

| 13.4 | 14.8 | 17.0 | 12.9 |
| 21.3 | 21.5 | 16.8 | 14.8 |
| 15.2 | 18.7 | 18.6 | 17.2 |
| 18.5 |  9.4 | 19.4 | 15.7 |
| 14.5 |  9.5 | 25.4 | 21.5 |
| 17.3 | 19.1 | 11.0 | 12.4 |
| 20.4 | 13.6 | 17.5 | 18.5 |
| 21.5 | 14.0 | 13.9 | 19.0 |

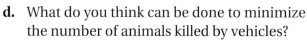

**COMMON CORE**

**Data Displays**
In this lesson, you will
- choose appropriate data displays.
- identify and analyze misleading data displays.

Learning Standard
S.ID.1

d. What do you think can be done to minimize the number of animals killed by vehicles?

### 2 ACTIVITY: Statistics Project

**Math Practice 4**

**Use a Graph**
How can you use a graph to represent the data you have gathered for your report? What does the graph tell you about the data?

**ENDANGERED SPECIES PROJECT** Use the Internet or some other reference to write a report about an animal species that is (or has been) endangered. Include graphical displays of the data you have gathered.

**Sample:** Florida Key Deer

In 1939, Florida banned the hunting of Key deer. The numbers of Key deer fell to about 100 in the 1940s.

In 1947, public sentiment was stirred by 11-year-old Glenn Allen from Miami. Allen organized Boy Scouts and others in a letter-writing campaign that led to the establishment of the National Key Deer Refuge in 1957. The approximately 8600-acre refuge includes 2280 acres of designated wilderness.

The Key Deer Refuge has increased the population of Key deer. A recent study estimated the total Key deer population to be approximately 800.

About half of Key deer deaths are due to vehicles.

One of two Key deer wildlife underpasses on Big Pine Key

### What Is Your Answer?

3. **IN YOUR OWN WORDS** How can you display data in a way that helps you make decisions? Use the Internet or some other reference to find examples of the following types of data displays.

    - Bar graph
    - Circle graph
    - Scatter plot
    - Stem-and-leaf plot
    - Box-and-whisker plot

Use what you learned about choosing data displays to complete Exercise 3 on page 662.

## 12.8 Lesson

Check It Out
Lesson Tutorials
BigIdeasMath.com

### Key Idea

| Data Display | What does it do? |
|---|---|
| Pictograph | shows data using pictures |
| Bar Graph | shows data in specific categories |
| Circle Graph | shows data as parts of a whole |
| Line Graph | shows how data change over time |
| Histogram | shows frequencies of data values in intervals of the same size |
| Stem-and-Leaf Plot | orders numerical data and shows how they are distributed |
| Box-and-Whisker Plot | shows the variability of a data set using quartiles |
| Dot Plot | shows the number of times each value occurs in a data set |
| Scatter Plot | shows the relationship between two data sets using ordered pairs in a coordinate plane |

### EXAMPLE 1  Choosing an Appropriate Data Display

**Choose an appropriate data display for the situation. Explain your reasoning.**

a. the number of students in a marching band each year

   • A line graph shows change over time. So, a line graph is an appropriate data display.

b. comparison of people's shoe sizes and their heights

   • You want to compare two different data sets. So, a scatter plot is an appropriate data display.

### On Your Own

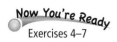
Now You're Ready
Exercises 4–7

**Choose an appropriate data display for the situation. Explain your reasoning.**

1. the population of the United States divided into age groups

2. the percents of students in your school who speak Spanish, French, or Haitian Creole

660  Chapter 12  Data Analysis and Displays

**EXAMPLE 2** **Identifying a Misleading Data Display**

**Which line graph is misleading? Explain.**

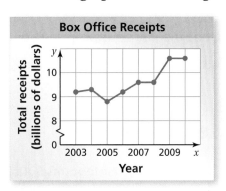
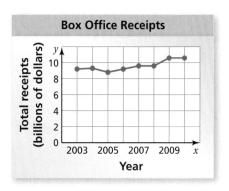

The vertical axis of the line graph on the left has a break (⌇) and begins at 8. This graph makes it appear that the total receipts increased rapidly from 2005 to 2010. The graph on the right has an unbroken axis. It is more honest and shows that the total receipts increased slowly.

∴ So, the graph on the left is misleading.

**EXAMPLE 3** **Analyzing a Misleading Data Display**

**A volunteer concludes that the numbers of cans of food and boxes of food donated were about the same. Is this conclusion accurate? Explain.**

Each icon represents the same number of items. Because the box icon is larger than the can icon, it looks like the number of boxes is about the same as the number of cans, but the number of boxes is actually about half of the number of cans.

∴ So, the conclusion is not accurate.

### On Your Own

*Now You're Ready*
Exercises 9–12

**Explain why the data display is misleading.**

3.
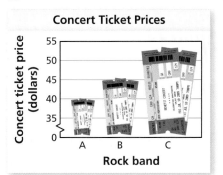

4.
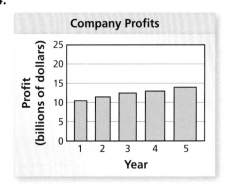

Section 12.8   Choosing a Data Display   **661**

# 12.8 Exercises

**Check It Out**
Help with Homework
BigIdeasMath✓com

## ✓ Vocabulary and Concept Check

1. **REASONING** Can more than one display be appropriate for a data set? Explain.
2. **OPEN-ENDED** Describe how a histogram can be misleading.

## Practice and Problem Solving

3. Analyze and display the data in a way that best describes the data. Explain your choice of display.

| Notebooks Sold in One Week | | | | |
|---|---|---|---|---|
| 192 red | 170 green | 203 black | 183 pink | 230 blue |
| 165 yellow | 210 purple | 250 orange | 179 white | 218 other |

**Choose an appropriate data display for the situation. Explain your reasoning.**

4. a student's test scores and how the scores are spread out
5. the distance a person drives each month
6. the outcome of rolling a number cube
7. homework problems assigned each day
8. **WRITING** When would you choose a histogram instead of a bar graph to display data?

**Explain why the data display is misleading.**

9.

10.

11.

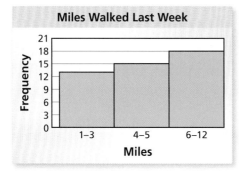

12.

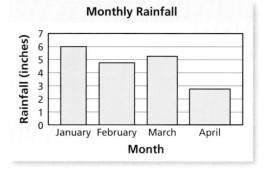

662   Chapter 12   Data Analysis and Displays

13. **VEGETABLES** A nutritionist wants to use a data display to show the favorite vegetables of the students at a school. Choose an appropriate data display for the situation. Explain your reasoning.

14. **CHEMICALS** A scientist gathers data about a decaying chemical compound. The results are shown in the scatter plot. Is the data display misleading? Explain.

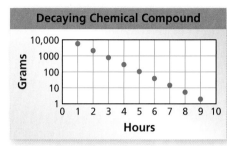

15. **REASONING** What type of data display is appropriate for showing the mode of a data set?

16. **SPORTS** A survey asked 100 students to choose their favorite sports. The results are shown in the circle graph.

    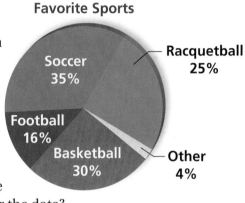

    a. Explain why the graph is misleading.
    b. What type of data display would be more appropriate for the data? Explain.

17. **Structure** With the help of computers, mathematicians have computed and analyzed billions of digits of the irrational number $\pi$. One of the things they analyze is the frequency of each of the numbers 0 through 9. The table shows the frequency of each number in the first 100,000 digits of $\pi$.

    a. Display the data in a bar graph.
    b. Display the data in a circle graph.
    c. Which data display is more appropriate? Explain.
    d. Describe the distribution.

| Number | 0 | 1 | 2 | 3 | 4 | 5 | 6 | 7 | 8 | 9 |
|---|---|---|---|---|---|---|---|---|---|---|
| Frequency | 9999 | 10,137 | 9908 | 10,025 | 9971 | 10,026 | 10,029 | 10,025 | 9978 | 9902 |

## Fair Game Review  *What you learned in previous grades & lessons*

**Write the verbal statement as an equation.** *(Skills Review Handbook)*

18. A number plus 3 is 5.
19. 8 times a number is 24.

20. **MULTIPLE CHOICE** What is 20% of 25% of 400? *(Skills Review Handbook)*

    Ⓐ 20    Ⓑ 200    Ⓒ 240    Ⓓ 380

# 12.5–12.8 Quiz

1. The scatter plot shows the amounts of money donated to a charity from 2005 to 2010. *(Section 12.5)*

   a. In what year did the charity receive $150,000?

   b. How much did the charity receive in 2008?

   c. Describe the relationship shown by the data.

2. Use a graphing calculator to find the equation of the line of best fit for the data in the table below. Identify and interpret the correlation coefficient. Make a scatter plot of the residuals and interpret the results. *(Section 12.6)*

| x | 0 | 1 | 2 | 3 | 4 | 5 | 6 | 7 |
|---|---|---|---|---|---|---|---|---|
| y | 12 | 16 | 15 | 14 | 18 | 22 | 20 | 25 |

3. The results of a recycling survey are shown in the two-way table. Find and interpret the marginal frequencies. *(Section 12.7)*

|  |  | Recycle | |
|---|---|---|---|
|  |  | Yes | No |
| Gender | Female | 28 | 9 |
|  | Male | 24 | 14 |

**Choose an appropriate data display for the situation. Explain your reasoning.** *(Section 12.8)*

4. percent of band students in each section of instruments

5. company's profit for each week

6. **CATS** The table shows the number of cats adopted from an animal shelter each month. *(Section 12.5)*

| Month | 1 | 2 | 3 | 4 | 5 | 6 | 7 | 8 | 9 |
|---|---|---|---|---|---|---|---|---|---|
| Cats | 3 | 6 | 7 | 11 | 13 | 14 | 15 | 18 | 19 |

   a. Make a scatter plot of the data and draw a line of fit.
   b. Write an equation of the line that fits the data.
   c. Interpret the slope of the line.
   d. Predict how many cats will be adopted in month 10.

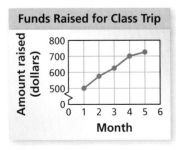

7. **FUNDRAISER** The line graph shows the amount of money that the eighth-grade students at a school raised each month to pay for a class trip. Is the graph misleading? Explain. *(Section 12.8)*

# 12 Chapter Review

## Review Key Vocabulary

measure of central tendency, *p. 608*
measure of dispersion, *p. 614*
range, *p. 614*
standard deviation, *p. 615*
box-and-whisker plot, *p. 620*
quartile, *p. 620*
five-number summary, *p. 620*
interquartile range, *p. 621*
scatter plot, *p. 638*
line of fit, *p. 640*
residual, *p. 646*
linear regression, *p. 647*
line of best fit, *p. 647*
correlation coefficient, *p. 647*
causation, *p. 648*
two-way table, *p. 654*
joint frequency, *p. 654*
marginal frequency, *p. 654*

## Review Examples and Exercises

### 12.1 Measures of Central Tendency (pp. 606–611)

The table shows the number of kilometers you ran each day for the past 10 days. Find the mean, median, and mode of the distances.

| Kilometers Run | |
|---|---|
| 3.5 | 4.1 |
| 4.0 | 4.3 |
| 4.4 | 4.5 |
| 3.9 | 2.0 |
| 4.3 | 5.0 |

**Mean:** (sum of the data / number of values) → $\frac{40}{10} = 4$

**Median:** 2.0, 3.5, 3.9, 4.0, 4.1, 4.3, 4.3, 4.4, 4.5, 5.0    Order the data.

$\frac{8.4}{2} = 4.2$    Mean of two middle values

**Mode:** 2.0, 3.5, 3.9, 4.0, 4.1, 4.3, 4.3, 4.4, 4.5, 5.0

The value 4.3 occurs most often.

∴ The mean is 4 kilometers, the median is 4.2 kilometers, and the mode is 4.3 kilometers.

### Exercises

1. Use the data in the example above. You run 4.0 kilometers on day 11. How does this additional value affect the mean, median, and mode? Explain.

Find the mean, median, and mode of the data.

2.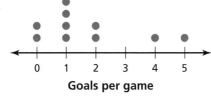
Goals per game

3. 

| Ski Resort Temperatures (°F) | | |
|---|---|---|
| 11 | 3 | 3 |
| 0 | −9 | −2 |
| 10 | 10 | 10 |

## 12.2 Measures of Dispersion (pp. 612–617)

Find the mean, range, and standard deviation of the bowling scores for each person. Then compare the data sets.

| Ryan | |
|---|---|
| 205 | 190 |
| 185 | 200 |
| 210 | 219 |
| 174 | 203 |
| 194 | 230 |

| Emma | |
|---|---|
| 228 | 205 |
| 172 | 181 |
| 154 | 240 |
| 235 | 235 |
| 168 | 192 |

**Ryan:** mean = $\frac{2010}{10} = 201$

174, 185, 190, 194, 200, 203, 205, 210, 219, 230   Order the data.

The range is $230 - 174 = 56$.

$$\sqrt{\frac{(205-201)^2 + (185-201)^2 + \cdots + (230-201)^2}{10}} = \sqrt{242.2} \approx 15.6$$

The standard deviation is 15.6.

**Emma:** mean = $\frac{2010}{10} = 201$

154, 168, 172, 181, 192, 205, 228, 235, 235, 240   Order the data.

The range is $240 - 154 = 86$.

$$\sqrt{\frac{(228-201)^2 + (172-201)^2 + \cdots + (192-201)^2}{10}} = \sqrt{919.8} \approx 30.3$$

The standard deviation is 30.3.

- The mean, 201, is the same for each data set. The range for Ryan's scores is 56 and the standard deviation is 15.6. The range for Emma's scores is 86 and the standard deviation is 30.3. So, Emma's scores are more spread out than Ryan's scores.

### Exercises

4. Find the mean, range, and standard deviation of the prices (in dollars) of portable keyboards at each store. Then compare the data sets.

| Store A | |
|---|---|
| 130 | 180 |
| 200 | 250 |
| 150 | 190 |
| 250 | 160 |

| Store B | |
|---|---|
| 225 | 310 |
| 260 | 190 |
| 200 | 285 |
| 210 | 230 |

## 12.3 Box-and-Whisker Plots *(pp. 618–625)*

**Make a box-and-whisker plot for the weights (in pounds) of pumpkins sold at a market. Identify the shape of the distribution.**

16, 20, 11, 15, 10, 8, 8, 19, 11, 9, 9, 16

**Step 1:** Order the data. Find the median and the quartiles.

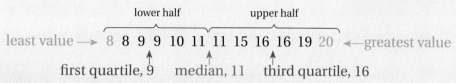

**Step 2:** Draw a number line that includes the least and greatest values. Graph points above the number line for the five-number summary.

**Step 3:** Draw a box using the quartiles. Draw a line through the median. Draw whiskers from the box to the least and greatest values.

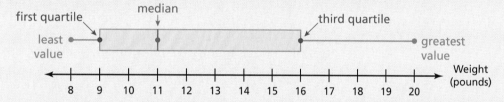

• The right whisker is longer than the left whisker, and most of the data are on the left side of the display. So, the distribution is skewed right.

### Exercises

**Make a box-and-whisker plot for the data. Identify the shape of the distribution.**

5. Ages of volunteers at a hospital:
   14, 17, 20, 16, 17, 14, 21, 18

6. Masses (in kilograms) of lions:
   120, 230, 180, 210, 200, 200, 230, 160

## 12.4 Shapes of Distributions *(pp. 626–633)*

**The histogram shows the numbers of words spelled correctly by students at a spelling bee. Describe the shape of the distribution. Which measures of central tendency and dispersion would best represent the data?**

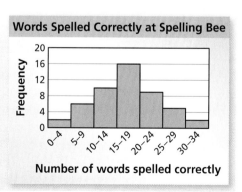

• The distribution is symmetric. So, use the mean to describe the center and the standard deviation to describe the spread.

Chapter Review  667

### Exercises

The frequency table shows the amounts of money the students in a class have in their pockets.

| Amount | Frequency |
|---|---|
| 0–0.99 | 9 |
| 1.00–1.99 | 10 |
| 2.00–2.99 | 9 |
| 3.00–3.99 | 7 |
| 4.00–4.99 | 4 |
| 5–5.99 | 1 |

7. Display the data in a histogram. Describe the shape of the distribution.

8. Which measures of central tendency and dispersion best represent the data?

## 12.5 Scatter Plots and Lines of Fit (pp. 636–643)

Your school is ordering custom T-shirts. The scatter plot shows the costs per T-shirt for various numbers of T-shirts ordered. What type of relationship do the data show?

The plotted points go down from left to right. As the number of T-shirts ordered increases, the cost per T-shirt decreases.

So, the scatter plot shows a negative relationship.

### Exercises

9. Use the scatter plot above. Write an equation of a line of fit. Predict the cost per T-shirt when you order 275 T-shirts.

## 12.6 Analyzing Lines of Fit (pp. 644–651)

The table shows the heights $x$ (in inches) and shoe sizes $y$ of several students. Use a graphing calculator to find an equation of the line of best fit. Identify and interpret the correlation coefficient.

| x | 65 | 62 | 70 | 72 | 68 | 67 | 70 | 67 | 64 | 63 |
|---|---|---|---|---|---|---|---|---|---|---|
| y | 8.5 | 7 | 10.5 | 12 | 10 | 10.5 | 11 | 10 | 8 | 7.5 |

**Step 1:** Enter the data from the table into your calculator.

**Step 2:** Use the *linear regression* feature of your calculator to find an equation of the line of best fit and the correlation coefficient.

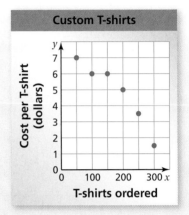

LinReg
y=ax+b
a=.4866803279
b=-23.0102459
r²=.943563901
r=.9713721743

An equation of the line of best fit is $y = 0.49x - 23.0$. The correlation coefficient is about 0.971. This means that the relationship between the heights and shoe sizes is a strong positive correlation and the equation closely models the data.

### Exercises

**10.** Use the data in the example. You take height and shoe size measurements of three more students: (64, 7), (65, 9), and (71, 11). Find a new equation of the line of best fit. Did the correlation coefficient change? Explain.

## 12.7 Two-Way Tables (pp. 652–657)

You randomly survey students in your school about whether they liked a recent school play. The results are shown. Make a two-way table that includes the marginal frequencies. What percent of the students surveyed liked the play?

*Male students*
*48 likes, 12 dislikes*

*Female students*
*56 likes, 14 dislikes*

| | | Student | | |
|---|---|---|---|---|
| | | Liked the Play | Did Not Like the Play | Total |
| **Gender** | Male | 48 | 12 | 60 |
| | Female | 56 | 14 | 70 |
| | Total | 104 | 26 | 130 |

Of the 130 students surveyed, 104 students liked the play.

Because $\frac{104}{130} = 0.8$, 80% of the students in the survey liked the play.

### Exercises

**11.** You randomly survey people at a mall about whether they like the new food court. The results are shown. Make a two-way table that includes the marginal frequencies. What percent of the adults surveyed dislike the new food court?

*Adults*
*21 likes, 79 dislikes*

*Teenagers*
*96 likes, 4 dislikes*

## 12.8 Choosing a Data Display (pp. 658–663)

**Choose an appropriate data display for the situation. Explain your reasoning.**

**a.** the percent of votes that each candidate received in an election

A circle graph shows data as parts of a whole. So, a circle graph is an appropriate data display.

**b.** the distribution of the ages of U.S. presidents at their inauguration(s)

A stem-and-leaf plot orders numerical data and shows how they are distributed. So, a stem-and-leaf plot is an appropriate data display.

### Exercises

**12.** A principal wants to use a data display to compare the number of cans of food donated by each eighth-grade class. Choose an appropriate data display for the situation. Explain your reasoning.

# 12 Chapter Test

**Find the mean, median, and mode of the data.**

1. 
| Distances (feet) Above or Below Water Level in Pool | | |
|---|---|---|
| −3 | 0 | −3 |
| 3 | 10 | 0 |
| 11 | −6 | −3 |

2. **Cooking Times (minutes)**

| Stem | Leaf |
|---|---|
| 3 | 5 8 |
| 4 | 0 1 8 |
| 5 | 0 4 4 4 5 9 |
| 6 | 0 |

Key: $4 \mid 1 = 41$ minutes

3. **TURTLES** The tables show the weights (in pounds) of turtles caught in two ponds. Find the mean, range, and standard deviation of the weights of the turtles in each pond. Then compare the data sets.

| Pond A | | | |
|---|---|---|---|
| 12 | 13 | 15 | 6 |
| 7 | 8 | 12 | 7 |

| Pond B | | | |
|---|---|---|---|
| 9 | 12 | 5 | 8 |
| 12 | 15 | 16 | 19 |

4. Which type of data display would you use for the information in Exercise 3? Explain.

5. **SWIMMING** The table shows the numbers of hours you swam for several weeks.

| Hours Swimming | | | |
|---|---|---|---|
| 7 | 3.5 | 8 | 3.5 |
| 6 | 7 | 7 | 2 |
| 5.5 | 7.5 | 7.5 | 7.5 |

   a. Make a box-and-whisker plot for the data.
   b. Find the range and interquartile range.
   c. Which measures of central tendency and dispersion best represent the data?

6. **NEWBORNS** The table shows the lengths and weights of several newborn babies.

| Length (inches) | Weight (pounds) |
|---|---|
| 19 | 6 |
| 19.5 | 7 |
| 20 | 7.75 |
| 20.25 | 8.5 |
| 20.5 | 8.5 |
| 22.5 | 11 |

   a. Write an equation of a line that fits the data.
   b. Use a graphing calculator to find an equation of the line of best fit. Identify and interpret the correlation coefficient.
   c. Predict the weight of a newborn that is 21 inches long using the equations from parts (a) and (b). Compare the results.

7. **SAT** The table shows the numbers $y$ of students (in thousands) who took the SAT from 2003 to 2010, where $x = 3$ represents the year 2003. Use a graphing calculator to find an equation of the line of best fit. Then make a scatter plot of the residuals to tell whether the line of best fit models the data well.

| x | 3 | 4 | 5 | 6 | 7 | 8 | 9 | 10 |
|---|---|---|---|---|---|---|---|---|
| y | 1406 | 1419 | 1476 | 1466 | 1495 | 1519 | 1530 | 1548 |

8. **RECYCLING** You randomly survey shoppers at a supermarket about whether they use reusable bags. Of 60 male shoppers, 15 use reusable bags. Of 110 female shoppers, 60 use reusable bags. Organize your results in a two-way table. Include the marginal frequencies.

# 12 Standards Assessment

1. What is the value of $x$ when the mean of the video game scores is 45.5? *(S.ID.2)*

   | Video Game Scores | | | |
   |---|---|---|---|
   | 36 | 28 | $x$ | 48 |
   | 42 | 57 | 63 | 52 |

   **A.** 35

   **B.** 38

   **C.** 45.5

   **D.** 46.57

**Test-Taking Strategy**
**Read Question Before Answering**

"Of 2048 cats, how many of them will NOT answer to 'Here, kitty kitty'?
Ⓐ 100%  Ⓑ 1  Ⓒ $\frac{4098}{2}$  Ⓓ 2048
Hey, it means 'free food.'"

"Be sure to read the question before choosing your answer. You may find a word that changes the meaning."

2. Which graph represents $y = 3x^2 - 1$? *(F.BF.3)*

   **F.** 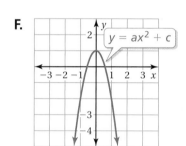   $y = ax^2 + c$

   **H.** 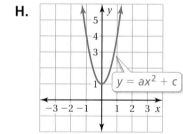   $y = ax^2 + c$

   **G.** 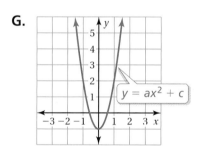   $y = ax^2 + c$

   **I.** 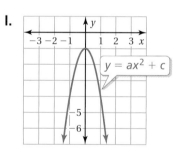   $y = ax^2 + c$

3. What is the solution of the equation $-6 + \sqrt{x} = -2$? *(N.RN.2)*

4. What is the product $(3z - 2)(2z + 4)$? *(A.APR.1)*

   **A.** $6z + 2$

   **C.** $6z^2 + 8z - 8$

   **B.** $6z^2 + 16z + 8$

   **D.** $6z^2 - 4z - 8$

5. The box-and-whisker plot represents the lengths of project presentations (in minutes) at a science fair. Find the interquartile range of the data. What does this represent in the context of the situation? *(S.ID.2)*

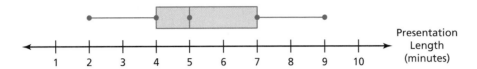

F. 7; The middle half of the presentation lengths vary by no more than 7 minutes.

G. 3; The presentation lengths vary by no more than 3 minutes.

H. 3; The middle half of the presentation lengths vary by no more than 3 minutes.

I. 7; The presentation lengths vary by no more than 7 minutes.

6. What is the simplified form of the expression? *(N.RN.2)*

$$\left(\frac{4}{5x}\right)^{-2}$$

A. $\dfrac{25}{16x^2}$  

B. $\dfrac{25x^2}{16}$

C. $\dfrac{16}{25x^2}$

D. $\dfrac{16x^2}{25}$

7. Which equation shows inverse variation? *(A.REI.10)*

F. $y = -3x + 7$

G. $2y = x$

H. $y = \dfrac{1}{5}x$

I. $y = \dfrac{4}{x}$

8. You randomly survey students in your school. You ask whether they have jobs. You display your results in the two-way table. How many male students do *not* have a job? *(S.ID.5)*

|  |  | Job | |
|---|---|---|---|
|  |  | Yes | No |
| Gender | Male | 27 | 12 |
|  | Female | 31 | 17 |

9. What is an equation for the $n$th term of the arithmetic sequence? *(F.LE.2)*

$$-\frac{3}{4}, -\frac{1}{4}, \frac{1}{4}, \frac{3}{4}, \ldots$$

A. $a_n = -\frac{3}{4}n + \frac{1}{2}$

B. $a_n = -\frac{3}{4}n + \frac{5}{4}$

C. $a_n = \frac{1}{2}n - \frac{1}{4}$

D. $a_n = \frac{1}{2}n - \frac{5}{4}$

10. A football field is 40 yards wide and 120 yards long. Find the distance between opposite corners of the football field. Show your work and explain your reasoning. *(8.G.7)*

    **Think Solve Explain**

11. Which scatter plot shows a negative relationship between $x$ and $y$? *(8.SP.1)*

    F.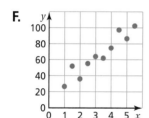

    H.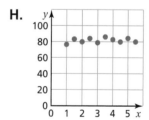

    G.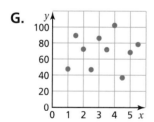

    I.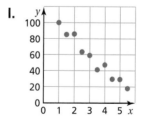

12. What is the solution of the equation? *(A.REI.3)*

    $$0.22(x + 6) = 0.2x + 1.8$$

    A. $x = 2.4$

    B. $x = 15.6$

    C. $x = 24$

    D. $x = 156$

Standards Assessment

# Appendix A
# My Big Ideas Projects

**A.1** **Literature Project**
   The Jungle Book

**A.2** **History Project**
   Mathematics in Medieval Islam

# My Big Ideas Projects

- **A.3** **Art Project**
  **Symmetry in Photographic Art**

- **A.4** **Science Project**
  **Tornadoes**

# A.1 Literature Project

## The Jungle Book

### 1 Getting Started

*The Jungle Book* is a collection of stories about an orphaned boy, Mowgli, who is raised by jungle animals in India. Written by Rudyard Kipling, the book was first published in 1894.

**Essential Question** How does the knowledge of mathematics help to visualize life in the jungle?

Read *The Jungle Book*. Look for descriptions of objects and places in the stories. Use the descriptions and mathematical formulas to estimate their size or distance.

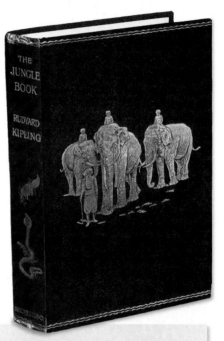

**Sample:** In the first story, Mowgli almost walks into a trap, described as a square box with a drop gate. Estimate the possible volume of the trap using inequalities.

The formula for the volume of a box is $V = \ell w h$. By letting $\ell = w = h$, you can rewrite the formula as $V = h \cdot h \cdot h = h^3$.

Mowgli is between 1 and 11 years old in the story. The median height of a 1-year-old is about 2.5 feet. The median height of an 11-year-old is about 4.5 feet. So, the inequality $2.5 \leq h \leq 4.5$ describes Mowgli's height during this time.

If the trap is as tall as Mowgli, the possible volume of the trap would be $2.5^3 \leq h^3 \leq 4.5^3$, or $15.625 \leq V \leq 91.125$ cubic feet.

A trap this size could ensnare a large animal, such as a wolf or a bear.

## 2  Things to Include

- Choose an object or place from each story. Write descriptions of each.
- State a formula you can use to estimate the size of each object or place.
- Estimate the value of each variable in the formula and state the units of measure.
- Calculate the measurement.
- Summarize your findings. Consider revising or refining your estimates based on the other estimates you made.

## 3  Things to Remember

- You can download each part of the book at *BigIdeasMath.com*.
- Add your own illustrations to your project. Label the dimensions of the objects and places in your illustrations.
- Include as many different formulas as possible.
- Organize your report in a folder, and think of a title for your report.

Section A.1  Literature Project  **A3**

# A.2 History Project

## Mathematics in Medieval Islam

### 1 Getting Started

Algebraic techniques date back thousands of years. In fact, they were discovered, applied, and debated long before they received the name *algebra*.

Mohammed ibn-Musa al-Khowarizmi, an Arab mathematician, wrote a book titled, "Al-jabr wa'l muqubalah." The book was written about 825 A.D. in Baghdad.

**Essential Question** How did the work of al-Khowarizmi and other Arabic and Islamic mathematicians influence modern-day mathematics?

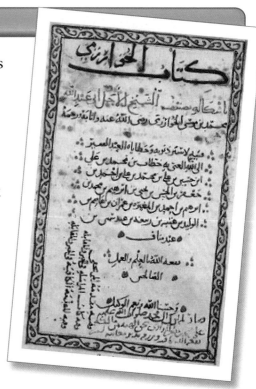

**Sample:** The word *algebra* is derived from the Arabic word *al-jabr*.

Al-Khowarizmi used the word *jabr* to describe moving a subtracted term to the other side of the equation.

$$x - 3 = 5$$
$$x = 8$$

## 2  Things to Include

- In addition to *jabr*, al-Khowarizmi used the word *muqubalah*. State the definitions and explain how al-Khowarizmi used *jabr* and *muqubalah* in solving equations.

- Give an example of how al-Khowarizmi solved an equation. Then demonstrate how you would solve the same equation.

- Discuss how Greek and Hindu civilizations influenced the development of algebra in medieval Islam.

- Islamic mathematicians and scientists contributed to another branch of mathematics called *trigonometry*. Discuss the development of trigonometry.

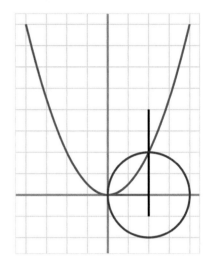

Omar Khayyám's geometric solution to a cubic equation

Medieval Islamic mosaics were produced using geometry that mathematicians did not fully understand until the 1970s.

## 3  Things to Remember

- Add your own illustrations to your project.

- Include as many different math concepts as possible.

- Organize your math stories in a folder, and think of a title for your report.

# A.3 Art Project

## Symmetry in Photographic Art

### 1 Getting Started

Symmetry is all around us, in nature and in manmade structures. A photographer chooses a perspective that either emphasizes the symmetry of the subject or focuses the attention on a specific part of the subject.

**Essential Question** How do the symmetry of the subject and the symmetry of the photograph affect a person's preference for a photograph?

Find two photographs with different perspectives of a subject having line and/or rotational symmetry. Discuss the symmetry of the subject and of the photograph. Conduct a survey asking which photograph is preferred.

**Sample:** In the first photograph, the bridge appears to have a vertical line of symmetry in the center of the bridge. In the second photograph, the bridge appears to have a vertical line of symmetry in the center of the bridge. There may be vertical lines of symmetry about the two vertical structures where the suspension lines are attached.

The first photograph emphasizes the symmetry of the bridge. The second photograph focuses on one of the vertical structures. The bridge in the second photograph may be symmetrical, but the photographer chose an angle that does not show the symmetry.

Of ten people surveyed, nine preferred the second photograph.

## 2 Things to Include

- Find two photographs of each of the following subjects. The subjects should have different symmetries, if possible. The photographs should have different perspectives.
    - flowers
    - houses or buildings
    - paintings or drawings by the same artist
    - sea creatures
    - a subject of your choice

- Discuss the types of symmetry (line and/or rotational) of the subject in each picture.

- Discuss whether the photographer's perspective emphasizes the symmetry of the subject in each photograph.

- Survey at least 10 people as to their visual preference of each pair.

This photograph has line symmetry and rotational symmetry. Would you prefer to view this photograph as displayed, or would you prefer to rotate it 90°?

## 3 Things to Remember

- Add your photographs to your project.
- Include as many different types of symmetry as possible.
- Organize your report in a folder, and think of a title for your report.

Section A.3   Art Project   **A7**

# A.4 Science Project

## Tornadoes

### 1 Getting Started

Tornadoes are considered some of nature's most violent storms. Tornadoes have strong rotating winds, which can range in speed from 40 to 318 miles per hour.

**Essential Question** Is there a relationship, that can be described by a mathematical model, between the occurrence of tornadoes and the month of the year?

Go to the National Oceanic and Atmospheric Administration website, *noaa.gov*, and enter the search words "storm event" to find the database containing statistics by state, county, and type of storm.

**Sample:** The table and scatter plot show the number of tornadoes $y$ in a given month $x$, for the years 1951 to 2010, in St. Lucie County, Florida.

| Month, $x$ | Number of Tornadoes, $y$ |
|---|---|
| 1 | 1 |
| 2 | 1 |
| 3 | 2 |
| 4 | 3 |
| 5 | 3 |
| 6 | 5 |
| 7 | 5 |
| 8 | 12 |
| 9 | 2 |
| 10 | 1 |
| 11 | 0 |
| 12 | 1 |

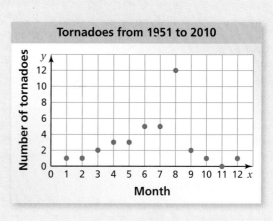

Summarize the data in the table and determine the pattern of the scatter plot (linear, quadratic, or exponential). Write an equation that roughly models the data and justify your work.

## 2 Things to Include

- Go to *noaa.gov* to find the dates on which tornadoes occurred in the county where you live.

- Organize the data by months in a table. Discuss the data in the table.

- Create a scatter plot of the data. Discuss the scatter plot.

- Determine if there is a linear, quadratic, or exponential relationship between the number of tornadoes and the month. Write an equation that roughly describes this relationship.

- Repeat the steps for a county in another state that has a different climate. Compare the scatter plots and equations for the two counties.

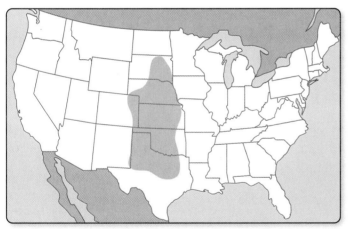

Tornado Alley in the central U.S. has a disproportionately high frequency of tornadoes during the late spring.

## 3 Things to Remember

- Add your own images to your project.

- Organize your report in a folder, and think of a title for your report.

Section A.4  Science Project  **A9**

# Selected Answers

## Section 1.1  Solving Simple Equations
(pages 7–9)

1. $+$ and $-$ are inverses. $\times$ and $\div$ are inverses.

3. $x - 3 = 6$; It is the only equation that does not have $x = 6$ as a solution.

5. $x = 57$      7. $x = -5$      9. $p = 21$      11. $x = 9\pi$      13. $d = \dfrac{1}{2}$      15. $n = -4.9$

17. **a.** $105 = x + 14$; $x = 91$

  **b.** no; Because $82 + 9 = 91$, you did not knock down the last pin with the second ball of the frame.

19. $n = -5$      21. $m = 7.3\pi$      23. $k = 1\dfrac{2}{3}$      25. $p = -2\dfrac{1}{3}$

27. They should have added 1.5 to each side.

  $-1.5 + k = 8.2$
  $k = 8.2 + 1.5$
  $k = 9.7$

29. $6.5x = 42.25$; $6.50 per hour

31. $420 = \dfrac{7}{6}b$, $b = 360$; $60

33. $h = -7$      35. $q = 3.2$      37. $x = -1\dfrac{4}{9}$

39. greater than; Because a negative number divided by a negative number is a positive number.

41. 3 mg      43. 8 in.      45. $7x - 4$      47. $\dfrac{25}{4}g - \dfrac{2}{3}$

## Section 1.2  Solving Multi-Step Equations
(pages 14 and 15)

1. $2 + 3x = 17$; $x = 5$      3. $k = 45$; $45°, 45°, 90°$      5. $b = 90$; $90°, 135°, 90°, 90°, 135°$

7. $c = 0.5$      9. $h = -9$      11. $x = -\dfrac{2}{9}$      13. 20 watches

15. $4(b + 3) = 24$; 3 in.      17. $\dfrac{2580 + 2920 + x}{3} = 3000$; 3500 people

19. $x = \dfrac{-7}{b}$      21. $x = \dfrac{3b}{2c}$      23. $x = \dfrac{c - b}{a}$

25. $<$      27. $>$

## Section 1.3  Solving Equations with Variables on Both Sides
(pages 22 and 23)

1. no; When 3 is substituted for $x$, the left side simplifies to 4 and the right side simplifies to 3.

3. $x = 13.2$ in.      5. $x = 7.5$ in.      7. $k = -0.75$

9. $p = -48$      11. no solution      13. $x = -4$

**A10**   Selected Answers

**15.** The 4 should have been added to the right side.
$$3x - 4 = 2x + 1$$
$$3x - 2x - 4 = 2x + 1 - 2x$$
$$x - 4 = 1$$
$$x - 4 + 4 = 1 + 4$$
$$x = 5$$

**17.** $15 + 0.5m = 25 + 0.25m$; 40 mi

**19.** 7.5 units

**21.** Sample answer: **a.** $5x = 5x - 3$   **b.** $x + 4 = 2x - x + 4$

**23.** fractions; Because $\frac{1}{3}$ is hard to perform operations with when written as a decimal.

**25.** 25 grams   **27.** 15.75 cm³   **29.** about 153.86 ft³

## Extension 1.3 — Solving Absolute Value Equations
(pages 24 and 25)

**1.** $x = 10$ or $x = -10$

**3.** no solution

**5.** $x = 6$ or $x = -2$

**7.** $x = 1$ or $x = -\frac{3}{5}$

**9.** Sample answer: $|x - 24| = 8$

## Section 1.4 — Rewriting Equations and Formulas
(pages 30 and 31)

**1.** no; The equation only contains one variable.

**3. a.** $A = \frac{1}{2}bh$   **b.** $b = \frac{2A}{h}$   **c.** $b = 12$ mm

**5.** $y = 4 - \frac{1}{3}x$

**7.** $y = \frac{2}{3} - \frac{4}{9}x$

**9.** $y = 3x - 1.5$

**11.** The $y$ should have a negative sign in front of it.
$$2x - y = 5$$
$$-y = -2x + 5$$
$$y = 2x - 5$$

**13. a.** $t = \frac{I}{Pr}$   **b.** $t = 3$ yr

**15.** $m = \frac{e}{c^2}$

**17.** $\ell = \dfrac{A - \frac{1}{2}\pi w^2}{2w}$

**19.** $w = 6g - 40$

**21. a.** $F = 32 + \frac{9}{5}(K - 273.15)$
**b.** 32°F
**c.** liquid nitrogen

**23.** $r^3 = \frac{3V}{4\pi}$; $r = 4.5$ in.

**25.** $6\frac{2}{5}$

**27.** $1\frac{1}{4}$

# Section 2.1 Graphing Linear Equations
(pages 46 and 47)

**1.** a line

**3.** Sample answer:

| x | 0 | 1 |
|---|---|---|
| y = 3x − 1 | −1 | 2 |

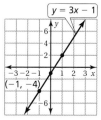

**5.**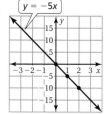

**7.**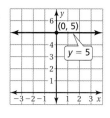

**9.**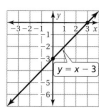

**11.**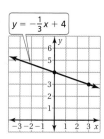

**13.**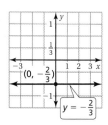

**15.**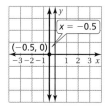

**17.** The equation $x = 4$ is graphed, not $y = 4$.

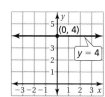

**19. a.**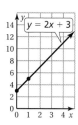

**b.** about $5

**c.** $5.25

**21.** $y = -\frac{5}{2}x + 2$

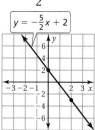

**23.** $y = -2x + 3$

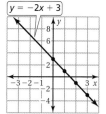

**25.** Begin this exercise by listing all of the given information.

**27. a.** Sample answer:

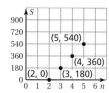

Yes, the points lie on a line.

**b.** No, $n = 3.5$ does not make sense because a polygon cannot have half a side.

**29.** $(-6, 6)$

**31.** $(-4, -3)$

A12 Selected Answers

## Section 2.2 Slope of a Line
(pages 53–55)

1. **a.** B and C
   **b.** A
   **c.** no; None of the lines are vertical.

3. The line is horizontal.

5.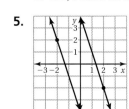
   The lines are parallel.

7. $\dfrac{3}{4}$

9. $-\dfrac{3}{5}$

11. 0

13. 0

15. undefined

17. $-\dfrac{11}{6}$

19. The denominator should be $2 - 4$.
    slope $= -1$

21. Choose any two points from the table and use the slope formula.
    slope $= 4$

23. Choose any two points from the table and use the slope formula.
    slope $= -\dfrac{3}{4}$

25. $\dfrac{1}{3}$

27. $k = 11$

29. $k = -5$

31. **a.** $\dfrac{3}{40}$
    **b.** The cost increases by $3 for every 40 miles you drive, or the cost increases by $0.075 for every mile you drive.

33. yes; The slopes are the same between the points.

35. yes; When you switch the coordinates, the differences in the numerator and denominator are the opposite of the numbers when using the slope formula. You still get the same slope.

37.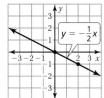

39.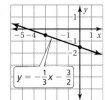

# Extension 2.2 Slopes of Parallel and Perpendicular Lines (pages 56 and 57)

1. blue and red; They both have a slope of $-3$.
3. yes; Both lines are horizontal and have a slope of 0.
5. yes; Both lines are vertical and have an undefined slope.
7. blue and green; The blue line has a slope of 6. The green line has a slope of $-\frac{1}{6}$. The product of their slopes is $6 \cdot \left(-\frac{1}{6}\right) = -1$.
9. yes; The line $x = -2$ is vertical. The line $y = 8$ is horizontal. A vertical line is perpendicular to a horizontal line.
11. yes; The line $x = 0$ is vertical. The line $y = 0$ is horizontal. A vertical line is perpendicular to a horizontal line.

# Section 2.3 Graphing Linear Equations in Slope-Intercept Form (pages 62 and 63)

1. Find the $x$-coordinate of the point where the graph crosses the $x$-axis.
3. *Sample answer:* The amount of gasoline $y$ (in gallons) left in your tank after you travel $x$ miles is $y = -\frac{1}{20}x + 20$. The slope of $-\frac{1}{20}$ means the car uses 1 gallon of gas for every 20 miles driven. The $y$-intercept of 20 means there is originally 20 gallons of gas in the tank.
5. A; slope: $\frac{1}{3}$; $y$-intercept: $-2$
7. slope: 4; $y$-intercept: $-5$
9. slope: $-\frac{4}{5}$; $y$-intercept: $-2$
11. slope: $\frac{4}{3}$; $y$-intercept: $-1$
13. slope: $-2$; $y$-intercept: 3.5
15. slope: 1.5; $y$-intercept: 11

17. a. [graph: $y = -10x + 3000$, Height (feet) vs Time (seconds)]
    b. The $x$-intercept of 300 means the skydiver lands on the ground after 300 seconds. The slope of $-10$ means that the skydiver falls to the ground at a rate of 10 feet per second.

19.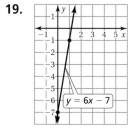
    $x$-intercept: $\frac{7}{6}$

21. $x$-intercept: $-\frac{5}{7}$

23.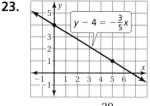
    $x$-intercept: $\frac{20}{3}$

**25.** $y = 0.75x + 5$

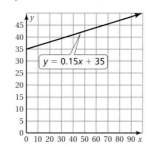

**27.** $y = 0.15x + 35$

**29.** $y = 2x + 3$

**31.** $y = \dfrac{2}{3}x - 2$

**33.** B

## Section 2.4 — Graphing Linear Equations in Standard Form
*(pages 68 and 69)*

**1.** no; The equation is in slope-intercept form.

**3.** $x =$ pounds of peaches
$y =$ pounds of apples
$y = -\dfrac{4}{3}x + 10$

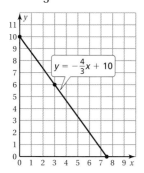

**5.** $y = -2x + 17$

**7.** $y = \dfrac{1}{2}x + 10$

**11.** $x$-intercept: $-6$
$y$-intercept: $3$

**13.** $x$-intercept: none
$y$-intercept: $-3$

**15. a.** $-25x + y = 65$
**b.** $390$

**9.**

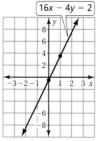

**17.**

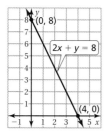

**19.** $x$-intercept: $9$
$y$-intercept: $7$

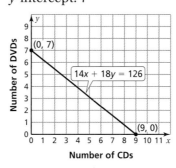

**21. a.** $9.45x + 7.65y = 160.65$
**b.**

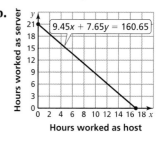

**23. a.** $y = 40x + 70$
**b.** $x$-intercept: $-\dfrac{7}{4}$; It will not be on the graph because you cannot have a negative time.
**c.**

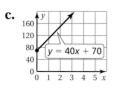

**25.**

| x | −2 | −1 | 0 | 1 | 2 |
|---|---|---|---|---|---|
| −5 − 3x | 1 | −2 | −5 | −8 | −11 |

Selected Answers **A15**

# Section 2.5
## Writing Equations in Slope-Intercept Form
(pages 76 and 77)

1. *Sample answer:* Find the ratio of the rise to the run between the intercepts.
3. $y = 3x + 2$; $y = 3x - 10$; $y = 5$; $y = -1$
5. $y = x + 4$
7. $y = \dfrac{1}{4}x + 1$
9. $y = \dfrac{1}{3}x - 3$
11. The *x*-intercept was used instead of the *y*-intercept. $y = \dfrac{1}{2}x - 2$
13. $y = 5$
15. $y = -2$
17. a–b. 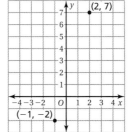 (0, 60) represents the speed of the automobile before braking. (6, 0) represents the amount of time it takes to stop. The line represents the speed *y* of the automobile after *x* seconds of braking.

    c. $y = -10x + 60$

19. Be sure to check that your rate of growth will not lead to a 0-year-old tree with a negative height.

21–23.

# Section 2.6
## Writing Equations in Point-Slope Form
(pages 82 and 83)

1. slope $= -2$; $(-1, 3)$
3. $y - 0 = \dfrac{1}{2}(x + 2)$
5. $y + 1 = -3(x - 3)$
7. $y - 8 = \dfrac{3}{4}(x - 4)$
9. $y + 5 = -\dfrac{1}{7}(x - 7)$
11. $y + 4 = -2(x + 1)$
13. $y = 2x$
15. $y = \dfrac{1}{4}x$
17. $y = x + 1$
19. a. $V = -4000x + 30,000$

    b. $30,000

21. The rate of change is 0.25 degree per chirp.
23. a. $y = 14x - 108.5$

    b. 4 meters

25. 175
27. D

A16 Selected Answers

## Extension 2.6 Writing Equations of Parallel and Perpendicular Lines (pages 84 and 85)

**1.** $y = 3x + 7$  **3.** $y = -4x - 21$  **5.** $y = 2x$  **7.** $y = 2x - 5$
**9.** $y = -x - 1$  **11.** $y = \frac{1}{5}x + \frac{19}{5}$  **13.** $y = \frac{1}{5}x - 1$

**15.** Example 1

$y - y_1 = m(x - x_1)$

$y + 2 = \frac{1}{2}(x - 6)$

$y + 2 = \frac{1}{2}x - 3$

$y = \frac{1}{2}x - 5$

Example 2

$y = mx + b$

$1 = \frac{1}{4}(-3) + b$

$1 = -\frac{3}{4} + b$

$\frac{7}{4} = b$

So, $y = \frac{1}{4}x + \frac{7}{4}$.

*Sample answer:* Point-slope form; The point-slope form gives you the equation of the line directly.

## Section 2.7 Solving Real-Life Problems (pages 90 and 91)

**1.** The $y$-intercept is $-6$ because the line crosses the $y$-axis at the point $(0, -6)$. The $x$-intercept is 2 because the line crosses the $x$-axis at the point $(2, 0)$. You can use these two points to find the slope.

Slope = $\frac{\text{change in } y}{\text{change in } x} = \frac{6}{2} = 3$

**3.** *Sample answer:* the rate at which something is happening

**5.** *Sample answer:* The temperature outside is falling 3°F every hour. After 7 hours, the temperature is 0°F.

**7. a.** slope: $-3.6$; $y$-intercept: 59  **b.** $y = -3.6x + 59$
 **c.** 59°F

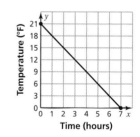

**9. a.** Antananarivo: 19°S, 47°E;  Denver: 39°N, 105°W;
 Brasilia: 16°S, 48°W;  London: 51°N, 0°W;  Beijing: 40°N, 116°E
 **b.** $y = \frac{1}{221}x + \frac{8724}{221}$
 **c.** a place that is on the prime meridian

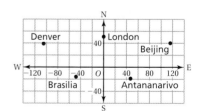

**11.** $h = \frac{5}{4}$  **13.** $q = -2.3$

## Section 3.1  Writing and Graphing Inequalities (pages 108 and 109)

1. An open circle would be used because 250 is not a solution.
3. no; $x \geq -9$ is all values of $x$ greater than or equal to $-9$. $-9 \geq x$ is all values of $x$ less than or equal to $-9$.
5. $x < -3$; all values of $x$ less than $-3$
7. $y + 5.2 < 23$
9. $k - 8.3 > 48$
11. yes
13. yes
15. no
17. [number line from $-7$ to $-1$ with closed circle at $-6$, shaded left]
19. [number line from $10\frac{1}{2}$ to $12$ with open circle at $11\frac{1}{4}$, shaded right]
21. $x \geq 21$
23. yes
25. a. $a \geq 10$; [number line 0 to 40]
    $s \geq 200$; [number line 0 to 400]
    $t \geq 10$; [number line 0 to 16]
    b. yes; You satisfy the swimming requirement of the course because $10(25) = 250$ and $250 \geq 200$.
27. a. $m < n$; $n \leq p$  b. $m < p$
    c. no; Because $n$ is no more than $p$ and $m$ is less than $n$, $m$ cannot be equal to $p$.
29. $-1.7$
31. D

## Section 3.2  Solving Inequalities Using Addition or Subtraction (pages 114 and 115)

1. no; The solution of $r - 5 \leq 8$ is $r \leq 13$ and the solution of $8 \leq r - 5$ is $r \geq 13$.
3. *Sample answer:* $A = 350$, $C = 275$, $Y = 3105$, $T = 50$, $N = 2$
5. *Sample answer:* $A = 400$, $C = 380$, $Y = 6510$, $T = 83$, $N = 0$
7. $t > 4$; [number line 0 to 6, open circle at 4, shaded right]
9. $a > -8$; [number line $-11$ to $-5$, open circle at $-8$, shaded right]
11. $-\dfrac{3}{5} > d$; [number line $-\frac{6}{5}$ to $0$, open circle at $-\frac{3}{5}$, shaded left]
13. $m \leq 1$; [number line $-3$ to $3$, closed circle at $1$, shaded left]
15. $h < -1.5$; [number line $-2.5$ to $0.5$, open circle at $-1.5$, shaded left]
17. $9.5 \geq u$; [number line 7.5 to 10.5, closed circle at 9.5, shaded left]
19. a. $100 + V \leq 700$; $V \leq 600$ in.$^3$   b. $V \leq \dfrac{700}{3}$ in.$^3$
21. $x + 2 > 10$; $x > 8$
23. 5
25. a. $4500 + x \geq 12{,}000$; $x \geq 7500$ points
    b. This changes the number added to $x$ by 60%, so the inequality becomes $7200 + x \geq 12{,}000$. So, you need less points to advance to the next level.
27. $2\pi h + 2\pi \leq 15\pi$; $h \leq 6.5$ mm
29. 10
31. 12
33. 0.5
35. $2\sqrt{3}$

# Section 3.3 Solving Inequalities Using Multiplication or Division (pages 121–123)

1. Multiply each side of the inequality by 6.
3. Sample answer: $-3x < 6$
5. $x \geq -1$
7. $x \leq -3$
9. $x \leq \dfrac{3}{2}$
11. $c \leq -36$;
13. $x < -28$;
15. $k > 2$;
17. $y \leq -4$;
19. The inequality sign should not have been reversed.

$$\dfrac{x}{2} < -5$$
$$2 \cdot \dfrac{x}{2} < 2 \cdot (-5)$$
$$x < -10$$

21. $\dfrac{x}{8} < -2$; $x < -16$
23. $5x > 20$; $x > 4$
25. $0.25x \leq 3.65$; $x \leq 14.6$; You can make at most 14 copies.
27. $n \geq -5$;
29. $h \leq -42$;
31. $y > \dfrac{11}{2}$;
33. $m > -12$;
35. $b > 4$;
37. no; You need to solve the inequality for $x$. The solution is $x < 0$. Therefore, numbers greater than 0 are not solutions.
39. $12x \geq 102$; $x \geq 8.5$ cm
41. $\dfrac{x}{4} < 80$; $x < \$320$
43. *Answer should include, but is not limited to:* Using the correct number of months that the CD has been out.
45. $n \geq -6$ and $n \leq -4$;
47. $m < 20$;
49. $8\dfrac{1}{4}$
51. 84

## Section 3.4 — Solving Multi-Step Inequalities
(pages 130 and 131)

1. *Sample answer:* They use the same techniques, but when solving an inequality, you must be careful to reverse the inequality symbol when you multiply or divide by a negative number.

3. $k > 0$ and $k \leq 16$ units

5. $b \geq 1$;

7. $m \geq -15$;

9. $p < -1$;

11. They did not perform the operations in proper order.
$$\frac{x}{4} + 6 \geq 3$$
$$\frac{x}{4} \geq -3$$
$$x \geq -12$$

13. all real numbers

15. $u < -17$     17. $z > -0.9$     19. no solutions

21. $20x + 100 \leq 320$; $x \leq 11$ $20 bills     23. $b < 3$;

25. $500 - 20x \geq 100$; $x > \$0$ and $x \leq \$20$ per hour

27. a. $3.5x + 350 \geq 500$; $x \geq 42\frac{6}{7}$; at least 43 more cars, so at least 143 cars total

    b. Because each car will pay $1 more, fewer cars will be needed for the theater to earn $500.

29. 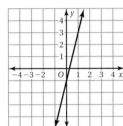     31. 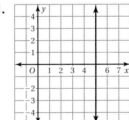     33. A

## Extension 3.4 — Solving Compound Inequalities
(pages 132–135)

1. $3 < k < 9$; 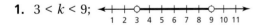     3. $w < -10$ or $w \geq -6$;

5. $-4 \leq x < -1$;     7. $9 < x < 12$;

9. $-2 < x \leq 3$;     11. $x > -6$ or $x \leq -7$;

13. $x > -8$ and $x < -6$;

15. no solution

17. $x > 6$ and $x < 14$;

19. $|x - 44| \leq 3$; The least percent of voters who will vote for the new mayor is 41%. The greatest percent of voters who will vote for the new mayor is 47%.

A20    Selected Answers

# Section 3.5 Graphing Linear Inequalities in Two Variables
(pages 141–143)

1. An ordered pair is a solution of an inequality if it makes the inequality true.

3. The graph of a linear equation in two variables will be a solid line. The graph of a linear inequality in two variables could be a solid or dashed line, and half of the coordinate plane will be shaded.

5. C

7. B

9. All the points on or above the line $y = -x + 5$.

11. yes

13. no

15. yes

17. no

19. yes

21. no

23. no

25. $2x + 3y \leq 60$; no, you cannot buy 12 yards of cotton lace and 15 yards of linen lace because (12, 15) is not a solution of the inequality.

27. The boundary line will be dashed.

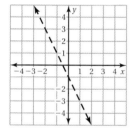

29. C

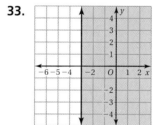

31. B

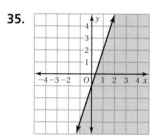

37.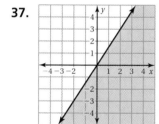

39.

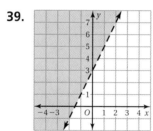

41. The boundary line should be dashed instead of solid.

43. *Sample answer:* Choosing a test point on the boundary line will not help you determine which half-plane to shade. The test point must be in one of the half-planes.

45. yes

47. no

49. $y > 2x + 1$

51. $y \leq -\dfrac{1}{2}x - 2$

53. a. $0.75x + 2.25y \leq 9$

b. Two possible solutions are (1, 3) and (12, 0). So, you can play 1 arcade game and buy 3 drinks, or you can play 12 arcade games and buy 0 drinks.

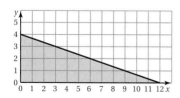

55. 256

57. 243

## Section 4.1 Solving Systems of Linear Equations by Graphing (pages 158 and 159)

1. yes; The equations are linear and in the same variables.
3. Check whether (3, 4) is a solution of each equation.
5. (4, 176)
7. B; (6, 7)
9. C; (3, −1)
11. (−5, 1)
13. (12, 15)
15. (8, 1)
17. (5, 1.5)
19. (−6, 2)
21. no; Two lines cannot intersect in exactly two points.
23. Make a table to compare your distance to your friend's distance.
25. $c = 8$
27. $x = 11$

## Section 4.2 Solving Systems of Linear Equations by Substitution (pages 164 and 165)

1. **Step 1:** Solve one of the equations for one of the variables.
   **Step 2:** Substitute the expression from Step 1 into the other equation and solve.
   **Step 3:** Substitute the value from Step 2 into one of the original equations and solve.
3. sometimes; A solution obtained by graphing may not be exact.
5. *Sample answer:* $x + 2y = 6$
   $x - y = 3$
7. $4x - y = 3$; The coefficient of $y$ is $-1$.
9. $2x + 10y = 14$; Dividing by 2 to solve for $x$ yields integers.
11. (6, 17)
13. (4, 1)
15. $\left(\dfrac{1}{4}, 6\right)$
17. **a.** $x = 2y$
      $64x - 132y = 1040$
    **b.** adult tickets: $8; student tickets: $4
19. (−2, 4)
21. The expression for $y$ was substituted back into the same equation; solution: (2, 1)
23. 30 cats, 35 dogs
25. Make a diagram to help visualize the problem.
27. $2x - 5y = -8$
29. B

## Section 4.3 Solving Systems of Linear Equations by Elimination (pages 173–175)

1. **Step 1:** Multiply, if necessary, one or both equations by a constant so at least one pair of like terms has the same or opposite coefficients.
   **Step 2:** Add or subtract the equations to eliminate one of the variables.
   **Step 3:** Solve the resulting equation for the remaining variable.
   **Step 4:** Substitute the value from Step 3 into one of the original equations and solve.

3. $2x + 3y = 11$ You have to use multiplication to solve the
   $3x - 2y = 10$; system by elimination.

5. (6, 2)   7. (2, 1)   9. (1, −3)   11. (3, 2)

13. The student added $y$-terms, but subtracted $x$-terms and constants; solution (1, 2)

15. **a.** $2x + y = 10$   17. (5, −1)   19. (−2, −1)   21. (4, 3)
    $2x + 3y = 22$
    **b.** 6 minutes

23. **a.** ±4   25. **a.** $23x + 10y = 86$
    **b.** ±7          $28x + 5y = 76$
                  **b.** Multiple choice: 2 points each
                         Short response: 4 points each

27. $95

29. 5 grams of 90% gold alloy, 3 grams of 50% gold alloy

31. (−1, 2, 1)   33. yes   35. D

## Section 4.4  Solving Special Systems of Linear Equations
*(pages 180 and 181)*

1. The graph of a system with no solution is two parallel lines, and the graph of a system with infinitely many solutions is one line.

3. infinitely many solutions; all points on the line $y = 4x + \dfrac{1}{3}$

5. no solution; The lines have the same slope and different $y$-intercepts.

7. infinitely many solutions; The lines are identical.

9. (−1, −2)

11. infinitely many solutions; all points on the line $y = -\dfrac{1}{6}x + 5$

13. (−2.4, −3.5)

15. no; because they are running at the same speed and your pig had a head start

17. When the slopes are different, there is one solution. When the slopes are the same, there is no solution if the $y$-intercepts are different and infinitely many solutions if the $y$-intercepts are the same.

19. $y = 0.99x + 10$
    $y = 0.99x$
    no; Because you paid $10 before buying the same number of songs at the same price, you spend $10 more.

21. Try using the Guess-and-Test method to help you answer this question.

23.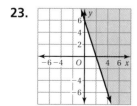

25.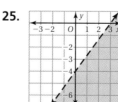

# Extension 4.4  Solving Linear Equations by Graphing
(pages 182 and 183)

1. $x = \dfrac{1}{2}$
3. no solution
5. $x = 2$
7. *Sample answer:* $6x - 3 = 6x$; Subtract 3 from the right side.
9. $x = \dfrac{21}{2}$
11. 6 mo

# Section 4.5  Systems of Linear Inequalities
(pages 189–191)

1. Substitute its coordinates for $x$ and $y$ in each inequality of the system and simplify each side. When both resulting inequalities are true, the ordered pair is a solution. Otherwise, it is not a solution.

3. no; The point is not part of the solution of the inequality bordered by the dashed line.

5. no

7.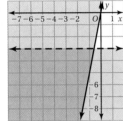
9.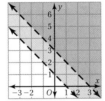
11.

13.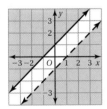
15.
17. The solid line is incorrect for the graph of $y < -x - 2$; Change it to a dashed line.

19. B
21. C

23. a. $x + y \leq 20$
    $12x + 10y \geq 110$

   b. *Sample answer:* (10, 8); Work 10 hours at the grocery store and 8 hours as a coach.

25. $y > -2x - 1$
    $y < -2x - 3$
27.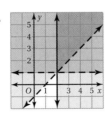
29.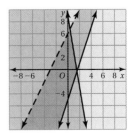

A24   Selected Answers

31. **a.** *Sample answer:* $y < 2x - 3$    **b.** *Sample answer:* $y > 2x - 3$

33. *Sample answer:* You drive 6 hours and your friend drives 8 hours for a total of 14 hours and a distance of 900 miles each day.

35. $-5$      37. $1$

## Section 5.1 — Domain and Range of a Function
*(pages 206 and 207)*

1. *Sample answer:* An independent variable represents an input value, and a dependent variable represents an output value.

3. **a.** $y = 6 - 2x$    **b.** domain: 0, 1, 2, 3; range: 6, 4, 2, 0

   **c.** $x = 6$ is not in the domain because it would make $y$ negative, and it is not possible to buy a negative number of headbands.

5. domain: $-2, -1, 0, 1, 2$; range: $-2, 0, 2$

7. The domain and range are switched. The domain is $-3, -1, 1$, and $3$. The range is $-2, 0, 2$, and $4$.

9. 
| x | −1 | 0 | 1 | 2 |
|---|----|---|---|---|
| y | −4 | 2 | 8 | 14 |

domain: $-1, 0, 1, 2$
range: $-4, 2, 8, 14$

11. 
| x | −1 | 0 | 1 | 2 |
|---|----|---|---|---|
| y | 1.5 | 3 | 4.5 | 6 |

domain: $-1, 0, 1, 2$
range: $1.5, 3, 4.5, 6$

13. Rewrite the percent as a fraction or decimal before writing an equation.

15.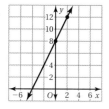

17.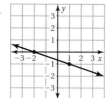

19. D

## Extension 5.1 — Relations and Functions
*(pages 208 and 209)*

1. not a function

3. not a function

5. false; A relation is not a function when an input has more than one output.

7. no; In any two games, if the winning teams had the same numbers of runs while the losing teams had different numbers of runs, then the relation is not a function.

9. function

11. no; All linear equations represent functions except for those representing vertical lines.

## Section 5.2 — Discrete and Continuous Domains
(pages 214 and 215)

1. A discrete domain consists of only certain numbers in an interval, whereas a continuous domain consists of all numbers in an interval.

3. domain: $0 \le x \le 6$
   range: $0 \le y \le 6$;
   continuous

5. 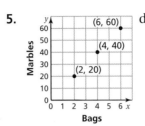 discrete

7. continuous

9. 2.5 is *not* in the domain, because the domain is discrete and consists only of the integers 1, 2, 3, and 4.

11. a. independent variable: $c$; dependent variable: $t$   b. discrete

13. no; A height can be any positive number.

15. a. *Sample answer:* the elevation of a sinking ship relative to sea level
    b. *Sample answer:* an overdraw leaves a negative checking account balance

17. $-\dfrac{5}{2}$

19. C

## Section 5.3 — Linear Function Patterns
(pages 220 and 221)

1. words, equation, table, graph

3. yes; All nonvertical lines intersect the $y$-axis.

5. $y = 2x$; $x$ is the base of the triangle; $y$ is the area of the triangle.

7. $y = -4x - 2$

9. $y = 2x$

11. $y = \dfrac{2}{3}x + 5$

13. a.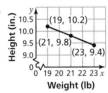
    linear

    b. $y = -0.2x + 14$
    c. 9.7 in.

15. Substitute 8 for $t$ in the equation.

17. $-4$; $-2$; $1$

19. $-1.25$; $-0.25$; $1.25$

## Section 5.4 — Function Notation
(pages 229–231)

1. Function notation assigns a name such as $f$ to a function and $f(x)$ represents the value of the function at $x$. Example: $y = 3x + 1$ can be written as $f(x) = 3x + 1$.

3. a line; Changing $b$ translates the graph vertically.

5. $g(-2) = -8$; $g(0) = -2$; $g(5) = 13$

**7.** $h(-2) = -5;\ h(0) = -7;\ h(5) = -12$

**9.** $f(-2) = 12;\ f(0) = 2;\ f(5) = -23$

**11.** $x = 1$

**13.** $x = -2$

**15.** $x = -6$

**17. a.** $198   **b.** 25 hours

**19.** C

**21.** A

**23.**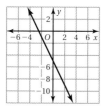

**25.**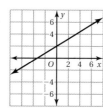

**27.**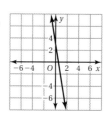

**29.** 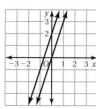 The graph of $g$ is a translation 2 units up of the graph of $f$.

**31.** 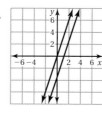 The graph of $v$ is a translation $\frac{7}{2}$ units down of the graph of $f$.

**33.** 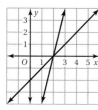 $(2, 0)$

**35.** 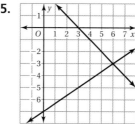 $(6, -3)$

**37.** $k = -1$

**39.** $k = -4$

**41.** Translate the graph of $y = x$ four units to the left.

**43.** $y = x$

**45.** $y = 1$

## Extension 5.4 Special Functions
*(pages 232–235)*

**1.**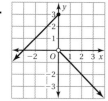
domain: all real numbers;
range: $y \leq 3$

**3.**
domain: all real numbers;
range: $y \geq -5$

**5.**
domain: all real numbers;
range: $-4 \leq y \leq 2$

**7.** no; It fails the Vertical Line Test.

**9.** $f(x) = \begin{cases} -4, & \text{if } x < 0 \\ -x, & \text{if } x \geq 0 \end{cases}$

**11.** $f(x) = \begin{cases} 100, & \text{if } 0 < x \leq 1 \\ 150, & \text{if } 1 < x \leq 2 \\ 200, & \text{if } 2 < x \leq 3 \\ 250, & \text{if } 3 < x \leq 4 \end{cases}$

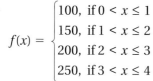

**13.** 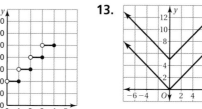 translation 5 units up; domain: all real numbers; range: $y \geq 5$

# Extension 5.4 Special Functions (continued)
(pages 232–235)

**15.**
translation 3 units right; domain: all real numbers; range: $y \geq 0$

**17.**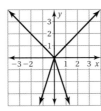
opens down and is narrower than $y = |x|$; domain: all real numbers; range: $y \leq 0$

**19.**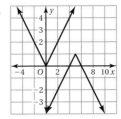
opens down and is a translation 5 units right and 1 unit up of the graph $y = |x|$; domain: all real numbers; range: $y \leq 1$

**21.** $y = |x| - 7$

**23.** $y = |x - 5| - 1$

**25. a.** positive $k$: translation up; negative $k$: translation down

**b.** positive $h$: translation to the right; negative $h$: translation to the left

**c.** positive $a$: Graph is narrower when $a > 1$, wider when $a < 1$; negative $a$: opens down, graph is narrower when $a < -1$, wider when $a > -1$.

**27.** $x = -7, 3$

**29.** Sample answer: $y = \begin{cases} -x - 4, & \text{if } x < -4 \\ x + 4, & \text{if } x \geq -4 \end{cases}$

# Section 5.5 Comparing Linear and Nonlinear Functions
(pages 240 and 241)

**1.** A linear function has a constant rate of change. A nonlinear function does not have a constant rate of change.

**3.** 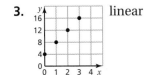 linear

**5.** nonlinear

**7.** linear; The graph is a line.

**9.** linear; As $x$ increases by 6, $y$ increases by 4.

**11.** nonlinear; As $x$ increases by 1, $V$ increases by different amounts.

**13.** linear; You can rewrite the equation in slope-intercept form.

**15.** nonlinear; As $x$ decreases by 65, $y$ increases by different amounts.

**17. a.** linear; The equation that represents the function is $h = 1.6x$.

**b.** tree B; It is growing at a rate of 1.6 feet per year.

**19. a.** 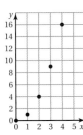 The points curve upward from left to right; nonlinear

**b.** $y = x^2$

**21.** $-6$

**23.** C

## Section 5.6 Arithmetic Sequences
(pages 247–249)

**1.** subtract a term from the following term

**3.**

| n | 1 | 2 | 3 | 4 |
|---|---|---|---|---|
| y | 25 | 50 | 75 | 100 |

As $n$ increases by 1, $y$ increases by 25.

**5.** 21.5, 25, 28.5  **7.** 5  **9.** 25  **11.** $-1.5$

**13.** 22, 25, 28  **15.** 36, 41, 46  **17.** 0.1, $-0.2$, $-0.5$

**19.** 4, 6, 8, 10, . . . ; yes, There is a common difference of 2.

**21.** no  **23.** yes; 6

**25.** 4.6, 4, 3.4  **27.** 1, $1\frac{1}{8}$, $1\frac{1}{4}$  **29.** 7.5, 12, 16.5

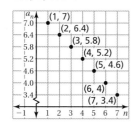
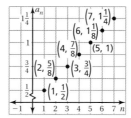
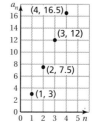

**31. a.**

| Number of Visits in One Year | Cost |
|---|---|
| 1 | $8 |
| 2 | $16 |
| 3 | $24 |
| 4 | $32 |

**b.** yes    **c.** $48

**d.** 4 times; The cost for the family to visit the zoo 3 times is $120, which is less than the $130 pass. The cost for the family to visit the zoo 4 times is $160, which is more than the $130 pass.

**33.** $a_n = n - 6$; $a_{10} = 4$    **35.** $a_n = \frac{1}{2}n$; $a_{10} = 5$

**37.** $a_n = -10n$; $a_{10} = -100$    **39.** $a_n = -20n + 120$

**41.** Make a table to show your speed at each second.

**43.** discrete; A sequence of $n$ terms has a domain of the integers 1 through $n$.

**45. a.** $a_n = 5n$

**b.** discrete

**c.** Substitute the number of minutes in a day, 1440, for $n$.

**47.** $(4, -2)$    **49.** D

## Section 6.1 — Properties of Square Roots
(pages 264 and 265)

1. when the radicand has no perfect square factors other than 1

3. $s = 8$ ft

5. $s = \dfrac{3}{4}$ cm

7. $-10\sqrt{2}$

9. $4\sqrt{3}$

11. $-\dfrac{\sqrt{23}}{8}$

13. $\dfrac{3\sqrt{2}}{7}$

15. $2\sqrt{2}$

17. $2\sqrt{13}$

19. $\dfrac{7\sqrt{3}}{2} \approx 6.1$ amperes

21. $3 + \sqrt{11}$

23. $2 + 2\sqrt{3}$

25. $\dfrac{1+\sqrt{7}}{2}$

27. $3\sqrt{10} \approx 9.5$ ft³

29. $xy\sqrt{42}$

31. $3xy\sqrt{2xz}$

33. 243

35. 125

## Section 6.2 — Properties of Exponents
(pages 273–275)

1. D

3. B

5. Simplify $3^{6-3}$; $3^3$; $3^9$

7. $x^2$

9. $64b^3$

11. $64a^4$

13. $b^{11}$

15. $\dfrac{1}{d^3}$

17. $\dfrac{1}{x^6}$

19. The exponents were added instead of multiplied; $(m^3)^4 = m^{3 \cdot 4} = m^{12}$

21. $38.44y^2$

23. $\dfrac{d^2}{36}$

25. $-3125x^5$

27. The exponent was not applied to the denominator.
$\left(\dfrac{x^3}{3}\right)^2 = \dfrac{(x^3)^2}{3^2} = \dfrac{x^6}{9}$

29. no; $(a^4)^2 = a^8$ by the Power of a Power Property. $a^{4^2} = a^{16}$ because $4^2 = 16$.

31. $V = \dfrac{32\pi m^6}{3}$

33. $5.1 \times 10^{-3}$

35. $3.456 \times 10^{-9}$

37. $4 \times 10^{-2}$

39. $\dfrac{y^{12}}{216x^6}$

41. $\dfrac{5b^5 c^{14}}{12}$

43. $14a^2 b$ microns

45. 4.4 on the Richter Scale

47. $4\sqrt{3}$

49. $\dfrac{6\sqrt{5}}{11}$

## Section 6.3 — Radicals and Rational Exponents
(pages 280 and 281)

1. 3; Find the fourth root of 81.

3. $s = 4$ in.

5. $s = \dfrac{7}{8}$ ft

7. $4^{2/3}$

9. $\sqrt[3]{15}$

11. $\left(\sqrt[5]{78}\right)^2$

13. 4

15. 4

17. 10

19. length: 3 ft, width: 2 ft

21. 25

23. 9

25. 2401

27. $a^{m/n}$; The $m$th power of the $n$th root of a positive number $a$ can be written as a power with base $a$ and an exponent of $m/n$.

29. always; $(x^{1/3})^3 = x^{3/3} = x^1 = x$

31. always; By definition, $x^{1/3} = \sqrt[3]{x}$.

33. always; By definition, $x^{(2/3) - (1/3)} = x^{1/3} = \sqrt[3]{x}$.

35.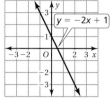

37.

## Section 6.4 Exponential Functions
(pages 289–291)

1. A linear function changes by a constant amount over equal intervals. An exponential function changes by equal factors over equal intervals.

3. $f(x) = (-3)^x$; It is not an exponential function.

5.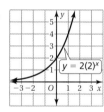

7. exponential; As $x$ increases by 1, $y$ is multiplied by 2.

9. linear; As $x$ increases by 3, $y$ decreases by 9.

11. 1.5   13. 8   15. 2   17. 6.25%   19. A

21.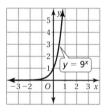
domain: all real numbers
range: all positive real numbers

23.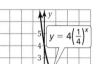
domain: all real numbers
range: all positive real numbers

25. When $b > 1$, the graph rises from left to right. When $0 < b < 1$, the graph falls from left to right.

27.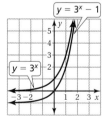
domain: all real numbers
range: all real numbers greater than $-1$
The graph is a translation 1 unit down of the graph of $y = 3^x$.

29.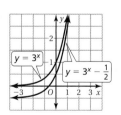
domain: all real numbers
range: all real numbers greater than $-\dfrac{1}{2}$
The graph is a translation $\dfrac{1}{2}$ unit down of the graph of $y = 3^x$.

## Section 6.4
### Exponential Functions (continued)
(pages 289–291)

**31.** An exponential function intersects the $x$-axis when it has the form $y = ab^x - c$, where $a$, $b$, and $c > 0$ or $y = ab^x + c$, where $a < 0$, $b > 0$, and $c > 0$; Example: $y = 2^x - 2$

**33.** $k = -1$

**35.** 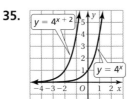 The graph is translated 2 units to the left.

**37.** $y = -8\left(\dfrac{1}{2}\right)^x$

**39.** $y = -3(5)^x$

**41.** You can begin by making a chart to show the amount of grills sold each year.

**43.** 0.23

**45.** 1.5

## Extension 6.4
### Solving Exponential Equations
(pages 292 and 293)

**1.** $x = 4$   **3.** $x = -2$   **5.** $x = 0$   **7.** $x = \dfrac{7}{2}$   **9.** $x = 3$

**11.** *Sample answer:* Because 1 raised to any power is equal to 1, $1^x = 1^y$ is true for all real numbers $x$ and $y$. The solution method fails in this case because it only gives the solution values for which $x = y$.

**13.** $x \approx 0.85$   **15.** $x \approx -10.00$   **17.** no solution

## Section 6.5
### Exponential Growth
(pages 298 and 299)

**1.** when $a > 0$ and $r > 0$

**3.** The terms are multiplied by a factor of about 1.2, so the pattern shows exponential growth.

**5.** $a = 25, r = 20\%; 62.2$   **7.** $a = 1500, r = 7.4\%; 2143.4$   **9.** $a = 6.7, r = 100\%; 214.4$

**11.** $y = 800(1.07)^t$

**13.** $y = 210{,}000(1.125)^t$

**15.** Because the growth rate is 150%, the growth factor should be 2.5, not 1.5.
$b(t) = 10(1 + 1.5)^t = 10(2.5)^t$
$b(8) = 10(2.5)^8 \approx 15{,}258.8$
After 8 hours, there are about 15,259 bacteria in the culture.

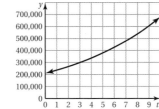

**17.** 468,000

**19.** $A = 6200(1.007)^{12t}$; $7029.46

**21.** $C(t) = 9000(1.003)^{12t} + 480t$; $C(t)$ represents the total amount of money you have saved after $t$ years.

**23.** yes; You can use the properties of exponents to transform the yearly balance function to the monthly balance function. You can use the compound interest formula with an annual interest rate of 2.4% compounded monthly to verify this.

**25.** $\dfrac{4}{9}$

**27.** $\dfrac{81}{625}$

## Section 6.6 — Exponential Decay
*(pages 304 and 305)*

1. When $b > 1$, the function represents exponential growth. When $0 < b < 1$, the function represents exponential decay.
3. The rate of change of the sequence is a constant $-2$, so it is a linear decay pattern.
5. exponential decay function
7. exponential growth function
9. neither
11. exponential decay function
13. exponential growth function
15. 20%
17. 25%
19. A
21. *Sample answer:* Graph an exponential function by hand when the value of $a$ is small enough to easily multiply by $1 - r$. Use a graphing calculator when the value of $a$ is too large or when $1 - r$ has many decimal places which would make multiplying by hand more difficult.

23. The time of day is an important piece of information to consider.
25. $a_n = 3n + 6$; 51
27. $a_n = -4n - 3$; $-63$

## Section 6.7 — Geometric Sequences
*(pages 310 and 311)*

1. *Sample answer:* You add a number to extend arithmetic sequences and you multiply by a number to extend geometric sequences.
3. When the common ratio is negative, $a_n$ will alternate between being positive and negative. These points do not lie on an exponential curve.
5. $-4$
7. 0.1
9. 9
11. 1250, 6250, 31,250
13. $1, -\dfrac{1}{3}, \dfrac{1}{9}$
15. $\dfrac{1}{36}, \dfrac{1}{216}, \dfrac{1}{1296}$

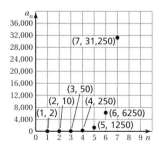
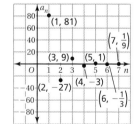
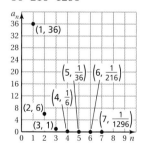

17. The ratio of the term to its previous term is $-\dfrac{1}{2}$, not 2. So, the next three terms are $-\dfrac{1}{2}, \dfrac{1}{4}$, and $-\dfrac{1}{8}$.

## Section 6.7
### Geometric Sequences (continued)
(pages 310 and 311)

19. arithmetic
21. neither
23. geometric
25. $a_n = (-5)^{n-1}$; 15,625
27. $a_n = 5(3)^{n-1}$; 3645
29. a. $a_n = (6)^{n-1}$  b. all positive integers; discrete
31. dependent; the terms are the output of the function; the position is the independent variable.
33. Remember that 1 gallon equals 128 fluid ounces.
35. $9x - 10$
37. C

## Extension 6.7
### Recursively Defined Sequences
(pages 312–315)

1. $0, -8, -16, -24, -32, -40$

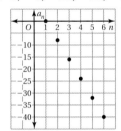

3. $-36, -18, -9, -4.5, -2.25, -1.125$

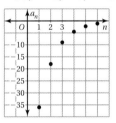

5. $a_1 = 8, a_n = a_{n-1} - 5$
7. $a_1 = 4, a_n = 5a_{n-1}$
9. $a_1 = 2, a_n = a_{n-1} + 1.5$
11. $a_n = 13(-3)^{n-1}$
13. $a_1 = -2.5, a_n = 2a_{n-1}$
15. $a_1 = -3, a_2 = -4,$
    $a_n = a_{n-2} + a_{n-1};$
    $-29, -47, -76$
17. $a_1 = 4, a_2 = 3,$
    $a_n = a_{n-2} - a_{n-1};$
    $7, -11, 18$
19. $a_1 = -2, a_2 = 2.5,$
    $a_n = a_{n-2} \cdot a_{n-1};$
    $-781.25, -48,828.125$

## Section 7.1
### Polynomials
(pages 332 and 333)

1. yes; It is a number.
3. Sample answer: $4x^5 - 2x^3 + 5$
5. [+][+][+][−][−][−][−]
7. 1
9. 9
11. 2
13. 0
15. $3p^2 + 7$; 2; binomial
17. $-4d^3 + 8d - 2$; 3; trinomial
19. $-v^{12} + 4v^{11}$; 12; binomial
21. $7.4z^5$; 5; monomial
23. $-\dfrac{5}{7}r^8 + 2r^5 + \pi r^2$; 8; trinomial

**25.** It is the product of two numbers and a variable with a whole number exponent; 3

**27.** *not* a polynomial

**29.** polynomial; 5; trinomial

**31.** $-16t^2 - 45t + 200$; 139 ft

**33.** This problem can be organized with a table.

**35.** $3x - 2y$      **37.** D

## Section 7.2 — Adding and Subtracting Polynomials
*(pages 338 and 339)*

**1.** Align like terms vertically and add; Group like terms and simplify.

**3.** $2x^2 + x + 1$   **5.** $3y + 10$   **7.** $n^2 - 8n + 5$   **9.** $4a^3 + 3a - 3$

**11.** Like terms are not aligned.

$$\begin{array}{r} -5x^2 \phantom{+2x} + 1 \\ +\phantom{-5x^2} 2x - 8 \\ \hline -5x^2 + 2x - 7 \end{array}$$

**13.** $-7k + 14$   **15.** $4r^2 - 2r - 17$

**17.** $\dfrac{1}{12}q^2 + 1$   **19.** $3b + 2$

**21.** $-2x^2 + 3xy + 8y^2$      **23.** $-3a^2 + 2ab + b^2$

**25.** The distance between your ball and your friend's ball is the difference between the two heights.

**27.** $2y + 7$      **29.** B

## Section 7.3 — Multiplying Polynomials
*(pages 345–347)*

**1.** *Sample answer:* Use the Distributive Property or the FOIL Method.

**3.** $x^2 - 4$   **5.** $x^2 + 4x + 3$   **7.** $z^2 - 2z - 15$   **9.** $g^2 - 9g + 14$

**11.** $3m^2 + 28m + 9$   **13.** $12 - 16s + 5s^2$   **15.** $(18x^2 + 12x + 2)$ in.$^2$   **17.** $y^2 + 5y - 50$

**19.** $8j^2 - 26j + 21$   **21.** $15d^2 - 71d + 84$   **23.** $w^2 + 15w + 54$   **25.** $x^2 + 4x - 32$

**27.** $z^2 - 14z + 45$   **29.** $10v^2 + 14v - 12$

**31.** The second term of the second binomial is $-7$, not 7.
$$(r + 6)(r - 7) = r(r) + r(-7) + 6(r) + 6(-7)$$
$$= r^2 - 7r + 6r - 42$$
$$= r^2 - r - 42$$

**33. a.** $x^2 - 20x - 300$      **35.** $p^2 - 2p - 3$      **37.** $2n^3 + 6n^2 + n + 3$

  **b.** 6000 yd$^2$

  **c.** 90 minutes

**39.** $2r^2 - 5rs - 3s^2$      **41.** $f^3 + 3f^2 - 12f + 8$      **43.** $t^3 - 8t^2 + 16t - 3$

## Section 7.3
### Multiplying Polynomials (continued)
(pages 345–347)

**45.** $18e^3 - 27e^2 + 37e + 7$  
**47. a.** $x^2 + 41x + 40$  **b.** $126$  
**49. a.** $m$ is negative and $n$ is positive; or $m$ is positive and $n$ is negative  
**b.** $m$ is positive and $n$ is positive; or $m$ is negative and $n$ is negative  
**51.** $4x^3 + 9x^2 - x - 6$  
**53.** $z^2 - \frac{5}{7}z$; 2; binomial  
**55.** B

## Section 7.4
### Special Products of Polynomials
(pages 352 and 353)

**1.** Sample answer: $x + 7, x - 7$   **3.** $x^2 - 36$   **5.** $4z^2 + 8z + 4$  
**7.** $g^2 - 25$   **9.** $b^2 - 144$   **11.** $9x^2 - 16$   **13.** $81 - c^2$  
**15.** $x - 4$ and $x + 4$; Rewrite the expression as $x^2 - 4^2$ and apply the sum and difference pattern.  
**17.** $y^2 + 16y + 64$   **19.** $d^2 - 20d + 100$   **21.** $25p^2 + 20p + 4$   **23.** $144 - 24x + x^2$  
**25.** The product should have a middle term.  
$(k + 4)^2 = k^2 + 2(k)(4) + 4^2$  
$= k^2 + 8k + 16$  
**27. a.** $x^2 + 100x + 2500$  
**b.** 4225 ft$^2$; 1725 ft$^2$  
**29.** $4x^2 + 28x + 49$   **31.** $x^2 - y^2$  
**33. a.** 75%  
**b.** $0.25N^2 + 0.5Na + 0.25a^2$  
**35.** $(x + 1)^3 = x^3 + 3x^2 + 3x + 1$  
$(x + 2)^3 = x^3 + 6x^2 + 12x + 8$  
$(a + b)^3 = a^3 + 3a^2b + 3ab^2 + b^3$  
**37.** $y^2 - 4y - 21$  
**39.** D

## Section 7.5
### Solving Polynomial Equations in Factored Form
(pages 360 and 361)

**1.** yes; When $x = 3$, the factor $(x - 3)$ equals 0.  
**3.** $x^2 + 4x$ does not belong because it is not in factored form.  
**5.** $t = 0, t = 5$   **7.** $q = -3, q = 2$   **9.** $m = -4$   **11.** $g = 3, g = 7$  
**13.** $z = 3$   **15.** $d = 6, d = -6$   **17.** $x = -8, x = 8$   **19.** $x = -22, x = 15$  
**21.** $z = 0, z = -2, z = 1$   **23.** $r = 4, r = -4, r = -8$  
**25.** To find the width of the arch at ground level, find the distance between the $x$-intercepts.  
**27.** 21   **29.** 15

## Section 7.6 Factoring Polynomials Using the GCF
*(pages 366 and 367)*

**1.** $6y$     **3.** $4(x + 2)$     **5.** $x(x - 4)$     **7.** $4m(2m + 1)$

**9.** $10x^2(2x + 3)$     **11.** $5(t^2 + 4t + 10)$     **13.** $100(1 + rt)$     **15.** $x = -\dfrac{3}{2}$

**17.** $m = 0, m = -2$     **19.** $r = 0, r = -7$     **21.** $k = 0, k = -\dfrac{13}{2}$

**23.** 3 should be factored from 15.
$$3x^2 = 15x$$
$$3x^2 - 15x = 0$$
$$3x(x - 5) = 0$$
$$3x = 0 \text{ or } x - 5 = 0$$
$$x = 0 \text{ or } \quad x = 5$$

**25.** $b = 0, b = 5$     **27.** $s = 0, s = -6$

**29.** Sample answer: $6x^2 + 3\pi x$

**31.** Begin by finding the time $t$ when the height $y = 0$.

**33.** $y^2 + 10y + 24$

**35.** $4k^2 - 4k - 3$

## Section 7.7 Factoring $x^2 + bx + c$
*(pages 373–375)*

**1.** Because $c$ is negative, $p$ and $q$ have different signs. Because $b$ is positive, the factor with the greater absolute value is positive.

**3.** $(x + 1)(x + 7)$     **5.** $(n + 2)(n + 6)$     **7.** $(h + 2)(h + 9)$

**9.** The factors of 48 (4 and 12) do not add up to 14.
$$t^2 + 14t + 48 = (t + 6)(t + 8)$$

**11.** $(x - 4)(x - 5)$     **13.** $(k - 4)(k - 6)$     **15.** $(j - 6)(j - 7)$     **17.** $x = -4, x = -7$

**19.** year 1 and year 5     **21.** $(x - 1)(x + 4)$     **23.** $(n - 2)(n + 6)$     **25.** $(h - 3)(h + 9)$

**27.** $(m - 7)(m + 1)$     **29.** $(t - 8)(t + 2)$     **31.** $x = 2, x = -7$

**33.** All of the terms were not collected on one side of the equation before factoring.
$$x^2 - 2x - 15 = 20$$
$$x^2 - 2x - 35 = 0$$
$$(x - 7)(x + 5) = 0$$
$$x - 7 = 0 \text{ or } x + 5 = 0$$
$$x = 7 \text{ or } \quad x = -5$$

**35.** length: 11 ft, width: 4 ft

**37.** length: 20 ft, width: 6 ft

**39. a.** 6     **b.** length: 13 in., width: 8 in.

**41.** Remember, area of a rectangle is length times width.

**43.** $2(y - 9)$

**45.** $4z^2(2z + 7)$

## Section 7.8

### Factoring $ax^2 + bx + c$
(pages 380 and 381)

1. First, factor out the GCF of the terms, if possible. Then, consider the possible factors of $a$ and $c$, list the possible products, and multiply.
3. $(2x - 1)(x - 1)$
5. $(4x + 3)(x + 2)$
7. $8(v - 2)(v + 3)$
9. $6(y - 3)(y - 1)$
11. $7(d - 5)(d - 4)$
13. $(3h + 2)(h + 3)$
15. $(4m + 1)(2m + 7)$
17. $(2n + 1)(n - 3)$
19. $2(g - 2)(4g + 3)$
21. $(7d - 2)(2d + 1)$
23. $(4x + 1)$ ft by $(2x + 5)$ ft
25. $k = -2$ or $k = \dfrac{9}{2}$
27. $-(3w - 4)(w + 2)$
29. $-5(4n - 1)(2n - 3)$
31. Remember that "$+ tx$" does not mean that $t$ is positive.

33. 4 ft
35. a. $24 or $36
   b. $30; *Sample answer:* By using a table, you can see that the maximum daily revenue occurs when $x = 5$. When $x = 5$, the price of the bobblehead is $30.
37. $(2x - y)(3x + 4y)$
39. $4x^2 - 49$
41. $9b^2 - 24b + 16$

## Section 7.9

### Factoring Special Products
(pages 386 and 387)

1. Use the difference of two squares pattern to factor $x^2 - 16$. Find the product $(x + 4)(x - 4)$.
3. $k^2 + 25$ does not belong because it cannot be factored.
5. $(3 + r)(3 - r)$
7. $(9d + 8)(9d - 8)$
9. $(h + 6)^2$
11. $(w - 7)^2$
13. The wrong sign is used.
   $n^2 - 16n + 64 = (n - 8)^2$
15. $s = -10$
17. $x = -\dfrac{7}{2}, x = \dfrac{7}{2}$
19. $y = 6$
21. $d + 4$
23. $2m(m + 5)(m - 5)$
25. $5f(f - 2)^2$
27. $x = -2, x = 2$; Solve by factoring or using square roots.
29. $(2y + 1)^2$
31. $9(m + 2)^2$
33. $(w + 4)(w - 3)$
35. $(d - 10)(d + 6)$

## Extension 7.9

### Factoring Polynomials Completely
(pages 388 and 389)

1. $(n + 2)(n^2 + 5)$
3. $(y + 4)(2y^2 + 3)$
5. $(8v - 5)(v^2 + 6)$
7. $(x + 3)(x + y)$
9. $(4y + 3)(x + 5)$
11. $5z^2(z + 1)(z - 1)$
13. cannot be factored
15. $3n^2(n + 4)(n - 4)$
17. $x = -6, x = 0, x = 4$

# Section 8.1

## Graphing $y = ax^2$
(pages 407–409)

1. The vertex is the lowest or highest point on a parabola. The vertical line that divides the parabola into two symmetric parts is the axis of symmetry.

3. The graph of $y = \frac{1}{6}x^2$ is wider because when $0 < a < 1$, the graph of $y = ax^2$ is wider than the graph of $y = x^2$. When $a > 1$, the graph of $y = ax^2$ is narrower than the graph of $y = x^2$.

5. Both graphs open down and have a U-shape. The graph of $y = -0.4x^2$ is wider than the graph of $y = -4x^2$.

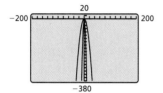

7. Both graphs open down and have a U-shape. The graph of $y = -0.004x^2$ is much wider than the graph of $y = -4x^2$.

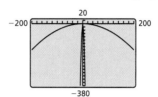

9. vertex: $(-2, 4)$; axis of symmetry: $x = -2$; domain: all real numbers; range: $y \geq 4$; When $x < -2$, $y$ increases as $x$ decreases. When $x > -2$, $y$ increases as $x$ increases.

11. Both graphs open up and have the same vertex, $(0, 0)$, and the same axis of symmetry, $x = 0$. The graph of $y = 6x^2$ is narrower than the graph of $y = x^2$.

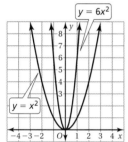

13. Both graphs open up and have the same vertex, $(0, 0)$, and the same axis of symmetry, $x = 0$. The graph of $y = \frac{1}{4}x^2$ is wider than the graph of $y = x^2$.

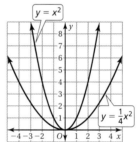

15. Both graphs open up and have the same vertex, $(0, 0)$, and the same axis of symmetry, $x = 0$. The graph of $y = \frac{5}{2}x^2$ is narrower than the graph of $y = x^2$.

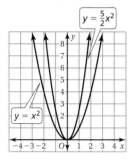

17. **a.** domain: $0 \leq t \leq 2$; range: $0 \leq y \leq 64$

**b.**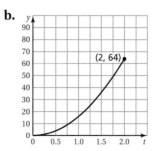

**c.** 2 seconds

Selected Answers **A39**

## Section 8.1  Graphing $y = ax^2$ (continued)
(pages 407–409)

**19.** Both graphs have the same vertex, (0, 0), and the same axis of symmetry, $x = 0$, but the graph of $y = -7x^2$ opens down. The graph of $y = -7x^2$ is narrower than the graph of $y = x^2$.

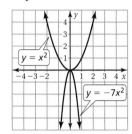

**21.** Both graphs have the same vertex, (0, 0), and the same axis of symmetry, $x = 0$, but the graph of $y = -\frac{5}{8}x^2$ opens down. The graph of $y = -\frac{5}{8}x^2$ is wider than the graph of $y = x^2$.

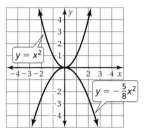

**23.** Both graphs have the same vertex, (0, 0), and the same axis of symmetry, $x = 0$, but the graph of $y = -\frac{9}{4}x^2$ opens down. The graph of $y = -\frac{9}{4}x^2$ is narrower than the graph of $y = x^2$.

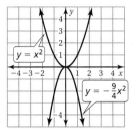

**25.** $a > 1$

**27.** $a < -1$

**29. a.** When $a > 0$, the domain is all real numbers and the range is $y \geq 0$.

**b.** When $a < 0$, the domain is all real numbers and the range is $y \leq 0$.

**31.** always; When $a > 1$ or $a < -1$, the graph of $y = ax^2$ is narrower than the graph of $y = x^2$.

**33.** never; As $|a|$ increases, the graph of $y = ax^2$ gets narrower.

**35.** When $a > 0$, the graph of $y = ax^2$ is increasing from left to right when $x > 0$. When $a < 0$, the graph of $y = ax^2$ is increasing from left to right when $x < 0$.

**37.** $\frac{3}{4}$

**39. a.** The domain includes the integers 0 through 10. It is discrete because you can only have a whole number of workers. It is positive because you cannot have a negative number of workers.

**b.**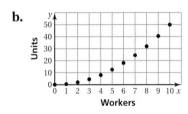

**c.** The company is better off running one assembly line at full capacity.

**39. c.** *(continued)*

*Sample answer:*

**(1) From a production standpoint:** Ten workers on one assembly line (full capacity) will produce 50 units per hour. Ten workers split on two assembly lines (half capacity) will produce a total of 25 units per hour. One assembly line at full capacity produces twice as many units as two assembly lines at half capacity.

**(2) From a financial standpoint:** The number of units produced is increasing at an increasing rate for each additional worker on an assembly line. This means production (i.e. revenue) is increasing faster than the labor cost when more employees work on the same assembly line. The company is getting more "bang for the buck."

**41.** 39  **43.** 14  **45.** B

## Section 8.2 — Focus of a Parabola
*(pages 414 and 415)*

1. The focus is a coordinate point which lies on the axis of symmetry. If the axis of symmetry is $x = k$, then the $x$-coordinate of the focus is $k$.

3. Sample answer: $y = -4x^2$

5. parabolic

7. focus: $\left(0, \dfrac{1}{4}\right)$

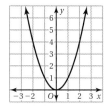

9. focus: $\left(0, -\dfrac{1}{48}\right)$

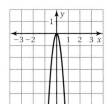

11. focus: $\left(0, \dfrac{1}{2}\right)$

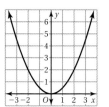

13. The student did not multiply 2 by 4. The focus is $\left(0, \dfrac{1}{8}\right)$.

15. $y = -\dfrac{1}{8}x^2$  17. $y = -x^2$  19. $y = \dfrac{1}{2}x^2$  21. B  23. C

25. always; The vertex of the graph of a function of the form $y = ax^2$ is always at the origin. So if the parabola opens down, then the focus will always lie below the $x$-axis.

27. As the distance between the vertex and focus increases, the graph of $y = ax^2$ gets wider.

29. 1.5 inches  31. 16  33. 1

## Section 8.3 — Graphing $y = ax^2 + c$
*(pages 420 and 421)*

1. The vertex is on the $y$-axis. The $y$-axis is the axis of symmetry. The value of $c$ is the $y$-coordinate of the vertex.

3. It is a translation $c$ units up or down of the graph of $y = ax^2$.

5. C

## Section 8.3 Graphing $y = ax^2 + c$ (continued)
(pages 420 and 421)

**7.** Both graphs open up and have the same axis of symmetry, $x = 0$. The graph of $y = x^2 + 4$ is a translation 4 units up of the graph of $y = x^2$.

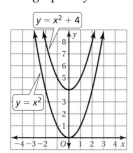

**9.** Both graphs have the same axis of symmetry, $x = 0$. The graph of $y = -x^2 + 5$ opens down. The vertex of the graph of $y = -x^2 + 5$ is a translation 5 units up of the vertex of the graph of $y = x^2$.

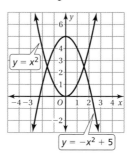

**11.** Both graphs open up and have the same axis of symmetry, $x = 0$. The graph of $y = 2x^2 - 4$ is narrower than the graph of $y = x^2$. The vertex of the graph of $y = 2x^2 - 4$ is a translation 4 units down of the vertex of the graph of $y = x^2$.

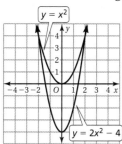

**13.** Both graphs open up and have the same axis of symmetry, $x = 0$. The graph of $y = 3x^2 + 4$ is narrower than the graph of $y = x^2$. The vertex of the graph of $y = 3x^2 + 4$ is a translation 4 units up of the vertex of the graph of $y = x^2$.

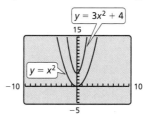

**15.** Both graphs have the same axis of symmetry, $x = 0$. The graph of $y = -\frac{1}{4}x^2 - \frac{1}{2}$ opens down, and is wider than the graph of $y = x^2$. The vertex of the graph of $y = -\frac{1}{4}x^2 - \frac{1}{2}$ is a translation 0.5 unit down of the vertex of the graph of $y = x^2$.

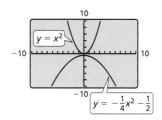

**17.** translate 3 units down

**19.** The student incorrectly states that the graph of $y = x^2 + 3$ is a translation 3 units down of the graph of $y = x^2$. The graph is actually a translation 3 units up.

**21.** after 1 second    **23.** $-1, 1$    **25.** $-3, 3$

**27.** *Sample answer:*

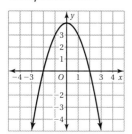

**29.** *Sample answer:*

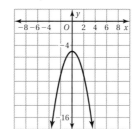

**31.** yes; $y = x^2 - 1$ has a vertex at $(0, -1)$ and a focus at $\left(0, \frac{1}{4}\right)$

**33. a.** When $k < 0$, the graph of $f(x) + k$ will translate the graph of $f(x)$ down. When $k > 0$, the graph of $f(x) + k$ will translate the graph of $f(x)$ up.

**b.** When $k > 1$, the graph of $k \cdot f(x)$ is narrower than the graph of $f(x)$. When $0 < k < 1$, the graph of $k \cdot f(x)$ is wider than the graph of $f(x)$. When $-1 < k < 0$, the graph of $k \cdot f(x)$ opens down and is wider than the graph of $f(x)$. When $k = -1$, the graph of $k \cdot f(x)$ opens down. When $k < -1$, the graph of $k \cdot f(x)$ opens down and is narrower than the graph of $f(x)$.

**35.** $(2x - 1)(x - 3)$    **37.** B

## Section 8.4  Graphing $y = ax^2 + bx + c$
(pages 429–431)

**1.** For a quadratic function $y = ax^2 + bx + c$ where $a \neq 0$, if $a > 0$, then the function has a minimum value and if $a < 0$, then the function has a maximum value.

**3.** vertex: $(2, -1)$
axis of symmetry: $x = 2$
$y$-intercept: 1

**5.** vertex: $(-2, -3)$
axis of symmetry: $x = -2$
$y$-intercept: 4

**7. a.** $x = -2$
**b.** $(-2, -12)$

**9. a.** $x = 2$
**b.** $(2, 4)$

**11. a.** $x = 4$
**b.** $(4, -6)$

**13.** domain: all real numbers
range: $y \geq -4$

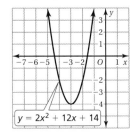

**15.** domain: all real numbers
range: $y \leq -1$

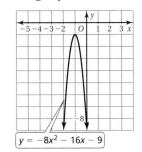

**17.** domain: all real numbers
range: $y \geq -5$

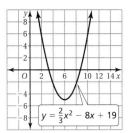

**19. a.** 4 seconds after it is launched at a height of 256 feet

**b.** The domain is $0 \leq t \leq 8$. The range is $0 \leq h \leq 256$.

**21.** minimum; $-12$    **23.** maximum; $-1$    **25.** maximum; 52

**27.** $(0, 8)$; Use the axis of symmetry to find the reflection of $(6, 8)$.

**29. a.** 0.5 second after it jumps

**b.** yes; 0.5 ft

**c.** 1 sec

**31.** vertex $\approx (-1.4, -4)$

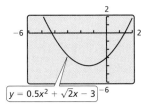

**33.** opens down; *Sample answer:* The $y$-values are increasing at a decreasing rate. This means the points are on the left side of a parabola that faces down.

## Section 8.4 Graphing $y = ax^2 + bx + c$ (continued)
(pages 429–431)

**35.** Use the wording in the exercise as a clue for defining the independent variable.

**37.** $\dfrac{k^2}{8}$ ft$^2$

**39.**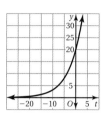

**41.** C

## Extension 8.4 Graphing $y = a(x - h)^2 + k$
(pages 432 and 433)

**1.**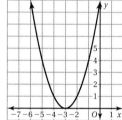

The graph of $y = (x + 3)^2$ is a translation 3 units to the left of the graph of $y = x^2$.

**3.**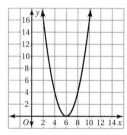

The graph of $y = (x - 6)^2$ is a translation 6 units to the right of the graph of $y = x^2$.

**5.**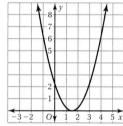

The graph of $y = (x - 1.5)^2$ is a translation 1.5 units to the right of the graph of $y = x^2$.

**7.** *Sample answer:* $x^2 + 6x + 9$ can be factored as $(x + 3)^2$. In this form, you can see that the graph of $y = (x + 3)^2$ is a translation 3 units to the left of the graph of $y = x^2$.

**9.**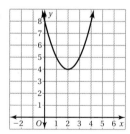

The graph of $y = (x - 2)^2 + 4$ is a translation 2 units to the right and 4 units up of the graph of $y = x^2$.

**11.**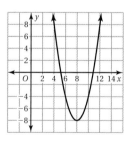

The graph of $y = (x - 8)^2 - 8$ is a translation 8 units to the right and 8 units down of the graph of $y = x^2$.

**13.**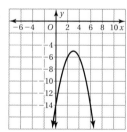

The graph of $y = -(x - 3)^2 - 5$ opens down. The vertex of the graph of $y = -(x - 3)^2 - 5$ is a translation 3 units to the right and 5 units down of the vertex of the graph of $y = x^2$.

**15.** $g(x)$ is a vertical translation 7 units down of $f(x)$.

**17.** $g(x)$ is narrower than $f(x)$.

**19.** $y = (x - 2)^2 - 5$; $y = x^2 - 4x - 1$

## Section 8.5 — Comparing Linear, Exponential, and Quadratic Functions (pages 439–441)

1. linear function: $y$-values have a common difference; exponential function: $y$-values have a common ratio; quadratic function: second differences are constant

3. $x > 1$ and $x < 0$

5. $x > -\dfrac{1}{2}$

7. C

9. 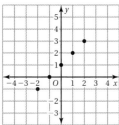 linear

11. quadratic

13. linear

15. exponential

17. quadratic

19. quadratic; $y = 2x^2$

21. linear; $y = -3x - 2$

23. quadratic; $y = -3x^2$

25. The student incorrectly substitutes the $x$-coordinate in for $y$ and the $y$-coordinate in for $x$. The equation should be $y = 4x^2$.

27. in general, no; It depends on the number of points and where they are located on the graph.

29. a.

| Round, x | 1 | 2 | 3 | 4 | 5 | 6 |
|---|---|---|---|---|---|---|
| Teams Remaining, y | 32 | 16 | 8 | 4 | 2 | 1 |

    b. exponential

    c. $y = 64(0.5)^x$

    d. sixth round

31. $-1, 1$

33. $-2, 2$

## Section 9.1 — Solving Quadratic Equations by Graphing (pages 459–461)

1. It is a nonlinear equation that can be written in the form $ax^2 + bx + c = 0$, where $a \neq 0$.

3. Use the graph to find the $x$-intercepts.

5. $x = 4, x = 6$

7. $x = -6$

9. $x = 3$

11. no real solutions

13. $x = -2$

15. $x = 7$

17. a. The $x$-intercepts give the horizontal positions of the ball where it is struck and where it lands.

    b. 5 yards

19. $x^2 - 6x + 8 = 0$; $x = 2, x = 4$

21. $x^2 + x + 3 = 0$; no real solutions

23. $x^2 - 5x + 6 = 0$; $x = 2, x = 3$

25. $x = -5, x = 2$

27. no real solutions

29. $x = -1$

31. *Sample answer:* Method 2; You do not have to rewrite the equation.

# Section 9.1
## Solving Quadratic Equations by Graphing
*(continued)* *(pages 459–461)*

33. **a.** about 1.7 seconds  **b.** 1.5 seconds
35. $x \approx -5.8, x \approx -0.2$
37. $x \approx -5.7, x \approx 0.7$
39. $x \approx 0.6, x \approx 3.4$
41. **a.** $h = -16t^2 + 20t + 8$
    **b.** about 1.6 seconds
43. sometimes; There are 2 $x$-intercepts when $c$ is positive.
45. never; A quadratic equation can never have more than 2 solutions.
47. 81
49. $25\sqrt{2}$

# Section 9.2
## Solving Quadratic Equations Using Square Roots *(pages 466 and 467)*

1. 2; 1; 0
3. 2; $x \approx 3.317, x \approx -3.317$
5. 2; $x \approx 1.225, x \approx -1.225$
7. 0; no real solutions
9. 2; $x = \sqrt{21}, x = -\sqrt{21}$
11. 2; $x = 13, x = -13$
13. no real solutions
15. $x = \sqrt{61}, x = -\sqrt{61}$
17. $x = 0$
19. $x = 0$
21. When taking the square root of each side, the student forgot the negative root; $x = 6, x = -6$
23. $x = -3$
25. $x = -4, x = 5$
27. $x = -\dfrac{7}{3}, x = \dfrac{1}{3}$
29. 8 in. by 8 in.
31. 12 ft
33. You can solve this problem by recalling a property of similar figures.

35. **a.** two solutions: $a$ is positive and $c$ is negative, or $c$ is positive and $a$ is negative.
    **b.** one solution: $c = 0$
    **c.** no solutions: $a$ and $c$ are both positive or both negative.
37. $x^2 + 10x + 25$
39. $4y^2 - 12y + 9$

# Section 9.3
## Solving Quadratic Equations by Completing the Square *(pages 472 and 473)*

1. Add $\left(\dfrac{b}{2}\right)^2$.

3. $x^2 - 4x + 4$

5. $x^2 - 6x + 9$

7. 4
9. 49
11. 81
13. $x^2 + 16x + 64 = (x + 8)^2$
15. $x^2 - 40x + 400 = (x - 20)^2$
17. $x^2 + 5x + \dfrac{25}{4} = \left(x + \dfrac{5}{2}\right)^2$
19. $x = -2, x = 8$
21. no real solutions
23. $x = \dfrac{1}{2}, x = 1$
25. **a.** $216 = x(x + 6)$  **b.** 12 ft by 18 ft
27. Divide each side by 3.

**29. a.** after about 4.4 seconds
  **b.** 96 ft; The vertex is (2, 96). The maximum is the y-coordinate of the vertex.
**31.** $x(x + 1) = 42$; 6, 7     **33.** $2\sqrt{3}$     **35.** $2\sqrt{5}$

## Section 9.4 — Solving Quadratic Equations Using the Quadratic Formula (pages 481–483)

**1.** $x = \dfrac{-b \pm \sqrt{b^2 - 4ac}}{2a}$

**3.** $x^2 - 7x = 0$; $a = 1, b = -7, c = 0$

**5.** $-2x^2 - 5x + 1 = 0$; $a = -2, b = -5, c = 1$

**7.** $x^2 - 6x + 4 = 0$; $a = 1, b = -6, c = 4$

**9.** $x = 6$

**11.** $x = -1, x = 11$

**13.** no real solutions

**15.** $x = \dfrac{2}{3}, x = \dfrac{3}{2}$

**17.** $x = \dfrac{1}{4}$

**19.** $x \approx -4.2, x \approx 2.2$

**21.** Used $-7$ for $-b$ instead of $-(-7) = 7$; $x = -\dfrac{2}{3}, x = 3$

**23.** 6 ft     **25.** A     **27.** 0     **29.** 1     **31.** 2

**33. a.** yes; When the solutions $m$ and $n$ are integers, the standard form can be factored as $(x - m)(x - n) = 0$.
  **b.** yes; When the solutions $\dfrac{m}{n}$ and $\dfrac{h}{k}$ are fractions, the standard form can be factored as $(nx - m)(kx - h) = 0$.
  **c.** Any quadratic equation with rational solutions can be solved by factoring.

**35. a.** 1994 and 2006
  **b.** no; The model predicts negative numbers of fish caught after 2015.

**37.** 2     **39.** 0

**41.** Sample answer: **a.** $c = 8$   **b.** $c = 2$

**43.** 2; When $a$ and $c$ have different signs, $b^2 - 4ac$ is positive.

**45.** Begin by writing an equation that represents the amount of fencing needed for both pastures and one that represents the area of each pasture.

**47.** $(1, -1)$     **49.** no solution

## Extension 9.4 — Choosing a Solution Method (pages 484 and 485)

**1.** $x = -7 + \sqrt{41}, x = -7 - \sqrt{41}$
**3.** $x = 6, x = -6$
**5.** $x = 1, x = -1$; square roots, because no $x$-term
**7.** $x = -5, x = 8$; factors easily
**9.** no real solutions; completing the square, because $a = 1$ and $b$ is even
**11.** $x = -4, x = 3$; factors easily

# Section 9.5 Solving Systems of Linear and Quadratic Equations (pages 490 and 491)

1. A solution of a system of linear and quadratic equations is an ordered pair that is a solution of each equation in the system.

3. B; (0, 1), (3, 4)   5. A; (0, −1)   7. (−1, −2)   9. (−3, −2), (1, 6)

11. $\left(-\dfrac{3}{2}, -4\right)$, (2, 10)   13. no real solutions   15. (0, 0), (−5, 10)   17. (2, −6), $\left(\dfrac{5}{2}, -\dfrac{23}{4}\right)$

19. (−1, −6), (−2, −9)   21. (2, 2)

23. *Sample answer:* The viewing window of the graphing calculator is too small. By increasing the window you can see that (5, 14) is also a solution.

25. a. 2   b. 0   27. $x = -1, x = 1$   29. no real solutions

# Section 10.1 Graphing Square Root Functions (pages 506 and 507)

1. no; It is a linear function. A square root function contains a square root with the independent variable in the radicand.

3. B   5. C

7. domain: $x \geq 0$   9. $x \geq 0$

11. $x \geq 2$

13. $x \geq -4$

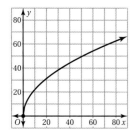

15. a. 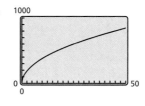   b. about 6 pounds per square inch

17. domain: $x \geq 0$; range: $y \geq 4$; The graph of $y = \sqrt{x} + 4$ is a translation 4 units up of the graph of $y = \sqrt{x}$.

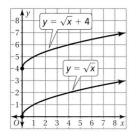

19. domain: $x \geq -2$; range: $y \geq -2$; The graph of $y = \sqrt{x + 2} - 2$ is a translation 2 units to the left and 2 units down of the graph of $y = \sqrt{x}$.

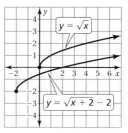

21. domain: $x \geq 1$; range: $y \leq 3$;
The graph of $y = -\sqrt{x-1} + 3$ is a reflection in the x-axis of the graph of $y = \sqrt{x}$, and then a translation 1 unit to the right, and a translation 3 units up.

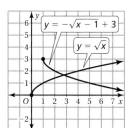

23. a. Sample answer: $y = \sqrt{x} + 1$
    b. Sample answer: $y = -\sqrt{x}$

25. a. domain: $A > 0$

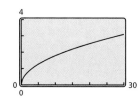

   b. about 28 square inches

27. Choose perfect cubes for $x$ to make evaluating $g(x)$ easier.

29. $x = -3$    31. B

## Extension 10.1 Rationalizing the Denominator
(pages 508 and 509)

1. $\dfrac{\sqrt{10}}{10}$    3. $\dfrac{3\sqrt{2}}{2}$    5. $\dfrac{\sqrt{10}}{6}$    7. $\sqrt{5}$

9. $\dfrac{4\sqrt{15} - 5}{15}$    11. $-3 + 3\sqrt{3}$    13. $-2\sqrt{2} + 2\sqrt{7}$

## Section 10.2 Solving Square Root Equations
(pages 515–517)

1. no; The radicand does not contain a variable.

3. 4;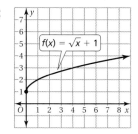

5. 24;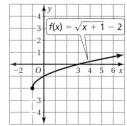

7. $x = 144$    9. $x = 121$    11. $x = 225$    13. $x = 19$

15. $x = 45$    17. $x = 7$    19. 96 in.²

21. Add $\sqrt{3m}$ to each side. Square each side. Subtract $m$ from each side. Divide each side by 2. Check your solution.

23. $x = 3$    25. $x = 7$    27. $x = \dfrac{2}{3}$    29. $x = 8$

31. $x = 1, x = 4$    33. $x = 5$    35. $x = 2$

## Section 10.2 Solving Radical Equations (continued)
*(pages 515–517)*

**37.** $x = -6$ does not check in the original equation, so it is extraneous; $x = 2$ is the only solution.

**39.** true  **41.** false  **43.** 14.4 ohms  **45.** about 29.2 ft

**47.** $\sqrt{x+2}$ has one term and $\sqrt{x} + 2$ has two terms. When you square $\sqrt{x+2}$, the result is the radicand, $x + 2$. When you square $\sqrt{x} + 2$, the result is an expression with 3 terms that includes a radical, $x + 4\sqrt{x} + 4$.

**49.** $x = -\dfrac{7}{16}$  **51.** 92°  **53.** 90°

## Section 10.3 The Pythagorean Theorem
*(pages 524 and 525)*

**1.** Use the Pythagorean Theorem to find the missing side length.

**3.** 12 yd  **5.** $\sqrt{799}$ ft  **7.** 6 in.

**9.** 20 should have been substituted for $c$, not $b$. The missing length is $2\sqrt{91}$ inches.

**11.** about 60 in.  **13.** The direct distance from the tee to the hole is 145 yards.

**15.** **a.** $28^2 + 21^2 = (5x)^2$

  **b.** Factoring: Subtract 1225 from each side. Factor out the GCF and factor the difference of squares. Set the factors equal to zero and solve. Divide each side by 25. Then take the square root of each side. *Sample answer:* taking square roots; Taking square roots is easier than factoring.

  **c.** 7

**17.** Both graphs have the same axis of symmetry, $x = 0$. The graph of $y = -2x^2 + 4$ opens down and is narrower than the graph of $y = x^2$. The vertex of the graph $y = -2x^2 + 4$ is a translation 4 units up of the vertex of the graph of $y = x^2$.

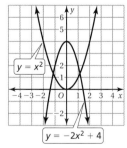

**19.** Both graphs open up and have the same axis of symmetry, $x = 0$. The graph of $y = 3x^2 + 8$ is narrower than the graph of $y = x^2$. The vertex of the graph of $y = 3x^2 + 8$ is a translation 8 units up of the vertex of the graph of $y = x^2$.

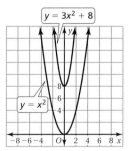

## Section 10.4 Using the Pythagorean Theorem
*(pages 530 and 531)*

**1.** the Pythagorean Theorem and the distance formula

**3.** If $a^2$ is even, then $a$ is an even number; true

**5.** yes  **7.** yes  **9.** no  **11.** $2\sqrt{10}$  **13.** $3\sqrt{10}$  **15.** $\sqrt{53}$

**17.** The squared quantities under the radical should be added, not subtracted; $2\sqrt{10}$

**19.** yes

**21.** yes

**23.** You can use what you learned about slopes of perpendicular lines to solve this problem.

**25.** Plane B; Plane A is about 8.35 kilometers away and Plane B is about 8.27 kilometers away.

**27.** $\frac{3}{2}$, 1

**29.** no real solutions

## Section 11.1 Direct and Inverse Variation
(pages 547–549)

**1.** In direct variation, $y$ is the product of $x$ and the constant $k$. In inverse variation, $y$ is the quotient of the constant $k$ and $x$.

**3.** direct variation; The ratio $\frac{y}{x}$ is constant.

**5.** direct variation; The equation can be written as $y = kx$.

**7.** $y = 3x$

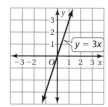

**9.** $y = \frac{1}{6}x$

**11.** $y = \frac{45}{x}$

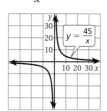

**13.** inverse variation; The product $hr$ is constant.

**15.** Both the domain and range are all real numbers except for 0.

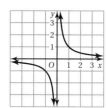

**17.** Both the domain and range are all real numbers except for 0.

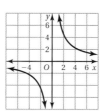

**19.** $y = \frac{12}{x}$; $x = 12$

**21.** $y = \frac{10}{x}$; $y = 2.5$

**23.** inverse variation;

| Number of Hats, x | 5 | 10 |
|---|---|---|
| Cost per Hat, y | 10 | 5 |

The product $xy$ is constant.

**25.** inverse variation; Suppose the swim team earns $1000.

| Number of Members, x | 1 | 5 | 10 | 20 |
|---|---|---|---|---|
| Earnings per Member, y | 1000 | 200 | 100 | 50 |

The product $xy$ is constant.

Selected Answers **A51**

## Section 11.1 Direct and Inverse Variation (continued)
(pages 547–549)

**27. a.** $v = 18t$

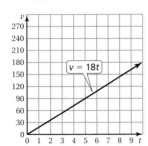

**b.** 90 hours

**29. a.** $t = \dfrac{5000}{p}$

**b.** 200 hours

**31.** odd

**33.** even

**35.** even

**37. a.** symmetric with respect to the $y$-axis

**b.** symmetric with respect to the origin (reflection in the $y$-axis followed by a reflection in the $x$-axis)

**39.** The graph of $y = 3x^2$ is narrower than the graph of $y = x^2$.

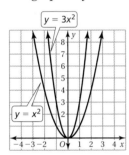

**41.** The graph of $y = x^2 - 1$ is a translation 1 unit down of the graph of $y = x^2$.

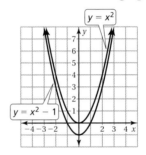

## Section 11.2 Graphing Rational Functions
(pages 555–557)

**1.** no; The denominator is not a polynomial.

**3.** The graph of a rational function approaches but never intersects the asymptotes.

**5.** The graph is two smooth curves. The domain appears to be all real numbers except 10. The range appears to be all real numbers except 10. As $x$ gets closer to 10, the graph approaches the vertical line $x = 10$. As $x$ increases and decreases, the graph approaches the horizontal line $y = 10$.

**7.** $x = -4$

**9.** $x = 9$

**11.** $x = -\dfrac{1}{2}$

**13.** The domain is all real numbers except 0. The range is all real numbers except 0.

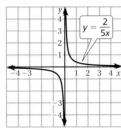

**15.** The domain is all real numbers except 3. The range is all real numbers except 0.

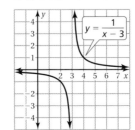

**17.** The domain is all real numbers except 2. The range is all real numbers except 0.

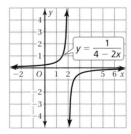

**19.** $x = 0$; $y = 0$; The domain is all real numbers except 0. The range is all real numbers except 0.

**21.** $x = 2$; $y = 7$; The domain is all real numbers except 2. The range is all real numbers except 7.

**23.** $x = 5$; $y = -2$; The domain is all real numbers except 5. The range is all real numbers except $-2$.

**25.** The wrong sign is used for the vertical asymptote; $x = -4$

**27.** Sample answer: $y = \dfrac{100}{x - 6} - 9$

**29.** The graph of $y = \dfrac{1}{x - 6}$ is a translation 6 units right of the graph of $y = \dfrac{1}{x}$.

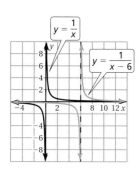

**31.** The graph of $y = \dfrac{1}{x + 7} + 3$ is a translation 7 units left and 3 units up of the graph of $y = \dfrac{1}{x}$.

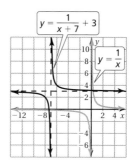

**33.** The graph of $y = 4 - \dfrac{1}{x + 8}$ is a reflection in the $x$-axis of the graph of $y = \dfrac{1}{x}$, and then a translation 8 units left and 4 units up.

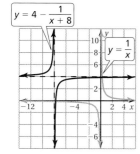

**35. a.** The domain is all real numbers greater than 0. The range is all real numbers greater than 0 and less than 15.

**b.** 12

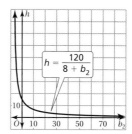

**37.** $y = \dfrac{1}{x - 5}$

**39.** $y = 2 - \dfrac{1}{x - 1}$

**41.** about 23°C

**43.** $x = -2$, $x = 2$, $y = 0$

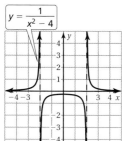

**45.** Read the exercise carefully before you begin.

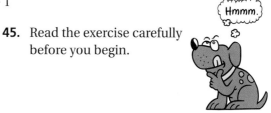

**47.** linear; This is a linear equation in standard form.

**49.** D

# Extension 11.2  Inverse of a Function
(pages 558 and 559)

1. (8, −5), (6, −5), (0, 0), (6, 5), (8, 10)

3. 
| Input | 4 | 1 | 0 | 1 | 4 |
|---|---|---|---|---|---|
| Output | −2 | −1 | 0 | 1 | 2 |

5. The domain of a relation is the range of its inverse relation. The range of a relation is the domain of its inverse relation.

7. $f^{-1}(x) = \dfrac{1}{3}x + \dfrac{1}{3}$

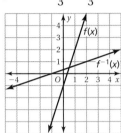

9. $f^{-1}(x) = \sqrt{0.5x}$

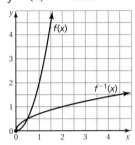

11. $f^{-1}(x) = \dfrac{1}{x}$

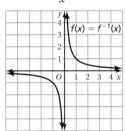

13. −2

15. Both equal $x$. For $f(g(x))$, $g(x)$ is the output value of $x$. So, then you have $f(g(x))$, which gives you back the input value, $x$, because they are inverses. For $g(f(x))$, $f(x)$ is the output value of $x$. So, then you have $g(f(x))$, which gives you back the input value, $x$, because they are inverses.

# Section 11.3  Simplifying Rational Expressions
(pages 564 and 565)

1. no; not a ratio of two polynomials

3. $\dfrac{1}{3x}$; $x = 0$

5. The expression is in simplest form; $n = -1$

7. $\dfrac{2t}{t+11}$; $t = -11, t = 0$

9. They did not list all of the excluded values; $x = 0, x = 3$

11. $-1$; $z = \dfrac{5}{2}$

13. $\dfrac{2-y}{y-5}$; $y = -2, y = 5$

15. $\dfrac{x+2}{2(x-2)}$; $x = 0, x = 2$

17. $\dfrac{x}{2}$ by $(x+3)$

19. $\dfrac{3(x+1)}{x(x+3)}$

21. Sample answer: $\dfrac{1}{x^2 + 8x + 15}$

23. $\left(\dfrac{3}{4}x + 3\right)$ in.

25. Begin by looking at the sum $6x^2 + 12x$ and its factors.

27. continuous

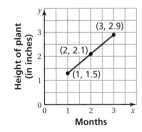

A54   Selected Answers

# Section 11.4
## Multiplying and Dividing Rational Expressions
*(pages 572 and 573)*

1. (1) Factor the numerators and denominators. (2) Multiply the numerators and denominators. (3) Divide out common factors. (4) Simplify.

3. $x = 0, x = 2; x = -1, x = 0, x = 2$

5. $\dfrac{n+3}{14n^4}$

7. $\dfrac{x}{6(x-5)}$

9. $\dfrac{-2}{r-6}$

11. $\dfrac{2t}{3}$

13. $\dfrac{p+4}{p-3}$

15. $\dfrac{3}{(z-6)^2}$

17. To divide rational expressions, you multiply by the reciprocal of the divisor, not the dividend;

$\dfrac{v-2}{4v} \div \dfrac{v-2}{6v^2} = \dfrac{v-2}{4v} \cdot \dfrac{6v^2}{v-2}$

$= \dfrac{6v^2(v-2)}{4v(v-2)}$

$= \dfrac{3v}{2}$

19. $3w(w+2)$

21. $-1$

23. To find the probability that your campsite has shade, compare the area of the green triangle to the area of the rectangle.

25. **a.** $T = \dfrac{50-x}{(1-0.05x)(0.05x^2+5)}$

   **b.** 2020; This will be 20 years after 2000 and 20 is the excluded value.

27. domain: all real numbers
    range: $y \geq -5$

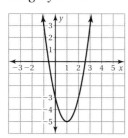

29. domain: all real numbers
    range: $y \leq 9$

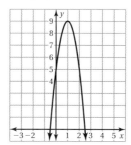

# Section 11.5
## Dividing Polynomials
*(pages 578 and 579)*

1. Monomial: Divide each term of the polynomial by the monomial.
   Binomial: You can use long division to divide a polynomial by a binomial.

3. When you divide the polynomial by the binomial, the remainder is 0.

5. $x - 2$

7. $\dfrac{n^2}{2} - \dfrac{2n}{3} + \dfrac{2}{n}$

9. $z + 7$

11. $3h - 1$

13. $8k - 1 + \dfrac{1}{k-1}$

15. $3g + 5 + \dfrac{8}{2g-3}$

## Section 11.5  Dividing Polynomials (continued)
(pages 578 and 579)

**17.** When subtracting the product from the dividend, the negative sign was not distributed to the second term; $(x^2 - x - 6) \div (x - 3) = x + 2$

**19.** $d - 3$

**21.** $4n + 2 + \dfrac{5}{2n - 1}$

**23.** The dividend is missing an $x$-term with a coefficient of 0 which resulted in like terms not being aligned; $(2x^2 - 5) \div (x + 2) = 2x - 4 + \dfrac{3}{x + 2}$

**25.** The sum of the degrees of the divisor and quotient are equal to the degree of the dividend.

**27.** The formula for the volume of a rectangular prism is $V = \ell wh$.

**29.**

| Quotient |
|---|
| $(x^2 - x + 1) \div (x + 1) = x - 2 + \dfrac{3}{x + 1}$ |
| $(x^3 - x^2 + x - 1) \div (x + 1) = x^2 - 2x + 3 - \dfrac{4}{x + 1}$ |
| $(x^4 - x^3 + x^2 - x + 1) \div (x + 1) = x^3 - 2x^2 + 3x - 4 + \dfrac{5}{x + 1}$ |
| $(x^5 - x^4 + x^3 - x^2 + x - 1) \div (x + 1) = x^4 - 2x^3 + 3x^2 - 4x + 5 - \dfrac{6}{x + 1}$ |

**31.** $x = -4 + \sqrt{23}, x = -4 - \sqrt{23}$

**33.** A

## Section 11.6  Adding and Subtracting Rational Expressions
(pages 585–587)

**1.** Find the LCM of the denominators in both cases.

**3.** $\dfrac{5s}{9}$

**5.** $\dfrac{7}{w}$

**7.** $\dfrac{3}{x + 2}$

**9.** $\dfrac{3t - 4}{t - 1}$

**11.** $\dfrac{1}{p - 1}$

**13.** $2x$

**15.** $(m + 5)(m - 4)$

**17.** $(h + 3)(h - 1)$

**19.** $\dfrac{S - 2\ell w}{2(\ell + w)}$

**21.** $\dfrac{7}{24}$

**23.** $\dfrac{2m - 4}{m - 7}$

**25.** $\dfrac{11}{(p - 4)(p - 3)}$

**27.** yes; not always; The product of the denominators is the product of *all* factors of both denominators. The LCM of the denominators is the product of the greatest power of each factor that appears in *either* denominator.

**29.** Sample answer:
a. $(x - 2)(x + 3)$
$\dfrac{1}{x - 2}, \dfrac{1}{x + 3}$
b. $\dfrac{2x + 1}{(x - 2)(x + 3)}$

A56  Selected Answers

**31.** $\dfrac{8p+7}{p+10}$  **33.** $\dfrac{2c+1}{c-1}$  **35.** $\dfrac{x(2x-1)}{(x+4)(x-5)}$  **37.** $\dfrac{33}{2b+20}$

**39.** Begin by subtracting the two rational expressions, $\dfrac{a}{b}$ and $\dfrac{c}{d}$.

**41.**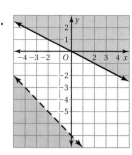

**43.** A

## Section 11.7 — Solving Rational Equations
*(pages 592 and 593)*

1. Use the Cross Products Property, or multiply each side by the LCD.
3. The solution may be extraneous.
5. $b = 1$
7. $m = -2, m = 6$
9. $k = -1, k = 5$
11. The solutions were not checked in the original equation. The solution $x = -1$ is extraneous because it is an excluded value. The only solution is $x = 2$.
13. $c = 5$
15. no solution
17. $a = -9, a = 6$
19. Rewrite the left side using a common denominator of $x$. Then use the Cross Products Property to solve.
21. Read through the problem and organize the given information.
23. 1.2 h (1 h 12 min)
25. $x = -4, x = 8$
27. $x = -2\dfrac{1}{5}, x = \dfrac{3}{5}$

## Section 12.1 — Measures of Central Tendency
*(pages 610 and 611)*

1. The mean is the sum of the data divided by the number of data values. The median of an odd number of values is the middle value. The median of an even number of values is the mean of the two middle values. The mode is the value or values that occur most often.
3. If the outlier is greater than the mean, removing it will decrease the mean. If the outlier is less than the mean, removing it will increase the mean.
5. mean: 1; median: 1; mode: $-1$
7. mean: $1\dfrac{29}{30}$ h; median: 2 h; modes: $1\dfrac{2}{3}$ h and 2 h
9. 4
11. 16
13. 465
15. Begin by ordering the data.
17. All measures increase by $k$.
19. All measures decrease by $k$.
21. bagel shop; Without seeing the data values, it is hard to tell for sure. With the information given, the 10% increase in wages at the bagel shop increases the mean and median hourly wage to $7.92, which is greater than these measures for the amusement park.
23. $-4.7, -2.8, -\dfrac{2}{3}, 1.2, \dfrac{3}{2}, 5.4$

## Section 12.2 — Measures of Dispersion
(pages 616 and 617)

1. A measure of central tendency is a value that represents a typical value in a data set. A measure of dispersion is a value that measures how spread out a data set is.

3. The mean is 85. The majority of the data are in clusters on each side, far from the mean.

5. Tigers: 69; 9; Centaurs: 69; 12; The means are equal but the range of the heights of the Centaurs is greater than the range of the heights of the Tigers. So, the heights of the Centaurs are more spread out.

7. 5; about 1.9     9. 21; about 1.6     11. 63; about 11.9

13. a. Kirsten: 90; 15; about 4.6
    Leah: 91; 8; about 2.5
    The means are about the same but Kirsten's range and standard deviation are much greater than Leah's range and standard deviation.
    b. Leah; Kirsten's scores are more spread out than Leah's scores.

15. 6.4; 8.2; about 3.0

17. no; Two data sets can have the same range and very different standard deviations due to outliers and/or the distribution of the data.
    Example: Data set 1: 1, 5, 6, 6, 6, 7, 11; range = 10, standard deviation = 2.7
    Data set 2: 1, 2, 2, 6, 10, 10, 11; range = 10, standard deviation = 4.0

19. a. *Sample answer:* height (in inches): 58, 58, 59, 60, 60, 61, 62, 62, 63, 65, 65, 65, 65, 66, 66, 68, 69, 71; mean = 63.5; range = 13; standard deviation ≈ 3.7
    b. The mean, range, and standard deviation should all increase; *Sample answer:* mean ≈ 64.6; range = 26; standard deviation ≈ 5.8

21. yes; no; The data set 4, 4, 4, 4, 4, 4, 4, has a standard deviation of 0. The formula for the standard deviation of a data set involves a square root, and standard deviation is a measure of how much a typical data value differs from the mean. So, it can never be negative.

23. [graph of $y = \dfrac{1}{x-6}$]    The domain is all real numbers except 6.    25. B
    The range is all real numbers except 0.

## Section 12.3 — Box-and-Whisker Plots
(pages 623–625)

1. 25%; 50%

3. The length gives the range of the data set and it tells how much the data vary.

5. 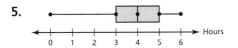 Hours

7. 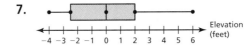 Elevation (feet)

A58    Selected Answers

**9.** 
  **a.** 11 in.
  **b.** about 8.5 in.; about 13.5 in.
  **c.** 

**11.** 
  **a.** 
  **b.** The middle half of the prices vary by no more than $4.38.
  **c.** One-quarter of the prices are $10.50 or less; One-half of the prices are between $10.50 and $14.88; One-quarter of the prices are $14.88 or more.
  **d.** 2.9; The typical price differs from the mean by about $2.90.

**13.** symmetric; The whiskers are about the same length and the median is in the middle of the data.

**15.** skewed left; The left whisker is longer than the right whisker and most of the data is on the right.

**17.** The number of data values on each side of the median is the same.

**19.** 
  **a.**   **b.** 944 calories   **c.**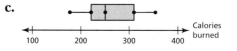
  **d.** The outlier makes the right whisker longer, increases the length of the box, increases the third quartile, and increases the median. In this case, the first quartile and the left whisker were not affected.

**21.** *Sample answer:* 0, 5, 10, 10, 10, 15, 20

**23.** *Sample answer:* 1, 7, 9, 10, 11, 11, 12

**25.** $y = 3x + 2$

**27.** $y = -\dfrac{1}{4}x$

**29.** B

## Section 12.4  Shapes of Distributions
*(pages 631–633)*

**1.** The shape of a skewed distribution will have a tail on one side. The shape of a symmetric distribution is even, or symmetrical, with respect to the mean.

**3.** about 95%

**5.**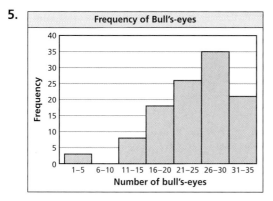

Most of the data are on the right.
So, the distribution is skewed left.

**7.**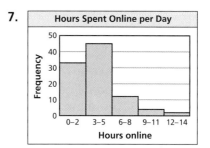

Most of the data are on the left.
So, the distribution is skewed right.

## Section 12.4 Shapes of Distributions (continued)
(pages 631–633)

9. Because the distribution is high on the right and the tail of the graph extends to the left, the distribution is skewed left. So, the median best represents the center of the data and the five-number summary best represents the spread of the data.

11. Because the distribution is high on the right and the tail of the graph extends to the left, the distribution is skewed left. So, the median best represents the center of the data and the five-number summary best represents the spread of the data.

13. A

15. *Sample answer:* histogram; A histogram can show a variety of intervals and easily show the distribution of large data sets. A stem-and-leaf plot uses every data value and can be very tedious for large data sets. Also, stem-and-leaf plots usually only show intervals that are multiples of 10.

17. Using too few or too many intervals may make it difficult to find a pattern or determine the shape of distribution.

19. $y = -x + 4$

21. $y = 2x - 7$

## Section 12.5 Scatter Plots and Lines of Fit
(pages 641–643)

1. They must be ordered pairs so there are equal amounts of $x$- and $y$-values.

3. a–b.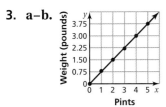
   c. *Sample answer:* $y = 0.75x$
   d. *Sample answer:* 7.5 lb
   e. *Sample answer:* $16.88

5. a. 3.5 h   b. $85
   c. There is a positive relationship between hours worked and earnings.

7. positive relationship

9. positive relationship

11. *Sample answer:* not a good representation; Too many points in the data set lie below the line.

13. *Sample answer:* good representation; The same number of points in the data set lie above and below the line.

15. a. positive relationship
    b. The more time spent studying, the better the test score.

17. The slope of the line of best fit should be close to 1.

19. *Sample answer:* The points follow a U-shaped pattern. There is a relationship between $x$ and $y$, but it is not linear.

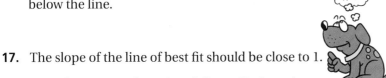

**21.** *Sample answer:* There appears to be two outliers. There appears to be a positive relationship between *x* and *y*.

**23.** 2         **25.** −4

## Section 12.6 Analyzing Lines of Fit
*(pages 649–651)*

**1.** A residual is positive when the actual value is greater than value from model. A residual is negative when the actual value is less than the value from model.

**3.** −0.98 because it is closer to −1 than 0.91 is to 1. $(|-0.98| > |0.91|)$

**5.** The points (*x*, residual) are all above the horizontal axis. So, the equation does not model the data well.

**7.** $y = 3.5$; $r = 0$; There is no correlation between *x* and *y*. The equation does not fit the data.

**9.** $y = 357.5x - 495$         **11.** A

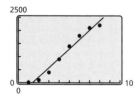

**13.** yes; yes; Talking longer causes life of battery to decrease.

**15.** no

**17. a.** $y = 0.2x + 25$; $r \approx 0.283$; The relationship between *x* and *y* is a weak positive correlation and the equation does not fit the data well.

**b.** The points (*x*, residual) form a ∩-shaped pattern. The equation does not fit the data well.

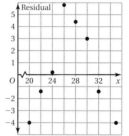

**19. a.** $y = -0.08x + 3.8$; $r \approx -0.965$; The relationship between *x* and *y* is a strong negative correlation and the equation closely models the data.

**b.** The slope is the change in grade point average per hour of television watched. The *y*-intercept is the grade point average of a student who does not watch television.

**c.** about 2.7

**d.** *Sample answer:* no; Watching television does not cause a lower grade point average, but there is a correlation. Studying less could be the cause.

**21.** Think back to quadratic functions and regression.

**23.** 15; 8; 2

# Section 12.7  Two-Way Tables
*(pages 656 and 657)*

1. The joint frequencies are the entries in the two-way table that differentiate the two categories of data collected. The marginal frequencies are the sums of the rows and columns of the two-way table.

3. total of females surveyed: 73;
   total of males surveyed: 59

5. 51

7. 71 students are juniors.   93 students are attending the school play.
   75 students are seniors.   53 students are not attending the school play.

9. **a.** 19; 42

   **b.** 72 9th-graders were surveyed.   112 students chose grades.
   74 10th-graders were surveyed.   40 students chose popularity.
   65 11th-graders were surveyed.   59 students chose sports.

   **c.** about 8.5%

11. **a.**

|  | Eye Color | | | |
|---|---|---|---|---|
| Gender | Green | Blue | Brown | Total |
| Male | 5 | 16 | 27 | 48 |
| Female | 3 | 19 | 18 | 40 |
| Total | 8 | 35 | 45 | 88 |

   **b.** 48 males were surveyed.
   40 females were surveyed.
   8 students have green eyes.
   35 students have blue eyes.
   45 students have brown eyes.

   **c.**

|  | Eye Color | | |
|---|---|---|---|
| Gender | Green | Blue | Brown |
| Male | 63% | 46% | 60% |
| Female | 38% | 54% | 40% |

   *Sample answer:* 62.5% represents that 62.5% of the students with green eyes are male. 40% represents that 40% of the students with brown eyes are female.

13. Be careful not to count the females with green eyes twice.

15. $\dfrac{x+2}{x+4}$; $x = 3$

17. B

A62  Selected Answers

# Section 12.8  Choosing a Data Display
(pages 662 and 663)

1. yes; Different displays may show different aspects of the data.

3. *Sample answer:*

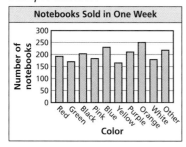

   A bar graph shows the data in different color categories.

5. *Sample answer:* Dot graph: shows changes over time.

7. *Sample answer:* Line graph: shows changes over time.

9. The pictures of the bikes are larger on Monday, which makes it seem like the distance is the same each day.

11. The intervals are not the same size.

13. *Sample answer:* bar graph; Each bar can represent a different vegetable.

15. *Sample answer:* dot plot

17. Does one display better show the differences in digits?

19. $8x = 24$

# Key Vocabulary Index

Mathematical terms are best understood when you see them used and defined *in context*. This index lists where you will find key vocabulary. A full glossary is available in your Record and Practice Journal and at *BigIdeasMath.com*.

absolute value equation, 24
absolute value function, 234
absolute value inequality, 134
arithmetic sequence, 244
asymptote, 553
axis of symmetry, 404
binomial, 331
box-and-whisker plot, 620
causation, 648
closed, 266
common difference, 244
common ratio, 308
completing the square, 470
compound inequality, 132
compound interest, 297
conjugates, 509
continuous domain, 212
correlation coefficient, 647
degree of a monomial, 330
degree of a polynomial, 331
dependent variable, 204
direct variation, 544
discrete domain, 212
discriminant, 480
distance formula, 528
domain, 204
excluded value, 552
exponential decay, 302
exponential decay function, 302
exponential function, 286
exponential growth, 296
exponential growth function, 296
extraneous solution, 513
factored completely, 389
factored form, 358
factoring by grouping, 388
five-number summary, 620
focus, 412
FOIL Method, 343
function, 204
function notation, 226
geometric sequence, 308
graph of an inequality, 107
graph of a linear inequality, 138

graph of a system of linear inequalities, 186
half-planes, 138
hypotenuse, 522
independent variable, 204
inequality, 106
interquartile range, 621
inverse function, 559
inverse relation, 558
inverse variation, 544
joint frequency, 654
least common denominator of a rational expression, 583
legs, 522
line of best fit, 647
line of fit, 640
linear equation, 44
linear function, 218
linear inequality in two variables, 138
linear regression, 647
literal equation, 28
marginal frequency, 654
maximum value, 427
measure of central tendency, 608
measure of dispersion, 614
minimum value, 427
monomial, 330
nonlinear function, 238
$n$th root, 278
parabola, 404
perpendicular lines, 57
piecewise function, 232
point-slope form, 80
polynomial, 331
prime polynomial, 389
Pythagorean Theorem, 522
quadratic equation, 456
quadratic formula, 478
quadratic function, 404
quartile, 620
range, 204
range (of a data set), 614
rational equation, 590

rational expression, 562
rational function, 552
rationalizing the denominator, 508
recursive rule, 312
relation, 208
residual, 646
rise, 50
root, 358
run, 50
scatter plot, 638
sequence, 244
simplest form of a radical expression, 508
simplest form of a rational expression, 562
slope, 50
slope-intercept form, 60
solution of an inequality, 106
solution of a linear equation, 44
solution of a linear inequality, 138
solution set, 106
solution of a system of linear equations, 156
solution of a system of linear inequalities, 186
square root equation, 512
square root function, 504
standard deviation, 615
standard form, 66
step function, 233
system of linear equations, 156
system of linear inequalities, 186
term, 244
theorem, 520
trinomial, 331
two-way table, 654
vertex, 404
vertex form, 432
Vertical Line Test, 209
$x$-intercept, 60
$y$-intercept, 60
zero, 419
Zero-Product Property, 358

# Student Index

This student-friendly index will help you find vocabulary, key ideas, and concepts. It is easily accessible and is designed to be a reference for you whether you are looking for a definition, a real-life application, or help with avoiding common errors.

## A

**Absolute value equation,** *See also*
  Equation(s)
    defined, 24
    real-life application, 25
    solving, 24
**Absolute value function,** *See also*
  Function(s)
    defined, 234
    graphing, 234–235
**Absolute value inequality,** *See also*
  Inequality
    defined, 134
    real-life application, 135
Addition
  as inverse of subtraction, 4
  polynomials, 334–339
    error analysis, 338
    real-life application, 337
  Property
    of Equality, 4
    of Inequality, 112
  rational expressions, 580–587
    error analysis, 585–586
    with like denominators, 582
    with unlike denominators, 583
  to solve equations, 4–5
Addition Property of Equality, 4
Addition Property of Inequality, 112
Algebra tiles
  polynomials
    adding, 334
    classifying, 328
    difference of squares, 382
    dividing, 574–575
    factoring with GCF, 362–363
    finding binomial factors, 368–369, 376–377
    multiplying, 340–341
    perfect square trinomials, 382
    square of binomial pattern, 349
    subtracting, 335
    sum and difference pattern, 348
  solving equations by completing the square, 468–469

Area formula
  of a rectangle, 28
  using, 26
**Arithmetic sequence(s),** 242–247
  defined, 244
  equation, 245
  error analysis, 248
  real-life application, 246
  recursive equation, 312
  writing, 247
**Asymptote(s)**
  defined, 553
  identifying, 553
  writing, 555
**Axis of symmetry,** defined, 404

## B

Bar graph, 660
**Binomial(s),** *See also* Polynomial(s)
  defined, 331
  square of binomial pattern, 348–353
    error analysis, 352
    real-life application, 351
  sum and difference pattern, 348–353
    error analysis, 352
**Box-and-whisker plot,** 618–625, 660
  error analysis, 623–624
  five-number summary of, 620
  interquartile range of, 621
  quartiles, 620
  skewed left, 622
  skewed right, 622
  symmetric, 622
  writing, 623–625

## C

**Causation,** defined, 648
Choose Tools, *Throughout. For example, see:*
  data displays, 632
  equations
    graphing linear equations, 43
    slope, 54
    solving, 7, 14
    square root, 517
    systems of linear and quadratic, 491
  exponential decay function, 305
  exponential functions, 305
  factoring polynomials, 381
  function notation, 231
  graphing quadratic functions, 431
  graphs
    linear equations, 43
    quadratic functions, 431
    scatter plots, 643
  polynomials
    dividing, 579
    factored form, 361
    factoring, 381
    scatter plots, 643
    slope of a line, 54
    square root equations, 517
    systems of linear and quadratic equations, 491
Circle graph, 660
**Closed set,** defined, 266
**Common difference,** defined, 244
Common Errors
  completing the square, 471
  graphing, 419
  inequalities, 119
  Pythagorean Theorem, 528
  quadratic equations, 471
  quadratic functions, 419
  subtracting rational expressions, 582
**Common ratio,** defined, 308
Comparing
  functions
    exponential, linear, and quadratic, 434–441
    graphs of, 442–443
    linear and nonlinear, 236–241
  graphs
    of exponential functions, 442–443
    of quadratic functions, 442–443
Comparison chart, 222
**Completing the square**
  to solve quadratic equations, 468–473
  defined, 470
  error analysis, 472
  real-life application, 471

A66  Student Index

**Compound inequalities,** *See also* Inequality
   defined, 132
   graphing, 132–135
   real-life application, 135
   solving, 132–133
**Compound interest,** defined, 297
Concept circle, 634
**Conjugates,** defined, 509
Constant of variation, 544
**Continuous domains,** *See also* Domain(s)
   defined, 210, 212
   error analysis, 214
   graphing, 213
Coordinate plane
   distance formula, 526–531
      error analysis, 530
      real-life application, 529
      writing, 530
**Correlation coefficient,** defined, 647
Critical thinking, *Throughout. For example, see:*
   comparing functions, 441
   data analysis, 617
      scatter plots, 643
   equations
      linear, 53–55, 63
      multi-step, 15
      quadratic, 483
      simple, 9
   exponential growth, 299
   function notation, 231
   functions
      comparing, 441
      exponential, 299
   geometric sequences, 310
   graphing, 53
      linear equations, 69
   inequalities, 109, 115, 122
      linear in two variables, 142–143
   irrational numbers, 267
   linear equations, 63, 69
      systems of, 159, 175
      writing, 76
   lines of fit, 643
   polynomials, dividing, 579
   quadratic equations, 483
   rational expressions, 565
   scatter plots, 643
   slope of a line, 53–55
   slope-intercept form, 62
   solving equations, 9, 15

   square root equations, 517
   standard deviation, 617
Cross Products Property, 590

## D

Data, *See* Data analysis and Data display(s)
Data analysis, *See also* Data display(s)
   causation, 648
   correlation coefficient, 647
   distribution
      skewed left, 622, 628–633
      skewed right, 622, 628–633
      symmetric, 622, 626–633
      uniform, 628
   linear regression, 647
      line of best fit, 647
   line of best fit, 647
   lines of fit
      correlation coefficient, 647
      linear regression, 647
      line of best fit, 647
      residuals, 646–647
      writing, 649
   mean, 606–611
      defined, 608
   measures of central tendency, 606–611
      defined, 608
      project, 617
   measures of dispersion, 612–617
      defined, 614
      range of, 614
      standard deviation, 615
   median, 606–611
      defined, 608
   mode, 606–611
      defined, 608
   outliers, 609, 629
   project, 617
   range of a data set, 614
   relationship of data, 639
   residuals, 646–647
   scatter plots
      lines of fit, 644–651
      relationship of data, 639
   standard deviation, 615
   writing, 610
Data display(s), *See also* Data analysis
   bar graphs, 660
   box-and-whisker plots, 618–625, 660
      error analysis, 623–624

      five-number summary of, 620
      interquartile range of, 621
      quartiles, 620
      skewed left, 622
      skewed right, 622
      symmetric, 622
      writing, 624–625
   choosing a, 658–663
   circle graphs, 660
   distributions, 626–633
   dot plots, 606, 660
   five-number summary
      defined, 620
   histograms, 660
   interquartile range
      defined, 621
   line graphs, 660
   misleading, 661
   pictographs, 660
   quartiles
      defined, 620
   scatter plots, 636–643, 660
      defined, 638
   stem-and-leaf plots, 660
   two-way tables, 652–657
      defined, 654
**Degree of a monomial,** defined, 330
**Degree of a polynomial,** defined, 331
**Dependent variable,** defined, 204
Difference of two squares pattern, 382–387
   real-life application, 385
   writing, 386
**Direct variation,** 542–549
   constant of variation for, 544
   defined, 544
   graphing, 546
   reading, 545
**Discrete domains,** *See also* Domain(s)
   defined, 210, 212
   error analysis, 214
   graphing, 212
**Discriminant**
   defined, 477, 480
   interpreting, 480
**Distance formula,** in a coordinate plane, 526–531
   error analysis, 530
   real-life application, 529
   writing, 530
Distribution, 626–633
   skewed left, 622, 628–633
   skewed right, 622, 628–633
   symmetric, 622, 626–633

uniform, 628
writing, 631
Distributive Property
  multiplying polynomials, 342
  solving equations with, 13
Division
  as inverse of multiplication, 5
  of polynomials 574–579
    error analysis, 578–579
    writing, 578
  Property
    of Equality, 5
    of Inequality, 116–123
  of rational expressions, 568–573
    error analysis, 572
    writing, 572
  to solve equations, 5
Division Property of Equality, 5
Division Property of Inequality, 116–123
  error analysis, 121–122
**Domain(s)**
  continuous
    defined, 210, 212
    graphing, 213
  discrete, 210–215
    defined, 210, 212
    graphing, 212
  error analysis, 214
  of a function, 202–207
    defined, 204
    error analysis, 206
    real-life application, 205
    square root, 504
  writing, 214
Dot plot, 660

# E

Equation(s)
  absolute value, 24–25
    defined, 24
    real-life application, 25
    solving, 24
    writing, 25
  arithmetic sequence, 245
  direct variation, 542–549
    graphing, 545
    reading, 545
  exponential
    solving, 292–293
  extraneous solutions, 513
  geometric sequences, 309
  graphing, 42–47
    error analysis, 46, 62
    real-life application, 45, 61
    slope, 48–55

    in slope-intercept form, 58–63
    solution points, 42
  inverse variation, 542–549
    error analysis, 548
    graphing, 545
    reading, 545
    real-life application, 546
  linear
    defined, 44
    error analysis, 46
    graphing, 42–47
    real-life applications, 45, 88–89
    slope, 48–55
  literal
    defined, 28
    error analysis, 30
    real-life application, 29
    rewriting, 26–31
  multi-step, 10–15
    error analysis, 14
    real-life application, 13
  parabola, 413
  parallel line, 84
  perpendicular line, 85
  in point-slope form, 78–83
    real-life application, 81
    writing, 82
  polynomial, *See* Polynomial equation(s)
  quadratic
    choosing a solution method, 484–485
    defined, 456
    error analysis, 460, 466, 472, 481–482
    real-life applications, 458, 465, 471, 479
    solutions, 456–457
    solving by completing the square, 468–473
    solving by graphing, 454–461
    solving by the quadratic formula, 476–483
    solving using square roots, 462–467
    writing, 459, 467
  rational
    defined, 590
    real-life application, 591
    solving, 588–593
    writing, 592
  rewriting, 26–31
    error analysis, 30
    real-life application, 29
  sequences
    arithmetic, 312

    geometric, 312
  slope
    defined, 48, 50
    error analysis, 54
    project, 54
  in slope-intercept form, 58–63
    error analysis, 62, 76
    real-life applications, 61, 75
  solving
    by addition, 4–5
    with Addition Property of Equality, 4,
    with Cross Products Property, 590
    with Distributive Property, 13
    by division, 5
    by graphing, 182–183
    with least common denominator, 591
    with like terms, 12
    multi-step, 10–15
    by multiplication, 5
    project, 91
    rational, 588–593
    real-life problems, 86–93
    simple, 2–9
    by subtraction, 4–5
    with Subtraction Property of Equality, 4
    with variables on both sides, 18–23
    writing, 592
  square root
    defined, 512
    error analysis, 515–516
    extraneous solutions of, 513
    real-life application, 514
    solving, 510–517
    writing, 515–516
  squaring both sides of, 512–517
  in standard form, 64–69
    real-life application, 67
  systems of linear
    defined, 154, 156
    error analysis, 159, 165, 173–174, 180
    number of solutions, 176–181
    reading, 156
    real-life applications, 157, 163, 172
    solving by elimination, 168–175
    solving by graphing, 154–159
    solving by substitution, 160–165
  with variables on both sides, 18–23

error analysis, 22
real-life application, 21
writing, 22–23
writing, 14
of parallel line, 84
of perpendicular line, 85
in point-slope-form, 78–83
real-life problems, 86–93
in slope-intercept form, 72–77
using two points, 81
Error analysis, *Throughout. For example, see:*
arithmetic sequences, 248
box-and-whisker plots, 623–624
common differences, 248
data displays
box-and-whisker plots, 623–624
distance formula, 530
domains, 214
equations
graphing, 46
inverse variation, 548
multi-step, 14
quadratic, 460, 466, 472, 481–482
rewriting, 30
simple, 8
slope, 54
in slope-intercept form, 62, 76
square root, 515–516
systems of linear and quadratic, 491
with variables on both sides, 22
writing, 76
evaluating a function, 229
exponential growth, 298
exponents, 273–274
rational, 280
expressions, 274
adding, 585–586
dividing, 572
simplifying, 564
factoring
perfect square trinomial, 386
polynomials, 360, 366, 373–374
trinomials, 373–374
function notation, 229
functions
arithmetic sequences, 248
domain and range, 206
domains, 214
exponential, 289, 298
quadratic, 408, 420, 429, 440
rational, 556
square root, 507

geometry
distance formula, 530
Pythagorean Theorem, 524
graphs
box-and-whisker plots, 623–624
inequalities, 121–122
graphing, 114
linear in two variables, 142
multi-step, 130
writing, 108
inverse variation, 548
linear equations, 159
in slope-intercept form, 62, 76
systems of, 165, 173
writing, 76
polynomials, 332
adding, 338
dividing, 578–579
factoring, 360, 366, 373–374, 380, 386
multiplying, 345–346
sum and difference pattern, 352
Pythagorean Theorem, 524
quadratic equations
solving by completing the square, 472
solving by graphing, 460
solving by the quadratic formula, 481–482
solving using square roots, 466
systems of linear and, 491
quadratic formula, 481–482
quadratic functions, 408
axis of symmetry, 429
difference and ratios of, 440
graphing, 408, 414, 420
rational expressions, 564
adding, 585–586
dividing, 572
rational functions, 556
slope, 54
square root equations, 515–516
square root functions, 507
square roots, 264
systems of linear equations
number of solutions, 180
solving by elimination, 173–174
solving by graphing, 159
solving by substitution, 165
systems of linear and quadratic equations, 491
systems of linear inequalities, 190

triangles
right, 524
trinomials, 373–374, 380
Example and non-example chart, 566
**Excluded value,** defined, 552
Exponent(s), 268–275
properties of, 268–275
error analysis, 273
Power of a Power, 270
Power of a Product, 271
Power of a Quotient, 271
Product of Powers, 270
Quotient of Powers, 270
real-life application, 272
using, 270–271
writing a rule for, 268–269
rational, 276–281
error analysis, 280
real-life application, 279
writing, 280–281
**Exponential decay,** 300–305
defined, 302
real-life application, 303
writing, 304
**Exponential decay function,** defined, 302
Exponential equations, 292–293
**Exponential function(s),** 284–291
compared with linear and quadratic, 434–441
defined, 286
differences and ratios of, 437
simplifying
error analysis, 289
geometric sequences, 306–311
equation for, 309
error analysis, 310
real-life application, 309
recursive equation for, 312
writing, 310
graphing
rates of change, 442–443
growth, 294–299
compound interest, 297
error analysis, 298
real-life application, 297
modeling, 285
rates of change, 442–443
real-life application, 288
**Exponential growth,** 294–299
compound interest, 297
defined, 296
error analysis, 298
function, 296–299
defined, 296
real-life application, 297

**Exponential growth function,** 296–299
  defined, 296
Expressions
  rational
    adding, 580–587
    defined, 562
    dividing, 568–573
    error analysis, 564, 572, 585–586
    multiplying, 568–573
    real-life applications, 563, 584
    simplest form of, 562
    simplifying, 560–565
    subtracting, 580–587
    writing, 564, 572, 585
  simplifying, 560–565
    error analysis, 274, 564
    real-life application, 563
    writing, 564
**Extraneous solution**
  defined, 513
  identifying, 514

## F

Factor(s)
  greatest common, 364
**Factored completely,** *See also* Polynomial(s), 389
**Factored form,** defined, 358
Factoring
  polynomials, 356–361, 368–381
    completely, 389
    difference of two squares, 382–387
    error analysis, 360, 366, 373–374, 380, 386
    by grouping, 388
    perfect square trinomials, 382–387
    prime, 389
    real-life applications, 359, 365, 372, 385
    trinomials, 368–381
    using greatest common factor, 362–367
    writing, 366
    Zero-Product Property, 358
**Factoring by grouping,** *See also* Polynomial(s), 388
**Five-number summary,** defined, 620
**Focus**
  of a parabola
    defined, 412

error analysis, 414
real-life application, 413
writing, 414
**FOIL Method,** defined, 343
Formula(s)
  for area
    of a rectangle, 28
  distance, 528
  period of a pendulum, 514
  rewriting, 26–31
    real-life application, 29
  simple interest, 28
  slope, 48, 50
  for surface area
    of a cone, 28
    of a cylinder, 28
  temperature conversion, 29
  using, 26–27
  for volume
    of a cylinder, 9
    of a sphere, 31
Four square, 124
**Function(s)**
  absolute value, 234–235
    defined, 234
  comparing, 434–441
    exponential, linear, and quadratic, 434–441
  comparing graphs of, 442–443
  defined, 204
  differences and ratios of, 437
    error analysis, 440
  direct variation, 542–549
    graphing, 545
    reading, 545
  domain, 202–207
    continuous, 210–215
    discrete, 210–215
    error analysis, 206, 214
    real-life application, 205
  even, 549
  exponential, 284–291
    defined, 286
    error analysis, 289
    geometric sequences, 306–311
    modeling, 285
    real-life applications, 288, 297, 303
    writing, 304
  exponential decay, 302–305
    defined, 302
  exponential growth, 294–299
    defined, 296
  graphing, 402–421, 424–443
    error analysis, 408, 414, 420, 429, 507, 556

properties of, 426
quadratic, 402–415, 424–443
rational, 550–557
real-life applications, 406, 413, 419, 505, 554
square root, 502–507
writing, 414, 420, 555
input-output tables and, 201–205
inverse, 558–559
  defined, 559
  reading, 559
inverse variation, 542–549
  error analysis, 548
  graphing, 545
  reading, 545
  real-life application, 546
linear
  arithmetic sequences, 242–249
  defined, 218
  patterns, 216–221
  real-life applications, 219, 239, 246
nonlinear
  compared to linear, 236–241
  defined, 238
  real-life application, 239
notation, 224–231
  error analysis, 229
odd, 549
piecewise, 232–233
  defined, 232
quadratic, 402–415
  defined, 404
  error analysis, 408, 414, 420, 429
  graphing, 402–415, 424–443
  maximum value of, 427–428
  minimum value of, 427–428
  parabola, 404
  real-life applications, 406, 413, 419, 428
  vertex form of, 432
  writing, 407, 420
  zero of, 419
range, 202–207
  error analysis, 206
  real-life application, 205
rates of change, 442–443
rational
  defined, 552
  error analysis, 556
  graphing, 550–557
  real-life application, 554

relations, 208–209
  defined, 208
square root
  defined, 504
  domain of, 504
  error analysis, 507
  graphing, 502–507
  real-life application, 505
step, 233
Vertical Line Test, 209
zero of, 419, 458
**Function notation,** 224–231
  defined, 224, 226
  error analysis, 229
  reading, 226

**Geometric sequences,** 306–311
  common ratio, 308
  equation for, 309
  error analysis, 310
  extending, 308
  graphing, 308
  real-life application, 309
  recursive equation for, 312
Geometry
  Pythagorean Theorem, 520–531
    converse of, 528
    defined, 522
    distance formula and, 526–531
    error analysis, 524, 530
    real-life applications, 523, 529
    using, 526–531
  triangles, See Triangle(s)
**Graph of an inequality,** See also Inequality
  defined, 107
  error analysis, 114
**Graph of a linear inequality,** See also Inequality
  defined, 136, 138
**Graph of a system of linear inequalities,** See also Systems of linear inequalities
  defined, 186
Graphic organizers
  comparison chart, 222
  concept circle, 634
  example and non-example chart, 566
  four square, 124
  idea and examples chart, 354
  information frame, 282
  information wheel, 474
  notetaking organizer, 166

  process diagram, 70
  summary triangle, 422
  word magnet, 518
  Y chart, 16
Graphing
  arithmetic sequences, 245
  direct variation, 545
  distance formula, 526–531
    error analysis, 530
    real-life application, 529
  equations
    direct variation, 545
    error analysis, 46, 62, 460
    exponential, 293
    inverse variation, 545
    real-life applications, 45, 61
    in slope-intercept form, 58–63
  exponential functions, 287
    real-life application, 288
  functions
    error analysis, 408, 414, 420, 429, 507, 556
    rational, 550–557
    real-life application, 554
    writing, 555
  geometric sequences, 308
  inequalities, 104–109
    compound, 132–133, 135
    half-plane, 138
    linear systems of, 186–187
    linear in two variables, 136–143
  inverse variations, 545–546
    real-life application, 546
  linear equations, 42–47, 182–183
    error analysis, 62, 68
    process diagram, 70
    real-life applications, 61, 67, 183
    in slope-intercept form, 58–63
    to solve, 182–183
    in standard form, 64–69
  linear functions, 216–221
  linear inequalities
    in one variable, 139
    in two variables, 139
  quadratic equations, 454–461
    error analysis, 460
    real-life application, 458
    writing, 459
  quadratic functions, 402–421, 424–443
    axis of symmetry, 404, 426
    error analysis, 408, 414, 420, 429
    focus, 412
    maximum value of, 427–428
    minimum value of, 427–428

    parabola, 404
    real-life applications, 406, 413, 419, 428
    vertex, 404, 426
    vertex form of, 432
    writing, 407, 414, 420
    zero of, 419
  rates of change
    of functions, 442–443
  rational functions, 550–557
    asymptotes, 553
    error analysis, 556
    real-life application, 554
    writing, 555
  slope, 48–55
    formula, 48, 50
    negative, 50, 52
    positive, 50, 52
    project, 54
    rise, 50
    run, 50
    undefined, 51–52
    zero, 51–52
  solution points, 42, 44
  square root functions, 502–507
    error analysis, 507
    real-life application, 505
  systems of linear equations, 154–159
    error analysis, 159
    real-life application, 157
  Vertical Line Test, 209
  vertical translations, 227, 287, 504
Graphs
  bar, 660
  box-and-whisker plot, 618–625
    error analysis, 623–624
    five-number summary of, 621
    interquartile range of, 621
    quartiles of, 620
    skewed left, 622
    skewed right, 622
    symmetric, 622
    writing, 623–625
  circle, 660
  histogram, 660
  line, 660
  pictograph, 660
  scatter plot, 636–643
    defined, 638
    lines of fit, 640, 644–651
**Greatest common factor (GCF),** 364
  factoring polynomials, 362–367
    error analysis, 366
    writing, 366

Student Index **A71**

## H

**Half-plane,** defined, 138
**Histogram,** 660
**Hypotenuse,** defined, 522

## I

Idea and examples chart, 354
**Independent variable,** defined, 204
**Inequality**
  absolute value, 134–135
    defined, 134
    real-life application, 135
  Addition Property of, 110–115
    reading, 113
  compound, 132–133
    defined, 132
    graphing, 132–133
    real-life application, 135
    writing, 132
  defined, 106
  Division Property of, 116–123
  graph of a linear, 136–143
    defined, 136, 138
    in two variables, 136–143
  graphing, 104–109
    defined, 107
    error analysis, 114
    real-life application, 140
  linear in one variable, 139
    graphing, 139
  linear in two variables, 136–143
    defined, 138
    error analysis, 142
    graphing, 139
    half-plane, 138
    real-life application, 140
  multi-step, 126–131
    error analysis, 130
    real-life application, 129
    writing, 130
  Multiplication Property of, 116–123
    error analysis, 121
  project, 123
  solution of, 106
  solution of linear, 138
  solution set, 106
  solving
    absolute value, 134–135
    compound, 132–133
    error analysis, 108, 114, 121–122, 130
    multi-step, 126–131
    reading, 113, 138
    real-life applications, 113, 129, 135, 140
    using addition and subtraction, 110–115
    using multiplication and division, 116–123
    using a table, 116–117
  Subtraction Property of, 110–115
  symbols of, 106
    reading, 106
  systems of linear, 184–191
    error analysis, 190
    graphing, 186–187
    real-life application, 188
    writing, 188
  Triangle inequality, 105
  writing, 104–109
    error analysis, 108
    project, 123
Information frame, 282
Information wheel, 474
**Interquartile range,** defined, 621
**Inverse function(s),** 558–559
  defined, 559
  reading, 559
Inverse operations, 4, 7
  addition and subtraction, 4
  multiplication and division, 5, 118
**Inverse relation,** defined, 558
**Inverse variation,** 542–549
  constant of variation for, 544
  defined, 544
  error analysis, 548
  graphing, 545
  reading, 545
  real-life application, 546
Irrational number(s), 266–267

## J

**Joint frequency,** defined, 654

## L

**Least common denominator**
  defined, 583
  solving rational equations with, 591
**Legs,** of a right triangle, 522
Like terms
  combining to solve equations, 12
Line(s)
  asymptote, 553
  graphing, 48–55
  parallel, 56, 84
    equation for, 84
  perpendicular, 85
    defined, 57
    equation for, 85
  slope
    defined, 48, 50
    error analysis, 54
    formula, 50
    project, 54
  Vertical Line Test, 209
  $x$-intercept of
    defined, 60
  $y$-intercept of
    defined, 60
**Line of best fit,** defined, 647
**Line of fit,** 636–651
  analyzing, 644–651
    correlation coefficient, 647
    defined, 640
    line of best fit, 647
    linear regression, 647
    residuals, 646–647
    writing, 641, 649
Line graph, 660
**Linear equation(s),** *See also* Equation(s)
  defined, 44
  graphing, 42–47, 182–183
    error analysis, 46, 62
    real-life application, 45
    slope, 48–55
    in slope-intercept form, 58–63
    solution points, 42
    in standard form, 64–69
  in point-slope form
    real-life application, 81
    writing, 78–83
  real-life problems, 88–89
    project, 91
  slope, 48–55
    defined, 48, 50
    error analysis, 54
    negative, 50, 52
    positive, 50, 52
    project, 54
    rise, 50
    run, 50
    undefined, 51–52
    zero, 51–52
  in slope-intercept form, 58–63
    error analysis, 62, 76
    real-life applications, 61, 75
    writing, 72–77
  solution of
    defined, 44
  solving by graphing, 182–183
    real-life application, 183

in standard form, 64–69
  error analysis, 68
  real-life application, 67
systems of, 154–159
  defined, 156
  error analysis, 159, 165, 173–174, 180
  number of solutions, 176–181
  reading, 156
  real-life applications, 157, 163, 172
  solving by elimination, 168–175
  solving by graphing, 154–159
  solving by substitution, 160–165
writing
  of parallel line, 84
  of perpendicular line, 85
  in point-slope form, 78–83
  real-life problems, 86–93
  in slope-intercept form, 72–77
  using two points, 81

**Linear function(s)**
arithmetic sequences, 242–249
  defined, 244
  error analysis, 248
  real-life application, 246
  recursive equation for, 312
  writing, 247
compared with exponential and quadratic, 434–441
defined, 218
differences and ratios of, 437
graphing
  rates of change, 442–443
nonlinear compared to, 236–241
  real-life application, 239
patterns, 216–221
  real-life application, 219
rates of change, 442–443

**Linear inequality in two variables,**
  *See also* Inequality
defined, 138
error analysis, 142
graphing, 139
reading, 138
real-life application, 140

**Linear regression,**
defined, 647
line of best fit, 647

**Literal equation(s)** *See also* Equation(s)
defined, 28
error analysis, 30
real-life application, 29
rewriting, 26–31

Logic, *Throughout. For example, see:*
  box-and-whisker plots, 624
  data displays
    box-and-whisker plots, 624
  domains, 215
  equations
    graphing, 42
    linear systems of, 175
    quadratic, 467
    simple, 9
    slope-intercept form, 59
    solving, 42
  exponential functions, 290
  exponents, 281
  factoring trinomials, 375
  formulas, 31
  functions
    exponential, 290
    quadratic, 409, 421
  graphing, 42
    quadratic functions, 409, 421
  graphs
    box-and-whisker plots, 624
  inequalities, 109
  linear equations
    in standard form, 69
    systems of, 175
  polynomials, 333
    factoring, 375
  quadratic equations, 467
  radicals, 281
  rational expressions, 587
  systems of linear equations, 175
    number of solutions, 181

## M

**Marginal frequency,** defined, 654
**Maximum value,** of a quadratic function, 427–428
  defined, 427
Mean, 606–611
  defined, 608
Meaning of a Word
  linear equations, 42
**Measures of central tendency,** 608–611
  defined, 608
  mean
    defined, 608
  median
    defined, 608
  mode
    defined, 608
**Measures of dispersion,** 612–617
  defined, 614

project, 617
range of a data set, 614
standard deviation, 615
Median, 606–611
  defined, 608
**Minimum value,** of a quadratic function, 427–428
  defined, 427
Mode, 606–611
  defined, 608
Modeling, *Throughout. For example, see:*
  box-and-whisker plots, 625
  data analysis
    lines of fit, 651
  data displays
    box-and-whisker plots, 625
    distribution, 633
  equations
    polynomial, 367
    quadratic, 461
    square root, 517
  functions
    linear *vs.* nonlinear, 241
    quadratic, 409
    rational, 557
  graphs
    box-and-whisker plots, 625
  inequalities, 115
    absolute value, 135
    linear in two variables, 142
  linear equations
    in standard form, 69
    systems of, 159
  lines of fit, 651
  polynomials, 339
    factors, 367
  quadratic equations, 461
  quadratic functions, 409
  rational functions, 557
  square root equations, 517
  square roots, 265
  systems of linear equations, 159
**Monomial(s),** *See also* Polynomial(s)
  defined, 330
  degree of, 330
Multi-step equation(s), *See also* Equation(s)
  error analysis, 14
  real-life application, 13
  solving, 10–15
    combining like terms, 12
    with Distributive Property, 13
  two-step, 12–13
Multi-step inequalities, *See* Inequality

Multiplication
   as inverse of division, 5
   polynomials, 340–347
      error analysis, 345–346
      real-life application, 344
   Property
      of Equality, 5
      of Inequality, 116–123
   rational expressions, 568–573
      writing, 572
   to solve equations, 5
Multiplication Property of
      Equality, 5
Multiplication Property of
      Inequality, 116–123

## N

**Nonlinear function(s)**
   defined, 238
   linear compared to, 236–241
      real-life application, 239
Notetaking organizer, 166
**$n$th root,** 276–278
   defined, 278
   writing, 280
Number(s)
   irrational, 266–267
   nonzero, 330
   rational, 266–267
   real, 266–267
   scientific notation, 272
   sets, closed, 266
Number Sense, *Throughout. For example, see:*
   absolute value inequality, 135
   arithmetic sequences, 248
   box-and-whisker plots, 623
   correlation coefficient, 649
   data displays
      box-and-whisker plots, 623
   exponential equations, 292
   exponential growth, 299
   expressions, 572
   factoring
      difference of two squares, 387
      trinomials, 375
   functions
      arithmetic sequences, 248
      exponential, 290, 299
      linear *vs.* nonlinear, 241
      quadratic, 420
   graphing
      quadratic functions, 420
   graphs
      box-and-whisker plots, 623
   lines of fit, 649

perfect square trinomial pattern, 473
polynomials, 333
   dividing, 578
   factoring, 375
   multiplying, 347
rational expressions, 572
systems of linear equations, 164–165
   solving by elimination, 173

## O

Open-Ended, *Throughout. For example, see:*
   arithmetic sequences, 249
   common differences, 249
   data analysis
      mode, 610
   data displays, 642
      choosing a, 662
      two-way tables, 656
   equations, 14
      rational, 592
      simple, 9
      slope, 53
      in slope-intercept form, 62
      solving, 592
      square root, 516
      with variables on both sides, 22
   exponential functions, 289
   exponents, 274
   functions
      quadratic, 414
      square root, 507
   graphing
      quadratic functions, 414
      rational functions, 556
      square root functions, 507
   histograms, 662
   inequalities, 121
      in two variables, 141
   mode, 610
   negative slope, 90
   polynomials, 332, 346
      factoring, 367
      sum and difference pattern, 352
      trinomial, 373
   rational equations, 592
   rational expressions, 565
      with like denominators, 586
   rational functions, 556
   simple equations, 9
   slope of a line, 53
   slope-intercept form, 62

square root equations, 516
square root functions, 507
trinomials, 332
two-way tables, 656
Operations
   closed set, 266
   inverse, 7
      addition and subtraction, 4
      multiplication and division, 5, 118
Ordered pairs, 44
   solution points, 42
   solution of a system of linear equations, 156
   solution of a system of linear inequalities, 186
Outlier, defined, 609

## P

**Parabola(s)**
   axis of symmetry, 404
   defined, 404
   equation, 413
   focus, 410–415
      defined, 412
      error analysis, 414
      real-life application, 413
      writing, 414
   properties, 426
   vertex, 404
Parallel line
   equation of, 84
   slope of, 56
Patterns
   difference of two squares, 382–387
   perfect square trinomial, 382–387
   square of a binomial, 348–353
   sum and difference, 348–353
Perfect square trinomial pattern, 382–387
   error analysis, 386
   writing, 472
Perimeter formulas, 26
**Perpendicular line**
   defined, 57
   equation of, 84–85
   slope of, 57
Pictograph, 660
**Piecewise function(s)**, *See also* Function(s)
   defined, 232
   graphing, 232–233
   writing, 233

**Point-slope form**
   defined, 80
   real-life application, 81
   writing equations in, 78–83
**Polynomial(s),** 328–333
   adding, 334–339
      error analysis, 338
      real-life application, 337
   binomial, 331
      error analysis, 352
      real-life application, 351
      square of binomial pattern, 348–353
      sum and difference pattern, 348–353
   classifying, 328, 331
   defined, 331
   degree of, 331
   difference of two squares pattern, 382–387
      writing, 386
   dividing, 574–579
      error analysis, 578–579
      writing, 578
   error analysis, 332, 338, 345–346, 352, 373–374
   factoring
      completely, 389
      difference of two squares, 382–387
      error analysis, 366, 373–374, 380, 386
      by grouping, 388
      perfect square trinomials, 382–387
      prime, 389
      real-life applications, 372, 385
      trinomials, 368–381
      using greatest common factor, 362–367
      writing, 366, 373, 380, 386
   FOIL Method, 343
   monomials, 330
   multiplying, 340–347
      error analysis, 345–346
      real-life application, 344
      using Distributive Property, 342
      using FOIL Method, 343
      perfect square trinomial pattern, 382–387
      error analysis, 386
   real-life applications, 331, 337, 344, 351
   square of binomial pattern, 348–353
      error analysis, 352
      real-life application, 351
   subtracting, 334–339
      error analysis, 338
      real-life application, 337
   sum and difference pattern, 348–353
      error analysis, 352
   trinomials
      defined, 331
      error analysis, 373–374
      factoring, 368–381
      real-life application, 372
      writing, 373, 380
   writing, 332, 338, 345, 373, 380
   Zero-Product Property, 358
**Polynomial equation(s),** *See also* Polynomial(s)
   factored form, 356–361
      defined, 358
      error analysis, 360, 366
      real-life applications, 359, 365
      using greatest common factor, 362–367
      writing, 360, 366
      Zero-Product Property, 358
**Power of a Power Property,** 270
**Power of a Product Property,** 271
**Power of a Quotient Property,** 271
**Precision,** *Throughout. For example, see:*
   data displays, 657
   direct and inverse variation, 549
   equations
      direct and inverse variation, 549
      graphing, 42, 46
      quadratic, 473
      solving, 23
      systems of linear, 181
      writing, 76
   exponents, 275
   FOIL Method, 347
   functions
      domains and range, 207
      linear *vs.* nonlinear, 241
      square root, 507
   geometric sequences, 311
   graphing
      quadratic functions, 430
      square root functions, 507
   graphs, 42
   inequalities, 123
   polynomials, 347
   Pythagorean Theorem, 525
   quadratic equations, 473
   right triangles, 525
   slope, 76
   systems of linear equations, 181
**Prime polynomial,** *See also* Polynomial(s), 389
**Problem Solving,** *Throughout. For example, see:*
   equations
      graphing, 47
      linear, 47, 175
      multi-step equations, 15
      in point-slope form, 83
      quadratic, 483
   exponents, 275, 281
   expressions, 565
   graphs, 47
   inequalities, 114
      linear in two variables, 143
   linear equations, 175
   linear functions, 221
   perfect square trinomial, 387
   polynomials, 339
      factoring, 387
   quadratic equations, 483
   quadratic functions, 421
   rational expressions, 565
   square root functions, 507
**Process diagram,** 70
**Product of Powers Property,** 270
**Product Property of Square Roots,** 262
**Properties**
   Addition Property of Equality, 4
   Addition Property of Inequality, 112
   Cross Products Property, 590
   Distributive Property, 13
   Division Property of Equality, 5
   Division Property of Inequality, 118
   Multiplication Property of Equality, 5
   Multiplication Property of Inequality, 118
   Power of a Power Property, 270
   Power of a Product Property, 271
   Power of a Quotient Property, 271
   Product of Powers Property, 270
   Product Property of Square Roots, 262
   Quotient of Powers Property, 270
   Quotient Property of Square Roots, 262
   Subtraction Property of Equality, 4
   Subtraction Property of Inequality, 112
   Zero-Product Property, 358

**Pythagorean Theorem,** 520–531
  converse of, 528
  defined, 522
  distance formula and, 526–531
    error analysis, 530
    writing, 530
  error analysis, 524, 530
  real-life applications, 523, 529
  using, 526–531
    distance formula, 526–531
    real-life application, 529

## Q

**Quadratic equation(s)**
  defined, 456
  roots, 456
  solutions of
    choosing a method, 484–485
    no real solutions, 457
    one real solution, 456
    two real solutions, 456
    using a discriminant, 480
  solving by completing the
    square, 468–473, 484
    defined, 470
    error analysis, 472
    real-life application, 471
  solving by factoring, 484
  solving by graphing, 454–461, 484
    error analysis, 460
    real-life application, 458
  solving by the quadratic
    formula, 476–484
    error analysis, 481–482
    real-life application, 479
  solving using square roots, 462–467, 484
    error analysis, 466
    real-life application, 465
  systems of linear and, 486–491
    error analysis, 491
    writing, 490
  writing, 459, 467
**Quadratic formula**
  defined, 478
  discriminant and, 480
    interpreting, 480
  using to solve quadratic
    equations, 476–483
    error analysis, 481–482
    real-life application, 479
**Quadratic function(s)**
  characteristics
    axis of symmetry, 404
    vertex, 404

  compared to linear and
    exponential, 434–441
  defined, 404
  differences and ratios of, 437
    error analysis, 440
  error analysis, 408, 414, 429
  graphing, 402–421, 424–443
    axis of symmetry, 404, 426
    error analysis, 408, 414, 420, 429
    focus, 412
    parabola, 404, 410–415
    rates of change, 442–443
    real-life applications, 406, 413, 419, 428
    vertex, 404, 426
    vertex form of, 432
    writing, 407, 414, 420
  maximum value of, 427–428
    real-life application, 428
  minimum value of, 427–428
  rates of change, 442–443
  real-life applications, 406, 428
  vertex form of, 432–433
  writing, 407
  zero of, 419, 458
**Quartile,** defined, 620
Quotient of Powers Property, 270
Quotient Property of Square Roots, 262

## R

Radical(s), 276–281
  $n$th root, 276–278
  reading, 278
Radical expression(s), 508–509
  rationalizing the denominator
    defined, 508
  simplest form of, 508
  simplifying, 508–509
    conjugates and, 509
    real-life application, 509
**Range**
  of a function, 202–207
    defined, 204
    error analysis, 206
    real-life application, 205
**Range (of a data set)**
  defined, 614
  project, 617
Rate of change, 442–443
**Rational equation(s)**
  defined, 590
  solving, 588–593
    real-life application, 591
    using Cross Products
      Property, 590

    using least common
      denominator, 591
    writing, 592
Rational exponent(s), *See also*
    Exponent(s)
  error analysis, 280
  real-life application, 279
  writing, 280–281
**Rational expression(s)**
  adding, 580–587
    error analysis, 585–586
    real-life application, 584
    with like denominators, 582
    with unlike denominators, 583
  defined, 562
  dividing, 568–573
    error analysis, 572
    writing, 572
  excluded values of, 552, 570
  least common denominator of, 583
    defined, 583
    writing, 585
  multiplying, 568–573
    writing, 572
  simplest form of, 562
    writing, 564–565
  simplifying, 560–565
    error analysis, 564
    real-life application, 563
  subtracting, 580–587
    with like denominators, 582
    with unlike denominators, 584
**Rational function(s)**
  defined, 552
  graphing, 550–557
    asymptotes, 553
    error analysis, 556
    excluded values, 552
    real-life application, 554
    writing, 555
Rational number(s), 266–267
**Rationalizing the denominator,**
    defined, 508
Reading
  direct variation, 545
  function notation, 226
  inequality, 113
    linear with two variables, 138
    symbols, 106
  inverse functions, 559
  inverse variation, 545
  radical sign, 278
  systems of linear equations, 156
Real number(s)
  operations, 266–267
  square of, 464

Real-life applications, *Throughout.*
*For example, see:*
arithmetic sequences, 246
equations
absolute value, 25
graphing, 45
inverse variation, 546
multi-step, 13
quadratic, 458, 465, 479
rational, 591
rewriting, 29
simple, 6
in slope-intercept form, 61, 75
square root, 514
in standard form, 67
systems of linear, 157
with variable(s) on both sides, 21
exponential decay, 303
exponential growth, 297
exponents, 272
rational, 279
expressions, 584
simplifying, 563
factoring
difference of two squares, 385
polynomials, 365
functions
arithmetic sequences, 246
domain and range, 205
exponential, 288, 297
linear, 219
linear *vs.* nonlinear, 239
maximum value of, 428
quadratic, 413, 419, 428
rational, 554
square root, 505
geometry
Pythagorean Theorem, 523, 529
graphing
exponential functions, 288
linear equations, 45
quadratic functions, 406, 413, 419
rational functions, 554
square root functions, 505
inequalities, 113
absolute value, 135
compound, 135
linear in two variables, 140
multi-step, 129
inverse variation, 546
linear equations
point-slope form, 81
slope-intercept form, 61, 75
solving by graphing, 183

standard form, 67
parabolas, 413
polynomials, 331, 337, 344
factoring, 359, 365, 385
square of binomial pattern, 351
Pythagorean Theorem, 523, 529
quadratic equations
solving by completing the square, 471
solving by graphing, 458
solving by the quadratic formula, 479
solving using square roots, 465
quadratic formula, 479
rational equations, 591
rational expressions, 563
adding, 584
rational functions, 554
simple equations, 6
slope-intercept form, 61
square root functions, 505
square roots, 263
systems of linear equations
solving by elimination, 172
solving by graphing, 157
solving by substitution, 163
systems of linear inequalities, 188
$x$-intercepts, 61
$y$-intercepts, 61
Real-life problems
solving, 86–93
writing, 86–93
Reasoning, *Throughout. For example, see*:
data analysis
distribution, 633
lines of fit, 643
measures of central tendency, 611
measures of dispersion, 616–617
standard deviation, 616–617
data displays
choosing, 662–663
scatter plots, 643
two-way tables, 657
direct variation, 548
distance formula in a coordinate plane, 531
domains, 215, 249
of a function, 408, 556
of a square root function, 506–507
equations
direct variation, 548–549

exponential, 292
inverse variation, 549
linear, 63, 68
point-slope form, 85
quadratic, 430, 460–461, 467, 473, 483, 491
rational, 593
real solutions of, 466
real-life problems, 91
rewriting, 31
slope, 53, 55
slope-intercept form, 63
square root, 516
exponential equations, 292
exponential functions, 290–291, 299
rates of change, 443
exponential growth, 299
exponents, 274–275
expressions
adding, 585
dividing, 573
multiplying, 586
factoring, 432
perfect square trinomial, 386
polynomials, 360, 366, 381
FOIL Method, 346
formulas, 31
functions, 209
absolute value, 235
arithmetic sequences, 249
differences and ratios of, 440
domains, 249, 506
exponential, 290–291, 299, 443
linear, 208, 220–221, 443
piecewise, 232
quadratic, 408, 415, 419–421
rates of change, 443
rational, 549, 556, 564
square root, 506–507
zero of, 421
graphs, 68, 290–291
comparing, 432
focus of a parabola, 415, 421
quadratic functions, 415, 419–421, 431
scatter plots, 643
indirect variation, 549
inequalities, 108, 114–115
linear in two variables, 141, 143
multi-step, 131
systems of linear, 189–190
intercepts, 90
inverse variation, 549
linear equations, 209
systems of, 159, 180–181

Student Index **A77**

linear functions, 219
    rates of change, 443
lines of fit, 643
    correlation coefficient, 651
measures of central tendency, 611
measures of dispersion, 616
mental math, 516
point-slope form, 85
polynomials
    dividing, 578–579
    factoring, 360, 366, 381, 386
    FOIL Method, 346
    multiplying, 346
    subtracting, 338
    sum and difference pattern, 352
Pythagorean Theorem, 531
quadratic equations, 460–461, 467, 483
    solving by completing the square, 473
    systems of linear and, 491
quadratic formula
    discriminant, 483
    solving by, 483
quadratic functions, 408, 431
    axis of symmetry, 430
    graphing, 415, 419–421
    vertex form of, 433
rational equations, 548, 593
rational functions, 556, 564
real-life problems, 90–91
right triangles, 531
scatter plots, 643
sequences
    arithmetic, 249
    geometric, 310–311
simple equations, 9
slope, 53, 55, 90
solutions, 466
square root equations, 516
square root functions, 507
    domain of, 506
standard deviation, 616–617
systems of linear equations, 159, 164, 174–175
    number of solutions, 180–181
systems of linear inequalities, 189–190
two-way tables, 657
**Recursive rule**
    defined, 312
    writing, 313
Recursively defined sequences, 312–315

recursive equation
    arithmetic sequence, 312
    geometric sequence, 312
recursive rule, 312
**Relation(s),**
    defined, 208
    functions and, 208–209
    inverse, 557
Repeated Reasoning, *Throughout. For example, see:*
    arithmetic sequences, 248
    equations
        linear systems of, 165
        quadratic, 482
    geometric sequences, 311
    graphing rational functions, 557
    inequalities, 109
        system of linear, 191
    polynomials
        dividing, 579
        multiplying, 353
    quadratic equations, 482
    quadratic formula, 482
    rewriting equations and formulas, 31
    systems of linear equations, 165
    systems of linear inequalities, 191
**Residual,** defined, 646
Right triangle, *See* Triangle(s)
**Rise,** defined, 50
**Root,** defined, 358
Roots of numbers
    cube, 276
    $n$th, 276–278
    square, 504
**Run,** defined, 50

**Ⓢ**

**Scatter plot,** 636–643, 660
    defined, 638
    lines of fit, 640–643
        analyzing, 644–651
        correlation coefficient, 647
        defined, 640
        line of best fit, 647
        linear regression, 647
        residuals, 646–647
        writing, 641
    relationship of data, 639
**Sequences**
    arithmetic, 242–247
        defined, 244
        equation for, 245
        error analysis, 248
        real-life application, 246

        recursive equation for, 312
        writing, 247, 310
    defined, 244
    geometric, 306–311
        common ratio, 308
        equation for, 309
        error analysis, 310
        extending, 308
        graphing, 308
        real-life application, 309
        recursive equation for, 312
        writing, 310
    recursively defined, 312–315
    recursive rule, 312
**Simplest form of a radical expression**
    conjugates and, 509
    defined, 508
    real-life application, 509
**Simplest form of a rational expression,** defined, 562
**Slope**
    defined, 48, 50
    error analysis, 54
    formula, 50
    graphing, 48–55
    negative, 50, 52
    parallel lines, 56, 84
    perpendicular lines, 57
    positive, 50, 52
    project, 54
    rise, 50
    run, 50
    undefined, 51–52
    zero, 51–52
**Slope-intercept form,** 58–63
    defined, 60
    graphing equations in
        error analysis, 62
        real-life application, 61
    writing equations in, 72–77
        error analysis, 76
        real-life application, 75
Solution(s)
    extraneous, 513
    of linear equations, 42, 44
    of quadratic equations
        choosing a method of, 484–485
        no real, 457
        one real, 456
        two real, 456
**Solution of an inequality,** defined, 106
**Solution of a linear inequality,** *See also* Inequality
    defined, 138

Solution points, *See* Solution(s)
**Solution set of an inequality,**
  defined, 106
**Solution of a system of linear equations,** *See also* Systems of linear equations
  defined, 156
**Solution of a system of linear inequalities,** *See also* Systems of linear inequalities
  defined, 186
Sphere, volume of, 31
**Square of binomial pattern,** 348–353
  error analysis, 352
  real-life application, 351
Square root(s), 260–265
  error analysis, 264
  evaluating, 262
  $n$th root, 276–278
    defined, 278
  operations with, 261
  Product Property of, 262
    writing, 264
  Quotient Property of, 262
    writing, 264
  real-life application, 263
  simplifying, 262
  symbol, 260
  to solve quadratic equations, 462–467
    error analysis, 466
    real-life application, 465
    writing, 467
  writing, 264
**Square root equation(s)**
  defined, 512
  solving, 510–517
    error analysis, 515–516
    extraneous solutions, 513
    real-life application, 514
    writing, 515–516
**Square root function(s)**
  defined, 504
  domain of, 504
  graphing, 502–507
    error analysis, 507
    real-life application, 505
**Standard deviation**
  defined, 615
  project, 617
**Standard form**
  of linear equations
    defined, 66
    error analysis, 68

real-life application, 67
Standardized Test Practice
  domains, 213
  equations
    inverse variation, 546
    quadratic, 489
    with variables on both sides, 21
    writing in slope-intercept form, 75
  exponents, 272
  expressions
    dividing, 571
  factoring
    polynomials, 359, 379
    trinomials, 379
  function notation, 228
  functions
    linear *vs.* nonlinear, 239
    quadratic, 419
  graphs
    of multi-step inequalities, 129
    quadratic functions, 419
  inequalities, 129
  inverse variation, 546
  polynomials, 337
    factoring, 359, 379
  quadratic equations
    systems of linear and, 489
  quadratic functions, 419
  rational expressions
    dividing, 571
  simple equations, 6
  slope-intercept form, 75
  trinomials
    factoring, 379
Stem-and-leaf plot, 660
**Step function(s),** *See also* Function(s)
  defined, 233
Structure, *Throughout. For example, see:*
  completing the square, 473
  data displays, 663
  distance formula in a coordinate plane, 531
  equations
    slope, 55
    solving, 23
    systems of linear, 165
  exponential functions, 291
    decay, 305
  expressions, dividing, 573
  factoring polynomials, 381
  functions
    domain and range, 235
    exponential, 291, 305

graphing, 415
notation, 231
piecewise, 235
inequalities, 123
  system of linear, 191
inverse variation, 549
polynomials. factoring, 381
Pythagorean Theorem, 525
  distance formula and, 531
quadratic equations
  solving by completing the square, 473
quadratic functions, 415
rational expressions, 573
rational numbers, 267
right triangles, 525
slope, 55
systems of linear equations, 165
systems of linear inequalities, 191
Study Tips
  absolute value function, 234
  arithmetic sequences, 246
  asymptotes, 554
  box-and-whisker plots, 620, 622
  completing the square, 470
  compound interest, 297
  conjugates, 509
  constant of variation, 544
  data displays
    box-and-whisker plots, 620, 622
    lines of fit, 640
    scatter plots, 639
  equations
    completing the square, 470
    constant of variation, 544
    factoring, 456
    inverse variation, 545
    linear, 66, 183
    of lines of fit, 647
    multi-step, 13
    quadratic, 480, 484
    slope, 50
    in standard form, 66
    writing, 74, 81
  exponential functions, 287–288
    compound interest, 297
    decay, 302
    geometric sequences, 309
    growth, 296
  exponents rational, 288
  expressions
    excluded values of, 562
    rational, 562
  factoring
    equations, 456
    polynomials, 364, 378–379

function notation, step, 233
functions, 438
   absolute value, 234
   arithmetic sequences, 246
   exponential, 287, 302
   inverse, 559
   linear, 437
   linear *vs.* nonlinear, 238–239
   notation, 227
   piecewise, 234
   quadratic, 428, 437
   square root, 505
geometric sequence, 309
graphing
   asymptotes, 554
   horizontal translation, 505
   square root function, 505
graphing calculator, 428
graphs
   box-and-whisker plots, 620, 622
   lines of fit, 640
   scatter plots, 639
horizontal translation, 505
inequalities, 112, 129
   absolute value, 135
   compound, 132–133
   graphing, 129
   systems of linear, 187
inverse functions, 559
inverse operations, 112
inverse variation, 545
linear functions, 437
lines of fit, 640, 648
   equations of, 647
parabolas, 428
period of a pendulum, 514
piecewise function, 234
polynomials, 336–337
   dividing, 576–577
   factoring, 364, 378–379
   sum and difference pattern, 350
Pythagorean triples, 528
quadratic equations
   roots of, 480
   solutions of, 480, 484
quadratic functions, 428, 437
   vertex form of, 433
radical expressions
   conjugates, 509
   rationalizing the denominator, 508
radical symbol, 278
rate of change, 238
rational exponents, 288
rational expressions, 562

rationalizing the denominator, 508
right triangles, 522
scatter plots, 639
   lines of fit, 640, 648
slope, 50
slope of a vertical line, 56
solutions
   checking for reasonableness, 465
   of quadratic equations, 480
systems of linear equations
   solving by elimination, 170–171
   solving by graphing, 157
   solving by substitution, 163
systems of linear inequalities, 187
triangles
   Pythagorean triples, 528
   right, 522
trinomials, factoring, 378–379
variation, constant of, 544
Subtraction
   as inverse of addition, 4
   polynomials, 334–339
      error analysis, 338
      real-life application, 337
   Property
      of Equality, 4
      of Inequality, 112
   rational expressions, 580–587
      least common denominator of, 583
      with like denominators, 582
      with unlike denominators, 584
   to solve equations, 4–5
Subtraction Property of Equality, 4
Subtraction Property of Inequality, 112
Sum and difference pattern, 348–353
   error analysis, 352
Summary triangle, 422
Surface area
   of a cone, 28
   of a cylinder, 28
Symbols
   of inequality, 106
      reading, 106
   radical sign, 278
   square root, 260
**Systems of linear equations**
   defined, 154, 156
   number of solutions, 176–181
      error analysis, 180

      infinitely many, 178–179
      no solution, 178
      one solution, 178
      writing, 180
   reading, 156
   solution of
      defined, 156
   solving by elimination, 168–175
      error analysis, 173–174
      real-life application, 172
      writing, 173
   solving by graphing, 154–159
      error analysis, 159
      real-life application, 157
   solving by substitution, 160–165
      error analysis, 165
      real-life application, 163
      writing, 164
   writing, 158, 164, 173
Systems of linear and quadratic equations
   solving
      by elimination, 486–487, 489–491
      by graphing, 486–487, 489–491
      by substitution, 486–488, 490–491
      error analysis, 491
      writing, 490
**Systems of linear inequalities,** 184–191
   defined, 186
   graph of
      defined, 186
   graphing, 186–187
   solution of
      defined, 186
   solving
      error analysis, 190
      real-life application, 188
      writing, 189
   writing, 188

**T**

Temperature conversion formula, 29
**Term of a sequence,** defined, 244
**Theorem,** defined, 520
Triangle(s)
   Pythagorean Theorem and, 520–531
      converse of, 528
      distance formula and, 526–531
      error analysis, 524

real-life application, 523, 529
  using, 526–531
 right, 520–531
  hypotenuse of, 522
  identifying, 528
  legs of, 522
  real-life application, 529
Triangle Inequality, 105
**Trinomial(s),** *See also*
  Polynomial(s)
 defined, 331
 factoring, 368–381
  error analysis, 373–374, 380
  real-life application, 372
  writing, 373, 380
**Two-way table(s),** 652–657
 defined, 654
 joint frequencies, 654
 marginal frequencies, 654
 relationships in, 655

## V

Variables
 dependent
  defined, 204
 independent
  defined, 204
  on both sides of an equation, 18–13
  error analysis, 22
  real-life application, 21
  writing, 22–23
Variation
 constant of, 544
 direct, *See* Direct variation
 inverse, *See* Inverse variation
**Vertex,** defined, 404
**Vertex form,** of a quadratic function, 432–433
**Vertical Line Test,** defined, 209
Volume, 26
 of a cylinder, 9
 of a sphere, 31

## W

Word magnet, 518
Writing, *Throughout. For example, see:*
 arithmetic sequences, 247, 310
 asymptotes, 555
 box-and-whisker plots, 623–625
 coordinate plane
  distance formula, 530

data analysis
 distribution, 631
 lines of fit, 641
 outliers, 610
data displays
 box-and-whisker plots, 624–625
 choosing a, 660
 quartiles, 623
distance formula, 530
domains, 214
equations, 14
 absolute value, 25
 quadratic, 459, 467
 rational, 592
 square root, 515–516
 systems of linear, 158
 with variables on both sides, 22–23
exponential decay, 304
exponents
 $n$th root, 280
 rational, 281
expressions
 dividing, 572
 least common denominator, 585
 multiplying, 572
 in simplest form, 564, 565
factoring
 difference of two squares, 386
 trinomials, 373
focus of a parabola, 414
function notation, 229
functions
 exponential, 304
 linear, 247
 quadratic, 407
geometric sequences, 310
graphing
 quadratic equations, 459
 rational functions, 555
graphs
 box-and-whisker plots, 623, 625
inequalities, 121, 130
 multi-step, 130
 systems of linear, 189
 in two variables, 141
linear equations, 76
lines of fit, 641, 649
monomials, 332
outliers, 610
perfect square trinomial pattern, 472

polynomials, 332, 338
 dividing, 578
 factoring, 360, 373, 386
 FOIL Method, 345
 with greatest common factor, 366
quadratic equations, 459, 467
quadratic functions, 407, 414, 420
rational equations, 592
residuals, 649
square root equations, 515–516
square roots, 264
systems of linear equations
 number of solutions, 180
 solving by elimination, 173
 solving by graphing, 158
 solving by substitution, 164
systems of linear and quadratic equations, 490–491
systems of linear inequalities, 189

## X

**$x$-intercept**
 defined, 60
 real-life application, 61

## Y

Y chart, 16
**$y$-intercept,** defined, 60

## Z

Zero
 of a function, 418, 458
**Zero-Product Property,** defined, 358

# Common Core State Standards

## Conceptual Category: Number and Quantity

### Domain: The Real Number System

**Extend the properties of exponents to rational exponents.**

**N.RN.1** Explain how the definition of the meaning of rational exponents follows from extending the properties of integer exponents to those values, allowing for a notation for radicals in terms of rational exponents.

**N.RN.2** Rewrite expressions involving radicals and rational exponents using the properties of exponents.

**Use properties of rational and irrational numbers.**

**N.RN.3** Explain why the sum or product of two rational numbers is rational; that the sum of a rational number and an irrational number is irrational; and that the product of a nonzero rational number and an irrational number is irrational.

### Domain: Quantities

**Reason quantitatively and use units to solve problems.**

**N.Q.1** Use units as a way to understand problems and to guide the solution of multi-step problems; choose and interpret units consistently in formulas; choose and interpret the scale and the origin in graphs and data displays.

**N.Q.2** Define appropriate quantities for the purpose of descriptive modeling.

**N.Q.3** Choose a level of accuracy appropriate to limitations on measurement when reporting quantities.

## Conceptual Category: Algebra

### Domain: Seeing Structure in Expressions

**Interpret the structure of expressions.**

**A.SSE.1** Interpret expressions that represent a quantity in terms of its context.
   a. Interpret parts of an expression, such as terms, factors, and coefficients.
   b. Interpret complicated expressions by viewing one or more of their parts as a single entity.

**A.SSE.2** Use the structure of an expression to identify ways to rewrite it.

**Write expressions in equivalent forms to solve problems.**

**A.SSE.3** Choose and produce an equivalent form of an expression to reveal and explain properties of the quantity represented by the expression.
   a. Factor a quadratic expression to reveal the zeros of the function it defines.
   b. Complete the square in a quadratic expression to reveal the maximum or minimum value of the function it defines.
   c. Use the properties of exponents to transform expressions for exponential functions.

## Domain: Arithmetic with Polynomials and Rational Expressions

**Perform arithmetic operations on polynomials.**

**A.APR.1** Understand that polynomials form a system analogous to the integers, namely, they are closed under the operations of addition, subtraction, and multiplication; add, subtract, and multiply polynomials.

## Domain: Creating Equations

**Create equations that describe numbers or relationships.**

**A.CED.1** Create equations and inequalities in one variable and use them to solve problems. *Include equations arising from linear and quadratic functions, and simple rational and exponential functions.*

**A.CED.2** Create equations in two or more variables to represent relationships between quantities; graph equations on coordinate axes with labels and scales.

**A.CED.3** Represent constraints by equations or inequalities, and by systems of equations and/or inequalities, and interpret solutions as viable or non-viable options in a modeling context.

**A.CED.4** Rearrange formulas to highlight a quantity of interest, using the same reasoning as in solving equations.

## Domain: Reasoning with Equations and Inequalities

**Analyze and solve linear equations and pairs of simultaneous linear equations.**

**8.EE.8** Analyze and solve pairs of simultaneous linear equations.

    **a.** Understand that solutions to a system of two linear equations in two variables correspond to points of intersection of their graphs, because points of intersection satisfy both equations simultaneously.

    **b.** Solve systems of two linear equations in two variables algebraically, and estimate solutions by graphing the equations. Solve simple cases by inspection.

    **c.** Solve real-world and mathematical problems leading to two linear equations in two variables.

**Understand solving equations as a process of reasoning and explain the reasoning.**

**A.REI.1** Explain each step in solving a simple equation as following from the equality of numbers asserted at the previous step, starting from the assumption that the original equation has a solution. Construct a viable argument to justify a solution method.

**Solve equations and inequalities in one variable.**

**A.REI.3** Solve linear equations and inequalities in one variable, including equations with coefficients represented by letters.

**A.REI.3.1** Solve one-variable equations and inequalities involving absolute value, graphing the solutions and interpreting them in context. (CA Standard)

**A.REI.4** Solve quadratic equations in one variable.

    **a.** Use the method of completing the square to transform any quadratic equation in $x$ into an equation of the form $(x - p)^2 = q$ that has the same solutions. Derive the quadratic formula from this form.

    **b.** Solve quadratic equations by inspection (e.g., for $x^2 = 49$), taking square roots, completing the square, the quadratic formula and factoring, as appropriate to the initial form of the equation. Recognize when the quadratic formula gives complex solutions [and write them as $a \pm bi$ for real numbers $a$ and $b$]*.

\* Items in brackets are not required until Algebra II.

## Solve systems of equations.

**A.REI.5**  Prove that, given a system of two equations in two variables, replacing one equation by the sum of that equation and a multiple of the other produces a system with the same solutions.

**A.REI.6**  Solve systems of linear equations exactly and approximately (e.g., with graphs), focusing on pairs of linear equations in two variables.

**A.REI.7**  Solve a simple system consisting of a linear equation and a quadratic equation in two variables algebraically and graphically.

## Represent and solve equations and inequalities graphically.

**A.REI.10**  Understand that the graph of an equation in two variables is the set of all its solutions plotted in the coordinate plane, often forming a curve (which could be a line).

**A.REI.11**  Explain why the $x$-coordinates of the points where the graphs of the equations $y = f(x)$ and $y = g(x)$ intersect are the solutions of the equation $f(x) = g(x)$; find the solutions approximately, e.g., using technology to graph the functions, make tables of values, or find successive approximations. Include cases where $f(x)$ and/or $g(x)$ are linear, [polynomial, rational, absolute value,] exponential, [and logarithmic functions].

**A.REI.12**  Graph the solutions to a linear inequality in two variables as a half-plane (excluding the boundary in the case of a strict inequality), and graph the solution set to a system of linear inequalities in two variables as the intersection of the corresponding half-planes.

# Conceptual Category: Functions
# Domain: Interpreting Functions

## Define, evaluate, and compare functions.

**8.F.1**  Understand that a function is a rule that assigns to each input exactly one output. The graph of a function is the set of ordered pairs consisting of an input and the corresponding output.

**8.F.2**  Compare properties of two functions each represented in a different way (algebraically, graphically, numerically in tables, or by verbal descriptions).

**8.F.3**  Interpret the equation $y = mx + b$ as defining a linear function, whose graph is a straight line; give examples of functions that are not linear.

## Use functions to model relationships between quantities.

**8.F.4**  Construct a function to model a linear relationship between two quantities. Determine the rate of change and initial value of the function from a description of a relationship or from two $(x, y)$ values, including reading these from a table or from a graph. Interpret the rate of change and initial value of a linear function in terms of the situation it models, and in terms of its graph or a table of values.

**8.F.5**  Describe qualitatively the functional relationship between two quantities by analyzing a graph (e.g., where the function is increasing or decreasing, linear or nonlinear). Sketch a graph that exhibits the qualitative features of a function that has been described verbally.

## Understand the concept of a function and use function notation.

**F.IF.1**  Understand that a function from one set (called the domain) to another set (called the range) assigns to each element of the domain exactly one element of the range. If $f$ is a function and $x$ is an element of its domain, then $f(x)$ denotes the output of $f$ corresponding to the input $x$. The graph of $f$ is the graph of the equation $y = f(x)$.

**F.IF.2**  Use function notation, evaluate functions for inputs in their domains, and interpret statements that use function notation in terms of a context.

**F.IF.3**  Recognize that sequences are functions, sometimes defined recursively, whose domain is a subset of the integers.

### Interpret functions that arise in applications in terms of the context.

**F.IF.4**  For a function that models a relationship between two quantities, interpret key features of graphs and tables in terms of the quantities, and sketch graphs showing key features given a verbal description of the relationship. *Key features include: intercepts; intervals where the function is increasing, decreasing, positive, or negative; relative maximums and minimums; symmetries; end behavior; [and periodicity].*

**F.IF.5**  Relate the domain of a function to its graph and, where applicable, to the quantitative relationship it describes.

**F.IF.6**  Calculate and interpret the average rate of change of a function (presented symbolically or as a table) over a specified interval. Estimate the rate of change from a graph.

### Analyze functions using different representations.

**F.IF.7**  Graph functions expressed symbolically and show key features of the graph, by hand in simple cases and using technology for more complicated cases.
   a. Graph linear and quadratic functions and show intercepts, maxima, and minima.
   b. Graph square root, [cube root,] and piecewise-defined functions, including step functions and absolute value functions.
   e. Graph exponential [and logarithmic] functions, showing intercepts and end behavior, {and trigonometric functions, showing period, midline, and amplitude].

**F.IF.8**  Write a function defined by an expression in different but equivalent forms to reveal and explain different properties of the function.
   a. Use the process of factoring and completing the square in a quadratic function to show zeros, extreme values, and symmetry of the graph, and interpret these in terms of a context.
   b. Use the properties of exponents to interpret expressions for exponential functions.

**F.IF.9**  Compare properties of two functions each represented in a different way (algebraically, graphically, numerically in tables, or by verbal descriptions).

## Domain: Building Functions

### Build a function that models a relationship between two quantities.

**F.BF.1**  Write a function that describes a relationship between two quantities.
   a. Determine an explicit expression, a recursive process, or steps for calculation from a context.
   b. Combine standard function types using arithmetic operations.

**F.BF.2**  Write arithmetic and geometric sequences both recursively and with an explicit formula, use them to model situations, and translate between the two forms.

### Build new functions from existing functions.

**F.BF.3**  Identify the effect on the graph of replacing $f(x)$ by $f(x) + k$, $k f(x)$, $f(kx)$, and $f(x + k)$ for specific values of $k$ (both positive and negative); find the value of $k$ given the graphs. Experiment with cases and illustrate an explanation of the effects on the graph using technology. [*Include recognizing even and odd functions from their graphs and algebraic expressions for them.*]

**F.BF.4**  Find inverse functions.
   a. Solve an equation of the form $f(x) = c$ for a simple function $f$ that has an inverse and write an expression for the inverse.

# Domain: Linear, Quadratic, and Exponential Models

**Construct and compare linear, quadratic, and exponential models and solve problems.**

**F.LE.1** Distinguish between situations that can be modeled with linear functions and with exponential functions.
- a. Prove that linear functions grow by equal differences over equal intervals; and that exponential functions grow by equal factors over equal intervals.
- b. Recognize situations in which one quantity changes at a constant rate per unit interval relative to another.
- c. Recognize situations in which a quantity grows or decays by a constant percent rate per unit interval relative to another.

**F.LE.2** Construct linear and exponential functions, including arithmetic and geometric sequences, given a graph, a description of a relationship, or two input-output pairs (include reading these from a table).

**F.LE.3** Observe using graphs and tables that a quantity increasing exponentially eventually exceeds a quantity increasing linearly, quadratically, [or (more generally) as a polynomial function].

**Interpret expressions for functions in terms of the situation they model.**

**F.LE.5** Interpret the parameters in a linear or exponential function in terms of a context.

**F.LE.6** Apply quadratic functions to physical problems, such as the motion of an object under the force of gravity. (CA Standard)

# Conceptual Category: Geometry

# Domain: Geometric Measurement and Dimension

**Understand and apply the Pythagorean Theorem.**

**8.G.6** Explain a proof of the Pythagorean Theorem and its converse.

**8.G.7** Apply the Pythagorean Theorem to determine unknown side lengths in right triangles in real-world and mathematical problems in two and three dimensions.

**8.G.8** Apply the Pythagorean Theorem to find the distance between two points in a coordinate system.

# Conceptual Category: Statistics and Probability

# Domain: Interpreting Categorical and Quantitative Data

**Investigate patterns of association in bivariate data.**

**8.SP.1** Construct and interpret scatter plots for bivariate measurement data to investigate patterns of association between two quantities. Describe patterns such as clustering, outliers, positive or negative association, linear association, and nonlinear association.

**8.SP.2** Know that straight lines are widely used to model relationships between two quantitative variables. For scatter plots that suggest a linear association, informally fit a straight line, and informally assess the model fit by judging the closeness of the data points to the line.

**8.SP.3** Use the equation of a linear model to solve problems in the context of bivariate measurement data, interpreting the slope and intercept.

**8.SP.4** Understand that patterns of association can also be seen in bivariate categorical data by displaying frequencies and relative frequencies in a two-way table. Construct and interpret a two-way table summarizing data on two categorical variables collected from the same subjects. Use relative frequencies calculated for rows or columns to describe possible association between the two variables.

## Summarize, represent, and interpret data on a single count or measurement variable.

**S.ID.1** Represent data with plots on the real number line (dot plots, histograms, and box plots).

**S.ID.2** Use statistics appropriate to the shape of the data distribution to compare center (median, mean) and spread (interquartile range, standard deviation) of two or more different data sets.

**S.ID.3** Interpret differences in shape, center, and spread in the context of the data sets, accounting for possible effects of extreme data points (outliers).

## Summarize, represent, and interpret data on two categorical and quantitative variables.

**S.ID.5** Summarize categorical data for two categories in two-way frequency tables. Interpret relative frequencies in the context of the data (including joint, marginal, and conditional relative frequencies). Recognize possible associations and trends in the data.

**S.ID.6** Represent data on two quantitative variables on a scatter plot, and describe how the variables are related.

  a. Fit a function to the data; use functions fitted to data to solve problems in the context of the data. *Use given functions or choose a function suggested by the context. Emphasize linear and exponential models.*

  b. Informally assess the fit of a function by plotting and analyzing residuals.

  c. Fit a linear function for a scatter plot that suggests a linear association.

## Interpret linear models.

**S.ID.7** Interpret the slope (rate of change) and the intercept (constant term) of a linear model in the context of the data.

**S.ID.8** Compute (using technology) and interpret the correlation coefficient of a linear fit.

**S.ID.9** Distinguish between correlation and causation.

# Photo Credits

**Cover**
Andrea Danti/Shutterstock.com, iurii/Shutterstock.com, valdis torms/Shutterstock.com

**Front matter**
**i** Andrea Danti /Shutterstock.com, iurii/Shutterstock.com, valdis torms/Shutterstock.com; **iv** Big Ideas Learning, LLC; **viii** *top* ©iStockphoto.com/Lisa Thornberg, ©iStockphoto.com/Ann Marie Kurtz; *bottom* ©iStockphoto.com/Jane norton; **ix** *top* ©iStockphoto.com/Jonathan Larsen; *bottom* wavebreakmedia ltd/Shutterstock.com; **x** *top* ©iStockphoto.com/ALEAIMAGE, ©iStockphoto.com/Ann Marie Kurtz; *bottom* Odua Images/Shutterstock.com; **xi** *top* stephan kerkhofs/Shutterstock.com, Cigdem Sean Cooper/Shutterstock.com, ©iStockphoto.com/Andreas Gradin; *bottom* James Flint/Shutterstock.com; **xii** *top* ©iStockphoto.com/sumnersgraphicsinc, ©iStockphoto.com/Ann Marie Kurtz; *bottom* william casey/Shutterstock.com; **xiii** *top* Chiyacat/Shutterstock.com, Zoom Team/Shutterstock.com; *bottom* Edyta Pawlowska/Shutterstock.com; **xiv** *top* Varina and Jay Patel/Shutterstock.com, ©iStockphoto.com/Ann Marie Kurtz; *bottom* PETER CLOSE/Shutterstock.com; **xv** *top* ©iStockphoto.com/Alistair Cotton; *bottom* Kharidehal Abhirama Ashwin/Shutterstock.com; **xvi** *top* ©iStockphoto.com/ALEAIMAGE, ©iStockphoto.com/Ann Marie Kurtz; *bottom* ©iStockphoto.com/Noraznen Azit; **xvii** *top* ©iStockphoto.com/Michael Flippo, ©iStockphoto.com/Ann Marie Kurtz; *bottom* ©iStockphoto.com/Thomas Perkins; **xviii** *top* Alexander Chaikin/Shutterstock.com, ©iStockphoto.com/Ann Marie Kurtz; *bottom* Evok20/Shutterstock.com; **xix** *top* Kasiap/Shutterstock.com, ©iStockphoto.com/Ann Marie Kurtz; *bottom* Sinisa Bobic/Shutterstock.com; **xx** Ljupco Smokovski/Shutterstock.com

**Chapter 1**
**1** ©iStockphoto.com/Lisa Thornberg, ©iStockphoto.com/Ann Marie Kurtz; **6** ©iStockphoto.com/David Freund; **7** ©iStockphoto.com/nicolas hansen; **8** NASA; **9** ©iStockphoto.com/Ryan Lane; **12** ©iStockphoto.com/Harley McCabe, **13** ©iStockphoto.com/Jacom Stephens; **14** ©iStockphoto.com/Harry Hu; **15** ©iStockphoto.com/Ralf Hettler, Vibrant Image Studio/Shutterstock.com **22** ©iStockphoto.com/Andrey Krasnov; **25** Chris Curtis/Shutterstock.com; **31** *top right* ©iStockphoto.com/Alan Crawford; *center left* ©iStockphoto.com/Julio Yeste; *bottom right* ©iStockphoto.com/Mark Stay; **36** *center right* Ljupco Smokovski/Shutterstock.com; *bottom left* emel82/Shutterstock.com

**Chapter 2**
**40** ©iStockphoto.com/Jonathan Larsen; **45** NASA; **46** ©iStockphoto.com/David Morgan; **47** *top right* NASA; *center left* ©iStockphoto.com/jsemeniuk, **54** ©iStockphoto.com/Amanda Rohde; **55** Julian Rovagnati/Shutterstock.com; **62** ©iStockphoto.com/Dreamframer; **63** *top right* Jerry Horbert/Shutterstock.com; *bottom left* ©iStockphoto.com/Chris Schmidt; **65** ©iStockphoto.com/biffspandex; **68** ©iStockphoto.com/Stephen Pothier; **69** *top left* Gina Smith/Shutterstock.com; *bottom left* Dewayne Flowers/Shutterstock.com; **71** Philip Lange/Shutterstock.com; **75** Photo courtesy of Herrenknecht AG; **76** ©iStockphoto.com/Adam Mattel; **77** *top left* ©iStockphoto.com/Gene Chutka; *center right* ©iStockphoto.com/marcellus2070, ©iStockphoto.com/beetle8; **81** ©iStockphoto.com/Connie Maher; **82** ©iStockphoto.com/Jacom Stephens; **83** *top right* ©iStockphoto.com/Petr Podzemny; *center left* ©iStockphoto.com/adrian beesley; **88** ©iStockphoto.com/Ryan Putnam; **89** ©iStockphoto.com/iLexx; **90** ©iStockphoto.com/Jeremy Edwards; **91** ©iStockphoto.com/Rober Kohlruber; **97** ©iStockphoto.com/Marcio Silva; **98** ©iStockphoto.com/Carmen Martinez Banús

**Chapter 3**
**102** ©iStockphoto.com/ALEAIMAGE, ©iStockphoto.com/Ann Marie Kurtz; **105** ©iStockphoto.com/Floortje; **109** *top right* The ESRB rating icons are registered trademarks of the Entertainment Software Association; *center left* ©iStockphoto.com/Kevin Panizza, ©iStockphoto.com/Island Effects, ©iStockphoto.com/Nico Smit; *center right* ©iStockphoto.com/Richard Goerg; **110** ©iStockphoto.com/George Peters; **111** ©iStockphoto.com/George Peters; **114** pandapaw/Shutterstock.com; **115** ©iStockphoto.com/Daniel Van Beek; **121** Alexander Kalina/Shutterstock.com; **122** *top right* ©iStockphoto.com/Peter Firus; *bottom* Robert Pernell/Shutterstock.com; **123** ©iStockphoto.com/Trevor Fisher; **130** ©iStockphoto.com/fotoVoyager; **144** *center right* Africa Studio/Shutterstock.com; *bottom right* lanitta/Shutterstock.com

**Chapter 4**
**152** stephan kerkhofs/Shutterstock.com, Cigdem Sean Cooper/Shutterstock.com, ©iStockphoto.com/Andreas Gradin; **154** Howard Sandler/Shutterstock.com, ©iStockphoto.com/Dori OConnell; **157** Richard Paul Kane/Shutterstock.com; **158** ©iStockphoto.com/pelicankate; **160** *top* YuriyZhuravov/Shutterstock.com; *bottom* Talvi/Shutterstock.com; **163** aguilarphoto/Shutterstock.com; **164** *Exercises 4–6* Talvi/Shutterstock.com; *bottom* Kiselev Andrey Valerevich/Shutterstock.com; **165** *center left* Susan Schmitz/Shutterstock.com; *center right* akva/Shutterstock.com; **167** Andrey Yurlov/Shutterstock.com; **168** Steve Cukrov/Shutterstock.com; **172** *top left* Le Do/Shutterstock.com; *center left* Quang Ho/Shutterstock.com, SergeyIT/Shutterstock.com, jon le-bon/Shutterstock.com; **173** Ariwasabi/Shutterstock.com; **174** Ewa/Shutterstock.com; **175** *top left* Gordana Sermek/Shutterstock.com; *center right* Rashevskyi Viacheslav/Shutterstock.com; **176** ©iStockphoto.com/walik; **180** ©iStockphoto.com/Corina Estepa; **181** ©iStockphoto.com/Tomislav Forgo; **183** Kateryna Larina/Shutterstock.com; **188** Sandra van der Steen/Shutterstock.com; **189** Nikola Bilic/Shutterstock.com; **190** Kristijan Zontar/Shutterstock.com; **192** Selena/Shutterstock.com; **196** Poznyakov/Shutterstock.com

**Chapter 5**
**200** ©iStockphoto.com/sumnersgraphicsinc, ©iStockphoto.com/Ann Marie Kurtz; **203** *bottom left* ©iStockphoto.com/Andrew Johnson; *bottom right* Image courtesy of Zappos.Com; **205** ©iStockphoto.com/alohaspirit; **206** ©iStockphoto.com/Timur Kulgarin; **207** *top right* Primo Ponies Photography; *center left* ©iStockphoto.com/Wayne Johnson; **211** ©Fotosearch.com; **214** ©iStockphoto.com/Hannu Liivaar; **215** ©iStockphoto.com/LoopAll; **219** ©iStockphoto.com/technotr; **221** *top right* ©iStockphoto.com/Mlenny Photography; *bottom right* ©iStockphoto.com/medobear; **224** Layland Masuda/Shutterstock.com; **229** Ferenc Szelepcsenyi/Shutterstock.com; **230** Todd Hackwelder/Shutterstock.com; **233** Pinkcandy/Shutterstock.com; **237** ©iStockphoto.com/PeskyMonkey; **241** ©iStockphoto.com/Tom Buttle; **246** Africa Studio/Shutterstock.com; **247** United States coin image from United States Mint; **248** ©iStockphoto.com/Victor Maffe; **249** *top right* Rafa Irusta/Shutterstock.com; *center left* violetkaipa/Shutterstock.com; **254** A. L. Spangler/Shutterstock.com

**Chapter 6**
**258** Chiyacat/Shutterstock.com, Zoom Team/Shutterstock.com; **265** ©iStockphoto.com/parema; **273** Vitaly Korovin/Shutterstock.com; **274** Fotana/Shutterstock.com; **275** SSSCCC/Shutterstock.com, **277** DM7/Shutterstock.com; **279** Glenda M. Powers/Shutterstock.com, siamionau pavel/Shutterstock.com; **280** ©iStockphoto.com/katkov, ©iStockphoto.com/Andrew Johnson; **283** ©iStockphoto.com/Mehrab Moghadasian; **284** *left* Kamira/Shutterstock.com; *center left* Matthew Jacques/Shutterstock.com; *center right* Rachelle Burnside/Shutterstock.com; *right* ©Norman Young; **290** Patsy Michaud/Shutterstock.com; **295** United States Dept. of the Interior, U.S. Fish and Wildlife Service, FloridaStock/Shutterstock.com; **296** locote/Shutterstock.com; **298** Viorel Sima/Shutterstock.com; **299** Elena Elisseeva/Shutterstock.com; **301** photomak/Shutterstock.com; **303** risteski goce/Shutterstock.com; **305** Maen Zayyad/Shutterstock.com; **310** Feng Yu/Shutterstock.com; **311** *top left* Alexander Lukin/Shutterstock.com; *center right* Andresr/Shutterstock.com; **313** Steshkin Yevgeniy/Shutterstock.com, mayer kleinostheim/Shutterstock.com, Matej Michelizza, Refat/Shutterstock.com; **315** Photo by Alvesgaspar/Joaquim Alves Gaspar, Lisboa, Portugal, modified by RDBury; **316** Faraways/Shutterstock.com

## Chapter 7
**326** Varina and Jay Patel/Shutterstock.com, ©iStockphoto.com/Ann Marie Kurtz; **328** ©iStockphoto.com/Derek Dammann; **331** Pinkcandy/Shutterstock.com, Dmitry Rukhlenko/Shutterstock.com; **332** ©iStockphoto.com/Mark Stay; **333** *Exercise 31* ©iStockphoto.com/edge69; *Exercise 32* Matt Antonino/Shutterstock.com; **337** ©iStockphoto.com/edge69; **338** Johanna Goodyear/Shutterstock.com; **339** ©iStockphoto.com/edge69; **344** Fejas/Shutterstock.com; **345** Rusian Ivantsov/Shutterstock.com; **347** Li Wa/Shutterstock.com; **357** ©iStockphoto.com/Aldo Murillo; **361** Bev Sykes; **366** ©iStockphoto.com/clu; **367** ©iStockphoto.com/Chris Grissom; **373** *bottom left* mmaxer/Shutterstock.com; *bottom right* ©iStockphoto.com/RonTech2000; **375** *top right* ©iStockphoto.com/Viktor Gmyria; *Exercise 41* Pichugin Dmitry/Shutterstock.com; **379** rattanapatphoto/Shutterstock.com; **380** Veronika Surovtseva/Shutterstock.com; **381** Martin Lehmann/Shutterstock.com; **385** G Tipene/Shutterstock.com; **386** Mark Stout Photography/Shutterstock.com; **390** ©iStockphoto.com/tirc83; **395** Elena Elisseeva/Shutterstock.com

## Chapter 8
**400** ©iStockphoto.com/Alistair Cotton; **407** Fred Hendriks/Shutterstock.com; **409** Wolna/Shutterstock.com, ©iStockphoto.com/PLAINVIEW; **410** Maxx-Studio/Shutterstock.com; **413** ©iStockphoto.com/Daniel Jensen; **414** *Exercise 4* Flashon Studio/Shutterstock.com; *Exercise 5* Yuri Kravchenko/Shutterstock.com; *Exercise 6* ARENA Creative/Shutterstock.com; *bottom right* Pavel Plakosh/Shutterstock.com; **415** ©iStockphoto.com/CAP53; **419** ©iStockphoto.com/edge69, L_amica/Shutterstock.com; **421** *top right* Dobrinya/Shutterstock.com; *center left* Ron Zmiri/Shutterstock.com; **423** Ilyashenko Oleksiy/Shutterstock.com; **430** *top right* James Thew/Shutterstock.com; *bottom right* VR Photos/Shutterstock.com; **431** Keith Lovett Photography; **447** ©iStockphoto.com/Geoffrey Holman, TyBy/Shutterstock.com; **448** cbpix/Shutterstock.com

## Chapter 9
**452** ©iStockphoto.com/ALEAIMAGE, ©iStockphoto.com/Ann Marie Kurtz; **458** RTimages/Shutterstock.com; **459** Ben Haslam/Haslam Photography/Shutterstock.com; **460** Golden Pixels LLC/Shutterstock.com, mikeledray/Shutterstock.com; **461** mobil11/Shutterstock.com; **463** ATurner/Shutterstock.com; **466** Digital Genetics/Shutterstock.com; **467** *center right* Zhukov Oleg/Shutterstock.com; *center left* Baker Alhashki/Shutterstock.com; **473** *top right* Maximus256/Shutterstock.com; *center left* Iakov Filimonov/Shutterstock.com, Shebeko/Shutterstock.com; **475** Ivan Bondarenko/Shutterstock.com; **477** Ra Studio/Shutterstock.com; **479** Maxim Kulko/Shutterstock.com; **481** Lukiyanova Natalia/frenta/Shutterstock.com; **482** Elena Elisseeva/Shutterstock.com; **483** NMorozova/Shutterstock.com; **486** Supri Suharjoto/Shutterstock.com; **490** ©iStockphoto.com/Scott Hirko; **491** Sinisa Bobic/Shutterstock.com; **492** Fotokostic/Shutterstock.com; **496** lpatov/Shutterstock.com

## Chapter 10
**500** ©iStockphoto.com/Michael Flippo, ©iStockphoto.com/Ann Marie Kurtz; **505** Zacarias Pereira da Mata/Shutterstock.com; **506** Johnny Habell/Shutterstock.com; **507** Eduard Härkönen/Shutterstock.com; **509** Radovan Spurny/Shutterstock.com, Worldpics/Shutterstock.com, wong yu liang/Shutterstock.com; **514** Sergej Razvodovskij/Shutterstock.com; **515** Pavel Burchenko/Shutterstock.com; **516** Poulsons Photography/Shutterstock.com; **517** eddtoro/Shutterstock.com; **519** BrendanReals/Shutterstock.com; **520** ©Oxford Science Archive/Heritage Images/Imagestate; **524** Mikateke/Shutterstock.com; **525** kavione/Shutterstock.com, Phase4Photography/Shutterstock.com; **527** Monkey Business Images/Shutterstock.com; **535** LoopAll/Shutterstock.com; **536** ©iStockphoto.com/Cathy Keifer

## Chapter 11
**540** Alexander Chaikin/Shutterstock.com, ©iStockphoto.com/Ann Marie Kurtz; **543** *bottom left* KellyNelson/Shutterstock.com; *bottom right* Mircea BEZERGHEANU/Shutterstock.com; **547** mangostock/Shutterstock.com; **548** Stephen Coburn/Shutterstock.com; **549** jean schweitzer/Shutterstock.com; **550** VectoriX/Shutterstock.com, Betacam-SP/Shutterstock.com; **551** wavebreakmedia ltd/Shutterstock.com; **555** ©iStockphoto.com/4x6; **556** Mark Herreid/Shutterstock.com; **557** *center left* Rene Hartmann/Shutterstock.com; *bottom right* Andy Dean Photography/Shutterstock.com; **560** Franco Volpato/Shutterstock.com; **564** Odua Images/Shutterstock.com, Jozsef Bagota/Shutterstock.com; **567** Dmitry Rukhlenko/Shutterstock.com; **573** qushe/Shutterstock.com; **578** Racheal Grazias/Shutterstock.com; **580** Suzanne Tucker/Shutterstock.com; **581** Alexandra Lande/Shutterstock.com; **585** Supri Suharjoto/Shutterstock.com; **586** Aspen Photo/Shutterstock.com; **587** Galina Barskaya/Shutterstock.com; **588** lsantilli/Shutterstock.com; **589** Richard Paul Kane/Shutterstock.com; **591** JCElv/Shutterstock.com; **592** altug/Shutterstock.com; **593** *top left* LoopAll/Shutterstock.com; *center right* Greg Epperson/Shutterstock.com; **594** ©iStockphoto.com/Colleen Butler; **599** *center right* ©iStockphoto.com/Mark Murphy; *center left* ©iStockphoto.com/Milorad Zaric

## Chapter 12
**604** Kasiap/Shutterstock.com, ©iStockphoto.com/Ann Marie Kurtz; **610** ©iStockphoto.com/Jan Will; **612** ©iStockphoto.com/George Peters; **613** Steve Byland/Shutterstock.com; **614** michaeljung/Shutterstock.com; **616** Stephen Coburn/Shutterstock.com; **617** Zinin Alexei/Shutterstock.com; **619** Laurence Gough/Shutterstock.com; **620** FREDERIC J. BROWN/Staff/AFP/Getty Images; **621** ©iStockphoto.com/4x6; **623** Krasowit/Shutterstock.com; **624** *top right* Fotofermer/Shutterstock.com; *center left* Komar Maria/Shutterstock.com; **625** ©iStockphoto.com/Neustockimages; **627** East/Shutterstock.com; **631** Otna Ydur/Shutterstock.com; **633** Adam Gregor/Shutterstock.com; **635** ©iStockphoto.com/David15; **636** Gina Brockett; **637** ©iStockphoto.com/Craig Dingle; **641** ©iStockphoto.com/Jill Fromer; **642** ©iStockphoto.com/Janis Litavnieks; **644** Oleksly Mark/Shutterstock.com; **648** Sashkin/Shutterstock.com; **649** koya979/Shutterstock.com; **650** Michael Shake/Shutterstock.com; **651** Pakhnyushcha/Shutterstock.com; **652** RTimages/Shutterstock.com; **657** *top right* Suponev Vladimir/Shutterstock.com; *bottom right* Alberto Zornetta/Shutterstock.com; **658** *center left* ©iStockphoto.com/Tony Campbell; *bottom right* ©iStockphoto.com/Eric Isselée; **659** *top right* Larry Korhnak; *bottom right* Photo by Andy Newman; **663** *center left* ©iStockphoto.com/Jane norton; *bottom right* ©iStockphoto.com/Krzysztof Zmij; **664** Dwight Smith/Shutterstock.com; **666** Tiplyashin Anatoly/Shutterstock.com; **667** Nikola Bilic/Shutterstock.com; **670** *center right* Iwona Grodzka/Shutterstock.com; *bottom right* Lim Yong Hian/Shutterstock.com

## Appendix A
**A1** *background* ©iStockphoto.com/Björn Kindler; *panther* Clipart deSIGN; *mosaic* Pentocelo; *fractal* ravl; *tornado* EmiliaU; *daisy* tr3gin; **A2** Elnur/Shutterstock.com; **A3** *top* Clipart deSIGN/Shutterstock.com; **A5** *bottom* Pentocelo; **A6** *top right* Taras Vyshnya/Shutterstock.com; *center right* topseller/Shutterstock.com; **A7** *top right* ravl/Shutterstock.com; *center left* tr3gin/Shutterstock.com; *bottom left* Artistas/Shutterstock.com; **A8** EmiliaU/Shutterstock.com; **A9** *top right* deepspacedave/Shutterstock.com; *bottom right* Delmas Lehman/Shutterstock.com

**Cartoon illustrations** Tyler Stout

# Common Core State Standards

## Kindergarten

| | |
|---|---|
| Counting and Cardinality | – Count to 100 by Ones and Tens; Compare Numbers |
| Operations and Algebraic Thinking | – Understand and Model Addition and Subtraction |
| Number and Operations in Base Ten | – Work with Numbers 11–19 to Gain Foundations for Place Value |
| Measurement and Data | – Describe and Compare Measurable Attributes; Classify Objects into Categories |
| Geometry | – Identify and Describe Shapes |

## Grade 1

| | |
|---|---|
| Operations and Algebraic Thinking | – Represent and Solve Addition and Subtraction Problems |
| Number and Operations in Base Ten | – Understand Place Value for Two-Digit Numbers; Use Place Value and Properties to Add and Subtract |
| Measurement and Data | – Measure Lengths Indirectly; Write and Tell Time; Represent and Interpret Data |
| Geometry | – Draw Shapes; Partition Circles and Rectangles into Two and Four Equal Shares |

## Grade 2

| | |
|---|---|
| Operations and Algebraic Thinking | – Solve One- and Two-Step Problems Involving Addition and Subtraction; Build a Foundation for Multiplication |
| Number and Operations in Base Ten | – Understand Place Value for Three-Digit Numbers; Use Place Value and Properties to Add and Subtract |
| Measurement and Data | – Measure and Estimate Lengths in Standard Units; Work with Time and Money |
| Geometry | – Draw and Identify Shapes; Partition Circles and Rectangles into Two, Three, and Four Equal Shares |

## Grade 3

| | |
|---|---|
| Operations and Algebraic Thinking | – Represent and Solve Problems Involving Multiplication and Division; Solve Two-Step Problems Involving Four Operations |
| Number and Operations in Base Ten | – Round Whole Numbers; Add, Subtract, and Multiply Multi-Digit Whole Numbers |
| Number and Operations—Fractions | – Understand Fractions as Numbers |
| Measurement and Data | – Solve Time, Liquid Volume, and Mass Problems; Understand Perimeter and Area |
| Geometry | – Reason with Shapes and Their Attributes |

## Grade 4

| | |
|---|---|
| Operations and Algebraic Thinking | – Use the Four Operations with Whole Numbers to Solve Problems; Understand Factors and Multiples |
| Number and Operations in Base Ten | – Generalize Place Value Understanding; Perform Multi-Digit Arithmetic |
| Number and Operations—Fractions | – Build Fractions from Unit Fractions; Understand Decimal Notation for Fractions |
| Measurement and Data | – Convert Measurements; Understand and Measure Angles |
| Geometry | – Draw and Identify Lines and Angles; Classify Shapes |

## Grade 5

| | |
|---|---|
| Operations and Algebraic Thinking | – Write and Interpret Numerical Expressions |
| Number and Operations in Base Ten | – Perform Operations with Multi-Digit Numbers and Decimals to Hundredths |
| Number and Operations—Fractions | – Add, Subtract, Multiply, and Divide Fractions |
| Measurement and Data | – Convert Measurements within a Measurement System; Understand Volume |
| Geometry | – Graph Points in the First Quadrant of the Coordinate Plane; Classify Two-Dimensional Figures |

# Mathematics Reference Sheet

## Conversions

**U.S. Customary to Metric**
1 inch = 2.54 centimeters
1 foot ≈ 0.30 meter
1 mile ≈ 1.61 kilometers
1 quart ≈ 0.95 liter
1 gallon ≈ 3.79 liters
1 cup ≈ 237 milliliters
1 pound ≈ 0.45 kilogram
1 ounce ≈ 28.3 grams
1 gallon ≈ 3785 cubic centimeters

**Metric to U.S. Customary**
1 centimeter ≈ 0.39 inch
1 meter ≈ 3.28 feet
1 kilometer ≈ 0.62 mile
1 liter ≈ 1.06 quarts
1 liter ≈ 0.26 gallon
1 kilogram ≈ 2.2 pounds
1 gram ≈ 0.035 ounce
1 cubic meter ≈ 264 gallons

**Temperature**

$$C = \frac{5}{9}(F - 32)$$

$$F = \frac{9}{5}C + 32$$

## Number Properties

Commutative Properties of Addition and Multiplication
$$a + b = b + a$$
$$a \cdot b = b \cdot a$$

Associative Properties of Addition and Multiplication
$$(a + b) + c = a + (b + c)$$
$$(a \cdot b) \cdot c = a \cdot (b \cdot c)$$

Addition Property of Zero
$$a + 0 = a$$

Multiplication Properties of Zero and One
$$a \cdot 0 = 0$$
$$a \cdot 1 = a$$

Distributive Property:
$$a(b + c) = ab + ac$$
$$a(b - c) = ab - ac$$

## Properties of Equality

Addition Property of Equality
If $a = b$, then $a + c = b + c$.

Subtraction Property of Equality
If $a = b$, then $a - c = b - c$.

Multiplication Property of Equality
If $a = b$, then $a \cdot c = b \cdot c$.

Division Property of Equality
If $a = b$, then $a \div c = b \div c$, $c \neq 0$.

Squaring both sides of an equation
If $a = b$, then $a^2 = b^2$.

## Properties of Exponents

Product of Powers Property: $a^m \cdot a^n = a^{m+n}$

Quotient of Powers Property: $\frac{a^m}{a^n} = a^{m-n}$, $a \neq 0$

Power of a Power Property: $(a^m)^n = a^{mn}$

Power of a Product Property: $(ab)^m = a^m b^m$

Power of a Quotient Property: $\left(\frac{a}{b}\right)^m = \frac{a^m}{b^m}$, $b \neq 0$

Zero Exponents: $a^0 = 1$, $a \neq 0$

Negative Exponents: $a^{-n} = \frac{1}{a^n}$, $a \neq 0$

Rational Exponents: $\sqrt[n]{a} = a^{1/n}$

## Properties of Square Roots

Product Property of Square Roots
$\sqrt{xy} = \sqrt{x} \cdot \sqrt{y}$, $x \geq 0$ and $y \geq 0$

Quotient Property of Square Roots
$\sqrt{\frac{x}{y}} = \frac{\sqrt{x}}{\sqrt{y}}$, $x \geq 0$ and $y > 0$

## Slope

$$m = \frac{\text{rise}}{\text{run}}$$

$$= \frac{\text{change in } y}{\text{change in } x}$$

$$= \frac{y_2 - y_1}{x_2 - x_1}$$

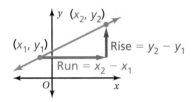

## Factoring

**Difference of Two Squares Pattern**

$$a^2 - b^2 = (a + b)(a - b)$$

**Perfect Square Trinomial Pattern**

$$a^2 + 2ab + b^2 = (a + b)^2$$
$$a^2 - 2ab + b^2 = (a - b)^2$$

## Pythagorean Theorem

$a^2 + b^2 = c^2$

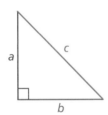

## Distance Formula

$$d = \sqrt{(x_2 - x_1)^2 + (y_2 - y_1)^2}$$

## Equations of Lines

Slope-intercept form
$$y = mx + b$$

Standard form
$$ax + by = c, a, b \neq 0$$

Point-slope form
$$y - y_1 = m(x - x_1)$$

## Forms of Quadratic Functions

Standard form
$$y = ax^2 + bx + c, a \neq 0$$

Vertex form
$$y = a(x - h)^2 + k, a \neq 0$$

## Quadratic Formula

$$x = \frac{-b \pm \sqrt{b^2 - 4ac}}{2a} \quad \leftarrow \text{discriminant}$$

## Sequences

**Arithmetic**

$a_n = a_1 + (n - 1)d$    Explicit equation

$a_n = a_{n-1} + d$    Recursive equation

**Geometric**

$a_n = a_1 r^{n-1}$    Explicit equation

$a_n = r \cdot a_{n-1}$    Recursive equation

## Volume

**Prism**

$V = Bh = \ell wh$

**Cylinder**

$V = Bh = \pi r^2 h$

**Cone**

$V = \frac{1}{3}Bh = \frac{1}{3}\pi r^2 h$

**Sphere**

$V = \frac{4}{3}\pi r^3$